Reagents for Organic Synthesis

INDEXER
Honor Ho

Fiesers'

Reagents for Organic Synthesis

VOLUME TWENTY ONE

Tse-Lok Ho

National Chiao Tung University
Republic of China

WILEY-INTERSCIENCE

A JOHN WILEY & SONS, INC., PUBLICATION

Published by John Wiley & Sons, Inc., Hoboken, New Jersey.
Published simultaneously in Canada.

Library of Congress Cataloging in Publication Data:

ISBN 0-471-21393-4
ISSN 0271-616X

Printed in the United States of America.

10 9 8 7 6 5 4 3 2 1

FOREWORD

When my friend and colleague Professor Tse-Lok Ho invited me to write a Foreword for Volume 21 of the Fieser/Ho Reagents for Organic Synthesis, I felt honored since this series had been, and is, of immense value in my research. In addition, my memories of Mary and Louis Fieser and their friendship over the years are very precious.

The very first synthetic organic experiment that I carried out (as an MIT sophomore in 1946) was the preparation of *n*-butyl bromide from *n*-butyl alcohol using the first edition of Louis Fieser's Experiments in Organic Chemistry (1935). I went on to carry out most of the other experiments in the manual, and added the later editions of the books to my personal library. The third edition, which appeared in 1955, contained an innovation: A section at the end of some 50 pages that provided valuable information on a collection of useful organic reagents and solvents. In the early 1960s during lunch with Mary and Louis, I mentioned my indebtedness to them for having provided that concise summary on reagents and suggested that they consider expanding it into a larger reference work. I think it likely that other colleagues may have made similar proposals. In any event, Mary and Louis in due course went on to produce the first volume of Reagents for Organic Synthesis (1967), a large (1457 pp) work, going far beyond what I had hoped for, that achieved immediate success. I believe that this volume was the first full documentation of the "organic reagent" revolution that occurred in the 1945–1965 period, the result of which was an enormous increase in the ability of chemists to construct complex organic molecules. That increase in the power of synthesis, which probably was closer exponential than linear with respect to the number of available reagents, was just the beginning. The continuing series of volumes, 21 with the appearance of this book, testifies to the continued vigor of reagent discovery and development and to the importance of this area of chemistry.

The Fiesers deserve enormous credit for their role in advancing research in the steroid field during the period 1940–1960 through their three landmark monographs on this important class of compounds. It is remarkable that their impact on the community of synthetic organic chemists through the series Reagents for Organic Synthesis is also large beyond calculation. But then, Mary and Louis Fieser were remarkable people. To Tse-Lok Ho has fallen their mantle and the responsibility to continue this classic series in the service of synthetic chemistry, and science in general. It is clear from Ho–Fieser Volumes 18–21 that the Reagents series is in strong hands. I recall the sage advice that Tse-Lok included in the Preface of Volume 20: "For a task perfectly done, first have the tool sharpened." Each new volume adds greatly to the toolbox and to the luster of the whole series. Tse-Lok Ho deserves our deepest thanks for his heroic effort, including preparation of all the text and formulas, proofreading and, of course, reading a massive amount of original literature. On behalf of the synthetic community, I convey our best wishes for future success.

E. J. Corey

PREFACE

學而時習之
不亦悅乎 ?
--論語

Learn, then apply constantly
Does it impart a joyful feeling?

-Analects

My quoting of another segment of the teachings of Confucius here is, by my interpretation, due to its relevance to the work of synthetic organic chemists. Learning from experience is essential to efficiently achieve set goals that give people intellectual satisfaction. Of course, experiences accumulated by others are also of enormous value. As is well known to readers of Reagents for Organic Synthesis, this series serves to facilitate the role of knowledge assimulation.

Volume 21, which covers the chemical literature from the period 1999–2000, is the first of the series to appear in the twenty-first century. The dawning of this new century brings many unexpected events. Fortunately, among the more pleasant constants human beings enjoy is scientific progress. For example, biotechnology that is primarily based on organic chemistry seems ready to blossom.

To be sure, organic synthesis has yet to attain perfection. The enormous achievements in this field are well recognized, at least by people in the field. While "green chemistry" attracts much current attention, it must be emphasized that active participation of the synthetic chemists is required. Conducting difficult chemistry in aqueous media, in the fluorous phase, in supercritical fluids, or without any solvent, is a great yet profitable challenge, and it behooves us to pursue these areas vigorously. Accordingly, pertinent information is included in volumes of the Reagents for Organic Synthesis series.

Every chemist is interested in the Nobel Prize. Organic synthesis received its share of limelight in the year 2001 as a result of the award going to three distinguished pioneers of asymmetric synthesis: William Knowles, Barry Sharpless, and Ryoji Noyori. Their contributions and profound influences are apparent from pages in this series.

I am deeply grateful to my respected mentor, Professor E. J. Corey of Harvard University, who succeeds Professor Louis F. Fieser to occupy the Sheldon Emery chair of Organic Chemistry, for writing a Foreword for this volume. His unsurpassed innovations in the realm of chemical synthesis, which include the creation of innumerable and invaluable reagents, are the major source of nourishment to sustain the growth of our science.

TSE-LOK HO

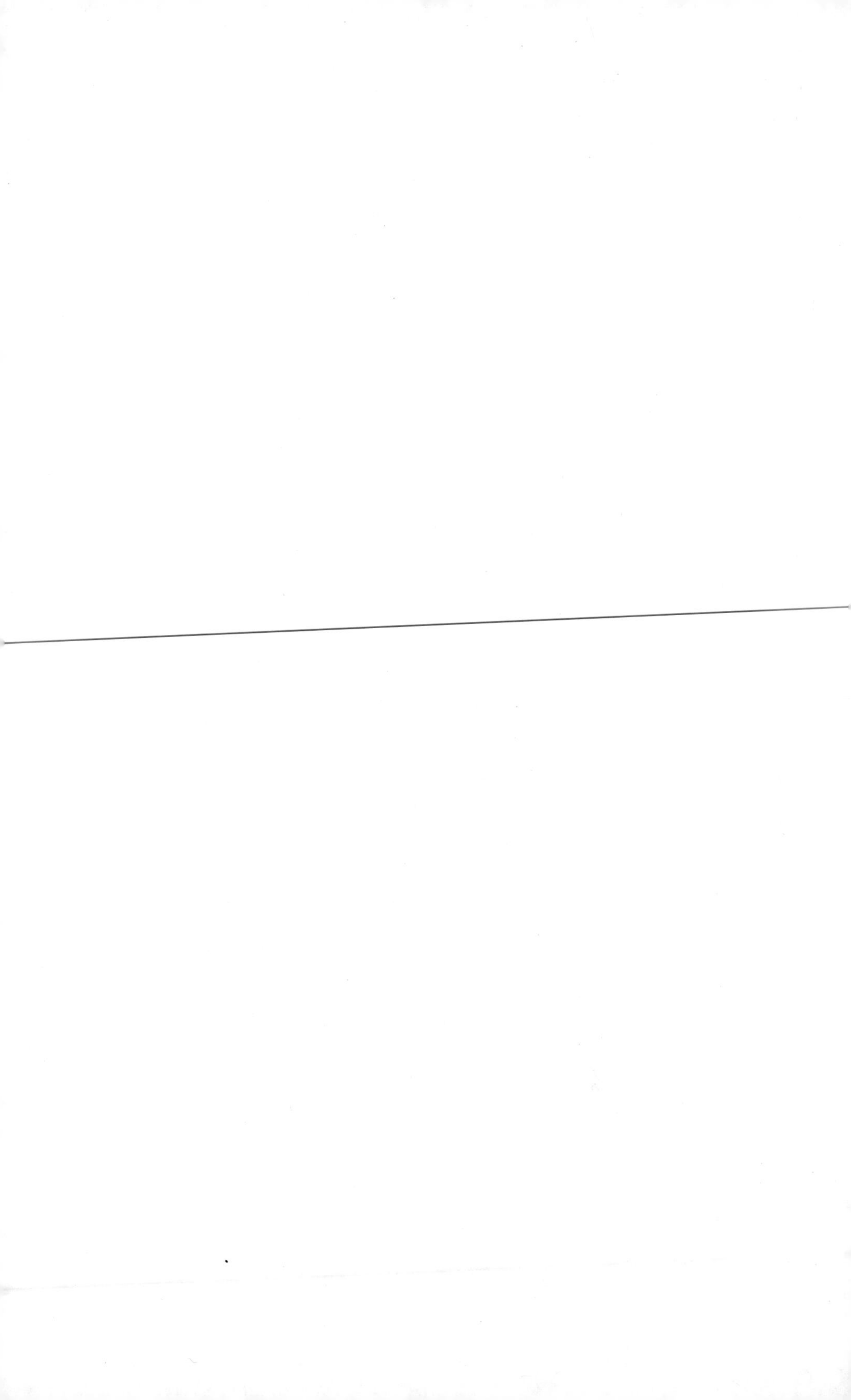

CONTENTS

GENERAL ABBREVIATIONS

Ac	acetyl
acac	acetylacetonate
ADDP	1,1′-(azodicarbonyl)dipiperidine
AIBN	2,2′-azobisisobutyronitrile
An	*p*-anisyl
aq	aqueous
Ar	aryl
ATPH	aluminum tris(2,6-diphenylphenoxide)
9-BBN	9-borabicyclo[3.3.1]nonane
BINOL	1,1′-binaphthalene-2,2′-diol
Bn	benzyl
Boc	*t*-butoxycarbonyl
bpy	2,2′-bipyridyl
Bt	benzotriazol-1-yl
Bu	*n*-butyl
Bz	benzoyl
18-c-6	18-crown-6
c-	cyclo
CAN	cerium(IV)ammonium nitrate
cat	catalytic
Cbz	benzyloxycarbonyl
cod	1,5-cyclooctadiene
cot	1,3,5-cyclooctatriene
Cp	cyclopentadienyl
Cp*	1,2,3,4,5-pentamethylcyclopentadienyl
CSA	10-camphorsulfonic acid
Cy	cyclohexyl
cyclam	1,4,8,11-tetraazacyclotetradecane
DABCO	1,4-diazobicyclo[2.2.2]octane
DAST	(diethylamino)sulfur trifluoride
dba	dibenzylideneacetone
DBN	1,5-diazobicyclo[4.3.0]non-5-ene
DBU	1,8-diazobicyclo[5.4.0]undec-7-ene
DCC	*N,N′*-dicyclohexylcarbodiimide
DDQ	2,3-dichloro-5,6-dicyano-1,4-benzoquinone
de	diastereomer excess
DEAD	diethyl azodicarboxylate

DIAD	diisopropyl azodicarboxylate
Dibal-H	diisobutylaluminum hydride
DMA	*N,N*-dimethylacetamide
DMAP	4-dimethylaminopyridine
DMD	dimethyldioxirane
DME	1,2-dimethoxyethane
DMF	*N,N*-dimethylformamide
DMPU	*N,N'*-dimethylpropyleneurea
DMSO	dimethyl sulfoxide
dppb	1,4-bis(diphenylphosphino)butane
dppe	1,2-bis(diphenylphosphino)ethane
dppf	1,2-bis(diphenylphosphino)ferrocene
dppp	1,3-bis(diphenylphosphino)propane
DTTB	4,4'-di-*t*-butylbiphenyl
E	COOMe
ee	enantiomer excess
en	ethylenediamine
Et	ethyl
EVE	ethyl vinyl ether
Fc	ferrocenyl
Fmoc	9-fluorenylmethoxycarbonyl
Fu	furanyl
HMDS	hexamethyldisilazane
HMPA	hexamethylphosphoric amide
hv	light
Hx	*n*-hexyl
i	iso
Ipc	isopinocampheyl
kbar	kilobar
L	ligand
LAH	lithium aluminum hydride
LDA	lithium diisopropylamide
LHMDS	lithium hexamethyldisilazide
LTMP	lithium 2,2,6,6-tetramethylpiperidide
LN	lithium naphthalenide
lut	2,6-lutidine
M	metal
MAD	methylaluminum bis(2,6-di-*t*-butyl-4-methylphenoxide)
MCPBA	*m*-chloroperoxybenzoic acid
Me	methyl
MEM	methoxyethoxymethyl
Men	menthyl

Mes	mesityl
MOM	methoxymethyl
Ms	methanesulfonyl (mesyl)
MS	molecular sieves
MTO	methyltrioxorhodium
MVK	methyl vinyl ketone
NBS	*N*-bromosuccinimide
NCS	*N*-chlorosuccinimide
NIS	*N*-iodosuccinimide
NMO	*N*-methylmorpholine *N*-oxide
NMP	*N*-methylpyrrolidone
Np	naphthyl
Ns	*p*-nitrobenzenesulfonyl
Nu	nucleophile
Oc	octyl
PCC	pyridinium chlorochromate
PDC	pyridinium dichromate
PEG	poly(ethylene glycol)
Ph	phenyl
phen	1,10-phenenthroline
Pht	phthaloyl
Piv	pivaloyl
PMB	*p*-methoxybenzyloxymethyl
PMHS	poly(methylhydrosiloxane)
Pr	*n*-propyl
py	pyridine
Q^+	quaternary onium ion
RAMP	(*R*)-1-amino-2-methoxymethylpyrrolidine
RaNi	Raney nickel
RCM	ring closure metathesis
R^f	perfluoroalkyl
s-	secondary
(s)	solid
salen	*N,N′*-ethylenebis(salicylideneiminato)
SAMP	(*S*)-1-amino-2-methoxymethylpyrrolidine
sens.	sensitizer
SEM	2-(trimethylsilyl)ethoxymethyl
SES	2-[(trimethylsilyl)ethyl]sulfonyl
TASF	tris(dimethylamino)sulfur(trimethylsilyl)difluoride
TBAF	tetrabutylammonium fluoride
TBDPS	*t*-butyldiphenylsilyl
TBDMS	*t*-butyldimethylsilyl

TBS	*t*-butyldimethylsilyl
Tf	trifluoromethanesulfonyl
THF	tetrahydrofuran
THP	tetrahydropyranyl
Thx	*t*-hexyl
TIPS	triisopropylsilyl
TMEDA	*N,N,N′,N′*-tetramethylethylenediamine
TMS	trimethylsilyl
Tol	*p*-tolyl
Ts	tosyl (*p*-toluenesulfonyl)
TSE	2-(trimethylsilyl)ethyl
TTN	thallium trinitrate
Z	benzyloxycarbonyl
Δ	heat
))))	microwave

REFERENCE ABBREVIATIONS

ACIEE	Angew. Chem. Int. Ed. Engl.
ACR	Acc. Chem. Res.
ACS	Acta Chem. Scand.
AJC	Aust. J. Chem.
AOMC	Appl. Organomet. Chem.
BC	Bioorg. Chem.
BCSJ	Bull. Chem. Soc. Jpn.
BRAS	Bull. Russ. Acad. Sci.
CC	Chem. Commun.
CEJ	Chem. Eur. J.
CJC	Can. J. Chem.
CL	Chem. Lett.
CPB	Chem. Pharm. Bull.
CR	Carbohydr. Res.
EJIC	Eur. J. Inorg. Chem.
EJOC	Eur. J. Org. Chem.
H	Heterocycles
HCA	Helv. Chim. Acta
JACS	J. Am. Chem. Soc.
JCC	J. Carbohydr. Chem.
JCCS(T)	J. Chin. Chem. Soc. (Taipei)
JCR(S)	J. Chem. Res. (Synopsis)
JCS(P1)	J. Chem. Soc. Perkin Trans. 1
JFC	J. Fluorine Chem.
JHC	J. Heterocycl. Chem.
JOC	J. Org. Chem.
JOCU	J. Org. Chem. USSR (Engl. Trans.)
JOMC	J. Organomet. Chem.
MC	Mendeleev Commun.
NJC	New J. Chem.
OL	Organic Letters
OM	Organometallics
PAC	Pure Appl. Chem.
PSS	Phosphorus Sulfur Silicon
RCB	Russian Chem. Bull.
RJGC	Russ. J. Gen. Chem.
RJOC	Russian J. Org. Chem.

S	Synthesis
SC	Synth. Commun.
SL	Synlett
SOC	Synth. Org. Chem. (Jpn.)
T	Tetrahedron
TA	Tetrahedron: Asymmetry
TL	Tetrahedron Lett.

Acetic acid.

Deacylation.[1] Peracetylated sugars can be regioselectively deacetylated at the anomeric center.

[1]Zhang, J., Kovac, P. *JCC* **18**, 461 (1999).

Acetic anhydride. 20, 1

Enolacetylation.[1] Assisted by microwave ketones readily undergo enolacetylation with Ac_2O-I_2.

Acetoxy isothiocyanates.[2] Amino alcohols are converted to isothiocyanatoalkyl acetates on treatment with Et_3N, CS_2, and then Ac_2O-DABCO at room temperature.

[1]Kalita, D.J., Borah, R., Sarma, J.C. *JCR(S)* 404 (1999).
[2]Hara, K., Tajima, H. *SC* **30**, 141 (2000).

Acetone oxime *O*-mesitylenesulfonate.

Arylamines.[1] Arylzinc compounds are converted into arylamines on Cu-catalyzed reaction with the reagent.

[1]Erdik, E., Daskapan, T. *SC* **29**, 3989 (1999).

Acetonyltriphenylphosphonium bromide.

Alcohol protection and deprotection.[1] The salt is an excellent catalyst. For example, THP ethers are formed within minutes in its presence, and deprotection in methanol can be achieved readily at room temperature.

[1]Hon, Y.-S., Lee, C.-F. *TL* **40**, 2389 (1999).

4-Acetoxybenzyl trichloroacetimidate.

Ether formation.[1] By using TfOH as catalyst, alcohols are derivatized with the title reagent to form 4-acetoxybenzyl ethers. These ethers can be cleaved with NaOMe, a valuable feature differentiating them from the unsubstituted benzyl ethers.

[1]Jobron, L., Hindsgaul, O. *JACS* **121**, 5835 (1999).

Acetylacetonato(dicarbonyl)rhodium.

Chain elongation. A stereoselective synthesis of 1,3-diols is by tandem silylformylation–allylation.[1] The potential for iterative operation further enhances the value of this method.

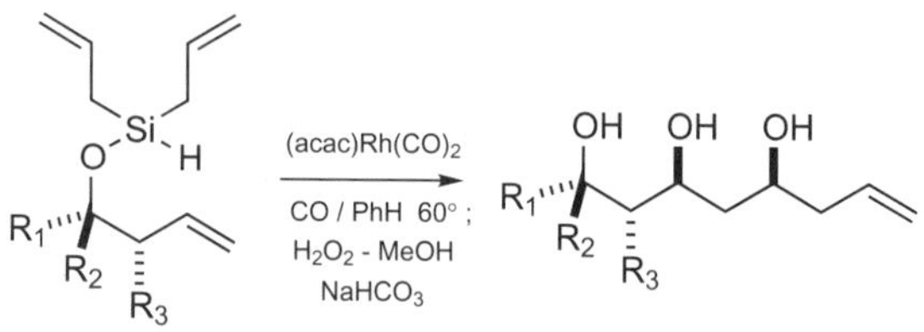

Conversion of RHgCl to RCHO is readily accomplished under H_2–CO.[2] Promotion of this reaction by DABCO indicates its possible role as ligand for the mercury atom.

[1]Zacuto, M.J., Leighton, J.L. *JACS* **122**, 8587 (2000).
[2]Sarraf, S.T., Leighton, J.L. *OL* **2**, 3205 (2000).

Acetylacetonato(diolefin)rhodium.

Aryl transfers. Transfer reactions involving arylboronic acids and aldehydes are catalyzed by (acac)Rh(cyclooctene)$_2$. A large acceleration effect by *t*-Bu$_3$P is evident.[1]

Michael addition. Asymmetric arylation of 1-alkenylphosphonates[2] and nitroalkenes[3] occurs in the presence of (acac)Rh(ethylene)$_2$ and (*S*)-BINAP.

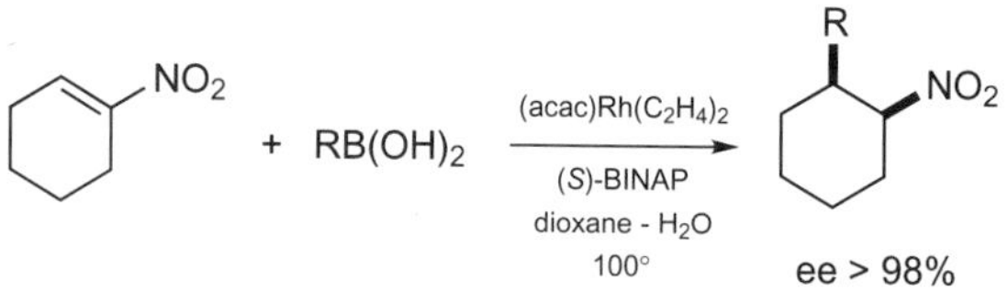

[1]Ueda, M., Miyaura, N. *JOC* **65**, 4450 (2000).
[2]Hayashi, T., Senda, T., Takaya, Y., Ogasawara, M. *JACS* **121**, 11591 (1999).
[3]Hayashi, T., Senda, T., Ogasawara, M. *JACS* **121**, 10716 (1999).

N-Acetylcysteine.

Amidines.[1] The title compound catalyzes the addition of ammonia to nitriles. The reaction conditions are compatible with many functional groups.

[1]Lange, U.E.W., Schäfer, B., Baucke, D., Buschmann, E., Mack, H. *TL* **40**, 7067 (1999).

1-Acylbenzotriazoles.

β-Dicarbonyl compounds.[1] *C*-Acylation of ketones with these reagents are reported (18 examples, 36–92%).

[1]Katritzky, A.R., Pastor, A. *JOC* **65**, 3679 (2000).

N-Acyl-5,5-dimethyloxazolidin-2-ones.

As latent aldehydes.[1] Dibal-H reduction affords aldehydes for the Emmons–Wadsworth condensation.

[1]Bach, J., Bull, S.D., Davies, S.G., Nicholson, R.L., Sanganee, H.J., Smith, A.D. *TL* **40**, 6677 (1999).

Acylhydrazones.

Imine surrogates. In the Mannich reaction and cyanide addition, the use of these hydrazones instead of unstable imines is advantageous.[1] Allylation may also be performed with allyltrichlorosilanes.[2]

Ph–CH₂–CH=N–NHBz + Sn(CH₂CH=CH₂)₄ —[Sc(OTf)₄, MeCN]→ Ph–CH₂–CH(NH–NHBz)–CH₂CH=CH₂

[1]Manabe, K., Oyamada, H., Sugita, K., Kobayashi, S. *JOC* **64**, 8054 (1999).
[2]Kobayashi, S., Hirabayashi, R. *JACS* **121**, 6942 (1999).

Acyl nitrates.

Nitration.[1] When immobilized, the acyl nitrates effect mononitration of arenes with good regioselectivity (6 examples, 86–100%).

[1]Rodrigues, J.A.R., Filho, A.P.O., Moran, P.J.S. *SC* **29**, 2169 (1999).

2-Acyloxy-4,6-dimethoxy-1,3,5-triazines.

Esters.[1] Prepared from 2-chloro-4,6-dimethoxy-1,3,5-triazine and RCOOH these compounds undergo transacylation to alcohols in the presence of $MgBr_2$.

[1]Kaminska, J.E., Kaminski, Z.J., Gora, J. *S* 593 (1999).

N-Acyl-N-(pentafluorophenyl)methanesulfonamides.

N-Acylation.[1] Amides are formed on reaction of these reagents with amines (36 examples, 62–99%).

[1]Kondo, K., Sekimoto, E., Nakao, J., Murakami, Y. *T* **56**, 5843 (2000).

Acylzirconocene chlorides.

Acyl anion equivalents.[1] Addition of the title reagents to enones occurs in the 1,2-mode. This Pd-catalyzed reaction provides unsaturated hydroxy ketones.

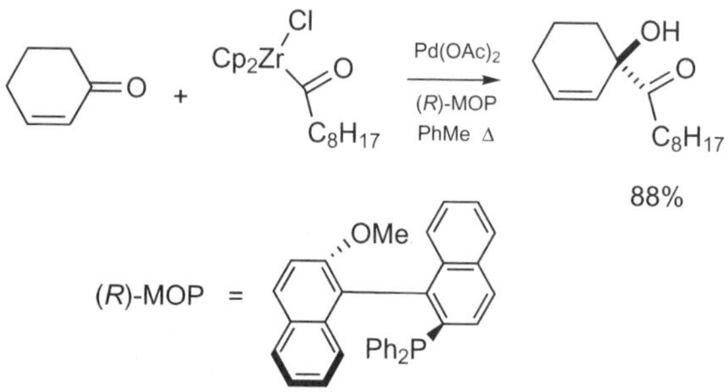

(R)-MOP =

[1]Hanzawa, Y., Tabuchi, N., Saito, K., Noguchi, S., Taguchi, T. *ACIEE* **38**, 2395 (1999).

Alane. 20, 2

Reduction of phosphine oxides.[1] Chemoselective reduction of phosphine oxides in the presence of sulfoxides are observed with AlH_3 in refluxing THF.

[1]Bootle-Wilbraham, A., Head, S., Longstaff, J., Wyatt, P. *TL* **40**, 5267 (1999).

Alkenylboronic acids and esters. 20, 3

2H-Chromenes.[1] Condensation of alkenylboronic acids with salicylaldehydes constitutes a method for the synthesis of 2*H*-chromenes.

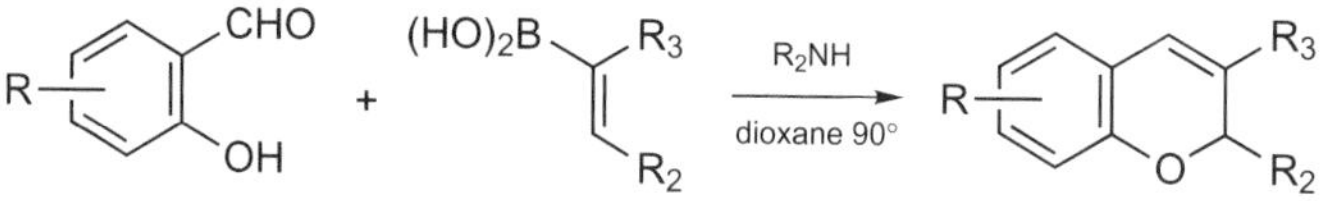

Allylamines.[2] Acylium ion precursors such as carbinolamines alkenylboronates react with in the presence of $BF_3·OEt_2$. Benzylic amines are similarly obtained by replacing alkenylboronates with the aryl analogues.

Diels–Alder reactions.[3] A synthesis of 2,6-disubstituted (also more highly substituted) 3-cyclohexenols is based on cycloaddition with preassemblage of alkadienols and alkenylboronates. Oxidation of the cycloadducts with Me_3NO completes the process.

[1]Wang, Q., Finn, M.G. *OL* **2**, 4063 (2000).
[2]Batey, R.A., Mackay, D.B., Santhakumar, V. *JACS* **121**, 5075 (1999).
[3]Batey, R.A., Thadani, A.N., Lough, A.J. *JACS* **121**, 450 (1999).

1-Alkenyldichloroboranes. 20, 3

As dienophiles.[1] These reagents are active in the Diels–Alder reaction at relatively low temperatures (e.g., $-10°$). The adducts afford alcohols readily.

[1]Zaidlewicz, M., Binkul, J.R., Sokol, W. *JOMC* **580**, 354 (1999).

Alkenylsulfones.

Alkenylation.[1] Free radicals such as those generated from alkyl iodides are intercepted by these reagents.

t-Bu_2O_2, PhCl Δ

57% (*exo* : *endo* = 8 : 2)

[1]Bertrand, F., Quiclet-Sire, B., Zard, S.Z. *ACIEE* **38**, 1943 (1999).

***N*-Alkoxycarbonyl (triflyl)anilides.**

N-Alkoxycarbonylation.[1] Excellent leaving group properties of triflylanilides make the transfer of the alkoxycarbonyl residue to amines very efficient. Both Boc and Cbz derivatives are readily prepared.

[1]Yasuhara, T., Nagaoka, Y., Tomioka, K. *JCS(P1)* 2233 (1999).

(α-Alkoxyalkyl)trimethylsilanes.

Alkoxymethyl anion equivalents.[1] Photoinduced radical electron-transfer conditions induce fission of alkoxymethylsilanes and the addition of $ROCH_2$ groups to electron-deficient alkenes.

BnO⌄SiMe$_3$ + COOMe/CN alkene; hv, Ph-Ph / MeCN - MeOH, 9,10-dicyanoanthracene → BnO…COOMe/CN

64%

Alkoxycarbenium ions.[2] Low-temperature electrooxidation of the reagents (together with Bu_4NBF_4–CH_2Cl_2) generates the carbenium species that can be intercepted with allylsilanes and silyl enol ethers.

[1]Gutenberger, G., Steckhan, E., Blechert, S. *ACIEE* **37**, 660 (1998).
[2]Suga, S., Suzuki, S., Yamamoto, A., Yoshida, J. *JACS* **122**, 10244 (2000).

Alkylaluminum chlorides.

Cyclization of alkenylsilanes.[1] Stereoselective intramolecular vinylsilyation of alkynes is effected by $EtAlCl_2$.

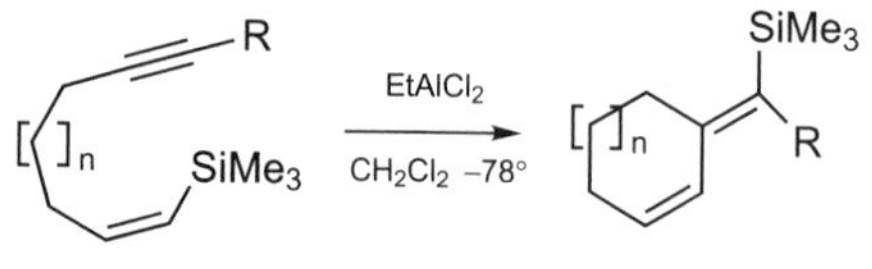

$EtAlCl_2$
CH_2Cl_2 −78°

SiMe3 R n

Cycloadditions. More examples of the $RAlCl_2$-catalyzed Diels–Alder reactions[2–4] and [5 + 2]cycloaddition[5] are shown in the following equations:

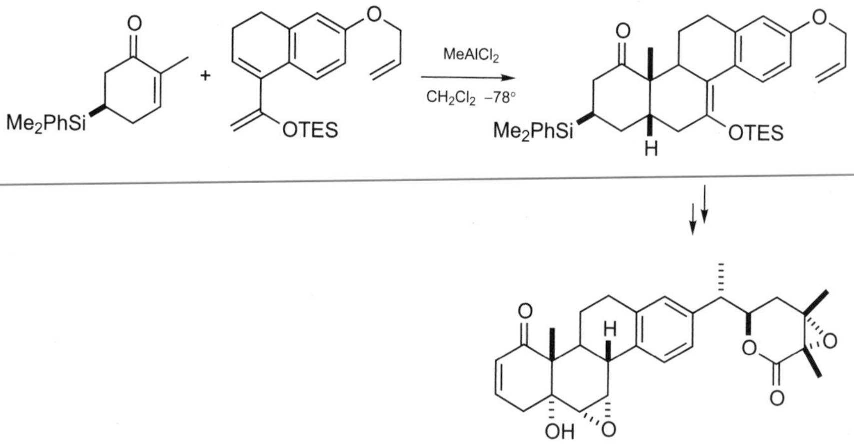

Me2PhSi OTES $MeAlCl_2$ CH_2Cl_2 −78° Me2PhSi H OTES

OH O

NIC-1 lactone

MeOOC CN R O $MeAlCl_2$ CH_2Cl_2 23° MeOOC CN HO R H HO OH H

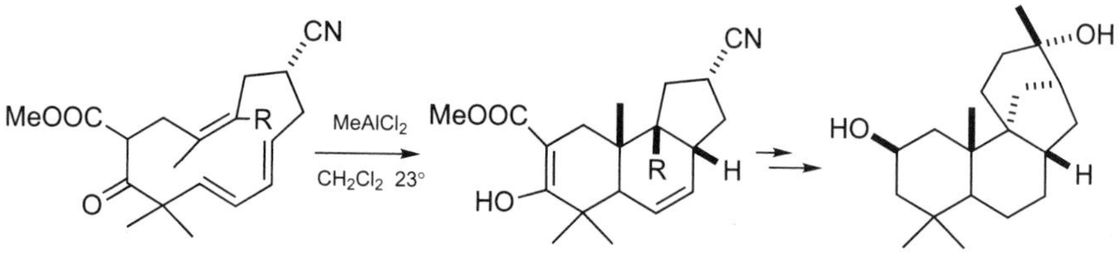

R = CHO 75% (+)-maritimol

TpMo(CO)2 O + R O $EtAlCl_2$ CH_2Cl_2 TpMo(CO)2 O O R hv HCl O O R

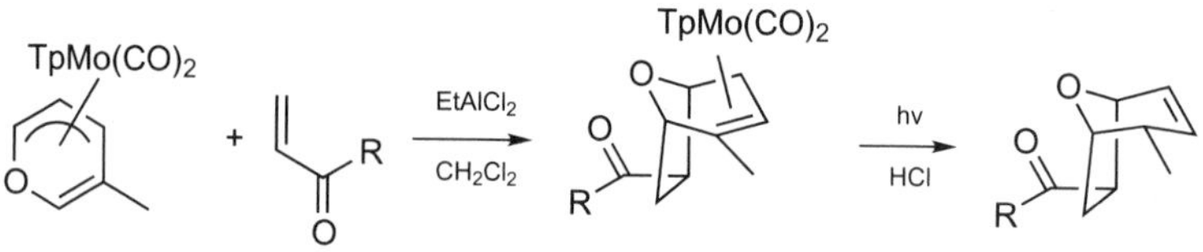

[1]Asao, N., Shimada, T., Yamamoto, Y. *JACS* **121**, 3797 (1999).
[2]Stoltz, B.M., Kano, T., Corey, E.J. *JACS* **122**, 9044 (2000).
[3]Ge, M., Stoltz, B.M., Corey, E.J. *OL* **2**, 1927 (2000).
[4]Toro, A., Nowak, P., Deslongchamps, P. *JACS* **122**, 4526 (2000).
[5]Yin, J., Liebeskind, L.S. *JACS* **121**, 5811 (1999).

Alkyl chloroformates.

Phenols.[1] Cyclization of 3,5-alkadienoic acids is readily effected, particularly in the context of a synthesis of *C*-aryl glycosides.

[1]Fuganti, C., Serra, S. *SL* 1241 (1999).

Alkyldibromoboranes.

Reductive bromination.[1] Aromatic aldehydes are converted into $ArCH_2Br$ by these reagents (e.g., isopinocampheylboron dibromide) (12 examples, 65–87%).

[1]Kabalka, G.W., Wu, Z., Yu, J. *TL* **41**, 5161 (2000).

Alkyl (diphenylphosphono)acetates.

(Z)-α,β-Unsaturated esters.[1] The condensation with aldehydes (Emmons–Wadsworth reaction) using these esters is stereoselective. Two convenient preparative methods for the reagents are (1) reaction of $(PhO)_2P(O)H$ with bromoacetic esters in dichloromethane, (2) alkoxycarbonylation of $(PhO)_2P(O)Me$ with ClCOOR. In the latter protocol, because anions of the methylphosphonate are very unstable, the acylating agents must be present before addition of base to $(PhO)_2P(O)Me$.

[1]Ando, K. *JOC* **64**, 8406 (1999).

Alkyl(3-hydroxyphenyl)diphenylphosphonium salts.

Wittig reactions.[1] These reagents are water soluble. Product isolation from their reaction with carbonyl compounds is also facilitated as the phosphine oxide can be removed by base.

[1]Russell, M.G., Warren, S. *JCS(P1)* 505 (2000).

Alkyliminotris(dimethylamino)phosphoranes.

Esters.[1] Transesterification using enol esters as acyl donors is catalyzed by the iminophosphoranes.

[1]Ilankumaran, P., Verkade, J.G. *JOC* **64**, 9063 (1999).

Alkyl methyl carbonates.

Methoxyarenes.[1] Phenols undergo *O*-methylation with methyl carbonates (K_2CO_3–DMF, 150°). Comparing to dimethyl carbonate the unsymmetrical carbonates with the other alkyl group larger than propyl can be used in an open flask.

[1]Perosa, A., Selva, M., Tundo, P., Zordan, F. *SL* 272 (2000).

1-Alkyl-3-methylimidazolium salts.

Reaction media. The low-melting salts (ionic liquids) are now routinely employed as reaction media. 1-Butyl-3-methylimidazolium tetrafluoroborate and hexafluorophosphate are particularly popular. Besides those described in various sections of this volume, their use in the $Ni(acac)_2$-catalyzed autoxidation of aromatic aldehydes,[1] reduction of aldehydes with Bu_3B,[2] Wittig reactions,[3] Diels–Alder reactions[4] may be mentioned. Many reactions are accelerated and the workup procedure for most of them is facilitated.

By using $(PhH)_4Ru_4BF_4$ as catalyst, the hydrogenation of arenes in reaction media containing ionic liquid and water has been described.[5]

Friedel–Crafts and related reactions conducted in a dialkylimidazolium haloaluminate appear in increasing frequency. The splitting of ethers with simultaneous acylation to afford alkyl benzoates is a recent example.[6]

[1]Howarth, J. *TL* **41**, 6627 (2000).
[2]Kabalka, G.W., Malladi, R.R. *CC* 2191 (2000).
[3]Le Boulaire, V., Gree, R. *CC* 2195 (2000).
[4]Fischer, T., Sethi, A., Welton, T., Woolf, J. *TL* **40**, 793 (1999).
[5]Dyson, P.J., Ellis, D.J., Parker, D.G., Welton, T. *CC* 25 (1999).
[6]Green, L., Hemeon, I., Singer, R.D. *TL* **41**, 1343 (2000).

Alkyl methyl sulfides.

Pauson–Khand reaction.[1] The sulfides have catalytic effects on both intramolecular and intermolecular version of the reaction.

[1]Sugihara, T., Yamada, M., Yamaguchi, M., Nishizawa, M. *SL* 771 (1999).

***N*-Alkyl trifluoroacetaldehyde imines.**

Friedel–Crafts reaction.[1] 1-Aryl-2,2,2-trifluoroethylamines are readily formed from the Lewis acid-catalyzed reaction of $CF_3CH{=}NR$ with activated arenes (e.g., indole).

[1]Gong, Y., Kato, K., Kimoto, H. *SL* 1058 (2000).

Allylbarium reagents.

Preparation.[1] The suitable form of BaI_2 for reaction with allyllithiums is obtained from stirring Ba with 1,2-diiodoethane in ether containing $Na_2S_2O_4$.

[1]Duval, E., Zoltobroda, G., Langlois, Y. *TL* **41**, 337 (2000).

1-Allylbenzotriazole.

2-Alkyl-1,3-butadienes.[1] This reagent undergoes alkylation at the α-position. By reaction with trimethylsilyl chloride, a valuable four-carbon building block is obtained. The latter compound on further alkylation and pyrolysis gives 2-alkyl-1,3-butadienes.

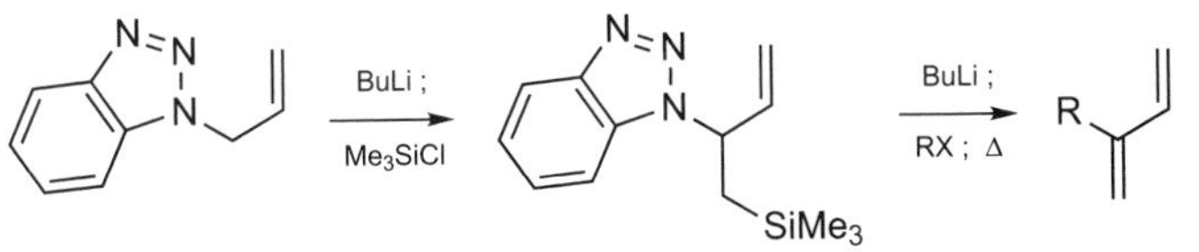

BuLi ; Me_3SiCl — SiMe3 — BuLi ; RX ; Δ — R

[1]Katritzky, A.R., Serdyuk, L., Toader, D., Wang, X. *JOC* **64**, 1888 (1999).

Allylboron reagents.

Homoallylic amines.[1] Water is critical in the asymmetric allylboration of *N*-trimethylsilylbenzaldimines with *B*-allyldiisopinocampheylborane.

Reaction with alkenyl epoxides.[2] The Lewis acid nature of the boranes is felt by the substrates so that rearrangement often precedes allylation.

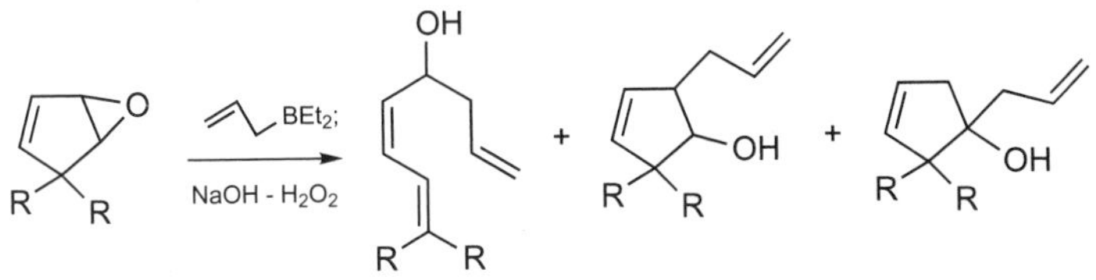

BEt_2; NaOH - H_2O_2

R = H (78%)	90	:	4	:	6
R = Me (81%)	93	:	4	:	3

4-Pentenols.[3] Alkenes undergo allylboration with allyldibromoborane. Oxidation of the adducts gives 2-substituted 4-pentenols.

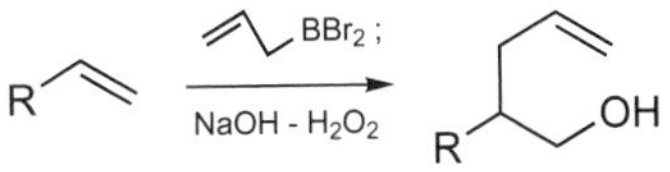

BBr_2 ; NaOH - H_2O_2

[1]Chen, G.-M., Ramachandran, P.V., Brown, H.C. *ACIEE* **38**, 825 (1999).
[2]Zaidlewicz, M., Krzeminski, M.P. *OL* **2**, 3897 (2000).
[3]Frantz, D.E., Singleton, D.A. *OL* **1**, 485 (1999).

(π-Allyl)bromotricarbonylruthenium(I).

Cyclopentenones.[1] A [2 + 2 + 1]cycloaddition involving an alkene, an allylic carbonate, and carbon monoxide occurs under the influence of the Ru(I) complex.

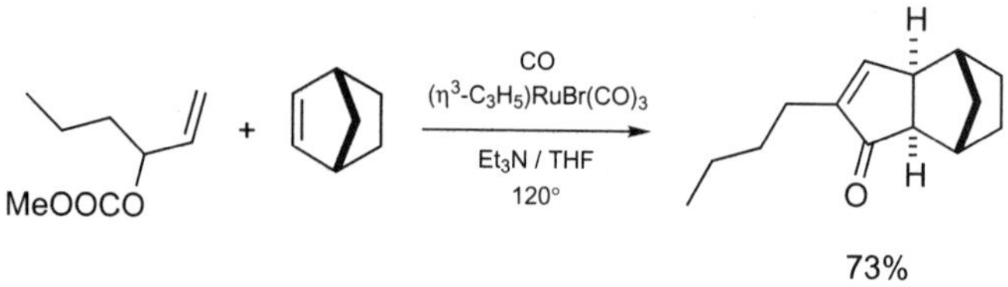

[1]Morisaki, Y., Kondo, T., Mitsudo, T. *OL* **2**, 949 (2000).

Allyl *N*-hydroxymethylcarbamate.

Protection of thiols.[1] Formation of $RSCH_2NHCOOCH_2CH{=}CH_2$ from RSH and the reagent uses CF_3COOH as catalyst. Cleavage of the protecting group is by a catalyzed deallylation [Bu_3SnH–$(Ph_3P)_3PdCl_2$-HOAc] in dichloromethane. (Thiols are oxidized to the disulfides for easy isolation.)

[1]Kimbonguila, A.M., Merzouk, A., Guibe, F., Loffet, A. *T* **55**, 6931 (1999).

Allylnickel bromide.

Cyclization of dienes.[1] Carbocycles and heterocycles are formed.

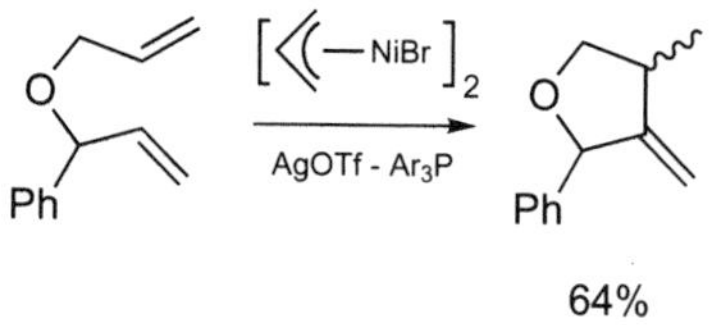

64%

[1]Radetich, B., RajanBabu, T.V. *JACS* **120**, 8007 (1998).

Allylpalladium complexes.

Cyclopropylation.[1] The complex formed from bis(allylpalladium chloride) by treatment with $AgBF_4$ and 2-(2-pyridyl)imidazole mediates cyclopropylation of ketene silyl acetals with allyl acetates.

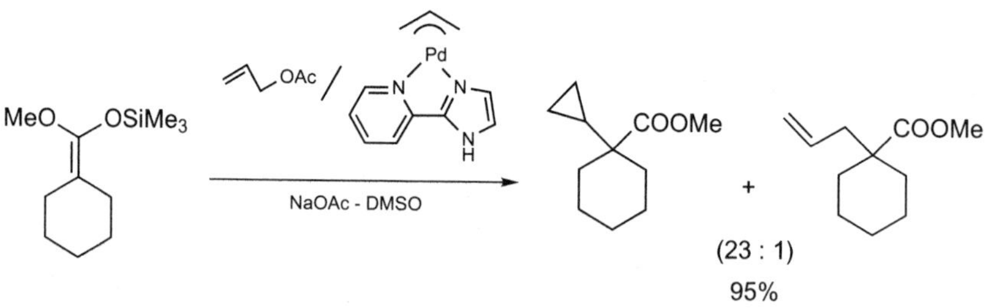

Biaryls.[2] Unsymmetrical biaryls are formed from a cross-coupling reaction involving ArI and tetrabutylammonium triaryldifluorosilicates. The catalyst is bis(allylpalladium chloride).

[1] Satake, A., Koshino, H., Nakata, T. *CL* 49 (1999).
[2] Mowery, M.E., DeShong, P . *JOC* **64**, 3266 (1999).

Allylsamarium halides.

Allyl selenides.[1] Allylsamarium bromide reacts with alkyl selenocyanates to give $RSeCH_2CH{=}CH_2$ (7 examples, 87–95%).

Conjugate addition to nitroalkenes.[2] δ,ε-Unsaturated nitro compounds are formed.

[1] Huang, Y., Chen, R. *SC* **30**, 3775 (2000).
[2] Bao, W., Zheng, Y., Zhang, Y . *JCR(S)* 732 (1999).

Allylsilanes.

Allylations. Opening of unsymmetrical acetals proceeds regioselectively according to the order of addition of the reagent and catalyst.[1]

C_6H_{13} ; CH_2Cl_2 ; HO ; O ; C_6H_{13} ; + ; C_6H_{13} ; O ; OH

$TiCl_4$; allyl-$SiMe_3$	91 : 9	
allyl-$SiMe_3$; $TiCl_4$	1 : 91	

Allylation of nascent *N*-tosylimines is observed when the carbonyl compounds are mixed with $TsNH_2$, $SnCl_2$, and NCS,[2] while chiral homoallylic amines can be obtained from imines in the presence of a π-allylpalladium catalyst, the advantage being the replacement of allylstannanes with allylsilanes.[3] When $YbCl_3$ is used as catalyst the allylation of aldehydes affords homoallyl silyl ethers.[4]

3,3-Bis(trimethylsilyl)propene, which is available from semihydrogenation of the corresponding propyne, is useful for the preparation of (*E*)-β-hydroxyvinyltrimethylsilanes.[5]

Me_3Si ; $SiMe_3$; + ; CHO ; $TiCl_4$; CH_2Cl_2 ; −60° ~ −20° ; Me_3Si ; OH ; 54%

For allylation using allyltrichlorosilanes, a new protocol consists of AgOTs and DMPU.[6] The chlorinated silanes react with benzoylhydrazones and tosylhydrazones readily in DMF, thereby opening up an expedient route to homoallylic amines.[7]

Reaction with oxygen heterocycles. 1,2-Dioxolanes and tetrahydrofurans are formed from reaction with ozonides[8] and epoxides,[9] respectively.

Me_3Si + ozonide ($SnCl_4$, CH_2Cl_2, −78° ~ 0°) → 61%

Reaction with benzotriazol-1-ylmethylamines. 4-Silylmethyl-1,2,3,4-tetrahydroquinolines[10] and 4-chloropiperidines[11] are the products from Lewis acid-catalyzed reactions of allylsilanes with *N*-aryl-*N*-(benzotriazol-1-yl)methylamines and the *N,N*-bis(benzotriazol-1-yl)methylamines, respectively.

Me_3Si + (N, N, N, NH) ($SnCl_4$, CH_2Cl_2 −78°) → Me_3Si (NH) 85%

[3 + 2]Cycloadditions.[12] Certain allylsilanes behave as 2-silylated 1,3-dipoles in the presence of Lewis acids. Their reaction with unsaturated compounds leads to five-membered rings.

$Si(CHPh_2)Me_2$ ($ClSO_2N{=}C{=}O$; Na_2SO_3) → $Me_2(Ph_2HC)Si$ (N H, O)

Hydroxypentenylation.[13] Sequential delivery of two substituents from Si to an acetal function accomplishes this chain extension.

Ph(OMe)OMe + (O, Si) ($BF_3.OEt_2$ / CH_2Cl_2 ; KF / H_2O) → Ph (OMe OH)

77% (*syn* : *anti* 62 : 38)

[1]Egami, Y., Takayanagi, M., Tanino, K., Kuwajima, I. *H* **52**, 583 (2000).
[2]Masuyama, Y., Tosa, J., Kurusu, Y. *CC* 1075 (1999).
[3]Nakamura, K., Nakamura, H., Yamamoto, Y. *JOC* **64**, 2614 (1999).
[4]Fang, X., Watkin, J.G., Warner, B.P. *TL* **41**, 447 (2000).
[5]Princet, B., Anselme, G., Pornet, J. *SC* **29**, 3329 (1999).
[6]Chataigner, I., Piarulli, U., Gennari, C. *TL* **40**, 3633 (1999).
[7]Kobayashi, S., Hirabayashi, R. *JACS* **121**, 6942 (1999).
[8]Dussault, P.H., Liu, X. *TL* **40**, 6553 (1999).
[9]Sugita, Y., Kimura, Y., Yokoe, I. *TL* **40**, 5877 (1999).
[10]Katritzky, A.R., Cui, X., Long, Q. *SC* **36**, 371 (1999).
[11]Katritzky, A.R., Luo, Z., Cui, X. *JOC* **64**, 3328 (1999).
[12]Peng, Z.-H., Woerpel, K.A. *OL* **2**, 1379 (2000).
[13]Frost, L.M., Smith, J.D., Berrisford, D.J. *TL* **40**, 2183 (1999).

Allylstannanes.

Allylations. Allylation of aldehydes with tetraallyltin can be conducted in ionic liquids.[1] The reaction of bifunctional reagents containing allylstannyl and silyl groups with 1,2-diketones via a photoinduced electron-transfer pathway takes place at the tin end.[2]

PhCOCOPh + Bu3Sn~C(=CH2)~SiMe3 → (hν, PhH - MeCN) PhCO–C(Ph)(OH)–CH2C(=CH2)CH2SiMe3 90%

Homolytic allylation of alkenyl iodides with allylstannanes[3] is catalyzed by either AIBN or Et_3B.

α-Ketoximes. A three-component assemblage of α-ketoximes via a free radical pathway employing RI, CO, and $RSO_2CH{=}NOBz$ is initiated by allyltributylstannane and AIBN. Actually, introduction of another alkyl iodide at the end changes the products from the α-ketoaldoxime benzoates to α-diketone monoxime derivatives.[4]

RI + CO + PhSO2CH=NOBz → (allyl-SnBu3, AIBN / PhH Δ) RCOCH=NOBz

Ts-N(CH2CH2I)(CH2CH=CH2) + CO + PhSO2CH=NOBz → (allyl-SnBu3, AIBN / PhH Δ) N-Ts-pyrrolidin-3-yl-CH2COCH=NOBz

Lactones.[5] Carbonylation of radicals derived from (γ-, δ-, ε-) iodoalkanols is followed by a back-transfer of the iodine atom (net insertion of CO to the C—I bond). The subsequent lactonization is an ionic process, therefore Et_3N is required.

n = 1, 2, 3 + CO → ($CH_2{=}CHCH_2SnBu_3$, AIBN / Et_3N, hexane 80°) 47–91%

[1]Gordon, C.M., McCluskey, A. *CC* 1431 (1999).
[2]Takuwa, A., Saito, H., Nishigaichi, Y. *CC* 1963 (1999).
[3]Miura, K., Saito, H., Itoh, D., Hosomi, A. *TL* **40**, 8841 (1999).
[4]Ryu, I., Kuriyama, H., Minakata, S., Komatsu, M., Yoon, J.-Y., Kim, S. *JACS* **121**, 12190 (1999).
[5]Kreimerman, S., Ryu, I., Minakata, S., Komatsu, M. *OL* **2**, 389 (2000).

Allyltriethylgermane.

Homoallylamines.[1] A chemoselective reaction with imines without affecting aldehydes can be achieved with $CH_2{=}CHCH_2GeEt_3$ in the presence of $BF_3{\cdot}OEt_2$ and HOAc in MeCN at 0°. Bronsted and Lewis acids activate the imine substrates.

[1]Akiyama, T., Iwai, J., Onuma, Y., Kagoshima, H. *CC* 2191 (1999).

Allyltris(2-methoxymethoxyphenyl)phosphonium bromide.

1,3-Dienes.[1] The Wittig reaction with this reagent with aldehydes affords conjugated dienes with high *cis*-selectivity (averaging 96:4). However, yields are moderate due to steric congestion of the transition state.

CHO + [2-(MeOCH₂O)C₆H₄]₃P⁺CH₂CH=CH₂ Br⁻ → (BuLi / THF, 0°) 98% (*Z* : *E* = 96 : 4)

[1]Wang, Q., El Khoury, M., Schlosser, M. *CEJ* **6**, 420 (2000).

Alumina.

N-Alkylation.[1] Gas-phase alkylation of amines with alcohols over γ-alumina has been reported.

Oxidations. Dramatic improvement has been claimed in handling alumina-supported MnO_2 for oxidation.[2] Potassium ferrate deposited on alumina can be used to remove a terminal $[CH_2O]$ unit from propargylic alcohols.[3]

Nonaqueous procedures for oxidation of mandelic esters[4] and the cleavage of *p*-nitrophenylhydrazones and semicarbazones involves treatment with ammonium chlorochromate adsorbed on alumina.[5]

[1] Valot, F., Fache, F., Jacquot, R., Spagnol, M., Lemaire, M. *TL* **40**, 3689 (1999).
[2] Stavrescu, R., Kimura, T., Fujita, M., Vinatoru, M., Ando, T. *SC* **29**, 1719 (1999).
[3] Caddick, S., Murtagh, L., Weaving, R. *TL* **40**, 3655 (1999).
[4] Zhang, G.-S., Gong, H. *SC* **29**, 3149 (1999).
[5] Zhang, G.-S., Gong, H., Yang, D.-H., Chen, M.-F. *SC* **29**, 1165 (1999).

Aluminum.

Reductions. Aluminum with NH_4Cl in methanol reduces nitroarenes to arylamines[1] on ultrasound irradiation. On the other hand, Al–KOH induces hydrazoarene formation.[2] Under similar conditions (Al–NaOH) reductive dimerization of araldehydes occurs.[3]

Concerning reductive dimerization in aqueous media, it is interesting to note that changing Al–KF to Al–FeF_2 system causes reduction to $ArCH_2OH$ only.[4]

Reduction of arenes by Al in ionic liquids is efficient. For example, pyrene is fully saturated (84% yield) and 9,10-dimethylanthracene gives the 9,10-dihydro derivative (81%).[5]

Aldehydes from nitroalkenes.[6] By using Al–$NiCl_2$·$6H_2O$ in THF, nitroalkenes are converted into aldehydes (9 examples, 60–88%).

Epoxide opening. Aluminum and an organomercury chloride mediate reaction of epoxides with acyl chlorides to furnish esters of chlorohydrins.[7]

[1] Nagaraja, D., Pasha, M.A. *TL* **40**, 7855 (1999).
[2] Khurana, J. M., Singh, S. *JCS(P1)* 1893 (1999).
[3] Sahade, D.A., Kawaji, T., Sawada, T., Mataka, S., Thiemann, T., Tsukinoki, T., Tashiro, M. *JCR(S)* 210 (1999).
[4] Li, L.-H., Chan, T.H. *OL* **2**, 1129 (2000).
[5] Adams, C.J., Earle, M.J., Seddon, K.R. *CC* 1043 (1999).
[6] Bezbarua, M.S., Bez, G., Barua, N.C. *CL* 325 (1999).
[7] Luzzio, F.A., Bobb, R.A. *T* **55**, 1851 (1999).

Aluminum chloride. 20, 12–13

Deprotection. Ethers are cleaved with $AlCl_3$–NaI without solvent.[1]

Friedel–Crafts reactions. Phenylsulfenylation is conveniently carried out using *N*-phenylthiophthalimide.[2] Acylation of benzodioxin derivatives[3] in the presence of $AlCl_3$-DMA without solvent is regioselective. This complex behaves similarly to $AlCl_3$-DMSO and $AlCl_3$–DMF.

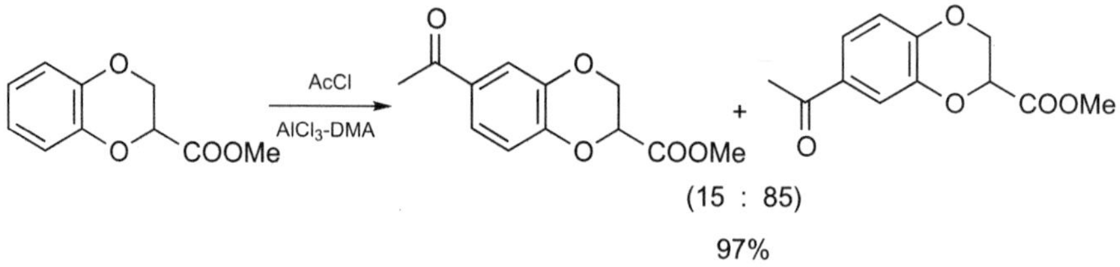

(15 : 85)

97%

Formylation of ferrocene can be accomplished by reaction with triethyl orthoformate in bromobenzene at room temperature (92% yield).[4] Somewhat lower yields are obtained in benzene or dichloromethane.

Reaction of arenes with $PhCCl_3$ in an ionic liquid (*N*-butylpyridinium chloroaluminate) gives diaryl ketones.[5]

Rearrangements. Aryl sulfonates undergo Fries-type rearrangement when exposed to $AlCl_3$–$ZnCl_2$ under microwave irradiation.[6]

A formal O -> C migration of an oxymethyl group to afford spirocyclic products[7] is realized by exposing dibenzodioxepins to $AlCl_3$. Other Lewis acids are less efficient.

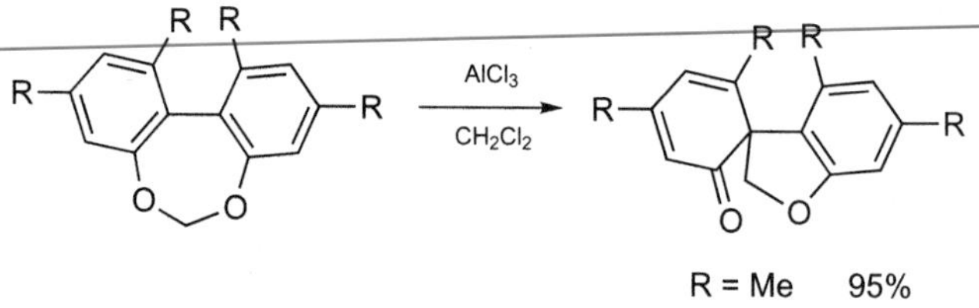

R = Me 95%

2-Chloroallyl sulfoxides.[8] Allenyl sulfoxides undergo addition of HCl in the presence of $AlCl_3$–H_2O.

Hydrosilylation.[9,10] Traditional methods of hydrosilylation involve the use of transition metal catalysts. However, Lewis acids such as $AlCl_3$ also show such reactivity.

Diels–Alder reactions.[11] Addition of $AlCl_3$ to 1-ethyl-3-methylimidazolium chloride forms a chloroaluminate ionic liquid. This substance accelerates and enhances the selectivity of Diels–Alder reactions.

Cyclopentadienes.[12] Replacement of the metal moiety of zirconacyclopentadienes that are readily available from alkynes by the [RCH] unit of an aldehyde is accomplished by an $AlCl_3$ catalyzed reaction.

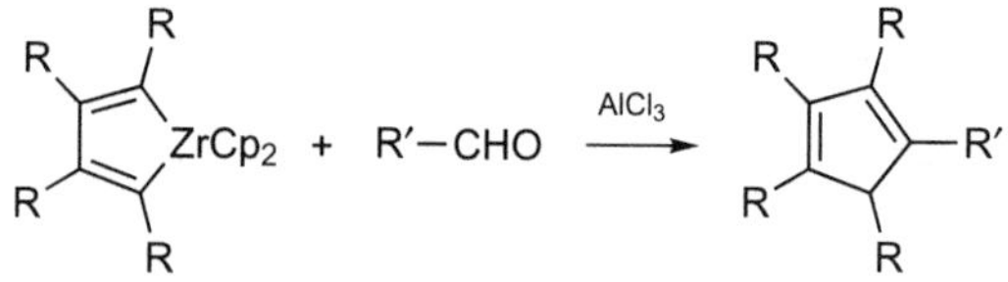

[1]Ghiaci, M., Asghari, J. *SC* **29**, 973 (1999).
[2]Suwa, S., Sakamoto, T., Kikugawa, Y. *CPB* **47**, 980 (1999).
[3]Suarez, A.G. *TL* **40**, 3523 (1999).
[4]Tang, J., Liu, X.-F., Zhang, L.-Y., Xu, X.-L., Zhang, P.-R. *SC* **30**, 1657 (2000).
[5]Rebeiro, G.L., Khadilkar, B.M. *SC* **30**, 1605 (2000).
[6]Moghaddam, F.M., Dakamin, M.G. *TL* **41**, 3479 (2000).
[7]Coleman, R.S., Guernon, J.M., Roland, J.T. *OL* **2**, 277 (2000).
[8]Ma, S., Wei, Q. *EJOC* 1939 (2000).
[9]Song, Y.-S., Yoo, B.R., Lee, G.-H., Jung, I.N. *OM* **18**, 3109 (1999).
[10]Sudo, T., Asao, N., Gevorgyan, V., Yamamoto, Y. *JOC* **64**, 2494 (1999).
[11]Lee, C.W. *TL* **40**, 2461 (1999).
[12]Xi, Z., Li, P. *ACIEE* **39**, 2950 (2000).

Aluminum hexafluoroantimonate.

β-Lactones.[1] Acid chlorides condense with aldehydes to give β-lactones.

$Al(SbF_6)_3$ / i-Pr_2NEt, CH_2Cl_2 −25¡

93%

[1]Nelson, S.G., Wan, Z., Peelen, T.J., Spencer, K.L. *TL* **40**, 5635 (1999).

Aluminum tris(2,6-diphenylphenoxide), ATPH. 20,14–15

Aldol reaction. α,β-Unsaturated carbonyl compounds show regioselectivity in the condensation with PhCHO at the γ-position after enol alumination with ATPH.[1]

ATPH ; LDA, PhCHO — 63%

ATPH ; LDA, PhCHO — 84%

α-Alkoxybutylation.[2] A relay attack of epoxides has been observed in the reaction of enolates in THF–toluene mixture. The coordinated THF is involved.

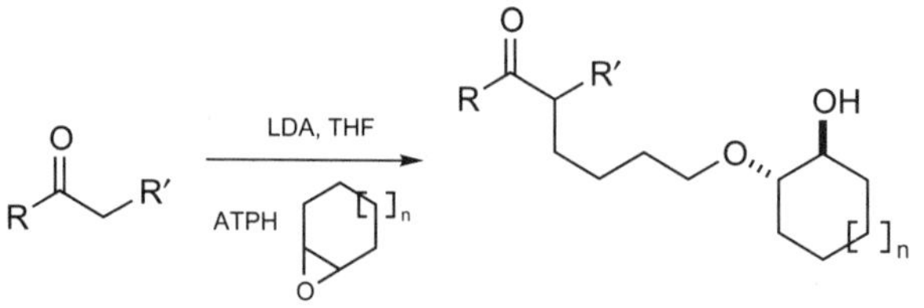

Addition to aromatic nuclei.[3] Complexes formed upon admixture of ATPH with ArCOCl are susceptible to attack by nucleophiles (e.g., enolates, *t*-BuLi) at a nuclear position.

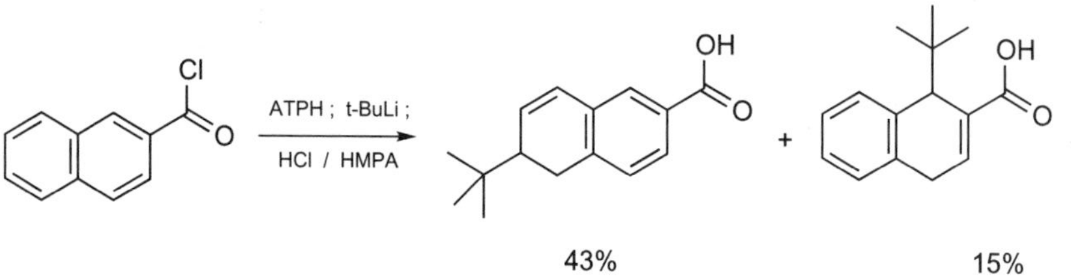

[1] Saito, S., Shiozawa, M., Nagahara, T., Nakadai, M., Yamamoto, H. *JACS* **122**, 7847 (2000).
[2] Saito, S., Yamazaki, S., Shiozawa, M., Yamamoto, H. *SL* 581 (1999).
[3] Saito, S., Sone, T., Murase, M., Yamamoto, H. *JACS* **122**, 10216 (2000).

Aluminum tris(2,6-diphenylphenoxide)–alkyllithium.

Fragmentation. γ-Iodo[1] and γ-stannyl ketones[2] derived from cycloalkenones are converted into unsaturated ketones. The reagent combinations are amphiphilic.

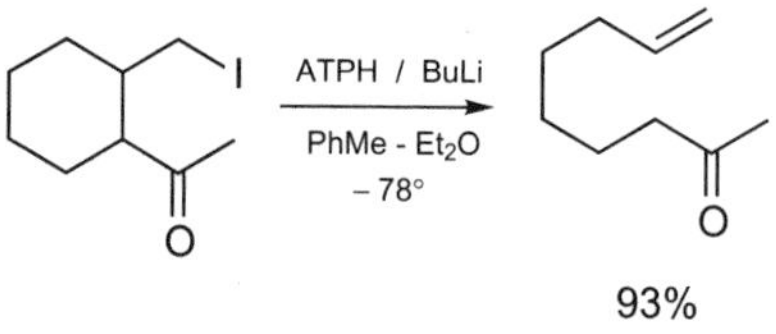

[1] Kondo, Y., Kon-i, K., Ooi, T., Maruoka, K. *TL* **40**, 9041 (1999).
[2] Kondo, Y., Kon-i, K., Iwasaki, A., Ooi, T., Maruoka, K. *ACIEE* **39**, 414 (2000).

1-[α-(Amino)arylmethyl]benzotriazoles

Aminobenzylation of phenols.[1] Sodium phenolates react with these reagents to furnish Mannich-type products.

[1] Katritzky, A.R., Abdel-Fattah, A.A.A., Tymoshenko, D.O., Belyakov, S.A., Ghivirigia, I., Steel, P.J. *JOC* **64**, 6071 (1999).

Ammonium molybdate.

α-Ketols.[1] Epoxides are converted to ketols by ammonium molybdate tetrahydrate at room temperature (11 examples, 92–96%).

[1] Ismail, N., Rao, R.N. *CL* 844 (2000).

Antimony.

Homoallylic alcohols.[1] Allylation of carbonyl compounds in aqueous media is mediated by activated Sb–KF.

[1] Li, L.-H., Chan, T.H. *TL* **41**, 5009 (2000).

Antimony(V) fluoride.

Carbonyl chloride fluoride.[1] Phosgene exchanges one of its chlorine atoms on treatment with SbF_5.

[1] Hoge, B., Christe, K.O. *JFC* **94**, 107 (1999).

Arenediazonium *o*-benzenedisullfonimides.

Aryl halides.[1] Decomposition of these salts **(1)** in the presence of a quaternary ammonium halide furnishes aryl halides.

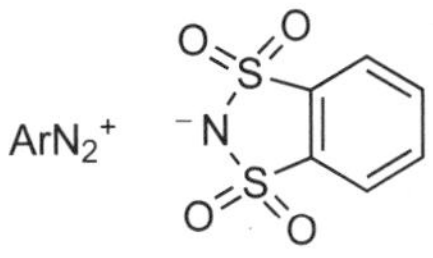

ArN_2^+

(1)

[1] Barbero, M., Degani, I., Dughera, S., Fochi, R. *JOC* **64**, 3448 (1999).

Arenesulfonic acids.

Mannich-type reaction.[1] Sulfonic acids bearing a long chain (e.g., dodecyl) act as excellent catalysts for the condensation in water.

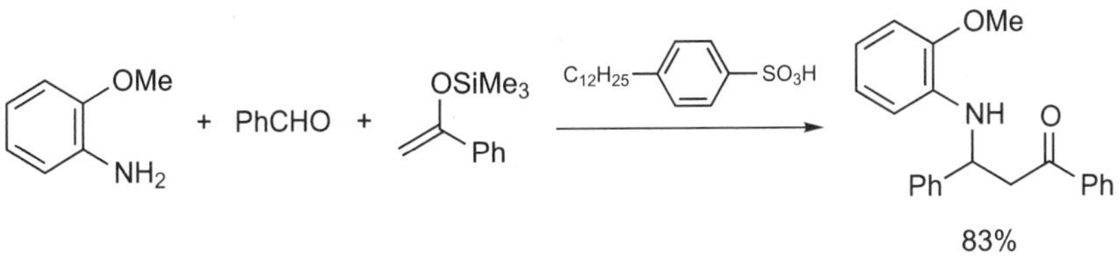

OMe, NH_2 + PhCHO + $OSiMe_3$, Ph → ($C_{12}H_{25}$–C_6H_4–SO_3H) OMe, NH, O, Ph, Ph

83%

[1] Manabe, K., Mori, Y., Kobayashi, S. *SL* 1401 (1999).

Arylboronic acids.

Glycosylation.[1] The glycosylation of unprotected sugars can be accomplished when they are activated by arylboronic acids such as **1**.

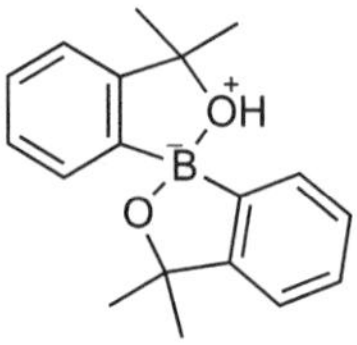

(1)

N-Arylimidazoles.[2] Arylation of imidazoles is efficiently catalyzed by a copper complex in the air.

[1]Oshima, K., Aoyama, Y. *JACS* **121**, 2315 (1999).
[2]Collman, J.P., Zhong, M. *OL* **2**, 1233 (2000).

Aryl *p*-nitrobenzenesulfonates.

Aryl esters.[1] Protected amino acids are converted to their aryl esters by these reagents.

[1]Pudhom, K., Vilaivan, T. *TL* **40**, 5939 (1999).

2,3-Azetidinediones.

Peptides.[1] Reaction of the heterocyclic compounds with amines gives peptides.

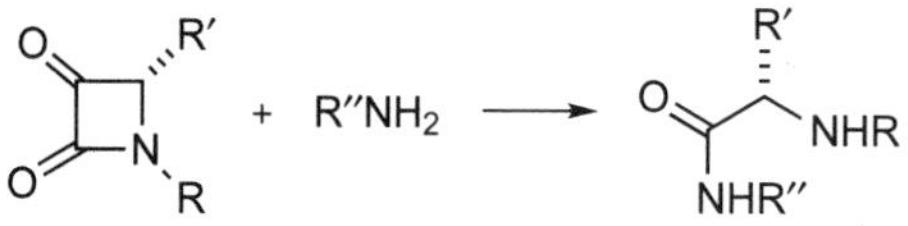

[1]Alcaide, B., Almendros, P., Aragoncillo, C. *CC* 757 (2000).

***B*-(2-Azido-2-propenyl)-1,3,2-dioxoborinane.**

Allylation.[1] The reagent reacts smoothly with aldehydes to afford azidohomoallyl alcohols.

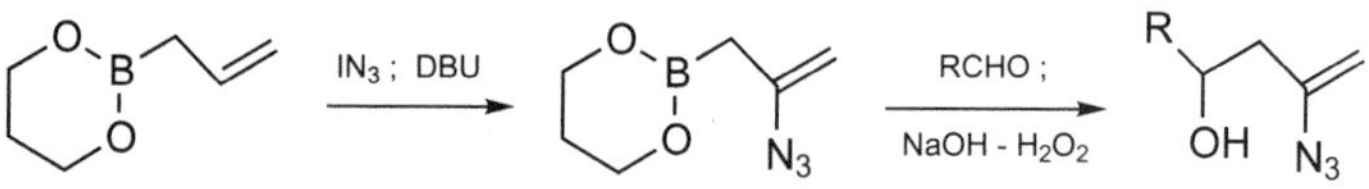

[1]Salunkhe, A.M., Ramachandran, P.V., Brown, H.C. *TL* **40**, 1433 (1999).

B

Barium permanganate.

Oxidation.[1] Solvent-free oxidation of allylic and benzylic alcohols is possible with this reagent or MnO_2.

[1]Firouzabadi, H., Karimi, B., Abbassi, M. *JCR(S)* 236 (1999).

Benzenesulfenyl chloride.

Cyclopropane ring opening.[1] Reaction conditions have great influence on the regioselectivity of ring opening.

PhSCl / CCl_4	10%	55%
PhSCl - py / CH_2Cl_2	75%	0%

[1]Graziano, M.L., Iesce, M.R., Cermola, F. *S* 1944 (1999).

3-Benzenesulfenyl-2-(*N*-cyanoimino)thiazolidine.

Sulfenylation.[1] Ketones and amines are sulfenylated with the reagent **1** under very mild conditions.

(**1**)

[1]Tanaka, T., Azuma, T., Fang, X., Uchida, S., Iwata, C., Ishida, T., In, Y., Maesaki, N. *SL* 33 (2000).

***N*-Benzenesulfenylsuccinimide.**

Sulfenylation.[1] Acylsilanes and aldehydes undergo acid-catalyzed α-sulfenylation. Ketones give low yields under the same condsitions.

[1]Huang, C.-H., Liao, K.-S., De, S.K., Tsai, Y.-M. *TL* **41**, 3911 (2000).

Benzenesulfonamide.

N-Benzenesulfonyl aldimines.[1] Aldehydes are readily converted to α-tosyl sulfonamides by reaction with $PhSO_2NH_2$, elimination of TsH is accomplished with $NaHCO_3$.

R−CHO → ($TolSO_2Na$, $PhSO_2NH_2$ / HCOOH / H_2O) → R–CH($NHSO_2Ph$)(SO_2Tol) → ($NaHCO_3$) → R–CH=N–SO_2Ph

[1]Chemla, F., Hebbe, V., Normant, J.-F. *S* 75 (2000).

Benzenethiol. 16, 327–329; **19,** 19; **20,** 20–21

Thioimidic esters.[1] These esters are convenient precursors of amidines and they are prepared from nitriles and PhSH in the presence of HBr.

Hydration of 3-aryl-2-propynols.[2] Alkenyl sulfides are formed as intermediates and these undergo hydrolysis with sulfuric acid in EtOH to afford α-ketols.

Michael–aldol reaction tandem.[3] Addition of lithium benzenethiolate to conjugated esters in the presence of aldehydes is followed by an aldol reaction in a stereoselective manner.

PhCHO + t-butyl acrylate → (PhSH / CH_2Cl_2, − 50°) → PhCH(OH)CH(CH_2SPh)CO_2t-Bu

80% (*syn* : *anti* 92 : 8)

Dechlorothiolation.[4] Activated trichloromethyl groups (e.g., in $RCOCCl_3$) are converted to the benzenethiomethyl residues using PhSH–PhSNa. Both substitution and hydrodechlorination occur on such treatment. However, chloroform and α,α,α-trichlorotoluene do not react in the same manner.

2-pyrrolyl–$COCCl_3$ → (PhSH - PhSNa / THF) → 2-pyrrolyl–$COCH_2SPh$

100%

α-Thioaldehydes.[5] A method for homologation of aldehydes accompanied by simultaneous introduction of an α-PhS group involves reaction with the lithiated chloromethyl phenyl sulfoxide and subsequent treatment with PhSK (PhSH + *t*-BuOK).

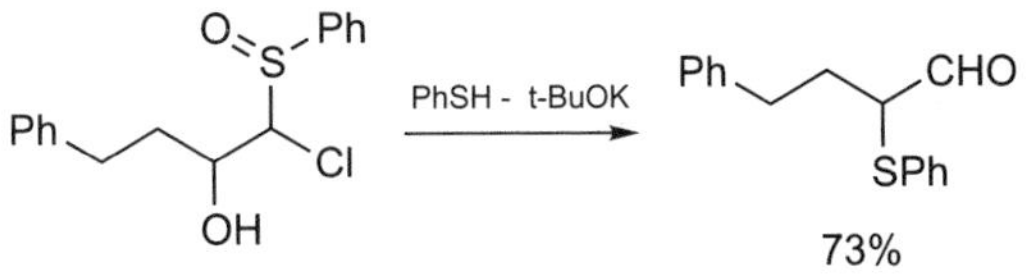

Deoxygenation of silyl epoxides.[6] Alkenylsilanes are generated from the epoxy silanes with retention of configuration, when they are exposed to PhSLi in THF at −78°. The reagent attacks the silylated carbon atom regioselectively unless the silyl group is very bulky (e.g., *t*-butyldiphenylsilyl group).

Cleavage of aryl esters. Aryl acetates, benzoates, pivalates, and tosylates are cleaved under nonhydrolytic conditions (PhSH, K_2CO_3, NMP reflux); alkyl esters are less reactive.[7] *o*-Aminobenzenethiol is also useful for this purpose.

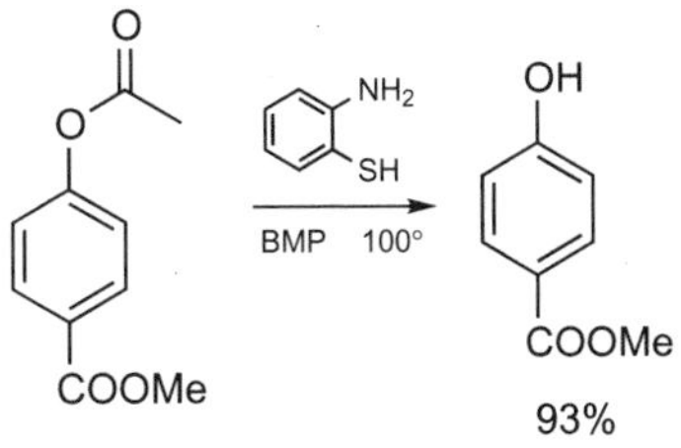

p-Methoxybenzyl ether cleavage.[8] The chemoselective cleavage with PhSH–$SnCl_4$ is valuable when oxidative reagents (DDQ, CAN, etc.) need to be avoided. However, it had been reported previously that such ethers can be cleaved by treatment with $SnCl_4$ alone.

[1]Baati, R., Gouverneur, V., Mioskowski, C. *S* 929 (1999).
[2]Waters, M.S., Cowen, J.A., McWilliams, J.C., Maligres, P.E., Askin, D. *TL* **41**, 141 (2000).
[3]Kamimura, A., Mitsudera, H., Asano, S., Kidera, S., Kakehi, A. *JOC* **64**, 6353 (1999).
[4]Romero-Ortega, M., Fuentes, A., Gonzalez, C., Morales, D., Cruz, R. *S* 225 (1999).
[5]Satoh, T., Kubota, K. *TL* **41**, 2121 (2000).
[6]Cuadrado, P., Gonzalez-Nogal, A.M. *TL* **41**, 1111 (2000).
[7]Chakraborti, A.K., Nayak, M.K., Sharma, L. *JOC* **64**, 8027 (1999).
[8]Yu, W., Su, M., Gao, X., Yang, Z., Jin, Z. *TL* **41**, 4015 (2000).

Benzimidazolium bromochromate.

Bromination and oxidation.[1] This reagent brominates arenes. Deactivated rings do not react, therefore *p*-nitroacetophenone gives ω-bromo- *p*-nitroacetophenone.

Oxidation of alcohols to carbonyl products is also observed in moderate yields.

[1]Ozgun, B., Degirmenbasi, N. *SC* **29**, 763 (1999).

Benzotriazole.

1,5-Disubstituted pyrrolidinones.[1] 1-Substituted 5-(1-benzotriazolyl)pyrrolidinones are available in one step from primary amines, benzotriazole, and 2,5-dimethoxy-2,5-dihydrofuran. The benzotriazolyl group is readily replaced by nucleophiles.

MeO–(2,5-dihydrofuran)–OMe + RNH_2 + benzotriazole $\xrightarrow{HOAc}$ 1-R-5-(benzotriazol-1-yl)pyrrolidin-2-one

70%

[1]Katritzky, A. R., Mehta, S., He, H.-Y., Cui, X. *JOC* **65**, 4364 (2000).

1 *H*-Benzotriazol-1-yl mesylate.

N-Mesylation.[1] Mesylation of an amine in the presence of hydroxyl groups is possible with this reagent. Primary amines are more reactive than secondary amines.

[1]Kim, S.Y., Sung, N.-D., Choi, J.-K., Kim, S.S. *TL* **40**, 117 (1999).

(1 *H*-Benzotriazol-1-ylmethyl)trimethylsilane.

Homologation of carboxylic acids.[1] A reaction sequence for the homologation starts from a reaction of $BtCH_2SiMe_3$ with RCOCl. After *O*-triflation, treatment with either TsOH and then Bu_4NF (aliphatic series) or NaOMe then HCl (aromatic series) completes the transformation.

[1]Katritzky, A. R., Zhang, S., Fang, Y. *OL* **2**, 3789 (2000).

***O*-(Benzotriazol-1-yl)- *N,N,N′,N′*-tetramethyluronium tetrafluoroborate.**

Ether cleavage.[1] This reagent catalyzes cleavage of THP, silyl, and 4,4′-dimethoxytrityl ethers.

[1]Ramasamy, K.S., Averett, D. *SL* 709 (1999).

Benzotriazol-1-yl alkyl carbonate.

Amides.[1] Reaction with a carboxylic acid followed by aminolysis leads to amides.

[1]Lee, J.S., Oh, Y.S., Lim, J.K., Yang, W.Y., Kim, I.H., Lee, C.W., Chung, Y.H., Yoon, S.J. *SC* **29**, 2547 (1999).

4-Benzyloxybutanal.

Acetals.[1] 1,3-Diols can be protected as substituted 1,3-dioxanes. Cleavage of the cyclic acetals is by catalytic hydrogenation in which an intramolecular exchange reaction is instigated by the released primary hydroxyl group.

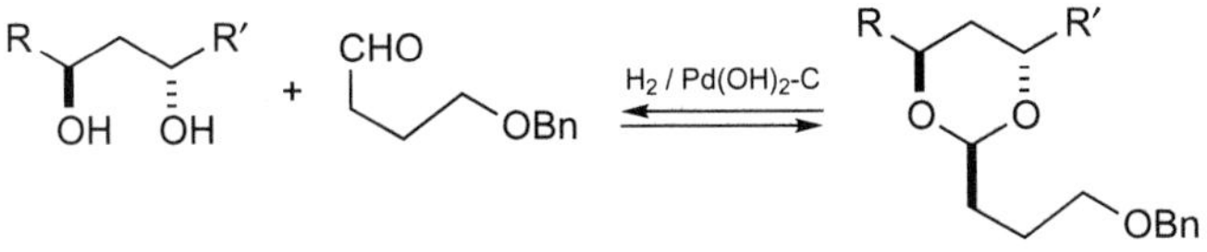

[1]Powell, N.A., Rychnovsky, S.D. *JOC* **64**, 2026 (1999).

Benzyltriethylammonium tetrathiomolybdate.

Disulfides. Alcohols are converted to disulfides on activation with DCC (CuCl catalyzed) and reaction with the reagent (9 examples, 45–88%).[1] On the other hand, organic disulfides are cleaved by the title reagent and the resulting thiolates can be trapped by Michael acceptors.[2]

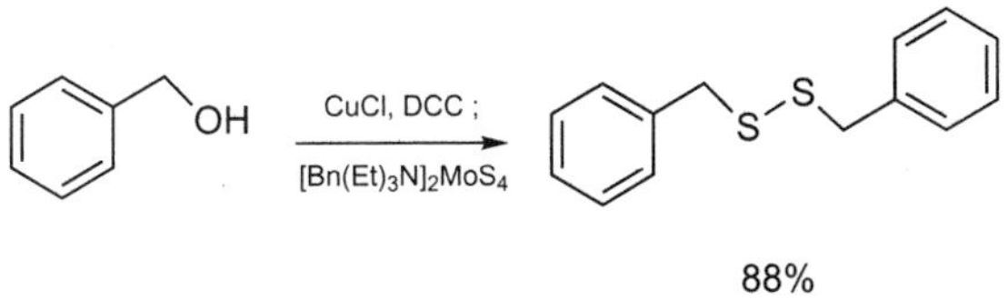

Cleavage of propargyl carbamates.[3] The amino-protecting group is removed on treatment with 1 equiv of the tetrathiomolybdate salt in MeCN with ultrasound irradiation.

[1]Sinha, S., Ilankumaran, P., Chandrasekaran, S. *T* **55**, 14769 (1999).
[2] Prabhu, K.R., Sivanand, P.S., Chandrasekaran, S. *ACIEE* **39**, 4316 (2000).
[3] Sinha, S., Ilankumaran, P., Chandrasekaran, S. *TL* **40**, 771 (1999).

Benzyltrimethylammonium diphenylphosphinate.

Stille coupling.[1] The title compound is an effective scavenger of organotin residues. On precipitating $R_3SnOP(O)Ph_2$ the Cu-catalyzed Stille coupling reactions are facilitated.

[1]Zhang, S., Marshall, D., Liebeskind, L.S. *JOC* **64**, 2796 (1999).

Benzyltrimethylammonium tetrachloroiodate.

Hydroximoyl chlorides.[1] Chlorination of aldoximes by this reagent at room temperature is rapid.

[1]Kanemasa, S., Matsuda, H., Kamimura, A., Kakinami, T. *T* **56**, 1057 (2000).

Benzyltriphenylphosphonium peroxodisulfate.

Oxidation.[1] Primary alcohols are oxidized to aldehydes under solvent-free conditions with this reagent and aluminum chloride as catalyst. Reaction in the presence of other Lewis acids including $FeCl_3$, $BiCl_3$, and $ZnCl_2$ are less efficient.

[1]Hajipour, A.R., Mallakpour, S.E., Adibi, H. *CL* 460 (2000).

Benzyl *N*-vinylcarbamate.

Arylethylamines.[1] The title compound is transformed into an alkylboronic acid via hydroboration and its Suzuki coupling with ArI affords $ArCH_2CH_2NHCbz$.

[1]Kamatani, A., Overman, L.E. *JOC* **64**, 8743 (1999).

Binaphane.

Asymmetric hydrogenation.[1] (*R,R*)-Binaphane (**1**) is prepared from 1,1′-bi-2,2′-naphthol in five steps. Its Ru complex is useful for asymmetric hydrogenation of trisubstituted enamides.

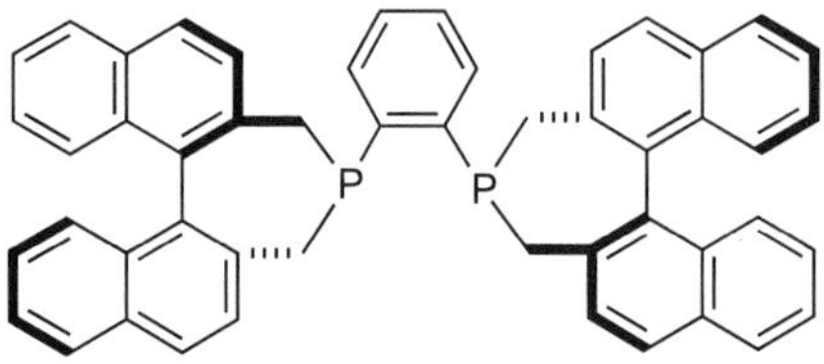

(**1**)

[1]Xiao, D., Zhang, Z., Zhang, X. *OL* **1**, 1679 (1999).

1,1′-Binaphthalene-2,2′-diol, BINOL.

Resolution.[1] The cyclic boronate ester derived from racemic BINOL and $BH_3 \cdot SMe_2$ preferentially forms a diastereoisomer with L-proline, thereby (*R*)-BINOL can be isolated by crystallization. On decomposition of the complex (*S*)-BINOL is recovered.

[1]Shan, Z., Xiong, Y., Zhao, D. *T* **55**, 3893 (1999).

1,1′-Binaphthalene-2,2′-diol (modified)–aluminum complexes.

Cyanations. Aluminum complexes of BINOLs (**1**) that are armed at C-3 and C-3′ with diarylphosphine oxide groups possess both Lewis acid and base centers. Asymmetric cyanation of aldehydes[1] and imines[2] with Me_3SiCN, and of quinolines and isoquinolines[3] in a manner analogous to the Reissert reaction is successful (ee ~ 70–90%). The asymmetric Strecker synthesis is applicable to conjugated aldimines and the higher reactivity of Me_3SiCN than HCN in the presence of 10 mol% of PhOH enables its use in catalytic amount while supplying stoichiometric HCN as the cyanide source.

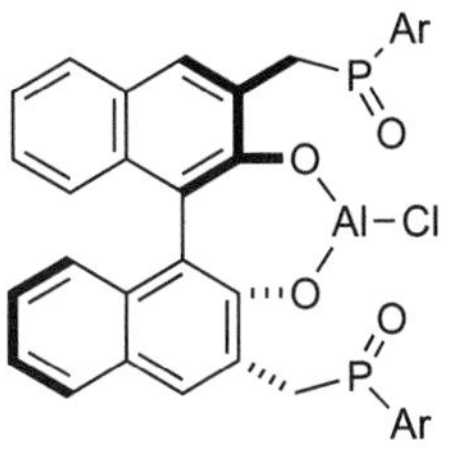

(**1**)

Aldol transfer.[4] An aluminum complex of BINOL is capable of catalyzing the replacement of the carbinol moiety of an aldol. Thus, adding the complex to a mixture of the aldol and an aldehyde in dichloromethane at room temperature completes the transformation.

[1]Hamashima, Y., Sawada, D., Kanai, M., Shibasaki, M. *JACS* **121**, 2641 (1999).
[2]Takamura, M., Hamashima, Y., Usuda, H., Kanai, M., Shibasaki, M. *ACIEE* **39**, 1650 (2000).
[3]Takamura, M., Funabashi, K., Kanai, M., Shibasaki, M. *JACS* **122**, 6327 (2000).
[4]Simpura, I., Nevalainen, V. *ACIEE* **39**, 3422 (2000).

1,1′-Binaphthalene-2,2′-diol–gallium/lithium complexes. 20, 24–25

2-Aryloxy alcohols.[1] *meso*-Epoxides are opened by phenols in the presence of the Ga—Li linked BINOL complex with high enantioselectivity.

[1]Matsunaga, S., Das, J., Roels, J., Vogl, E.M., Yamamoto, N., Iida, T., Yamaguchi, K., Shibasaki, M. *JACS* **122**, 2252 (2000).

1,1′-Binaphthalene-2,2′-diol–potassium/ytterbium complexes.

Addition to imines.[1] The heterobimetallic complex derived from BINOL, $(i\text{-PrO})_3Yb$ and *t*-BuOK catalyzes the asymmetric reaction between nitroalkanes and *N*-phosphonimines.

[1]Yamada, K., Harwood, S.J., Gröger, H., Shibasaki, M. *ACIEE* **38**, 3504 (1999).

1,1′-Binaphthalene-2,2′-diol–titanium complexes. 15, 26–27; **16,** 24–25; **17,** 28–30; **18,** 43–44; **19,** 25; **20,** 25–27

Enantioselective isomerization.[1] Epoxides of enol esters provide chiral α-acyloxy ketones and chiral starting materials.

4-Substituted butenolides. 2-Siloxyfurans react with aldehydes[2] and with imines[3] at C-5 to give chiral products. The hydroxyalkylbutenolides formed can be incorporated into the catalyst system thus rendering the aldol reaction autoinductive.

Chiral sulfoxides.[4] A catalytic oxidation of organothiomethylphosphonates enables the preparation of enantiomerically pure sulfoxides.

α,α-Disubstituted amines.[5] Addition of two Grignard reagents to a nitrile leads to the products. If a chiral BINOL–Ti complex is introduced in between the additions, the products are enantioenriched.

BnO⌄CN —EtMgBr ; [BINOL–Ti(Cl)(O-i-Pr)]→ —MeMgBr→ BnO⌄C*(NH$_2$)(Me)Et

Mukaiyama aldol reaction. Reasonably good enantioselection is observed when the catalyzed reaction of ketene silyl acetals derived from thioesters with aldehydes is conducted in supercritical fluoroform.[6]

The homoaldol version involving 1-ethoxy-1-trimethylsiloxycyclopropane as donor employs a catalyst prepared from (i-PrO)$_2$Ti(binol) and Me_3SiOTf.[7]

2-(α-Hydroxyalkyl)-1,3-butadienes.[8] A synthetic process for these compounds involves reaction of aldehydes with 2,3-butadienylstannane. A BINOL–Ti complex combined with synergistic reagent such as i-PrSBEt$_2$ constitutes a superior catalytic system.

PhCHO + Bu_3SnCH$_2$CH=C=CH$_2$ —i-PrSBEt$_2$, (i-PrO)$_4$Ti / (S)-BINOL PhCF$_3$ / – 20°→ PhCH(OH)C(=CH$_2$)CH=CH$_2$

81% (98% ee)

(R)-1-Phenylpropanol.[9] In this preparative example, the repeated usage of a 6,6′-bis(tridecafluorooctyl)silylated BINOL in catalyzing the reaction of PhCHO with Et_2Zn has been demonstrated.

[1]Feng, X., Shu, L., Shi, Y. *JACS* **121**, 11002 (1999).
[2]Szlosek, M., Figadere, B. *ACIEE* **39**, 1799 (2000).

[3]Martin, S.F., Lopez, O.D. *TL* **40**, 8949 (1999).
[4]Capozzi, M.A.M., Cardellicchio, C., Fracchiolla, G., Naso, F., Tortorella, P. *JACS* **121**, 4708 (1999).
[5]Charette, A.B., Gagnon, A. *TA* **10**, 1961 (1999).
[6]Mikami, K., Matsukawa, S., Kayaki, Y., Ikariya, T. *TL* **41**, 1931 (2000).
[7]Martins, E.O., Gleason, J.L. *OL* **1**, 1643 (1999).
[8]Yu, C.-M., Lee, S.-J., Jeon, M. *JCS(P1)* 3557 (1999).
[9]Nakamura, Y., Takeuchi, S., Ohgo, Y., Curran, D.P. *TL* **41**, 57 (2000).

1,1′-Binaphthalene-2,2′-diol (modified)–zinc complexes.

Chiral benzylic alcohols.[1] The addition of diphenylzinc to aldehydes is rendered enantioselectively by 3,3′-bis(2,5-dihexyloxyphenyl)-1,1′-bi-2-naphthol.

[1]Huang, W.-S., Pu, L. *JOC* **64**, 4222 (1999).

1,1′-Binaphthalene-2,2′-diol (modified)–zirconium complexes. 19, 25–26

Aldol reaction.[1] A catalytic asymmetric version using a Zr complex of 3,3′-diiodo-BINOL has been developed.

Hetero-Diels–Alder reaction.[2] The zirconium complexes of 3,3′-diaryl-BINOL are effective chiral catalysts for the smooth condensation of aldimines with a Danishefsky diene. Linkage of the BINOL moiety to polymer support also provides viable catalysts.

[1]Ishitani, H., Yamashita, Y., Shimizu, H., Kobayashi, S. *JACS* **122**, 5403 (2000).
[2]Kobayashi, S., Kusakabe, K., Ishitani, H. *OL* **2**, 1225 (2000).

1,1′-Binaphthalene-2,2′-diol derived 20-crown-6.

Mukaiyama aldol reaction.[1] The crown ether **1** on complexation with $Pb(OTf)_2$ serves as a catalyst for the aldol reaction in aq EtOH at 0°. High diastereoselectivity (*syn*-selective) is observed.

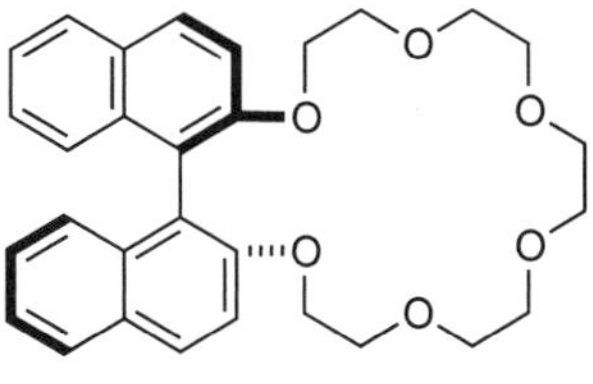

(**1**)

[1]Nagayama, S., Kobayashi, S. *JACS* **122**, 11531 (2000).

Bis[(1,1′-binaphthalene-2,2′-diol)-3-methyl] ether.

Michael reaction.[1] Lanthanide–alkali metal complexes of this "dimeric" BINOL derivative are stable, storable, and reusable chiral catalysts.

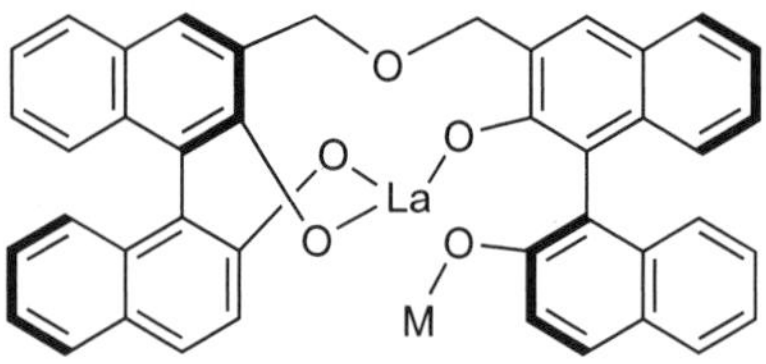

(1)

[1]Kim, Y.S., Matsunaga, S., Das, J., Sekine, A., Ohshima, T., Shibasaki, M. *JACS* **122**, 6506 (2000).

1,1′-Binaphthalene-2,2′-diyl heteroesters.

4-Alkynones.[1] The *B*-alkynylboronates act as alkynyl group donors of an enone.

Hydrogenation.[2] Easily prepared monodentate phosphoramidite ligands containing a BINOL residue are employed in conjunction with $(cod)_2RhBF_4$ in asymmetric hydrogenation of dehydroamino acid derivatives and α-substituted acrylic esters (ee > 99% reachable).

Conjugate addition. A phosphoramidite ligand (**1**) is useful for chiral induction during reaction of organozinc reagents with enones[3] and 4,4-disubstituted 1,5-cyclohexadienones[4] in the presence of $Cu(OTf)_2$.

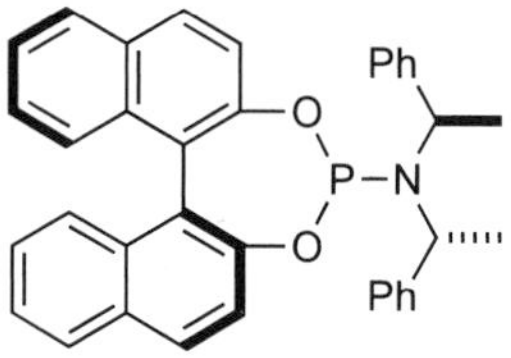

(**1**)

Allylic displacements.[5] The mixed phosphite of BINOL and phenol serves as a ligand to make up a catalyst with $[(cod)IrCl]_2$ for the allylic substitution. Remarkable rate enhancement and increase in yield and ee are realized when the reaction is performed with BuLi–$ZnCl_2$ also.

Hetero-Diels–Alder reaction.[6] In the cyclocondensation of the Danishefsky diene and aldehydes in the presence of **2**, a remarkably high asymmetric amplification is realized.

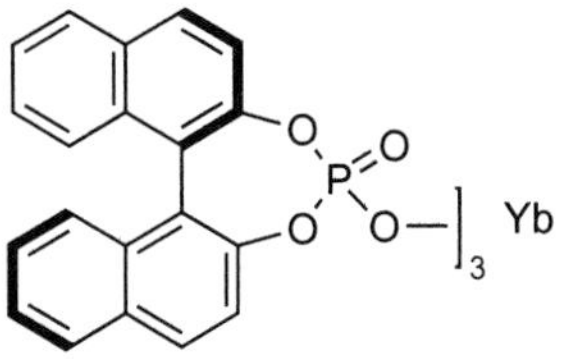

(2)

Hydroformylation.[7] A polymer-supported (*R,S*)-BINAPHOS complex **3** serves as catalyst for asymmetric hydroformylation of gaseous substrates (e.g., 3,3,3-trifluoropropene) to afford branched aldehydes.

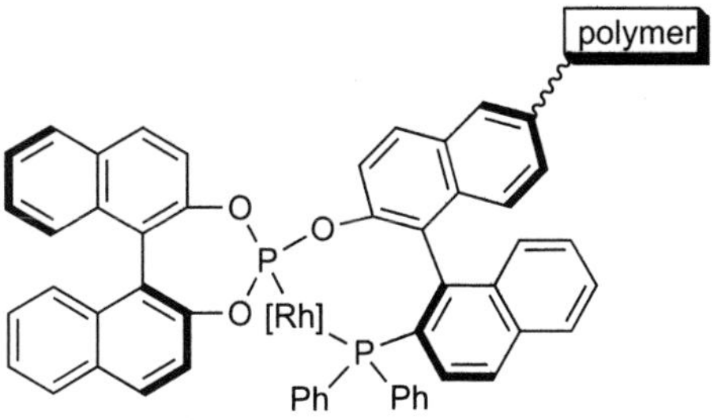

(3)

[1]Chong, J.M., Shen, L., Taylor, N.J. *JACS* **122**, 1822 (2000).
[2]van der Berg, M., Minnaard, A.J., Schudde, E.P., van Esch, J., de Vries, A.H.M., de Vries, J.G., Feringa, B.L. *JACS* **122**, 11539 (2000).
[3]Naasz, R., Arnold, L.A., Pineschi, M., Keller, E., Feringa, B.L. *JACS* **121**, 1104 (1999).
[4]Imbos, R., Brilman, M.H.G., Pineschi, M., Feringa, B.L. *OL* **1**, 623 (1999).
[5]Fuji, K., Kinoshita, N., Tanaka, K., Kawabata, T. *CC* 2289 (1999).
[6]Furuno, H., Hanamoto, T., Sugimoto, Y., Inanaga, J. *OL* **2**, 49 (2000).
[7]Nozaki, K., Shibahara, F., Hiyama, T. *CL* 694 (2000).

1,1′-Binaphthalene-2,2′-diyl siloxane.

Cyclosilylation.[1] Yttrocene **1** catalyzes the conversion of polyenes to cyclic compounds while incorporating a silyl group from hydrosilane reagents. However, asymmetric induction is poor.

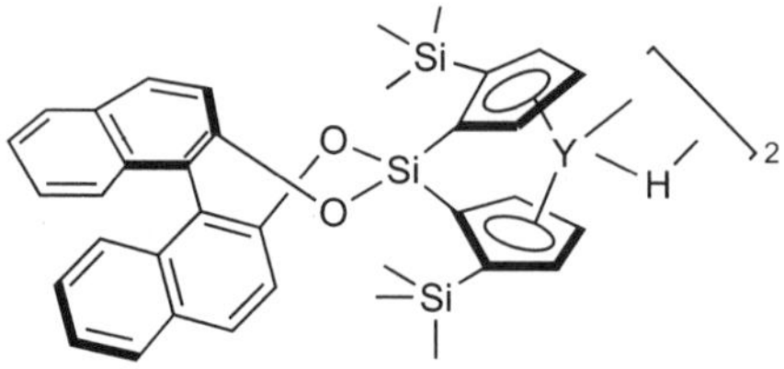
Si
O
Si
O
Y
H
2
Si

(**1**)

[1] Muci, A.R., Bercaw, J.E. *TL* **41**, 7609 (2000).

Bis[1-(1,1′-Binaphthalene-2,2′-methyl)-3-methylimidazol-2,2′-ylidene]palladium(II) iodide.

Heck reaction.[1] The carbene complex **1** is very stable to air, water, and high temperature. It also survives silica gel chromatography. Preliminary investigations of **1** indicate its catalytic activity in the Heck reaction.

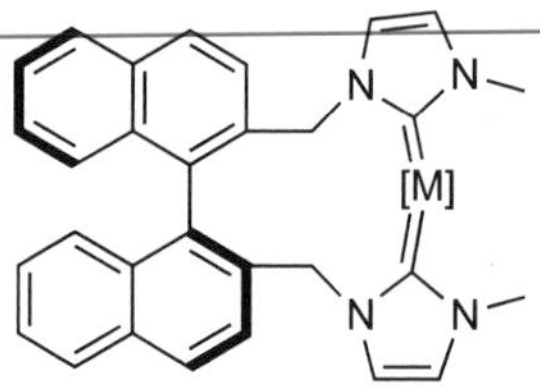
N
N
[M]
N
N

(**1**)

[1] Clyne, D.S., Jin, J., Genest, E., Gallucci, J.C., RajanBabu, T.V. *OL* **2**, 1125 (2000).

Bis(1,1′-binaphthalene-2,2′-methyl)tin dibromide.

Enantioselective benzoylation.[1] The title compound **1** catalyzes selective benzoylation of racemic 1,2-diols.

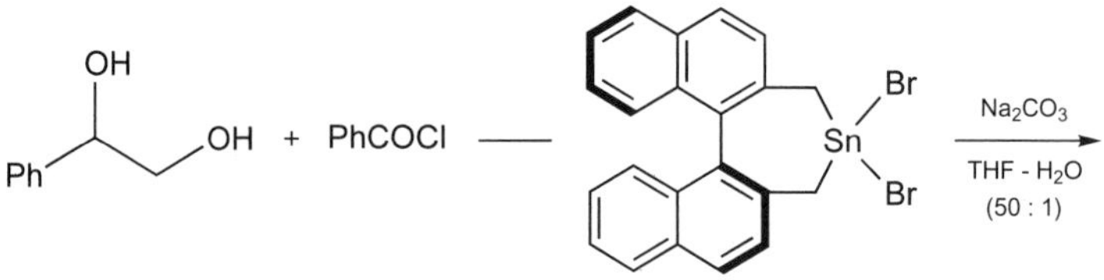
OH
Ph
OH
+ PhCOCl
Br
Sn
Br
Na_2CO_3
THF - H_2O
(50 : 1)

(**1**)

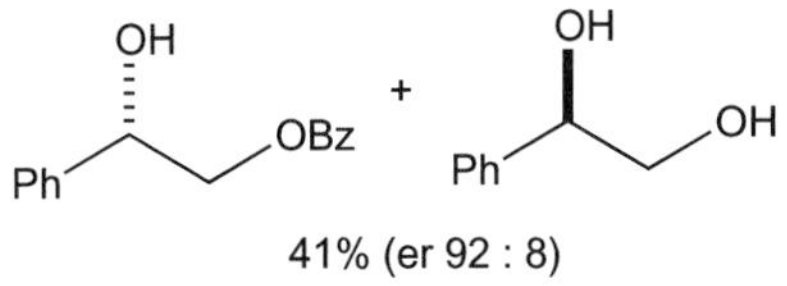

Ph / OH (hashed) / OBz + Ph / OH / OH

41% (er 92 : 8)

[1]Iwasaki, F., Maki, T., Onomura, O., Nakashima, W., Matsumura, Y. *JOC* **65**, 996 (2000).

Bis(acetonitrile)(1,5-cyclooctadiene)rhodium(I) tetrafluoroborate.

Aryl transfer. The aryl group of $ArSnMe_3$ is transferred to *N*-sulfonylimines[1] under the influence of the ionic Rh complex.

[1] Oi, S., Moro, M., Fukuhara, H., Kawanishi, T., Inoue, Y. *TL* **40**, 9259 (1999).

Bis(acetonitrile)cyclopentadienyl(triorganophosphine)ruthenium(I) tetrafluoroborate.

Isomerization. Efficiency for the transformation of allyl alcohols to saturated ketones by intramolecular redox isomerization is improved by using these complexes [vs. $CpRu(PPh_3)_2Cl–NH_4PF_6$], but there is the disadvantage of limited substitution patterns on the substrates. The catalyst is readily prepared from $RuCl_3 \cdot 3H_2O$.

[1] Slugovc, C., Ruba, E., Schmid, R., Kirchner, K. *OM* **18**, 4230 (1999).

Bis(acetonitrile)dichloropalladium(II). 13, 33, 211, 236; **14,** 35–36; **15,** 28–29; **16,** 25–26; **17,** 30–31; **18,** 44–45; **19,** 26

Thio-Claisen rearrangement.[1] Introduction of a side chain to the α-position of chiral bicyclic thiolactams via a thio-Claisen rearrangement at room temperature is effected with assistance of $(MeCN)_2PdCl_2$.

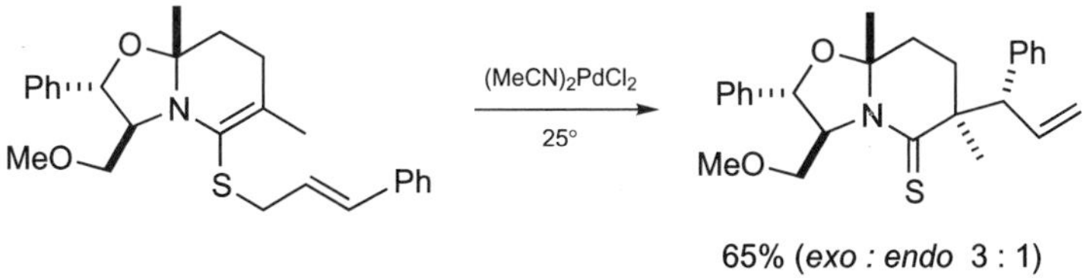

Ph, O, N, MeO, S, Ph —$(MeCN)_2PdCl_2$, 25°→ Ph, O, N, MeO, S, Ph

65% (*exo* : *endo* 3 : 1)

Allylic displacements. For achieving dehydrative cyclization of *N*-Boc-7-amino-2-alkenols to form 2-vinylpiperidine derivatives,[2] only a catalytic amount of $(MeCN)_2PdCl_2$ is needed because the Pd(II) species maintains its oxidation state throughout the reaction.

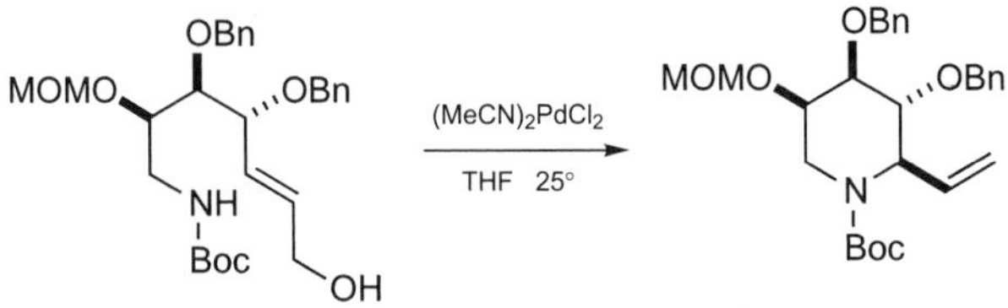

Ring opening of 1,4-oxa-1,4-dihydronaphthalene with R_2Zn proceeds in the presence of $(MeCN)_2PdCl_2$ to afford *cis*-2-alkyl-1-hydroxy-1,2- dihydronaphthalenes.[3]

Acylboration. Allenes are functionalized to give 2-acylallylboronates on reaction with bis(pinacolato)diboron and acyl chlorides.[4]

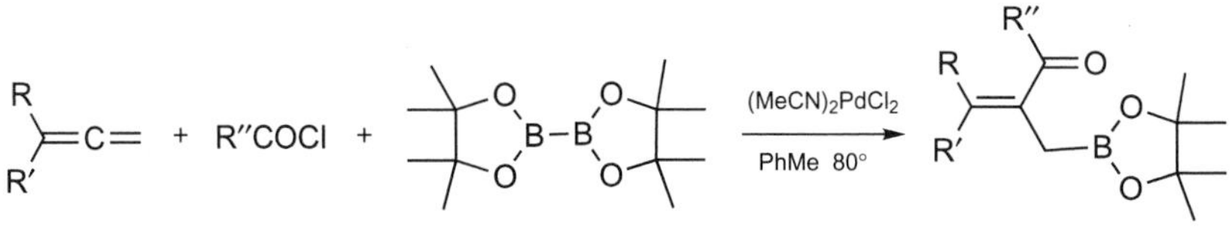

Stille coupling.[5] An expedient synthesis of [6]dendralene involves coupling of 2,3-bis(trimethylstannyl)-1,3-butadiene with 3-iodosulfolene and pyrolysis. A lesser amount of [8]dendralene is also produced.

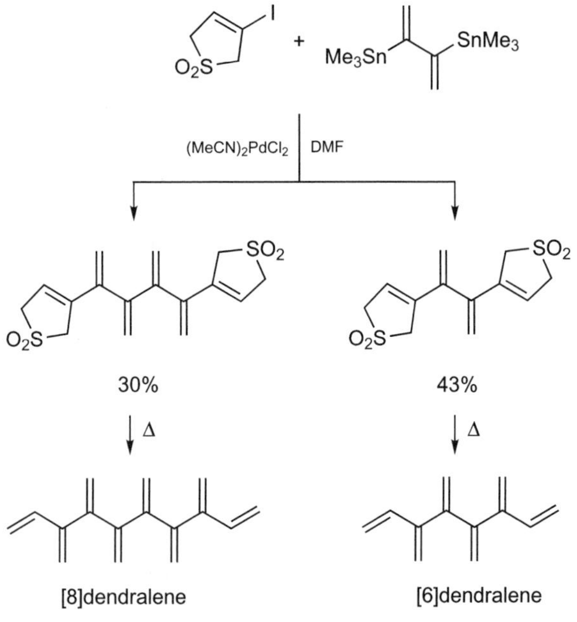

[1]Watson, D.J., Devine, P.N., Meyers, A.I. *TL* **41**, 1363 (2000).
[2]Yokoyama, H., Otaya, K., Kobayashi, H., Miyazawa, M., Yamaguchi, S., Hirai, Y. *OL* **2**, 2427 (2000).
[3]Lautens, M., Renaud, J.-L., Hiebert, S. *JACS* **122**, 1804 (2000).
[4]Yang, F.-Y., Wu, M.-Y., Cheng, C.-H. *JACS* **122**, 7122 (2000).
[5]Fielder, S., Rowan, D.R., Sherburn, M.S. *ACIEE* **39**, 4331 (2000).

Bis(allyl)dichlorodipalladium. 20, 29

Allylic displacements. Glycine derivatives are *C*-allylated via Zn-chelate enolates with an allyl carbonate.[1] BINAP mediated asymmetric allylation assisted by microwaves[2] is rapid.

The reaction with 2-bromo-1,3-dienes to form allene products[3] is more remarkable. Similarly, 3-bromo-3-alken-1-ynes are susceptible to transpositional substitution, forming butatriene products.[4] In this reaction, 2,2′-bis(diphenylphosphino)-1,1′-biphenyl is the added ligand.[4]

Ph, Br, MeO, O⁻, Na⁺, O, MeO, PdCl, 2, dpbp / THF, Ph, C, C, COOMe, COOMe, 45%

In contrast to the other Pd catalysts, the $[(C_3H_5)PdCl]_2$–Cy_3P system promotes the regioselective formation of branched products from 1-alken-3-yl acetates.[5]

Alkylative cyclization. A Pd-catalyzed Michael reaction product that still contains a C—Pd bond can undergo cyclization with an allenyl group.[6]

C, NC, CN, CN, Ph, CN, PdCl, 2, dppf / THF, NC, NC, CN, Ph, CN, 71%

The cationic π-allylpalladium that is ligated to 2-imidazolyl-3-methylpyridine catalyzes cyclopropanation with cinnamyl acetate.[7]

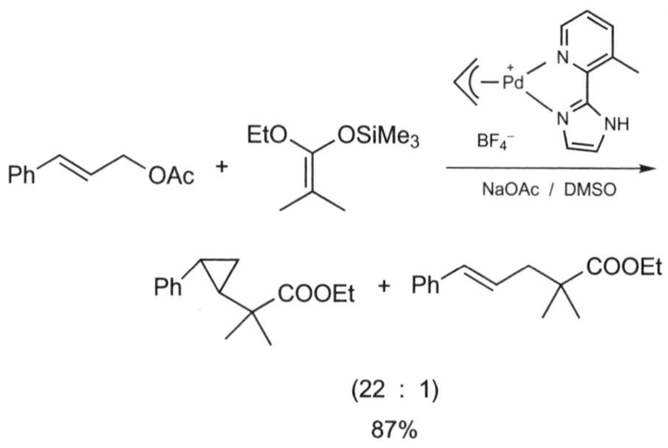

Coupling reactions. Carbostannylation of the palladacyclopentadiene intermediates derived from two molecules of alkynoic esters leads to dienoic esters that also bear a reactive stannyl residue.[8]

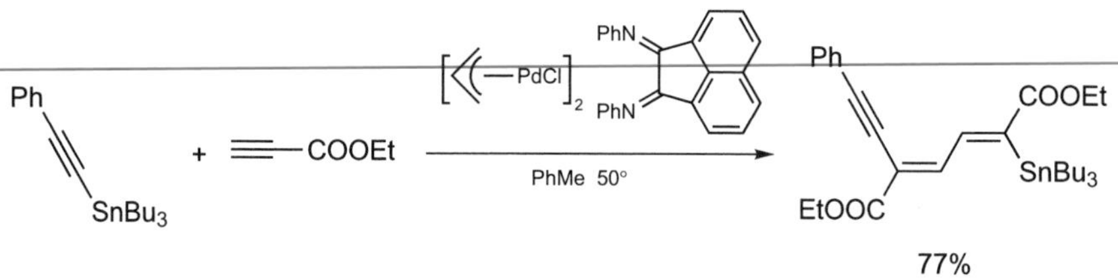

Homocoupling of aryl iodides to form biaryls with this complex in the presence of tetrabutylammonium fluoride has been carried out in DMSO at 120°.[9]

α-Arylalkenyl ethers are easily prepared in a coupling reaction involving ArI and α-alkoxyalkenyl(hydro)silanes or α-alkoxyalkenyl(hydroxy)silanes.[10] The reaction conditions are compatible with a wide array of functional groups.

Many 1,2,7,8-nonatetraenes couple with $R_3M—MR_3$ to give 1,2-bis[1-(triorganometallo)vinyl]cyclopentanes that can be transformed into ring-fused 1,2-dimethylenecyclobutanes via demetalative coupling.[11]

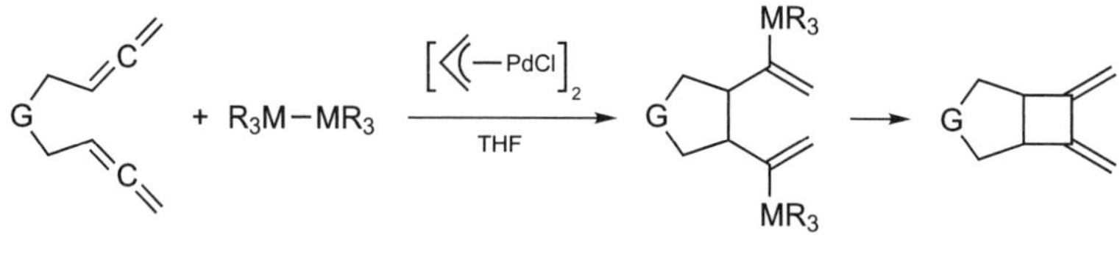

G = NTs, C(COOEt)$_2$, CPh$_2$. CHOBz, CHOBn

M = Si, R = Me
M = Sn, R = Bu

[2 + 2 + 2]Cyclotrimerization.[12] An unusual trimerization of 1-alkynes furnishes substituted fulvenes.

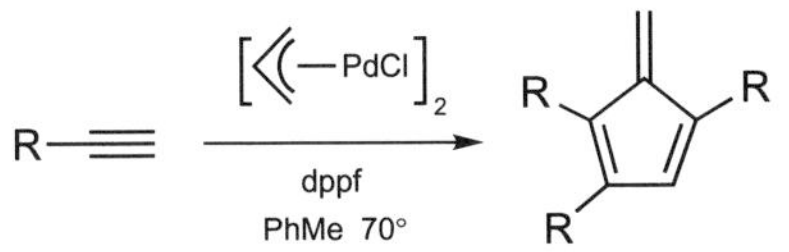

Arylsilanes.[13] The conversion of ArBr into $ArSiMe_3$ on reaction with hexamethyldisilane is catalyzed by the Pd-complex in the presence of (2-hydroxyphenyl)diphenylphosphine. This ligand activates both Pd and Si by providing both a soft and a hard donor.

[1]Kazmaier, U., Zumpe, F.L. *ACIEE* **38**, 1468 (1999); **39**, 802 (2000).
[2]Bremberg, U., Larhed, M., Moberg, C., Hallberg, A. *JOC* **64**, 1082 (1999).
[3]Ogasawara, M., Ikeda, H., Hayashi, T. *ACIEE* **39**, 1042 (2000).
[4]Ogasawara, M., Ikeda, H., Ohtsuk, K., Hayashi, T. *CL* 776 (2000).
[5]Blacker, A.J., Clarke, M.L., Loft, M.S., Williams, J.M.J. *OL* **1**, 1969 (1999).
[6]Meguro, M., Yamamoto, Y. *JOC* **64**, 694 (1999).
[7]Satake, A., Koshino, H., Nakata, T. *CL* 49 (1999).
[8]Shirakawa, E., Yoshida, H., Nakao, Y., Hiyama, T. *JACS* **121**, 4290 (1999).
[9]Albanese, D., Landini, D., Penso, M., Petricci, S. *SL* 199 (1999).
[10]Denmark, S.E., Neuville, L. *OL* **2**, 3221 (2000).
[11]Kang, S.-K., Baik, T.-G., Kulak, A.N., Ha, Y.-H., Lim, Y., Park, J. *JACS* **122**, 115290 (2000).
[12]Radhakrishnan, U., Gevorgyan, V., Yamamoto, Y. *CC* 1971 (2000).
[13]Shirakawa, E., Kurahashi, T., Yoshida, H., Hiyama, T. *CC* 1895 (2000).

1,2-Bis(benzenesulfonyl)ethylene.

Cyclic acetals. Derivatization of diols with this disulfone under basic conditions leads to 2-benzenesulfonylmethyl-1,3-dioxa cycles.[1] In protection of carbohydrates, these cyclic acetals resist acid hydrolysis but can be cleaved by reducing agents.

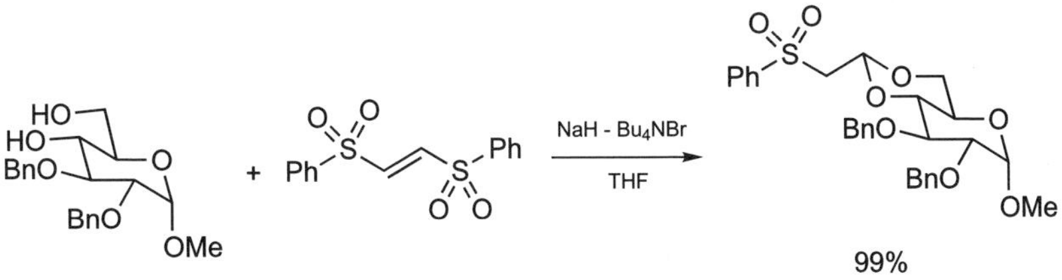

[1]Chery, F., Rollin, P., De Lucchi, O., Cossu, S. *TL* **41**, 2357 (2000).

Bis(benzonitrile)[bis(1,4-diphenylphosphino)butane]palladium(0).

β-Lactones.[1] With the Pd-complex, [2 + 2]cycloaddition between aldehydes and ketene forms β-substituted β-lactones in very high yields. The adduct from crotonaldehyde is easily transformed into sorbic acid.

sorbic acid

[1] Hattori, T., Suzuki, Y., Uesugi, O., Oi, S., Miyano, S. *CC* 73 (2000).

Bis(benzonitrile)dichloropalladium(II). 13, 34; **15,** 29; **18,** 46–47; **19,** 27; **20,** 30–31

Claisen rearrangement.[1] The catalyzed rearrangement of methoxycarbonylmethyl allyl ethers is stereoselective, in contrast to the thermal version.

Coupling reactions. The catalyst system for the Sonogashira coupling that includes t-Bu_3P is useful for room temperature reaction with aryl bromides.[2]

$$\text{Ph}-\equiv \;+\; \text{Br}-\text{Ph} \xrightarrow[\text{CuI, i-Pr}_2\text{NH / dioxane}]{(\text{PhCN})_2\text{PdCl}_2\text{ - (t-Bu)}_3\text{P}} \text{Ph}-\equiv-\text{Ph} \quad 94\%$$

Carboxylative coupling of allylstannanes and allyl halides under carbon dioxide (50 atm) affords unsaturated esters.[3]

$$\text{CH}_2{=}\text{CHCH}_2\text{SnBu}_3 \;+\; \text{ClCH}_2\text{CH}{=}\text{CH}_2 \xrightarrow[\text{CO}_2\text{ / THF}]{(\text{PhCN})_2\text{PdCl}_2\text{ - Bu}_3\text{P}} \text{CH}_2{=}\text{CHCH}_2\text{CO}_2\text{CH}_2\text{CH}{=}\text{CH}_2 \quad 97\%$$

Dehalogenation.[4] The catalyst supported on mesoporous silica is useful for dehalogenation of aryl halides under hydrogen.

3-Allylfurans.[5] Concurrent allylation and cyclization involving allenyl ketones and allyl bromides is catalyzed by $(PhCN)_2PdCl_2$.

Ph Br + =C= R ; $(PhCN)_2PdCl_2$, K_2CO_3 / MeCN → Ph, R

[1] Hiersemann, M. *SL* 1823 (1999).
[2] Hundertmark, T., Littke, A.F., Buchwald, S.L., Fu, G.C. *OL* **2,** 1729 (2000).

[3]Franks, R.J., Nicholas, K.M. *OM* **19**, 1458 (2000).
[4]Kantam, M.L., Rahman, A., Bandyopadhyay, T., Haritha, Y. *SC* **29**, 691 (1999).
[5]Ma, S., Li, L. *OL* **2**, 941 (2000).

Bis(1-benzyl-3,5,7-triaza-1-azoniatricyclo[3.3.1.1^{3,7}]decane peroxodisulfate.

Oxidations.[1] Various substances are oxidized by the title reagent: alcohols, oximes, and hydrazones to carbonyl compounds and thiols to disulfides.

[1] Wu, M., Yang, G., Chen, Z. *SC* **30**, 3127 (2000).

Bis[1,1′-bi-2,2′-hydroxynaphth-3-ylmethyl] ether–gallium/lithium complex.

2-Aryloxy alcohols.[1] These glycol monoethers are formed by epoxide opening. From *meso*-epoxides, chiral products are obtained in the presence of the linked BINOL complex. Reaction with simple BINOLs is complicated by erosion of enantioselectivity due to phenolate exchange (i.e., BINOL and phenolate nucleophile).

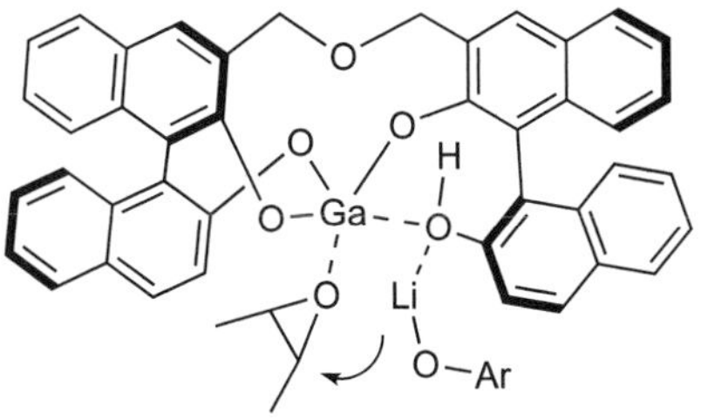

[1]Matsunaga, S., Das, J., Roels, J., Vogl, E.M., Yamamoto, N., Iida, T., Yamaguchi, K., Shibasaki, M. *JACS* **122**, 2252 (2000).

Bis[chloro(1,5-cyclooctadiene)iridium(I)].

Aziridines.[1] Aziridines are obtained in one step from aliphatic amines, aldehydes, and ethyl α-diazoacetate in the presence of $[(cod)IrCl]_2$.

Enynes.[2] 1-Alkynes are dimerized. The (*E/Z*) ratio of the products is strongly dependent on the phosphine ligand that is added. *tert*-Alkylalkynes afford 1,2,3-butatienes.

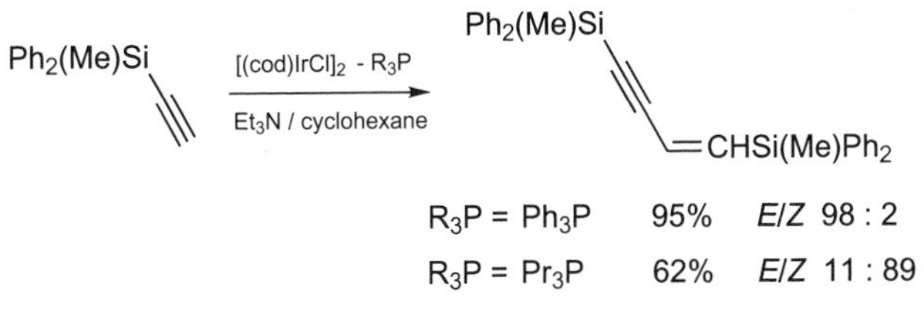

$Ph_2(Me)Si$–C≡CH $\xrightarrow[Et_3N\ /\ cyclohexane]{[(cod)IrCl]_2\ -\ R_3P}$ $Ph_2(Me)Si$–C≡C–CH=CHSi(Me)Ph_2

$R_3P = Ph_3P$ 95% *E/Z* 98 : 2

$R_3P = Pr_3P$ 62% *E/Z* 11 : 89

N,N-Disubstituted hydroxylamines.[3] The Ir complex, together with BINAP, forms a homogeneous hydrogenation catalyst for nitrones.

[1]Kubo, T., Sakaguchi, S., Ishii, Y. *CC* 625 (2000).
[2]Ohmura, T., Yorozuya, S., Yamamoto, Y., Miyaura, N. *OM* **19**, 365 (2000).
[3]Murahashi, S.-I., Tsuji, T., Ito, S. *CC* 409 (2000).

Bis[chloro(1,5-cyclooctadiene)rhodium(I)].

Functionalization of unsaturated compounds. Hydroacylation of alkynes, allenes, and alkenes using 2-hydroxyaraldehydes affords aryl ketones.[1]

[(cod)RhCl]$_2$ / dppf, Na_2CO_3 / PhH Δ

(83 : 17)

72%

cis-1-Alkenylboronates are obtained from the reaction of 1-alkynes with pinacolatoborane and catecholborane, which is catalyzed by [(cod)RhCl]$_2$–*i*-Pr_3P (Et_3N) in cyclohexane.[2]

A merry-go-round alkylation leading up to four contiguous norbornyl groups in a benzene ring has been achieved.[3]

$B(OH)_2$ + ; [(cod)RhCl]$_2$ / DPPP, CsF / PhMe 100°

Racemization of N-acyl α-amino acids.[4] A phosphine is included in the reaction mixture. The mild conditions of the racemization have implications in kinetic resolution as well as low pressure asymmetric hydrogenation at high conversion. Under such conditions, the occurrence of racemization results in a decrease of ee.

Acylation. Acylation at an α-position of *N*-(2-pyridyl)amines by a combination of CO and an alkene (e.g., ethylene) is conducted with the aid of [(cod)RhCl]$_2$. The pyridyl substituent in the substrates is essential.[5]

[(cod)RhCl]$_2$

CO / ethylene

i - PrOH 160°

Aldol reactions. Reductive aldolization of methyl acrylate furnishes predominantly methyl *syn*-2-methyl-3-hydroxyalkanoates in moderate yields.[6] The catalyst system consists of [(cod)RhCl]$_2$, Me—DuPhos and $Cl_2Si(H)Me$. For this reaction, significant interdependence of the metal, ligand, and hydride source for reactivity and selectivity has been witnessed.

With CO–H_2 under pressure, the Rh complex mediates hydroformylation of alkenes and then cyclization involving remote silyl enol ethers.[7]

OSiMe$_3$ [(cod)RhCl]$_2$ CO - H_2 / CH_2Cl_2 90° R — OSiMe$_3$ H R

Methylation.[8] On promotion by the Rh complex, arylzinc compounds undergo methylation with MeX to afford ArMe.

[1]Kokubo, K., Matsumasa, K., Nishinaka, Y., Miura, M., Nomura, M. *BCSJ* **72**, 303 (1999).
[2]Ohmura, T., Yamamoto, Y., Miyaura, N. *JACS* **122**, 4990 (2000).
[3]Oguma, K., Miura, M., Satoh, T., Nomura, M. *JACS* **122**, 10464 (2000).
[4]Hateley, M.J., Schichl, D.A., Kreuzfeld, H.-J., Beller, M. *TL* **41**, 3821 (2000).
[5]Chatani, N., Asaumi, T., Ikeda, T., Yorimitsu, S., Ishii, Y., Kakiuchi, F., Murai, S. *JACS* **122**, 12882 (2000).
[6]Taylor, S.J., Morken, J.P. *JACS* **121**, 12202 (1999).
[7]Hollmann, C., Eilbracht, P. *T* **56,** 1685 (2000).
[8]Hossain, K.M., Takagi, K. *CL* 1241 (1999).

Bis[chloro(diphenylphosphinobutane)rhodium(I)].

Intramolecular ene reaction.[1] Internal 1,6-enynes are induced by the Rh-complex to undergo an intramolecular ene reaction (cycloisomerization).

[(dppb)RhCl]$_2$ - AgSbF$_6$

85%

[1]Cao, P., Wang, B., Zhang, X. *JACS* **122**, 6490 (2000).

Bis[chloro(pentamethylcyclopentadienyl)methylthioruthenium].

Substitution reactions.[1] Replacement of the hydroxyl group of 1-alkyn-3-ols on reaction with alcohols, thiols, and amines is accomplished at room temperature in the presence of binuclear complex **1** and NH_4BF_4.

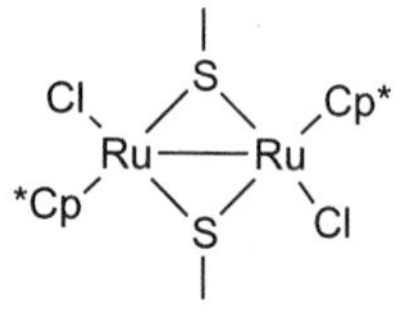

(1)

Cycloisomerization.[2] 1,ω-Diynes undergo cyclization to provide conjugated enynes on exposure to the title complex.

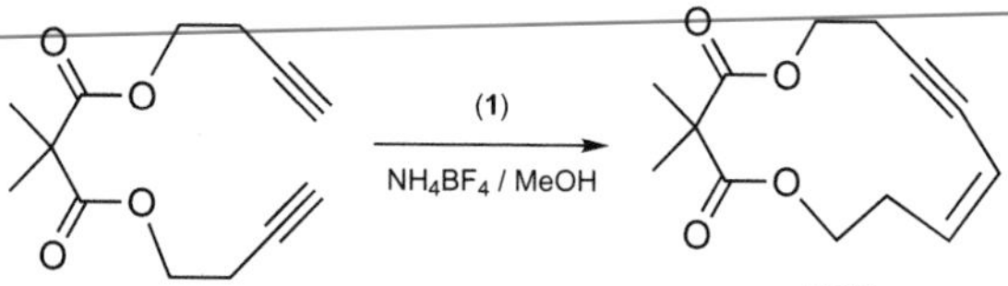

(1)

NH_4BF_4 / MeOH

72%

[1]Nishibayashi, Y., Wakiji, I., Hidai, M. *JACS* **122**, 11019 (2000).
[2]Nishibayashi, Y., Yamanashi, M., Wakiji, I., Hidai, M. *ACIEE* **39,** 2909 (2000).

Bis(*sym*-collidine)halogen(I) hexafluorophosphate. 15, 30; **17,** 155; **18,** 49; **19,** 29; **20,** 33

Bromoalkenes.[1] This reagent transforms α-diazo esters into α-bromo-α, β-unsaturated esters at low temperatures.

Halolactonization.[2] A method for the preparation of all-*cis* 2,3-dihydroxycyclohexanemethanol via a bicyclic bromolactone is as follows:

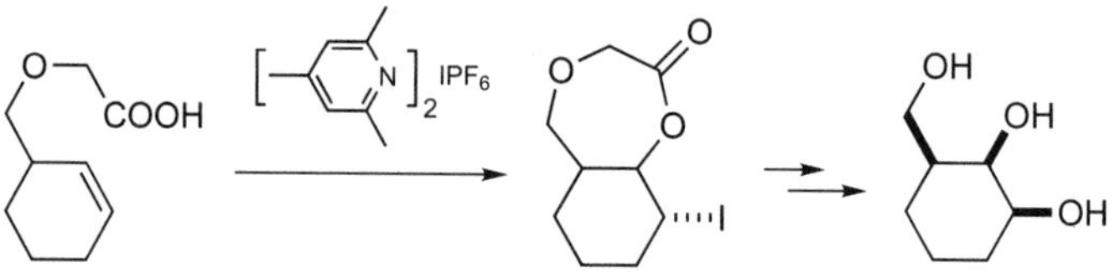

COOH

IPF_6

OH

OH

OH

1-Haloalkynes and 1-haloalkenes.[3] The Hunsdiecker reaction occurs when the conjugated acids are treated with these positive halogen salts. There is a stricter limitation for such a reaction with alkenoic acids as they must be α-unsubstituted, and the β-carbon supports a positive charge well (e.g., cinnamic acid and analogues).

Oxidation.[4] Alcohols are oxidized to carbonyl compounds in good yields. Activated benzylic alcohols undergo oxidative cleavage and ipso-bromination.

OH; [2,4,6-trimethylpyridine]$_2$ $BrPF_6$; CH_2Cl_2 25°; O; 82%

OH; MeO; [2,4,6-trimethylpyridine]$_2$ $BrPF_6$; CH_2Cl_2 25°; MeO; Br; CHO; 67%

[1]Rousseau, G., Marie, J.-X. *SC* **29**, 3705 (1999).
[2]Clausen, R.P., Bols, M. *JOC* **65**, 2797 (2000).
[3]Homsi, F., Rousseau, G. *TL* **40**, 1495 (1999).
[4]Rousseau, G., Robin, S. *TL* **41**, 8881 (2000).

Bis(1,5-cyclooctadiene)nickel(0).

Addition to alkynes. Carboxylation[1] and carbostannylation[2] of alkynes are readily achieved via Ni(0)-catalyzed addition. In the latter process, stannanes bearing allyl, alkynyl, and acyl groups can be used. Therefore a great variety of trisubstituted alkenylstannanes can be synthesized.

$SnBu_3$ + R—≡—Ph; R = Me, $SiMe_3$; $(cod)_2Ni$; PhMe 80°; $SnBu_3$; R; Ph

A synthesis of allylic alcohols from alkynes and aldehydes is conducted in the presence of $Ni(cod)_2$, Bu_3P, and Et_3B.[3]

Cyclopentenones.[4] π-Allylnickel complexes derived from enals undergo carbonylative cycloaddition with CO and alkynes.

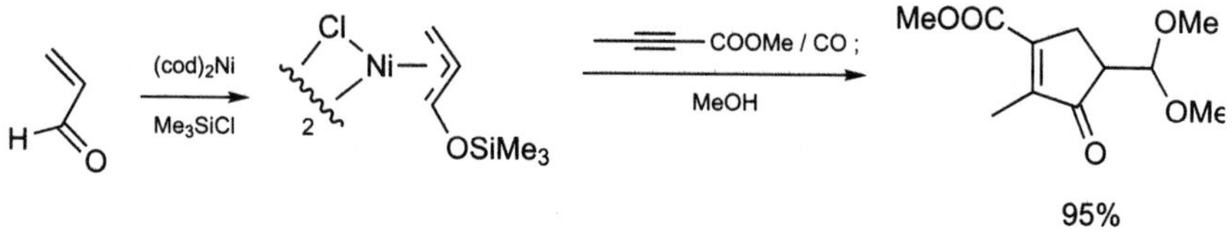

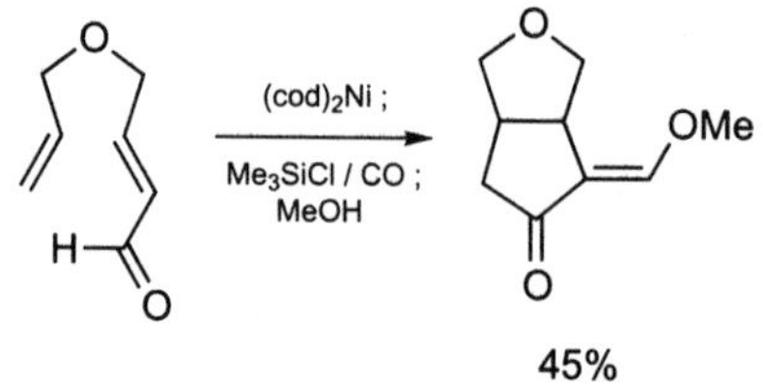

Cycloadditions and cyclization. 2-Alken-7-ynones undergo a [2 + 2 + 2]cycloaddition with electron-deficient alkenes to give cyclohexene derivatives.[5] This process is complementary to the Diels–Alder reaction.

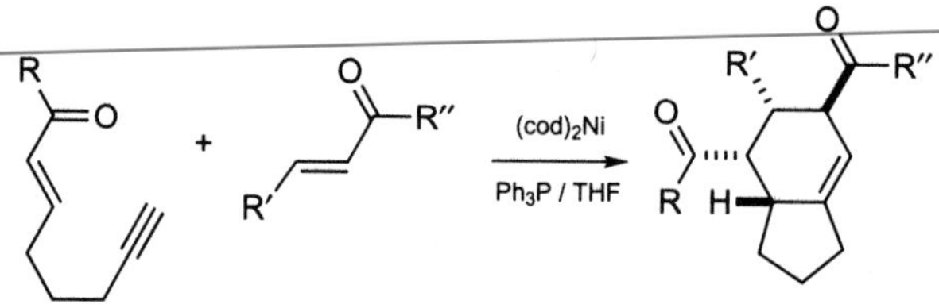

In the presence of $Ni(cod)_2$–TMEDA, certain 2-alken-7-ynones are transformed into bicyclo[3.3.0]octenes or bicyclo[3.1.0]hexanes, depending on whether an electrophile is added.[6]

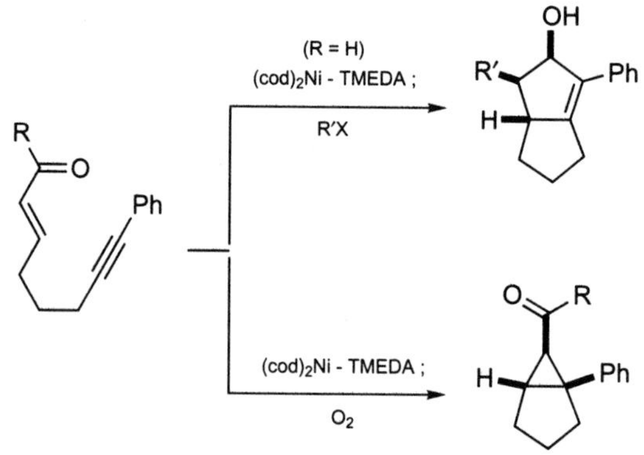

A combination of a hydrosilane and $Ni(cod)_2$ directs reductive cyclization of alkynals. Valuable applications to the synthesis of pyrrolidines, indolizidines, and quinolizidines are found.[7]

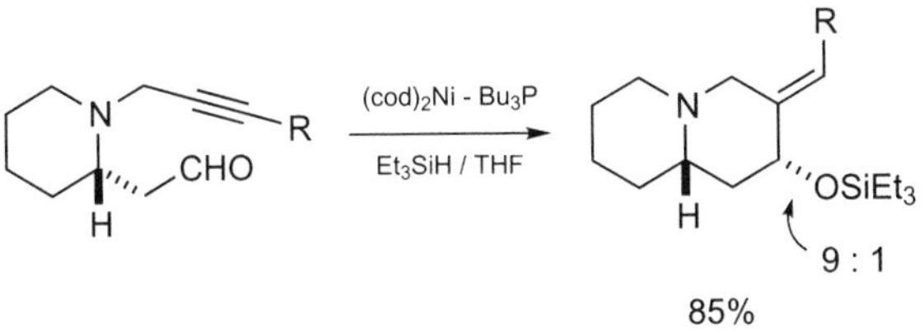

[1]Saito, S., Nakagawa, S., Koizumi, T., Hirayama, K., Yamamoto, Y. *JOC* **64**, 3975 (1999).
[2]Shirakawa, E., Yamasaki, K., Yoshida, H., Hiyama, T. *JACS* **121**, 10221 (1999).
[3]Huang, W.-S., Chan, J., Jamison, T.F. *OL* **2**, 4221 (2000).
[4]Garcia-Gomez, G., Moreto, J.M. *JACS* **121**, 878 (1999).
[5]Seo, J., Chui, H.M.P., Heeg, M.J., Montgomery, J. *JACS* **121**, 476 (1999).
[6]Choudhury, S.K., Amarasinghe, K.K.D., Heeg, M.J., Montgomery, J. *JACS* **121**, 6775 (1999).
[7]Tang, X.-Q., Montgomery, J. *JACS* **121**, 6098 (1999).

Bis(1,5-cyclooctadiene)rhodium(I) tetrafluoroborate.

Cyclobutanone cleavage.[1] Benzannulated lactones are obtained from 2-(*o*-hydroxyphenyl)cyclobutanones as a result of Rh insertion. Whether the benzylic position is fully substituted is a determinant factor for giving rise to γ-lactones or ε-lactones.

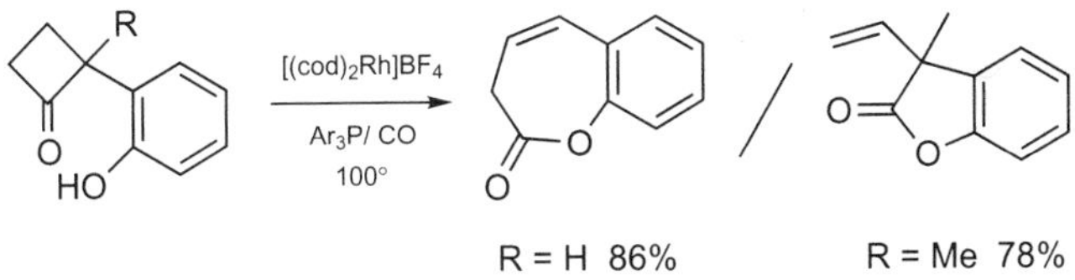

Stannylations.[2] Rh-catalyzed reaction of R_3SnH with propargyl ethers follows different courses depending on the organic residue R.

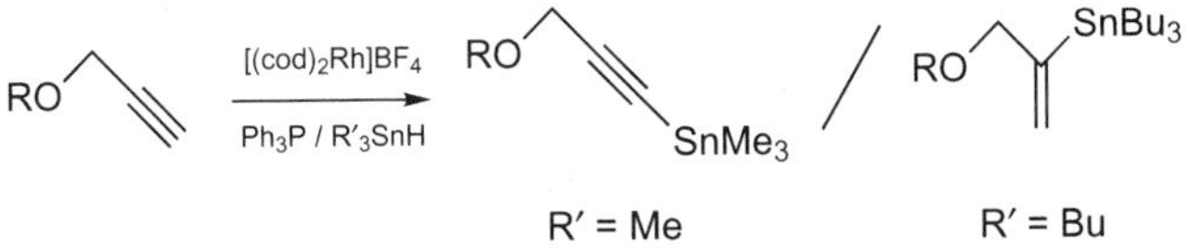

Benzyl alcohols.[3] Catalyzed by $(cod)_2RhBF_4$–NaF, phenyltrialkylstannanes transfer the phenyl group to aldehydes (11 examples, 52–92%).

Reduction.[4] Ketones are reduced enantioselectively by hydrosilanes in the presence of this Rh complex and a chiral diphosphinyl ferrocene ligand.

Isomerization.[5] The isomerization of allylic alcohols to β,β-disubstituted aldehydes by the Rh complex is rendered enantioselective (ee 72–86% in 6 examples) by the ligand **1**.

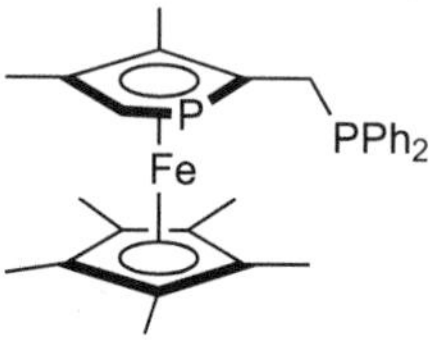

P PPh2 Fe

(**1**)

[1]Murakami, M., Tsuruta, T., Ito, Y. *ACIEE* **39**, 2484 (2000).
[2]Mitchell, T.N., Moschref, S.-N. *SL* 1259 (1999).
[3]Li, C.-J., Meng, Y. *JACS* **122**, 9538 (2000).
[4]Kuwano, R., Sawamura, M., Shirai, J., Takahashi, M., Ito, Y. *BCSJ* **73**, 485 (1999).
[5]Tanaka, K., Qiao, S., Tobisu, M., Lo, M.M.-C., Fu, G.C. *JACS* **122**, 9870 (2000).

Bis(dibenzylideneacetone)palladium(0).

Arylamines and diaryl ethers. Polyhaloarenes have been employed in *N*-arylation.[1] By using certain carbene ligands, rapid reaction with chloroarenes at room temperature has been demonstrated.[2] The $(dba)_2Pd$–t-Bu_3P system has a wide scope of arylation including that of indole and carbamates.[3] (Formation of *N*-dimethylaminoindole from 2-chlorophenylacetaldehyde *N,N*-dimethylhydrazone is novel.[4]) A more complicated ligand for arylation of amines by aryl chlorides is **1**.[5]

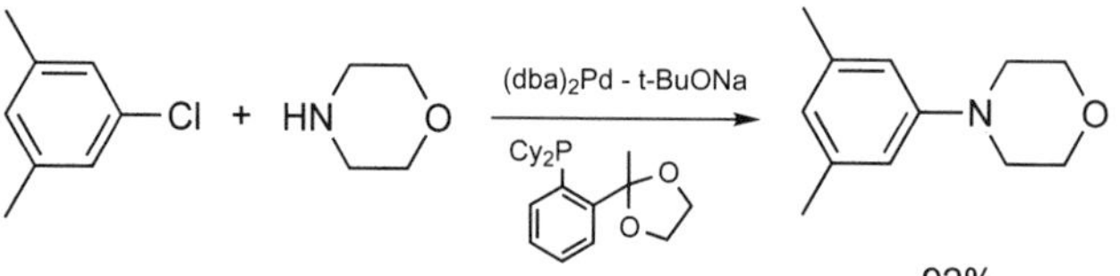

Cl + HN O → (dba)2Pd - t-BuONa, Cy2P (1) → N O 92%

(**1**)

Coupling involving unactivated aryl halides and ArONa is successful under similar conditions (but *t*-BuONa is unnecessary).[6,7]

Cross-couplings. The coupling of aryl iodides and iodoalkenes with alkenylsilanols[8] or alkenylsilacyclobutanes,[9] is very efficient in furnishing styrenes and conjugated dienes.

The process involving aryltrialkoxysilanes is an alternative to Stille and Suzuki couplings.[10] Change of reaction conditions permits the synthesis of either styrenes or β-silylstyrenes from vinyltrimethylsilane.[11]

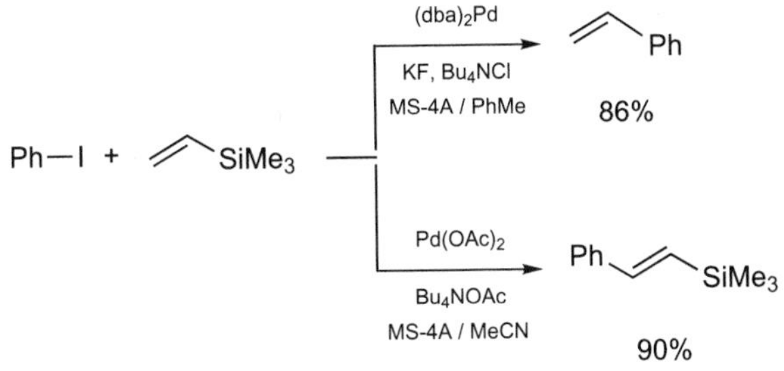

A facile synthesis of 2-aryl-1,3-dienes has been realized from aryl halides and allenes.[12] Ionic liquids prove to be a superior media for the Heck reaction with aryl chlorides.[13]

Cyclopropanol cleavage.[14] 1-Substituted cyclopropanols undergoes oxidative opening to afford 1-alken-3-ones. Although many other Pd reagents are also useful for this transformation, $PdCl_2$ without phosphine ligand is an exception.

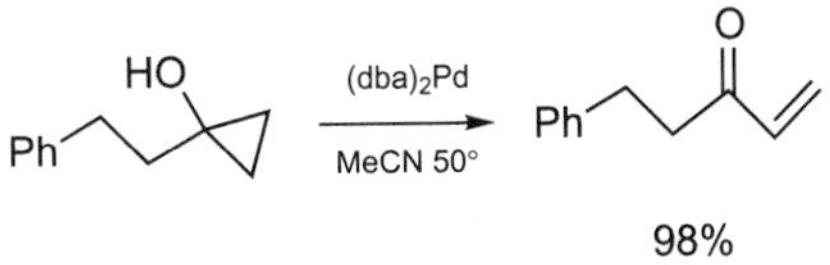

Annulation.[15] *o*-Iodoaryl cyanides and the homologous arylacetonitriles are united with unsaturated compounds. Ring formation results.

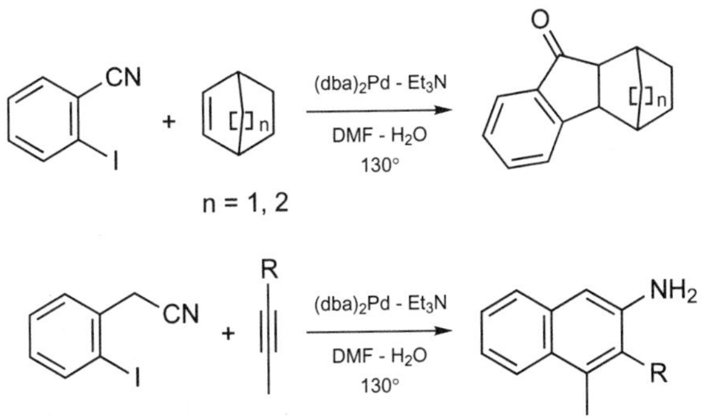

Heterocycle formation. 4-Alkenyl-1,3-oxazolidines are readily prepared from vinylepoxides and *N*-tosylaldimines in the presence of $(dba)_2Pd$–dppe.[16] A synthesis of isoquinolines and pyridines from allenes is similarly achieved.[17]

R′ = Bn, t-Bu, CH_2CH_2CN

$(dba)_2Pd$ - dppp, MeCN - Na_2CO_3, 100°

Carbostannylation of allenes.[18] Allylstannanes are obtained from aryl iodide and hexamethylditin.

Arylation.[19] Ketones and malonate esters are arylated. Bis(1,1′-di-*t*-butyl)ferrocene is a ligand that is used.

[1]Beletskaya, I.P., Bessmertnykh, A.G., Guilard, R. *SL* 1459 (1999).
[2]Stauffer, S.R., Lee, S., Stambuli, J.P., Hauck, S.I., Hartwig, J.F. *OL* **2**, 1423 (2000).
[3]Hartwig, J. F., Kawatsura, M., Hauck, S.I., Shaughnessy, K.H., Alcazar-Roman, L.M. *JOC* **64**, 5575 (1999).
[4]Watanabe, M., Yamamoto, T., Nishiyama, M. *ACIEE* **39**, 2501 (2000).
[5]Bei, X., Guram, A.S., Turner, H.W., Weinberg, W.H. *TL* **40**, 1237 (1999).
[6]Mann, G., Incarvito, C., Rheingold, A.L., Hartwig, J. F. *JACS* **121**, 3224 (1999).
[7]Shelby, Q., Kataoka, N., Mann, G., Hartwig, J. F. *JACS* **122**, 10718 (2000).
[8]Denmark, S.E., Wehrli, D. *OL* **2**, 565 (2000).
[9]Denmark, S.E., Choi, J.Y. *JACS* **121**, 5821 (1999).
[10]Mowery, M.E., DeShong, P. *JOC* **64**, 1684 (1999).
[11]Jeffery, T. *TL* **40**, 1673 (1999); **41**, 8445 (2000).
[12]Chang, H.-M., Cheng, C.-H. *JOC* **65**, 1767 (2000).
[13]Bohm, V.P.W., Herrmann, W.A. *CEJ* **6**, 1017 (2000).
[14]Okumoto, H., Jinnai, T., Shimizu, H., Harada, Y., Mishima, H., Suzuki, A. *SL* 629 (2000).
[15]Larock, R.C., Tian, Q., Pletner, A.A. *JACS* **121**, 3238 (1999).
[16]Shim, J.-G., Yamamoto, Y. *H* **52**, 885 (2000).
[17]Diederen, J.J.H., Sinkeldam, R.W., Fr*u*ha*u*f, H.-W., Hiemstra, H., Vrieze, K. *TL* **40**, 4255 (1999).
[18]Yang, F.-Y., Wu, M.-Y., Cheng, C.-H. *TL* **40**, 6055 (1999).
[19]Kawatsura, M., Hartwig, J. F. *JACS* **121**, 1473 (1999).

Bis(dibenzylideneacetone)platinum(0).

Hydroboration and silylboration. Hydroboration of allenes with $(dba)_2Pt$ as the catalyst shows regioselectivity dependence on the phosphine ligands.[1]

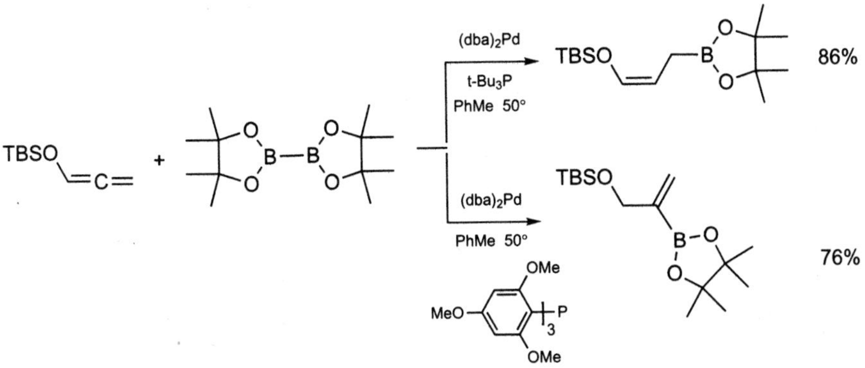

Silyboration of alkylidenecyclopropanes involves different C—C bonds using $(dba)_2Pt$ and $(dba)_2Pd$.[2]

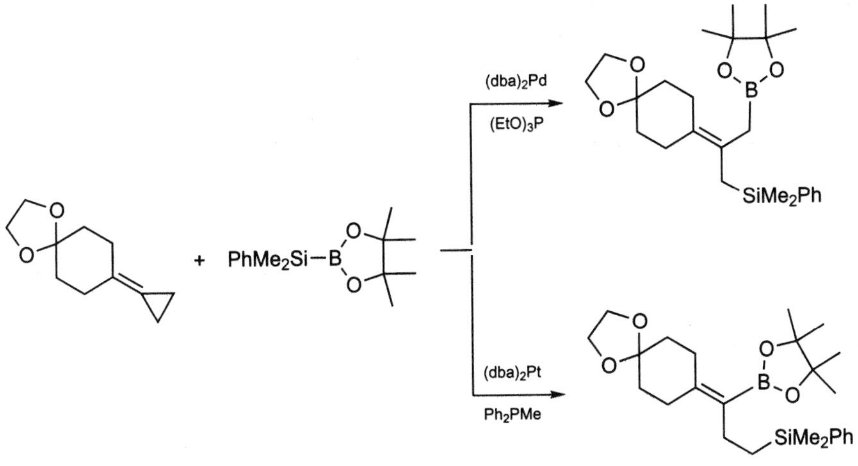

Allylboranes.[3] Displacement of allyl halides with pinacolborane is promoted by $(dba)_2Pt–Ph_3As–Et_3N$.

[1]Yamamoto, Y., Fujikawa, R., Yamada, A., Miyaura, N. *CL* 1069 (1999).
[2]Suginmoe, M., Matsuda, T., Ito, Y. *JACS* **122**, 11015(2000).
[3]Murata, M., Watanabe, S., Masuda, Y. *TL* **41**, 5877 (2000).

Bis[di-*t*-butyl(hydroxy)tin chloride].

Deacylation. The reagent catalyzes cleavage of esters in methanol.

[1]Orita, A., Sakamoto, K., Hamada, Y., Otera, J. *SL* 140 (2000).

Bis[dicarbonylchlororhodium(I)] and bis[(1,5-cyclooctadiene)chlororhodium(I)].

Epoxide opening.[1] The monoepoxide of a diene is regioselectively attacked by nucleophile (ArNHR, ROH) at the allylic position using $[Rh(CO)_2Cl]_2$ as catalyst, affording *anti*-1,2-amino alcohols and alkoxy alcohols. The results are apparently complementary to those obtained from the Pd-catalyzed process.

Addition to multiple bonds. Hydroboration of 1-alkynes with catecholborane (when i-Pr_3P and Et_3N are also present) gives (*Z*)-1-alkenylboronates.[2] On the other hand, dehydrogenative coupling[3] between styrenes and pinacolborane is observed.

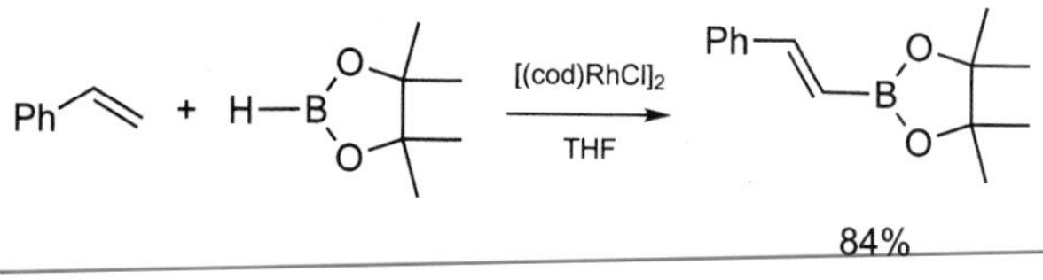

[(cod)RhCl]2

THF

84%

Under hydroformylation conditions, amines effect hydroaminomethylation to alkenes in an anti-Markovnikov sense.[4,5]

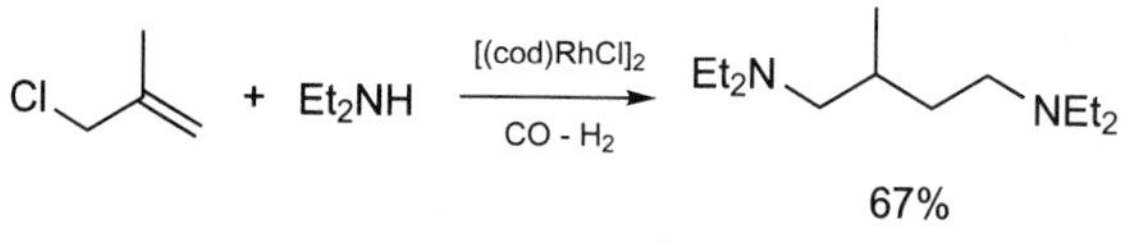

[(cod)RhCl]2

CO - H2

67%

Cycloadditions. Cycloheptenones are prepared from 1-alkenylcyclopropane derivatives via a [5 + 2]cycloaddition.[6,7] Improvement of stereoselectivity (retention) by the use of $[Rh(CO)_2Cl]_2$ is recognized. A synthesis of (+)-aphanamol I has been realized based on such a cycloaddition involving an allenic double bond as the two-carbon component.[8]

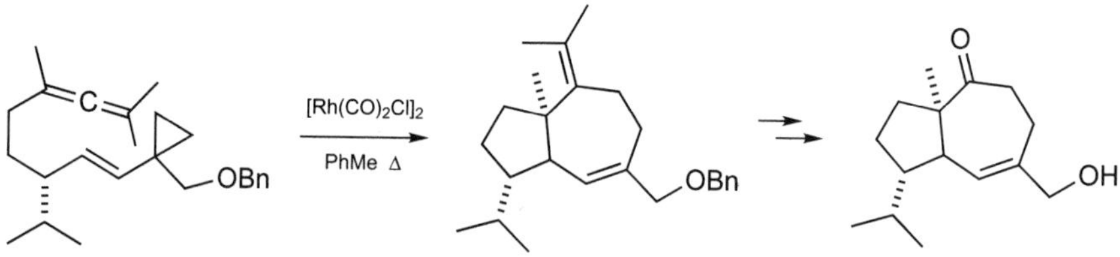

[Rh(CO)2Cl]2

PhMe Δ

(+)-aphanamol-I

Regarding substitution effects, a heteroatom substituent in a 1,1-disubstituted cyclopropane moiety significantly enhances reaction rates. 1,1-Disubstituted cyclopropyl derivatives afford only one regioisomer.[9] Of synthetic importance is the obervation of regioselectivity dependence on catalyst for certain substrates.[10]

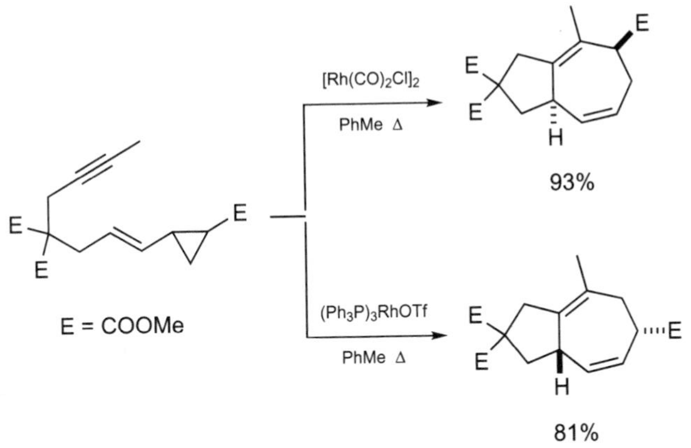

By employing substrates possessing a 2-alkenylcyclobutanone unit instead of the alkenylcyclopropane, the cycloaddition leads to 4-cyclooctenones.[11]

Intramolecular Pauson–Khand reaction.[12] Fused cyclopentenones are formed. The reaction exhibits modest asymmetric induction when a chiral BINAP is present.

[1]Fagnou, K., Lautens, M. *OL* **2**, 2319 (2000).
[2]Ohmura, T., Yamamoto, Y., Miyaura, N. *JACS* **122**, 4990 (2000).
[3]Murata, M., Watanabe, S., Masuda, Y. *TL* **40**, 2585 (1999).
[4]Eilbracht, P., Kranemann, C.L., Barfacker, L. *EJOC* 1907 (1999).
[5]Rosche, T., Eilbracht, P. *T* **55**, 3917 (1999).
[6]Wender, P.A., Dyckman, A.J., Husfeld, C.O., Scanio, M.J.C. *OL* **2**, 1609 (2000).
[7]Wender, P.A., Glorius, F., Husfeld, C.O., Langkopf, E., Love, J.A. *JACS* **121**, 5348 (1999).
[8]Wender, P.A., Zhang, L. *OL* **2**, 2323 (2000).
[9]Wender, P.A., Dyckman, A.J., Husfeld, C.O., Love, J.A., Rieck, H. *JACS* **121**, 10442 (1999).
[10]Wender, P.A., Dyckman, A.J. *OL* **1**, 2089 (1999).
[11]Wender, P.A., Correa, A.G., Sato, Y., Sun, R. *JACS* **122**, 7815 (2000).
[12]Jeong, N., Sung, B.K., Choi, Y.K. *JACS* **122**, 6771 (2000).

Bis[dicarbonyl(formyloxy)triphenylphosphineruthenium(I)].

Conjugate addition.[1] Under neutral conditions, various carbon pronucleophiles (β-ketoesters, 1-alkynes, etc) are induced by the title complex to add to enones in the Michael fashion.

$[(Ph_3P)Ru(CO)_2P(OCHO)]_2$

67%

$[(Ph_3P)Ru(CO)_2P(OCHO)]_2$

74%

Enol esters.[2] The regioselective addition of carboxylic acids to alkynes is extendable to diynes.

[1]Picquet, M., Bruneau, C., Dixneuf, P.H. *T* **55**, 3937 (1999).
[2]Kabouche, A., Kabouche, Z., Bruneau, C., Dixneuf, P.H. *JCR(S)* 1247 (1999).

Bis[dichloro(*p*-cymene)ruthenium].

Hydrosilylation.[1] Efficient synthesis of (*Z*)-1-silylalkenes from 1-alkynes by this Ru complex is reported.

Ph_3SiH + ≡—Ph → $[(cymene)RuCl_2]_2$, CH_2Cl_2 45° → Ph_3Si / Ph

94%

Silanols.[2] Autoxidation of silanes in the air is catalyzed by $[(p\text{-cymene})RuCl_2]_2$. This oxidation, which gives silanols of inverted configuration, is amenable to large-scale production.

Ring-closing metathesis.[3] Cyclic alkenes are obtained by heating dienes with $[(p\text{-cymene})RuCl_2]_2$ and Cy_3P in dichloromethane under neon light.

[1]Na, Y., Chang, S. *OL* **2**, 1887 (2000).
[2]Lee, M., Ko, S., Chang, S. *JACS* **122**, 12011 (2000).
[3]Fürstner, A., Ackermann, L. *CC* 95 (1999).

Bis(2,4-dichlorophenyl)chlorophosphate.

Hydroxyl activation.[1] Alcohols become activated on reaction with this reagent and it is possible to use them in situ to form alkyl azides. In other words, it requires only admixture of ROH, $ClPO(OAr)_2$, DMAP, and NaN_3 in DMF to complete the transformation.

[1]Yu, C., Liu, B., Hu, L. *OL* **2**, 1959 (2000).

Bis(dimethylamino)methane.

Isoflavanones.[1] *o*-Hydroxydiphenylethanones condense with the reagent to form isoflavanones directly.

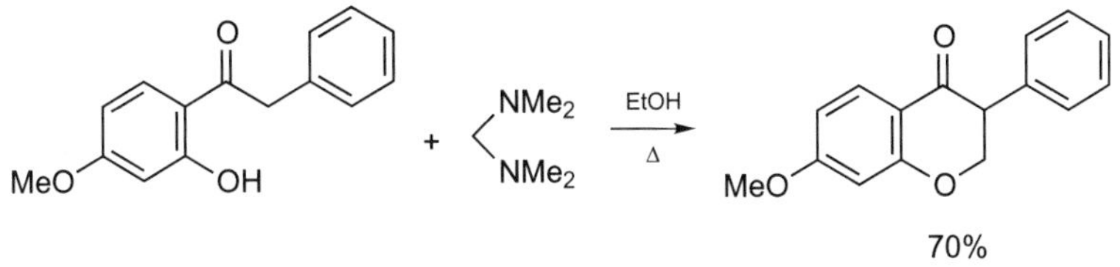

[1]Valenti, P., Belluti, F., Rampa, A., Bisi, A. *SC* **29**, 3895 (1999).

2,2′-Bis(diphenylphosphino)-1,1′-binaphthyl. 13, 36–37; **14,** 38–44; **15,** 34; **16,** 32–36; **17,** 34–38; **18,** 39–41; **19,** 33–55; **20,** 41–44

Iridium(I) complexes

Carbonylcyclization.[1] The use of $[(cod)IrCl]_2$ in conjunction with BINAP and CO to promote formation of 2-cyclopentenones provides an alternative method to the Pauson–Khand reaction. Moreover, chiral products are obtained (9 examples, 82–97% ee).

[1] Shibata, T., Takagi, K. *JACS* **122**, 9852 (2000).

Palladium(II) complexes

N-Arylations. Aryl bromides containing electron-withdrawing group(s) undergo displacement with $BocNHNH_2$ to afford $ArN(Boc)NH_2$ [1]and sulfoximines to *N*-aryl derivatives,[2] both by catalysis with $Pd(OAc)_2$–BINAP. Introducting an aryl group to hydrazones by this method provides a new route to indoles.[3] Chiral diamines such as 1,2-diphenyl-1,2-ethanediamine can be arylated once at each nitrogen atom.[4]

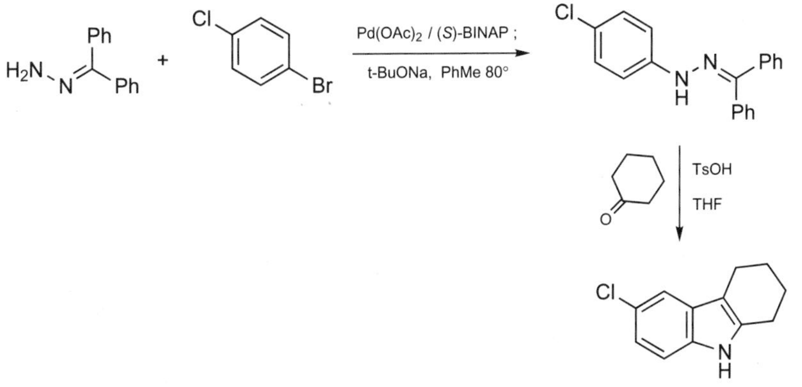

A synthesis of 2-arylaminotropones from 2-tosyloxytropone extends the scope of this method.[5]

Hydroamination. The addition of arylamines to styrenes to afford *N*-arylbenzylamines is promoted by $Pd(OTf)_2$–BINAP, although catalytic systems constituting other Pd(II) salts with TfOH are also effective.[6] Faster reactions are observed with styrenes that are relatively electron poor, and those bearing an ortho-substituent furnish products in high yields.

Allylation. Chiral products are obtained from α-acetamido-β-ketoesters.[7]

Hetero-Diels–Alder reaction. High enantioselectivity is observed for this process using chiral cationic Pd(II) complexes.[8]

Allylic displacement. The asymmetric version of the conventional displacement is improved (in ee) by using diethylzinc as a base to deprotonate the pronucleophiles.[9]

Ene reaction.[10] An expedient route to chiral α-hydroxy esters is based on the ene reaction of alkenes with ethyl glyoxylate in the presence of (*S*)-BINAP-(bisacetonitrile)-palladium hexafluoroantimonate. The tolyl–BINAP analogue is an even better ligand.

[1]Wang, Z., Skerlj, R.T., Bridger, G.J. *TL* **40**, 3543 (1999).
[2]Bolm, C., Hildebrand, J.P. *JOC* **65**, 169 (2000).
[3]Wagaw, S., Yang, B.H., Buchwald, S.L. *JACS* **121**, 10251 (1999).
[4]Cabanal-Duvillard, I., Mangeney, P. *TL* **40**, 3877 (1999).
[5]Hicks, F.A., Brookhart, M. *OL* **2**, 219 (2000).
[6]Kawatsura, M., Hartwig, J.F. *JACS* **122**, 9546 (2000).
[7]Kuwano, R., Ito, Y. *JACS* **121**, 3236 (1999).
[8]Oi, S., Terada, E., Ohuchi, K., Kato, T., Tachibaba, Y., Inoue, Y. *JOC* **64**, 8660 (1999).
[9]Fuji, K., Kinoshita, N., Tanaka, K. *CC* 1895 (1999).
[10]Hao, J., Hatano, M., Mikami, K. *OL* **2**, 4059 (2000).

Platinum complexes

Baeyer–Villiger oxidation.[1] A BINAP–Pt complex has been used to catalyze the reaction of *meso*-cyclohexanones with H_2O_2.

Diels–Alder reaction.[2] BINAP–Pt(II) complexes with perchlorate and hexafluoroantimonate counterions are highly effective catalysts for the enantioselective Diels–Alder reaction. The corresponding Pd(II) complexes have comparable (and for certain substrates, somewhat inferior) reactivity.

[1]Paneghetti, C., Gavagnin, R., Pinna, F., Strukul, G. *OM* **18**, 5057 (1999).
[2]Ghosh, A.K., Matsuda, H. *OL* **1**, 2157 (1999).

Rhodium complexes

Conjugate additions. Arylborates generated in situ from ArBr are effective as asymmetric addends[1] to enones in the presence of a Rh-complex derived from BINAP and $(acac)_2Rh(C_2H_4)_2$. They give excellent chemical yields and ee (>91%, often 99%). In these processes (such as conjugate addition to alkenylphosphonates[2]), the use of arylboroxines $(ArBO)_3$ instead of arylboronic acids is advantageous because catalytic activity of the Rh complex decimated by large amount of water can be avoided.

[1]Takaya, Y., Ogasawara, M., Hayashi, T. *TL* **40**, 6957 (1999).
[2]Hayashi, T., Senda, T., Takaya, Y., Ogasawara, M. *JACS* **121**, 11591 (1999).

Ruthenium(II) complexes

Asymmetric hydrogenation. A precursor of (*S,S*)-CHIRAPHOS has been obtained from asymmetric hydrogenation of 2,3-bis(diphenylphosphinoyl)-1,3-butadiene.[1] Chiral 1,1,1-trifluoroalkan-2-ols can be synthesized from enol acetates of the trifluoromethyl ketones.[2] A general method for accessing secondary alcohols from aromatic[3] and heteroaromatic ketones[4] is by asymmetric hydrogenation with the complex **1**.

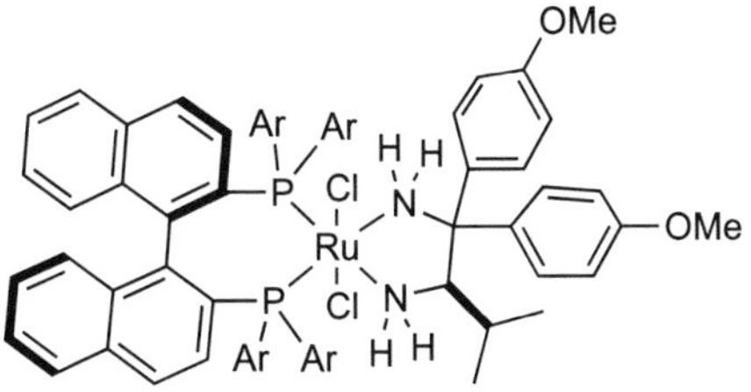

(**1**)

Catalysts containing chiral 1,2-diamine ligands in addition to (modified) BINAP have been developed for the hydrogenation of α-amino ketones to furnish chiral 1,2-amino alcohols.[5]

6,6′-Bis(aminomethyl)–BINAP is a precursor of a polymer-bound ligand.[6]

[1]Beghetto, V., Matteoli, U., Scrivanti, A. *CC* 155 (2000).
[2]Kuroki, Y., Asada, D., Sakamaki, Y., Iseki, K. *TL* **41**, 4603 (2000).
[3]Ohkuma, T., Koizumi, M., Ikehira, H., Yokozawa, I., Noyori, R. *OL* **2**, 659 (2000).
[4]Ohkuma, T., Koizumi, M., Yoshida, M., Noyori, R. *OL* **2**, 1749 (2000).
[5]Ohkuma, T., Ishii, D., Takeno, H., Noyori, R. *JACS* **122**, 6510 (2000).
[6]ter Halle, R., Colasson, B., Schulz, E., Spagnol, M., Lemaire, M. *TL* **41**, 643 (2000).

Silver(I) complexes

Aldol reactions. A combination of AgOTf, (*R*)-BINAP, and R_3SnOMe promotes a diastereoselective–enantioselective aldol reaction of enol trichloroacetates with aldehydes.[1]

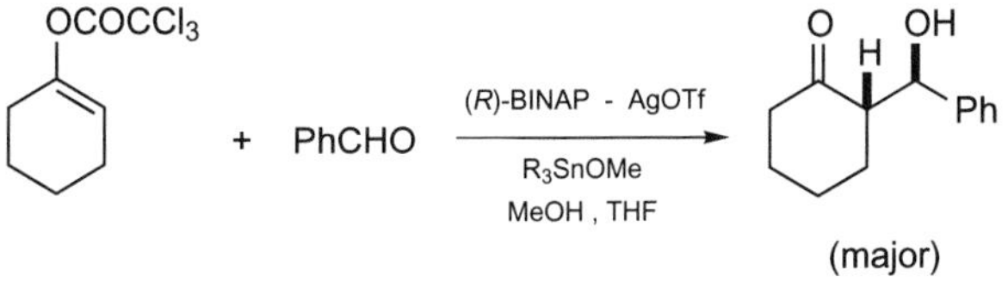

(major)

[1]Yanagisawa, A., Matsumoto, Y., Asakawa, K., Yamamoto, H. *JACS* **121**, 892 (1999).

2,2′-Bis(diphenylphosphino)-1,1′-binaphthyl + 1,1′-Bi-2,2′-naphthol copolymer.

Alkylation + hydrogenation.[1] The dual activity for the transformation of ketoaldehydes by complexes of (diamine)$RuCl_2$ with this copolymer is clearly demonstrated. Alkylation of the formyl group with organozinc reagents (e.g., Et_2Zn) followed by hydrogenation of the ketone is readily accomplished.

[1]Yu, H.-B., Hu, Q.-S., Pu, L. *JACS* **122**, 6500 (2000).

2,2′-Bis(di-*p*-tolylphosphino)-1,1′-binaphthyl.

Copper(I) complexes

Addition to N-tosylaldimines. Protected α-amino acids are obtained from the addition to ethyl glyoxylate under the influence of $CuPF_6$-[*p*-tolyl]BINAP. Indole[1] and allyltributylstannane[2] are typical addends. The α-hydroxyamine precursor of the imine can be used in the reaction with enol silyl ethers,[3] but a higher (room) temperature is needed.

EtOOC–CH(NHTs)–OH + CH_2=C(OSiMe$_3$)Ph + [(*p*-tolyl)BINAP: Ar$_2$P–binaphthyl–PAr$_2$]·$CuClO_4$ → EtOOC–CH(NHTs)–CH_2–C(=O)–Ph

Reduction. α,β-Unsaturated esters[4] and ketones[5] are reduced asymmetrically, using poly(methylhydrosiloxane) as the hydrogen source.

Conjugate reduction.[6] β,β-Disubstituted α,β-unsaturated esters undergo asymmetric reduction by polymethylhydrosiloxane in the presence of *t*-BuONa and a complex derived from CuCl and a slightly modified (i.e., *p*-tolyl) (*S*)-BINAP.

[1]Johannsen, M. *CC* 2233 (1999).
[2]Fang, X., Johannsen, M., Yao, S., Gathergood, N., Hazell, R. G., Jorgensen, K. A. *JOC* **64**, 4844 (1999).
[3]Ferraris, D., Dudding, T., Young, B., Drury, W.J., Leckta, T. *JOC* **64**, 2168 (1999).
[4]Apella, D.H., Moritani, Y., Shintani, R., Ferreira, E.M., Buchwald, S.L. *JACS* **121**, 9473 (1999).
[5]Moritani, Y., Apella, D.H., Jurkauskas, V., Buchwald, S.L. *JACS* **122**, 6797 (2000).
[6]Appella, D.H., Moritani, Y., Shintani, R., Ferreira, E.M., Buchwald, S.L. *JACS* **121**, 9473 (1999).

Silver(I) complexes

Allylation. Allyltrimethoxysilanes are useful donors[1] and the reaction of aldehydes with allylstannanes can be carried out in an aqueous medium.[2]

[1]Yanagisawa, A., Kageyama, H., Nakatsuka, Y., Asakawa, K., Matsumoto, Y., Yamamoto, H. *ACIEE* **38**, 3701 (1999).
[2]Loh, T.-P., Zhou, J.-R. *TL* **41**, 5261 (2000).

Tin(II) complexes

Aziridines.[1] The reaction of α-imino esters with diazoalkanes provides aziridinecarboxylic esters. A copper(I) complex can also be used.

[1]Juhl, K., Hazell, R. G., Jorgensen, K. A. *JCS(P1)* 2293 (1999).

N,O-Bis(ethoxycarbonyl)hydroxylamine.

Lossen rearrangement.[1] Condensation of a carboxylic acid with this reagent followed by mild base treatment leads to primary amines with one less carbon.

R−COOH + HN(OCOOEt)(COOEt) —DCC→ [R–C(=O)–N-dioxazolidinone] —Et_3N / H_2O→ R−NH_2

[1]Anilkumar, R., Chandrasekhar, S., Sridhar, M. *TL* **41**, 5291 (2000).

Bis(ethylenedichloro)platinum.

Ring opening. 1,2-Cyclopropano sugars afford 2-alkyl glycosides on reaction with alcohols in the presence of the Pt complex.

[1]Beyer, J., Madsen, R. *JACS* **120**, 12137 (1998).

Bis(iodozincio)methane.

1,2-Cyclopropanediols. Consecutive nucleophilic addition of the reagent[1] to α-diketones affords the *vic*-diols that can be trapped with Ac_2O and Me_3SiCl.[2]

R(C=O)(C=O)R′ + IZn⁄⁀ZnI —THF→ HO, R / R′, OH cyclopropane

Enol ethers.[3] The title compound is an alternative to the Tebbe reagent for methylenation of esters using $TiCl_2$ as catalyst in THF at room temperature.

Conjugate addition.[4] Addition of bis(iodozincio)methane to enals and enones in the presence of Me_3SiCl leads to reactive species that allow further functionalization or chain elongation.

R–CH=CH–C(=O)–R′ —(Me_3SiCl; $IZn\diagup\diagdown ZnI$)→ IZn–CH₂–CH(R)–CH=C(R′)OSiMe₃ —(allyl bromide)→ CH₂=CH–CH₂–CH₂–CH(R)–CH=C(R′)OSiMe₃

[1]Matsubara, S., Yamamoto, Y., Utimoto, K. *SL* 1471 (1998).
[2]Ukai, K., Oshima, K., Matsubara, S. *JACS* **122**, 12047 (2000).
[3]Matsubara, S., Ukai, K., Mizuno, T., Utimoto, K. *CL* 825 (1999).
[4]Matsubara, S., Arioka, D., Utimoto, K. *SL* 1253 (1999).

Bis(2-methoxyethyl)aminosulfur trifluoride.

Alkyl fluorides. This reagent $(MeOCH_2CH_2)_2NSF_3$, converts alcohols into fluorides,[1] and with $SbCl_3$ as catalyst, thiono compounds are converted into *gem*-difluorides.[2]

Acyl fluorides.[3] Carboxylic acids are readily transformed into acid fluorides by this reagent in the presence of *i*-Pr_2NEt. Weinreb amides can be prepared henceforth. *N*-Protected chiral α-amino acids are derivatized without racemization.

[1]Lal, G.S., Pez, G.P., Pesaresi, R.J., Prozonic, F.M., Cheng, H. *JOC* **64**, 7048 (1999).
[2]Lal, G.S., Lobach, E., Evans, A. *JOC* **65**, 4830 (2000).
[3]Tunoori, A.R., White, J.M., Georg, G.I. *OL* **2**, 4091 (2000).

Bismuth. 20, 44

Nitroarene reduction.[1] Either azoarenes or azoxyarenes can be obtained on the reaction of nitroarenes with Bi–KOH in methanol.

Allylbismuth reagents.[2] Formation and use of these reagents in situ for allylation of carbonyl compounds can be achieved. Thus, an allylic acohol, after conversion into the iodide, readily afford the reagent by treatment with Bi.

[1]Laskar, D.D., Prajapati, D., Sandhu, J. S. *JCS(P1)* 67 (2000).
[2]Miyoshi, N., Nishio, M., Murakami, S., Fukuma, T., Wada, M. *BCSJ* **73**, 689 (2000).

Bismuth(III) acetate.

Azlactones.[1] $Bi(OAc)_3$ is a useful catalyst for the preparation from α-amino acids.

[1]Monk, K.A., Sarapa, D., Mohan, R.S. *SC* **30**, 3167 (2000).

Bismuth(III) bromide. 20, 44–45

Benzylation.[1] Alcohols form benzyl ethers on standing with BnOH in the presence of $BiBr_3$ (solvent: CCl_4).

Selective desilylation.[2] In aq MeCN, this salt catalyzes hydrolysis of a *t*-butyldimethylsilyl alkyl ether while preserving aryl derivatives.

TBSO–C6H4–CH2–O–(CH2)3–OTBS —[BiBr3, MeCN - H2O]→ TBSO–C6H4–CH2–O–(CH2)3–OH 93%

[1]Boyer, B., Keramane, E.-M., Roque, J.-P., Pavia, A.A. *TL* **41**, 2891 (2000).
[2]Bajwa, J.S., Vivelo, J., Slade, J., Repic, O., Blacklock, T. *TL* **41**, 6021 (2000).

Bismuth(III) chloride. 15, 37; 18, 52; 19, 37; 20, 45–46

Beckmann rearrangement.[1] With microwave assistance, the catalyzed rearrangement is completed within minutes (10 examples, 70–90%).

Sulfonylation. Doping of TfOH with $BiCl_3$ affords a catalyst system for sulfonylation of arenes.[2] Apparently, the reactive species is $Bi(OTf)_3$.[3]

Halogen exchange.[4] An alkyl bromide can be changed to a chloride (and vice versa by using the proper BiX_3) with retention of configuration. The reactivity pattern of the halide substrates is primary < secondary < tertiary.

Br, Br —[$BiCl_3$]→ Cl, Br

[1]Thakur, A.J., Boruah, A., Prajapati, D., Sandhu, J. S. *SC* **30**, 2105 (2000).
[2]Repichet, S., Le Roux, C., Dubac, J. *TL* **40**, 9233 (1999).
[3]Repichet, S., Le Roux, C., Hernandez, P., Dubac, J., Desmurs, J.-R. *JOC* **64**, 6479 (1999).
[4]Boyer, B., Keramane, E.M., Arpin, S., Montero, J.-L., Roque, J.-P. *T* **55**, 1971 (1999).

Bismuth(III) triflate.

Preparation.[1] An improved method for the preparation of $Bi(OTf)_3$ consists of mixing Ph_3Bi with 3 equiv of TfOH in dichloromethane at −78° and allowing the mixture to warm to room temperature.

2,3-Dihydro-4-pyridones.[2] These are the hydrolytic products from the condensation of Danishefsky's diene and imines. The condensation is efficeintly achieved at room temperature in the presence of $Bi(OTf)_3$.

Sulfonylation.[3] Heating mixtures of arenes and arenesulfonyl chlorides generates diaryl sulfones.

Acylation.[4] $Bi(OTf)_3$ is an effective catalyst for acylation. Thus, in its presence tertiary propargylic alcohols are acetylated in good yields.

[1]Labrouillere, M., Le Roux, C., Gaspard, H., Dubac, J. *TL* **40**, 285 (1999).
[2]Laurent-Robert, H., Garrigues, B., Dubac, J. *SL* 1160 (2000).
[3]Repichet, S., Roux, C.L., Hernandez, P., Dubac, J., Desmurs, J.-R. *JOC* **64**, 6479 (1999).
[4]Orita, A., Tanahashi, C., Kakuda, A., Otera, J. *ACIEE* **39**, 2877 (2000).

Bis(pentamethylcyclopentadienyl)cerium tetraphenylborate.

2-Aryl-2,3-dihydro-4-pyranones.[1] The hetero-Diels–Alder reaction of Danishefsky's diene and aromatic aldehydes is catalyzed by the metallocenium complex. The adducts are readily hydrolyzed.

[1]Molander, G.A., Rzasa, R.M. *JOC* **64**, 1215 (2000).

Bis(phenylthio)methane.

Methane polyanion. Generation of organolithium species from this reagent can be achieved by either deprotonation and/or reductive cleavage of the C—S bond(s).[1]

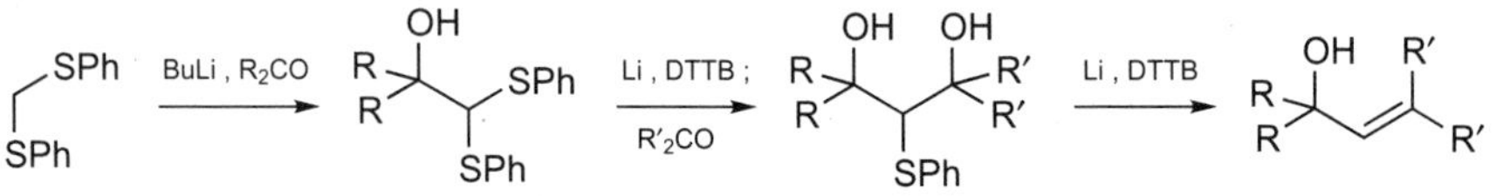

[1]Foubelo, F., Gutierrez, A., Yus, M. *TL* **40**, 8177 (1999).

[Bis(2-pyridyldimethylsilyl)methyl]lithium.

Alkenylsilanes. The precursor, bis(2-pyridyldimethylsilyl)methane, is available from 2-trimethylsilylpyridine via lithiation, reaction of the resulting 2-$PyMe_2SiCH_2Li$ with 2-$PyMe_2SiH$.

Lithiation of bis(2-pyridyldimethylsilyl)methane with BuLi in ether and treatment with carbonyl compounds furnish alkenylsilanes in good yields.[1]

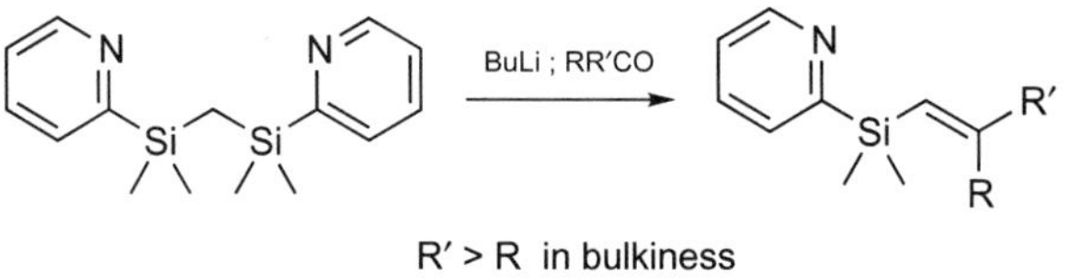

R′ > R in bulkiness

[1]Itami, K., Nokami, T. Yoshida, J. *OL* **2**, 1299 (2000).

Bis(2-tosyloxyethyl)diphenylsilane.

Protection of amines.[1] Reagent **1** is prepared from diphenyldichlorosilane by the following sequence: reaction with vinylmagnesium bromide, hydroboration of the resulting divinylsilane, and tosylation. The 4-diphenylsilapiperidines that are formed with primary amines can be cleaved by exposure to CsF, Bu_4NF in THF or DMF.

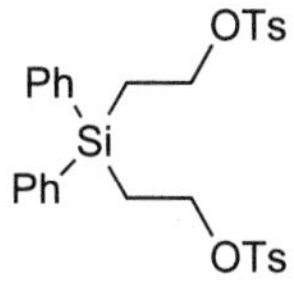

(1)

[1]Kim, B.M., Cho, J.H. *TL* **40**, 5333 (1999).

Bis(tributyltin) oxide. 13, 41–42; **15,** 39; **18,** 54; **19,** 40; **20,** 50

Oxepanes.[1] 2-(α-Hydroxyalkyl)oxepanes are obtained by an intramolecular substitution of 6,7-epoxy alcohols. An epoxide that is part of a glycidyl ether system reacts by mediation of $(Bu_3Sn)_2O$. The corresponding glycidyl esters afford related products in low yields.

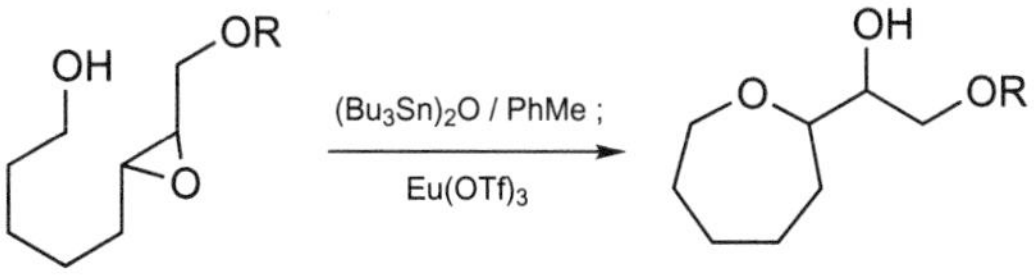

[1]Matsumura, R., Suzuki, T., Sato, K., Oku, K., Hagiwara, H., Hoshi, T., Ando, M., Kamat, V.P. *TL* **41**, 7701 (2000).

Bis(2,2,2-trifluoroethyl)acylmethylphosphonates.

Functionalized alkenes. (*Z*)-Enones are accessible from the acylmethylphosphonates by the Emmons–Wadsworth reaction.[1] Similarly, (*E*)-α-bromoacrylates have been prepared from the α-bromo-α-phosphonoacetate esters.[2]

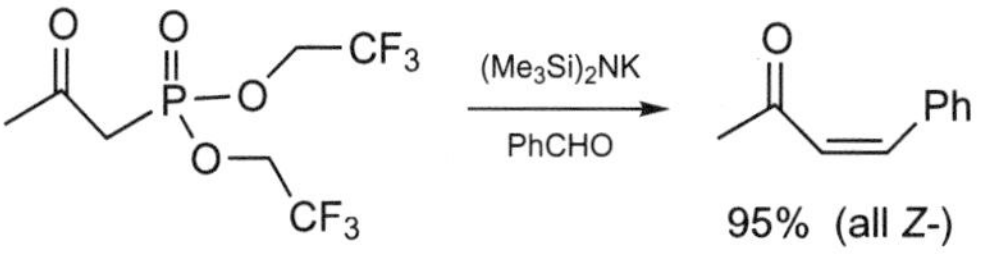

95% (all *Z*-)

[1]Yu, W., Su, M., Jin, J. *TL* **40**, 6725 (1999).
[2]Tago, K., Kogen, H. *OL* **2**, 1975 (2000).

2-Bis(trifluoromethylsulfonyl)amino-5-chloropyridine.

Triflation.[1] The stereochemical outcome in the *O*-triflation of β-keto esters (K salts) with this reagent is solvent dependent: (*E*)-selective in THF, (*Z*)-selective in DMF.

[1]Shao, Y., Eummer, J.T., Gibbs, R.A. *OL* **1**, 627 (1999).

N,O-Bis(trimethylsilyl)acetamide.

Proton acceptor.[1] In the displacement reaction of allyl carbonates with nitroalkanes, the presence of this silylated acetamide affords good diastereoselectivity.

OCOOMe + NO_2 → ($NSiMe_3$ / $OSiMe_3$ acetamide, Bu_4NCl, CH_2Cl_2) → H, H, NO_2 85%

Methyl β-nitrosoacrylate.[2] Methyl 3-nitropropionate is partially deoxygenated by *N,O*-bis(trimethylsilyl)acetamide. In the presence of a cyclic conjugated diene the conjugated nitroso compound is trapped as a Diels–Alder adduct.

O_2N … COOMe → ($NSiMe_3$ / $OSiMe_3$) → [O=N … COOMe] → (diene, X) → adduct COOMe

X = CH_2 57%
X = CH_2CH_2 21%

[1]Trost, B.M., Surivet, J.-P. *JACS* **122**, 6291 (2000).
[2]Tishkov, A.A., Lyapkalo, I.M., Ioffe, S.L., Strelenko, Y.A., Tartakovsky, V.A. *OL* **2**, 1323 (2000).

cis-2,3-Bis(trimethylsilyl)cyclopropanone.

Furans.[1] The reaction of this reagent with α-ketomethylenephosphoranes unexpectedly affords furans.

Me_3Si, Me_3Si cyclopropanone =O + Ph_3P= … R (O) → (Δ, PhMe) → Me_3Si, Me_3Si furan R + Me_3Si furan R

R = Me 70% (77 : 23)

[1]Takanami, T., Ogawa, A., Suda, K. *TL* **41**, 3399 (2000).

2,2′-Bis(2-trimethylsilylethoxymethoxy)-1,1′-binaphthyl.

2-(Trimethylsilyl)ethoxymethylation.[1] The ether **1** behaves as an electrophile toward silyl enol ethers in the presence of $SnCl_4$.

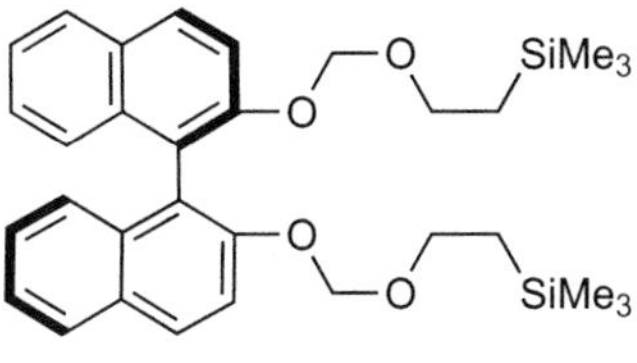

(**1**)

[1] Ishihara, K., Nakamura, H., Yamamoto, H. *JACS* **121**, 7720 (1999).

Bis(trimethylsilyl) peroxide. 20, 51

Nef reaction.[1] Nitronate ions are oxidized at room temperature.

Addition to double bond.[2] Alkenes are converted to either chlorohydrins or 1,2-acetoxy alcohols on reaction with Me_3SiX (X = Cl, OAc), $Me_3SiOOSiMe_3$, and with promotion by a Lewis acid.

Nucleotide phosphates.[3] Nucleotide H-phosphonates are oxidized to the phosphates in dichloromethane by bis(trimethylsilyl) peroxide in the presence of *N,O*-bis(trimethylsilyl)acetamide and a catalytic amount of Me_3SiOTf.

[1] Shahi, S.P., Vankar, Y.D. *SC* **29**, 4321 (1999).
[2] Sakurada, I., Yamasaki, S., Göttlich, R., Iida, T., Kanai, M., Shibasaki, M. *JACS* **122**, 1245 (2000).
[3] Kato, T., Hayakawa, Y. *SL* 1796 (1999).

3,3-Bis(trimethylsilyl)propene.

3-Piperideines.[1] Desilylative condensation of this reagent with HCHO and a primary amine (CF_3COOH promotion) results in the heterocycle formation.

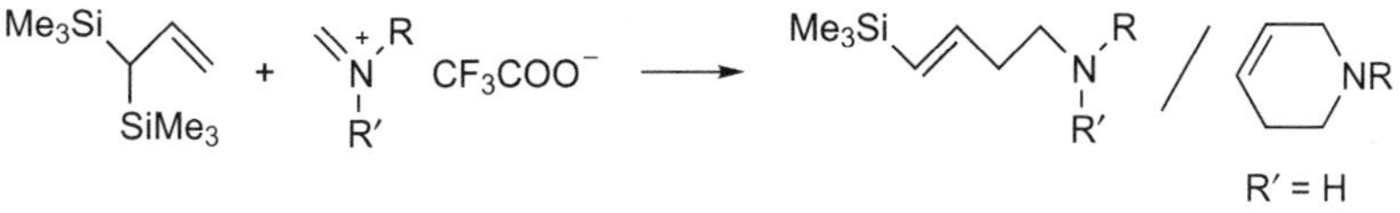

R′ = H

4-Amino-1-trimethylsilyl-1-butenes.[2] Iminium ions are readily trapped with this reagent. Silylated homoallylic amines are obtained.

[1] Princet, B., Anselme, G., Pornet, J. *JOMC* **592**, 34 (1999).
[2] Princet, B., Gardes-Gariglio, H., Pornet, J. *JOMC* **604**, 186 (2000).

Bis(trimethylsilyl) sulfide.

Thiols.[1] Preparation of thiols from alkyl halides is readily carried out with a combination of $(Me_3Si)_2S$ and Bu_4NF (9 examples, 68–94%).

[1]Hu, J., Fox, M.A. *JOC* **64**, 4959 (1999).

9-Borabicyclo[3.3.1]nonane, 9-BBN. 14, 52–53; **15,** 43–44; **17,** 49–50; **20,** 52

Aminoethyl extension.[1] Hydroboration of benzyl *N*-vinylcarbamate with 9-BBN followed by a Suzuki coupling with aryl or alkenyl substrates accomplishes a synthesis of these amines (still in protected form).

Cyclopropanes.[2] Aryl- and alkenylcyclopropanes are readily prepared from propargyl bromide via hydroboration and Pd-catalyzed coupling of the *B*-hydroxyborates with aryl and alkenyl halides.

H B + Br THF Δ ; NaOH → B OH PhBr $(Ph_3P)_4Pd$ →

Cyclization. A rare but documented C—C bond formation involving an organoborane moiety and a ketone is again revealed.[3]

O H OTBS MeOOC 9-BBN ; CrO_3 - HOAc → HO OTBS MeOOC

Reduction of lactams.[4] Tertiary lactams are transformed into tertiary amines. 9-BBN holds an advantage over borane because it does not complex with amines and therefore no excess reagent is required.

B-Methyl-BBN.[5] *B*-Methyl-BBN traps *N*-unsubstituted aldimines as stable adducts from their *N*-trimethylsilyl derivatives during methanolysis.

[1]Kametani, A., Overman, L.E. *JOC* **64**, 8743 (1999).
[2]Soderquist, J.A., Huertas, R., Leon-Colon, G. *TL* **41**, 4251 (2000).
[3]Boeckman, Jr., R.K., Mitchell, L.H., Shao, P., Lachicotte, R.J. *TL* **41**, 603 (2000).
[4]Collins, C.J., Lanz, M., Singaram, B. *TL* **40**, 3673 (2000).
[5]Chen, G.-M., Brown, H.C. *JACS* **122**, 4217 (2000).

Borane-amines. 13, 42; **18,** 58; **19,** 43–44; **20,** 53

Borane adducts with tertiary amines in which one of the alkyl groups is tertiary (e.g., **1**[1] and **2**[2]) have been prepared for hydroboration.

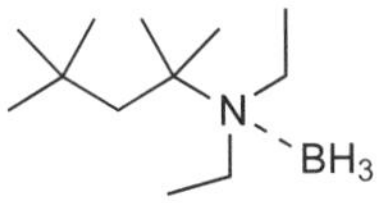

(**1**) 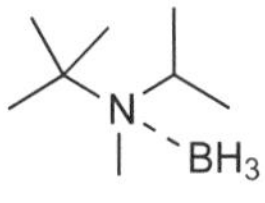 (**2**)

Reductions. *syn*-Selective reduction of 1,3-diketones[3] and α-alkyl-β-keto esters[4] is observed with borane–pyridine in the presence of $TiCl_4$. On the other hand, *anti*-products predominate in the reduction with $LiBHEt_3$–$CeCl_3$.

$(Ph_3P)_4Pd$ catalyzes removal of the R^fSO_3 group from aryl perfluorosulfonates[5] and the transformation of γ,δ-epoxy-α,β-unsaturated esters[6] to δ-hydroxy-α,β-unsaturated esters by the borane–dimethylamine complex.

Allyl group scavengers.[7] In the solid-phase peptide synthesis, the protection of nitrogen by an *N*-allyloxycarbonyl group can be removed by a Pd(0) catalyzed reaction in the presence of a borane–amine complex.

[1]Brown, H. C., Kanth, J.V.B., Dalvi, P. V., Zaidlewicz, M. *JOC* **65**, 4655 (2000).
[2]Brown, H. C., Kanth, J.V.B., Dalvi, P. V., Zaidlewicz, M. *JOC* **64**, 6263 (1999).
[3]Bartoli, G., Bosco, M., Bellucci, M.C., Dalpozzo, R., Marcantoni, E., Sambri, L. *OL* **2**, 45 (2000).
[4]Marcantoni, E., Alessandrini, S., Malavolta, M., Bartoli, G., Bellucci, M.C., Sambri, L., Dalpozzo, R. *JOC* **64**, 1986 (1999).
[5]Lipshutz, B.H., Buzard, D.J., Vivian, R.W. *TL* **40**, 6871 (1999).
[6]David, H., Dupuis, L., Guillerez, M.-G., Guibe, F. *TL* **41**, 3335 (2000).
[7]Gomez-Martinez, P., Dessolin, M., Guibe, F., Albericio, F. *JCS(P1)* 2871 (1999).

Boron tribromide. 13, 43; **14,** 53–54; **18,** 59; **19,** 45; **20,** 54

Ring contraction.[1] 3-Methoxypiperidines undergo ring contraction on exposure to BBr_3 in dichloromethane. This reaction is a rare transformation of piperidines.

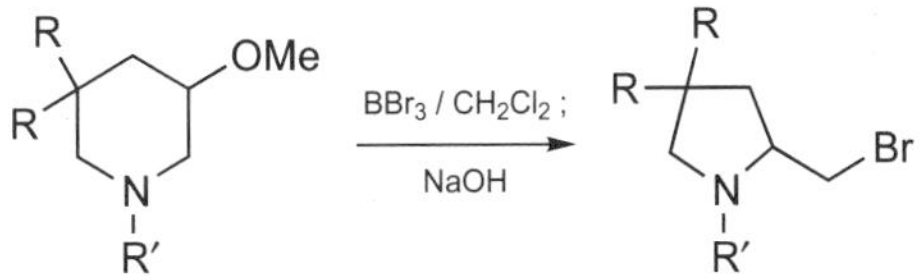

BBr_3 / CH_2Cl_2 ; NaOH

[1]Tehrani, K.A., Van Syngel, K., Boelens, M., Contreras, J., De Kimpe, N., Knight, D.W. *TL* **41**, 2507 (2000).

Boron trichloride. 13, 43; **14,** 54; **15,** 44; **18,** 59–60; **19,** 45–46; **20,** 54–55

Ether cleavage. Trityl ethers suffer cleavage selectively.[1] Phenols are released from $ArOCH_2R$ with BCl_3.[2]

Et3Si ... O–CPh3 → BBr3 / CH2Cl2 ; – 30° → Et3Si ... OH 91%

Benzal chlorides.[3] Chlorodeoxygenation of aromatic aldehydes is readily accomplished with BCl_3 in refluxing hexane (9 examples, 76–99%).

Alkenyldichloroboranes.[4] The species that are useful for Suzuki coupling can be prepared from alkenylsilanes with BCl_3 in dichloromethane.

[1]Jones, G.B., Hynd, G., Wright, J.M., Sharma, A. *JOC* **65**, 263 (2000).
[2]Brooks, P.R., Wirtz, M.C., Vetelino, M.G., Rescek, D.M., Woodworth, G.F., Morgan, B.P., Coe, J.W. *JOC* **64**, 9719 (1999).
[3]Kabalka, G.W., Wu, Z. *TL* **41**, 579 (2000).
[4]Babudri, F., Farinola, G.M., Naso, F., Panessa, D. *JOC* **65**, 1554 (2000).

Boron trifluoride etherate. 13, 43–47; **14,** 54–56; **15,** 45–47; **16,** 44–47; **17,** 52–53; **18,** 60–63; **19,** 46–48; **20,** 55–57

Heterocycles. Alkenyl and aryl boronates deliver their organic residues to the α-position of the nitrogen atom of cyclic carbinolamines to replace the hydroxyl group.[1] Synthesis of 2,3,5-trisubstituted furans,[2] from a catalyzed reaction of alkynyl borates and enones, is achieved in one step.

OH, OH, N–COOBn + dioxaborolane–CH=CH–Bu → BF3·OEt2 / CH2Cl2, – 78° ~ 25° → OH, N–COOBn, CH=CH–Bu 93%

Fused oxepane arrays are derived from regioselective oxacyclization of polyepoxides.[3]

BF3·OEt2 / CH2Cl2, – 40° → H, O, OH, H

Dealkylation.[4] Tertiary amines are converted into the acetamides of secondary amines on treatment with $Ac_2O–BF_3·OEt_2$.

γ-Ketoesters.[5] Fischer carbene complexes in which the oxygen atom is bonded to BF_2 serve as acyl anion equivalents in a conjugate addition.

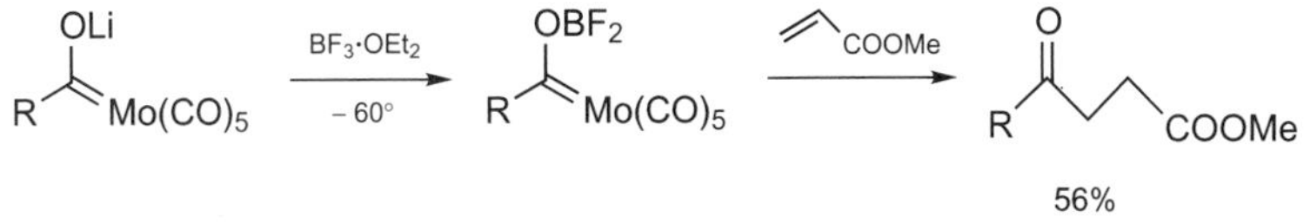

Reactions of alkyne–cobalt complexes. The hexacarbonyldicobalt complex of a propargylic alcohol containing a juxtaposed benzyloxy group suffers elimination of the propargylic hydroxyl on treatment with $BF_3·OEt_2$. An intramolecular hydride transfer is involved.[6] Complexes of *vic*-diols give deconjugated alkynones.[7]

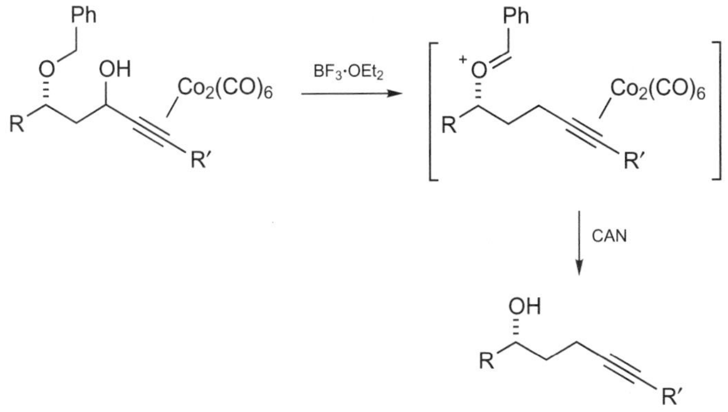

Rearrangements. Cyclic ketones undergo ring expansion on reaction with hydroxyalkyl azides.[8] Regioselectivity is dependent on steric and electronic factors. Beckmann rearrangement of ketoxime carbonates[9] has been observed at room temperature.

There is a change in the regiochemical course for Claisen rearrangement[10] of a tetrahydropyridyl allyl ether by $BF_3·OEt_2$. The problem in stereocontrol of the aldol reaction involving cyclic ketones and aldehydes can be solved by way of the rearrangement of hydroxyalkyl epoxides.[11] Ultimately, it rests on the proper choice of the haloalkenes.

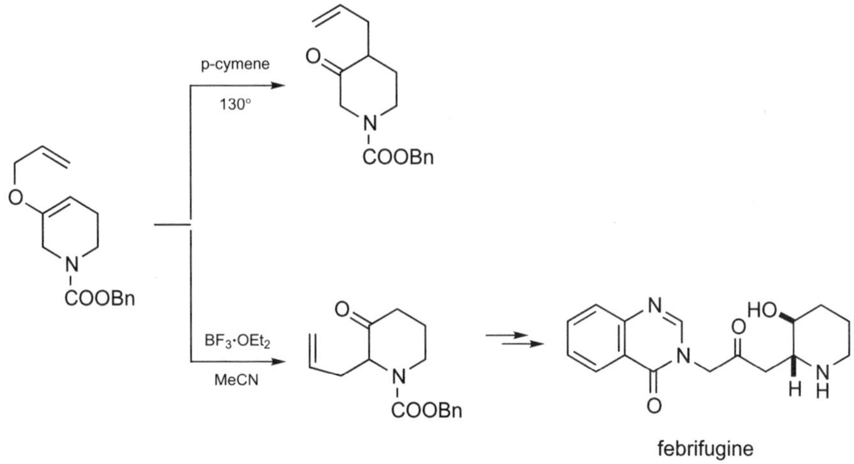

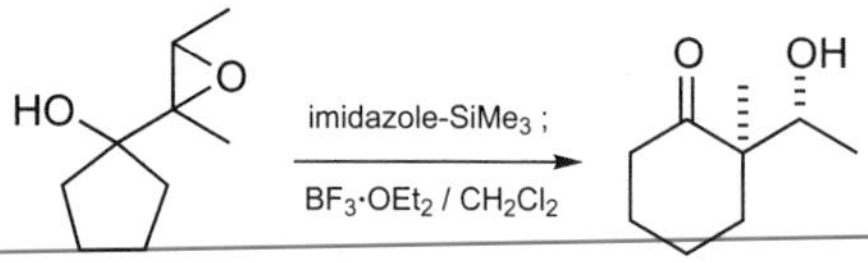

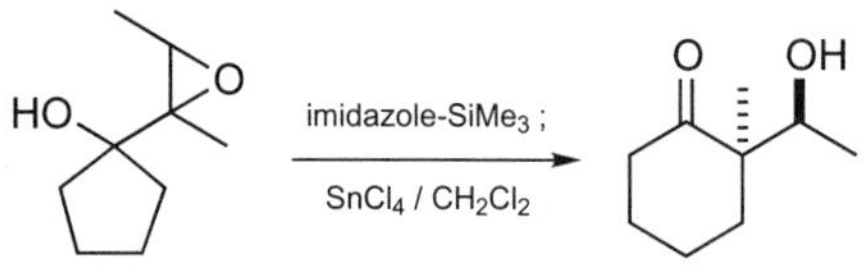

p-cymene 130°
COOBn
$BF_3·OEt_2$ MeCN
febrifugine
imidazole-$SiMe_3$; $BF_3·OEt_2$ / CH_2Cl_2
imidazole-$SiMe_3$; $SnCl_4$ / CH_2Cl_2

Alkenyl methyl sulfides.[12] $BF_3·OEt_2$ plays a role in promoting the Horner condensation.

Mannich-type reactions.[13] Aldimines are activated in preference to aldehydes by $BF_3·OEt_2$ (0.2 equiv + 10 equiv H_2O) toward condensation with silyl enol ethers.

(E)-Styrenes.[14] The $BF_3·OEt_2$-catalyzed reaction of an araldehyde with a symmetrical ketone leads to styrenes.

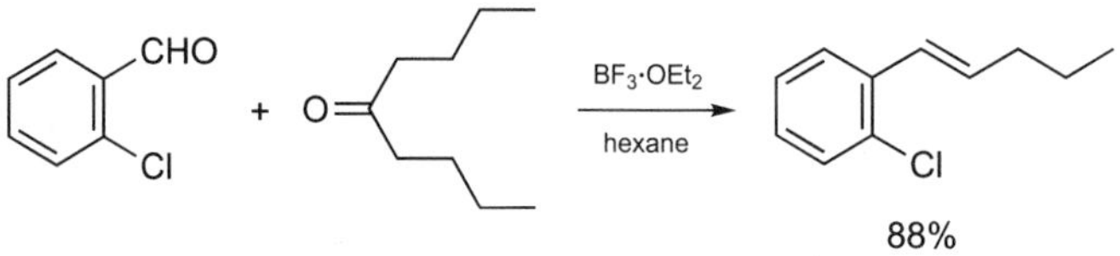

CHO Cl + O $BF_3·OEt_2$ hexane Cl 88%

syn-1,2-Amino alcohols.[15] It is possible to utilize γ-oxygenated allylstannanes to prepare enantioenriched *syn*-1,2-amino alcohol derivatives that are potential precursors of α-amino-β-hydroxy acids, β-amino-α-hydroxy acids, and aza sugars. This $BF_3{\cdot}OEt_2$-catalyzed process has advantages over other methods because it stereoselectively creates two stereocenters upon C—C bond formation.

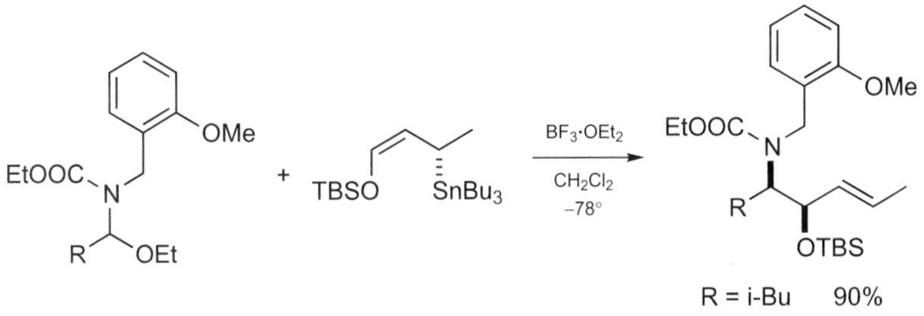

EtOOC–N(CH₂C₆H₄OMe)–CH(R)OEt + TBSO–CH=CH–CH(SnBu₃)CH₃ → ($BF_3{\cdot}OEt_2$, CH_2Cl_2, −78°) → EtOOC–N(CH₂C₆H₄OMe)–CH(R)–CH(OTBS)–CH=CH–CH₃

R = i-Bu 90%

Dehydration.[16] Propargylic alcohols are readily dehydrated after complexation with a dicobalt hexacarbonyl residue. Interestingly, renewed treatment of the diols (formed by dihydroxylation of the double bond) with $BF_3{\cdot}OEt_2$ leads to the deconjugated 3-alkynones.

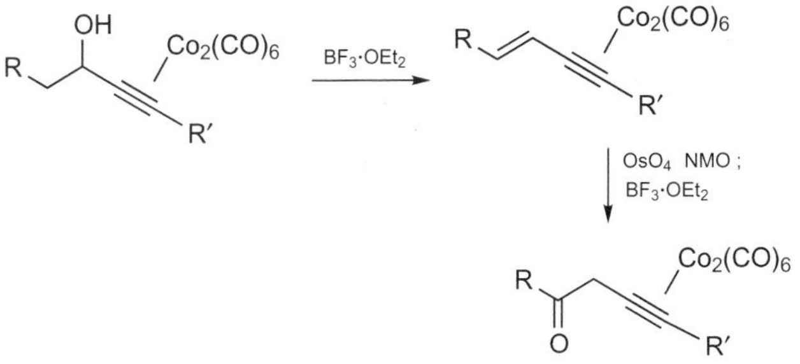

R–CH₂–CH(OH)–C≡C–R′ [$Co_2(CO)_6$] → ($BF_3{\cdot}OEt_2$) → R–CH=CH–C≡C–R′ [$Co_2(CO)_6$] → (OsO_4 NMO ; $BF_3{\cdot}OEt_2$) → R–C(=O)–CH₂–C≡C–R′ [$Co_2(CO)_6$]

Cyclic mutation. Intramolecular allylation leads to formation of a carbocycle in exchange for an aziridine when the heterocycle contains an allylsilane moiety in a side chain. The transformation is initiated by $BF_3{\cdot}OEt_2$.[17]

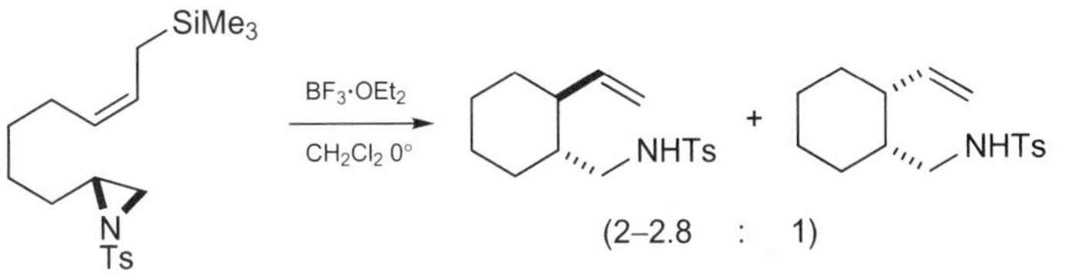

N-Ts aziridine with allylsilane ($SiMe_3$) side chain → ($BF_3{\cdot}OEt_2$, CH_2Cl_2 0°) → cyclohexane bearing vinyl and CH₂NHTs (cis) + cyclohexane bearing vinyl and CH₂NHTs (trans)

(2–2.8 : 1)

A twofold intramolecular Friedel–Crafts alkylation serves to create the tricyclic skeleton of the pseudopterosins.[18]

[1]Batey, R.A., MacKay, D.B., Santhakumar, V. *JACS* **121**, 5075 (1999).
[2]Brown, C.D., Chong, J.M., Shen, L. *T* **55**, 14233 (1999).
[3]McDonald, F.E., Wang, X., Do, B., Hardcastle, K.I. *OL* **2**, 2917 (2000).
[4]Dave, P.R., Kumar, K.A., Duddu, R., Axenrod, T., Dai, R., Das, K.K., Guan, X.-P., Sun, J., Trivedi, N.J., Gilardi, R.D. *JOC* **65**, 1207 (2000).
[5]Barluenga, J., Rodriguez, F., Fananas, F.J. *CEJ* **6**, 1930 (2000).
[6]Diaz, D., Martin, V.S. *TL* **41**, 743 (2000).
[7]Diaz, D., Martin, V.S. *TL* **40**, 2815 (1999).
[8]Smith, B.T., Gracias, V., Aube, J. *JOC* **65**, 3771 (2000).
[9]Anilkumar, R., Chandrasekhar, S. *TL* **41**, 5427 (2000).
[10]Takeuchi, Y., Hattori, M., Abe, H., Harayama, T. *S* 1814 (1999).
[11]Baldwin, S.W., Chen, P., Nikolic, N., Weinseimer, D.C. *OL* **2**, 1193 (2000).
[12]Stephan, E., Olaru, A., Jaouen, G. *TL* **40**, 8571 (1999).
[13]Akiyama, T., Takaya, J., Kagoshima, H. *CL* 947 (1999).
[14]Kabalka, G.W., Li, N.-S., Tejedor, D., Malladi, R.R., Trotman, S. *JOC* **64**, 3157 (1999).
[15]Marshall, J.A., Gill, K., Seletsky, B.M. *ACIEE* **39**, 953 (2000).
[16]Soler, M.A., Martin, V.S. *TL* **40**, 2815 (1999).
[17]Bergmeier, S.C., Seth, P.P. *JOC* **64**, 3237 (1999).
[18]Harrowven, D.C., Sibley, G.E.M. *TL* **40**, 8299 (1999).

Bromine. 13, 47; **14,** 56–57; **15,** 47; **18,** 64; **19,** 48; **20,** 57–58

Sulfoxides.[1] The oxidation of sulfides with the bromine–hexamethylenetetramine complex is a rapid and high-yielding process (7 examples, 91–98%).

Oxidation.[2] Bromine adsorbed on alumina (0.77 mmol/g) is useful as an oxidant for alcohols. It is stable in a stoppered bottle for several months. The oxidation is carried out in the solid state.

Cleavage of tertiary alcohols.[3] Bromine–K_2CO_3 in chloroform at 0° performs a retro-Barbier reaction. Thus, cyclic tertiary alcohols give ω-bromoalkyl ketones.

[1]Shaabani, A., Teimouri, M.B., Safaei, H.R. *SC* **30**, 265 (2000).
[2]Love, B.E., Nguyen, B.T. *SC* **30**, 963 (2000).
[3]Zhang, W.-C., Li, C.-J. *JOC* **650**, 5831 (2000).

Bromine trifluoride. 19, 48

Trifluoromethyl ethers.[1] These ethers are formed when alcohols are transformed into xanthates and then treated with BrF_3.

Halogen exchange. Replacement of other halogen atoms by fluorine using BrF_3 affords a simple approach to potentially anesthetic fluorinated ethers.[2] Examples include the synthesis of $CF_3CH_2OCF_3$ from $CF_3CH_2OCF_2Cl$.

[1]Ben-David, I., Rechavi, D., Mishani, E., Rozen, S. *JFC* **97**, 75 (1999).
[2]Hudlicky, T., Duan, C., Reed, J.W., Yan, F., Hudlicky, M., Endoma, M.A., Eger, E.I. *JFC* **102**, 363 (2000).

***B*-Bromocatecholborane–acetic acid.**

Cleavage of MOM ethers.[1] This combination of reagents enables smooth removal of the methoxymethyl group from protected 1,3-diols and 1,3-amino alcohols in dichloromethane. Complications arise if acetic acid is omitted. For example, 1,3-oxazines are formed in the case of protected 3-aminopropyl methoxymethyl ethers.

[1]Yu, C., Liu, B., Hu, L S. *TL* **41**, 819 (2000).

2-Bromoethyl acetate.

Amine protection.[1] Secondary amines are protected by alkylation with this acetate. The products are reverted to RR′NH on photolysis in the presence of 4,4′-dimethoxybenzophenone.

[1]Cossy, J., Rakotoarisoa, H. *TL* **41**, 2097 (2000).

Bromoform.

α-Methoxycarboxylic acids.[1] Treatment of ketones with $CHBr_3$, KOH in methanol leads to the acids.

[1]Yabuuchi, T., Kusumi, T. *CPB* **47**, 684 (1999).

Bromomethydiisopropoxyborane.

1,3-Diols.[1] Allylic alcohols are selectively transformed into 1,3-diols via the mixed borates. Radical cyclization followed by oxidative C—B bond cleavage completes the process.

OH — $(iPrO)_2BCH_2Br$ → O–B(OiPr)CH$_2$Br — $(Me_3Si)_3SiH$ - AIBN ; NaOH - H_2O_2 → OH, OH

74%

[1]Batey, R.A., Smil, D.V. *TL* **40**, 9183 (1999).

***N*-Bromosuccinimide.** **13,** 49; **14,** 57–58; **15,** 50–51; **16,** 49; **18,** 65–67; **19,** 50–51; **20,** 58–59

Brominations. Hydrochloric acid is a catalyst for arene bromination in acetone.[1] Deactivated arenes are brominated with NBS–CF_3COOH–H_2SO_4.[2]

Some hydroxylated heterocycles are readily brominated (e.g., 2-hydroxypyridine to 2-bromopyridine) by reaction with NBS–Ph_3P.[3]

Reaction with unsaturated acids and derivatives. Propiolic acids undergo halodecarboxylation at room temperature with NXS in the presence of a quaternary ammonium salt.[4] Analogous Hunsdiecker reaction products from alkenoic acids can be used in situ for Heck coupling.[5]

COOH

NBS , $Bu_4N(OCOCF_3)$

COOEt Pd(OAc)$_2$

Ph_3Sb - LiCl -Et_3N

COOEt

81%

Conjugated amides are cyclized to afford β-bromo-α-lactams.[6]

Acylsilanes.[7] 2-Silyl-1,3-dithianes undergo oxidative hydrolysis on exposure to NBS in aqueous acetone. Addition of a base (triethylamine or barium hydroxide) prevents further conversion of the products to carboxylic acids.

1-Alkyl-7-azaisatins.[8] 7-Azaindoles are brominated at C-3 with bromine. Treatment of the products with NBS in DMSO transforms them into the isatins.

Br

N N R

NBS , DMSO 60° – 80°

O O

N N R

87 – 90%

Acetalization. The use of NBS as a catalyst for acetalization with triethyl orthoformate[9] shows chemoselectivity in favor of aldehydes, although formation of ketone acetals is also feasible. 1,3-Dioxanes are formed when 1,3-propanediol is present in the reaction.[10]

[1]Andersh, B., Murphy, D.L., Olson, R.J. *SC* **30**, 2091 (2000).
[2]Duan, J., Zhang, L. H., Dolbier, W. R. *SL* 1245 (1999).

[3]Sugimoto, O., Mori, M., Tanji, K. *TL* **40**, 7477 (1999).
[4]Naskar, D., Roy, S. *JOC* **64**, 6896 (1999).
[5]Naskar, D., Roy, S. *T* **56**, 1369 (2000).
[6]Naskar, D., Roy, S. *JCS(P1)* 2435 (1999).
[7]Patrocinio, A.F., Moran, P.J.S. *JOMC* **603**, 220 (2000).
[8]Tatsugi, J., Tong, Z., Amano, T., Izawa, Y. *H* **53**, 1145 (2000).
[9]Karimi, B., Seradj, H., Ebrahimian, G.R. *SL* 1456 (1999).
[10]Karimi, B., Ebrahimian, G.R, Seradj, H. *OL* **1**, 1737 (1999).

P-Bromotri(pyrrol-1-yl)phosphonium bromide.

Formamidines.[1] This well-known reagent for peptide synthesis activates formamides for reaction with primary amines to furnish formamidines.

R−NH2 + H(C=O)NMe2 + (pyrrol-1-yl)3P+−Br Br− —(i-Pr2NEt)→ R−N=CH−NMe2

[1] Delarue, S., Sergherart, C. *TL* **40**, 5487 (1999).

(1,3-Butadien-1-yl)dimethylsulfonium tetrafluoroborate.

Alkenyl epoxides.[1] A three-component condensation reaction involves an anion, an aldehyde, and the dienylsulfonium salt.

PhCHO + CH2=CH−CH=CH−S+Me2 BF4− + Na+ −C(COOEt)2NHAc —(CH2Cl2, −40 ~ 23°)→ Ph-epoxide−CH=CH−CH2−C(COOEt)2NHAc

[1]Rowbottom, M.W., Mathews, N., Gallagher, T. *JCS(P1)* 3927 (1998).

(2,3-Butadien-1-yl)tributylstannane.

Preparation.[1] A simple route to this reagent is by reaction of 1,4-dichloro-2-butyne with Bu_3SnLi. Note that coupling of 1,3-butadien-2-ylmagnesium chloride with Ph_3SnCl leads to the triphenyl analogue but the coupling with Bu_3SnCl proceeds without skeletal change.

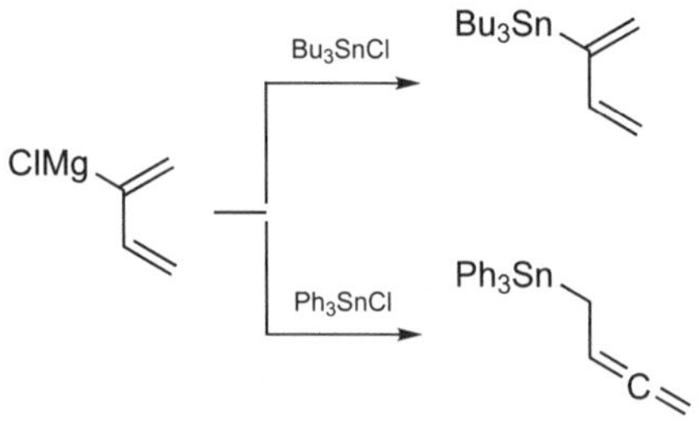

1,3-Butadien-2-ylation.[1] The nucleophilic site of the allenylmethylstannane reagent is the central carbon atom of the allene moiety. Its reaction [Ti(IV)-catalyzed] with aldehydes and acetals affords products possessing a conjugated diene unit.

[1]Luo, M., Iwabuchi, Y., Hatakeyama, S. *CC* 267 (1999).

N-*t*-Butoxycarbonyl-*O*-(2-pyridylsulfonyl)hydroxylamine.

Carbamoylation.[1] Hydroxamic esters can be synthesized via reaction of β-dicarbonyl compounds with reagent **1**, which is readily obtained.

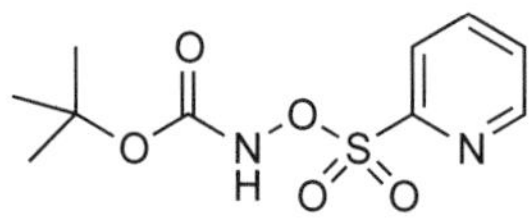

(**1**)

[1]Hanessian, S., Johnstone, S. *JOC* **64**, 5896 (1999).

1-*t*-Butoxycarbonyl-1-tosylhydrazine.

Substituted hydrazines.[1] Carbonyl compounds undergo reductive condensation with this reagent, which is prepared from 1,2-di-*t*-Boc-1-tosylhydrazine by selective hydrogenolysis.

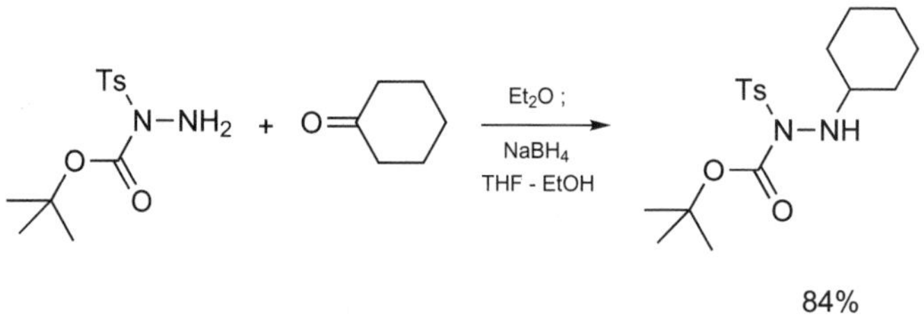

Et_2O ; $NaBH_4$ THF - EtOH

84%

[1]Grehn, L., Ragnarsson, U. *T* **55**, 4843 (1999).

2-*t*-Butoxycarbonyl-3-trichloromethyloxaziridine.

Hydroxylamines.[1] Various alcohols and carboxylic acids undergo *O*-amination with this reagent after deprotonation.

[1]Foot, O.F., Knight, D.W. *CC* 975 (2000).

***t*-Butyl azidoacetate.**

α-Azido-β-hydroxyalkanoates.[1] The condensation products with aldehydes are useful for the preparation of serine homologues. The adduct with PhCHO can be used in a synthesis of *t*-butyl indole-2-carboxylate.

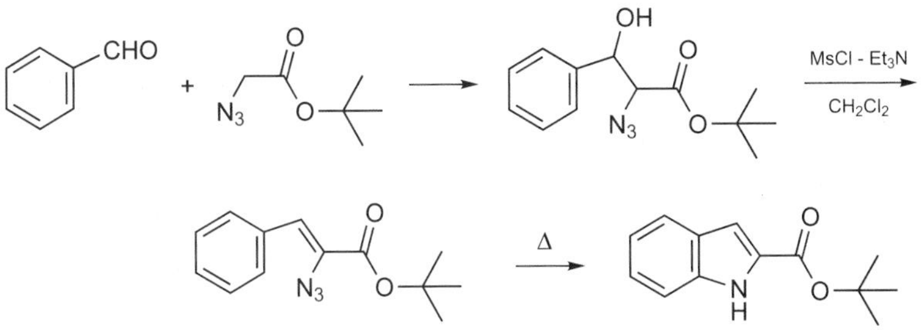

MsCl - Et_3N, CH_2Cl_2; Δ

[1]Kondo, K., Morohoshi, S., Mitsuhashi, M., Murakami, Y. *CPB* **47**, 1227 (1999).

***t*-Butyldimethylsiloxymalononitrile.**

α-Siloxy amides.[1] A rapid reaction of amines and $TBSOCH(CN)_2$ with carbonyl compounds in MeCN at 0° leads to α-siloxy amides. The title reagent behaves as an amphiphilic [CO] component.

[1]Nemoto, H., Ma, R., Suzuki, I., Shibuya, M. *OL* **2**, 4245 (2000).

6-(*t*-Butyldimethylsilyl)-1,5,6-trimethyl-1,4-cyclohexadiene.

Radical generation.[1] The title reagent serves as a surrogate of tributyltin hydride in initiating many free radical reactions, including reduction of organic halides.

[1]Studer, A., Amrein, S. *ACIEE* **39**, 3080 (2000).

***t*-Butyldiphenylsilyl chloride.**

Silylation.[1] Selective monosilylation of symmetrical primary 1,*n*-diols is possible.

[1]Yu, C., Liu, B., Hu, L.S. *TL* **41**, 4281 (2000).

***t*-Butyl hydroperoxide metal salts. 19,** 51–53; **20,** 61–62

Oxidations. Ketones are cleaved to afford carboxylic acids[1] in moderate yields by Re_2O_7–*t*-BuOOH in acetic acid at 100°. Vanadium-catalyzed epoxidation of unsaturated alcohols in liquid carbon dioxide is feasible.[2] A resin-bound cobalt–phosphine complex is also effective as a catalyst for oxidation.[3]

Interestingly, α-silylated allyl alcohols are transformed into α-silyl-β-hydroxy ketones[4] on oxidation with $(i\text{-PrO})_4Ti$–*t*-BuOOH.

(i-PrO)4Ti - t-BuOOH

Allyl alkyl ethers are converted to ketones[5] on reaction with *t*-BuOOH in the presence of a catalytic amount of CrO_3 in dichloromethane at room temperature. This oxidation does not affect THP, TBS, and MOM ethers.

Arylation of alkenes.[6] Coupling reaction of arenes with alkenes by $Pd(OAc)_2$ is carried out in the presence of benzoquinone and *t*-BuOOH as a reoxidant. Note, this is not a Heck reaction (involving ArH instead of ArX).

[1]Gurunath, S., Sudalai, A. *SL* 559 (1999).
[2]Pesiri, D.R., Morita, D.K., Walker, T., Tumas, W. *OM* **18**, 4916 (1999).
[3]Leadbeater, N.E., Scott, K.A. *JOC* **65**, 4770 (2000).
[4]Fässler, J., Enev, V., Bienz, S. *HCA* **82**, 561 (1999).
[5]Chandrasekhar, S., Mohanty, P.K., Ramachander, T. *SL* 1063 (1999).
[6]Jia, C., Lu, W., Kitamura, T., Fujiwara, Y. *OL* **1**, 2097 (1999).

***t*-Butyl hypochlorite. 13,** 55; **18,** 74; **19,** 54

Nitrosoarenes. *N*-Arylhydroxylamines undergo oxidation (5 examples, 88–92%).

[1]Davey, M.H., Lee, V.Y., Miller, R.D., Marks, T.J. *JOC* **64**, 4976 (1999).

2-*t*-Butylimino-2-diethylamino-3-methylperhydro-1,3,2-diazaphosphorine, polymer bound.

N-Acylation.[1] The polymer-supported base **1** adequately serves for use in the acylation of weakly nucleophilic amines.

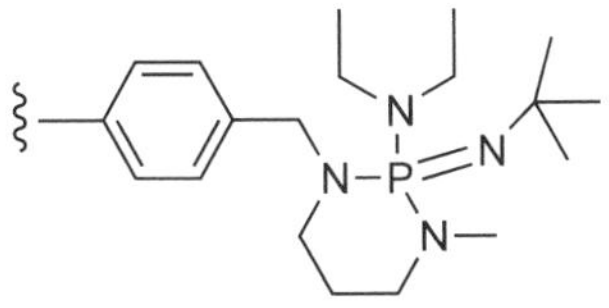

(**1**)

[1]Kim, K., Le. K. *SL* 1957 (1999).

t-Butyl isocyanide.

Reaction with nitro compounds.[1] Homologation to afford α-oximino amide derivatives is observed.

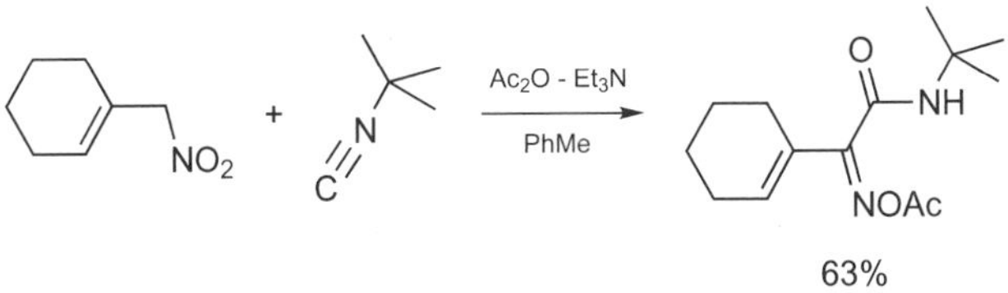

Ac_2O - Et_3N, PhMe; 63%

[1]Dumestre, P., El Kaim, L., Gregoire, A. *CC* 775 (1999).

Butyllithium. 13, 56; **14,** 63–68; **15,** 59–61; **17,** 59–60; **18,** 74–77; **19,** 54–59; **20,** 62–65

Lithiation. Various uses of BuLi for deprotonation include ethers that subsequently react with aldimines to provide precursors of pyrrolidin-3-ones.[1] 2-Alkylidene tetrahydrofurans are readily prepared via alkylation of 2-sulfonylmethylenetetrahydrofuran.[2]

Allenamides are functionalized at the α-position by virtue of the regioselective lithiation. Thus, on introduction of an alkynyl chain, valuable precursors of the Pauson–Khand reaction can be prepared.[3] Lithiation of the silyl ethers of propargyl or allenyl alcohols engenders O → C transsilylation.[4,5]

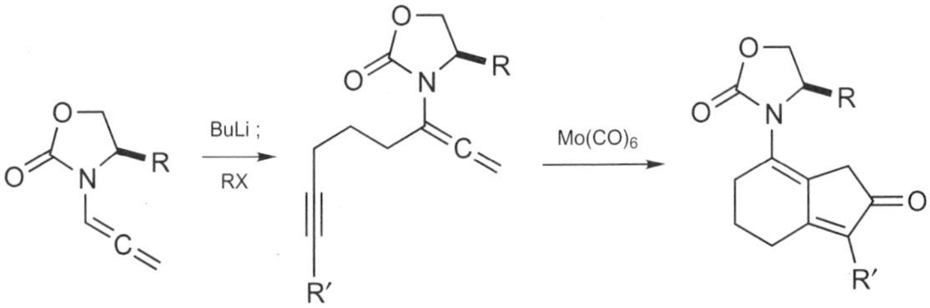

BuLi ; RX; $Mo(CO)_6$

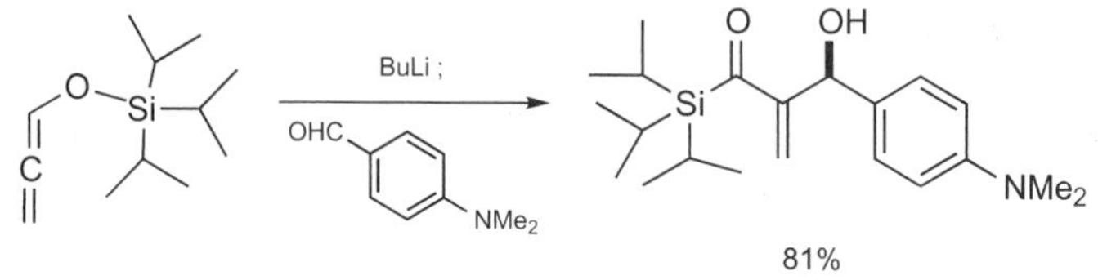

The above reverse Brook rearrangement also occurs during treatment of the silyl enol ether of *o*-bromoacetophenone with BuLi. Quenching the enolate with PhCHO gives an aldol product.[6]

Halogen–lithium exchange followed by intramolecular substitution creates a new ring. Such a method is the basis of a benzodihydrofuran synthesis. The presence of two exchangeable halogen atoms (2,6-disubstitution pattern) makes it possible to introduce another functionality at C-7.[7]

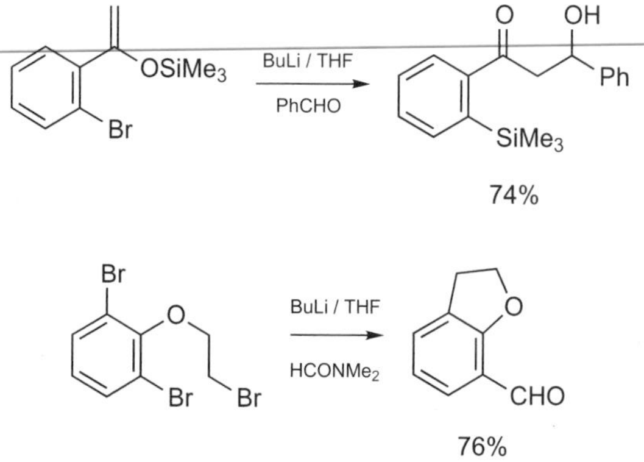

O,C$_\beta$-Dilithio enolates are generated from β-stannyl ketones by Sn–Li exchange and enolization.[8]

Alkenes. (Phenylsulfinylmethyl)diphenylphosphine oxide is readily deprotonated for condensation with aldehydes, affording *(E)*-alkenyl phenyl sulfoxides.[9]

A sulfoxide version of the Julia–Lythgoe olefination is terminated by elimination of vicinal hydroxyl and phenylsulfinyl groups on treatment with BuLi.[10]

Solvent dependence. Solvent dependence of the products arising from either isomerization or elimination of propargylic acetals[11] on treatment with BuLi has been witnessed.

The halogen–lithium exchange of 2,5-dibromopyridine also shows different regioselectivity in ether and toluene.[12]

N-Acylation.[13] Primary arylamines form amides on dilithiation with BuLi and subsequent exposure to an ester at low temperature.

Wittig rearrangements.[14] Solvent effects dominate the mode of Wittig rearrangement of benzyl propargyl ethers.

Et_2O, 0°	0%	55%	14%
Et_2O -THF - hexane −110°	62%	4%	8%

[1] Amombo, M.O., Hausherr, A., Reissig, H.-U. *SL* 1871 (1999).
[2] Edwards, G.L., Sinclair, D.J. *TL* **40**, 3933 (1999).
[3] Xiong, H., Hsung, R.P., Wei, L.-L., Berry, C.R., Mulder, J.A., Stockwell, B. *OL* **2**, 2869 (2000).
[4] Sakaguchi, K., Fujita, M., Suzuki, H., Higashino, M., Ohfune, Y. *TL* **41**, 6589 (2000).
[5] Stergiades, I.A., Tius, M.A. *JOC* **64**, 7547 (1999).
[6] Comanita, B.M., Woo, S., Fallis, A.G. *TL* **40**, 5283 (1999).
[7] Plotkin, M., Chen, S., Spoors, P.G. *TL* **41**, 2269 (2000).
[8] Ryu, I., Nakahira, H., Ikebe, M., Sonoda, N., Yamato, S., Komatsu, M. *JACS* **122**, 1219 (2000).
[9] van Steenis, J.H., van Es, J.J.G.S., van der Gen, A. *EJOC* 2787 (2000).
[10] Satoh, T., Hanaki, N., Yamada, N., Asano, T. *T* **56**, 6223 (2000).
[11] Le Strat, F., Maddaluno, J. *TL* **41**, 5367 (2000).
[12] Wang, X., Rabbat, P., O'Shea, P., Tillyer, R., Grabowski, E.J.J., Reider, P.J. *TL* **41**, 4335 (2000).
[13] Ooi, T., Tayama, E., Yamada, M., Maruoka, K. *SL* 729 (1999).
[14] Tomooka, K., Harada, M., Hanji, T., Nakai, T. *CL* 1394 (2000).

Butyllithium–2-(*N,N*-dimethylamino)ethanol.

Lithiation of pyridines. 3-Methylpyridine is regioselectively lithiated at C-6 with this base system,[1] which is markedly different from its behavior toward LDA (benzylic lithiation). 2-Chloropyridine behaves similarly.[2]

[1]Mathieu, J., Gros, P., Fort, Y. *CC* 951 (2000).
[2]Choppin, S., Gros, P., Fort, Y. *OL* **2**, 803 (2000).

Butyllithium–potassium *t*-butoxide.

Cyclization.[1] Fused 1-vinylcyclopropanes are formed on lithiation of allyl sulfides that bear an alkenyl chain at the α-position.

Lithiation. *m*-Methoxybenzoic acid and 3,5-dimethoxybenzoic acid are lithiated at C-4.[2]

[1]Cheng, D., Knox, K.R., Cohen, T. *JACS* **122**, 412 (2000).
[2]Sinha, S., Mandal, B., Chandrasekaran, S. *TL* **41**, 3157 (2000).

Butyllithium–*N,N,N′N′*-tetramethylethylenediamine. 19, 60; **20,** 66–67

Lithiation. The facile lithiation of β-silylalkanamides enables substitution at C-3.[1]

Oxazolinylepoxides under lithiation at −100°. Isomerization of the oxiranyllithiums to 2-acyloxazolines occurs on warming, therefore alkylation must be conducted at low temperature.[2]

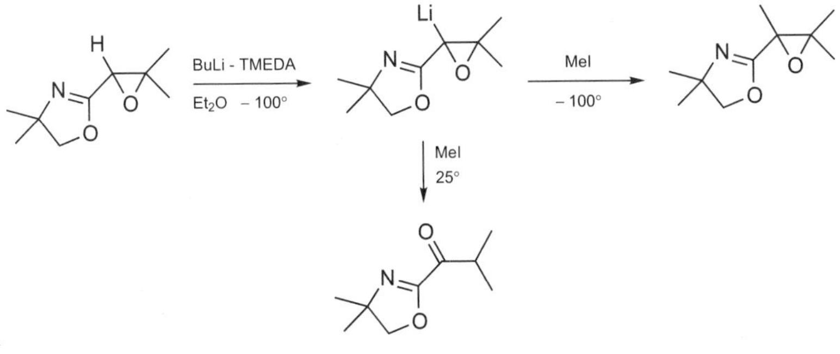

[1]Shinozuka, T., Utsumi, T., Inagaki, M., Asaoka, M. *JCS(P1)* 2433 (1999).
[2]Capriati, V., Florio, S., Luisi, R., Russo, V., Salomone, A. *TL* **41**, 8835 (2000).

s-Butyllithium. 14, 69; **16,** 56; **18,** 77–79; **19,** 60–61

Lithiation of aromatic compounds. Hydrazides of the constitution $ArCONHNMe_2$ are lithiated at the ortho position.[1]

Indole-2-carboxylic esters.[2] Benzylic lithiation of *t*-butyl *N*-(*o*-alkylphenyl)carbamates followed by addition of alkyl oxalates also prompts cyclization and completes an indole-2-carboxylic ester synthesis on treatment with CF_3COOH.

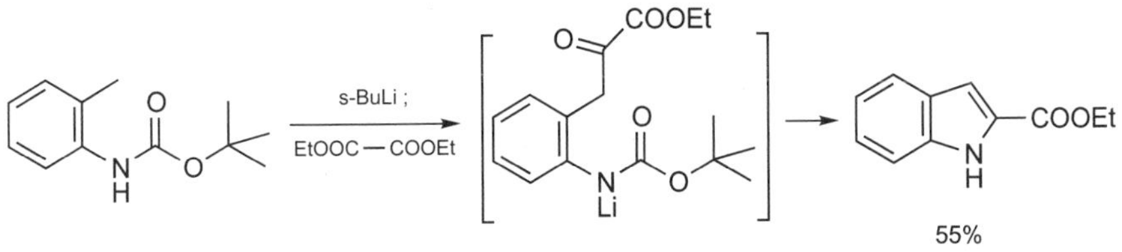

Alkylation of benzylic acetals.[3] It has been shown that *s*-BuLi is the best base for the second-staged allylation of a dimeric acetal. Note that *n*-BuLi functions well in the first-staged alkylation.

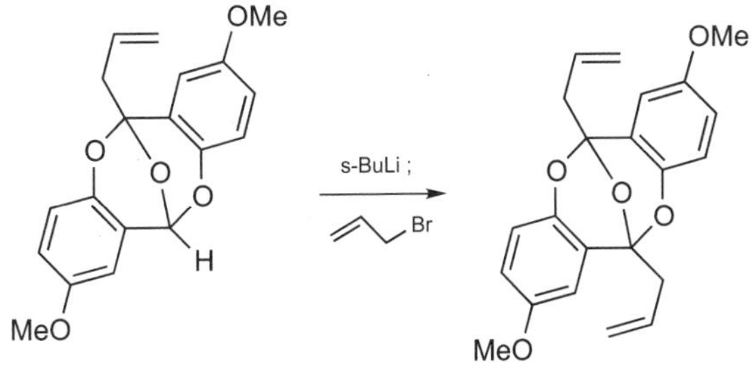

[1]McCombie, S.W., Lin, S.-I., Vice, S.F. *TL* **40**, 8767 (1999).
[2]Allen, D.A. *SC* **29**, 447 (1999).
[3]Ragot, J.P., Prime, M.E., Archibald, S.J., Taylor, R.J.K. *OL* **2**, 1613 (2000).

s-Butyllithium–*N,N,N′N′*-tetramethylethylenediamine. 18, 78–79; **20,** 67–68

Lithiation. Indoles are lithiated at C-7 when the nitrogen atom is protected by a 2,2-diethylbutanoyl group.[1] Aminoallenes are available by two routes: from reaction of lithiated 1-silylalkynes with oxazolidines[2] and reaction of propargylic substrates with amines,[3] after the latter undergo α-lithiation and Li–Cu exchange.

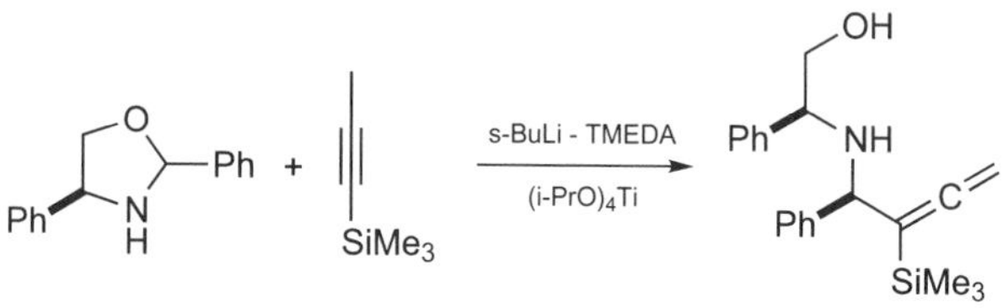

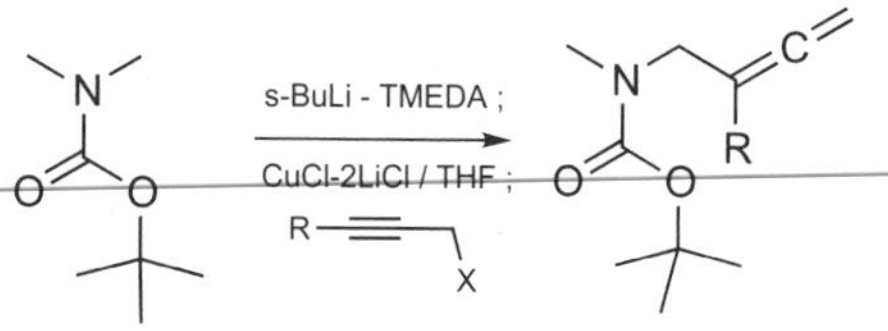

s-BuLi - TMEDA
(i-PrO)$_4$Ti

s-BuLi - TMEDA ;
CuCl-2LiCl / THF ;

Indolizidines.[4] The *t*-Boc derivative of azacyclononene oxide is deprotonated at an α-position of the epoxide ring by *s*-RLi–TMEDA. A transannular reaction follows.

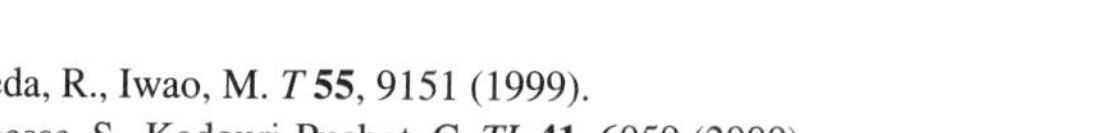

s-BuLi - TMEDA
Et$_2$O

74%

[1]Fukuda, T., Maeda, R., Iwao, M. *T* **55**, 9151 (1999).
[2]Agami, C., Comesse, S., Kadouri-Puchot, C. *TL* **41**, 6059 (2000).
[3]Dieter, R.K., Nice, L.E. *TL* **40**, 4283 (1999).
[4]Hodgson, D.M., Robinson, L.A. *CC* 309 (1999).

***t*-Butyllithium. 13,** 58; **15,** 64–65; **16,** 56–57; **18,** 79–81; **19,** 61–63; **20,** 68–69

Lithiation. The terminus of a polyene can be lithiated on treatment with *t*-BuLi, permitting chain extension by alkylation.[1]

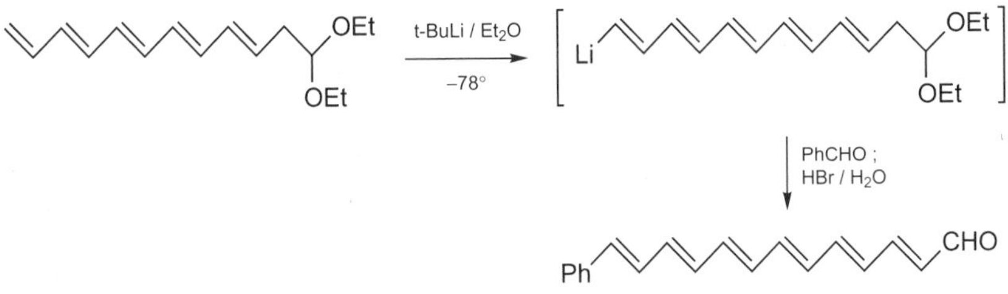

An interesting synthesis of kainic acid[2] starts from α-lithiation of an *N*-benzylbenzamide, which undergoes cyclization to form a dihydroisoindolone (dearomatization).

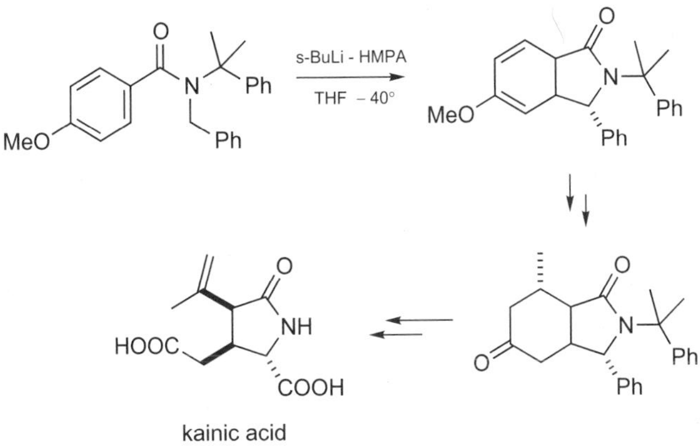

Halogen/lithium exchanges. After treatment with *t*-BuLi, 1-iodo-3-alkynes react with certain alkenes (vinyl sulfides, vinylsilanes, styrenes) to give alkylidenecyclopentanes.[3]

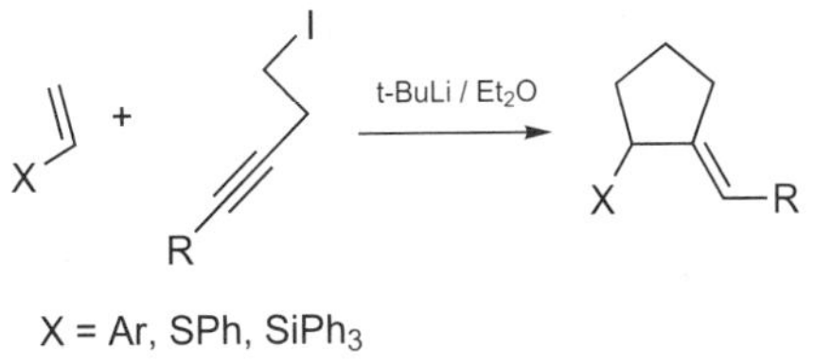

X = Ar, SPh, $SiPh_3$

Lithium alkynoxides are generated from ethyl 2,2-dibromoalkanoates with *t*-BuLi. These species react with keto esters in a remarkable manner, yielding cyclalkenones.[4]

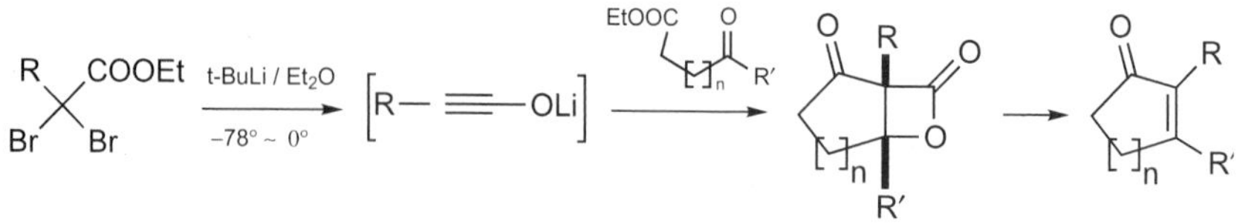

The elaboration of a polyunsaturated 5:9-fused ring system[5] based on transannular addition of an alkenyllithium species to a triple bond following the Br—Li exchange of a bromoalkene is clean and elegant.

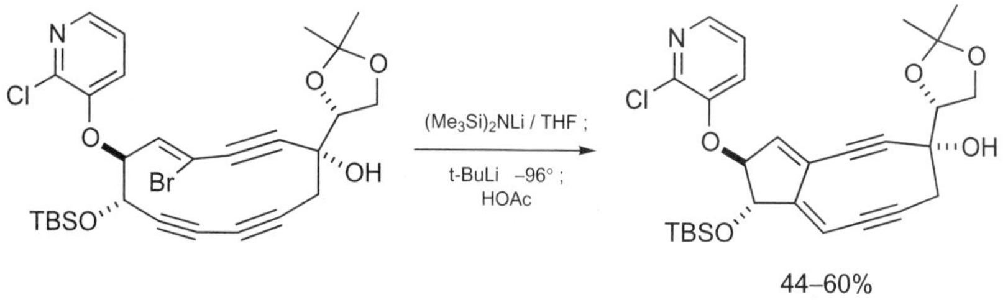

Deallylation. Allyl ethers are cleaved by *t*-BuLi via an S_N2' process.[6]

Addition to styrenes. *t*-Butyllithium (other alkyllithiums also) adds to the terminus of styrenes. Benzylic carboxylation is feasible.When the substrates contain an allyloxy or benzyloxy group the allyl or benzyl residue is transfer to the benzylic position.[7]

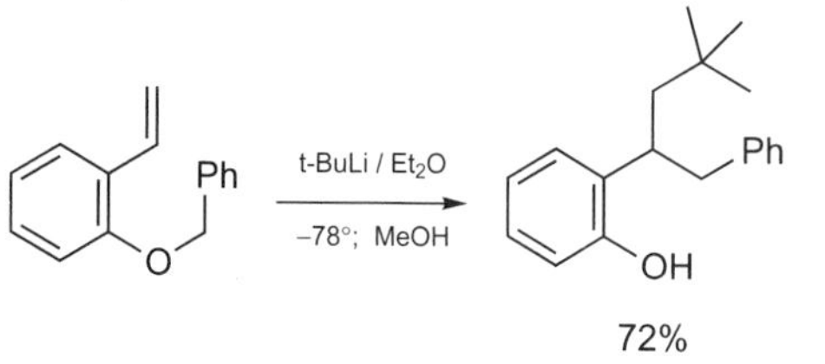

[1]Villiers, P., Soullez, D., Ramondenc, Y., Ple, G. *T* **55**, 14031 (1999).
[2]Clayden, J., Tchabanenko, K. *CC* 317 (2000).
[3]Wei, X., Taylor, R.J.K. *ACIEE* **39**, 409 (2000).
[4]Shindo, M., Sato, Y., Shishido, K. *JACS* **121**, 6507 (1999).
[5]Myers, A.G., Goldberg, S.D. *ACIEE* **39**, 2732 (2000).
[6]Bailey, W. F., England, M.D., Mealy, M.J., Thongsornkleeb, C., Teng, L. *OL* **2**, 489 (2000).
[7]Wei, X., Taylor, R.J.K. *JCS(P1)* 1109 (2000).

t-Butyl methanesulfonyloxycarbamate.

Lossen rearrangement.[1] A modified version of the rearrangment for degradation of carboxylic acids is via reaction of the corresponding acid chlorides RCOCl with *t*-BuOCONHOMs, and the subsequent treatment with zinc triflate, 2,6-di-*t*-butylpyridine, benzyl alcohol in MeCN. Benzyl carbamates RNHCOOBn are obtained.

[1]Stafford, J. A., Gonzales, S. S., Barrett, D. G., Suh, E. M., Feldman, P. L. *JOC* **63**, 10040 (1998).

1-Butyl-3-methylimidazolium salts. 20, 70

Allylic displacements.[1] Palladium-catalyzed allylic displacement reactions have been carried out in ionic liquids at room temperature.

[1]Chen, W., Xu, L., Chatterton, C., Xiao, J. *CC* 1247 (1999).

***N*-(*t*-Butyl)phenylsulfinimidoyl chloride.**

Oxidation of alcohols. This stable reagent,[1] prepared from *t*-BuNH$_2$ by *N,N*-dichlorination and then treatment with PhSAc, functions in the same fashion as the reactive species generated in the Swern and Corey–Kim oxidations. With which efficient oxidation of alcohols is performed in dichloromethane in the presence of DBU (11 examples, 90–99%).[2]

Conjugated ketones.[3] Dehydrogenation of ketones via their lithium enolates on treatment with *t*-BuN═S(Cl)Ph at −78° is very convenient.

[1]Markovskii, L.N., Dubinina, T.N., Levchenco, E.S., Kirsanov, A.V. *JOC(USSR)* **9**, 1435 (1973).
[2]Mukaiyama, T., Matsuo, J., Yanagisawa, M. *CL* 1072 (2000).
[3]Mukaiyama, T., Matsuo, J., Kitagawa, H. *CL* 1250 (2000).

***t*-Butyl propynoate.**

Protection of 1,2-diols.[1] Acetal formation with this ester (via a double Michael reaction) at room temperature in the presence of DMAP is limited to 1,2-diols (1,3- and 1,4-diols do not react in a similar manner). These acetals survive conditions that acetonides are hydrolyzed (e.g., aq HOAc) but they can be cleaved with bases.

[1]Ariza, X., Costa, A.M., Faja, M., Pineda, O., Vilarrasa, J. *OL* **2**, 2809 (2000).

***t*-Butylsulfinyl chloride.**

Amide radicals.[1] After *O*-sulfinylation, hydroxamic acids undergo homolysis at the N—O bond. Trapping of the *N*-radical by an alkene leads to cyclic structures. This method has been applied to a formal synthesis of peduncularine.

MeO, HO, N, O — t-BuSOCl / i-Pr$_2$NEt, CH$_2$Cl$_2$, Ph$_2$Se$_2$ → N, MeO, O, SePh → → N, NH

peduncularine

[1] Lin, X., Stien, D., Weinreb, S.M. *TL* **41**, 2333 (2000).

2-(*t*-Butylsulfonyl)allyl chloroformate.

Amine protection.[1] Carbamates are formed at room temperature or below. The removal of the protecting group, is most expediently achieved by treatment with a polymer-bound secondary amine. The released fragment is trapped and the isolation of amines is uncomplicated.

[1] Carpino, L.A., Philbin, M. *JOC* **64**, 4315 (1999).

2-(*t*-Butylsulfonyl)phenyliodine(III) diacetate.

Hypervalent iodine reagents.[1] This compound is obtained from *t*-butyl 2-iodophenyl sulfone by oxidation with H_2O_2-Ac_2O. It can be transformed into reagents such as **1** and **2**.

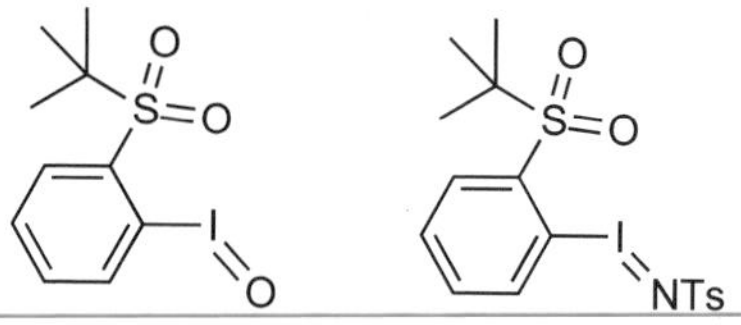

(**1**) (**2**)

[1] Macikenas, D., Skrzypczak-Jankun, E., Protasiewicz, J.D. *JACS* **121**, 7164 (1999).

1-*t*-Butyl-1*H*-tetrazol-5-yl sulfones.

(Z)-Alkenes.[1] These sulfones are very stable and participate in the Julia olefination with improved yields.

[1] Kocienski, P.J., Bell, A., Blakemore, P.R. *SL* 365 (2000).

(*t*-Butylthio)azoarenes.

Hydrazones.[1] Reaction of ketones with these reagents conveniently introduces an arylhydrazono group to ketones.

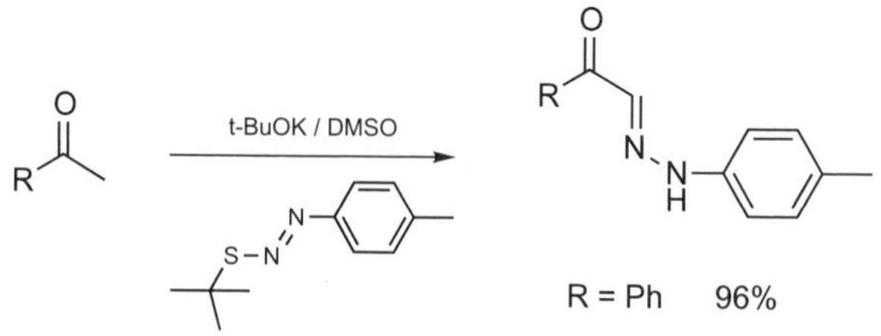

t-BuOK / DMSO

R = Ph 96%

[1] Caposcialli, N., Dell'Erba, C., Novi, M., Petrillo, G., Tavani, C. *T* **54**, 5315 (1998).

C

Calcium hypophosphite.

Reduction of arenediazonium ions.[1] This reagent is used in nonaqueous media (e.g., MeCN) to remove amino groups from aromatic rings after diazotization. Good yields are obtained from substrates that contain electron-withdrawing groups.

[1]Mitsuhashi, H., Kawakami, T., Suzuki, H. *TL* **41**, 5567 (2000).

Carbon dioxide.

Carbonates and carbamates. In the presence of phosphine–CBr_4 and base, alcohols combine with CO_2 to afford carbonates.[1] Mixed carbonates are obtained under different conditions[2], and with a Mg—Al mixed oxide as catalyst, epoxides formed cyclic carbonates.[3] The preparation of dimethyl carbonate from acetone dimethyl acetal and supercritical carbon dioxide in the presence of a metal catalyst (e.g., dibutyltin methoxide) is successfully carried out.[4]

Carbamates are formed by trapping the condensation products of amines and CO_2 with alkyl halides.[5] The initial step is promoted by electrogenerated superoxide ion. Heating amines and oxetanes at 40 atm leads to 3-hydroxyalkyl carbamates.[6]

$$\text{oxetane} + CO_2 + R_2NH \xrightarrow[\text{40 atm}]{120^\circ} R_2N\text{-C(=O)-O-(CH}_2)_3\text{-OH}$$

2,4-Dihydroxyquinazolines are obtained when 2-cyanoanilines are treated with CO_2 in the presence of an amine such as DBU.[7]

$$\text{2-}NH_2\text{-}C_6H_4\text{-CN} \xrightarrow[\text{DBU / DMF}]{CO_2} \text{2,4-(OH)}_2\text{-quinazoline}$$

Reaction medium. By virtue of its attributes supercritical carbon dioxide deserves its increasing applications as a reaction medium in synthesis. These include epoxidation[8] with oxygen and PhCHO, rhodium-catalyzed hydroboration,[9] Pd(0)-catalyzed 1,4-hydroarylation of enones[10] and Heck reaction in the presence of Pd–C,[11] carbonylation of aryl halides,[12] Glaser coupling of 1-alkynes (mediated by $CuCl_2$) using a solid base (NaOAc),[13] as well as scandium(III) perfluorooctanesulfonate-catalyzed Diels–Alder and hetero-Diels–Alder reactions.[14]

Investigations have shown that in Pd-catalyzed reactions in supercritical carbon dioxide, superior results are obtained with catalysts containing fluorinated ligands [e.g., $(CF_3COO)_2Pd$ and (F_6-acac)Pd vs. their respective hydrogen analogues].[15]

Continuous, selective hydroformylation in supercritical CO_2 using (acac)Rh$(CO)_2$ immobilized on silica as catalyst[16] shows certain advantages. A version of asymmetric hydroformylation in this medium has also been reported,.[17] (Subcritical CO_2 gas accelerates solventless synthesis involving solid reactants,[18] including hydrogenation and hydroformylation.) The regioselective and enantioselective nickel-catalyzed hydrovinylation of styrenes in supercritical CO_2 make 3-arylpropenes available in an optically active form.[19]

Improvement in the performance of the Pauson–Khand reaction in supercritical media is due to thorough dispersion of catalytic metal species. Besides CO_2,[20] supercritical ethylene has the same effect.[21]

Water and supercritical carbon dioxide form an excellent medium for hydrogenation of unsaturated aldehydes to allylic alcohols with ruthenium(III) chloride and a water-soluble triarylphosphine,[22] because the limitation pertaining to gas–liquid–liquid mass transfer is eliminated due to the very high solubility of reactant gas.

[1]Kadokawa, J., Habu, H., Fukamachi, S., Karasu, M., Tagaya, H., Chiba, K. *JCS(P1)* 2205 (1999).
[2]Kim, S.-I., Chu, F., Dueno, E.E., Jung, K.W. *JOC* **64**, 4578 (1999).
[3]Yamaguchi, K., Ebitani, K., Yoshida, T., Yoshida, H., Kaneda, K. *JACS* **121**, 4526 (1999).
[4]Sakakura, T., Choi, J.-C., Saito, Y., Masuda, T., Sako, T., Oriyama, T. *JOC* **64**, 4506 (1999).
[5]Casadei, M.A., Moracci, F.M., Zappia, G., Inesi, A., Rossi, L. *JOC* **62**, 6757 (1997).
[6]Ishii, S., Zhou, M., Yoshida, Y., Noguchi, H. *SC* **29**, 3207 (1999).
[7]Mizuno, T., Okamoto, N., Ito, T., Miyata, T. *TL* **41**, 1051 (2000).
[8]Loeker, F., Leitner, W. *CEJ* **6**, 2011 (2000).
[9]Carter, C.A.G., Baker, R.T., Nolan, S.P., Tumas, W. *CC* 347 (2000).
[10]Cacchi, S., Fabrizi, G., Gasparrini, F., Pace, P., Villani, C. *SL* 650 (2000).
[11]Cacchi, S., Fabrizi, G., Gasparrini, F., Villani, C. *SL* 345 (1999).
[12]Kayaki, Y., Noguchi, Y., Iwasa, S., Ikariya, T., Noyori, R. *CC* 1235 (1999).
[13]Li, J., Jiang, H. *CC* 2369 (1999).
[14]Matsuo, J., Tsuchiya, T., Odashima, K., Kobayashi, S. *CL* 178 (2000).
[15]Shezad, N., Oakes, R.S., Clifford, A.A., Rayner, C.M. *TL* **40**, 2221 (1999).
[16]Meehan, N.J., Sandee, A.J., Reek, J.N.H., Kamer, P.C.J., van Leeuwen, P.W.N.M., Poliakoff, M. *CC* 1497 (2000).
[17]Francio, G., Leitner, W. *CC* 1663 (1999).
[18]Jessop, P., Wynne, D.C., DeHaai, S., Nakawatase, D. *CC* 693 (2000).
[19]Wegner, A., Leitner, W. *CC* 1583 (1999).
[20]Jeong, N., Hwang, S.H., Lee, Y.W., Lim, J.S. *JACS* **119**, 10549 (1997).
[21]Jeong, N., Hwang, S.H. *ACIEE* **39**, 636 (2000).
[22]Bhanage, B.M., Ikushima, Y., Shirai, M., Arai, M. *CC* 1277 (1999).

Carbon disulfide.

1,3-Dithiacycles.[1] 1,3-Dithianes and 1,3-dithiepanes are formed from 1,3- and 1,4-dihalides on reaction with CS_2 and $NaBH_4$.

Isothiocyanates.[2] Sequential treatment of organic azides with Ph_3P and CS_2 gives isothiocyantes. This process can be applied to a synthesis of isocyanatoalkylphosphonates.

[1]Wan, Y., Kurchan, A.N., Barnhurst, L.A., Kutateladze, A.G. *OL* **2**, 1133 (2000).
[2]Sikora, D., Gajda, T. *PSS* **157**, 201 (2000).

Carbon monoxide. 20, 72

Diaryl diketones.[1] Highly hindered aryllithiums such as mesityllithium react with CO to provide the 1,2-diketones.

[1]Nudelman, N., Schulz, H. *JCR(S)* 422 (1999).

Carbonyl(chloro)hydridotris(tricyclohexylphosphine)ruthenium.

Hydrogenation.[1] The Ru complex is a homogeneous hydrogenation catalyst.

Silylation of alkenes.[2] Terminal alkenes give 1-silylalkenes using vinylsilanes as reagents and the Ru complex as catalyst.

Dehalogenation.[3] Haloarenes are reduced in basic media with a related complex that can be generated in situ from $[(cod)RuCl_2]_x$ and Cy_3P under hydrogen.

[1]Yi, C.S., Lee, D.W. *OM* **18**, 5152 (1999).
[2]Yi, C.S., He, Z., Lee, D.W., Rheingold, A.L., Lam, K.-C. *OM* **19**, 2036 (2000).
[3]Cucullu, M.E., Nolan, S.P., Belderrain, T.R., Grubbs, R.H. *OM* **18**, 1299 (1999).

Carbonyl(chloro)hydridotris(triphenylphosphine)ruthenium.

Double-bond migration.[1] Deconjugation of α,β-unsaturated esters is observed.

SnBu3
Bu3Sn COOMe
(Ph3P)3Ru(CO)ClH
PhMe Δ
SnBu3
Bu3Sn COOMe
80%

BnO COOMe
(Ph3P)3Ru(CO)ClH
PhMe Δ
BnO COOMe

Silyl group transfer.[2, 3] Allylsilanes and vinylsilanes serve as silyl group donors to alkenes such as styrenes and vinyl ethers. It is apparently a metathesis reaction.

EtO + SiMe3
(Ph3P)3Ru(CO)ClH
PhH Δ
EtO SiMe3

[1]Wakamatsu, H., Nishida, M., Adachi, N., Mori, M. *JOC* **65**, 3966 (2000).
[2]Kakiuchi, F., Yamada, A., Chatani, N., Murai, S., Furukawa, N., Seki, Y. *OM* **18**, 2033 (1999).
[3]Marciniec, B., Kujawa, M., Pietraszuk, C. *OM* **19**, 1677 (2000).

Carbonyldihydridotris(triphenylphosphine)ruthenium(II). 19, 65–66

Addition reactions.[1, 2] Electron-deficient arenes, such as aromatic nitriles and ketones, add to unsaturated silanes in the presence of the Ru catalyst. Carbon chain extension occurs at the ortho position of the CN or CO group.

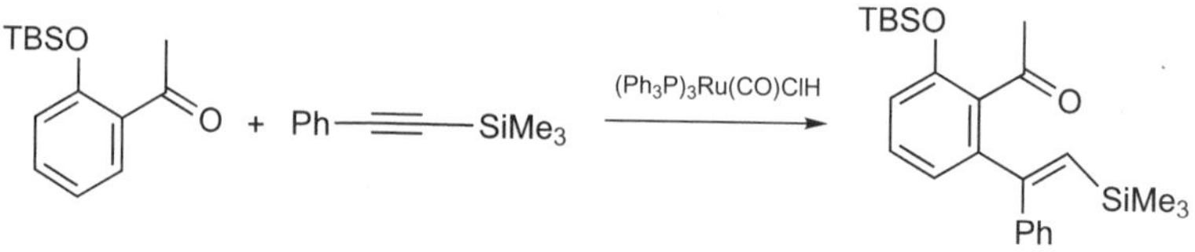

[1]Harris, P.W.R., Rickard, C.E.F., Woodgate, P.D. *JOMC* **589**, 168 (1999).
[2]Kakiuchi, F., Sonoda, M., Tsujimoto, T., Chatani, N., Murai, S. *CL* 1083 (1999).

1,1′-Carbonyldiimidazole. 13, 66; **16,** 64; **18,** 85; **20,** 73

Oxazole.[1] Ethyl isocyanoacetate is converted in ethyl oxazole-4-carboxylate on reaction with this reagent in the presence of HCOOH and Et_3N. The parent heterocycle, which is no longer commercially available, is easily obtained.

[1]Shafer, C.M., Molinski, T.F. *H* **53**, 1167 (2000).

Catecholborane.

Conjugate addition.[1] *B*-Alkylcatecholboranes, obtained from hydroboration of alkenes, serve as a source of radicals in the addition.

[1]Ollivier, C., Renaud, P. *CEJ* 5, 1468 (1999).

Cerium(IV) ammonium nitrate CAN. 13, 67–68; **14,** 74–75; **15,** 70–72; **16,** 66; **17,** 68; **18,** 85–87; **19,** 67–69; **20,** 73–75

Protection and deprotection. Tetrahydropyranyl ether cleavage is catalyzed by CAN at pH 8.[1] Acetonides are formed from diols under the influence of CAN.[2]

Cyclic acetals are deprotected. In certain cases, the method is superior to that using pyridinium tosylate–acetone.[3]

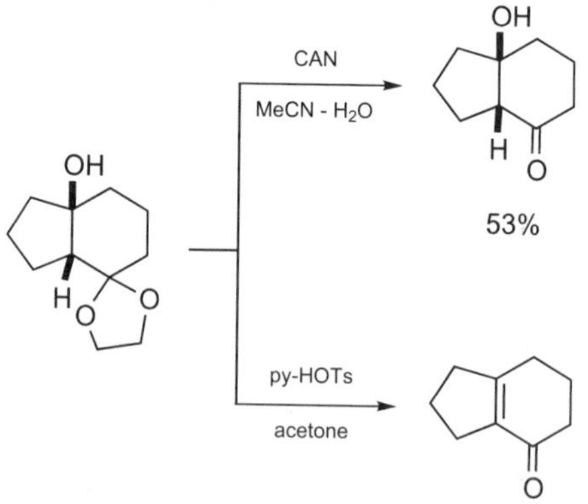

Debenzylation. Tertiary amines containing a benzyl group are converted to secondary amines.[4]

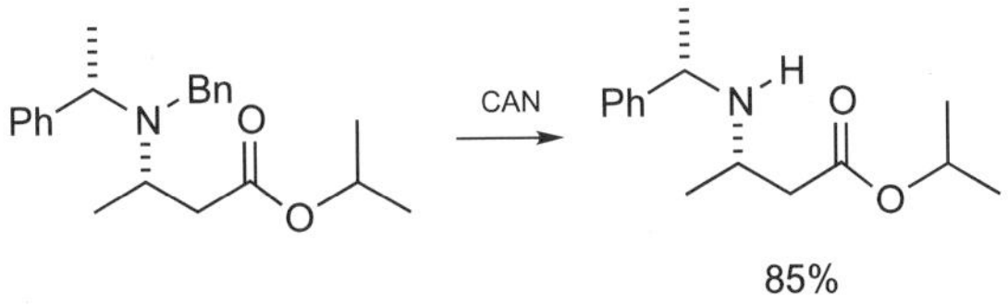

CAN

85%

Nitrile oxide formation.[5] Oxidative formation of these species from oxime derivatives is advantageous. Side reactions can be avoided.

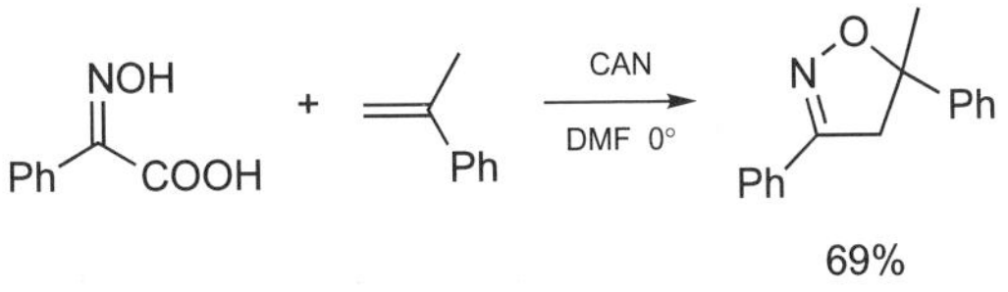

CAN, DMF 0°

69%

Oxidative cyclizations. Cyclopropyl sulfides undergo ring opening with participation of a juxtaposed hydroxyl group leading to a new oxycycle.[6]

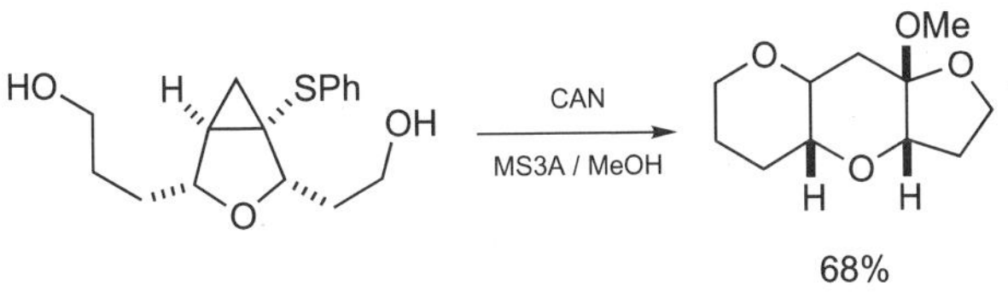

CAN, MS3A / MeOH

68%

3,6-Dihydroxyphthalic esters are obtained when bisenolsilylated β-ketoesters are treated with CAN.[7]

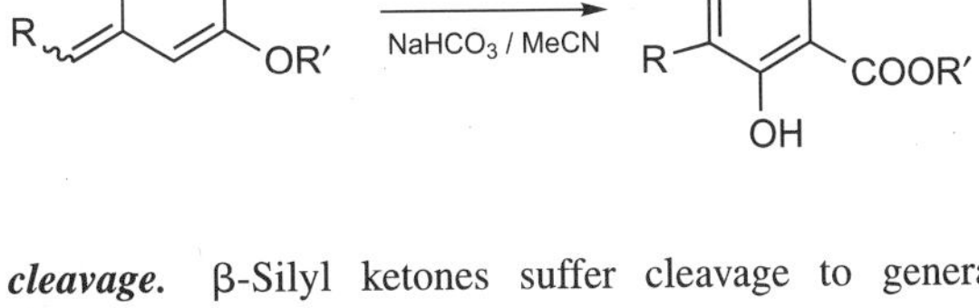

CAN, $NaHCO_3$ / MeCN

Oxidative cleavage. β-Silyl ketones suffer cleavage to generate unsaturated carboxylic acids.[8]

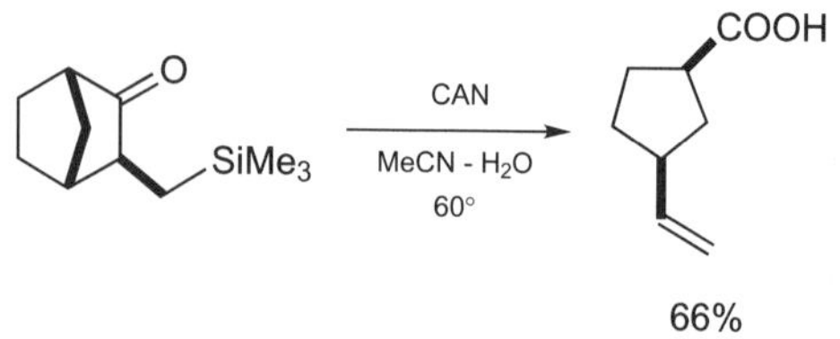

Unusual decomplexation. A highly strained cycloalkyne cannot be produced from its hexacarbonyldicobalt complex on reaction with CAN. Instead, carbonylation at the two nascent *sp*-hybridized carbon sites is observed.[9]

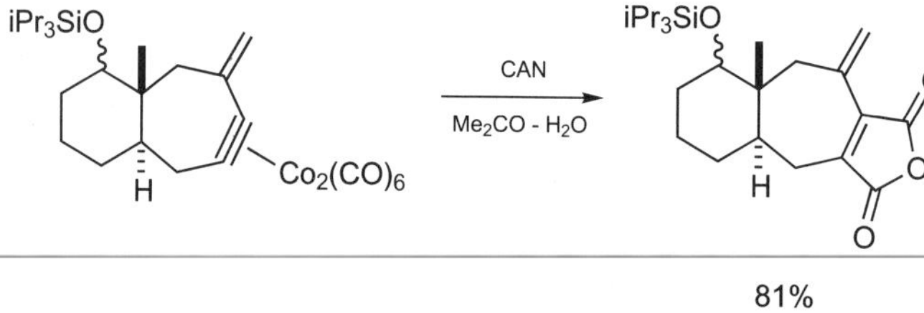

Note that heating the 1-alkyne complexes with amines in toluene alone leads 2-alkenamides.[10]

[1]Marko, I.E., Ates, A., Augostyns, B., Gautier, A., Quesnel, Y., Turet, L., Wiaux, M. *TL* **40**, 5613 (1999).
[2]Manzo, E., Barone, G., Parrilli, M. *SL* 887 (2000).
[3]Ates, A., Gautier, A., Leroy, B., Plancher, J.-M., Quesnel, Y., Marko, I.E. *TL* **40**, 1799 (1999).
[4]Bull, S.D., Davies, S.G., Fenton, G., Mulvaney, A.W., Prasad, R.S., Smith, A.D. *CC* 337 (2000).
[5]Arai, N., Iwakoshi, M., Tanabe, K., Narasaka, K. *BCSJ* **72**, 2277 (1999).
[6]Takemoto, Y., Furuse, S., Hayase, H., Echigo, T., Iwata, C., Tanaka, T., Ibuka, T. *CC* 2515 (1999).
[7]Langer, P., Kohler, V. *CC* 1653 (2000).
[8]Hwu, J.R., Shiao, S.-S., Tsay, S.-C. *JACS* **122**, 5899 (2000).
[9]Tanino, K., Shimizu, T., Miyama, M., Kuwajima, I. *JACS* **122**, 6116 (2000).
[10]Sugihara, T., Okada, Y., Yamaguchi, M., Nishizawa, M. *SL* 768 (1999).

Cerium(III) chloride heptahydrate. 14, 75–77; **15,** 72–73; **16,** 67–68; **18,** 87; **20,** 75

a-Chloro enones.[1] α,β-Epoxy ketones undergo ring opening and dehydration, but changing the solvent system from aq MeOH to MeCN stops the dehydration step. The behavior of 2,3-epoxycyclopentanone differs from that of the cyclohexanone homologue in that the product dehydrates much more readily.

The regioselectivity of epoxide opening mediated by $CeCl_3$ is opposite to that by $TiCl_4$.

[1]Montalban, A.G., Wittenberg, L.-O., McKillop, A. *TL* **40**, 5893 (1999).

Cerium(III) chloride heptahydrate–sodium iodide.

Hydrolysis of p-methoxybenzyl ethers. The cleavage of *p*-methoxybenzyl ethers[1] is performed in refluxing MeCN. Remarkably, alcohols are converted to iodides[2] under essentially the same conditions.

Dehydration.[3] β-Ketols are dehydrated in refluxing acetonitrile to enones by this system.

Dealkylation.[4] Regioselective removal of an alkyl group from an *o*-alkoxy substituent of aromatic carbonyl compounds is by treatment with the combined salts in refluxing MeCN.

[1]Cappa, A., Marcantoni, E., Torregiani, E., Bartoli, G., Bellucci, M.C., Bosco, M., Sambri, L. *JOC* **64**, 5696 (1999).
[2]Di Deo, M., Marcantoni, E., Torregiani, E., Bartoli, G., Bellucci, M.C., Bosco, M., Sambri, L. *JOC* **65**, 2830 (2000).
[3]Bartoli, G., Bellucci, M.C., Petrini, M., Marcantoni, E., Sambri, L., Torregiani, E. *OL* **2**, 1791 (2000).
[4]Yadav, J.S., Reddy, B.V.S., Madan, C., Hashim, S.R. *CL* 738 (2000).

Cerium(III) isopropoxide–diethylzinc.

Pinacol coupling.[1] Aldehydes selectively provide *anti*-pinacols (*anti/syn*~ 9:1 to 98:1).

$$R{-}CHO \xrightarrow[Me_3SiCl\ /\ THF]{(i\text{-}PrO)_3Ce\ -\ Et_2Zn}$$ (syn/anti bis-$OSiMe_3$ diol products, 9–49 : 1)

[1]Groth, U., Jeske, M. *ACIEE* **39**, 574 (2000).

Cerium(IV) sulfate-iodine.

Ring cleavage.[1] 2-Alkylcycloalkanones are cleaved to give keto esters.

$$\xrightarrow[EtOH]{Ce(SO_4)_2.4H_2O\ -\ I_2}$$ COOEt 61%

[1]He, L., Kanamori, M., Horiuchi, C.A. *JCR(S)* 122 (1999).

Cerium(IV) triflate. 20, 75

Iodoetherification.[1] $Ce(OTf)_4$ is a catalyst for functionalization of alkenes (e.g., with iodine in MeOH).

[1]Iranpoor, N., Shekarriz, M. *T* **56**, 5209 (2000).

Cesium carbonate. 13, 70; 14, 77–78; 15, 73–75; 18, 87–88; 19, 70; 20, 76

Aryl ethers.[1] With Cs_2CO_3 as base, aryl mesylates act as phenolate ions because desulfonylation occurs in situ.

OMs, Br + F, NO_2 $\xrightarrow[DMSO\ 80°]{Cs_2CO_3}$ O, NO_2, Br 95%

O-Alkylations. Carboxylic acids undergo *O*-alkylation in DMF at room temperature.[2] Replacement of the chlorine atoms of the Merrifield resin by carbonate or carbamate groups is efficiently accomplished by reaction with alcohols or amines and carbon dioxide in the presence of Cs_2CO_3 and tetrabutylammonium iodide.[3] There are related reports on etherification[4] and mixed carbonate formation.[5]

Hydroquinolines.[6] Reactive heterodienes are generated from *N*-mesyl-2-halomethylanilines by treatment with Cs_2CO_3 at low temperature. Interception of heterodienes readily affords adducts.

Cs_2CO_3, CH_2Cl_2 –78°; OEt; NH, Ms; N, Ms; 78%

Pyridines.[7] The intramolecular Diels–Alder adducts of alkadienoic esters of α-cyanooximies aromatize by a double elimination process on exposure to Cs_2CO_3 in DMF at room temperature.

X = CN, COOEt; Cs_2CO_3, DMF; X = CN, R = H 67%

Alkylation.[8] A modification described for the synthesis of amino acid derivatives via alkylation of the *N*-diphenylmethyleneglycine ester consists of the use of poly-(ethylene glycol) as solvent, promotion by microwave irradiation, and linking the substrate onto a polymer support (PEG ester).

[1]Dinsmore, C.J., Zartman, C.B. *TL* **40**, 3989 (1999).
[2]Parrish, J.P., Dueno, E.E., Kim, S.-I., Jung, K.W. *SC* **30**, 2687 (2000).
[3]Salvatore, R.N., Flanders, V.L., Ha, D., Jung, K.W. *OL* **2**, 2797 (2000).
[4]Dueno, E.E., Chu, F., Kim, S.I., Jung, K.W. *TL* **40**, 1843 (1999).
[5]Chu, F., Dueno, E.E., Jung, K.W. *TL* **40**, 1847 (1999).
[6]Steinhagen, H., Corey, E.J. *ACIEE* **38**, 1928 (1999).
[7]Bland, D.C., Raudenbush, B.C., Weinreb, S.M. *OL* **2**, 4007 (2000).
[8]Sauvagnat, B., Lamaty, F., Lazaro, R., Martinez, J. *TL* **41**, 6371 (2000).

Cesium fluoride. 13, 68; **14,** 79; **15,** 75–76; **16,** 69–70; **17,** 68; **18,** 88–89; **19,** 70–72; **20,** 77–78

Disulfides.[1] Organoboranes are sulfurized with sulfur in the presence of CsF. Organoborates are more reactive.

2-Cyanomethyl-1,2-dihydroquinolines.[2] A combination of Me_3SiCH_2CN and CsF conveniently effects alkylation at C-2 of quinolinium salts.

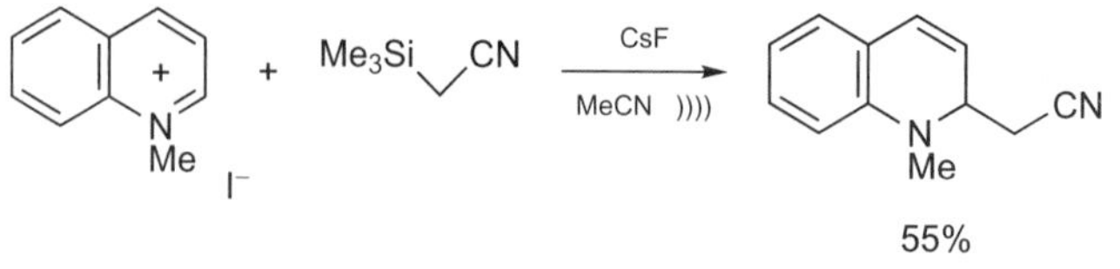

Annulation.[3] *trans*-3,4-Disubstituted tetrahydrothiophenes are obtained from conjugated carbonyl compounds and chloromethyl trimethylsilylmethyl sulfide in a reaction mediated by CsF. When the conjugated carbonyl component bears a chiral auxiliary, the process is amenable to synthesis of enantiopure derivatives.

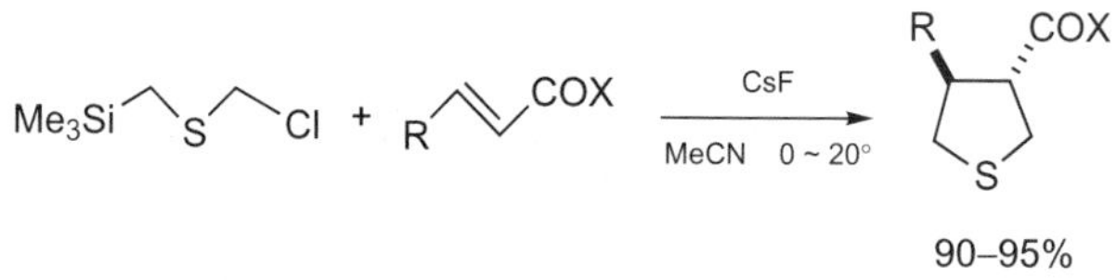

Organotin halide cleanup.[4] Residual tin halides are a nuissance of organic reactions that generate them as coproducts. A method using a 2:1 mixture of CsF and CsOH on silica to scavenge the tin halides (from dichloromethane or THF solutions) is effective.

[1]Kerverdo, S., Gingras, M. *TL* **41**, 6053 (2000).
[2]Diaba, F., Le Houerou, C., Grignon-Dubois, M., Gerval, P. *JOC* **65**, 907 (2000).
[3]Karlsson, S., Hogberg, H.-E. *OL* **1**, 1667 (1999).
[4]Edelson, B.S., Stoltz, B.M., Corey, E.J. *TL* **40**, 6729 (2000).

Cesium hydroxide.

Alkynylation. This base (monohydrate) is an excellent catalyst for alkynylation of carbonyl compounds[1] and selective *N*-monoalkylation of primary amines.[2] Cesium hydroxide suppresses overalkylation, and therefore is very useful in the exclusive formation of secondary amines from amino acids.

Alkylations. An ether synthesis[3] from alcohols and alkyl bromides is promoted by CsOH and Bu_4NI in the presence of molecular sieves 4Å in DMF at room temperature.

Carbanion generation.[4] A 1:1 mixture of CsOH and CsF suspended in dichloromethane or THF is very effective for generation of carbanions from 1-trimethylsilylalkynes, 2-trimethylsilyl-1,3-dithiane, trifluoromethyltrimethylsilane, and silyl enol ethers, all through desilylation.

[1]Tzalis, D., Knochel, P. *ACIEE* **38**, 1463 (1999).
[2]Salvatore, R.N., Nagle, A.S., Schmidt, S.E., Jung, K.W. *OL* **1**, 1893 (1999).
[3]Dueno, E.E., Chu, F., Kim, S.-I., Jung, K.W. *TL* **40**, 1843 (1999).
[4]Busch-Petersen, J., Bo, Y., Corey, E.J. *TL* **40**, 2065 (1999).

Chiral auxiliaries and catalysts. 18, 89–97; **19,** 72–93; **20,** 78–103

Chiral derivatizing agents. (*S*)-*O*-acetyllactyl chloride is prepared from lactic acid by treatment with AcCl and the $SOCl_2$.[1] Unsymmetrical 1,2-bis(phosphanyl)ethanes are obtained from chiral diols via displacement of the cyclic sulfates using R_2PLi in two stages.[2] C_2-symmetric bis-sulfoxides are accessible from reaction of bis-sulfinyl chlorides with diacetone-D-glucose and then Grignard reagents.[3] Useful ligands can be prepared from chiral *N,N*-dibenzyl α-aminoketones by Grignard reactions and reductive amination.[4] QUIPHOS (**1**) is readily prepared in large scale from L-glutamic acid.[5]

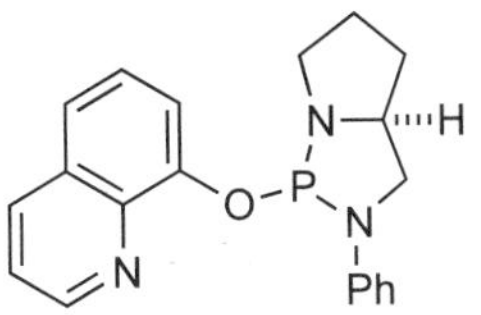

(**1**)

Kinetic resolutions. Baylis–Hillman adducts are deracemized by exploiting their reactivity toward Pd(0)-catalyzed substitution, using chiral ligand **2**.[6] Both the planar chiral DMAP derivative **3**[7–9] and the axially chiral analogue (**4**)[10] and **5**[11] have been developed as catalysts for enantioselective acylation. Benzylic alcohols undergo enantioselective acylation with the aid of **6**.[12] Methanolysis of *meso*-anhydrides in the presence of a cinchona alkaloids is a good way to desymmetrize such compounds.[13,14]

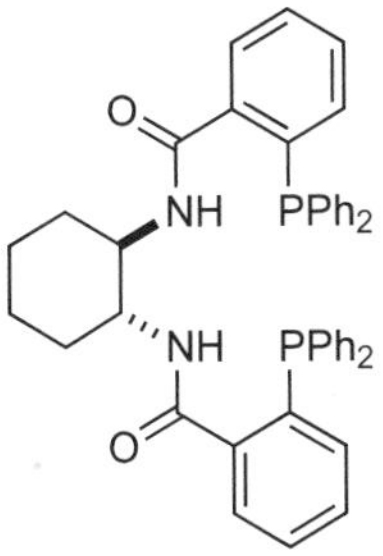

(**2**) 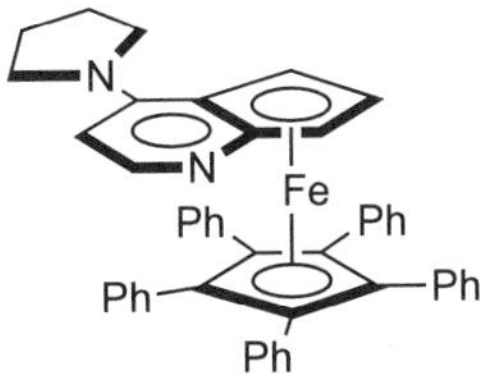 (**3**)

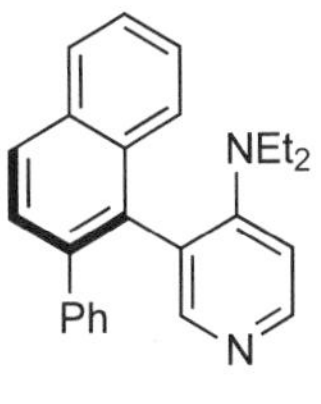

(**4**) 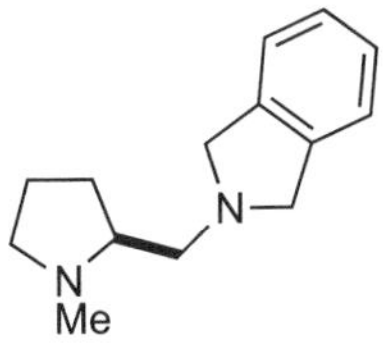 (**5**) 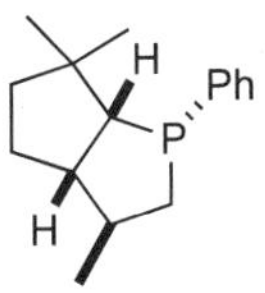 (**6**)

Halogenations. Fluorination of ketones can make use of **7**,[15] although the enantioselectivity is only moderate. Of the active methylene compounds, fluorination in the presence of dihydroquinine esters is adequate.[16] A synthesis of β-amino-α-hydroxy acids involves iodination of substrates such as **8**, which is attended by spontaneous cyclization.[17]

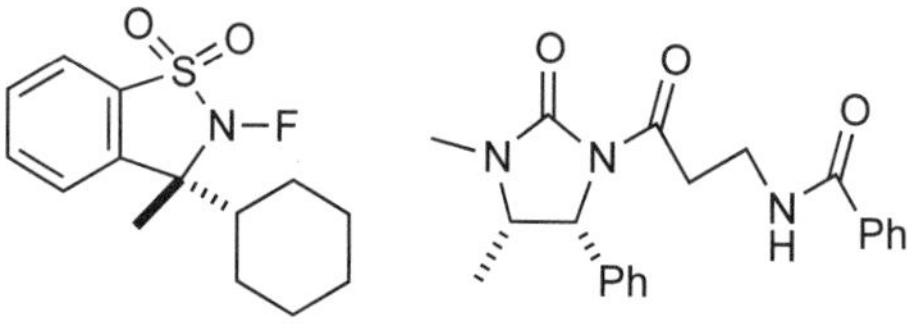

(**7**) (**8**)

Alkylations. Ketone alkylation has been mediated by **9**,[18] and the process also exhibits excellent regioselectivity in favor of the more highly substituted position. Using allyl esters as electrophiles and a Pd(0)-catalyst, ligand *ent*-**2**[19] furnishes suitable support (a similar alkylation of azlactones employs ligand **2**[20]).

The enantioselective alkylation of *N*-protected α-amino esters has been studied with many chiral catalysts, including spirocyclic ammonium salt **10A**,[21] while (**10B**) containing two binaphthyl components is an effective mediator for alkylation of protected glycine under phase-transfer conditions.[22] β-*t*-Boc-amino acid derivatized with (+)-pseudoephedrine enables enantioselective alkylation of the ensuing amides.[23] Note the enolate derived from **11** remains chiral, alkylation products are produced in high ee.[24]

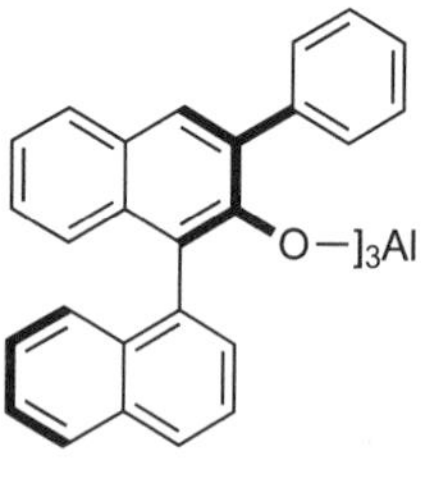

(**9**) 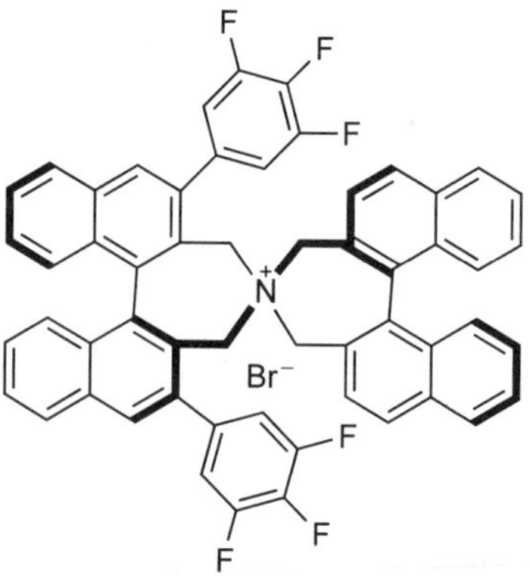 (**10A**)

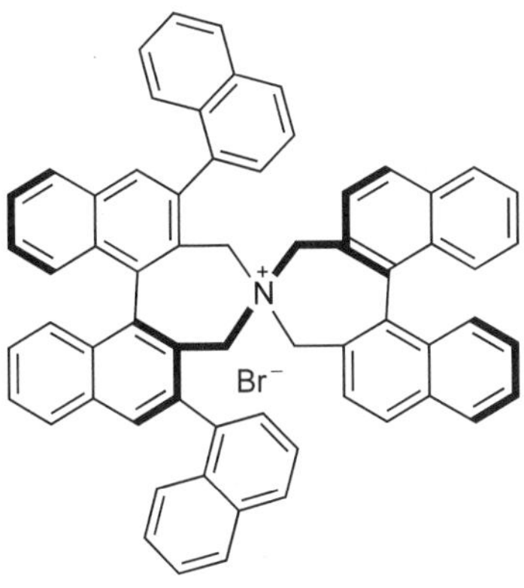

(10B)

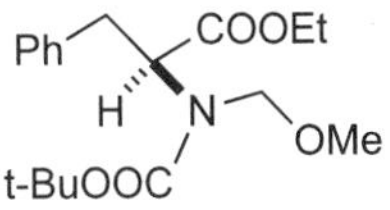

(11)

Esters of a chiral *trans*-2-(2-naphthalenesulfonyl)cyclohexanol undergo alkylation with excellent diastereofacial (hence enantiomeric) selectivity.[25] Glycolic acid is transformed into chiral homologues via formation of 1,3-dioxolanone with *N,N*-diisopropyl-10-camphorsulfonamide and subsequent alkylation.[26]

The alkylation products of SAMP-hydrazones with *N*-tosylaziridine can be convertied to chiral γ-amino nitriles and ketones.[27] 4,5,5-Trisubstituted oxazolidinones such as **12** have been studied in the context of their capacity for inducing enantioselectivity during alkylation of their *N*-acyl[28a] and *N*-methylthiomethyl derivatives.[28b]

Under phase-transfer conditions a quaternized cinchona alkaloid is used in the Darzens condensation.[29]

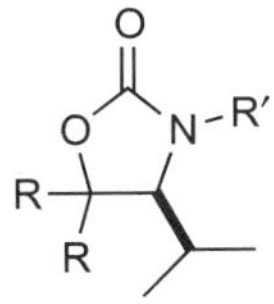

(12)

R = Me, R′ = COEt
R = Ph, R′ = CH_2Me

Displacements involving allylic systems. Allylic substitution reactions continue to receive considerable attention. Improvement by microwave is noted.[30] New ligands for the Pd-catalyzed reaction include iminophosphine **13**, which is derived from 1-mesitylethylamine,[31] pyridylphenylphosphine **14**,[32] and the phosphinite **15**, obtained from D-glucosamine.[33] Phospholanes such as **16**, **17**, **18**, prepared from mannitol, are excellent ligands.[34]

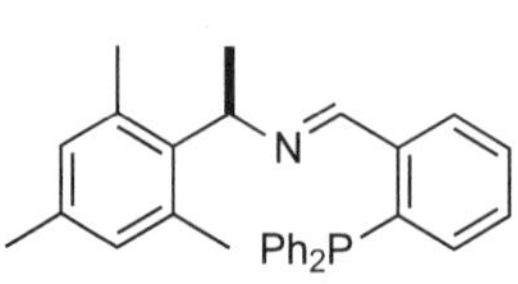

(13) 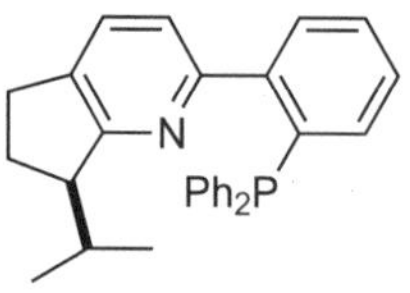 (14) 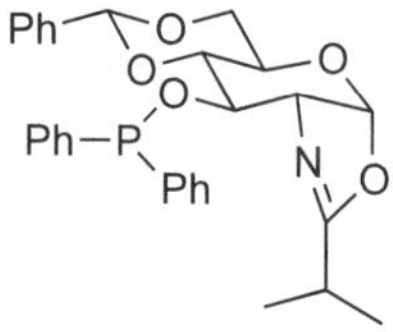 (15)

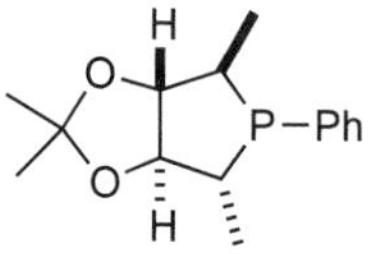

(16) 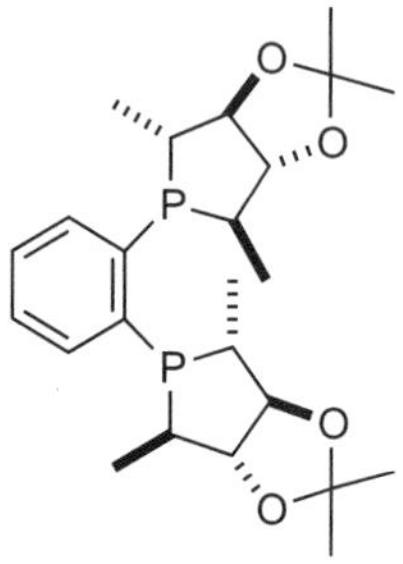 (17) 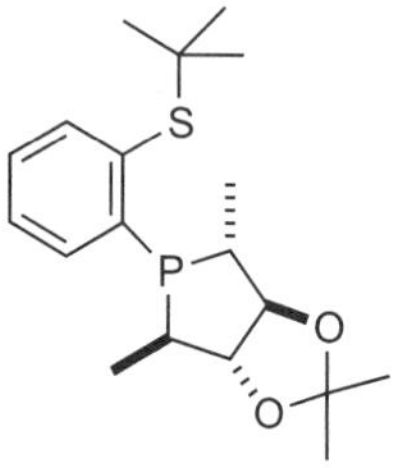 (18)

There is a remarkable reversal of enantioselectivity for reactions involving a bis(oxazoline) ligand **19** by merely changing the oxygen arms.[35] (4-Diphenylphosphinoethyl) oxazolines **20** represent a second generation of these class of ligands, being superior to the lower homologues (the phosphinomethyl compounds) and as effective as the well-known benzologue.[36] Placing the phosphorus atom at a bridgehead of the norbornadiene skeleton creates relatively rigid phosphinooxazolines (**21**).[37] The planar chirality of the modified ferrocene moiety in **22** is responsible for the stereochemical course of the allylic substitution, so in these cases the stereocenter in the oxazoline ring plays at most a minor role.[38] With ligand **23** derived from β-pinene, the reaction of cinnamyl 2-(benzenesulfonyl)vinyl ethers shows that vinylogous sulfonates are viable substrates.[39] A study on the acyclic *P,N*-ligands **24** manifests the importance of substituent effects (best with R = NMe).[40]

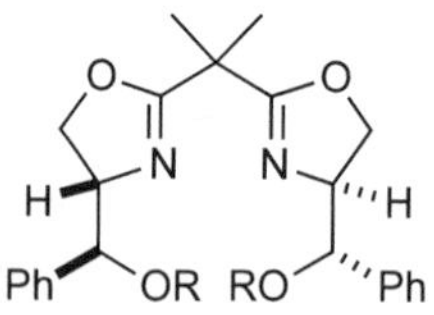

R = H, Bz

(19) (20) 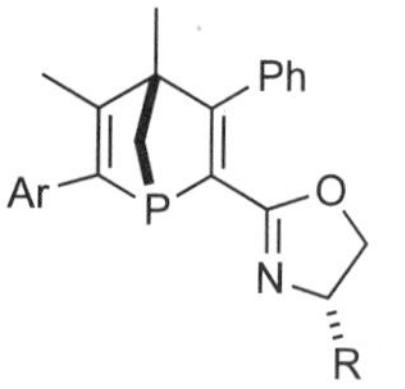 (21)

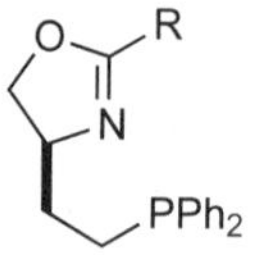

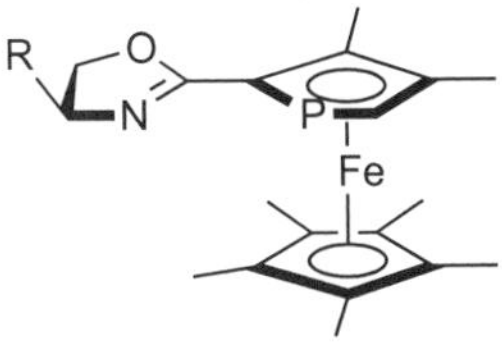

(22) 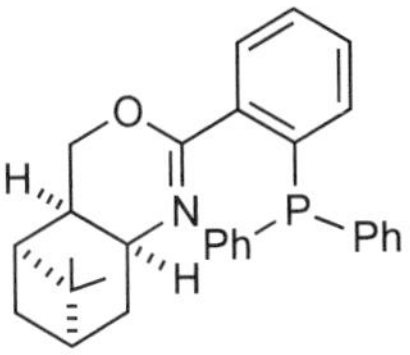 (23) 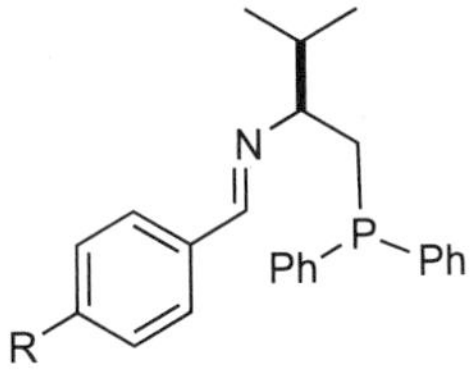 (24)

Metallocene-catalyzed ring opening of 7-oxabicyclo[2.2.1]heptenes with alcohols[41] and organometallic reagents[42] proceeds with very high ee. Novel *P,S*-ligands (e.g., **25**), based on the planar chiral ferrocenyl unit,[43] render good electronic control due to the presence of both the π-acceptor phosphorus group and the donor thioether. Another recently synthesized *P,S*-ligand is **26**.[44]

Chiral ligands related to BINAP and used in the Pd-catalyzed allylic displacement context are **27**[45] and **28**.[46] The aldimine representatives of the latter series are not effective.

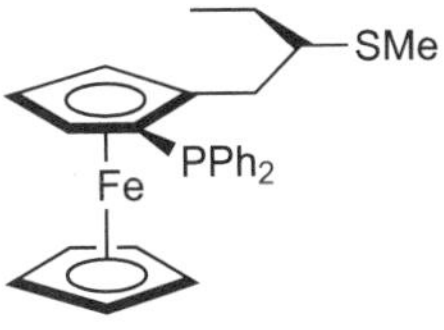

(25) 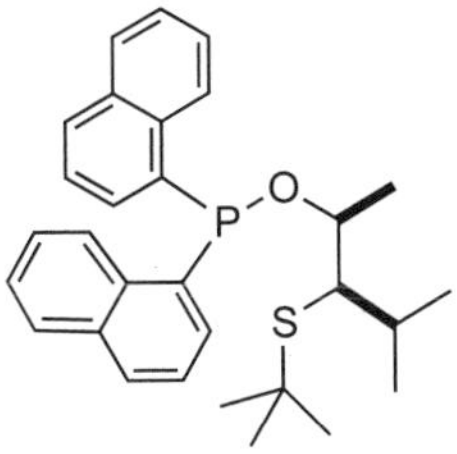 (26)

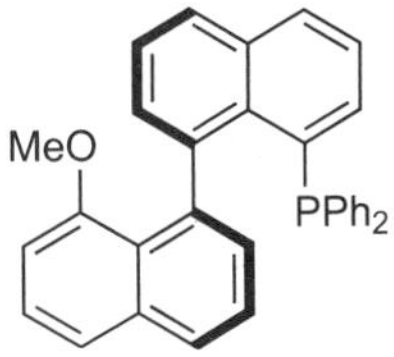

(27) 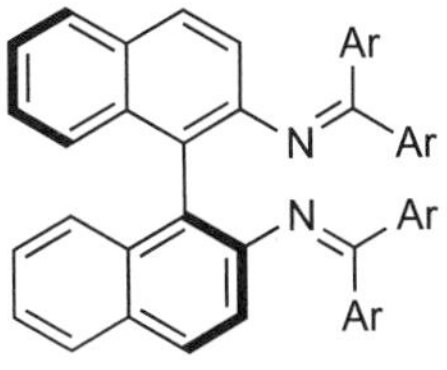 (28)

γ-(*t*-Butoxycarbonyloxy)butenolide is readily transformed into aryloxy analogues with chirality. In a formal synthesis of aflatoxin B, such a reaction precedes the elaboration of the furanobenzofuran subunit.[47] Allylic acylals also afford chiral sulfones in the reaction with $PhSO_2Na$, the products are "chiral aldehyde" equivalents.[48] The Mo(0)-catalyzed reaction leading to branched products with excellent ee is catalyzed by *trans-N,N'*-dipicolinoyl-1,2-cycloheanediamine[49] and the bis(oxazolin-4-carbonyl) analogue **29**.[50] It also has proven to be valuable in the preparation of 3-substituted 1,4,6-alkatrienes from 2,4,6-alkatrienyl carbonates. By using ligand **30**, products are obtained in good ee.[51]

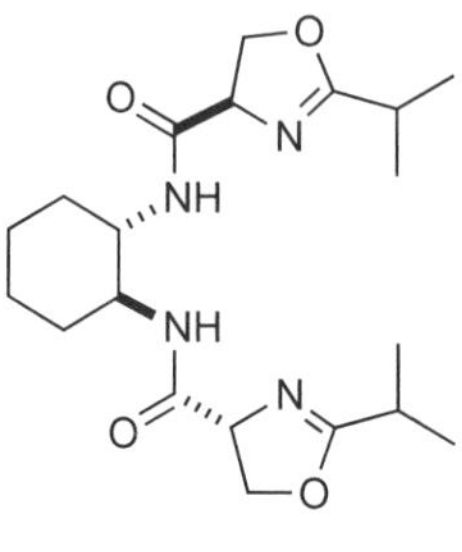

(29) 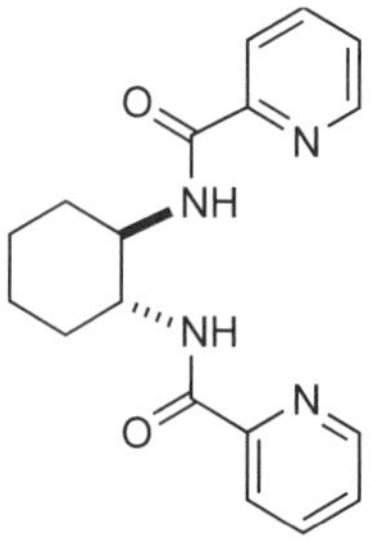 **(30)**

Opening of epoxides and aziridines. An observation that concerns the enantioselective opening of *meso*-epoxides by $SiCl_4$ in the presence of **31** indicates a beneficial effect of the *o*-methoxy group to form an octahedral complex.[52] β-Titanoxy radical intermediates are involved when using **32** to open epoxides.[53] A chiral Cr–salen complex has been used to mediate epoxide opening with KHF_2. Although the ee values are moderate, this is the first report of fluorohydrin formation by such a method.[54]

2,2-Disubstituted epoxides can be resolved on reaction with Me_3SiN_3 in the presence of a chiral (salen)CrN_3 complex.[55] Good enantioselectivity is associated with the enantioselective opening of *meso*-epoxides with Me_3SiCN mediated by (pybox)$YbCl_3$.[56] Simultaneous activation of both epoxide and cyanide is indicated.

Chiral vinylglycidols containing a quaternary carbon center are obtained from racemic vinyl epoxides via ring opening with *p*-methoxybenzyl alcohol. One such product has been used in a synthesis of (−)-malyngolide.[57]

Proximal Co–salen complexes that are part of a dendrimer show cooperative reactivity in catalyzing epoxide ring opening with water.[58a] Chromium-complex **33** is the optimal catalyst for converting *meso*-aziridines to chiral β-amino azides on reaction with Me_3SiN_3.[58b]

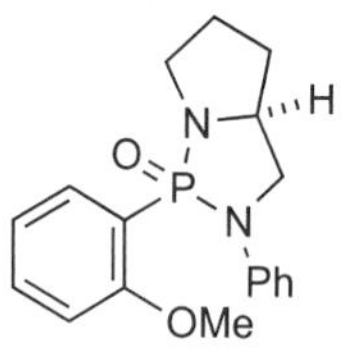

(31) 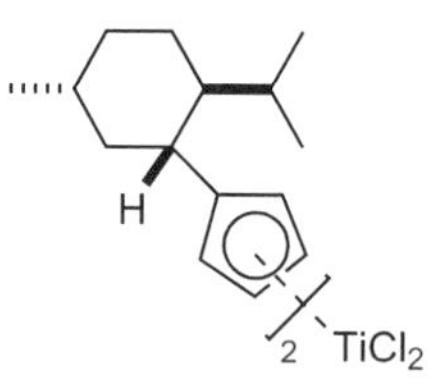 **(32)** 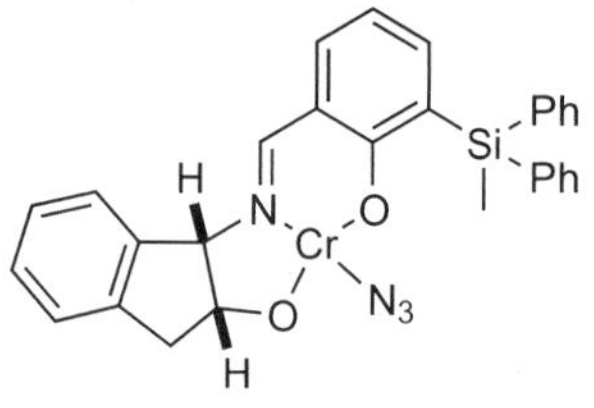 **(33)**

Addition to C═O. Several protocols have been developed for the preparation of chiral cyanohydrins or their trimethylsilyl ethers. Catalytic systems including vanadyl–salen complex,[59] (pybox)$YbCl_3$,[60] zirconium–TADDOLate,[61] and the **34**-titanate[62] are effective.

Variable enantioselectivities are observed during allylation of aldehydes in the presence of ligand **35**.[63] β-Hydroxyaldehydes can be allylated in the unprotected form with a chiral allyltitanium species **36**,[64] whereas α-hydroxyaldehydes are readily prepared from **37** by various organometallic reactions.[65] Cinchona alkaloids, being commercially available and inexpensive, are suitable chiral promoters for the indium-mediated allylation, even though they induce only moderate enantioselectivities.[66] Allylation such as that effected by **38** affords products useful for substrate-based alkylation.[67]

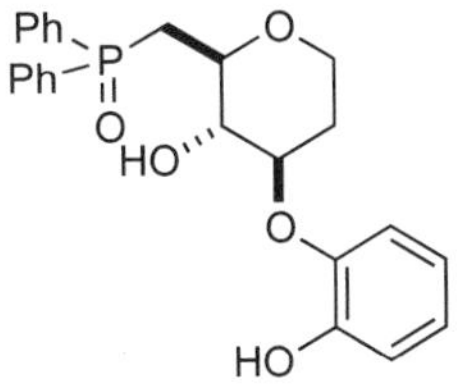

(**34**) 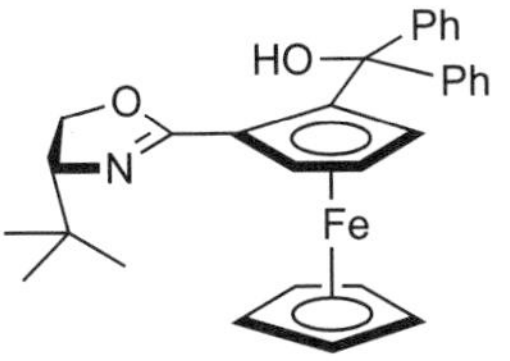 (**35**) 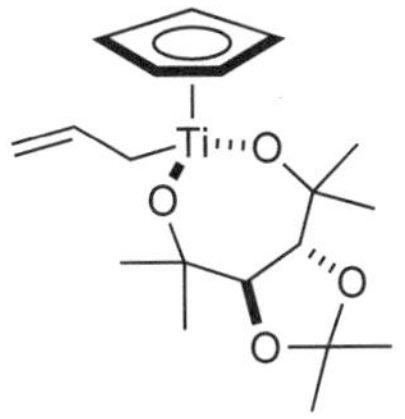 (**36**)

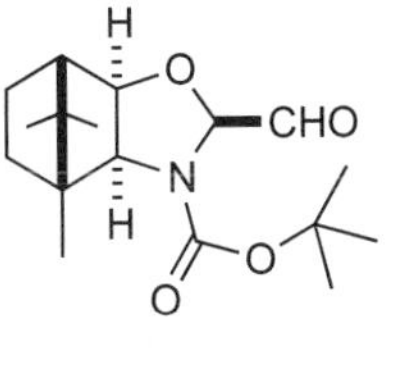

(**37**) 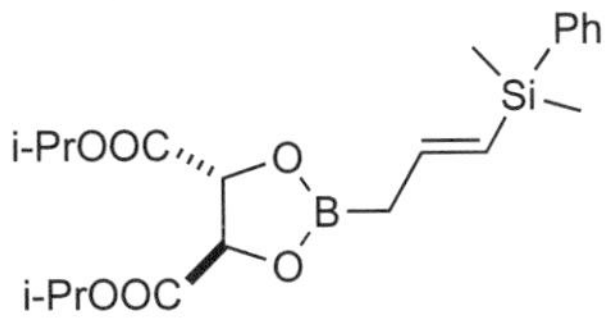 (**38**)

The chiral sulfoxide **39** is valuable for assemblage of γ-butenolides on lithiation and reaction with aldehydes (low ee with ketones).[68] The γ-selective alkylation of **40** with aldehydes has certain synthetic use.[69]

1,1-Bis(borylmethyl)ethylenes such as **41** are useful assembling platforms that allows bidirectional synthesis of 1,5-diols.[70] Diastereoselectivity and enantioselectivity are usually high.

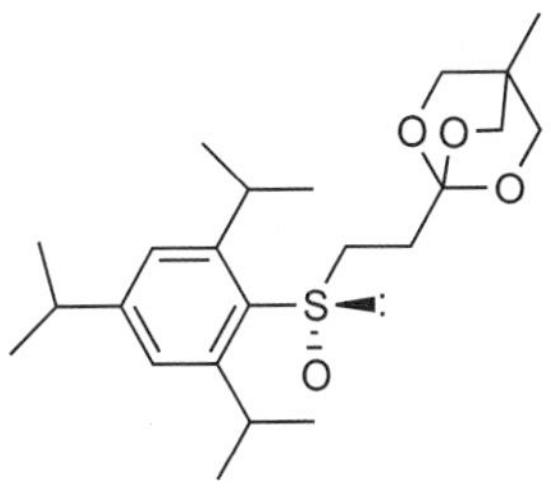

(**39**) 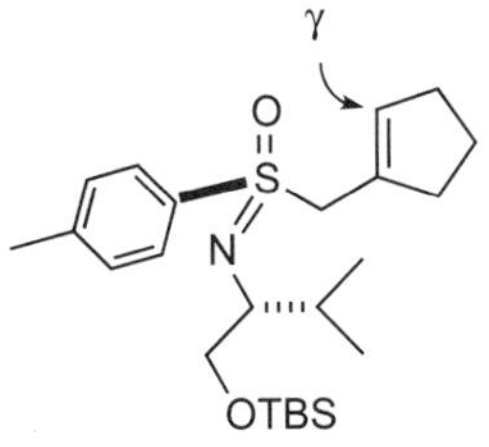 (**40**)

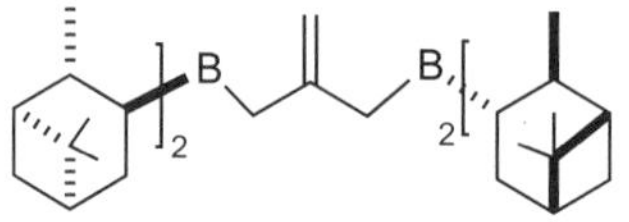

(**41**)

A direct synthesis of chiral propargylic alcohols from 1-alkynes and aldehydes in the presence of $Zn(OTf)_2$, Et_3N, and (+)-*N*-methylephedrine has a broad scope.[71] Several new ligands are found suitable for inducing asymmetric addition of R_2Zn (mostly diethylzinc) to aldehydes. These include **42**,[72] **43**,[73] **44**,[74] **45**,[75] and **46**.[76] Other β-amino alcohols that show desirable features are 3-*exo*-morpholinoisoborneol,[77] which is more stable in air than the dimethylamino analogue, (*S*)-2-(pyrrolidin-1-yl)-1,2,2-triphenylethanol,[78] and a polymer-supported *N*-alkyl-α,α-diphenyl-L-prolinols.[79] *N,N*-Dibutylnorephedrine is useful in a solvent-free reaction.[80]

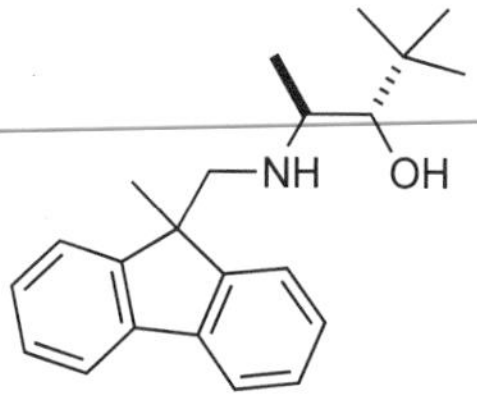

(**42**) 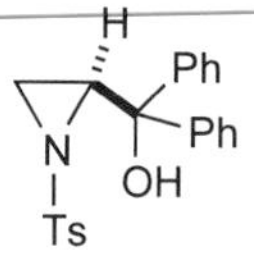 (**43**) 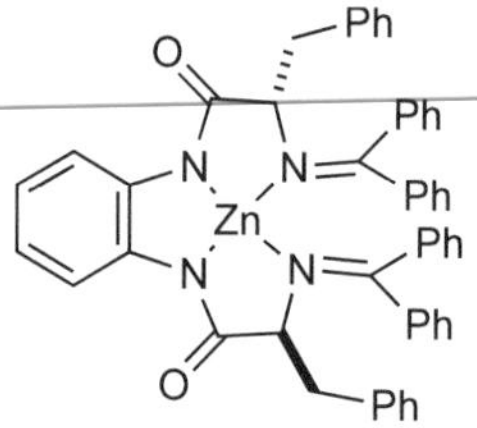 (**44**)

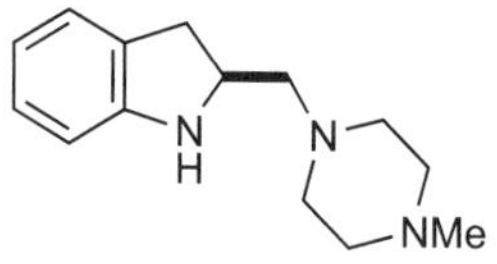

(**45**) 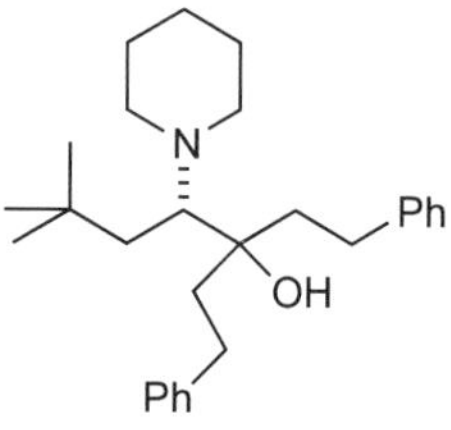 (**46**)

Practically perfect asymmetric autocatalysts of the (2-alkynyl-5-pyrimidyl)alkanol series have been identified.[81] A *t*-butylalkynyl residue fulfills the role of proper bulkiness and moderate electron-withdrawing power. Quartz also induces enantioselective addition of *i*-Pr_2Zn to this heterocyclic aldehyde,[82] by virtue of its morphological chirality and acidity, which enable differentiation of the enantiofaces of the aldehyde upon coordination with the oxygen and nitrogen atoms. Chiral sodium chlorate crystals have the same effect.[82a]

Phosphoramide ligands represented by **47** are valuable for Lewis base-catalyzed allylation and aldol reaction. Their preparations are detailed.[83]

Aldol reaction with L-proline as catalyst[84] has been extended to α-ketols thereby generating *anti*-diols.[85] Transition state **48** is consistent with the results of aldol condensation catalyzed by Et_2Zn–Ph_3PS in the presence of a bisprolinol.[86] α-Amino acid derived imidazolidinones[87] serve as chiral auxiliaries in the same manner as the corresponding oxazolidinones. In employing 4-*t*-butylthiazolidin-2-thione,[88] the presence of one or two equiv of a base leads to *syn* products of opposite enantiomeric series.

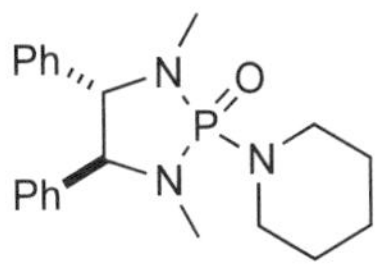

(47) 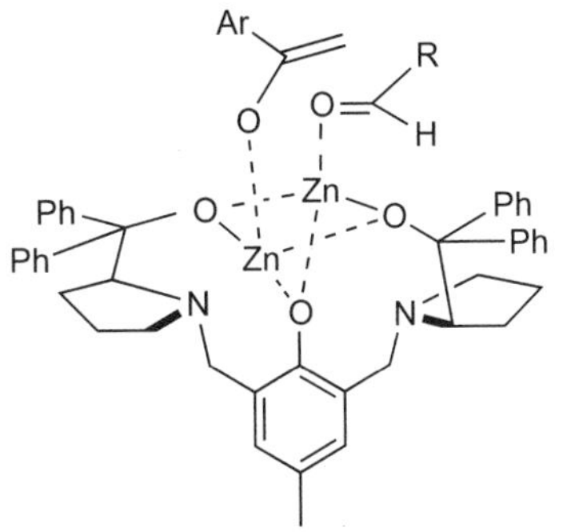 **(48)**

Aldol-type reaction of the trimethylsilyl–t-butyl keteneacetal of *N*-diphenylmethyleneglycine with aldehydes in the presence of an *N*-anthracen-9-ylmethyl-*O*-benzylcinchonidinium salt leads predominantly to *(2S,3R)*-β-hydroxy-α-amino esters.[89]

A prior derivatization of glycolic acid into chiral 5,6-diphenyl-1,4-dioxan-2-one paves the way to *anti*-2,3-dihydroxycarboxylic acids by an aldol-type process.[90]

Chiral δ-lactones are formed in the condensation of keteneacetals with α-diketones (2:1 stoichiometry) in the presence of the bis(oxazoline)–$Cu(OTf)_2$ complex **49**.[91] Aldol adducts derived from α-bromoenolates are precursors of chiral α,β-epoxyalkanoic esters.[92,93]

From a $TiCl_4$-catalyzed Baylis–Hillman reaction in the presence of 10-methylthioisoborneol, chiral adducts ensue.[94] The acryloyl derivative **50** is a particularly interesting addend because either (*R*)- or (*S*)-alcohols can be prepared by merely changing the solvent.[95] Another variation employs hexafluoroisopropyl acrylate and ligand **51**.[96]

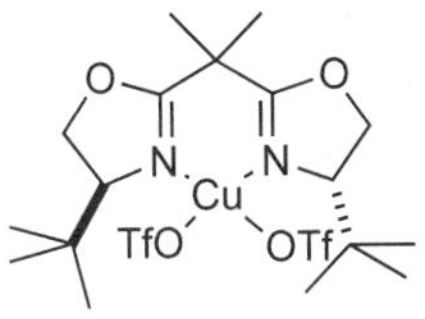

(49) 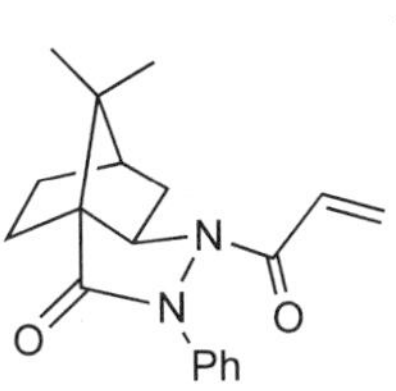 **(50)** 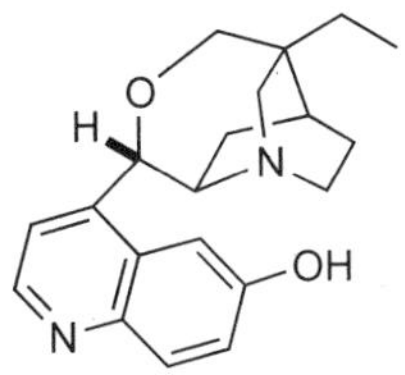 **(51)**

The Cu(II)-catalyzed reaction of silyl enol ethers with oxomalonic esters in the presence of a bis(oxazoline) ligand constitutes the first step of an access to chiral β-hydroxy acids.[97] Enantioselective Mukaiyama aldol reaction performed in the presence of **52**,[98] and that in aqueous ethanol has been accomplished to a certain degree of success (32–85% ee).[99]

Excellent diastereoselectivity (*anti*-selective) and enantioselectivity are observed in the reaction between phenylthioacetic esters and aldehydes. A *B*-bromodiazaborolidine (**53**) mediates the condensation.[100] (−)-Sparteine is useful in a synthesis of α-substituted serines from ethyl 2,5-diethoxy-3,6-dihydropyridine-3-carboxylate on reaction with aldehydes.[101]

A catalytic asymmetric cyanosilylation of carbonyl compounds with Me_3SiCN using either a carbohydrate-based phosphine oxide (**54**)[102] or the monolithium salt of a chiral salen ligand[103] has been studied. Comparing to the BINOL analogues, the reaction involving **54** does not require phosphine oxide additives to attain high levels of asymmetric induction, the catalytic activity is higher, and Me_3SiCN can be introduced rapidly.

Homoenolate species derived from α-substituted acrylic esters of isosorbide (via SmI_2 reduction) can be used to genrate chiral γ-butyrolactones.[104]

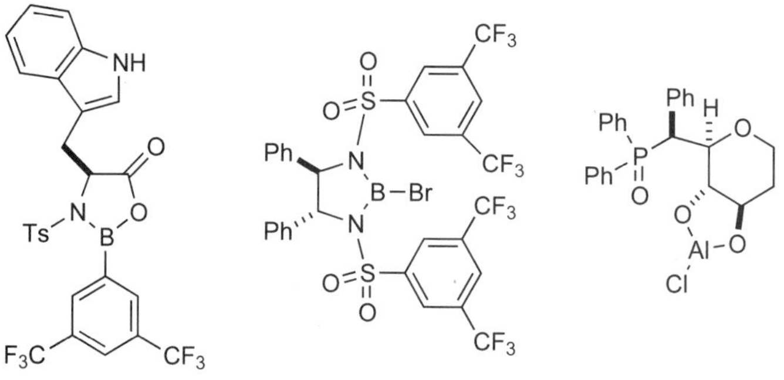

(**52**) (**53**) (**54**)

Addition to C=N. Substrate-based synthesis of chiral amines often relies on the availability of enantiopure imine derivatives. An improved preparation of chiral sulfinimine building blocks employs titanium(IV) ethoxide to promote condensation of chiral $TolS(O)NH_2$ with carbonyl compounds.[105] An asymmetric synthesis of β-amino esters involves reaction of chiral sulfinimines in which the sulfinyl group is linked to C-10 of an isobornyl system.[106] Generation of a dehydroglycinamide from 2,10-camphorsultam is readily achieved by treatment with Me_3Al and the *O*-benzyloxime derivative of methyl glyoxylate. The dehydroglycinamide is receptive to free radicals with high diastereofacial discrimination.[107]

Chiral Mannich reaction is also successful between *endo*-2-acetylisobornyl trimethylsilyl ether and *N*-alkylidenecarbamate precursors,[108] and with aldimines activated by (*S*)-*N,N*-phthaloyl-*t*-leucine chloride.[109]

During addition of Et_2Zn to *N*-sulfonylimines[110] and *N*-phosphonylimines,[111] asymmetric induction is rendered by **55** and **56**, respectively. *N*-Sulfonylimines also accept the aryl group from $ArSnMe_3$ asymmetrically in the presence of **57**.[112]

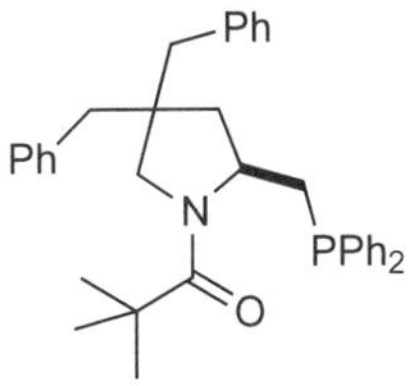

(**55**) 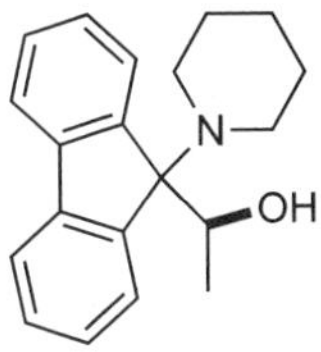 (**56**) 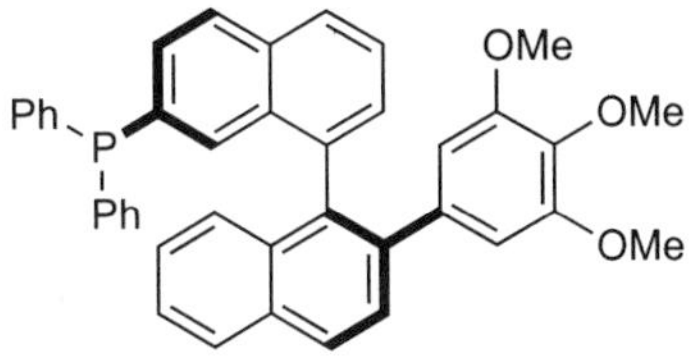 (**57**)

The addition of organolithiums to imines derived from an optically active α-naphthylethylamine forms mainly one diastereomer.[113] Hydrazones **58** undergo radical addition, therefrom chiral amines are available.[114]

α-Sulfinyl carbanions also add to chiral sulfinimines stereoselectively.[115] Allyldiisopinocampheylborane is an allyl transfer agent reactive toward *N*-silylaldimines,[116] whereas high-leveled 1,6-asymmetric induction is observed during the addition of (*R,R*)-4,5-dicyclohexyl-2-vinyl-1,3-dioxolane to aldimines.[117]

Strecker reaction to establish a new stereocenter is subject to asymmetric induction, capable of creating either a tertiary[118] or quaternary carbon atom[119] in the presence of **59**. The peptido-imine **60** proves to be an excellent ligand for the Ti(IV)-mediated cyanation of aldimines.[120] On catalysis of the bicyclic guanidine **61** the addition of HCN to *N*-benzhydrylaldimines affords α-amino nitrile derivatives with moderate to good ee.[121]

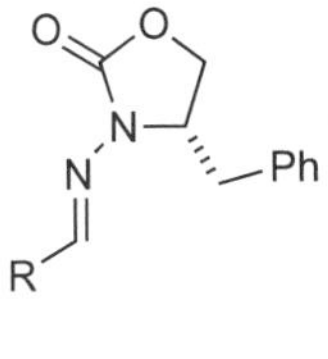

(**58**) 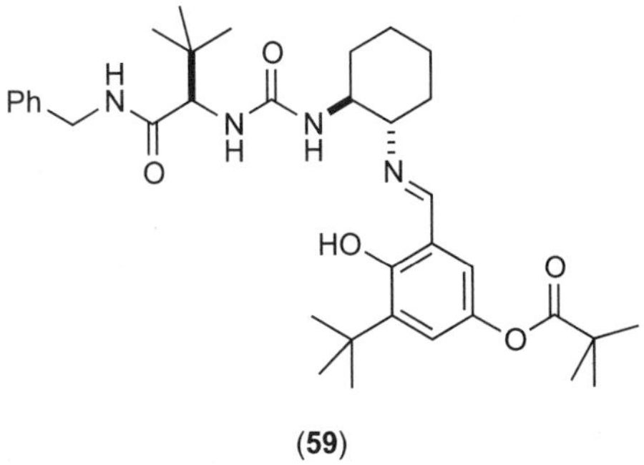 (**59**)

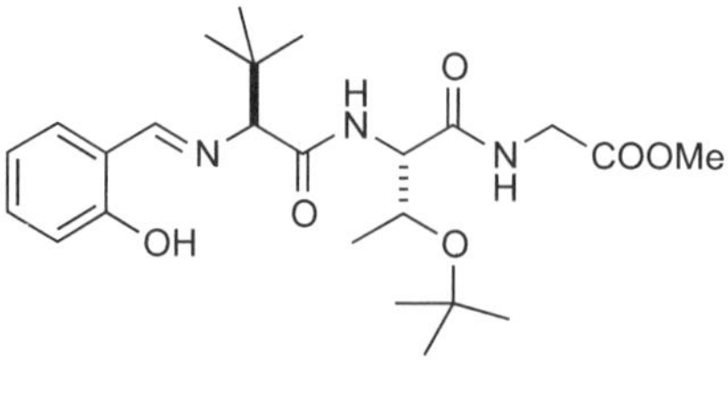

(60) 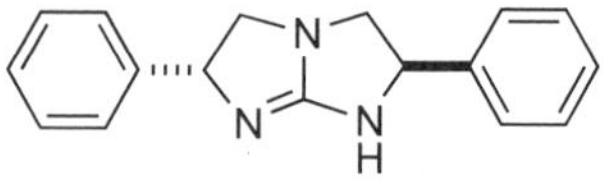 (61)

It is interesting to note that *anti*-selectivity of aldolization (with L-proline promotion) involving hydroxyacetone as the donor is switched in the Mannich reaction. Thus *syn*-2-hydroxy-3-amino ketones are obtained as major products.[122]

α-Branched amines can be synthesized from aldimines or ketimines. A chiral version involving hydrosilylation of *N*-aryl ketimines with PMHS adopts ethylenebis(η^3-tetrahydroindenyl)titanium difluoride as a precatalyst,[123] diarylmethylamines

Addition to C═C. Chiral α-arylpropanoic esters are generated by the union of arylketenes with alcohols in the presence of **62**.[124] An arylselenomethoxylating agent that depends on an *o*-(α-methylthioethyl group to exert its enantioselectivity is prepared in situ from the diaryl diselenide by treatment with bromine and AgOTf.[125] Another chiral reagent is *endo*-3-bornylselenyl triflate.[126]

Cyclopropanone acetals with a quaternary carbon atom in chiral form can be established by addition of bisoxazoline-ligated allylzinc reagents to the cyclopropenes.[127] The *t*-BuLi/(−)-sparteine combination favors Br–Li exchange and also promotes enantioselection in the intramolecular addition of aryllithium to an *o*-alkenyl side chain (e.g., indoline synthesis).[128,129] Moderate asymmetric induction is shown in the hydroarylation of norbornene in the presence of **63**.[130]

Cyclization of dienes and enynes to form five-membered rings proceeds enantioselectively in the presence of Pd complexes of chiral ligands such as **64**[131] and those of the bis(oxazoline) type.[132]

The intramolecular Pauson–Khand reaction finds another version in chiral titanocene catalysis.[133]

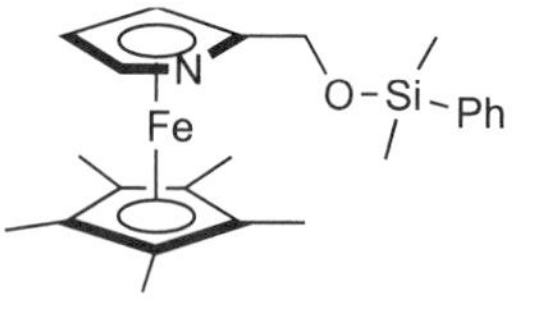

(62) 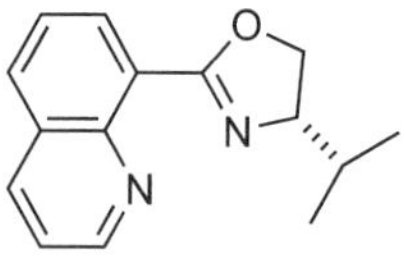 (63) 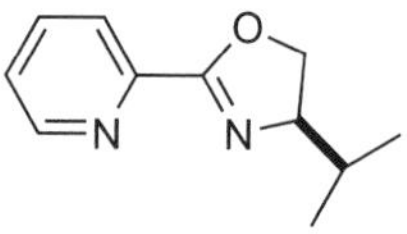 (64)

Michael reactions. A *P,N*-ligand (**65**) for the $Cu(OTf)_2$-catalyzed conjugate addition by diethylzinc has been developed.[134] Substrate-directed asymmetric induction in the addition of α-amino radicals is the basis of a necine base synthesis.[135] With a chiral

Al–salen complex, the addition of hydrazoic acid to mixed imides in which one of the acyl groups is α,β-unsaturated, excellent enantioselectivity results. Thus, this process is very valuable for the synthesis of β-amino acids.[136]

A synthesis of 2,8-dioxabicyclo[3.3.0]octan-3-one derivatives consists of conjugate addition of formaldehyde SAMP-hydrazone to α,β-unsaturated δ-valerolactone and α-alkylation, with hydrolytic reorganization of the functional groups.[137]

66 was discovered after screening a library of 100 ligands for the copper-catalyzed Michael reaction to enones compound.[138] The asymmetric reaction using the bis(oxazoline)–$Cu(SbF_6)_2$ catalyst between silyl enol ethers and *N*-alkenoyloxazolidin-2-ones establishes two adjacent stereocenters favoring the *anti* arrangement.[139] That between silyl keteneacetals and alkylidenemalonic esters is surprisingly effective despite the fact that the alkylidene group is quite far away from the chiral environment created on coordination.[140]

Another conjugate addition involving 1,3-dicarbonyl compounds and nitroalkenes is catalyzed by **67**.[141] For the addition of nitroalkanes to 2-cycloalkenones, L-proline is the most viable chiral auxiliary and *trans*-2,5-dimethylpiperazine is a most beneficial additive.[142]

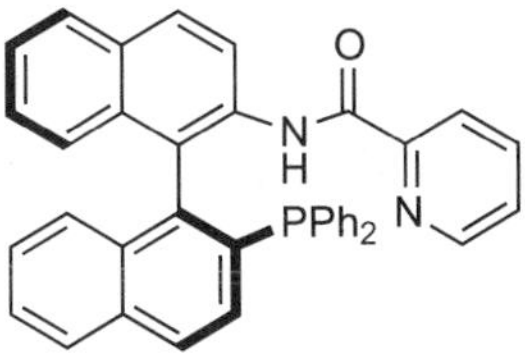

(65)

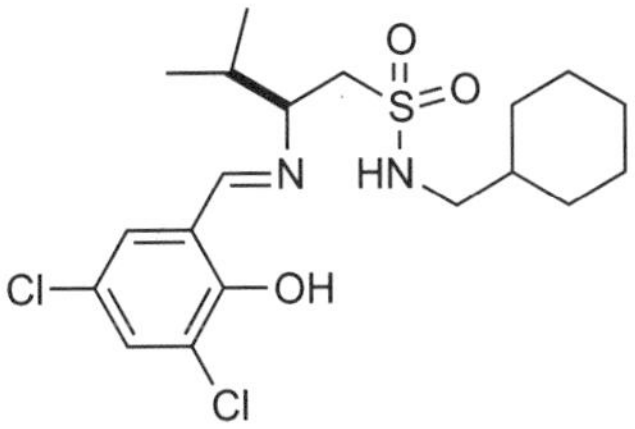

(66)

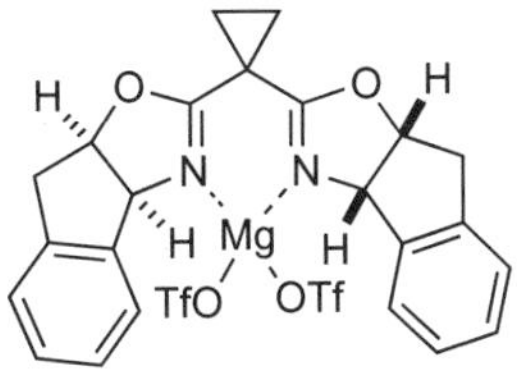

(67)

By using *N*-anthracen-9-ylmethyl-*O*-benzyldihydrocinchoninium bromide as catalyst, the conjugate addition of nitromethane to 1-phenyl-3-*p*-chlorophenyl-2-propen-1-one yields a mixture of (*R/S*) adducts in a 85:15 ratio.[143] After recrystallization, the product is suitable for conversion into (*R*)-baclofen hydrochloride, which is an important drug for treatment of spasms caused by spinal cord injury. Also worthy of mention is the application of a similar catalyst for the synthesis of methyl dihydrojasmonate under conditions of solid–liquid phase-transfer catalysis.[144] On the other hand, *N*-(2-alkenoyl)-4-phenyloxazolidin-2-ones are Michael acceptors with built-in diastereofacial control. Therefore a method for generating enantiopure β-substituted γ-butyrolactam-γ-carboxylic acids evolves to consist merely of using the chiral substrates to form adducts with glycine Schiff base–Ni(II) complexes.[145]

syn-2,3-Disubstituted 4-pentenoic esters of high optical purity have been synthesized using the Michael reaction and trapping protocol in the first stage. Sulfoxide **68** in serving a Michael donor role, is a chiral vinyl anion equivalent.[146]

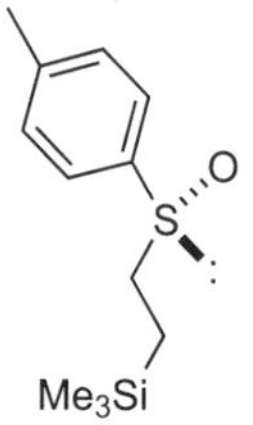

Me$_3$Si

(**68**)

Amination of silyl enol ethers with azodicarboxylic acid derivatives is usually quite efficient when conducted in the presence of C_2-symmetric bis(oxazoline)–Cu(OTf)$_2$ complex.[147]

Cycloadditions. A chiral 2,2′-bipyridine (**69**) in which each nucleus is fused to a ferrocene has found use in a CuOTf-catalyzed asymmetric cyclopropanation.[148]

The catalytic activities of chiral biaryl ligands **70** and **71** have been evaluated in aziridine formation. The former is treated with borane–THF before introduction of the addends (*N*-benzhydrylaldimines and ethyl diazoacetate),[149] while the latter is used in conjunction with (MeCN)$_4$CuBF$_4$ to promote the transfer of the TsN group from PhI=NTs to alkenes.[150] The best are those with 2,6-disubstituted Ar group because they clearly define the steric and electronic profile of the active site and discourage formation of inactive L$_2$M$_2$ species.

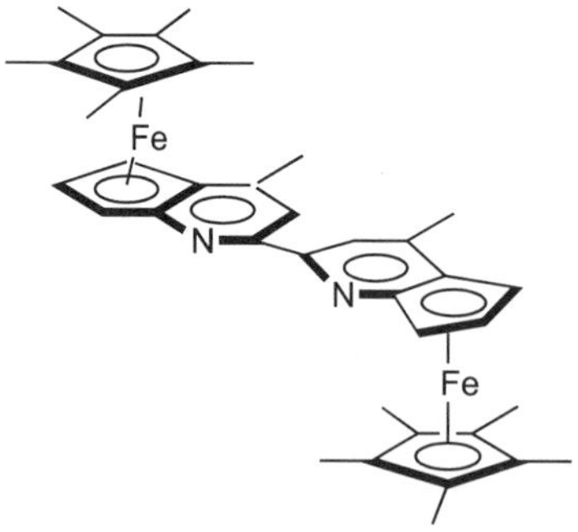

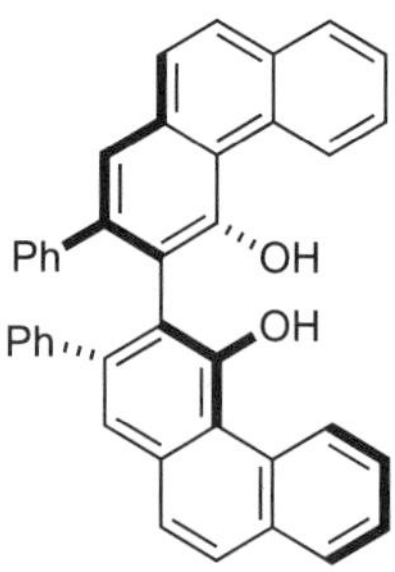

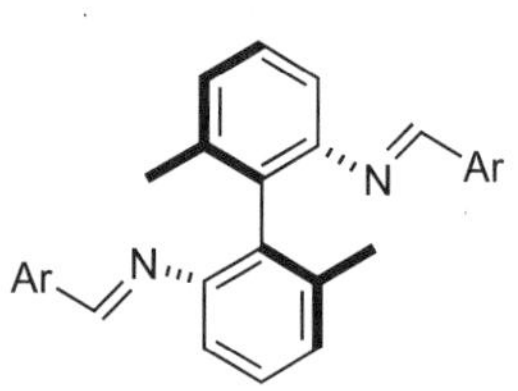

(**69**) (**70**) (**71**)

Optically active β-lactones are readily prepared by ketene–aldehyde cycloaddition using **72** as the chirality inducer.[151,152] The products are excellent precursors of β-amino acids. The [2 + 2]cycloaddition of ketenes and imines in the presence of *O*-benzoylquinine leads to numerous *cis*-2,3-disubstituted β-lactams.[153]

Reports on the advances of asymmetric 1,3-dipolar cycloadditions include the reaction of diazoalkanes to *N*-(2-alkenoyl)oxazolidin-2-ones catalyzed by Mg or Zn complexes of **73**, showing cooperative chiral control by the achiral oxazolidinone auxiliary and the chiral ligand.[154] An intramolecular cycloaddition of the same kind from substrates containing a chiral cyclic *N,N′*-dimethylaminal unit adjacent to the dipolarophilic double bond (i.e., **74**) proves very successful in the asymmetric sense, although the reaction of an analogous nitrone lacks stereoselectivity.[155]

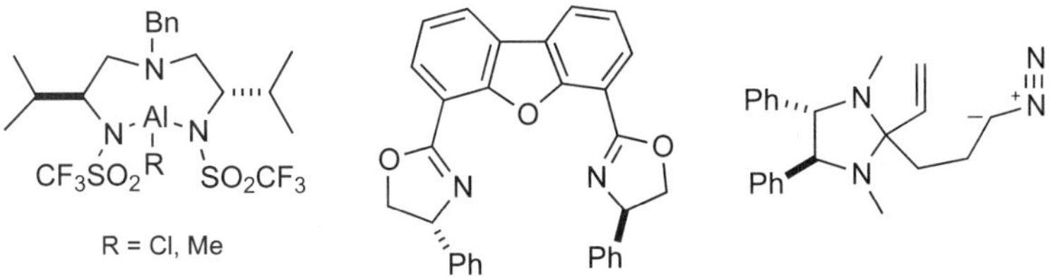

Bn N Al N N R CF_3SO_2 SO_2CF_3

R = Cl, Me

(72)

O O N N O Ph Ph

(73)

Ph N N Ph N N

(74)

Alkenoylation of a camphor-based cyclic acylhydrazide leads to substrate-defined dipolarophiles (for nitrile oxides).[156] On temporary (in situ) derivatization with **75**, enals are activated to undergo enantioselective cycloaddition with nitrones.[157]

The 2,2′-bis(benzamido-1,1′-binaphthyl) complex of $Yb(OTf)_3$ is capable of catalyzing the Diels–Alder reaction.[158] Complex **76** has a very high stability and therefore an attractive recycling possibility.[159]

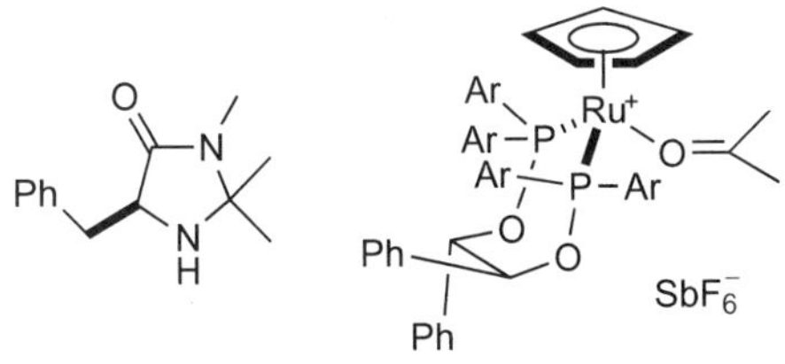

O N Ph N H

(75)

Ar Ru+ Ar P Ar P Ar O O O Ph Ph SbF_6^-

(76)

Asymmetric Diels–Alder reaction and its applications based on catalysis by bis(oxazoline)copper(II) complexes have been published in full details.[160] This method has been further extended to the synthesis of 2,3-dihydropyran-6-ylphosphonates,[161] 4-aminodihydropyrans,[162] and piperidones.[163] Note that both epimeric amines are accessible by merely changing the diene substrate, that is, (*E*)- versus (*Z*)-configuration. Diazolidinone **75** serves as a chiral catalyst that also activates enals (actual dienophiles being the iminium ions).[164]

Hetero-Diels–Alder reactions involving *N*-phenethyl-(2-pyridylmethylene)imine as the dienophile have been reported.[165] A Cr complex **77** is useful for the synthesis of trisubstituted tetrahydro-4-pyranones,[166] whereas a (salen)Cr(III) complex is involved in those reaction with 1-amino-3-siloxy-1,3-butadienes.[167]

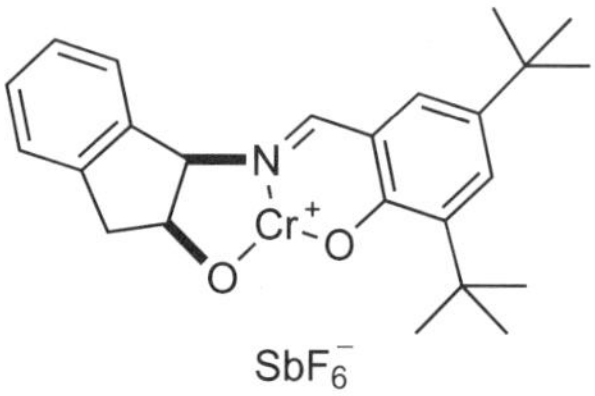

SbF_6^-

(77)

Compund **78** is among the new transition metal catalysts that have found good use in the decomposition of diazo compounds and delivery of the metal carbenoids to alkenes.[168] Iminodiazaphospholidine (**79**) possesses a stereogenic phosphorus center and its applicability to effect asymmetric cyclopropanation[169] is now known. The Zn chelate of **80** is effective for the Simmons–Smith reaction of allylic alcohols.[170]

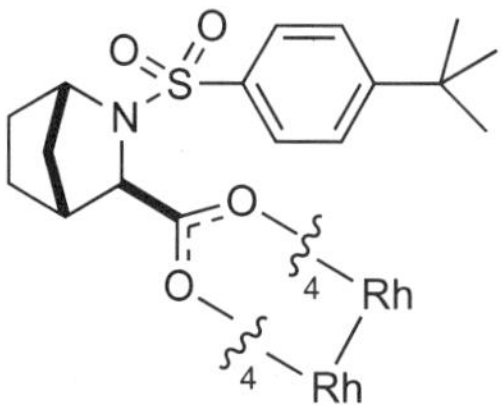

(78) 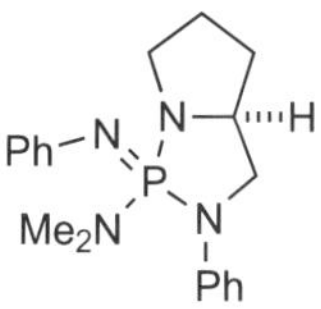 **(79)**

Epoxidations. Grafting tantalum onto silica to form a useful catalyst for the Sharpless asymmetric epoxidation of allyl alcohols is contrary to the ineffective titanium species on a similar support.[171] Vanadium-complexed chiral hydroxamic acids such as the [2.2]paracyclophane derivative **81**[172] and the crowded **82**[173] have been used.

Ketones represented by **83** are a new generation of catalysts for mediating asymmetric epoxidation of alkenes by Oxone®.[174] Ketone **84** exhibits desirable characteristics in the epoxidation of *cis*-alkenes.[175] For example, it does not cause isomerization of the double bond.

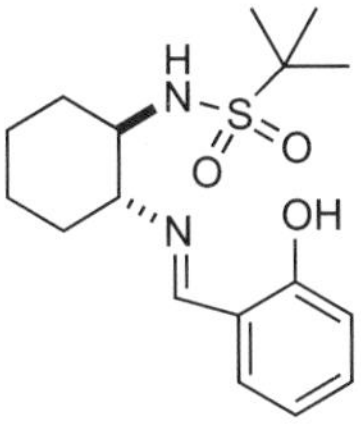

(**80**) 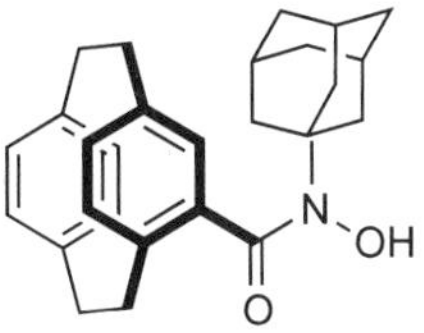 (**81**) 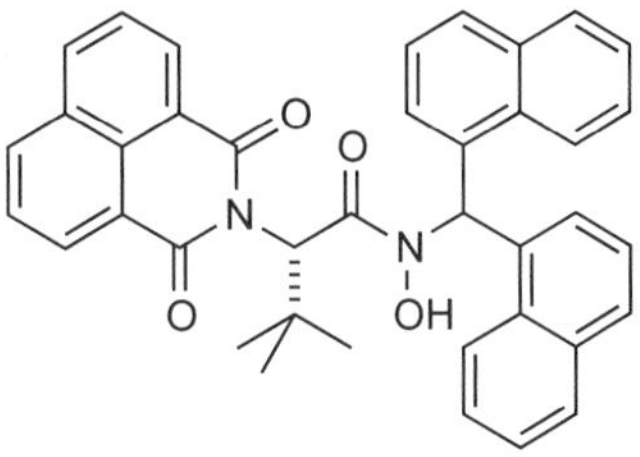 (**82**)

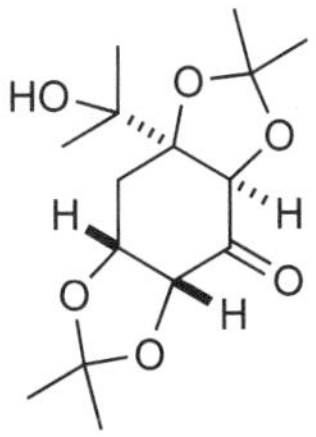

(**83**) 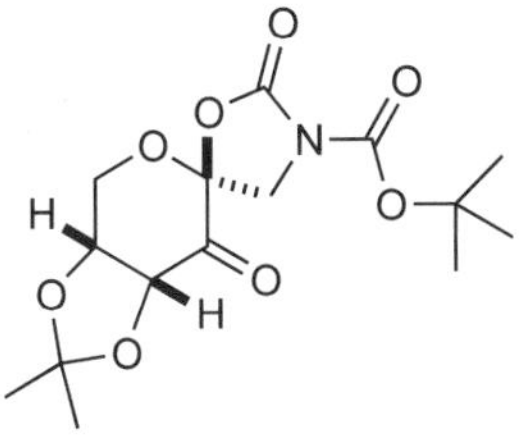 (**84**)

Poly-L-leucine[176] and a polymeric BINOL-zinc complex[177] have been evaluated for the epoxidation of enones. Poly-L-leucine immobilized on silica is also effective.[178]

An air- and moisture-stable ionic liquid in the reaction media facilitates recovery of chiral (salen)Mn(III) complexes that are used in asymmetric epoxidation.[179]

Ketone reductions. Corey's oxazaborolidine continues to be a popular catalyst for borane reduction. A cheap borane source for the reduction is LiH–$BF_3 \cdot OEt_2$.[180] There are also modifications, for example, using aluminum(III) ethoxide,[181] and the sulfonamide derivative **85**.[182]

The use of $LiBH_4$ to reduce an arylborane derivative **86** prepared from tartaric acid shows good results.[183] Ketone reduction with $NaBH_4$ in the presence of a dendrimer in which the polyhydroxy outer rim is derived from glucose results in secondary alcohols with ee values exceeding 95%.[184]

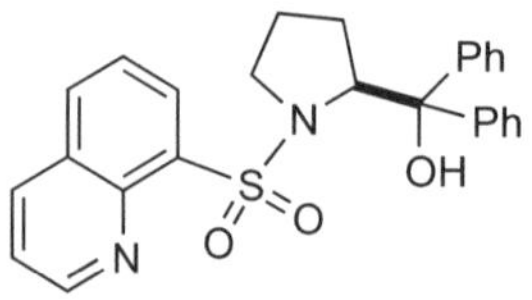

(**85**)

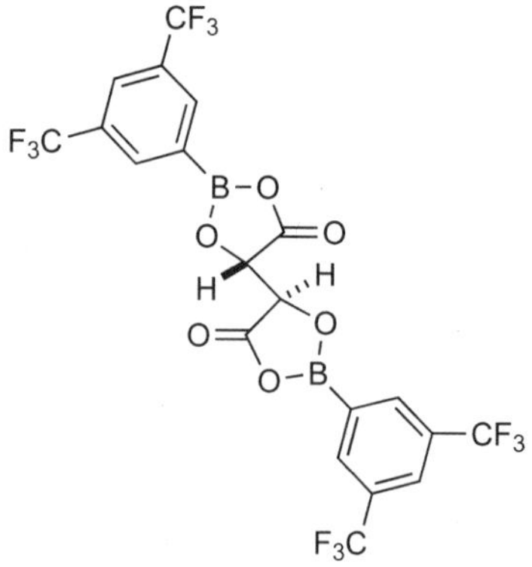

(**86**)

In free radical reduction mediated by chiral hydrostannanes, there is a remarkable enhancement of enantioselectivity by Lewis acids.[185] When using a hydrosilane in the ketone reduction, Rh complexes of **87**,[186] **88**,[187] and **89**[188] serve well as chiral catalysts.

A Lewis acid shows a remarkable enhancement of enantioselectivity during free radical reduction by some stannanes.[189] Accordingly, borohydride reduction of 1,3-diaryl-1,3-propanediones in the presence of the cobalt complex **90** conveniently provides chiral diols.[190]

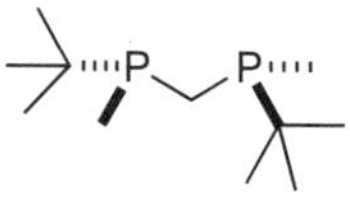

(**87**)

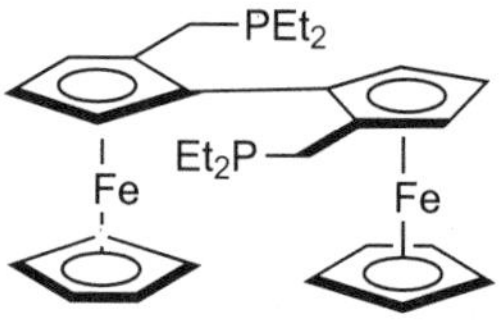

(**88**)

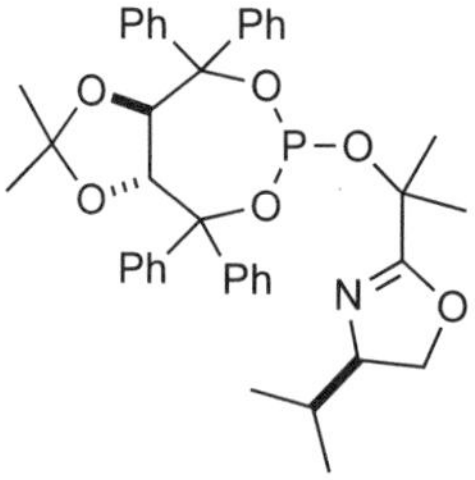

(**89**)

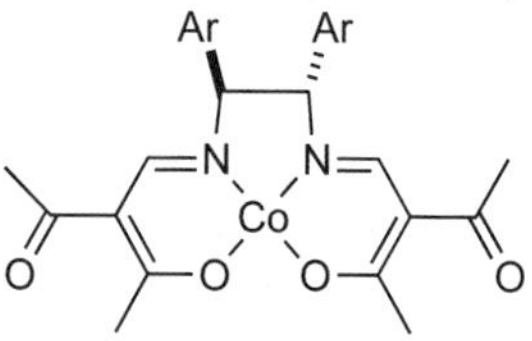

(**90**)

A route to chiral alcohols from enones relies on stereoselective hydride transfer (Meerwein–Ponndorf–Verley reduction) from the isoborneol moiety after the substrates are attached to the monoterpene skeleton through Michael reaction with a 10-thiol group.[191] Aromatic ketones are reduced by the system of *t*-BuOK–*i*-PrOH and **91**.[192]

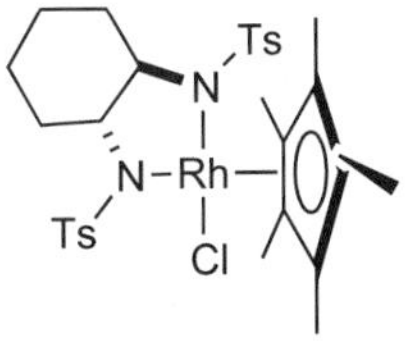

(**91**)

Asymmetric hydrogenation. With complex **92** to mediate the asymmetric transfer hydrogenation of 1-aryl-1,2-propanediones, the acetyl group is preferentially transformed, but the second reduction follows when the temperature is raised from 10° to 40°, furnishing *trans*-diols.[193]

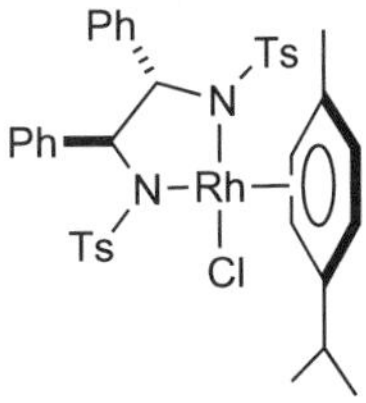

(**92**)

2-Azanorbornyl-3-methanol (**93**) shows its value as a chiral ligand in Ru-catalyzed transfer hydrogenation.[194] On binding to Ru, C_2-symmetric diphosphines (**94**, **95**, etc.)[195] and the ferrocene-based ligand **96**[196] are able to exert their steric influences during hydrogenation of 1,3-diketones.

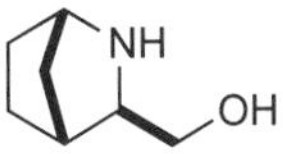

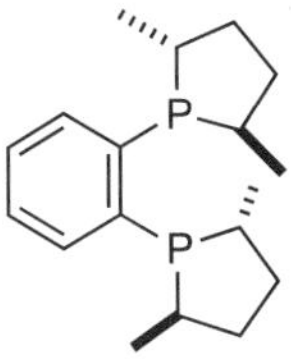

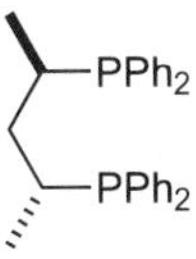

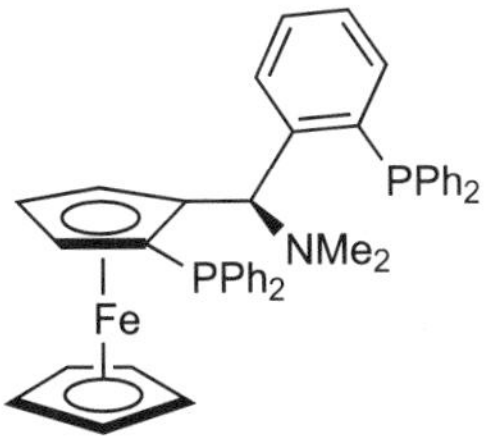

(**93**) (**94**) (**95**) (**96**)

Among the effective diphosphine–Ru–diamine complexes used for catalyzing carbonyl hydrogenation enantioselectively is one member in which the phosphorus atoms are attached to a [2,2]paracyclophane unit.[197]

The principle of asymmetric activation–deactivation has been demonstrated in the hydrogenation of ketones using a catalytic system consisting of a Ru complex of racemic dimethyl-BINAP and two chiral diamines.[198] The principle is depicted in the following diagram.

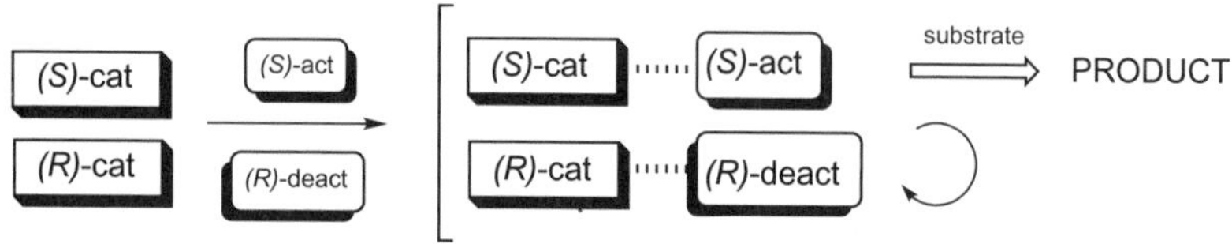

Platinum catalysts modified by cinchona alkaloids have been developed for asym metric hydrogenation of α-keto acetals[199] and α-keto esters,[200] and the Ru complex of chiral 2-(MeO)BIPHEP **97** has been developed for β-keto sulfone reduction.[201]

Ruthenium-based asymmetric hydrogenation of α-arylacrylic acids and β-keto esters uses the bipyridylbisphosphine **98** (and its enantiomer).[202]

MeO MeO PPh$_2$ PPh$_2$

(97)

OMe N MeO MeO PPh$_2$ PPh$_2$ N OMe

(98)

The ferrocenyldiphosphine ligand **99** constitutes a catalytic system with $(cod)_2RhBF_4$ for synthesizing α-alkylsuccinic acid derivatives from the alkylidene compounds.[203] Secondary acetates in chiral form are accessible from hydrogenation of enol acetates using the catalytic system of $(cod)_2RhBF_4$–**100**.[204]

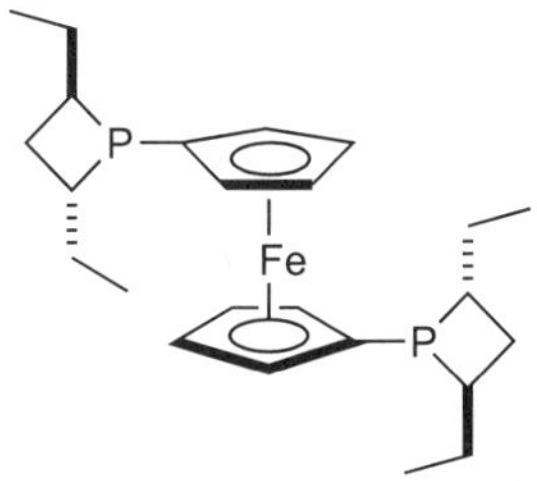

(**99**) 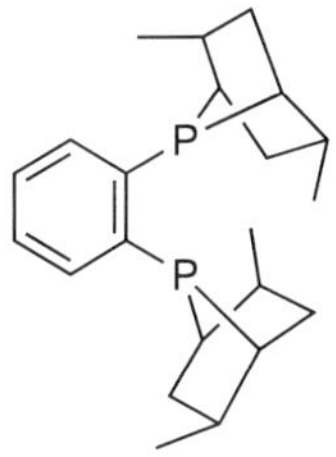 (**100**)

Good results have been obtained from hydrogenation of tetrasubstituted alkenes with a cationic zirconocene catalyst,[205] of indoles with a bis(ferrocenylphosphine)–Rh system, [206] and of 2-pyrones with a 2-(MeO)BIPHEP–Ru complex.[207]

Very efficient ligands that complement $(cod)_2Rh^+$ salts for hydrogenation of dehydroamino acids (and derivatives) are **101**,[208] **102**,[209] and **103**.[210] A combination of ligand **87** with Rh^+ has the versatility of catalyzing asymmetric hydrogenation, hydrosilylation, and conjugate addition.[211]

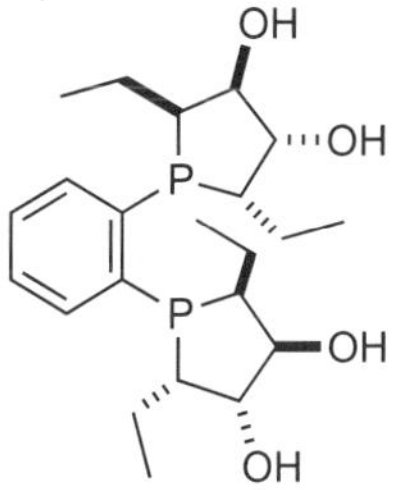

(**101**) 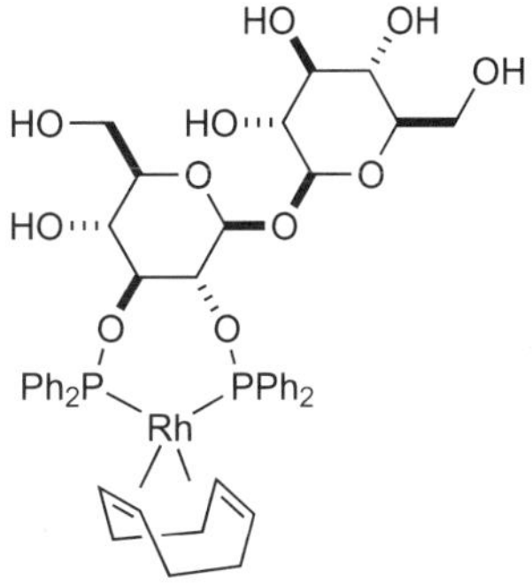 (**102**) 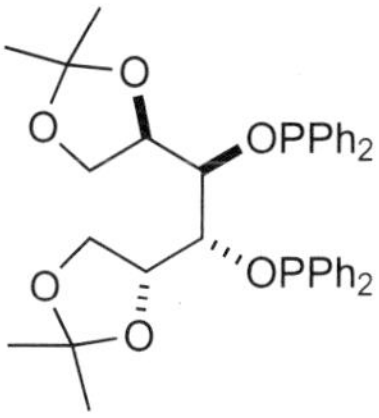 (**103**)

The paradox concerning the generation of opposite enantiomers in the hydrogenation of *N*-(1-phenylvinyl)acetamide and *N*-(1-*t*-butylvinyl)acetamide with **104** is now understood on the grounds that the mode of substrate docking at the migratory insertion step of the catalytic cycle determines the stereochemical outcome.[212]

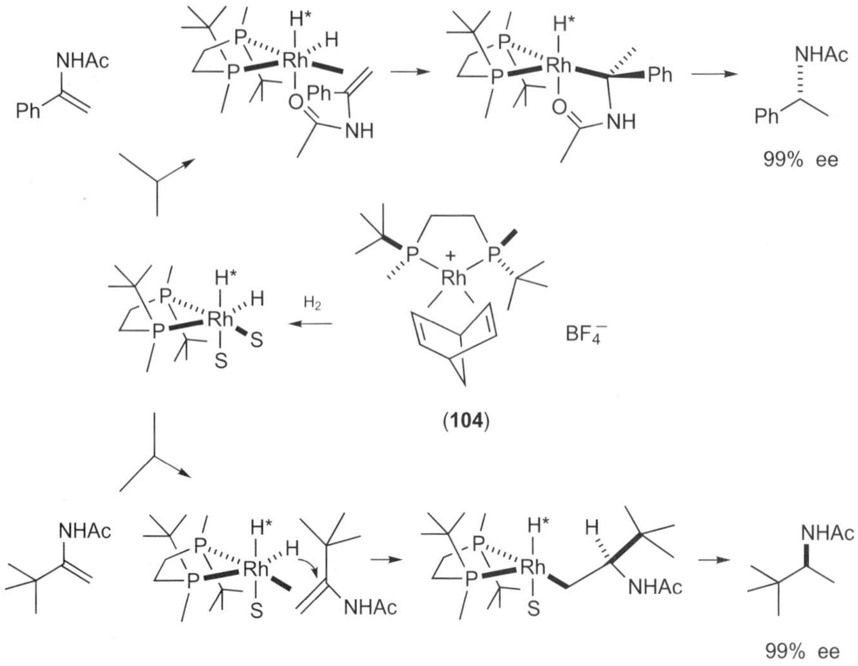

(104)

99% ee

99% ee

Other enantioselective reactions. Several asymmetric reactions worth mentioning are the Cu-catalyzed allylic oxidation in the presence of **105**,[213] **106**,[214] or **107**[215] with *t*-butyl perbenzoate, oxidation of sulfides (*t*-BuOOH–Ti^{IV}) in the presence of a 4,4′-dimer of B-aromatic 3-hydroxyestrane,[216] the reductive amination by chiral *t*-butylsulfinamide,[217] the glyoxylate ene reaction promoted by $Yb(OTf)_3$ and *ent*-**73**,[218] *C*-arylation of phenols with aryllead reagents under the influence of brucine,[219] and the C—H bond insertion by Rh-carbenoids.[220]

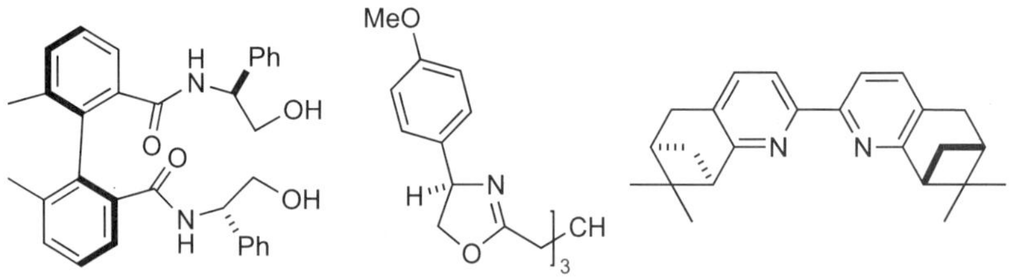

(105) (106) (107)

Heterocyclic analogues of BINAP, such as the bis(benzothiophene) **108** as well as the 2,2′,5,5′-tetramethyl-4,4′-bis(diphenylphosphino)-3,3′-bithiophene, have been prepared and tested in the Heck reaction.[221] Incorporation of a chiral *o*-(*N,N*-dimethylamino)phenylsulfinyl group to a double bond elicits enantioselectivity at the β-carbon during the intramolecular Heck reaction.[222] Optically active 5-arylcyclopenten-1-yl *t*-butyl sulfoxides are produced from (*R*)-1-*t*-butylsulfinylcyclopentene.[223]

Palladacycle (**109**) and its analogues induce chirality of *N*-allylalkanamides during rearrangement of allylic imidates.[224] For a synthesis of α-branched chiral 4-alkenethioamides, it is convenient to allylate thioamides of (+)-*trans*-2,5-diphenylpyrrolidine.[225] This process involves a thio-Claisen rearrangement.

Enantioselective deprotonation of ketones is achieved by using a homochiral magnesium bis[*N*-benzyl-*N*-(α-phenethyl)]amide.[226] On the other hand, **110** is excellent for protonation of enolates.[227]

Access to bicyclic enones from 1,6-enynes by the Pauson–Khand method is rendered enantioselective by installing a chiral *t*-butylsulfinyl group at C-1.[228] Cyclization of α-benzylidene-aroylacetamides to furnish 3-arylindanones is subject to 1,5-asymmetric induction when the amide moiety is derived from a bulky 4-substitulted oxazolidin-2-one.[229] Chiral ligands for the Pauson–Khand reaction have also been studied. Phosphine-borane (**111**) derived from (+)-pulegone is an example.[230]

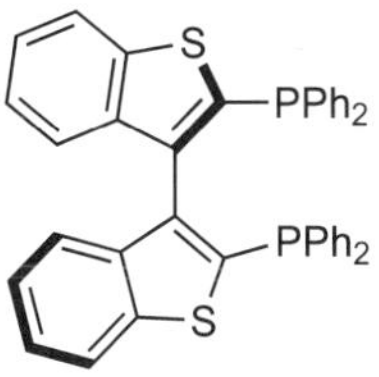

(**108**) 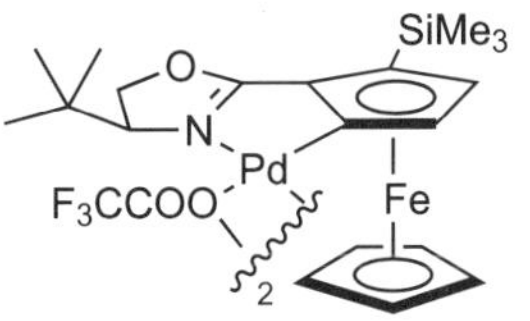 (**109**) 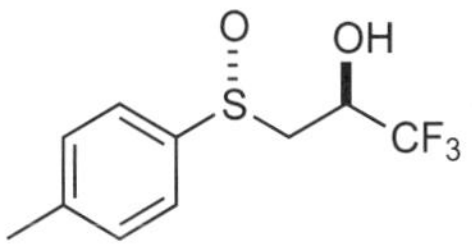 (**110**)

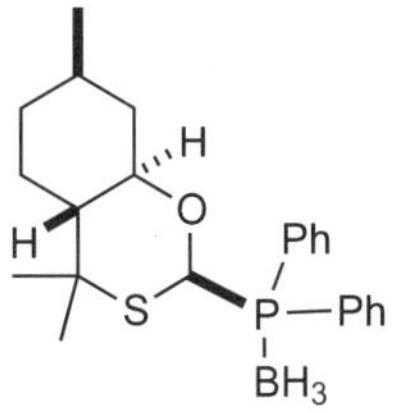

(**111**)

Enantioselective cyclization of carbanions derived from *N,N*-diisopropylcarbamates can lead to *cis*- or *trans*-1,2-disubstituted cyclopentane derivatives by treatment with *x*-BuLi–(−)-sparteine.[231,232]

A cyclization route to 2-substituted (e.g., cinnamyl) pyrrolidines and piperidines involves cyclization of *N*-Boc amines bearing a 3-chloropropyl or 4-chlorobutyl chain, respectively. Asymmetry is induced by (−)-sparteine.[233]

[1]Buisson, D., Azerad, R. *TA* **10**, 2997 (1999).
[2]Fries, G., Wolf, J., Pfeiffer, M., Stalke, D., Werner, H. *ACIEE* **39**, 564 (2000).
[3]Khiar, N., Alcudia, F., Espartero, J.-L., Rodriguez, L., Fernandez, I. *JACS* **122**, 7598 (2000).
[4]Reetz, M.T., Schmitz, A. *TL* **40**, 2737, 2741 (1999).
[5]Brunel, J.M., Constantieux, T., Buono, G. *JOC* **64**, 8940 (1999).
[6]Trost, B.M., Tsui, H.-C., Toste, F.D. *JACS* **122**, 3534 (2000).
[7]Ie, Y., Fu, G.C. *CC* 119 (2000).
[8]Tao, B., Ruble, J.C., Hoic, D.A., Fu, G.C. *JACS* **121**, 5091 (1999).
[9]Bellemin-Laponnaz, S., Tweddell, J., Ruble, J.C., Breitling, F.M., Fu, G.C. *CC* 1009 (2000).
[10]Spivey, A.C., Fekner, T., Spey, S.E. *JOC* **65**, 3154 (2000).
[11]Sano, T., Imai, K., Ohashi, K., Oriyama, T. *CL* 265 (1999).
[12]Vedejs, E., Daugulis, O. *JACS* **121**, 5813 (1999).
[13]Bolm, C., Gerlach, A., Dinter, C.L. *SL* 195 (1999).
[14]Chen, Y., Tian, S.-K., Deng, L. *JACS* **122**, 9542 (2000).
[15]Takeuchi, Y., Suzuki, T., Satoh, A., Shiragami, T., Shibata, N. *JOC* **64**, 5708 (1999).
[16]Shibata, N., Suzuki, E., Takeuchi, Y. *JACS* **122**, 10728 (2000).
[17]Cardillo, G., Gentilucci, L., Tolomelli, A., Tomasini, C. *SL* 1727 (1999).
[18]Saito, S., Nakadai, M., Yamamoto, H. *SL* 1107 (2000).
[19]Trost, B.M., Schroeder, G.M. *JACS* **121**, 6759 (1999).
[20]Trost, B.M., Ariza, X. *JACS* **121**, 10727 (1999).
[21]Ooi, T., Takeuchi, M., Kameda, M., Maruoka, K. *JACS* **122**, 5228 (2000).
[22]Ooi, T., Kameda, M., Maruoka, K. *JACS* **121**, 6519 (1999).
[23]Nagula, G., Huber, V.J., Lum, C., Goodman, B.A. *OL* **2**, 3527 (2000).
[24]Kawabata, T., Suzuki, H., Nagae, Y., Fuji, K. *ACIEE* **39**, 2155 (2000).
[25]Sarakinos, G., Corey, E.J. *OL* **1**, 1741 (1999).
[26]Chang, J.-W., Jang, D.-P., Uang, B.-J., Liao, F.-L., Wang, S.-L. *OL* **1**, 2061 (1999).
[27]Enders, D., Janeck, C.F., Runsink, J. *SL* 641 (2000).
[28a]Bull, S.D., Davies, S.G., Jones, S., Sanganee, H.J. *JCS(P1)* 387 (1999).
[28b]Gaul, C., Seebach, D. *OL* **2**, 1501 (2000).
[29]Arai, S., Shirai, Y., Ishida, T., Shioiri, T. *CC*49 (1999).
[30]Kaiser, N.-F.K., Bremberg, U., Larhed, M., Moberg, C., Hallberg, A. *JOMC* **603**, 2 (2000).
[31]Kohara, T., Hashimoto, Y., Saigo, K. *SL* 517 (2000).
[32]Ito, K., Kashiwagi, R., Iwasaki, K., Katsuki, T. *SL* 1563 (1999).
[33]Yonehara, K., Hashizume, T., Mori, K., Ohe, K., Uemura, S. *CC* 415 (1999).
[34]Yan, Y.-Y., RajanBabu, T.V. *OL* **2**, 199 (2000).
[35]Hoarau, O., Ait-Haddou, H., Daran, J.-C., Cramailere, D., Balavoine, G.G.A. *OM* **18**, 4718 (1999).
[36]Hou, D.-R., Burgess, K. *OL* **1**, 1745 (1999).
[37]Gilbertson, S.R., Genov, D.G., Rheingold, A.L. *OL* **2**, 2885 (2000).
[38]Shintani, R., Lo, M.M.-C., Fu, G.C. *OL* **2**, 3695 (2000).
[39]Evans, P.A., Brandt, T.A. *OL* **1**, 1563 (1999).
[40]Saitoh, A., Misawa, M., Morimoto, T. *SL* 483 (1999).
[41]Lautens, M., Fagnou, K., Rovis, T. *JACS* **122**, 5650 (2000).
[42]Millward, D.B., Sammis, G., Waymouth, R.M. *JOC* **65**, 3902 (2000).
[43]Enders, D., Peters, R., Runsink, J., Bats, J.W. *OL* **1**, 1863 (1999).
[44]Evans, D.A., Campos, K.R., Tedrow, J.S., Michael, F.E., Gagne, M.R. *JOC* **64**, 2994 (1999).

[45]Fuji, K., Ohnishi, H., Moriyama, S., Tanaka, K., Kawabata, T., Tsubaki, K. *SL* 351 (2000).
[46]Reetz, M.T., Haderlein, G., Angermund, K. *JACS* **122**, 996 (2000).
[47]Trost, B.M., Toste, F.D. *JACS* **121**, 3543 (1999).
[48]Trost, B.M., Crawley, M.L., Lee, C.B. *JACS* **122**, 6120 (2000).
[49]Kaiser, N.-F.K., Bremberg, U., Larhed, M., Moberg, C., Hallberg, A. *ACIEE* **39**, 3596 (2000).
[50]Glorius, F., Pfaltz, A. *OL* **1**, 141 (1999).
[51]Trost, B.M., Hildbrand, S., Dogra, K. *JACS* **121**, 10416 (1999).
[52]Brunel, J.M., Legrand, O., Reymond, S., Buono, G. *ACIEE* **39**, 2554 (2000).
[53]Gansäuer, A., Lauterbach, T., Bluhm, H., Noltemeyer, M. *ACIEE* **38**, 2909 (1999).
[54]Bruns, S., Haufe, G. *JFC* **104**, 247 (2000).
[55]Lebel, H., Jacobsen, E.N. *TL* **40**, 7303 (1999).
[56]Schaus, S.E., Jacobsen, E.N. *OL* **2**, 1001 (2000).
[57]Trost, B.M., Tang, W., Schulte, J.L. *OL* **2**, 4013 (2000).
[58a]Breinbauer, R., Jacobsen, E.N. *ACIEE* **39**, 3604 (2000).
[58b]Li, Z., Fernandez, M., Jacobsen, E.N. *OL* **1**, 1611 (1999).
[59]Belokon, Y.N., North, M., Parsons, T. *OL* **2**, 1617 (2000).
[60]Aspinall, H.C., Greeves, N., Smith, P.M. *TL* **40**, 1763 (1999).
[61]Ooi, T., Takaya, K., Miura, T., Ichikawa, H., Maruoka, K. *SL* 1133 (2000).
[62]Hamashima, Y., Kanai, M., Shibasaki, M. *JACS* **122**, 7412 (2000).
[63]Bolm, C., Muniz, K. *CC* 1295 (1999).
[64]BouzBouz, S., Cossy, J. *OL* **2**, 501 (2000).
[65]Colombo, L., Di Giacomo, M. *TL* **40**, 1977 (1999).
[66]Loh, T.-P., Zhou, J.-R., Yin, Z. *OL* **1**, 1855 (1999).
[67]Micalizio, G.C., Roush, W.R. *OL* **2**, 461 (2000).
[68]Renard, M., Ghosez, L. *TL* **40**, 6237 (1999).
[69]Reggelin, M., Gerlach, M., Vogt, M. *EJOC* 1011 (1999).
[70]Barrett, A.G.M., Braddock, D.C., de Koning, P.D. *CC* 459 (1999).
[71]Frantz, D.E., Fässler, R., Carreira, E.M. *JACS* **122**, 1806 (2000).
[72]Paleo, M.R., Cabezza, I., Sardina, F.J. *JOC* **65**, 2108 (2000).
[73]Lawrence, C.F., Nayak, S.K., Thijs, L., Zwanenburg, B. *SL* 1571 (1999).
[74]Dangel, B.D., Polt, R. *OL* **2**, 3003 (2000).
[75]Asami, M., Watanabe, H., Honda, K., Inoue, S. *TA* **9**, 4165 (1998).
[76]Kawanami, Y., Mitsuie, T., Miki, M., Sakamoto, T., Nishitani, K. *T* **56**, 175 (2000).
[77]Nugent, W.A. *CC* 1369 (1999).
[78]Reddy, K.S., Sola, L., Moyano, A., Pericas, M.A., Riera, A. *JOC* **64**, 3969 (1999).
[79]Hodge, P., Kell, R.J., Ma, J., Morris, H. *AJC* **52**, 1041 (1999).
[80]Sato, I., Saito, T., Soai, K. *CC* 2471 (2000).
[81]Shibata, T., Yonekubo, S., Soai, K. *ACIEE* **38**, 659 (1999).
[82]Soai, K., Osanai, S., Kodowaki, K., Yonekubo, S., Shibata, T., Sato, I. *JACS* **121**, 11235 (1999).
[82a]Sato, I., Kodowaki, K., Soai, K. *ACIEE* **39**, 1510 (1999).
[83]Denmark, S.E., Su, X., Nishigaichi, Y., Coe, D.M., Wong, K.-T., Winter, S.B.D., Choi, J.Y. *JOC* **64**, 1958 (1999).
[84]List, B., Lerner, R.A., Barbas III, C.F. *JACS* **122**, 2395 (2000).
[85]Notz, W., List, B. *JACS* **122**, 7386 (2000).
[86]Trost, B.M., Ito, H. *JACS* **122**, 12003 (2000).
[87]Kim, T.H., Lee, G.-J. *TL* **41**, 1505 (2000).
[88]Crimmins, M.T., Chaudhary, K. *OL* **2**, 775 (2000).
[89]Horikawa, M., Busch-Petersen, J., Corey, E.J. *TL* **40**, 3843 (1999).
[90]Andrus, M.B., Sekhar, B.B.V.S., Meredith, E.L., Dalley, N.K. *OL* **2**, 3035 (2000).
[91]Audrain, H., Jorgensen, K.A. *JACS* **122**, 11543 (2000).

[92]Kiyooka, S., Shahid, K.A. *TA* **11**, 1537 (2000).
[93]Wang, Y.-C., Li, C.-L., Tseng, H.-L., Chuang, S.-C., Yan, T.-H. *TA* **10**, 3249 (1999).
[94]Iwama, T., Tsujiyama, S., Kinoshita, H., Kanematsu, K., Tsurukami, Y., Iwamura, T., Watanabe, S., Kataoka, T. *CPB* **47**, 956 (1999).
[95]Yang, K.-S., Chen, K. *OL* **2**, 729 (2000).
[96]Iwabuchi, Y., Nakatani, M., Yokoyama, N., Hatakeyama, S. *JACS* **121**, 10219 (1999).
[97]Reichel, F., Fang, X., Yao, S., Ricci, M., Jorgensen, K.A. *CC* 1505 (1999).
[98]Ishihara, K., Kondo, S., Yamamoto, H. *SL* 1283 (1999).
[99]Kobayashi, S., Nagayama, S., Busujima, T. *CL* 71 (1999).
[100]Corey, E.J., Choi, S. *TL* **41**, 2769 (2000).
[101]Sano, S., Ishii, T., Miwa, T., Nagao, Y. *TL* **40**, 3013 (1999).
[102]Kanai, M., Hamashima, Y., Shibasaki, M. *TL* **41**, 2405 (2000).
[103]Holmes, I.P., Kagan, H.B. *TL* **41**, 7457 (2000).
[104]Xu, M.-H., Wang, W., Lin, G.-Q. *OL* **2**, 2229 (2000).
[105]Davis, F.A., Zhang, Y., Andemichael, Y., Fang, T., Fanelli, D.L., Zhang, H. *JOC* **64**, 1403 (1999).
[106]Kawecki, R. *JOC* **64**, 8724 (1999).
[107]Miyabe, H., Ushiro, C., Ueda, M., Yamakawa, K., Naito, T. *JOC* **65**, 176 (2000).
[108]Palomo, C., Oiarbide, M., Gonzalez-Rego, M.C., Sharma, A.K., Garcia, J.`M., Gonzalez, A., Landa, C., Linden, A. *ACIEE* **39**, 1063 (2000).
[109]Müller, R., Goesmann, H., Waldmann, H. *ACIEE* **38**, 184 (1999).
[110]Fujihara, H., Nagai, K., Tomioka, K. *JACS* **122**, 12055 (2000).
[111]Jimeno, C., Reddy, K.S., Sola, L., Moyano, A., Pericas, M.A., Riera, A. *OL* **2**, 3157 (2000).
[112]Hayashi, T., Ishigedani, M. *JACS* **122**, 976 (2000).
[113]Yamada, H., Kawate, T., Nishida, A., Nakagawa, M. *JOC* **64**, 8821 (1999).
[114]Friestad, G.K., Qin, J. *JACS* **122**, 8329 (2000).
[115]Ruano, J.L.G., Alcudia, A., del Prado, M., Barros, D., Maestro, M.C., Fernandez, I. *JOC* **65**, 2856 (2000).
[116]Itsuno, S., Watanabe, K., Matsumoto, T., Kuroda, S., Yokoi, A., El-Shehawy, A. *JCS(P1)* 2011 (1999).
[117]Teng, X., Takayama, Y., Okamoto, S., Sato, F. *JACS* **121**, 11916 (1999).
[118]Sigman, M.S., Vachal, P., Jacobsen, E.N. *ACIEE* **39**, 1279 (2000).
[119]Vachal, P., Jacobsen, E.N. *OL* **2**, 867 (2000).
[120]Krueger, C.A., Kuntz, K.W., Dzierba, C.D., Wirschum, W.G., Gleason, J.D., Snapper, M.L., Hoveyda, A.H. *JACS* **121**, 4284 (1999).
[121]Corey, E.J., Grogan, M.J. *OL* **1**, 157 (1999).
[122]List, B. *JACS* **122**, 9336 (2000).
[123]Hansen, M.C., Buchwald, S.L. *OL* **2**, 713 (2000).
[124]Hodous, B.L., Ruble, J.C., Fu, G.C. *JACS* **121**, 2637 (1999).
[125]Tiecco, M., Testaferri, L., Bagnoli, L., Marini, F., Temperini, A., Tomassini, C., Santi, C. *TL* **41**, 3241 (2000).
[126]Back, T.G., Dyck, B.P., Nan, S. *T* **55**, 3191 (1999).
[127]Nakamura, M., Inoue, T., Sato, A., Nakamura, E. *OL* **2**, 2193 (2000).
[128]Bailey, W.F., Mealy, M.J. *JACS* **122**, 6787 (2000).
[129]Gil, G.S., Groth, U.M. *JACS* **122**, 6789 (2000).
[130]Wu, X.-Y., Xu, H.-D., Zhou, Q.-L., Chan, A.S.C. *TA* **11**, 1255 (2000).
[131]Perch, N.S., Pei, T., Widenhoefer, R.A. *JOC* **65**, 3836 (2000).
[132]Zhang, Q., Lu, X. *JACS* **122**, 7604 (2000).
[133]Hicks, F.A., Buchwald, S.L. *JACS* **121**, 7026 (1999).
[134]Hu, X., Chen, H., Zhang, X. *ACIEE* **38**, 3518 (1999).
[135]Bertrand, S., Hoffmann, N., Pete, J.-P. *EJOC* 2227 (2000).

[136]Myers, J.K., Jacobsen, E.N. *JACS* **121**, 8959 (1999).
[137]Enders, D., Vazquez, J., Raabe, G. *CC* 701 (1999).
[138]Chataigner, I., Gennari, C., Piarulli, U., Ceccarelli, S. *ACIEE* **39**, 916 (2000).
[139]Evans, D.A., Willis, M.C., Johnston, J.N. *OL* **1**, 865 (1999).
[140]Evans, D.A., Rovis, T., Kozlowski, M.C., Tedrow, J.S. *JACS* **121**, 1994 (1999).
[141]Ji, J., Barnes, D.M., Zhang, J., King, S.A., Wittenberger, S.J., Morton, H.E. *JACS* **121**, 10215 (1999).
[142]Hanessian, S., Pham, V. *OL* **2**, 2975 (2000).
[143]Corey, E.J., Zhang, F.-Y. *OL* **2**, 4257 (2000).
[144]Perrad, T., Plaquevent, J.-C., Desmurs, J.-R., Hebrault, D. *OL* **2**, 2959 (2000).
[145]Soloshonok, V.A., Cai, C., Hruby, V.J. *OL* **2**, 747 (2000).
[146]Nakamura, S., Watanabe, Y., Toru, T. *JOC* **65**, 1758 (2000).
[147]Evans, D.A., Johnson, D.S. *OL* **1**, 595 (1999).
[148]Rios, R., Liang, J., Lo, M.M.-C., Fu, G.C. *CC* 376 (2000).
[149]Antilla, J.C., Wulff, W.D. *JACS* **121**, 5099 (1999).
[150]Sanders, C.J., Gillespie, K.M., Bell, D., Scott, P. *JACS* **122**, 7132 (2000).
[151]Nelson, S.G., Peelen, T.J., Wan, Z. *JACS* **121**, 9742 (1999).
[152]Nelson, S.G., Spencer, K.L. *ACIEE* **39**, 1323 (2000).
[153]Taggi, A.E., Hafez, A.M., Wack, H., Young, B., Drury III, W.J., Lectka, T. *JACS* **122**, 7831 (2000).
[154]Kanemasa, S., Kanai, T. *JACS* **122**, 10710 (2000).
[155]Jung, M.E., Huang, A. *OL* **2**, 2659 (2000).
[156]Yang, K.-S., Lain, J.-C., Lin, C.-H., Chen, K. *TL* **41**, 1453 (2000).
[157]Jen, W.S., Wiener, J.J.M., MacMillan, D.W.C. *JACS* **122**, 9874 (2000).
[158]Nishida, A., Yamanaka, M., Nakagawa, M. *TL* **40**, 1555 (1999).
[159]Kündig, E.P., Saudan, C.M., Bernardinelli, G. *ACIEE* **38**, 1220 (1999).
[160]Evans, D.A., Miller, S.J., Lectka, T., von Matt, P. *JACS* **121**, 7559, 7582 (1999).
[161]Evans, D.A., Johnson, J.S., Olhava, E.J. *JACS* **122**, 1635 (2000).
[162]Zhuang, W., Thorhauge, J., Jorgensen, K.A. *CC* 459 (2000).
[163]Jnoff, E., Ghosez, L. *JACS* **121**, 2617 (1999).
[164]Ahrendt, K.A., Borths, C.J., MacMillan, D.W.C. *JACS* **122**, 4243 (2000).
[165]Hedberg, C., Pinho, P., Roth, P., Andersson, P.G. *JOC* **65**, 2810 (2000).
[166]Dossetter, A.G., Jamison, T.F., Jacobsen, E.N. *ACIEE* **38**, 2398 (1999).
[167]Huang, Y., Iwama, T., Rawal, V.H. *JACS* **122**, 7843 (2000).
[168]Bertilsson, S.K., Andersson, P.G. *JOMC* **603**, 13 (2000).
[169]Brunel, J.M., Legrand, O., Reymond, S., Buono, G. *JACS* **121**, 5807 (1999).
[170]Balsells, J., Walsh, P.J. *JOC* **65**, 5005 (2000).
[171]Meunier, D., Piechaczyk, A., de Mallmann, A., Basset, J.-M. *ACIEE* **38**, 3540 (1999).
[172]Bolm, C., Kuhn, T. *SL* 899 (2000).
[173]Hoshino, Y., Yamamoto, H. *JACS* **122**, 10452 (2000).
[174]Wang, Z.-X., Miller, S.M., Anderson, O.P., Shi, Y. *JOC* **64**, 6443 (1999).
[175]Tian, H., She, X., Shu, L., Yu, H., Shi, Y. *JACS* **122**, 11551 (2000).
[176]Allen, J.V., Drauz, K.-H., Flood, R.W., Roberts, S.M., Skidmore, J. *TL* **40**, 5417 (1999).
[177]Yu, H.-B., Zheng, X.-F., Lin, Z.-M., Hu, Q.-S., Huang, W.-S., Pu, L. *JOC* **64**, 8149 (1999).
[178]Carde, L., Davies, H., Geller, T.P., Roberts, S.M. *TL* **40**, 5421 (1999).
[179]Song, C.E., Roh, E.J. *CC* 837 (2000).
[180]Ford, A., Woodward, S. *SC* **29**, 189 (1999).
[181]Yanagi, T., Kikuchi, K., Takeuchi, H., Ishikawa, T., Nishimura, T., Kamijo, T. *CL*1203 (1999).
[182]Yang, G.-S., Hu, J.-B., Zhao, G., Ding, Y., Tang, M.-H. *TA* **10**, 4307 (1999).
[183]Nozaki, K., Kobori, K., Uemura, T., Tsutsumi, T., Takaya, H., Hiyama, T. *BCSJ* **72**, 1109 (1999).
[184]Schmitzer, A., Perez, E., Rico-Lattes, I., Lattes, A. *TL* **40**, 2947 (1999).

[185]Dakternieks, D., Dunn, K., Perchyonok, V.T., Schiesser, C.H. *CC* 1665 (1999).
[186]Yamanoi, Y., Imamoto, T. *JOC* **64**, 2988 (1999).
[187]Kuwano, R., Uemura, T., Saitoh, M., Ito, Y. *TL* **40**, 1327 (1999).
[188]Heldmann, D.K., Seebach, D. *HCA* **82**, 1096 (1999).
[189]Dakternieks, D., Dunn, K., Perchyonok, V.T., Schiesser, C.H. *CC* 1665 (1999).
[190]Ohtsuka, Y., Kubota, T., Ikeno, T., Nagata, T., Yamada, T. *SL* 535 (2000).
[191]Node, M., Nishide, K., Shigeta, Y., Shiraki, H., Obata, K. *JACS* **122**, 1927 (2000).
[192]Murata, K., Ikariya, T., Noyori, R. *JOC* **64**, 2186 (1999).
[193]Koike, T., Murata, K., Ikariya, T. *OL* **2**, 3833 (2000).
[194]Alonso, D.A., Nordini, S.J.M., Roth, P., Tarnai, T., Andersson, P.G., Thommen, M., Pittelkow, U. *JOC* **65**, 3116 (2000).
[195]Blanc, D., Ratovelomanana-Vidal, V., Marinetti, A., Genet, J.-P. *SL* 480 (1999).
[196]Ireland, T., Grossheimann, G., Wieser-Jeunesse, C., Knochel, P. *ACIEE* **38**, 3212 (1999).
[197]Burk, M.J., Hems, W., Herzberg, D., Malan, C., Zanotti-Gerosa, A. *OL* **2**, 4173 (2000).
[198]Mikami, K., Korenaga, T., Ohkuma, T., Noyori, R. *ACIEE* **39**, 3707 (2000).
[199]Studer, M., Burkhardt, S., Blaser, H.-U. *CC* 1727 (1999).
[200]LeBlond, C., Wang, J., Liu, J., Andrews, A.T., Sun, Y.-K. *JACS* **121**, 4920 (1999).
[201]Bertus, P., Phansavath, P., Ratovelomanana-Vidal, V., Genet, J.-P., Touati, A.R., Homri, T., Hassine, B.B. *TA* **10**, 1369 (1999).
[202]Pai, C.-C., Lin, C.-W., Lin, C.-C., Chen, C.-C., Chan, A.S.C., Wong, W.T. *JACS* **122**, 11513 (2000).
[203]Berens, U., Burk, M.J., Gerlach, A., Hems, W. *ACIEE* **39**, 1981 (2000).
[204]Jiang, Q., Xiao, D., Zhang, Z., Cao, P., Zhang, X. *ACIEE* **38**, 516 (1999).
[205]Troutman, M.V., Apella, D.H., Buchwald, S.L. *JACS* **121**, 4916 (1999).
[206]Kuwano, R., Sato, K., Kurokawa, T., Karube, D., Ito, Y. *JACS* **122**, 7614 (2000).
[207]Fehr, M.F., Consiglio, G., Scalone, M., Schmid, R. *JOC* **64**, 5768 (1999).
[208]Li, W., Zhang, Z., Xiao, D., Zhang, X. *JOC* **65**, 3489 (2000).
[209]Yonehara, K., Ohe, K., Uemura, S. *JOC* **64**, 9381 (1999).
[210]Chen, Y., Li, X., Tong, S., Choi, M.C.K., Chan, A.S.C. *TL* **40**, 957 (1999).
[211]Yamano, Y., Imamoto, T. *JOC* **64**, 2988 (1999).
[212]Gridnev, I.D., Higashi, N., Imamoto, T. *JACS* **122**, 10486 (2000).
[213]Andrus, M.B., Asgari, D. *T* **56**, 5775 (2000).
[214]Kohmura, Y., Katsuki, T. *TL* **41**, 3941 (2000).
[215]Malkov, A.V., Bella, M., Langer, V., Kocovsky, P. *OL* **2**, 3047 (2000).
[216]Bolm, C., Dabard, O.A.G. *SL* 360 (1999).
[217]Borg, G., Cogan, D.A., Ellman, J.A. *TL* **40**, 6709 (1999).
[218]Qian, C., Wang, L. *TA* **11**, 2347 (2000).
[219]Saito, S., Kano, T., Muto, H., Nakadai, M., Yamamoto, H. *JACS* **121**, 8943 (1999).
[220]Davies, H.M.L., Hansen, T., Churchill, M.R. *JACS* **122**, 3063 (2000).
[221]Tietze, L.F., Thede, K., Schimpf, R., Sannicolo, F. *CC* 583 (2000).
[222]Buezo, N.D., Mancheno, O.G., Carretero, J.C. *OL* **2**, 1451 (2000).
[223]Priego, J., Carretero, J.C. *SL* 1603 (1999).
[224]Donde, Y., Overman, L.E. *JACS* **121**, 2933 (1999).
[225]He, S., Kozmin, S.A., Rawal, V.H. *JACS* **122**, 190 (2000).
[226]Henderson, K.W., Kerr, W.J., Moir, J.H. *CC* 479 (2000).
[227]Asensio, G., Cuenca, A., Medio-Simon, M., Gavina, P. *TL* **40**, 3939 (1999).
[228]Adrio, J., Carreyero, C. *JACS* **121**, 7411 (1999).
[229]Pridgen, L.N., Huang, K., Shilcrat, S., Tickner-Eldridge, A., DeBrosse, C., Haltiwanger, R.C. *SL* 1612 (1999).
[230]Verdaguer, X., Moyano, A., Pericas, M.A., Riera, A., Maestro, M.A., Mahia, J. *JACS* **122**, 10242 (2000).

[231]Deiters, A., Hoppe, D. *ACIEE* **38**, 546 (1999).
[232]Tomooka, K., Komine, N., Sasaki, T., Shimizu, H., Nakai, T. *TL* **39**, 9715 (1998).
[233]Serino, C., Stehle, N., Park, Y.S., Florio, S., Beak, P. *JOC* **64**, 1160 (1999).

Chloramine-M.

Epoxidation.[1] A combination of chloramine-M, PhCHO, and benzyltriethylammonium yields an epoxidation agent.

[1]Yang, D., Zhang, C., Wang, X.-C. *JACS* **122**, 4039 (2000).

Chloramine-T. 20, 103

Chlorination.[1] Heterocyclic ketene aminals undergo chlorination with this reagent (10 examples, 80–92%).

chloramine T
MeOH , $CHCl_3$
0 ~ 25°
85%

[1]Liu, B., Wang, M.-X., Huang, Z.-T. *SC* **29**, 4241 (1999).

Chlorine dioxide.

Alcohols.[1] Oxidation of trialkylaluminums with ClO_2 in ether at room temperature affords alcohols (3 examples, 85–90%).

[1]Kuchin, A.V., Dvornikova, I.A., Nalimova, I.Yu. *RCB* **48**, 2001 (1999).

Chloroborane.

Hydroboration.[1] The dioxane complex of this borane is a highly reactive hydroboration agent.

[1]Kanth, J.V.B., Brown, H.C. *OL* **1**, 315 (1999).

Chloro(1,5-cyclooctadiene)cyclopentadienylruthenium(I). 20, 104

Unsaturated ketones. A three-component coupling that linearly combines a 1-alkyne and an enone while introducing a chlorine atom at C-2 of the original alkyne, giving *(E)*-5-chloro-4-alkenyl ketones[1] is achieved. With 3-hexynol as activator, a 1,5-diene synthesis from allenes and enones is effected by Cp(cod)RuCl and $CeCl_3 \cdot 7H_2O$ in DMF.[2]

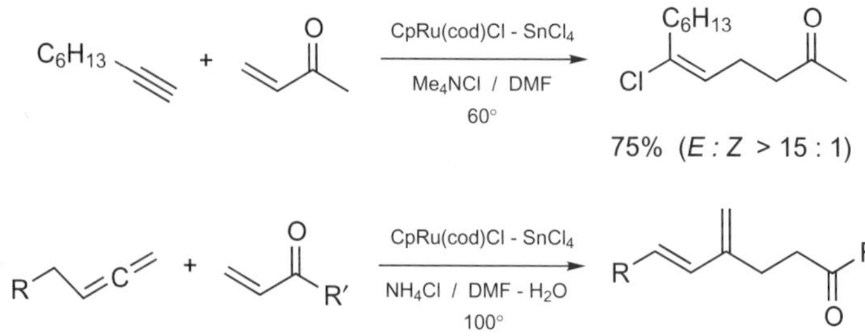

Allylic substitutions.[3] Successful substitution of allylic carbonates with retention of configurations using carbon and nitrogen nucleophiles has been carried out in the presence of Cp(cod)RuCl–NH_4PF_6.

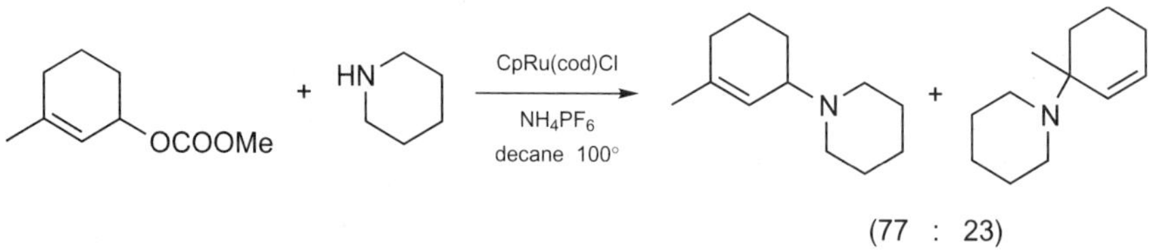

γ-Butyrolactones.[4] Homopropargyl alcohols undergo catalyzed oxidation–isomerization. The oxidizing agent is *N*-hydroxysuccinimide.

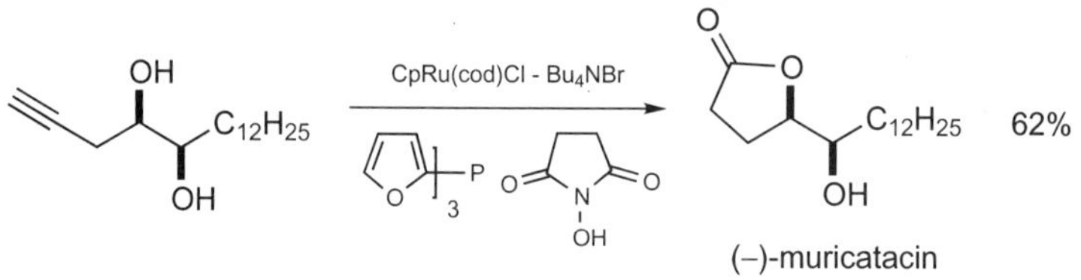

[1]Trost, B.M., Pinkerton, A.B. *JACS* **121**, 1988 (1999).
[2]Trost, B.M., Pinkerton, A.B. *JACS* **121**, 4068 (1999).
[3]Morisaki, Y., Kondo, T., Mitsudo, T. *OM* **18**, 4742 (1999).
[4]Trost, B.M., Rhee, Y.H. *JACS* **121**, 11680 (1999).

Chloro(1,5-cyclooctadiene)pentamethylcyclopentadienylruthenium(I).

1,3-Dienes.[1,2] Two types of reactions involving 1-alkynes leading to 1,3-dienes are illustrated in the following.

Ph—≡ + Ph—COOH —Cp*Ru(cod)Cl, dioxane 23°→ Ph–CH=CH–CH=C(Ph)–OBz

98%

Ph—≡ + $Me_3Si-CHN_2$ —Cp*Ru(cod)Cl, dioxane 60°→ Me_3Si–CH=C(Ph)–CH=CH–$SiMe_3$

72% (*E* : *Z* 70 : 30)

5-Methylene-2-oxytetrahydropyrans.[3] Regioselective coupling of propargyl alcohols and allyl alcohols furnishes the cyclic hemiacetals (at room temperature) or acetals (at 80°).

OH + OH —Cp*Ru(cod)Cl, 23°→ OR, O

R = H , allyl

(Z)-1,5-Hexadienyl carboxylates.[4] Codimerization of vinyl carboxylates with 2-substituted 1,3-butadienes proceeds in high yields with excellent regioselectivity. The reaction involves the disubstituted double bond of the diene.

Sulfur compounds. Displacement of allylic carbonates with thiols takes place at room temperature. Linear products also predominate from reaction with secondary carbonates.[5] 1,2-Bis(phenylthio)alkanes are formed when 1-alkenes are treated with diphenyl disulfide under catalytic conditions ([Ru] catalyst, toluene, 100°).[6] Various functional groups (ester, silyl, etc.) directly bonded to the double bond are tolerated.

[1]Le Paih, J., Derien, S., Dixneuf, P.H. *CC* 1437 (1999).
[2]Le Paih, J., Derien, S., Ozdemir, I., Dixneuf, P.H. *JACS* **122**, 7400 (2000).
[3]Derien, S., Ropartz, L., Le Paih, J., Dixneuf, P.H. *JOC* **64**, 3524 (1999).
[4]Fujiwhara, M., Nishikawa, T., Hori, Y. *OL* **1**, 1635 (1999).
[5]Kondo, T., Morisaki, Y., Uenoyama, S., Wada, K., Mitsudo, T. *JACS* **121**, 8657 (1999).
[6]Kondo, T., Uenoyama, S., Fujita, K., Mitsudo, T. *JACS* **121**, 482 (1999).

Chloro(1,5-cyclooctadiene)triisopropylphosphinerhodium(I).

Diaryl ketones.[1] *N*-Pyrazylaldimines couple with aryl halides and the products are readily hydrolyzed to the ketones.

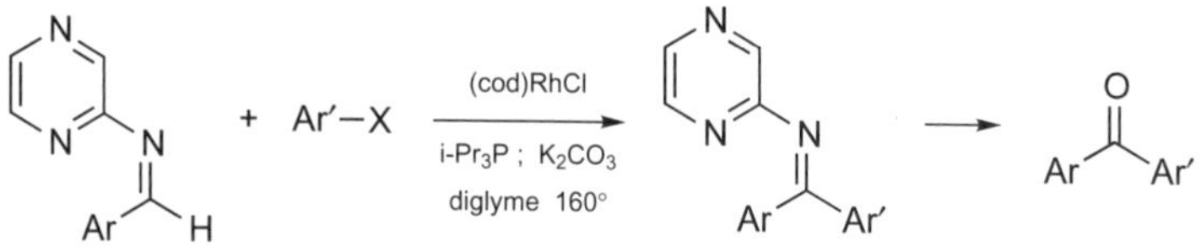

(cod)RhCl
i-Pr$_3$P ; K_2CO_3
diglyme 160°

[1]Ishiyama, T., Hartwig, J. *JACS* **122**, 12043 (2000).

Chloro(cyclopentadienyl)bis(triphenylphosphine)ruthenium(I).

(Z)-Enediones.[1] α-Diazo ketones dimerize with decomposition on exposure to the Ru-complex. As more than statistical quantitites of cross-reaction products are obtained from an α-diazo ketone with either ethyl diazoacetate or trimethylsilyldiazomethane, the method has a preparative value for the corresponding unsymmetrical (*Z*)-enones.

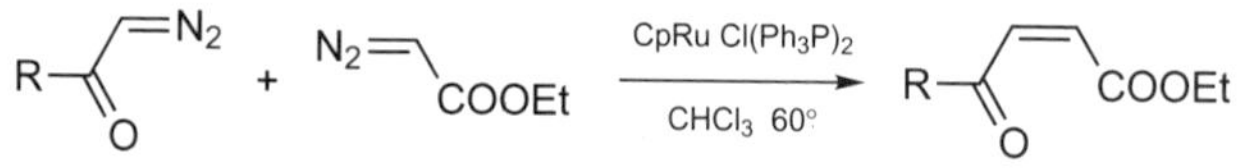

CpRu Cl(Ph_3P)$_2$
$CHCl_3$ 60°

α-Heterosubstituted ketones.[2] When α-diazo ketones are treated with the Ru catalyst in the presence of amines or thiols, α-amino ketones and α-thio ketones are formed in excellent yields.

[1]Del Zotto, A., Baratta, W., Verardo, G., Rigo, P. *EJOC* 2795 (2000).
[2]Del Zotto, A., Baratta, W., Rigo, P. *JCS(P1)* 3079 (1999).

***B*-Chlorodicyclohexylborane.**

Enolboration.[1] Reaction of α-bromoketones with this reagent and triethylamine (enolboration) followed by addition of aldehydes and oxidative workup leads to *anti*-α - bromo-β-hydroxyketones.

[1]Brown, H.C., Zou, M.-F., Ramachandran, P.V. *TL* **40**, 7875 (1999).

2-Chloro-4,6-dimethoxy-1,3,5-triazine.

Esterification.[1] Activated esters derived from carboxylic acids with the title reagent react with alcohols (including tertiary alcohols) to give esters.

[1]Kaminski, J.E., Kaminski, Z.J., Gora, J. *S* 593 (1999).

2-Chloro-1,3-dimethylimidazolinium chloride. 20, 105

Ketenimines.[1] Elimination of hydrogen sulfide from secondary *N*-arylthioamides is readily accomplished with the imidazolinium salt (and triethylamine).

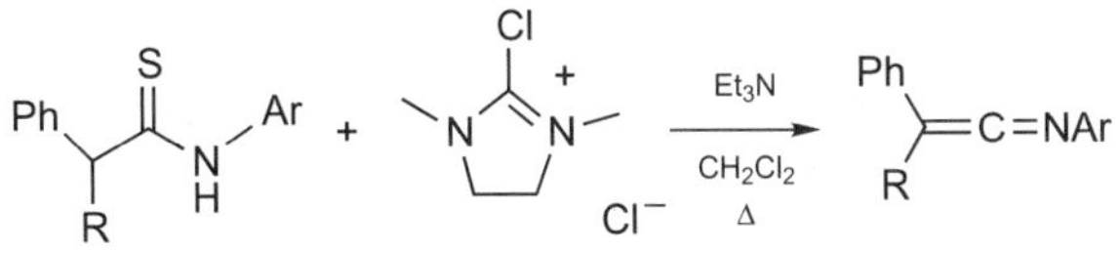

Assorted transformations.[2] The function of this reagent includes the conversion of primary alcohols to chlorides, oxidation of secondary and primary alcohols in the presence of hexamethylenetetramine, chlorination of 1,3-diketones, reduction of sulfoxides, preparation of ureas and carbamates from hydroxamic acids, and of carbamides from oximes.

[1]Shimizu, M., Gama, Y., Takagi, T., Shibakami, M., Shibuya, I. *S* 517 (2000).
[2]Isobe, T., Ishikawa, T. *JOC* **64**, 5832 (1999).

Chloromethyl(dimethyl)silyl chloride.

N-Methylation of amides.[1] This method involves formation of silaoxazolines and decomposition with CsF. (9 examples, 72–85%).

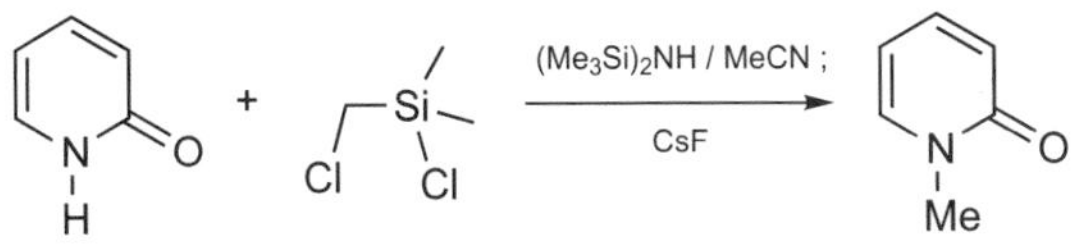

[1]Bassindale, A.R., Parker, D.J., Patel, P., Taylor, P.G. *TL* **41**, 4933 (2000).

1-Chloromethyl-4-fluoro-1,4-diazoniabicyclo[2.2.2]octane salts. 18, 100; **20,** 106

2-Fluoro sugars.[1] Glycals undergo fluoroalkoxylation by reaction with one of these salts followed by alcoholysis.

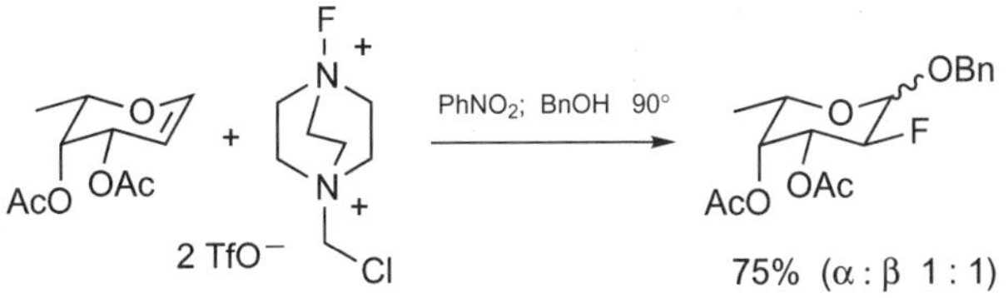

Fluorination of aromatic compounds. The reaction of various indoles takes place at room temperature in aq MeCN to afford 3-fluoroxindoles.[2] 4-Substituted phenols are converted to 4-fluoro-2,5-cyclohexadienones.[3]

[1]Vincent, S.P., Burkart, M.D., Tsai, C.-Y., Zhang, Z., Wong, C.-H. *JOC* **64**, 5264 (1999).
[2]Takeuchi, Y., Tarui, T., Shibata, N. *OL* **2**, 639 (2000).
[3]Stavber, S., Jereb, M., Zupan, M. *SL* 1375 (1999).

Chloromethyllithium.

Allylamines.[1] *N,N*-Disubstituted α-aminoaldehydes are transformed to the allylamines in one step without racemization.

[1]Concellon, J.M., Baragana, B., Riego, E. *TL* **41**, 4361 (2000).

Chloromethylsulfonyl chloride.

Sulfonates.[1] The sulfonates derived from alcohols and the title reagent are good electrophiles toward azide and cyanide ions.

[1]Shimizu, T., Ohzeki, T., Hiramoto, K., Hori, N., Nakata, T. *S* 1373 (2000).

Chloromethyl trimethylsilylmethyl sulfide.

Tetrahydrothiophenes.[1] Thiocarbonyl ylides are generated from the title reagent on contact with CsF. Trapping with 1,3-dipolarophiles leads to substituted tetrahydrothiophenes. Chiral adduct formation by this reaction is possible when the trapping agents contain a chiral auxiliary.

[1]Karlsson, S., Hogberg, H.-E. *OL* **1**, 1667 (1999).

***m*-Chloroperoxybenzoic acid, MCPBA. 13,** 76–79; **14,** 84–87; **15,** 86; **16,** 80–83; **17,** 76; **18,** 101; **19,** 94–95; **20,** 106–108

Oxidation of nitrogen and sulfur compounds. These oxidations serve to complete transformation of amines to hydroxylamines,[1] hydrazones to nitriles,[2] and β-amino esters to conjugated esters.[3]

OTBS
MCPBA - NaHCO$_3$
OTBS
N
—OMe
CH_2Cl_2
N
CN
63% (93% ee)

Dithioesters are oxidized at the thiono group.[4]

Benzylic oxidation.[5] $ArCH_2R$ are oxidized to ArCOR with MCPBA–O_2 ($NaHCO_3$) in dichloromethane at room temperature.

Baeyer–Villiger oxidations. The oxidation with MCPBA is catalyzed by TfOH or $Sc(OTf)_3$.[6] 1,3-Dioxenes give ring contraction products (aldehydes) if direct distillation is applied.[7] The rearrangement is caused by *m*-chlorobenzoic acid.

A preparation of 3-hydroxyindole-2-carboxylic esters (yields 80–90%) involves Vilsmeier–Haack formylation and the Baeyer–Villiger reaction.[8]

R–indole–COOR′ → ($POCl_3$ - PhN(CHO)Me ; MCPBA - TsOH / CH_2Cl_2) → R–3-hydroxyindole–COOR′ (OH, NH)

80–90%

Epoxidation.[9] Alkenes are epoxidized at low temperatures (e.g., −78°) when catalyzed by $(MeCN)_4CuPF_6$.

Dihydroxylation.[10] The conventional method of alkene dihydroxylation with OsO_4 and *N*-methylmorpholine *N*-oxide (NMO) has been modified such that the latter reagent is replaced by substoichiometric *N*-methylmorpholine and 1.4 equiv of MCPBA.

[1]Tokuyama, H., Kuboyama, T., Amano, A., Yamashita, T., Fukuyama, T. *S* 1299 (2000).
[2]Diez, E., Fernandez, R., Martin-Zamora, E., Pareja, C., Prieto, A., Lassaletta, J.M. *TA* **10**, 1145 (1999).
[3]Davies, S.G., Smethurst, C.A.P., Smith, A.D., Smyth, G.D. *TA* **11**, 2437 (2000).
[4]Corbin, F., Alayrac, C., Metzner, P. *TL* **40**, 2319 (1999).
[5]Ma, D., Xia, C., Tian, H. *TL* **40**, 8915 (1999).
[6]Kotsuki, H., Arimura, K., Araki, T., Shinohara, T. *SL* 462 (1999).
[7]Wattenbach, C., Maurer, M., Frauenrath, H. *SL* 303 (1999).
[8]Hickman, Z., Sturino, C.F., Lachance, N. *TL* **41**, 8217 (2000).
[9]Andrus, M.B., Poehlein, B.W. *TL* **41**, 1013 (2000).
[10]Bergstad, K., Piet, J.J.N., Bäckvall, J.-E. *JOC* **64**, 2545 (1999).

[Chloro(phenylthio)methylene]dimethylammonium chloride.

Alkyl halides.[1] Primary alcohols are converted by this reagent, $[Me_2N{=}C(Cl)SPh]Cl$, to the chlorides in excellent yields without affecting unprotected secondary hydroxyl groups. In the presence of Bu_4NBr, the bromides are obtained.

[1]Gomez, L., Gellibert, F., Wagner, A., Mioskowski, C. *TL* **41**, 6049 (2000).

***N*-Chlorosuccinimide NCS. 13,** 79–80; **15,** 86–88; **18,** 101–102; **19,** 95–96; **20,** 108

(E)-Bromoalkenes.[1] (*Z*)-1-Dialkylbora-1-bromoalkenes are induced to decompose stereoselectively by NCS. A special solvent effect is exerted by *N*, *N*-dimethylformamide, and is superior to DMSO.

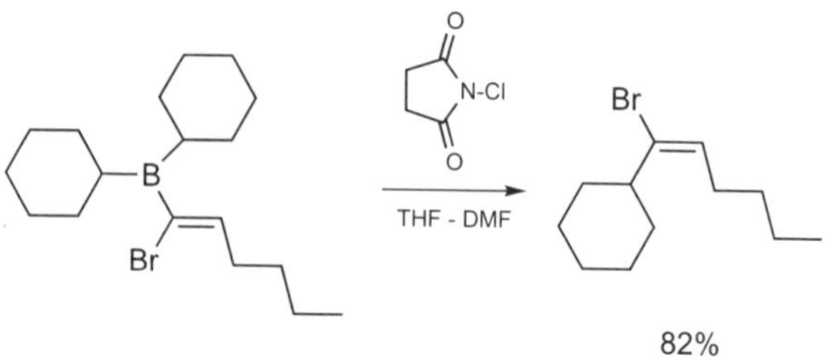

THF - DMF

82%

Methylthiomethylation.[2] The Corey–Kim reagent (NCS–dimethyl sulfide) induces cyclization of tryptamine derivatives while introducing a methylthiomethyl group at C-3. An efficient route to physostigmine is based on this process.

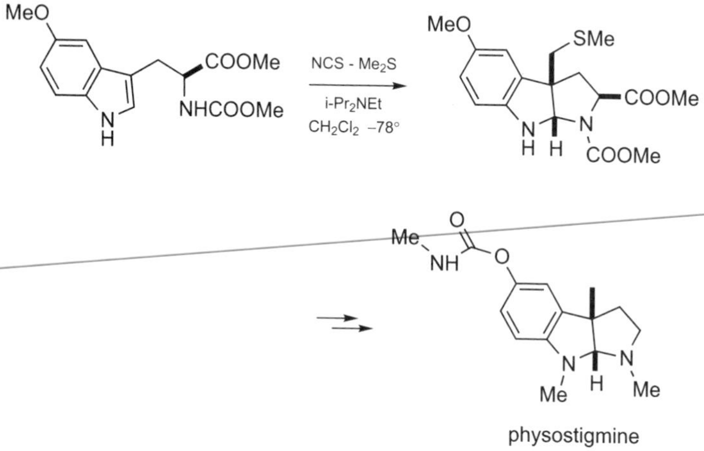

NCS - Me_2S / i-Pr_2NEt / CH_2Cl_2 −78°

physostigmine

Chlorination.[3] Anilides and deactivated anilines are chlorinated with NCS–2-propanol.

Biaryls.[4] Homocoupling of ArZnI employs NCS as an oxidant in the presence of a Pd(0) catalyst. Yields are good.

[1]Hoshi, M., Shirakawa, K. *TL* **41**, 2595 (2000).
[2]Kawahara, M., Nishida, A., Nakagawa, M. *OL* **2**, 675 (2000).
[3]Zanka, A., Kubota, A. *SL* 1984 (1999).
[4]Hossain, K.M., Shibata, T., Takagi, K. *SL* 1137 (2000).

Chlorosulfonyl isocyanate. 13, 80–81; **18,** 102

N-Allyl carbamates.[1] Preparation of the carbamates from allyl ethers may incur 1,3-transposition.

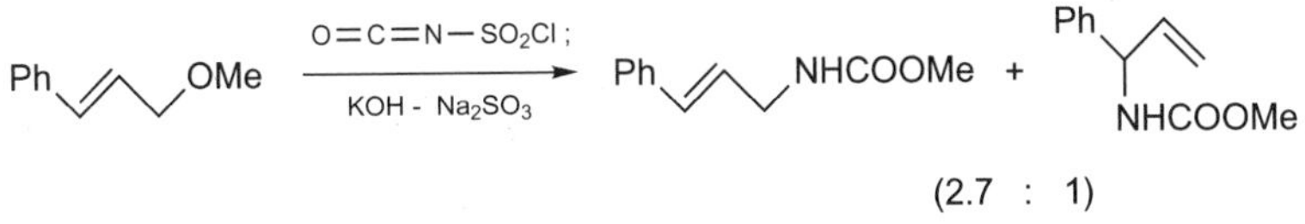

(2.7 : 1)

[3 + 2]Cycloaddition. γ-Lactams are formed on reaction with allylsilanes.[2] Of particular interest is the formation of a bridged ring representative that serves as a precursor of peduncularine.[3]

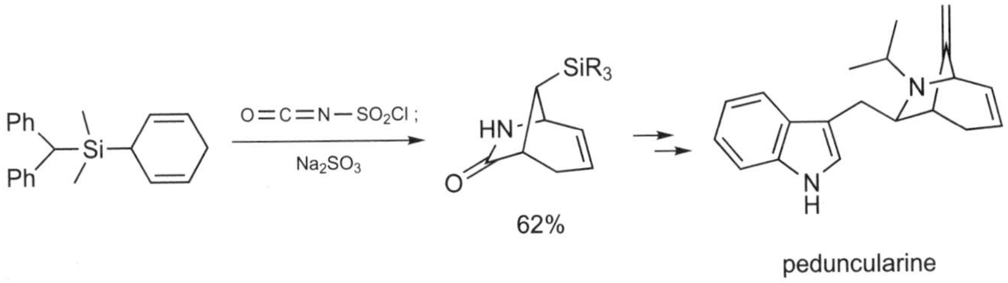

62%

peduncularine

[1]Kim, J.D., Lee, M.H., Lee, M.J., Jung, Y.H. *TL* **41**, 5073 (2000).
[2]Isaka, M., Williard, P.G., Nakamura, E. *BCSJ* **72**, 2115 (1999).
[3]Roberson, C.W., Woerpel, K.A. *OL* **2**, 621 (2000).

Chloro(triphenylphosphine)gold(I).

Hexaalkylditins.[1] Oxidative dimerization of R_3SnH (e.g., Bu_3SnH) is mediated by $(Ph_3P)AuCl$.

Reductive silylations.[2] In the presence of $(Ph_3P)AuCl$ and Bu_3P, aldehydes and imines undergo reductive silylation by a hydrosilane ($PhMe_2SiH$). Ketones do not react under these conditions.

[1]Ito, H., Yajima, T., Tateiwa, J., Hosomi, A. *TL* **40**, 7807 (1999).
[2]Ito, H., Yajima, T., Tateiwa, J., Hosomi, A. *CC* 981 (2000).

Chlorotris(triphenylphosphine)rhodium(I). 19, 96–98; 20, 108–109

Allylic alkylations.[1–3] Highly regioselective alkylation of both hard and soft nucleophiles (phenolates, sulfonamides, and phenylsulfonylacetic esters, respectively) is also possible with the Rh catalyst modified by the added $(MeO)_3P$. The countercation seems to play an important role in the displacement with alkali phenolates to afford branched allylic ethers; thus, reaction with Li salts shows the highest regioselectivity but the product yields are low. The best compromise is to use Na phenolates.

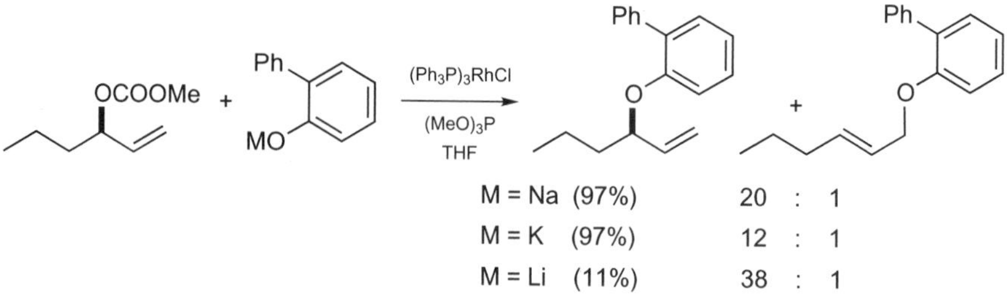

M = Na	(97%)	20 : 1
M = K	(97%)	12 : 1
M = Li	(11%)	38 : 1

Reformatsky-type reaction.[4] This procedure involves treatment of the α-bromo esters with Et_2Zn and the Rh(I) complex. This mild and efficient reaction is applicable to both inter- and intramolecular versions.

Cycloadditions. Substituted indolines are formed from *N*-functionalized 1-alkynyl amides via a [2 + 2 + 2]cycloaddition process.[5] A study on the regioselectivity and stereoselectivity of the [5 + 2]cycloaddition promoted by the Wilkinson catalyst together with AgOTf has been delineated.[6]

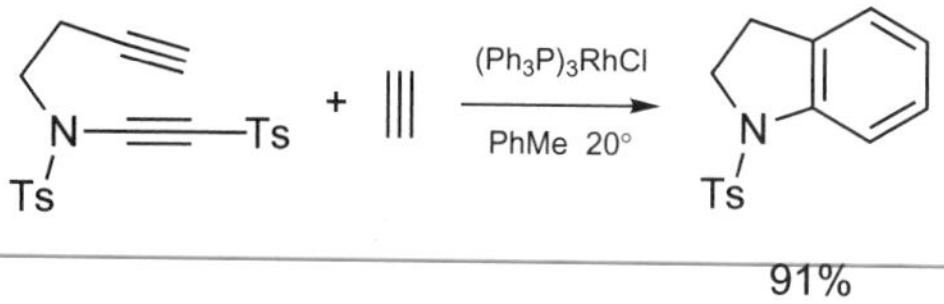

91%

Addition to vinyl aminopropyl ethers.[7] Boranes add to the double bond in the anti-Markovnikov sense under the influence of the Rh catalyst. However, intramolecular hydroamination to give tetrahydro-2-methyloxazine occurs in the presence of Pd or Pt complexes.

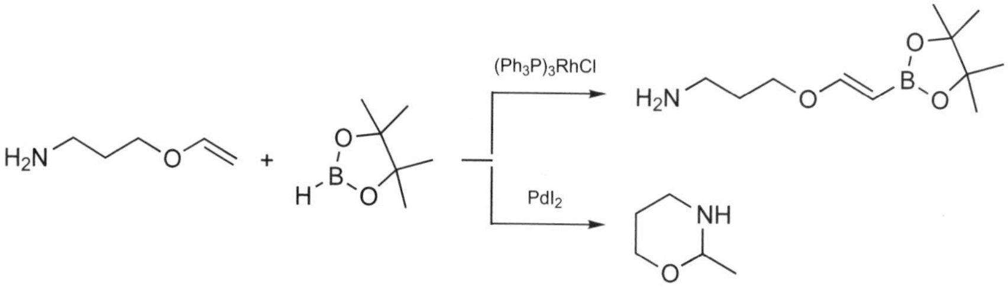

[1]Evans, P. A., Leahy, D.K. *JACS* **122**, 5012 (2000).
[2]Evans, P. A., Robinson, J.E., Nelson, J.D. *JACS* **121**, 6761 (1999).
[3]Evans, P.A., Kennedy, L.J. *OL* **2**, 2213 (2000).
[4]Kanai, K., Wakabayashi, H., Honda, T. *OL* **2**, 2549 (2000).

[5]Witulski, B., Stengel, T. *ACIEE* **38**, 2426 (1999).
[6]Wender, P.A., Dyckman, A.J., Husfeld, C.O., Kadereit, D., Love, J.A., Rieck, H. *JACS* **121**, 10442 (1999).
[7]Vogels, C.M., Hayes, P.G., Shaver, M.P., Westcott, S.A. *CC* 51 (2000).

Chlorotris(triphenylphosphine)rhodium(I)–2-Amino-3-picoline.

Hydroacylation. α-Cleavage of a phenylethyl ketone and delivery of the acyl group (after elimination of styrene) to an alkene occurs around Rh. The reaction starts from imine formation of the ketone with 2-amino-3-picoline. After cleavage, the [Rh]-H species combines with the alkene and then undergoes reductive elimination.[1]

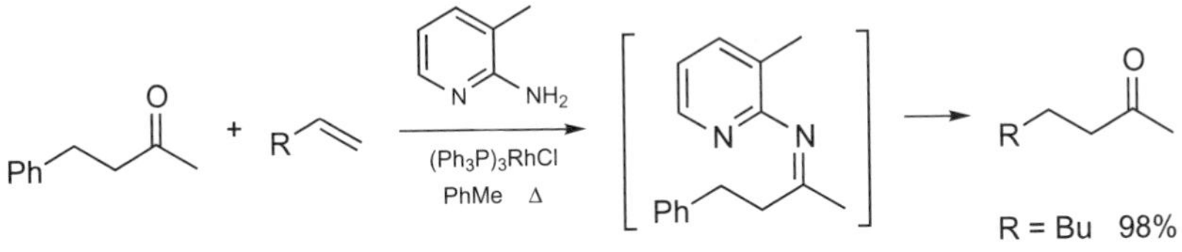

$(Ph_3P)_3RhCl$, PhMe Δ

R = Bu 98%

Further extension of the method results in a direct synthesis of ketones from aldehydes and alkenes[2] and, in the case of an aromatic aldehyde, *o*-substituted aryl ketones.[3] Imines of aryl ketones undergo *o*-alkylation.

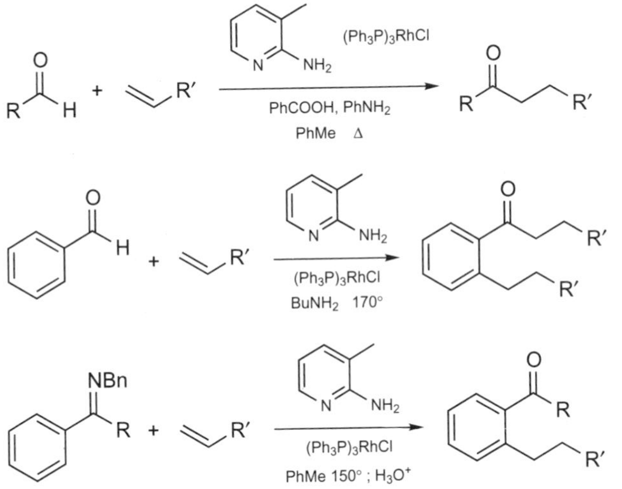

$(Ph_3P)_3RhCl$, PhCOOH, $PhNH_2$, PhMe Δ

$(Ph_3P)_3RhCl$, $BuNH_2$ 170°

$(Ph_3P)_3RhCl$, PhMe 150° ; H_3O^+

[1]Jun, C.-H., Lee, H. *JACS* **121**, 880 (1999).
[2]Jun, C.-H., Lee, D.-Y., Lee, H., Hong, J.-B. *ACIEE* **39**, 3070 (2000).
[3]Jun, C.-H., Hong, J.-B., Kim, Y.-H., Chung, K.-Y. *ACIEE* **39**, 3440 (2000).

Chromium–carbene complexes. 13, 82–83; **14,** 91–93; **15,** 93–95; **16,** 88–92; **17,** 80–84; **18,** 103–104; **19,** 98–101; **20,** 110–111

Rearrangement. Allyloxy(aryl)carbene complexes are converted to allyl aryl ketones on Pd(0) catalysis.[1]

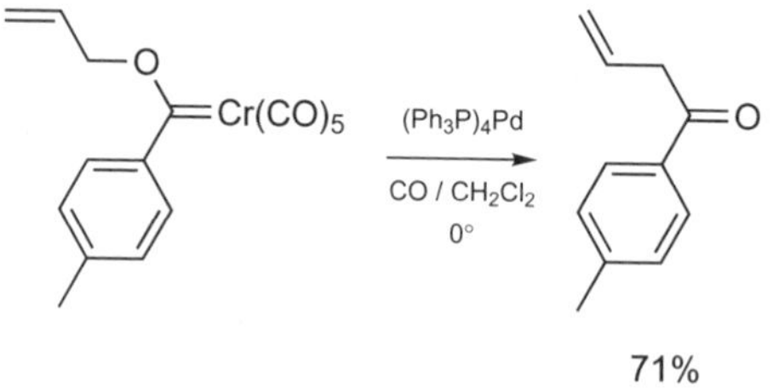

=Cr(CO)5 (Ph3P)4Pd CO / CH2Cl2 0° =O

71%

Cycloadditions. Pyrolysis of *O*-alkynyl cyclopropylcarbene complexes generates fused cyclopentenones. This transformation has been exploited in a synthesis of a vitamin D_3 synthon.[2]

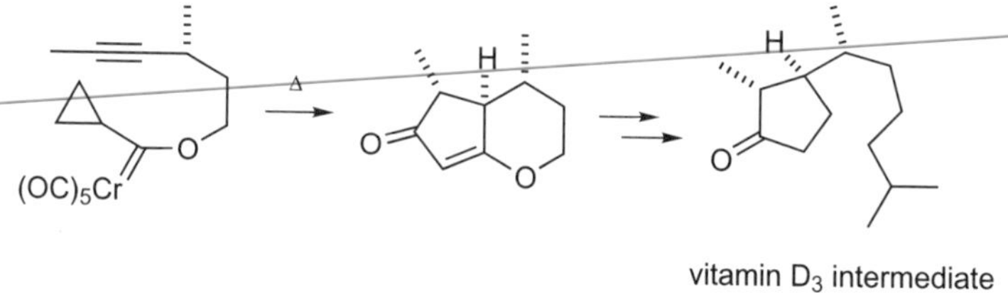

(OC)5Cr O Δ H O O H O

vitamin D_3 intermediate

3-Substituted 1,2-naphthoquinones are formed by an intramolecular reaction of *o*-alkenylarylcarbene complexes. With an electron-rich aromatic nucleus, the photoinduced benzannulation is sluggish. The use of *t*-butyl isocyanide instead of CO circumvents such problems.[3]

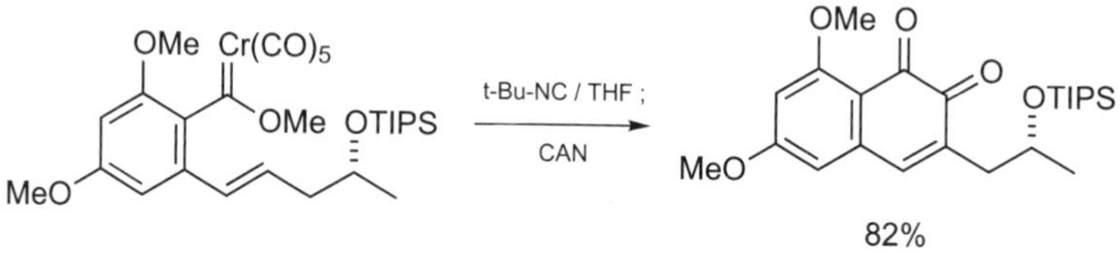

OMe Cr(CO)5 OMe OTIPS MeO t-Bu-NC / THF ; CAN OMe O O OTIPS MeO

82%

A chromane synthesis is readily accomplished from pentacarbonyltetrahydropyranylidenechromium via alkylidenation and photoinduced cycloaddition.[4]

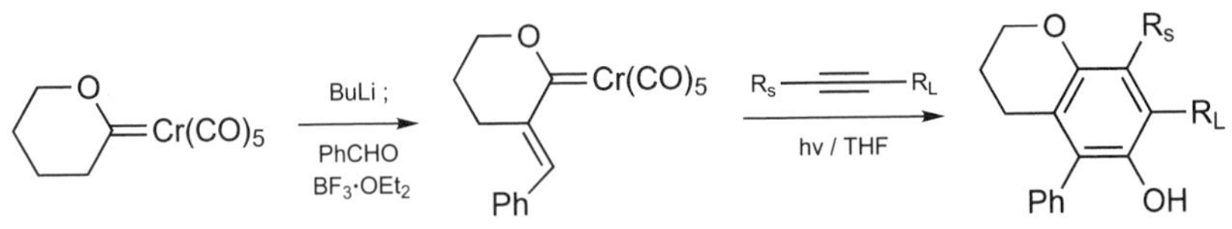

Reductive cyclization of chromium–carbene complexes that contain a triple bond in the carbon chain leads to bicyclic butenolides.[5] Five- and seven-membered enol ethers are assembled when conjugated carbene complexes react with lithium enolates and dienolates, respectively.[6]

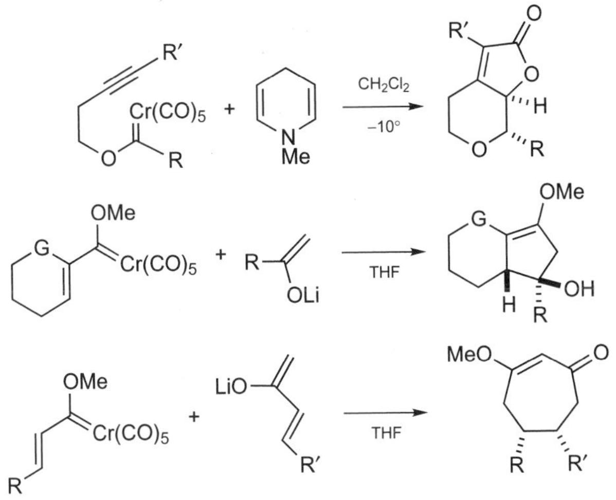

Highly functionalized diaryl ethers are accessible from reaction of aryloxy-substituted Fischer carbene complexes with alkynes.[7]

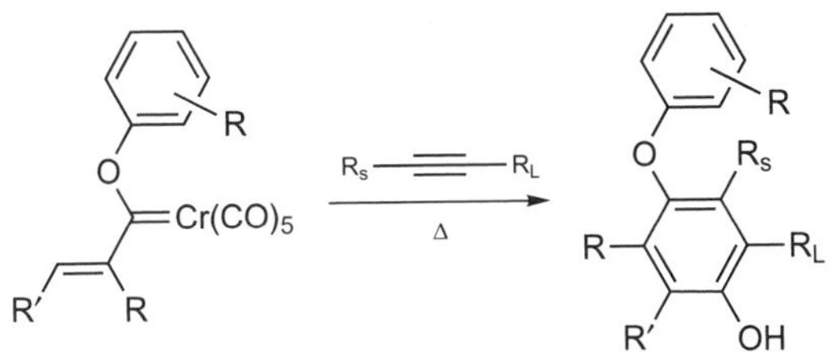

Diketones. Pd(0)-catalyzed coupling reactions involving $ArC(ONMe_4){=}Cr(CO)_5$ and ArI–CO give α-diketones,[8] whereas a Pd(II) catalyst mediates diacylation of alkenes to furnish 1,4-diketones.[9]

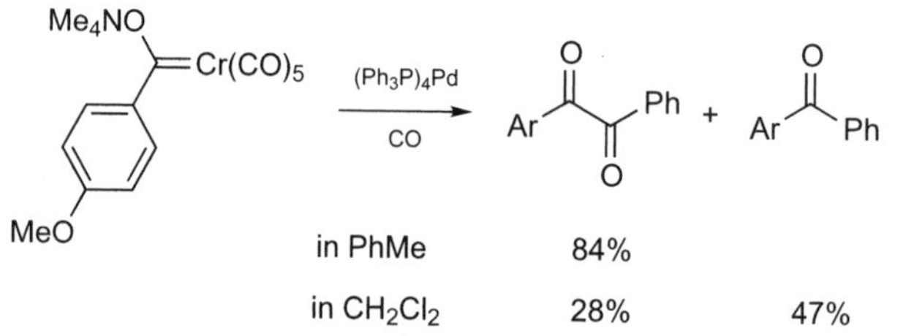

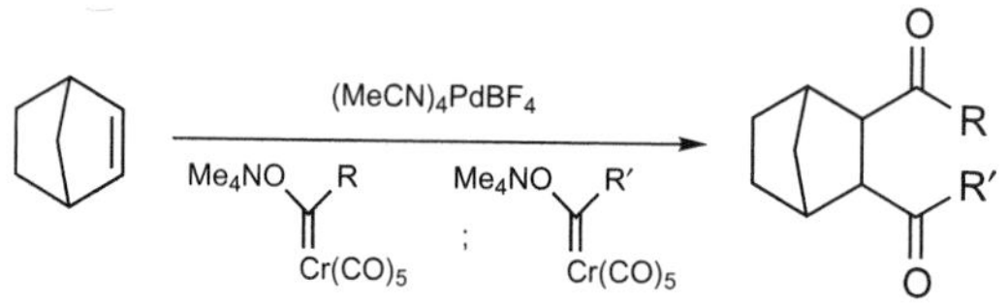

in PhMe 84%

in CH_2Cl_2 28% 47%

[1]Sakurai, H., Tanabe, K., Narasaka, K. *CL* 75, 309 (1999).
[2]Yan, J., Herndon, J.W. *JOC* **63**, 2325 (1998).
[3]Merlic, C.A., Aldrich, C.C., Albaneze-Walker, J., Saghatelian, A. *JACS* **122**, 3224 (2000).
[4]Weyershausen, B., Dötz, K.-H. *SL* 231 (1999).
[5]Rudler, H., Parlier, A., Certal, V., Vaissermann, J. *ACIEE* **39**, 3417 (2000).
[6]Barluenga, J., Alonso, J., Rodriguez, F., Fananas, F.J. *ACIEE* **39**, 2460 (2000).
[7]Pulley, S.R., Sen, S., Vorogushin, A., Swanson, E. *OL* **1**, 1721 (1999).
[8]Sakurai, H., Tanabe, K., Narasaka, K., Yamane, M., Ishibashi, Y. *CL* 168 (2000).
[9]Yamane, M., Ishibashi, Y., Sakurai, H., Narasaka, K. *CL* 174 (2000).

Chromium(II) chloride. 13, 84; **14,** 94–97; **15,** 95–96; **16,** 93–94; **17,** 84–85; **18,** 104; **19,** 101; **20,** 111–113

Alcohol synthesis. Halogenated allylchromium reagents can be prepared and used in a reaction with carbonyl compounds.[1]

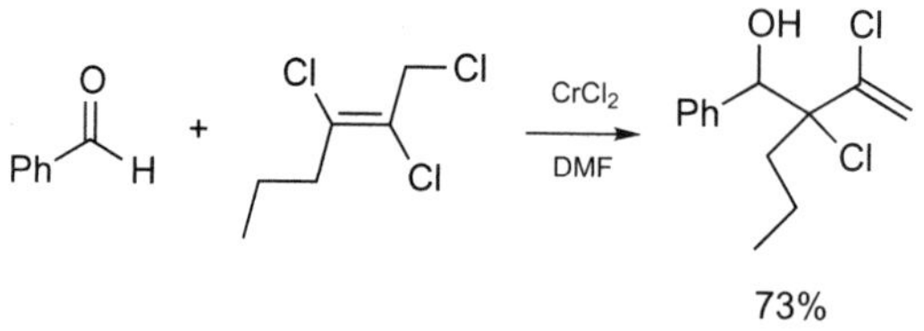

73%

2-Tetrahydrofuranyl ethers.[2] At room temperature alcohols are derivatized into the ethers with $CrCl_2$ and CCl_4 in THF under essentially neutral conditions.

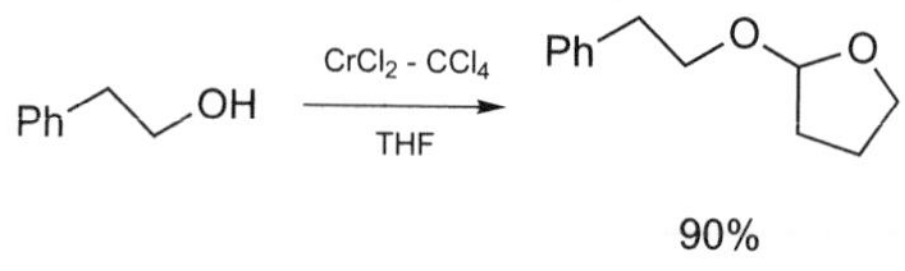

90%

Homologation. 2-Chloro-1-alken-3-ols are obtained from aldehydes on treatment with 1,1,1-trichloroethane in the presence of $CrCl_2$.[3] They are precursors of propargyl alcohols. Conversion of aldehydes to alkenylsilanes with one more carbon is accomplished by the $CrCl_2$-catalyzed reaction with MeI, Mn, and Me_3SiCl in THF.[4]

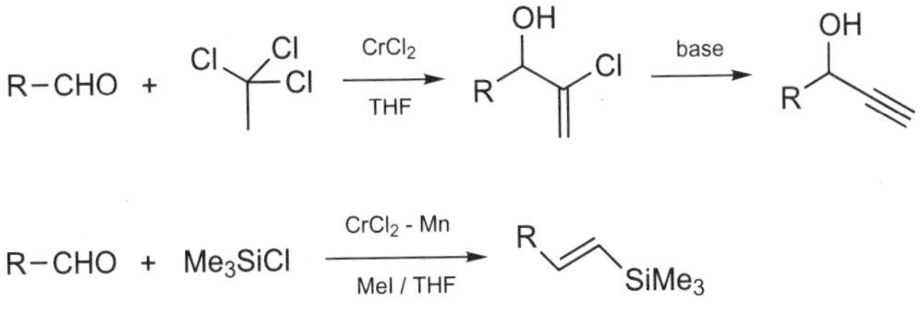

Reductive cyclization.[5] Azide reduction with $CrCl_2$ is followed by intramolecular addition of the resulting amine to a cross-conjugated dienone system. This efficient assemblage of fused-bridged tricycles has been incorporated as the final and key step in a synthesis of (−)-cylindricine.

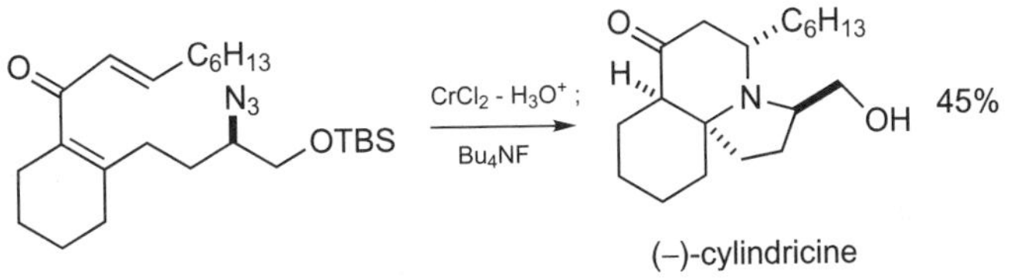

(–)-cylindricine

2-(1-Hydroxyalkyl)cyclopropanols.[6] 2-Alkoxycyclopropylchromium(III) species are generated from enones on treatment with $CrCl_2$. The subsequent reaction of these reagents with aldehydes afford diol products.

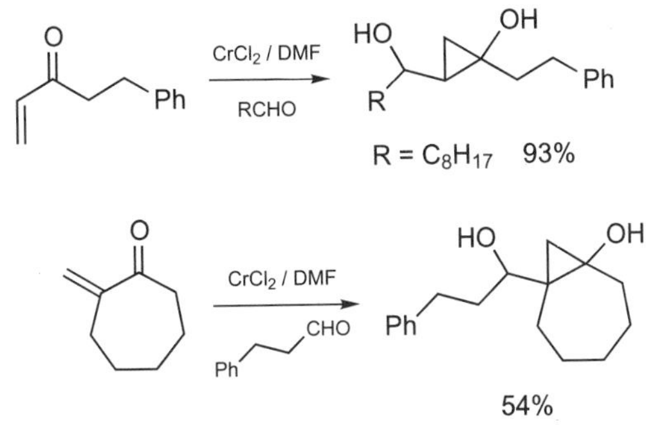

[1]Baati, R., Gouverneur, V., Mioskowski, C. *JOC* **65**, 1235 (2000).
[2]Baati, R., Valleix, A., Mioskowski, C., Barma, D.K., Falck, J.R. *OL* **2**, 485 (2000).
[3]Falck, J.R., Barma, D.K., Mioskowski, C. *TL* **40**, 2091 (1999).
[4]Takai, K., Hikasa, S., Ichiguchi, T., Sumino, N. *SL* 1769 (1999).
[5]Molander, G.A., Rönn, M. *JOC* **64**, 5183 (1999).
[6]Toratsu, C., Fujii, T., Suzuki, T., Takai, K. *ACIEE* **39**, 2725 (2000).

Chromium(II) chloride–nickel(II) halide. **14,** 97–98; **15,** 96–97; **17,** 86; **18,** 105; **19,** 102; **20,** 113–114

Alkenylation of aldehydes. Alkenylchromium(III) species are generated by the electrochemical reaction of $CrCl_2$-$NiBr_2$ with alkenyl bromides in DMF. Their reaction with ArCHO in the presence of Me_3SiCl gives silyl ethers of allylic alcohols.[1] Instead of electrochemical reduction, aluminum can be employed as the electron source.[2] An alternative promoter is tetrakis(dimethylamino)ethylene.[3]

[1]Kuroboshi, M., Tanaka, M., Kishimoto, S., Tanaka, H., Torii, S. *SL* 69 (1999).
[2]Kuroboshi, M., Tanaka, M., Kishimoto, S., Goto, K., Tanaka, H., Torii, S. *TL* **40**, 2785 (1999).
[3]Kuroboshi, M., Tanaka, M., Kishimoto, S., Goto, K., Mochizuki, M., Tanaka, H. *TL* **41**, 81 (2000).

Chromium(III) chloride.

Benzhydrols and diaryl ketones.[1] Mediation of the ArZnI reaction with ArCHO by $CrCl_3$ leads to benzhydrols. In the case of *o*-zinciobenzoic esters, phthalides are formed.

When PhCHO (or any other ArCHO) is added at the termination of the reaction as a hydrogen acceptor to the chromium(III) diarylmethanoates, an Oppenauer oxidation occurs and the products are diaryl ketones.

Ar–ZnI + Ar′CHO $\xrightarrow{CrCl_3}$ [Ar–CH(OCr^{III})–Ar′] $\xrightarrow{PhCHO}$ Ar–CO–Ar′ (70–80%)

Allylations.[2] Allylating agents prepared from allyl halides, $CrCl_3$, $NiBr_2$, and Me_3SiCl react with carbonyl compounds, provided that tetrakis(dimethylamino)ethylene is used as electron source.

(E)-Iodoalkenes. Aldehydes are homologated with iodoform, $(thf)_3CrCl_3$, Zn, and Me_3SiCl in dioxane to afford (*E*)-iodoalkenes[3] at room temperature. On replacing Zn with Mn the reaction products are (*E*)-alkenylsilanes.[4]

Conjugate additions.[5] Organochromium reagents derived from halides such as BnBr and $CrCl_3$–Mn add to acrylonitrile (water is an additive in the reaction medium).

[1]Ogawa, Y., Saiga, A., Mori, M., Shibata, T., Takagi, K. *JOC* **65**, 1031 (2000).
[2]Kuroboshi, M., Goto, K., Mochizuki, M., Tanaka, H. *SL* 1930 (1999).

[3]Takai, K., Ichiguchi, T., Hikasa, S. *SL* 1268 (1999).
[4]Takai, K., Hikasa, S., Ichiguchi, T., Sumino, N. *SL* 1769 (1999).
[5]Auge, J., Gil, R., Kalsey, S. *TL* **40**, 67 (1999).

Chromium(IV) oxide. 20, 114

Reaction of acetals.[1] Dimethyl acetals of aromatic aldehydes (including cinnamaldehyde) are hydrolyzed in aqueous chloroform with CrO_2 as catalyst. However, aliphatic aldehydes so generated undergo oxidation to afford carboxylic acids.

[1]Ko, K.-Y., Park, S.T. *TL* **40**, 6025 (1999).

Chromium(VI) oxide–hydrochloric acid.

Aryliodine(III) dichlorides.[1] Oxidative chlorination of ArI is completed by using CrO_3–HCl in aq HOAc at room temperature.

[1]Kazmierczak, P., Skulski, L., Obeid, N. *JCR(S)* 64 (1999).

Chromium(VI) oxide–periodic acid.

Benzylic oxidation.[1] This combination oxidizes substituted toluenes to the corresponding acids and diarylmethanes (including fluorene) into ketones at room temperature in MeCN. Benzyl ethers such as phthalan and isochroman are converted to phthalide and 3,4-dihydroisocoumarin in quantitative yield.

[1]Yamazaki, S. *OL* **1**, 2129 (1999).

Chromium peroxide.

Oxidation of alcohols.[1] Polymer-supported CrO_5 selectively oxidizes allylic and benzylic alcohols.

[1]Lakouraj, M.M., Keyvan, A. *JCR(S)* 206 (1999).

Cobalt.

Pauson–Khand reaction. Cobalt is deposited by decomposing dicobalt octacarbonyl in refluxing toluene in the presence of mesoporous silica. The substance is active as a heterogeneous catalyst for the Pauson–Khand reaction.[1] A related catalyst is cobalt-on-carbon, which can be used at least 10 times with yields maintaining at the 95% level.[2]

[1]Kim, S.-W., Son, S.U., Lee, S.I., Hyeon, T., Chung, Y.K. *JACS* **122**, 1550 (2000).
[2]Son, S.U., Lee, S.I., Chung, Y.K. *ACIEE* **39**, 4158 (2000).

Cobalt(II) bromide. 19, 104

Conjugate additions.[1] Cobalt bromide catalyzes electrochemical addition of aryl halides to activated alkenes.

[1]Gomes, P., Gosmini, C., Nedelec, J.-Y., Perichon, J. *TL* **41**, 3385 (2000).

Cobalt(II) chloride. 14, 99; **15,** 97–98; **18,** 107–108; **19,** 104–105; **20,** 115–116;

Hydrosilylation.[1] A new catalyst for regioselective hydrosilylation of acrylonitrile is $CoCl_2$. Only the β-silylated nitriles are obtained.

α-Hydroxy-β-ketoesters.[2] The hydroxylation of β-ketoesters with molecular oxygen is carried out under neutral conditions.

[1]Chauhan, M., Chauhan, B.P.S., Boudjouk, P. *TL* **40**, 4127 (1999).
[2]Baucherel, X., Levoirier, E., Uziel, J., Juge, S. *TL* **41**, 1385 (2000).

Cobalt(III) fluoride.

Quinones.[1] Hydroquinone dimethyl ethers undergo oxidative demethylation on contact with CoF_3. The reagent is comparable to AgO or CAN in efficiency.

Fluorination.[2] Fluorination of unsaturated compounds with CoF_3 is carried out in a stainless steel vessel from −196° to room temperature. The method transforms methyl trifluorovinyl ether to methyl pentafluoroethyl ether in 70% yield.

[1]Tomatsu, A., Takemura, S., Hashimoto, K., Nakata, M. *SL* 1474 (1999).
[2]Tamura, M., Takubo, S., Quan, H., Sekiya, A. *SL* 343 (2000).

Cobalt(III) halochromate, ammine complexes.

Oxidation.[1] The cobalt complexes are mild and efficient oxidants with which alcohols are oxidized to carbonyl compounds.

[1]Kooti, M., Esm-Hosseini, M. *SC* **30**, 651 (2000).

Copper(II) acetate. 18, 109–110; **19,** 106; **20,** 117

Arylations. Copper(II) acetate catalyzes the reaction of arylboronic acids with thiols[1] and of amines with hypervalent arylsiloxanes.[2] 4-Substituted imidazoles react with aryllead(IV) reagents (in dichloromethane at room temperature) to afford N-1 arylated derivatives regioselectively under catalysis of $Cu(OAc)_2$.[3]

Radical cyclizations. Copper(II) acetate has a great influence on the Ni–HOAc promoted 5-*endo* and 5-*exo* cyclizations.[4]

Aroylformic acids.[5] 1-Aryl-2-nitroethanols are oxidized by $Cu(OAc)_2$ in HOAc–MeOH at 90° to afford the ketoacids. Nitroaldols derived from aliphatic aldehydes only give the corresponding methyl ethers under these conditions.

Cleavage of picolinic acid esters.[6] One advantage of the Mitsunobu reaction protocol using picolinic acid as the nucleophile is that the configurationally inverted alcohols are readily recovered, that is, after treatment with $Cu(OAc)_2$–MeOH.

[1]Herradura, P.S., Pendola, K.A., Guy, R.K. *OL* **2**, 2019 (2000).
[2]Lam, P. Y. S., Deudon, S., Averell, K.M., Li, R., He, Y., DeShong, P., Clark, C.G. *JACS* **122**, 7600 (2000).
[3]Elliott, G.I., Konopelski, J.P. *OL* **2**, 3055 (2000).
[4]Cassayre, J., Dauge, D., Zard, S.Z. *SL* 471 (2000).
[5]Nikalje, M.D., Ali, I.S., Dewkar, G.K., Sudalai, A. *TL* **41**, 959 (2000).
[6]Sammakia, T., Jacobs, J.S. *TL* **40**, 2685 (1999).

Copper(I) bromide.

Cyclization.[1] Cyclization of *N*-allyl α-bromoamides induced by ligated CuBr at room temperature involves atom transfer. 2-Pyridylformaldimines are better ligands than 2,2′-bipyridine.

[1]Clark, A.J., Duncalf, D.J., Filik, R.P., Haddleton, D.M, Thomas, G.H., Wongtap, H. *TL* **40**, 3807 (1999).

Copper(II) bromide. 14, 100; 15, 100; 18, 111; 19, 106

4-Halo-5-hydroxypyrrol-2(5H)-ones.[1] Treatment of 2,3-alleneamides with $CuBr_2$ (slightly less efficiently, $CuCl_2$) furnishes the title compounds. Halolactamization is followed by oxidation at C-5 when the allenyl moiety is not fully substituted.

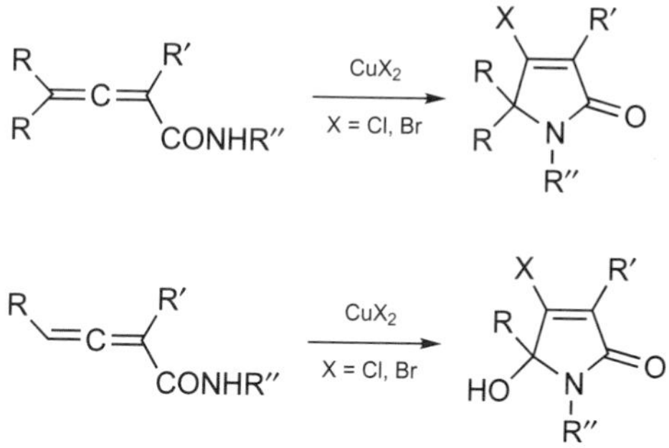

Insertion into silacyclopropanes.[2] Remarkably regioselective reactions of silacyclopropanes with methyl formate are mediated by $CuBr_2$ and $ZnBr_2$. The regioselectivity is totally switched from one of these two salts to the other.

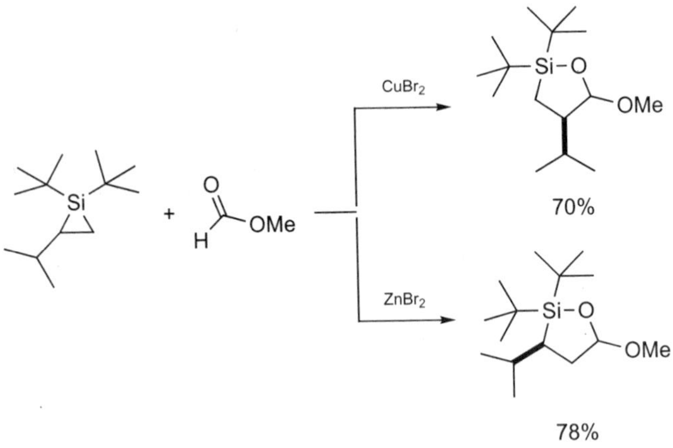

CuBr$_2$ 70%

ZnBr$_2$ 78%

[1]Ma, S., Xie, H. *OL* **2**, 3801 (2000).
[2]Franz, A.K., Woerpel, K.A. *ACIEE* **39** 4295 (2000).

Copper(I) chloride. 13, 85; **15,** 101; **18,** 112–113; **19,** 107–108; **20,** 118–120

Oxidations. Carbonyl compounds are obtained from autoxidation of alkyl halides and tosylates promoted by CuCl on Kieselguhr (15 examples, 81–99%).[1]

Coupling reactions. The Stille coupling is accelerated by CuCl. Sterically congested substrates are readily coupled in the presence of $(Ph_3)_4Pd$, LiCl, CuCl in DMSO.[2]

Allylfurans and allylthiophenes can be synthesized from the corresponding stannanes by coupling with allyl halides.[3] The coupling of alkynylsilanes and 1-chloroalkynes to furnish conjugated diynes tolerates many sensitive functional groups such as C=O.[4] Benzotropylidenes and benzotropones are accessible from the reaction of zirconacyclopentadienes with 2-iodobenzyl halide and 2-iodobenzoyl halide, respectively.[5]

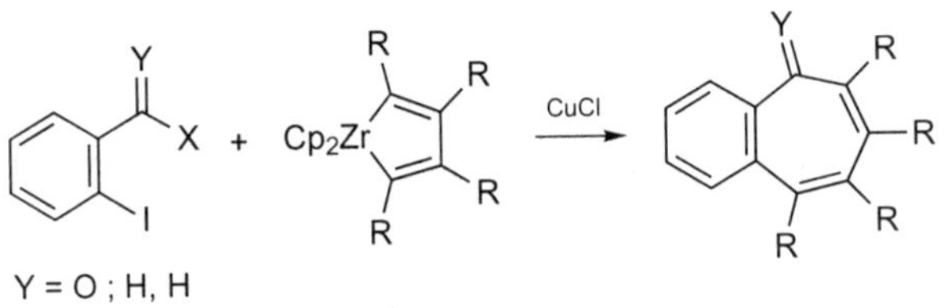

Y = O ; H, H

CuCl

2,2′-Bridged biaryls are readily obtained from short-chain 1,ω-bis(2-trimethylstannylaryl)alkanes by treatment with CuCl in DMF at room temperature.[6]

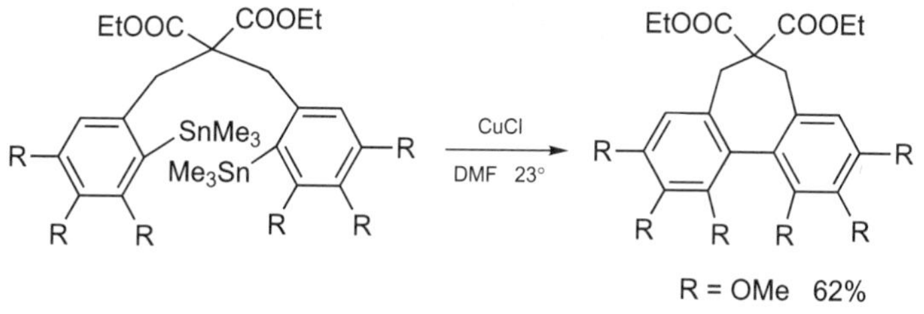
EtOOC
COOEt
SnMe3
Me3Sn
R
CuCl
DMF 23°
R = OMe 62%

Coupling of zirconacyclopentadienes with (*Z,Z*)-1,4-diiodo-1,3-dienes by CuCl results in cyclooctatetraenes.[7]

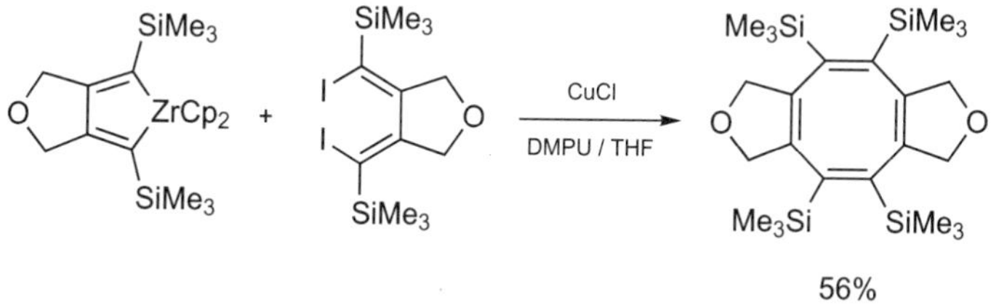
SiMe3
ZrCp2
+
I
O
CuCl
DMPU / THF
Me3Si
56%

Dehydration.[8] Tertiary and benzylic alcohols undergo dehydration with CuCl–DCC via pseudourea intermediates.

Aryl radicals.[9] A synthetic application of the arenediazonium ion reduction by CuCl is radical generation at an α-position of amines. Thus, diazotization of the *N*-(2-aminobenzoyl) derivatives in the presence of CuCl in MeOH leads to benzamides in which the α-position becomes chlorinated and/or methoxylated. Hydrogen abstraction by the nascent aryl radicals is evident.

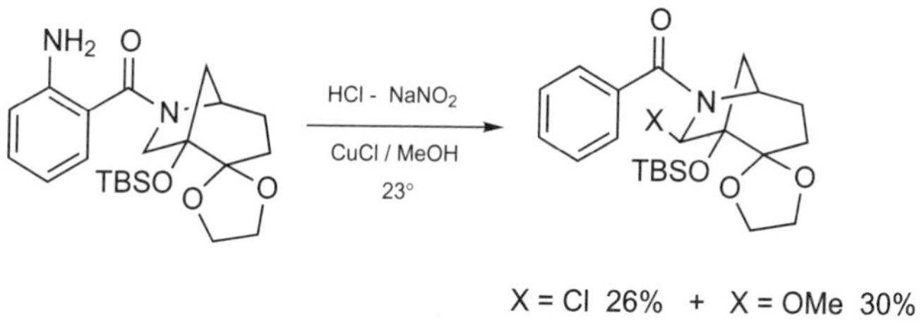
NH2
O
N
TBSO
HCl - NaNO2
CuCl / MeOH
23°
X
X = Cl 26% + X = OMe 30%

Pinacolatoboration.[10] Hydroboration of 1-alkynes to give 2-pinacolato-1-alkenes is effected with bis(pinacolato)diboron in the presence of CuCl and KOAc. Perhaps a borylcopper species is involved in the reaction. The same reagent is useful as a nucelophile toward allylic halides and Michael acceptors.

CuCl - LiCl - KOAc
DMF 20°

[1]Hashemi, M.M., Beni, Y.A. *JCR(S)* 434 (1999).
[2]Han, X., Stoltz, B.M., Corey, E.J. *JACS* **121**, 7600 (1999).
[3]Nudelman, N.S., Carro, C. *SL* 1942 (1999).
[4]Nishihara, Y., Ikegashira, K., Hirabayashi, K., Ando, J.-I., Mori, A., Hiyama, T. *JOC* **65**, 1780 (2000).
[5]Takahashi, T., Sun, W.-H., Duan, Z., Shen, B. *OL* **2**, 1197 (2000).
[6]Piers, E., Yee, J.G.K., Gladstone, P.L. *OL* **2**, 481 (2000).
[7]Yamamoto, Y., Ohno, T., Itoh, K. *CC* 1543 (1999).
[8]Majetich, G., Hicks, R., Okha, F. *NJC* **23**, 129 (1999).
[9]Han, G., LePorte, M.G., Folmer, J.J., Werner, K.M., Weinreb, S.M *ACIEE* **39**, 237 (2000); Han, G., LePorte, M.G., McIntosh, M.C., Weinreb, S.M., Parvez, M. *JOC* **61**, 9483 (1996).
[10]Takahashi, K., Ishiyama, T., Miyaura, N. *CL* 982 (2000).

Copper(II) chloride. 14, 100; **18,** 113–114; **19,** 108; **20,** 120

Halolactonization.[1] Allenic acids give β-halo-γ-butenolides on reaction with CuX_2 (X = Cl, Br).

Coupling of organometallics. The demetallative dimerization of RLi by $CuCl_2$ is different from the conjugate addition catalyzed by CuI.[2] Organostannanes $RSnBu_3$ give R—R when R is an alkynyl, alkenyl, or aryl group.[3]

Hydrolysis of SAMP-hydrazones.[4] After asymmetric alkylation of the SAMP-hydrazones, it is critical to generate the chiral ketones without racemization. Treatment of the products with CuX_2 in THF or MeCN followed by aqueous ammonia provides a solution.

MeO
$CuCl_2$
THF ; NH_3
(99% ee)

Cyclooctatetraenes.[5] (*Z,Z*)-1,3-Dienyl-1,4-dicopper species are formed on treatment of zirconacyclopentadienes with $CuCl_2$. Further reaction with NBS results in the substituted cyclooctatetraenes.

[1]Ma, S., Wu, S. *JOC* **64**, 9314 (1999).
[2]Pastor, I.M., Yus, M. *TL* **41**, 1589 (2000).
[3]Kang, S.-K., Baik, T.-G., Jiao, X.H., Lee, Y.-T. *TL* **40**, 2383 (1999).

[4]Enders, D., Hundertmark, T., Lazny, R. *SC* **29**, 27 (1999).
[5]Takahashi, T., Sun, W.-H., Nakajima, K. *CC* 1595 (1999).

Copper(I) iodide. 16, 98; **18,** 114–115; **19,** 109–110; **20,** 120–121

Alkynyl ketones.[1] The reaction of 1-alkynes with acyl halides is promoted by CuI–Et_3N.

Cyclization. Alkynes in which the triple bond is separated by three skeletal atoms from a pronucleophlic site undergo cyclization in the presence of catalytic amounts of CuI and *t*-BuOK.[2] The triple bond is activated on coordination with the copper salt.

X COOMe
CuI - t-BuOK
THF
X COOMe
X = CN, COMe,...

Coupling between an imino chloride and 2,2-dimethyl-4-alkynamide is accompanied by cyclization.[3] This Pd-catalyzed process is aided by CuI, for without which considerable decomposition occurs and the reaction also becomes sluggish.

N N CN Cl + NH_2 O
CuI - Pd
N N CN O HN
46%

Either pyrroles or 3-pyrrolines are formed when *N*-propargylanilines are heated with formaldehyde and CuI.[4]

R HN Ar
i-Pr_2NEt - HCHO aq.
CuI Δ
R Ar N   R Ar N
in dioxane   in ethanediol

Desilylallylation. 2-(1-Trimethylsilylalkenyl)thiopyridines undergo allylation. Regioselectivity differences are noted for silylallyl- and silylvinyl-type substrates.[5]

Me_3Si N S Cl CuI - KF THF / Me_2CO N S 58%

Me_3Si N S Cl CuI - KF THF / Me_2CO N S 78%

[1]Chowdhury, C., Kundu, N.G. *T* **55**, 7011 (1999).
[2] Bouyssi, D., Monteiro, N., Balme, G. *TL* **40**, 1297 (1999).
[3]Jacobi, P.A., Liu, H. *JACS* **121**, 1958 (1999).
[4]Jayaprakash, K., Venkatachalam, C.S., Balasubramanian, K.K. *TL* **40**, 6493 (1999).
[5]Takeda, T., Uruga, T., Gohroku, K., Fujiwara, T. *CL* 821 (1999).

Copper(II) nitrate. 15, 101; **18,** 115–116; **19,** 110; **20,** 121

Biaryls.[1] Diarylstannanes including heteroaromatic analogues (furan and thiophene series) undergo reductive elimination on treatment with copper(II) nitrate trihydrate in THF at room temperature.

Azidolysis of glycidic acids.[2] In the presence of copper(II) nitrate trihydrate, regioselective ring opening of the epoxide by sodium azide in water (pH 4) to provide 3-azido-2-hydroxy carboxylic acids is observed.

N-Nitro-1-methyluracil.[3] 1-Methyluracil undergoes *N*-nitration with a mixture of copper(II) nitrate trihydrate and acetic anhydride, while nitration occurs at C-5 in fuming nitric acid.

[1]Harada, G., Yoshida, M., Iyoda, M. *CL* 160 (2000).
[2]Fringuelli, F., Pizzo, F., Vaccaro, L. *SL* 311 (2000).
[3]Giziewicz, J., Wnuk, S.F., Robins, M.J. *JOC* **64**, 2149 (1999).

Copper(I) oxide. 16, 99

Perfluoroalkylation.[1] By using Cu_2O as the catalyst, anilines undergo perfluoroalkylation with R^fI in DMSO at 130°. The perfluoroalkyl groups enter at ortho- and para-positions that are open. *N,N*-Dimethylaniline suffers demethylation in the process.

$$C_6H_5NH_2 \xrightarrow[\text{DMSO } 130^\circ]{Cu_2O - C_8F_{17}I} 2,4,6\text{-}(C_8F_{17})_3C_6H_2NH_2$$

73%

[1]Moreno-Manas, M., Plexixata, R., Villarroya, S. *SL* 1996 (1999).

Copper(II) tetrafluoroborate.

β-Nitrostyrenes.[1] Styrenes undergo nitration with sodium nitrite in the presence of $Cu(BF_4)_2$ and I_2 in MeCN.

[1]Campos, P.J., Garcia, B., Rodriguez, M.A. *TL* **41**, 979 (2000).

Copper(I) 2-thiophenecarboxylate. 19, 112; 20, 122

Enamides.[1] Enamides are prepared by coupling alkenyl iodides with amides in NMP with Cs_2CO_3 as base.

[1]Shen, R., Porco, J.A. *OL* **2**, 1333 (2000).

Copper(II) triflate. 19, 112; 20, 122–123

Cleavage of aziridines.[1] Copper(II) triflate is a good catalyst for cleavage of aziridines by arylamines.

Mukaiyama aldol reaction.[2] With $Cu(OTf)_2$ as catalyst, the condensation of silyl enol ethers with aldehydes can be carried out in aqueous ethanol.

Acetylation. Alcohols, thiols, and amines are acetylated by a $Cu(OTf)_2$-catalyzed reaction with acetic anhydride at room temperature.[3] Various aldehydes (but not ketones) are similarly transformed into *gem*-diacetates.[4]

N-Arylimidazoles.[5] Together with 1,10-phenanthroline and dibenzylideneacetone as additives, $Cu(OTf)_2$ and cesium carbonate promote formation of *N*-arylimidazoles.

[1]Sekar, G., Singh, V.K. *JOC* **64**, 2537 (1999).
[2]Kobayashi, S., Nagayama, S., Busujima, T. *CL* 71 (1999).
[3]Saravanan, P., Singh, V.K. *TL* **40**, 2611 (1999).
[4]Chandra, K.L., Saravanan, P., Singh, V.K. *SL* 359 (2000).
[5]Kiyomori, A., Marcoux, J.-F., Buchwald, S.L. *TL* **40**, 2657 (1999).

Copper(II) trifluoromethylthiolate. 19, 112; 20, 122–123

Trifluoromethylthioarenes.[1] Arylamines are converted to $ArSCF_3$ by diazotization in the presence of the title reagent.

[1]Adams, D.J., Goddard, A., Clark, J.H., Macquarrie, D.J. *CC* 987 (2000).

1-Cyanobenzotriazole.

Aryl cyanides.[1] Cyanation of ArLi by the title compound is expedient.

[1]Hughes, T.V., Cava, M.P. *JOC* **64**, 313 (1999).

***N*-(2-Cyanoethoxycarbonyloxy)succinimide.**

Amine protection.[1] Amino groups present in oligonucleotides are readily protected with this activated carbonate **1** in the form of carbamates.

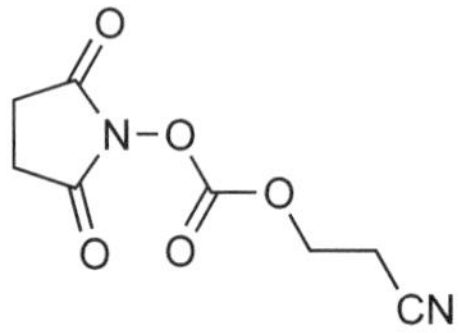

(**1**)

[1]Manoharan, M., Prakash, T.P., Barber-Peoc'h, I., Bhat, B., Vasquez, G., Ross, B.S., Cook, P.D. *JOC* **64**, 6468 (1999).

1-Cyanoimidazole.

Cyanation.[1] This reagent is prepared from imidazole and cyanogen bromide. It donates the cyano group to various nucelophiles such as amines, thiols, and RMgX (or RLi).

[1]Wu, Y., Limburg, D.C., Wilkinson, D.E., Hamilton, G.S. *OL* **2**, 795 (2000).

Cyanomethylenetriorganophosphoranes.

Cyanomethylenation of carbonyl compounds. This powerful Wittig reagent $Me_3P{=}CHCN$ reacts with esters, lactones, and imides.[1] The reaction is valuable for synthesis of *C*-glycosides from glyconolactones and $Ph_3P{=}CHCN$. Microwave assistance shortens reaction time to minutes.[2]

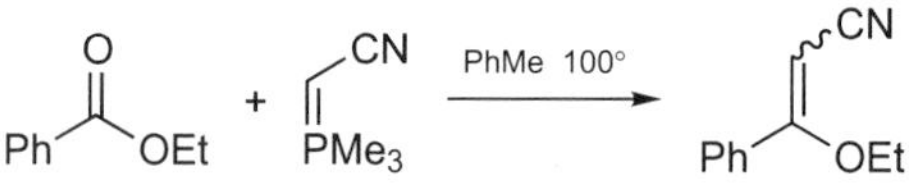

PhMe 100°

[1]Tsunoda, T., Takagi, H., Takaba, D., Kaku, H., Ito, S. *TL* **41**, 235 (2000).
[2]Lakhrissi, Y., Taillefumier, C., Lakhrissi, M., Chapleur, Y. *TA* **11**, 417 (2000).

β-Cyclodextrin.

2-Amino alcohols. Epoxides are opened regioselectively with arylamines[1] and trimethylsilyl azide[2] in the presence of β-cyclodextrin.

Terephthalic acid. Benzene is selectively converted to tetrephthalic acid (46 mol% yield) in a Cu mediated reaction with CCl_4 in the presence of NaOH and β-cyclodextrin.[3] The C—C bond formation occurs when benzene is trapped in the cavity of β-cyclodextrin.

[1]Reddy, L.R., Reddy, M.A., Bhanumathi, N., Rao, K.R. *SL* 339 (2000).
[2]Kamal, A., Arifuddin, M., Rao, M.V. *TA* **10**, 4261 (1999).
[3]Shiraishi, Y., Tashiro, S., Toshima, N. *CL* 828 (2000).

Carbonylhydridotris(triphenylphosphine)rhodium.

Hydroformylation + Wittig reaction.[1] The tandem process on alkenes is accomplished.

PPh2, EtOOC, O, O; H_2 / CO - $(Ph_3P)_3Rh(CO)H$; Ph_3P=CH–CO–CH$_3$, PhMe 90°

60% (*syn* : *anti* 9 : 1)

[1]Breit, B., Zahn, S.K. *ACIEE* **38**, 969 (1999).

(1,5-Cyclooctadiene)(1,3,5-cyclooctatriene)ruthenium(0).

Isomerization.[1] 2-Allylphenol is isomerized to 2-propenylphenol in 95% yield with (cod)Ru(cot) in methanol at room temperature. The (*Z/E*) ratio of the products is 94:6. Triethylphosphine is also added.

[1]Sato, T., Komine, N., Hirano, M., Komiya, S. *CL* 441 (1999).

(1,5-Cyclooctadiene)cyclopentadienylcobalt.

[2 + 2 + 2]Cycloaddition.[1] This method of pyridine formation, which unites two alkynes and a nitrile, has been extended to the synthesis of spiroannulated analogues.

NC, CN, +, R′, R; Cp(cod)Co, PhMe; R, N, R′, R′, N, R

[1]Varela, J.A., Castedo, L., Saa, C. *OL* **1**, 2141 (1999).

(1,5-Cyclooctadiene)(η^6-naphthalene)rhodium(I) tetrafluoroborate.

[4 + 2]Cycloaddition. This cationic rhodium complex is useful for inducing the cycloaddition of conjugated dienes to unactivated alkynes in dichloromethane (15°, 15 min), forming 1,4-cyclohexadienes.[1]

[1]Paik, S.-J., Son, S.U., Chung, Y.K. *OL* **1**, 2045 (1999).

(1,5-Cyclooctadiene)(η^6-tetraphenylborato)rhodium(I).

Hydroformylation. Alkynes undergo hydroformylation to afford branched-chain aldehydes[1] or lactones.[2]

CO - H_2
$(PhO)_3P$ / CH_2Cl_2
CHO
55%

B
$^+Rh(cod)$
O
O
91%

[1]van den Hoven, B.G., Alper, H. *JOC* **64**, 3964, 9640 (1999).
[2]van den Hoven, B.G., El Ali, B., Alper, H. *JOC* **65**, 4131 (1999).

Cyclopentadienylbis(ethylene)cobalt.

[2 + 2 + 2]Cycloaddition. Extension of this cycloaddition method to annulation of benzofuran successfully elaborates four rings of the morphinoids.[1] A strychnine synthesis[2] has been completed based on an analogous elaboration.

X
MeO
O
+
$SiMe_3$
$SiMe_3$
$CpCo(C_2H_4)_2$
X
H
CoCp
$SiMe_3$
MeO
O
H
$SiMe_3$

NHAc

$CpCo(C_2H_4)_2$

C_2H_4 / THF

NHAc CoCp

47%

[1] Perez, D., Siesel, B.A., Malaska, M.J., David, E., Vollhardt, K.P.C. *SL* 306 (2000).
[2] Eichberg, M.J., Dorta, R.L., Lamottke, K., Vollhardt, K.P.C. *OL* **2**, 2479 (2000).

Cyclopentadienylindium(I).

Cyclopentadienylcarbinols.[1] The title reagent reacts with carbonyl compounds in water.

In + RCHO → (H_2O / THF) HO R

[1] Yang, Y., Chan, T.H. *JACS* **122**, 402 (2000).

D

Decaborane.

Reductive etherification.[1] Benzyl ethers are generated from ArCHO on treatment with decaborane in an alcohol (e.g., MeOH, EtOH).

[1]Lee, S.H., Park, Y.J., Yoon, C.M. *TL* **40**, 6049 (1999).

Dess–Martin periodinane.

Deoximation.[1,2] A method for the regeneration of carbonyl compounds from oximes calls for oxidation with the Dess–Martin periodinane.

A related application of the reagent is in the preparation of *N*-acyl nitroso compounds from hydroxamic acids.[3]

Heterocycle formation.[4] When *o*-iodination of acylanilines in which the acyl group contains a double bond at a proper distance is followed by several steps, benzomorpholines result. Unsaturated carbamates and amides form oxazolidinones and lactams under such reaction conditions.

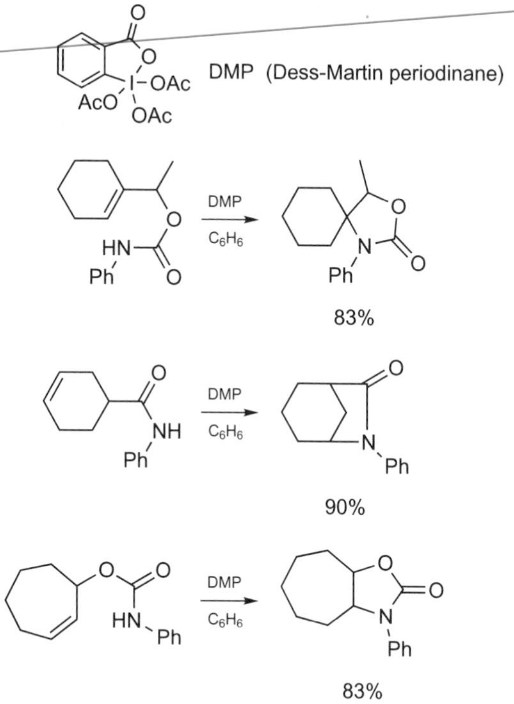

DMP (Dess-Martin periodinane)

DMP, C_6H_6 — 83%

DMP, C_6H_6 — 90%

DMP, C_6H_6 — 83%

[1]Bose, D.S., Narsaiah, A.V. *SC* **29**, 937 (1999).
[2]Chaudhari, S.S., Akamanchi, K.G. *S* 760 (1999).
[3]Jenkins, N.E., Ware, R.W., Atkinson, R.N., King, S.B. *SC* **30**, 947 (2000).
[4]Nicolaou, K.C., Zhong, Y.-L., Baran, P.S. *ACIEE* **39**, 623, 625 (2000).

Dialkylaluminum chloride. 20, 126–127

Addition to imines. The adducts from organocuprate reagents can be used in situ in the addition to *N*-sulfonylimines, which is promoted by Et_2AlCl.[1]

COOEt, Me_2CuLi / Et_2O ; Tol–S(O)–N=CHPh, Et_2AlCl → Tol–S(O)–NH, Ph, COOEt

77%

Propargylic alcohols.[2] When catalyzed by Me_2AlCl, the addition of alkynylstannanes to aldehydes is subject to chelation control by a β-alkoxy or siloxy group in the substrate.

3-Acylindoles.[3] Friedel–Crafts acylation of indoles at C-3 with acid chlorides is catalyzed by R_2AlCl (R = Me, Et).

Michael and aldol reaction tandem. Michael reaction products arising from dicarbonyl compounds that contain one conjugated double bond undergo cyclization (aldol reaction) if the second carbonyl group is properly juxtaposed. For example, a synthesis of 2-hydroxycyclohexanecarboxylic acid derivatives is accomplishable in one step from 7-keto-2-alkenamides.[4] The tandem reaction is promoted by Me_2AlCl.

R, O, O, N, O → R′M - Me_2AlCl, THF → R′, H, O, N, R, OH, O

[1]Li, G., Wei, H.-X., Whittlesey, B.R., Batrice, N.N. *JOC* **64**, 1061 (1999).
[2]Evans, D.A., Halstead, D.P., Allison, B.D. *TL* **40**, 4461 (1999).
[3]Okauchi, T., Itonaga, M., Minami, T., Owa, T., Kitoh, K., Yoshino, H. *OL* **2**, 1485 (2000).
[4]Schneider, C., Reese, O. *ACIEE* **39**, 2948 (2000).

Dialkylaluminum cyanide.

Cyanohydrins.[1] Selective addition of Et_2AlCN to unhindered aldehydes gives α-cyanohydrins en route to homologous nitriles. Ketones and enones are not homologated.

β-Fluoro α-amino acids.[2] Asymmetric addition to chiral sulfinilimines paves the way to the fluorinated amino acids.

Tol S(O) N=CH–CH(F)–CH₂Ph → (Et_2AlCN, i-PrOH) → NC–CH(NH-S(O)Tol)–CH(F)–CH₂Ph 78%

[1]Nicolaou, K.C., Vassllikogiannakis, G., Kranich, R., Baran, P.S., Zhong, Y.-L., Natarajan, S. *OL* **2**, 1895 (2000).
[2]Davis, F.A., Srirajan, V., Titus, D.D. *JOC* **64**, 6931 (1999).

Dialkylaluminum triflamides.

Condensation reactions.[1] The title compounds ($R_2Al–NTf_2$) are excellent catalysts for Mukaiyama aldol reactions, Michael reactions, and allylation. The allylation is sensitive to steric environments of the substrates, enabling selective attack on a less crowded formyl group.

R–CHO + Me_3Si–CH₂CH=CHCH=CH₂ → (Me_2AlNTf_2) → R–CH(OH)–CH₂CH=CHCH=CH₂

[1]Marx, A., Yamamoto, H. *ACIEE* **39**, 178 (2000).

Dialkyl α-nitroalkylphosphonates.

Emmons–Wadsworth reaction.[1] Very mild bases (e.g., K_2CO_3) are required for the synthesis of nitroalkenes using these reagents.

[1]Franklin, A.S. *SL* 1154 (2000).

Diallylmercury.

Allylations. This reagent or allylmercury bromide react with aldehydes in aqueous media.[1]

[1]Chan, T.H., Yang, Y. *TL* **40**, 3863 (1999).

Diallyltin dibromide.

Allylation. Formation of homoallylic alcohols from aldehydes and the title reagent is possible in aqueous media.[1]

[1]Chan, T.H., Yang, Y., Li, C.J. *JOC* **64**, 4452 (1999).

Diaryliodonium salts.

Arylations. Arylation of malonic esters is achieved in good yields. The less electron-rich aryl group is transferred from unsymmetrical iodonium salts.[1] *N*-Arylation of amines, amides, and lactams is under catalysis by CuI,[2] whereas pyrimidinones undergo *N*-arylation using *t*-BuOK as base (no copper salt).[3]

TfO⁻ + EtOOC, COOEt, Na+ ; DMF, 0°

EtOOC COOEt OMe 5.6% + EtOOC COOEt 74%

Biaryls.[4] The coupling of diaryliodonium salts with aryllead triacetates to afford unsymmetrical biaryls is effected by $(dba)_3Pd_2.CHCl_3$. For example,4-$MeOC_6H_4Ph$ is obtained from the reaction of 4-anisyl(phenyl)iodonium tetrafluoroborate with $PhPb(OAc)_3$.

[1]Oh, C.H., Kim, J.S., Jung, H.H. *JOC* **64**, 1338 (1999).
[2]Kang, S.-K., Lee, S.-H., Lee, D. *SL* 1022 (2000).
[3]Jacobsen, S.A., Rodbotten, S., Benneche, T. *JCS(P1)* 3265 (1999).
[4]Kang, S.-K., Choi, S.-C., Baik, T.-G. *SC* **29**, 2493 (1999).

1,4-Diazabicyclo[2.2.2]octane, DABCO. 13, 92; **15,** 109; **18,** 120; **19,** 116–117; **20,** 128–129

Baylis–Hillman reaction.[1] One application of the reaction is for a convenient synthesis of 3-acylchromenes from salicylaldehydes and 1-alken-3-ones.

Dechloroacetylation. Selective cleavage of chloroacetates (e.g., in the presence of benzoates, and acetates.) by transacylation to EtOH is readily performed.[2]

[1]Kaye, P.T., Nocanda, X.W. *JCS(P1)* 1331 (2000).
[2]Lefeber, D.J., Kamerling, J.P., Vliegenthart, J.F.G. *OL* **2**, 701 (2000).

1,8-Diazabicyclo[5.4.0]undec-7-ene, DBU. 13, 92; **14,** 109; **15,** 109–110; **16,** 105–106; **17,** 99–100; **18,** 120–121; **19,** 117; **20,** 129–130

Alkylations. There are reports on the alkylation of *N*-acylhydroxylamines,[1] 1,2,4-triazole,[2] and glycine methyl ester[3] using DBU as base. In the last case, an α,*N*,*N*-triallyl derivative results because of a [2.3]Stevens rearrangement.

NH2 COOMe + Br → (Bu4NI - K2CO3, DBU / DMF, 40°) COOMe 80%

Eliminations. Protected dehydro-α-amino acids are formed from serine-type derivatives by elimination[4] or from condensation of nitroalkanes with α-tosylglycines.[5]

There is a similar process for the transformation of enediones to 2-alkylidene-1,4-diones.[6]

HN EtOOC Ts + O2N → (DBU / THF) 68%

An elimination of toluenesulfinic acid is via double bond migration.[7]

HO R R′ Ts COOMe → (DBU / THF, 23°) R R′ OH COOMe

γ-Alkylidenation of 1,3-dicarbonyl compounds.[8] The reaction gives rise to unsaturated products with enals.

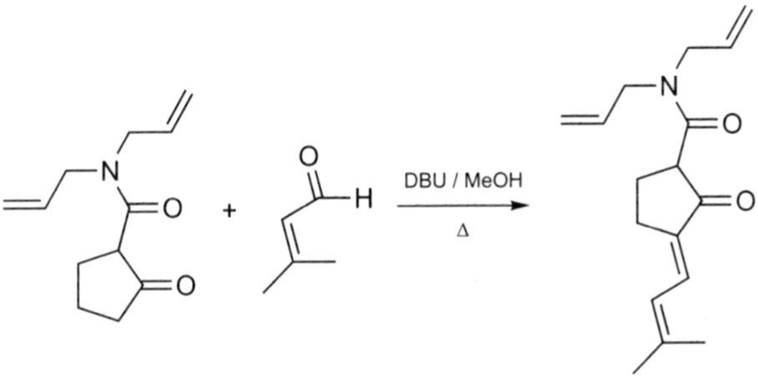

DBU / MeOH
Δ

44%

Removal of protecting groups. A combination of DBU with a thiol can be used in the removal of an Fmoc group[9] in a large scale process. Regeneration of alcohols from trichloroacetimidates is accomplished by treatment with DBU in MeOH (other methods involve acid-catalyzed hydrolysis and Zn dust reduction).[10] Alcohols temporarily protected as trichloroacetimidates permit their differentiation from others that are masked in acetonide, ester (acetate, benzoate, etc.), and ether (allyl, TBS, etc.) forms.

Ramberg–Bäcklund rearrangement.[11] An expedient access to 1,1-dichloro-1,3-dienes is by treatment of allylic trichloromethyl sulfones with DBU.

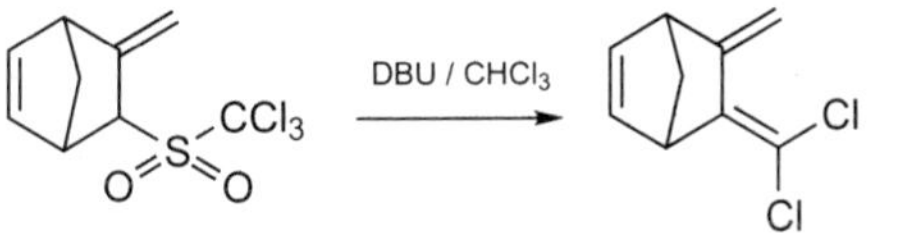

DBU / $CHCl_3$

92%

Baylis–Hillman reaction.[12] DBU is an excellent promoter (better than DBN, DMAP, etc.) because it significantly enhances the reaction rate, giving good yields of the adducts.

Isocyanates.[13] In DMSO, trihaloacetamides undergo elimination of CHX_3 (X = Cl, Br) on treatment with DBU to afford RN═C═O.

[1]Jones, D.S., Hammaker, J.R., Tedder, M.E. *TL* **41**, 1531 (2000).
[2]Bulger, P.G., Cottrell, I.F., Cowden, C.J., Davies, A.J., Dolling, U.-H. *TL* **41**, 1297 (2000).
[3]Arbore, A.P.A., Cane-Honeysett, D.J., Coldham, I., Middleton, M.L. *SL* 236 (2000).
[4]Stohlmeyer, M.M., Tanaka, H., Wandless, T.J. *JACS* **121**, 6100 (1999).
[5]Nagano, T., Kinoshita, H. *BCSJ* **73**, 1605 (2000).
[6]Ballini, R., Bosica, G., Petrelli, L., Petrini, M. *S* 1236 (1999).
[7]Caturla, F., Najera, C., Varea, M. *TL* **40**, 5957 (1999).
[8]Charonnet, E., Filippini, M.-H., Rodriguez, J. *SL* 1951 (1999).
[9]Sheppeck, J.E., Kar, H., Hong, H. *TL* **41**, 5329 (2000).
[10]Yu, B., Yu, H., Hui, T., Han, X. *SL* 753 (1999).

[11]Raj, C.P., Pichnit, T., Braverman, S. *TL* **41**, 1501 (2000).
[12]Agggarwal, V.K., Mereu, A. *CC* 2311 (1999).
[13]Braverman, S., Cherkinsky, M., Kedrova, L. *TL* **40**, 3235 (1999).

N,N-Dibenzylformamide dimethyl acetal.

Amine protection.[1] Primary amines are converted to *N,N*-dibenzylformamidines with this reagent or *N,N*-dibenzyl chloroformimidinium chloride.

EtO–CH(OEt)–(CH₂)₃–NH_2 → (MeO)(OMe)CH–NBn_2, MeCN [or $ClCH{=}NBn_2^+\ Cl^-$, CH_2Cl_2] → EtO–CH(OEt)–(CH₂)₃–N=CH–NBn_2 75% [99%]

[1]Vincent, S., Lebeau, L., Mioskowski, C. *SC* **29**, 167 (1999).

Diborane.

Hydroboration.[1] In chlorohydrocarbon solvents (dichloromethane, chloroform, 1,2-dichloroethane, etc), diborane hydroborates unsaturated compounds rapidly at –16°.

[1]Kanth, J.V.B., Brown, H.C. *TL* **41**, 9361 (2000).

N,N'-Dibromo-*N,N'*-ethylenebis(2,5-xylenesulfonamide).

Bromination.[1] Monobromination of arenes is achieved with this reagent.

[1] Ardeshir, K., Abbas, S. *SC* **29**, 4079 (1999).

1-Dibutylamino-2,2,2-trifluoroethanol.

Trifluoromethylation.[1] The adduct of dibutylamine (or diethylamine) and trifluoroacetaldehyde decomposes in situ and transfers the CF_3 group to carbonyl compounds. (So far only PhCHO has been reported.)

$F_3C{-}CHO + R_2NH \rightarrow F_3C{-}CH(OH){-}NR_2$ (R = Et, Bu) $\xrightarrow{\text{PhCHO - t-BuONa; HOAc}}$ $F_3C{-}CH(OH){-}Ph$ 48%

[1]Mispelaere, C., Roques, N. *TL* **40**, 6411 (1999).

Dibutylboron triflate. 20, 132

Aldol reaction.[1] Predominant *syn*-aldol products from the reaction of methyl phenylacetate with aldehydes are obtained, in contrast to that promoted by LDA.

Oxazolinyl epoxides.[2] Condensation of 2-chloromethyloxazolines with ketones gives the epoxides. High stereoselectivity is induced by a chiral oxazoline.

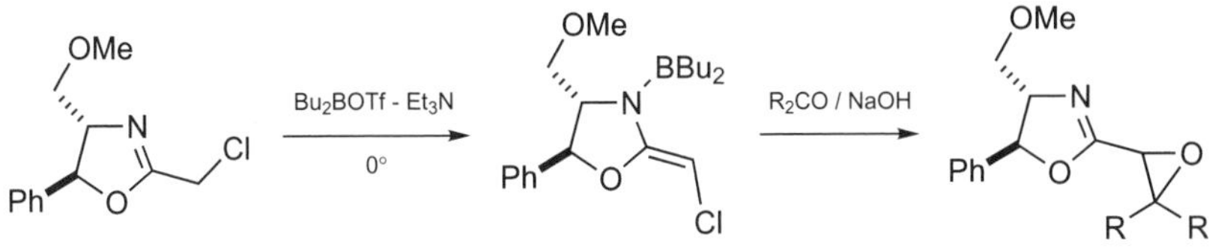

[1]Pinheiro, S., Lima, M.B., Goncalves, C.B.S.S., Pedraza, S.F., de Farias, F.M.C. *TL* **41**, 4033 (2000).
[2]Florio, S., Capriati, V., Luisi, R. *TL* **41**, 5295 (2000).

Dibutylchlorostannane.

Reductive amination.[1] The $Bu_2Sn(H)Cl$–HMPA complex converts carbonyl compounds to secondary amines.

Hydrostannylation.[2] The *trans*-hydrostannylation of propargyl ethers to give (*Z*)-1-dibutylchlorostannyl-3-hydroxy-1-alkenes is probably directed by the ethereal oxygen atom.

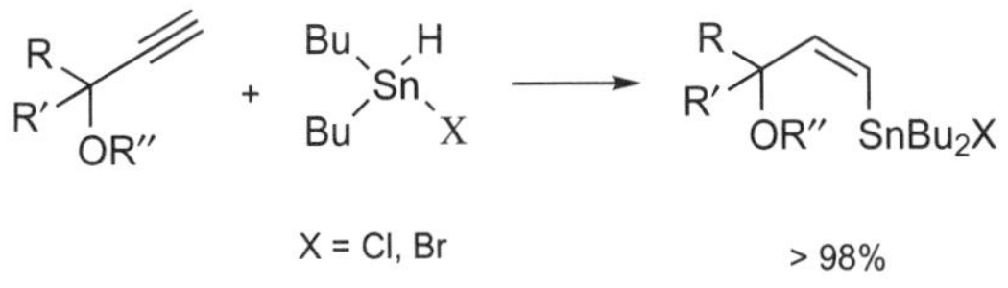

[1]Suwa, T., Sugiyama, E., Shibata, I., Baba, A. *S* 789 (2000).
[2]Mitchell, T.N., Moschref, S.-N. *CC* 1201 (1998).

Dibutylchlorostannyl oxide.

Esters.[1] The catalyst promotes transesterification, for example, the acetyl group from EtOAc to alcohols. Acetic anhydride can be used as acylating agent.

[1]Orita, A., Sakamoto, K., Hamada, Y., Mitsutome, A., Otera, A. *T* **55**, 2899 (1999).

2,7-Di-*t*-butylfluoren-9-ylmethoxycarbonyl chloride.

Amine protection.[1] Compared to simple *O*-(9-fluorenylmethyl) carbamates, the *t*-butyl substituents make them more soluble. On deprotection by treatment with 20%

piperidine in DMF, the formation of a highly lipophilic byproduct enables its removal by hexane extraction.

[1]Stigers, K.D., Koutroulis, M.R., Chung, D.M., Nowick, J.S. *JOC* **65**, 3858 (2000).

Dibutyliodostannane.

Reductive amination. Aldehydes form secondary amines on treatment with $ArNH_2$ and $Bu_2Sn(H)I$. When a proximal Michael acceptor is present, cyclization follows.[1] However, a reduction–aldolization sequence prevails without the amine.[2]

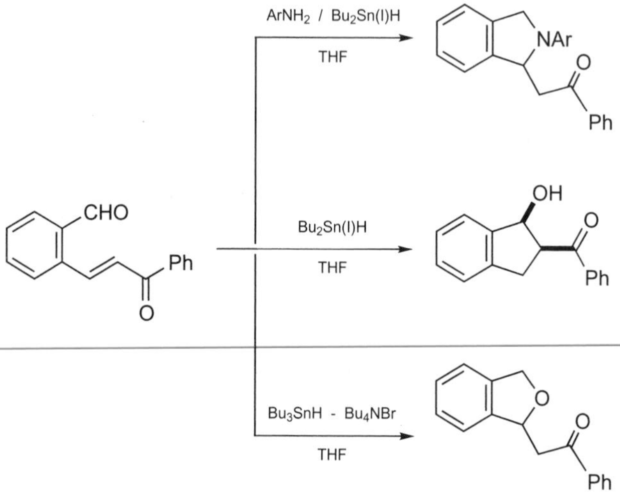

CHO, Ph, O; $ArNH_2$ / $Bu_2Sn(I)H$, THF → NAr, O, Ph; $Bu_2Sn(I)H$, THF → OH, O, Ph; Bu_3SnH - Bu_4NBr, THF → O, O, Ph

[1]Suwa, T., Shibata, I., Nishino, K., Baba, A. *OL* **1**, 1579 (1999).
[2]Suwa, T., Nishino, K., Miyatake, M., Shibata, I., Baba, A. *TL* **41**, 3403 (2000).

Di-*t*-butyl pyrocarbonate. 20, 133

Dehydration.[1] Serine analogues undergo dehydration on treatment with $(Boc)_2O$ and DMAP to afford dehydroamino acid derivatives. Peptides containing serine residues are similarly affected.

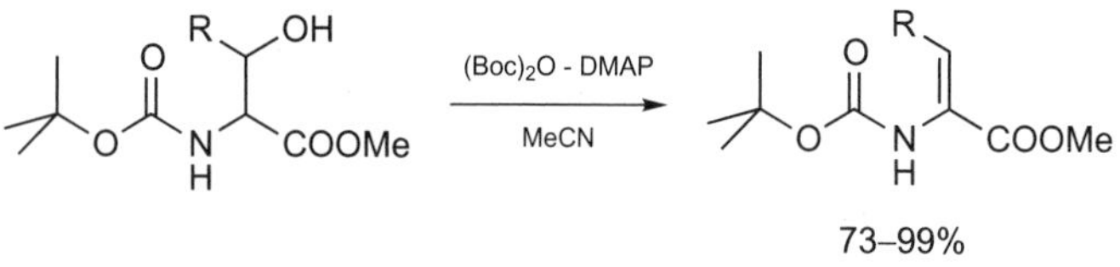

R, OH, O, O, N, H, COOMe; $(Boc)_2O$ - DMAP, MeCN → R, O, O, N, H, COOMe

73–99%

Carbamate exchange.[2] Benzyl carbamates are converted to *t*-butyl carbamates on hydrogenolysis (Pd/C, polymethylhydrosiloxane in EtOH) in the presence of $(Boc)_2O$.

[1]Ferreira, P.M.T., Maia, H.L.S., Monteiro, L.S., Sacramento, J. *JCS(P1)* 3697 (1999).
[2]Chandrasekhar, S., Chandraiah, L., Reddy, C.R., Reddy, M.V. *CL* 780 (2000).

Dibutyltin oxide. 13, 95–96; **15,** 116–117; **16,** 112; **18,** 125; **20,** 133–134

syn-Amino alcohols.[1] *syn*-1,2-Diols, after forming dibutylstannylene derivatives, are converted to cyclic carbamates on reaction with BzN═C═S.

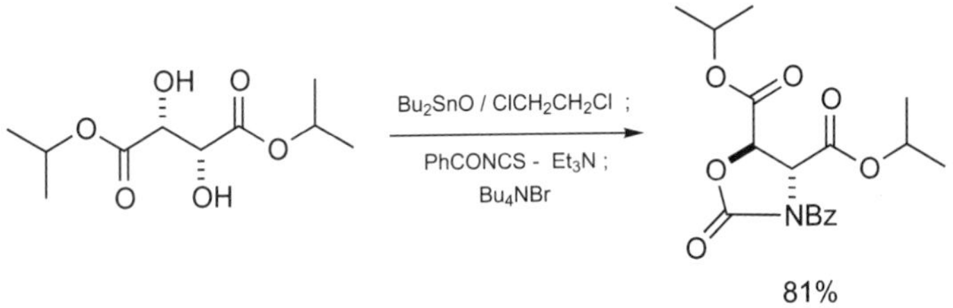

Bu_2SnO / $ClCH_2CH_2Cl$; PhCONCS - Et_3N ; Bu_4NBr

81%

Monotosylation.[2] 1,2-Diols undergo selective tosylation with this catalyst (without which yields are poor). In some cases the efficiency is comparable to that involving stoichiometric formation of the cyclic stannylene derivatives.

[1]Cho, G.Y., Ko, S.Y. *JOC* **64**, 8745 (1999).
[2]Martinelli, M.J., Vaidyanathan, R., Khau, V.V. *TL* **41**, 3773 (2000).

Dicarbonylcyclopentadienylcobalt.

Cycloisomerization.[1] β-Keto esters that contain an alkynyl chain at the α-position undergo ene reaction.

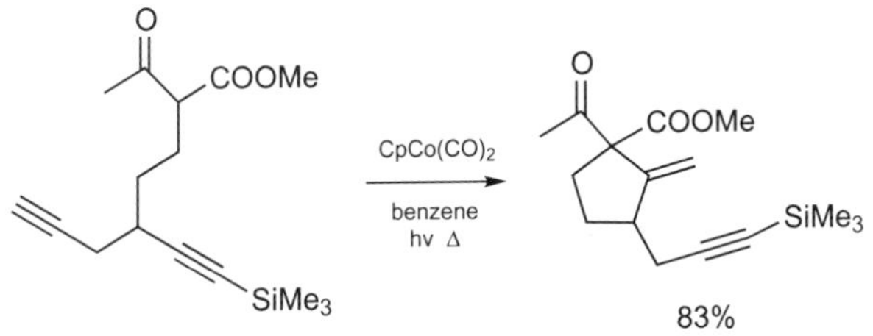

COOMe, $SiMe_3$ → $CpCo(CO)_2$ / benzene / hν Δ → COOMe, $SiMe_3$

83%

[1]Renaud, J.-L., Aubert, C., Malacria, M. *T* **55**, 5113 (1999).

Dichloroalane.

1,1-Diiodoalkanes.[1] Reaction of 1-alkynes with $HAlCl_2$ followed by quenching the *gem*-dimetallic species with iodine leads to a twofold hydroiodination. The reagent is prepared in situ from $LiAlH_4$ and $AlCl_3$ (ratio 1:3) in ether with subsequent exchange of the solvent.

Ph–C≡CH $\xrightarrow[\text{PhMe } 90°]{HAlCl_2}$ $PhCH_2CH(AlCl_2)CH_2AlCl_2$ (> 90%) $\xrightarrow[\text{THF}]{I_2}$ $PhCH_2CH(I)CH_2I$

[1]Aufauvre, L., Knochel, P., Marek, I. *CC* 2207 (1999).

Dichloro(1,5-cyclooctadiene)ruthenium(II).

Cyclization.[1] An efficient synthesis of methylenecyclopentanes from 1,6-dienes is by Ru catalysis.

[1]Yamamoto, Y., Ohkoshi, N., Kameda, M., Itoh, K. *JOC* **64**, 2178 (1999).

Dichlorobis(triphenylphosphine)ruthenium(II). 19, 123–125; **20,** 136–137

Isomerization. The isomerization of allyl alcohols to ketones by this complex has been applied to Baylis–Hillman adducts, therefore a new way to acquire α-methyl-β-ketoesters is developed.[1]

2-Thiazolines.[2] Dehydrogenation of thiazolidines by *t*-BuOOH is catalyzed by $(Ph_3P)_2RuCl_2$.

Reduction of ketones.[3] Extremely highly enantioselective reduction by *i*-PrONa/ *i*-PrOH is observed in the presence of the Ru complex and a chiral (oxazolinylferrocenyl)phosphine ligand. The same catalytic system effects oxidation of racemic secondary alcohols with acetone via kinetic resolution, again with ee of ~ 99.9%.

[1]Basavaiah, D., Muthukumaran, K. *SC* **29**, 713 (1999).
[2]Fernandez, X., Fellous, R., Dunach, E. *TL* **41**, 3381 (2000).
[3]Nishibayashi, Y., Takei, I., Uemura, S., Hidai, M. *OM* **18**, 2291 (1999).

2,3-Dichloro-5,6-dicyano-1,4-benzoquinone, DDQ. 13, 104–105; **14,** 126–127; **15,** 125–126; **16,** 120; **18,** 130; **19,** 121–122; **20,** 137–138

Deprotection. Removal of the 2-naphthylmethyl group from 2-NpCH_2OR[1] and oxidative cleavage of oximes and tosylhydrazones with DDQ proceed under mild conditions.[2] Smooth and selective transformation of $RNBn_2$ to RNHBn at room temperature by DDQ in dichloromethane containing some water is observed.[3]

Acetalization. Diethyl acetals are formed from carbonyl compounds by functional group exchange with triethyl orthoformate (EtOH also present) under the influence of DDQ.[4]

Benzylic activation. Hydride abstraction by DDQ from benzyl ethers, where the benzylic position is also activated by a nuclear substituent (e.g., methoxy group), prepares such compounds to be attacked by nucleophiles. The reaction constitutes an important step in a synthesis of deoxyfrenolicin.[5]

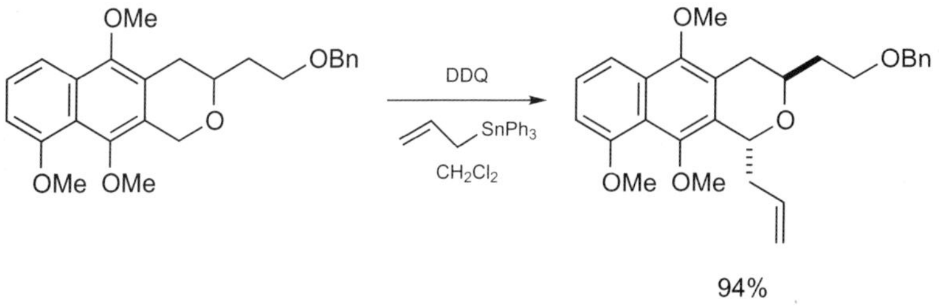

[1]Xia, J., Abbas, S.A., Locke, R.D., Piskorz, C.F., Alderfer, J.L., Matta, K.L. *TL* **41**, 169 (2000).
[2]Chandrasekhar, S., Reddy, C.R., Reddy, M.V. *CL* 430 (2000).
[3]Hungerhoff, B., Samanta, S.S., Roels, J., Metz, P. *SL* 77 (2000).
[4]Karimi, B., Ashtiani, A.M. *CL* 1199 (1999).
[5]Xu, Y.-C., Kohlman, D.T., Liang, S.X., Erikkson, C. *OL* **1**, 1599 (1999).

1,1-Dichloroethyllithium.

2,2-Dichloro-3-alkanones.[1] The title reagent, which is generated from 1,1-dichloroethane, reacts with esters at low temperatures to give mixed alkyl–silyl acetals of 2,2-dichloro-3-alkanones. The dichloroketones are obtained directly from reaction of MOM esters.

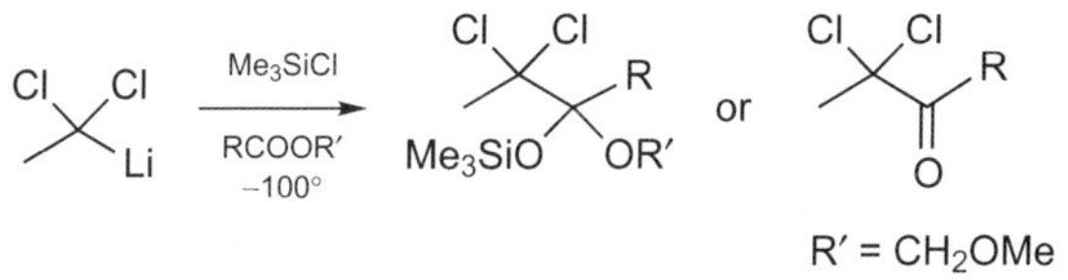

[1]Shiina, I., Imai, Y., Suzuki, M., Yanagisawa, M., Mukaiyama, T. *CL* 1062 (2000).

[1-(2,5-Dichlorophenyl)-2,2-bismethylthio]vinyl benzoate.

N-Acylation.[1] Reagent **1** is obtained by enolbenzoylation of α,α-bismethylthioacetophenone. It is a highly efficient acylating agent for amines. Thus, only the benzamides are formed from amino alcohols and amino thiols, using limited amount of **1**. A higher reactivity is observed toward aliphatic OH than phenolic OH.

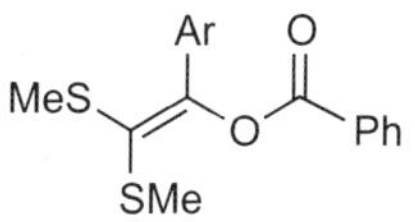

[1]Degani, I., Dughera, S., Fochi, R., Serra, E. *S* 1200 (1999).

_N,N_-Dichloro-2-nitrobenzenesulfonamide.

Aminochlorination.[1] The dichloro compound effects addition to cinnamic esters in combination with the sodium salt of 2-nitrobenzenesulfonamide.

NsN(Na)Cl - $NsNCl_2$; CuOTf · C_6H_6 / MeCN ; Na_2SO_3

76% (*syn* : *anti* 1 : 30)

[1]Li, G., Wei, H.-X., Kim, S.H. *OL* **2**, 2249 (2000).

Dichlorotin oxide.

Alkene functionalization.[1] Vicinal addition of the [HO/N_3] and [HO/RCOO] to alkenes is effected by Me_3SiX (X = N_3, OCOR) in the presence of bis(trimethylsilyl) peroxide and catalyzed by $(Cl_2SnO)_n$.

[1]Sakurada, I., Yamasaki, S., Kanai, M., Shibasaki, M. *TL* **41**, 2415 (2000).

Dicobalt octacarbonyl. 13, 99–101; **14,** 117–119; **15,** 117–118; **16,** 113–115; **17,** 102–105; **18,** 132; **19,** 125–126; **20,** 139–141

Hydrosilylation.[1] The cobalt-catalyzed hydrosilylation of oxygen-containing alkenes has been reported.

Pauson–Khand reaction. Many modifications, particularly regarding the use of substoichiometric quantities of $Co_2(CO)_8$, have been studied. Addition of cyclohexylamine makes catalytic reaction practical, and there is no need to purify $Co_2(CO)_8$, which is somewhat air sensitive.[2,3] Various alkyne–$Co_2(CO)_6$ complexes are useful catalyst [instead of $Co_2(CO)_8$] for carrying out the reaction under carbon monoxide.[4]

Other reports include reactions in supercritical ethylene,[5] promotion by molecular sieves,[6] and the effect of a phosphine sulfide.[7] Clusters such as methylidynetricobalt nonacarbonyl are also effective catalyst of the Pauson–Khand reaction.[8] Vinyl esters are identified as surrogates for ethylene.[9]

Conventional Pauson–Khand reaction conditions are employed in annulating a menthene derivative in an approach to (−)-dendrobine.[10]

(−)-dendrobine

Depropargylation.[11] The value of propargyl derivatives as protected forms of alcohols, carboxylic esters, and amines depends on their facile reversal. It can be achieved using $Co_2(CO)_8$ and trifluoroacetic acid in dichloromethane at room temperature. Alcohols and amines can also be masked as $RXCOOCH_2CCH$.

β-Hydroxy esters.[12] Regioselective ring opening and homologation (methoxycarbonylation) of epoxides are achieved on treatment with $Co_2(CO)_8$ and 3-hydroxypyridine.

β-Lactams.[13] Insertion of CO into aziridines is a new approach to β-lactam synthesis. Regioselectivity for such a reaction has been observed in certain substrates.

(92 : 8)
99.8%

[1]Chatani, N., Kodama, T., Kajikawa, Y., Murakami, H., Kakiuchi, F., Ikeda, S., Murai, S. *CL* 14 (2000).
[2]Krafft, M.E., Bonaga, L.V.R., Hirosawa, C. *TL* **40**, 9171 (1999).
[3]Krafft, M.E., Bonaga, L.V.R. *SL* 959 (2000).
[4]Krafft, M.E., Hirosawa, C., Bonaga, L.V.R. *TL* **40**, 9177 (1999).
[5]Jeong, N., Hwang, S.H. *ACIEE* **39**, 636 (2000).
[6]Perez-Serrano, L., Blanco-Urgoiti, J., Casarrubios, L., Dominguez, G., Perez-Castells, J. *JOC* **65**, 3513 (2000).
[7]Hayashi, M., Hashimoto, Y., Yamamoto, Y., Usuki, J., Saigo, K. *ACIEE* **39**, 631 (2000).
[8]Sugihara, T., Yamaguchi, M. *JACS* **120**, 10782 (1998).
[9]Kerr, W.J., McLaughlin, M., Pauson, P.L., Robertson, S.M. *CC* 2171 (1999).
[10]Cassayre, J., Zard, S.Z. *JACS* **121**, 6072 (1999).
[11]Fukase, Y., Fukase, K., Kusumoto, S. *TL* **40**, 1169 (1999).
[12]Hinterding, K., Jacobsen, E.N. *JOC* **64**, 2164 (1999).
[13]Davoli, P., Moretti, I., Prati, F., Alper. H. *JOC* **64**, 518 (1999).

Dicyanoketene ethyleneacetal.

Mannich reaction.[1] This reagent and similar π-acids on polymer support effect chemoselective condensation between silyl enol ethers and imines in refluxing MeCN; aldehydes do not participate in the reaction.

Ether cleavage.[2] The polymer-bound reagent catalyzes hydrolysis of TBS ether and acetals in aqueous acetonitrile. Methoxymethyl and tetrahydropyranyl ethers are stable under such conditions.

[1]Tanaka, N., Masaki, Y. *SL* 406 (2000).
[2]Tanaka, N., Masaki, Y. *SL* 1960 (1999).

Dicyclohexylborane. 20, 141–142

Hydroboration. Hydroboration of alk-3-en-1-ynes provides conjugated dienylboranes that are reactive toward unactivated dienophiles. Cyclohexenols are readily acquired from the cycloadducts.[1]

Cy_2BH ; Me_3NO ; allyl alcohol (OH) 190° ; Me_3NO → OH/OH products + OH/OH (6 : 4) 60%

[1]Batey, R.A., Thadani, A.N., Lough, A.J. *CC* 475 (1999).

Dicyclohexylboron chloride.

Alkylation. Reaction of the reagent with ArCHO followed by oxidation delivers secondary alcohols. There are hints that other secondary R_2BCl behave similarly, but the scope is severely limited to aldehydes without an α-hydrogen (otherwise enolization prevails).[1]

Aldol reactions. In conjunction with an amine, dicyclohexylboron chloride is used to generate kinetic enolates from ketones. Some interesting and useful information concerns the stereoselectivity of the aldol reactions.[2]

O, OR + Cy_2B-Cl → Et_3N / Et_2O –78°, RCHO → O OH, R, OR
→ Et_2NMe / pentane 25°, RCHO → O OH, R, OR

[1]Kabalka, G.W., Wu, Z., Trotman, S.E., Gao, X. *OL* **2**, 255 (2000).
[2]Galobardes, M., Gascon, M., Mena, M., Romea, P., Urpi, F., Vilarrasa, J. *OL* **2**, 2599 (2000).

Dicyclohexylboron triflate.

Double aldol reactions.[1] The consecutive (one-pot) aldol reaction of carbonyl compounds with two aldehydes is highly stereoselective. It is applicable to the synthesis of C_3-symmetric triols.

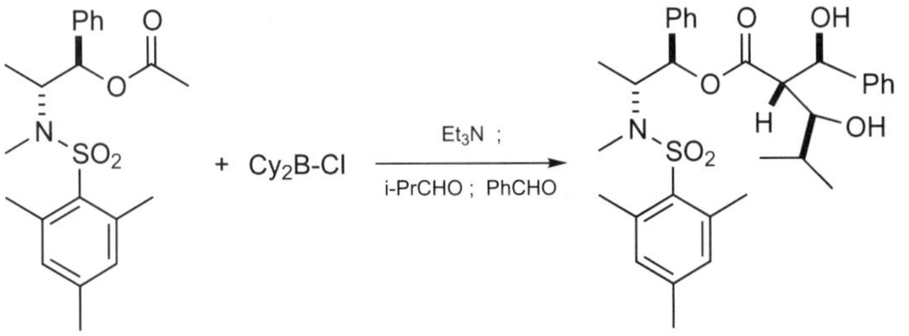
Ph
O
N
SO2
+ Cy2B-Cl
Et3N ;
i-PrCHO ; PhCHO
Ph
O
OH
Ph
H
OH
N
SO2

[1]Abiko, A., Liu, J.-F., Buske, D.C., Moriyama, S., Masamune, S. *JACS* **121**, 7168 (1999).

***N,N'*-Dicyclohexylcarbodiimide, DCC. 14,** 131–132; **16,** 128; **18,** 133–134

Nitroacetic esters.[1] Esterification of nitroacetic acid is not easy to achieve by conventional methods, but it can be performed with the aid of DCC. This procedure serves to anchor nitroacetyl groups to hydroxylated Merrifield resin.

4-O-Alkyltetronic acids.[2] Preparation of these compounds from the parent tetronic acids under mild conditions relies on the exchange reaction with isoureas derived from DCC.

[1]Sylvain, C., Wagner, A., Mioskowski, C. *TL* **40**, 875 (1999).
[2]Schobert, R., Siegfried, S. *SL* 686 (2000).

Diethylaluminum ethoxide.

Aldol reaction.[1] Enol esters are used as donors in the crossed-aldol reaction in THF at 0°. Product yields are in the 70% range.

Diels-Alder reaction.[2] At low temperatures, the cycloaddition occurs between enones and 1-acetoxy-1,3-dienes in the presence of Et_2AlOEt. The acetate suffers hydrolysis during the condensation.

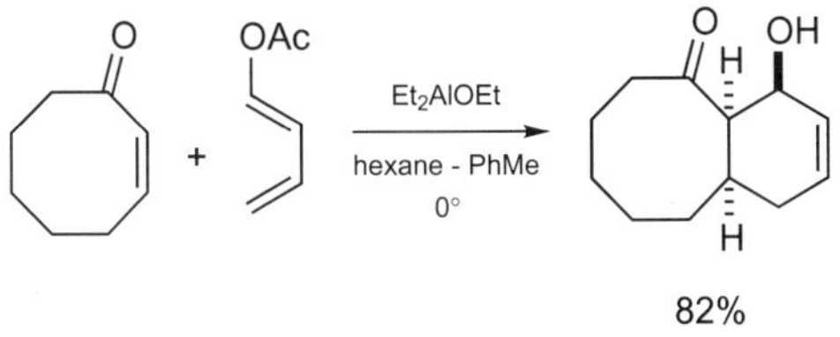
O
OAc
+
Et2AlOEt
hexane - PhMe
0°
O
H
OH
H
82%

[1]Mukaiyama, T., Shibata, J., Shimamura, T., Shiina, I. *CL* 951 (1999).
[2]Shibata, J., Shiina, I., Mukaiyama, T. *CL* 313 (1999).

(Diethylamino)sulfur trifluoride, DAST. 13, 110–112; **16,** 128–129; **18,** 135; **19,** 126–127; **20,** 142

Cyclizations. A one-step cyclization of *N*-hydroxyethyl amides to oxazoles and oxazolines[1] is mediated by DAST. β-Hydroxy carbamates usually undergo fluorination (OH → F), but the *N*-Boc derivatives afford oxazolidin-2-ones.[2]

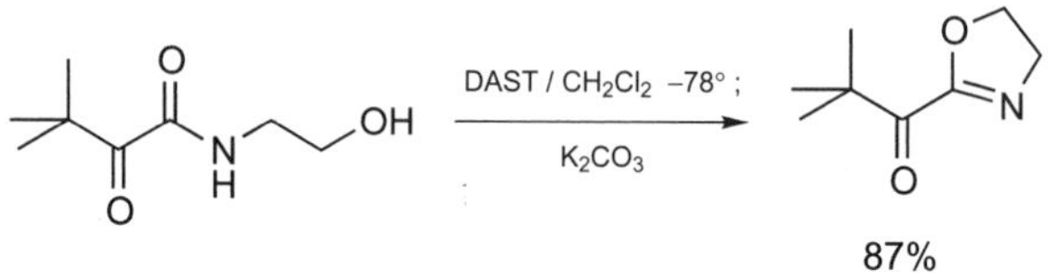

DAST / CH_2Cl_2 –78° ; K_2CO_3

87%

[1]Phillips, A.J., Uto, Y., Wipf, P., Reno, M.J., Williams, D.R. *OL* **2**, 1165 (2000).
[2]Zhao, H., Thurkauf, A. *SL* 1280 (1999).

1,3-Dihydro-1,3-diacetyl-2*H*-benzimidazol-2-one.

Acetylation.[1] Selective acetylation of amines (primary amines > secondary amines, arylamines) is readily accomplished.

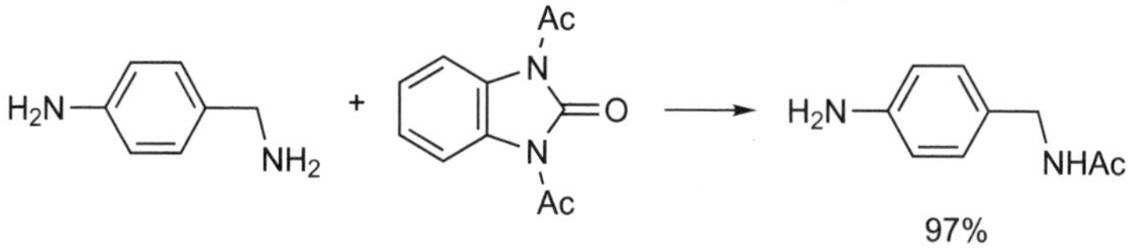

97%

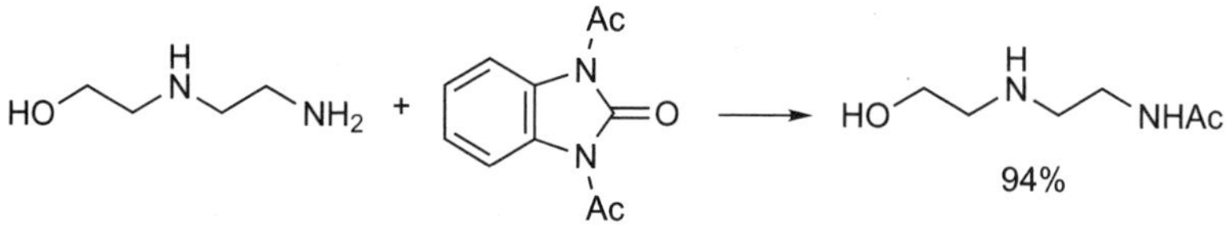

94%

[1]Chung, I.H., Cha, K.S., Seo, J.H., Kim, J.H., Chung, B.Y., Kim, C.S. *H* **53**, 529 (2000).

2,6-Dihydroxypyridine.

Sulfoxide reduction.[1] By using this reagent to reduce sulfoxides, many functional groups (e.g., esters, carbamates) are not affected.

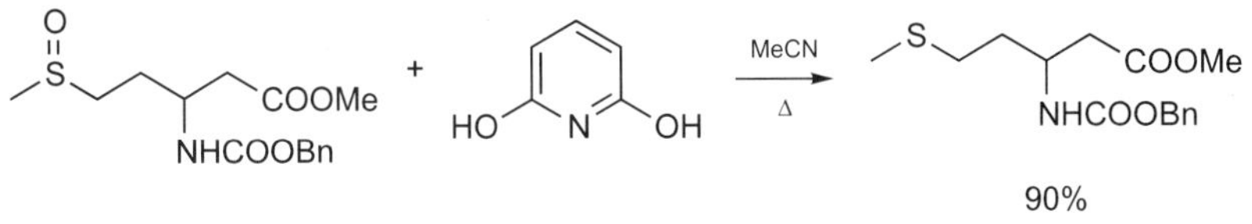

[1] Miller, S.J., Collier, T.R., Wu, W. *TL* **41**, 3781 (2000).

Diiodomethane. 13, 110–115; 275–276; **16,** 184–185; **17,** 155; **18,** 139–140; **19,** 128; **20,** 143–144

1,3-Diketones. Simmons–Smith reagent reacts with acid chlorides to afford the 1, 3-diones.[1] By addition of tetrahydrothiophene the monomeric form is stabilized.

[1] Matsubara, S., Yamamoto, Y., Utimoto, K. *SL* 1471 (1999).

Diiodosilane.

Ureas.[1] Carbamates are converted to isocyanates by H_2SiI_2. On exposing the isocyanates to amines, unsymmetrical ureas result.

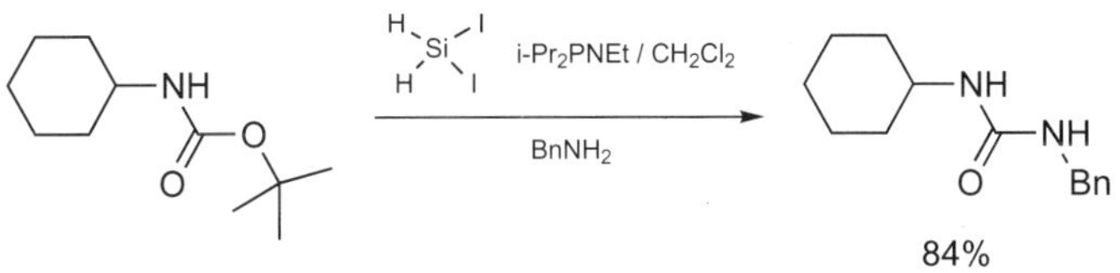

[1] Gastaldi, S., Weireb, S.M., Stien, D. *JOC* **65**, 3239 (2000).

Diisobutylaluminum hydride, Dibal-H. 13, 115–116; **15,** 137–138; **16,** 134–135; **17,** 123–125; **18,** 140–141; **19,** 128–129; **20,** 144–146

Conjugate reduction.[1] The double bond of α,β-unsaturated ketones, esters, nitriles, and tertiary amides are reduced by Dibal-H in the presence of $Co(acac)_2$.

Debenzylation.[2] Perbenzylated cyclodextrins (α-, β-, and γ-forms) are partially debenzylated by Dibal-H in toluene in a regioselective manner to provide the mono-6-*O*-debenzylated or AD-di-*O,O*-debenzylated derivative.

Claisen rearrangement.[3] Dibal-H serves as a Lewis acid catalyst for the Claisen rearrangement. Thus, allyl 4-*t*-butylphenyl ether is converted to the 2-allylphenol product at room temperature in 87% yield. $Yb(OTf)_3$ is a far less effective catalyst.

[1] Ikeno, T., Kimura, T., Ohtsuka, Y., Yamada, T. *SL* 96 (1999).
[2] Pearce, A.J., Sinay, P. *ACIEE* **39**, 3610 (2000).
[3] Sharma, G.V.M., Ilangovan, A., Sreenivas, P., Mahalingam, A.K. *SL* 615 (2000).

p-Diisocyanatobenzene.

Nitrile oxide generation.[1] Dehydration of nitroalkanes with a diisocyanate–triethylamine combination simplifies the product (of 1,3-dipolar cycloaddition) purification, because the urea byproduct is polymeric which can be filtered off.

[1] Kantorowski, E.J., Brown, S.P., Kurth, M.J. *JOC* **63**, 5272 (1998).

Diisopropyl azodicarboxylate, DIAD.

N-Debenzylation.[1] Benzyl ethers are unaffected when *N*-benzylamines undergo C—N bond scission with the title reagent.

PhMe, Δ

81%

[1] Kroutil, J., Trnka, T., Cerny, M. *OL* **2**, 1681 (2000).

Dilauroyl peroxide.

Radical addition.[1] The adducts of *S*-(α-ketoalkyl) xanthates to allylamines (from reaction catalyzed by lauroyl peroxide) are useful precursors of piperidines.

dilauroyl peroxide

89%

Azidation.[2] On heating with $EtSO_2N_3$ and dialuroyl peroxide in chlorobenzene at 100°, alkyl iodides or xanthates are transformed into azides.

Cyclizations. Indanes and indolines are readily formed by treatment of 3-butenylarenes and *N*-allylaniline derivatives with a stoichiometric amount of dilauroyl peroxide.[3] Formation of a seven-membered ring adjoining an existing aromatic nucleus by radical cyclization of xanthate precursors has a useful scope.[4] Several different functional groups inside the chain are tolerated.

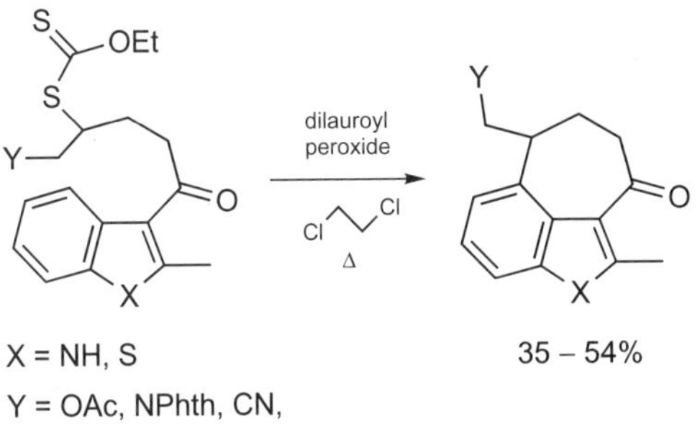

X = NH, S
Y = OAc, NPhth, CN,

35 – 54%

[1]Boivin, J., Pothier, J., Zard, S.Z. *TL* **40**, 3701 (1999).
[2]Ollivier, C., Renaud, P. *JACS* **122**, 6496 (2000).
[3]Ly, T.-M., Quiclet-Sire, B., Sortais, B., Zard, S.Z. *TL* **40**, 2533 (1999).
[4]Kaoudi, T., Quiclet-Sire, B., Seguin, S., Zard, S.Z. *ACIEE* **39**, 731 (2000).

Dimanganese decacarbonyl.

Radical reactions. Initiation of the cyclization to form five-membered rings[1] by a photochemical reaction in the presence of $Mn_2(CO)_{10}$ is shown in the following.

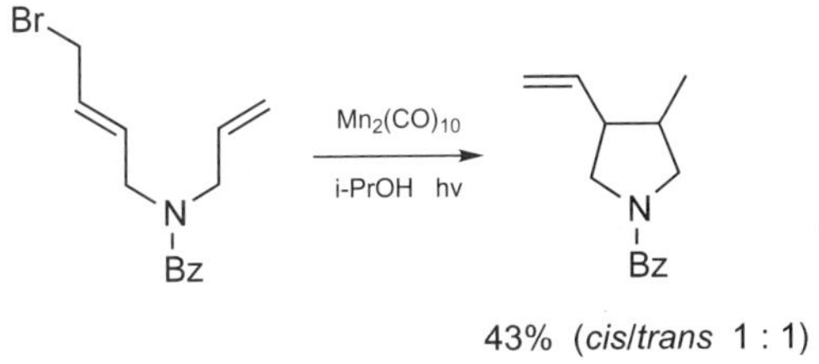

43% (*cis/trans* 1 : 1)

Homocoupling and cross-coupling of bromides[2] are efficiently achieved with $Mn_2(CO)_{10}$.

[1]Gilbert, B.C., Kalz, W., Lindsay, C.I., McGrail, P.T., Parsons, A.F., Whittaker, D.T.E. *JCS(P1)* 1187 (2000).
[2]Gilbert, B.C., Lindsay, C.I., McGrail, P.T., Parsons, A.F., Whittaker, D.T.E. *SC* **29**, 2711 (1999).

***N,N′*-Dimethoxy-*N,N′*-dimethyloxamide.**

Condensation.[1] Reaction of this reagent with dilithiated β-ketoesters results in functionalized butenolides.

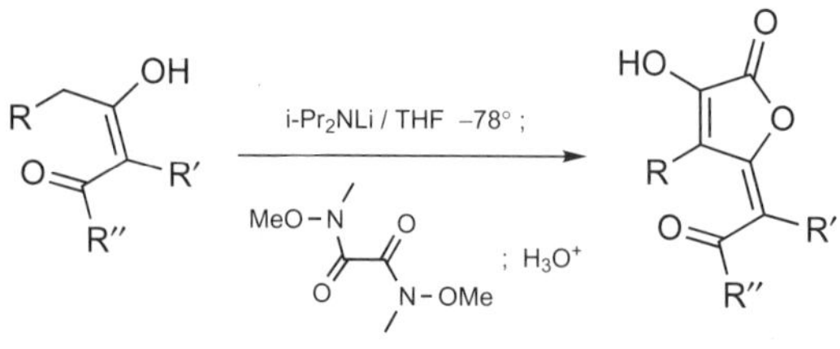

[1]Langer, P., Stoll, M. *ACIEE* **38**, 1803 (1999).

N-(4,6-Dimethoxy-1,3,5-triazin-2-yl)-N-methylmorpholinium chloride.

Activation of carboxylic acids. The coupling reagent **1**, derived from *N*-methylmorpholine and 2-chloro-4,6-dimethoxy-1,3,5-triazine, reacts with RCOOH to form heteroaryl esters that are susceptible to alcoholysis[1] and serve as transacylating agents in peptide synthesis.[2]

The esters can be converted to aldehydes or alcohols on hydrogenation over Pd–C. Longer reaction times favor production of the alcohols.[3]

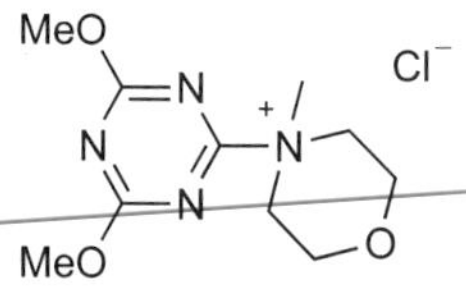

(**1**)

[1]Kurishima, M., Morita, J., Kawachi, C., Iwasaki, F., Terao, K., Tani, S. *SL* 1255 (2000).
[2]Falchi, A., Giacomelli, G., Porcheddu, A., Taddei, M. *SL* 275 (2000).
[3]Falorni, M., Giacomelli, G., Porcheddu, A., Taddei, M. *JOC* **64**, 8962 (1999).

N,N-Dimethylbenzotriazol-1-ylmethyleniminium chloride.

Preparation.[1] The stable salt **1** is obtained in 92% yield by heating *N*-trimethylsilylbenzotriazole with DMF and thionyl chloride in THF.

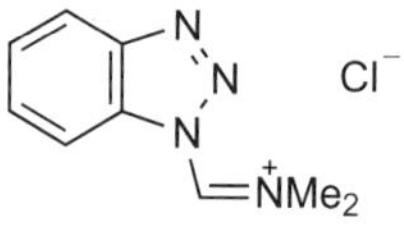

(**1**)

Heterocycles. Synthesis of substituted pyridines[2] and quinolines[3] from ketones, via dienamines and *N*-arylimines, respectively, is by extraction of the [CH] unit from reagent **1**.

N,N-Dimethylformamidrazones.[4] The reaction of hydrazines with **1** in refluxing THF delivers $R_2NN{=}CHNMe_2$.

[1]Katritzky, A.R., Cheng, D., Leeming, Ghiviriga, I., Hartshorn, C.M., Steel, P.J. *JHC* **33**, 1935 (1996).
[2]Katritzky, A.R., Denisenko, A., Arend, M. *JOC* **64**, 6076 (1999).
[3]Katritzky, A.R., Arend, M. *JOC* **63**, 9989 (1998).
[4]Katritzky, A.R., Huang, T.-B., Voronkov, M.V. *JOC* **65**, 2246 (2000).

2-Dimethylamino-2′-di-*t*-butylphosphino-1,1′-binaphthyl.

Diaryl ethers.[1] This electron-rich and bulky *P,N*-ligand (**1**) facilitates the Pd-catalyzed synthesis of diaryl ethers from phenols and aryl halides.

(**1**)

[1]Aranyos, A., Old, D.W., Kiyomori, A., Wolfe, J.P., Sadighi, J.P., Buchwald, S.L. *JACS* **121**, 4369 (1999).

4-Dimethylaminopyridine, DMAP.

Esters. With promotion by DMAP the methanolysis of *N*-acyloxazolidinethiones[1] and *N*-acylthiazolidinethiones[2] is achieved at room temperature.

DMAP has an ameliorating effect on the displacement of secondary alkyl sulfonates with inversion of configuration that employs CsOAc.[3]

[1]Su, D.-W., Wang, Y.-C., Yan, T.-H. *TL* **40**, 4197 (1999).
[2]Su, D.-W., Wang, Y.-C., Yan, T.-H. *CC* 545 (1999).
[3]Hawryluk, N.A., Snider, B.B. *JOC* **65**, 8379 (2000).

***N,N*-Dimethyl-(1-benzotriazolyloxy)methyleneiminium hexachloroantimonate.**

Peptide coupling.[1] Excellent yields are obtained with the coupling reagent **1**. Results are better than those of several other reagents.

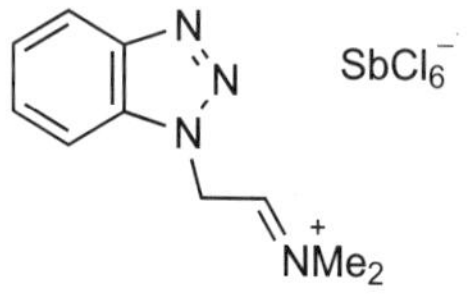

$SbCl_6^-$ $\overset{+}{N}Me_2$

(**1**)

[1]Li, P., Xu, J.C. *TL* **40**, 3605 (1999).

(*R*)-3,3′-Dimethyl-1,1′-binaphthalene-2,2′-diamine.

Asymmetric hydrogenation.[1] The ligand enables the Ru-catalyzed hydrogenation of ketones to afford optically active alcohols (7 examples, ~99% yield, ee 91–96%).

[1]Mikami, K., Korenaga, T., Ohkuma, T., Noyori, R. *ACIEE* **39**, 3707 (2000).

2,7-Dimethyl-1,8-biphenylenedioxybis(dialkylaluminum). 20, 149

Redox reactions. Reversal of oxidation states in hydroxyaldehydes results when Oppenaur oxidation and Meerwein–Ponndorf–Verley reduction operate at the responsive functional groups.[1]

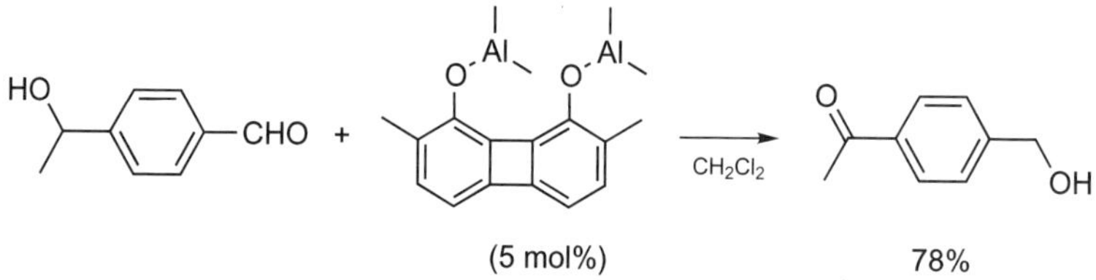

HO — CHO + (5 mol%) $\xrightarrow{CH_2Cl_2}$ O — OH 78%

A rapid Tishchenko reaction occurs when aldehydes are exposed to the bis(diisopropoxyaluminum) analogue.[2]

[1]Ooi, T., Itagaki, Y., Miura, T., Maruoka, K. *TL* **40**, 2137 (1999).
[2]Ooi, T., Miura, T., Takaya, K., Maruoka, K. *TL* **40**, 7695 (1999).

Dimethyl carbonate. 18, 144; **20,** 150

Carbamates and ureas. Stepwise displacement of the methoxy groups of dimethyl carbonate by amines leads to carbamates and then ureas.[1]

[1]Vauthey, I., Valot, F., Gozzi, C., Fache, F., Lemaire, M. *TL* **41**, 6347 (2000).

***N,N*-Dimethyl-2-(dimethylamino)aniline *N*-oxide.**

Carbonyl compounds.[1] An alternative to the Kornblum method for conversion of halides to carbonyl products is employment of this reagent.

[1]Chandrasekhar, S., Sridhar, M. *TL* **41**, 5423 (2000).

Dimethyldioxirane, DMD. 12, 413; **13,** 120; **14,** 148; **15,** 143–144; **16,** 142–144; **18,** 144–146; **19,** 135–136; **20,** 150–152

Epoxidation. A simple and efficient protocol for alkene epoxidation uses DMD in acetone.[1] *N,N'*-Dialkylalloxans are now found to be effective catalysts.[2]

Trifluoromethyl ketones that are frequently used for mediating alkene epoxidation with dioxiranes are subject to degradation. A stabilized form is made by covalent anchoring to silica.[3]

Iodohydrins.[4] A combination of DMD and MeI generates HOI, which readily converts alkenes to iodohydrins.

Oxidation of aromatic compounds. 3-Arylisocoumarins are obtained from 3-arylisochromanes[5] whereas flavanones are oxygenated in the aromatic ring.[6]

MeO, OMe, OMe O, O, + dioxirane, HCl, −30°, OH, MeO, HO, OMe O, OMe, ~ 100%

1,2-Dicarbonyl compounds. α-Diazoketones are readily oxidized.[7]

α-Bromo-α,β-enones.[8] When enones are treated with DMD and NaBr, introduction of a bromine atom at the α-position results.

[1]Ferraz, H.M.C., Muzzi, R.M., de O.Vieira, T., Viertler, H. *TL* **41**, 5021 (2000).
[2]Carnell, A.J., Johnstone, R.A.W., Parsy, C.C., Sanderson, W.R. *TL* **40**, 8029 (1999).
[3]Song, C.E., Lim, J.S., Kim, S.C., Lee, K.-J., Chi, D.Y. *CC* 2415 (2000).
[4]Asensio, G., Andreu, C., Boix-Bernardini, C., Mello, R., Gonzalez-Nunez, M.E. *OL* **2**, 2125 (1999).
[5]Bovicelli, P., Lupattelli, P., Crescenzi, B., Sanetti, A., Bernini, R. *T* **55**, 14719 (1999).
[6]Bernini, R., Mincioni, E., Sanetti, A., Mezzetti, M., Bovicelli, P. *TL* **41**, 1087 (2000).
[7]Darkins, P., Groarke, M., McKervey, M.A., Moncrieff, H.M., McCarthy, N., Nieuwenhuyzen, M. *JCS(P1)* 381 (2000).
[8]Righi, G., Bovliclli, P., Sperandio, A. *TL* **40**, 5889 (1999).

Dimethyl disulfide.

Unsymmetrical disulfides.[1] Disproportionation between a symmetrical disulfide with MeSSMe (NaOH) gives rise to RSSMe.

[1]Kamiyama, H., Noguchi, T., Onodera, A., Yokomori, S., Nakayama, J. *CL* 755 (1999).

Dimethyl 3,3′-dithiodipropionate.

Methyl acrylate precursor.[1] A heterocyclic NH is converted into the NCH_2CH_2COOMe moiety by the reagent under basic conditions.

[1]Hamel, R., Girard, M. *JOC* **65**, 3123 (2000).

***N,N*-Dimethylformamide–phosphoryl chloride. 18,** 146; **20,** 153

Formylations. Direct conversion of silyl ethers to formic esters[1] occurs at 0° with DMF–$POCl_3$. Silyl enol ethers give α-hydroxymethylene ketones.[2]

[1]Koeller, S., Lellouche, J.-P. *TL* **40**, 7043 (1999).
[2]Jameleddine, K., Bechir, B.H., Mustapha, M. *SC* **30**, 2759 (2000).

Dimethyl(methylthio)sulfonium tetrafluoroborate.

Furans.[1] The reagent activates thioacetal function by methylthiolation. It induces cyclization of thioacetals derived from carbonyl compounds (including lactones and lactams) and bis(methylthio)acetaldehyde via aldol reaction and acetylation.

OAc SMe SMe N R O + S⁺ S BF_4^- → N R O SMe

[1]Padwa, A., Ginn, J.D., McClure, M.S. *OL* **1**, 1559 (1999).

2,5-Dimethylphenacyl chloride.

Photoreleasable esters.[1] A method for carboxyl group protection is based on esters obtained by *O*-alkylation of acids with the phenacyl chloride. The photochemical decomposition that regenerates the acids does not require a photosensitizer.

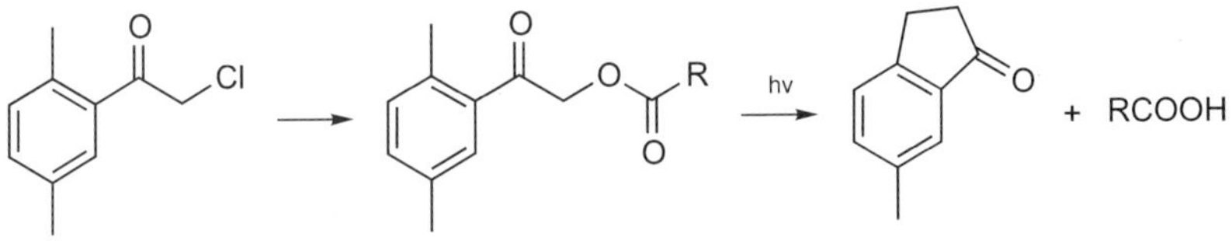

[1] Klan, P., Zabadal, M., Heger, D. *OL* **2**, 1569 (2000).

2,6-Dimethylphenylsulfinyl chloride.

Ketimine radicals.[1] Ketoximes are deoxygenated on treatment with sulfinyl chlorides via homolysis of the oxime sulfinates. When suppression of diradical recombination by steric hindrance is exploited, imine radical intermediates tend to undergo other reactions such as intramolecular addition.

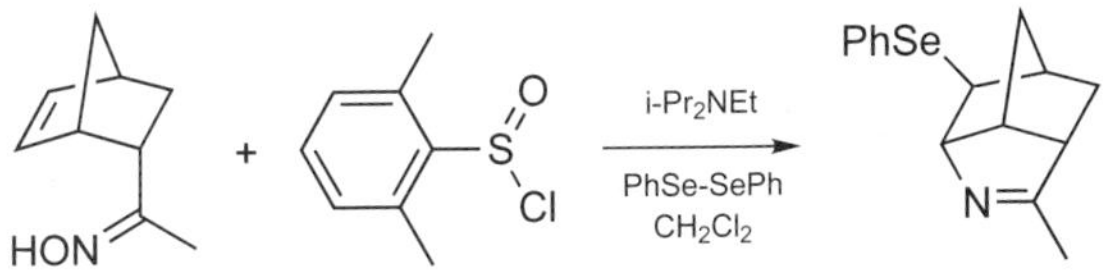

[1] Lin, X., Stien, D., Weinreb, S.M. *OL* **1**, 637 (1999).

Dimethyl phosphite.

Debromination.[1] 1,1-Dibromoalkenes undergo monodebromination by the action of $(MeO)_2POH–Et_3N$ to furnish the (*E*)-bromoalkenes.

[1] Abbas, S., Hayes, C.J., Worden, S. *TL* **41**, 3215 (2000).

Dimethyl sulfonium ylide. 18, 149

Allylic alcohols.[1] Epoxides are homologated, and this reaction can be used to construct polyhydroxylated carbon chains.

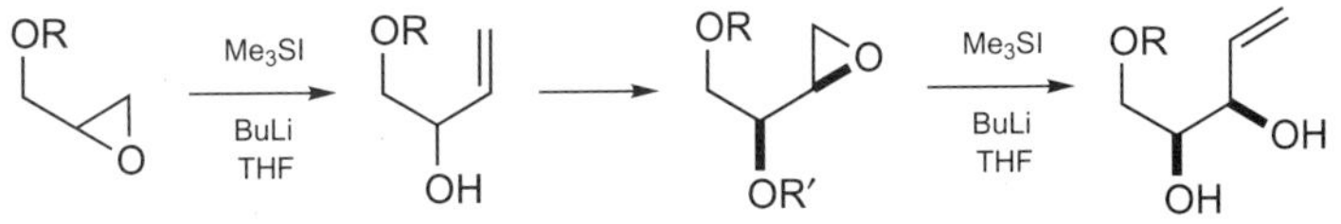

[1]Davoille, R.J., Rutherford, D.T., Christie, S.D.R. *TL* **41**, 1255 (2000).

Dimethyl sulfoxide. 13, 124; **16,** 149; **18,** 149; **20,** 154–155

Aldol and Michael reactions.[1] Spontaneous aldol and Michael reactions involving silyl ketene acetals are observed at room temperature in dry DMSO.

(E)-Bromoalkenes.[2] *gem*-Borylbromoalkenes are transformed to bromoalkenes by DMSO, which acts as an oxidizing agent.

Oxidative acetalization.[3] Secondary alcohols are converted to 1,3-dioxolanes directly in a Re-catalyzed oxidation in the presence of DMSO and 1,2-ethanediol.

[1]Genisson, Y., Gorrichon, L. *TL* **41**, 4881 (2000).
[2]Hoshi, M., Tanaka, H., Shirakawa, K., Arase, A. *CC* 627 (1999).
[3]Arterbum, J.B., Perry, M.C. *OL* **1**, 769 (1999).

Dimethyl sulfoxide–2-chloro-1,3-dimethylimidazolinium chloride.

Oxidation.[1] This combination together with Et_3N is comparable to the Swern reagent.

[1] Isobe, T., Ishikawa, T. *JOC* **64**, 5832 (1999).

Dimethyl sulfoxide–oxalyl chloride.

Desilylative oxidation.[1] Selective transformation of a primary silyl ether to an aldehyde without affecting secondary silyl ethers is achieved.

OSiEt3 / R / OSiEt3 / OSiEt3 —[DMSO - $(COCl)_2$; Et_3N / CH_2Cl_2, −70° ~ 25°]→ OSiEt3 / R / CHO / OSiEt3

[1]Rodriguez, A., Nomen, M., Spur, B.W., Godfroid, J.J. *TL* **40**, 5161 (1999).

Dimethyltin dichloride.

Monobenzoylation.[1] 1,2-Diols form monobenzoates on reaction with BzCl and Me_2SnCl_2.

[1]Iwasaki, F., Maki, T., Onomura, O., Nakashima, W., Matsumura, Y. *JOC* **65**, 996 (2000).

Dimethyltitanium dichloride.

Cyclization.[1] 4-Alkynylamines undergo cyclization to pyrrolines under the influence of Me_2TiCl_2. In some cases, chain extension at the alkyne terminal can be effected.

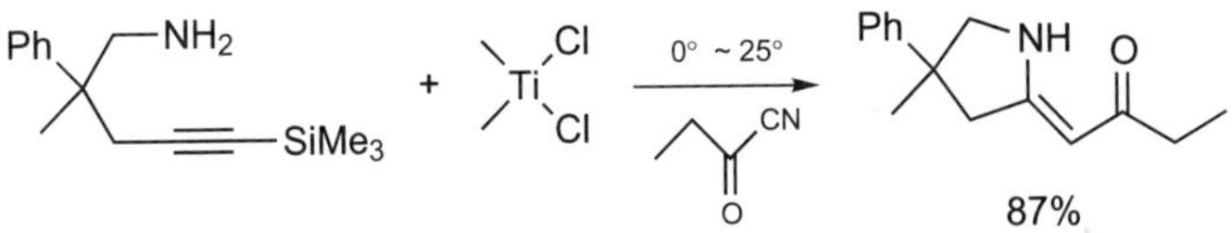

87%

[1]Duncan, D., Livinghouse, T. *OM* **18**, 4421 (1999).

Dimethyltitanocene.

Hydroamination. Alkynes are converted to enamines on addition of amines in the presence of Cp_2TiMe_2.[1,2] Immediate reduction provides stable products, whereas hydrolysis results in ketones.

An intramolecular version in which an acyl cyanide is added leads to the vinylogous lactam.[3]

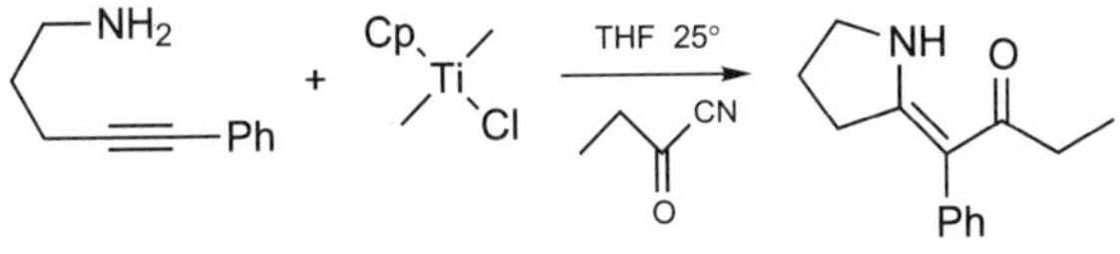

87%

Lactones. Methylenation (Cp_2TiMe_2 in refluxing toluene) of cyclic carbonates that bear a vinyl group at a carbinolated carbon is followed by a Claisen rearrangement to give lactones.[4]

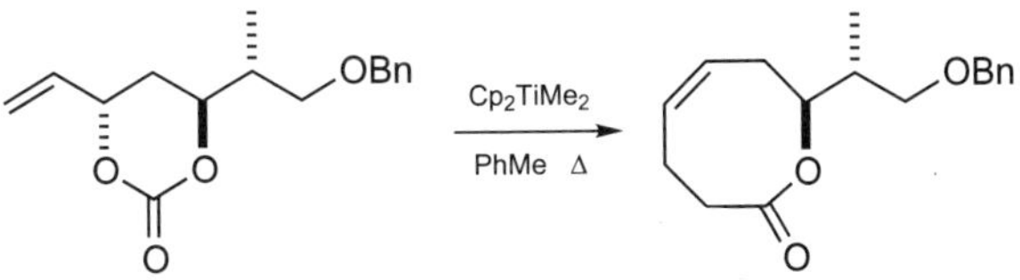

52%

2-Methyleneazetidines. Methylenation of β-lactams by the title reagent is efficient when the substrate bears a strongly deactivating *N*-substituent (e.g., *N*-Boc derivative).[5]

[1]Haak, E., Bytschkov, I., Doye, S. *ACIEE* **38**, 3389 (1999).
[2]Haak, E., Siebeneicher, H., Doye, S. *OL* **2**, 1935 (2000).
[3]Fairfax, D., Stein, M., Livinghouse, T., Jensen, M. *OM* **16**, 1523 (1997).
[4]Davidson, J.E.P., Anderson, E.A., Buhr, W., Harrison, J.R., O'Sullivan, P.T., Collins, I., Green, R.H., Holmes, A.B. *CC* 629 (2000).
[5]Martinez, I., Howell, A.M. *TL* **41**, 5607 (2000).

Diphenylammonium triflate.

Esterification.[1] Heating carboxylic acids with alcohols in toluene at 80° in the presence of Ph_2NH_2OTf (1 mol%) furnishes esters (12 examples, 78–96%). The same catalyst can be used in transesterification. Improved yields are obtained by adding Me_3SiCl as cocatalyst.

[1]Wakasugi, K., Misaki, T., Yamada, K., Tanabe, Y. *TL* **41**, 5249 (2000).

Diphenyl diselenide. 13, 125; **18,** 151–152; **19,** 140–141; **20,** 157–158

Coupling reaction.[1] Photoinduced three-component coupling combines an alkene (enol ether, silyl enol ether, 1,3-diene, etc.) and ethyl propiolate to form a carbon chain in which the two PhSe groups from PhSeSePh are separately located. The reaction starts from β-addition of PhSe radical to the ester.

BnO + EtOOC ——PhSe-SePh, hv——> BnO ... SePh, SePh COOEt

[1]Ogawa, A., Doi, M., Ogawa, I., Hirao, T. *ACIEE* **38**, 2027 (1999).

(Diphenylphosphinoethane)rhodium(I) perchlorate.

Cyclooctenones.[1] Intramolecular hydroacylation of δ-cyclopropyl-γ,δ-unsaturated aldehydes is effected by $(dppe)RhClO_4$. This reaction is carried out under ethylene to suppress the decarbonylation pathway.

O, H, Ph ——$(dppe)RhClO_4$, $H_2C{=}CH_2$, Cl⁀Cl——> O, Ph

65%

[1]Aloise, A.D., Layton, M.E., Shair, M.D. *JACS* **122**, 12610 (2000).

Diphenyl phosphorazidate.

β-Amino acids.[1] A method involving asymmetric alkylation of chiral *N*-acyloxazolidin-2-ones with *t*-butyl bromoacetate and saponification furnishes substituted succinic monoesters. On submission to degradation by $(PhO)_2PON_3$, properly protected β-amino acids are obtained.

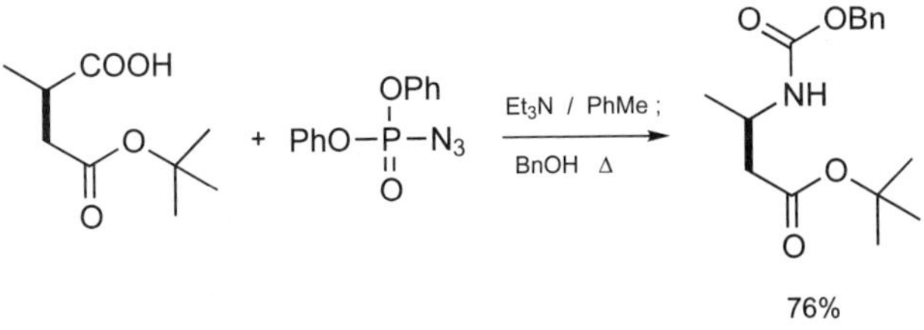

76%

[1]Evans, D.A., Wu, L.-D., Wiener, J.J.M., Johnson, J.S., Ripin, D.H.B., Tedrow, J.S. *JOC* **64**, 6411 (1999).

2,6-Diphenyl-4*H*-thiopyran-4-thione.

Baylis–Hillman reaction.[1] A useful catalytic system is formed when the thiopyranone is combined with $TiCl_4$.

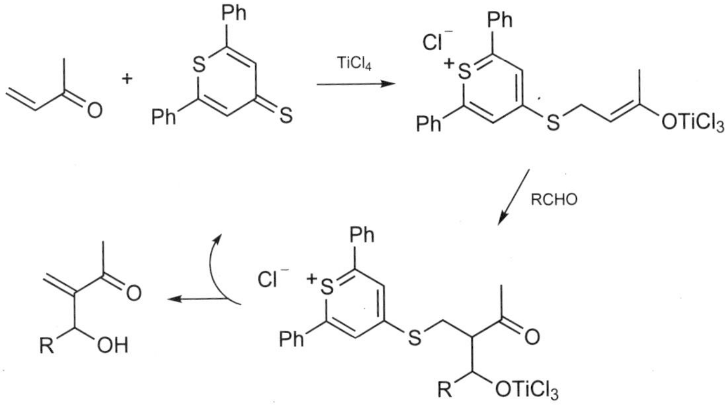

[1]Iwama, T., Kinoshita, H., Kataoka, T. *TL* **40**, 3741 (1999).

Dodecylbenzenesulfonic acid.

Mannich reaction. The sulfonic acid is both a Bronsted acid and a surfactant. It proves to be a most effective catalyst for promoting condensation of silyl enol ethers with aldehydes and *o*-anisidine.[1] Actually, other arylamines and ketones can be used.[2]

[1]Manabe, K., Mori, Y., Kobayashi, S. *SL* 1401 (1999).
[2]Manabe, K., Kobayashi, S. *OL* **1**, 1965 (1999).

Dysprosium.

Propargylation of carbonyl compounds.[1] This metal in the presence of $HgCl_2$ promotes formation of homopropagyl alcohols.

[1]Li, Z., Jia, Y., Zhou, J. *SC* **30**, 2515 (2000).

Dysprosium(II) iodide.

Alkylation of carbonyl compounds.[1] The Barbier-type reaction is mediated by DyI_2 in THF. The reagent, possessing a reactivity pattern similar to that of $(thf)_xTmI_2$, must be prepared just before use.

Reductions.[1] The complex $(dme)_3DyI_2$ is a reducing agent whose reactivity lies between Birch reducing agents and Sm(II) and Tm(II) salts. It can be used to convert tolane to (*Z*)-stilbene and naphthalene to the 1,4-dihydro derivative.

[1]Evans, W.J., Allen, N.T., Ziller, J.W. *JACS* **122**, 11749 (2000).

Dysprosium(III) triflate.

Cycloaddition.[1] In the presence of $Dy(OTf)_3$ aromatic aldehydes and arylamines form aldimines in situ, and then react as heterodienes toward enecarbamates.

R–C6H4–NH2 + BnOOC–N(pyrroline) + OHC–C6H4–R′ —Dy(OTf)3 / MeCN→ product (BnOOC, N, N–H, R, R′)

[1]Batey, R.A., Simoncic, P.D., Lin, D., Smyj, R.P., Lough, A.J. *CC* 651 (1999).

E

Ethanesulfonyl azide.

Azidoalkanes.[1] A preparation of RN_3 from RI involves initiation by dilauroyl peroxide to add a carbon radical to a nitrogen atom of the reagent. The adduct decomposes by extrusion of sulfur dioxide and an ethyl radical, the latter then carries the chain reaction on.

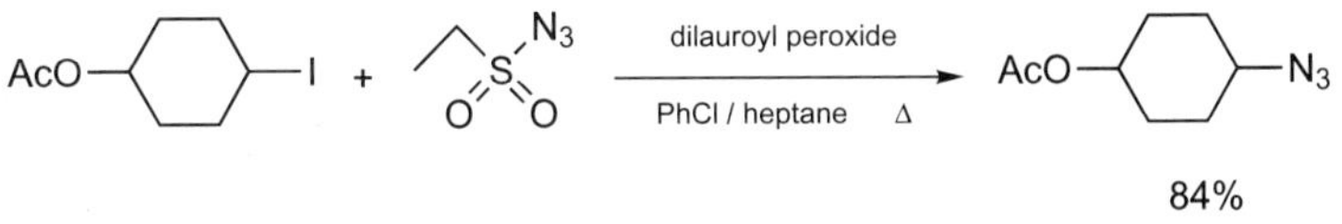

84%

[1]Ollivier, C., Renaud, P. *JACS* **122**, 6496 (2000).

2-Ethoxy-1-ethoxycarbonyl-1,2-dihydroquinoline.

Activation of α-hydroxy acids.[1] The derivatization is complete at room temperature. When an amine is present, the carboxyl group is transformed into an amide.

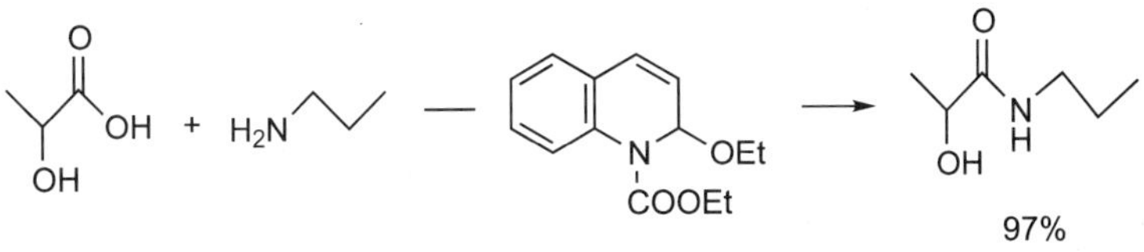

97%

[1]Hyun, M.H., Kang, M.H., Han, S.C. *TL* **40**, 3435 (1999).

Ethylaluminum sesquichloride.

Friedel–Crafts alkylation.[1] Under the influence of $Et_3Al_2Cl_3$ isopropyl chloroformate acts as an alkylating agent for alkenes.

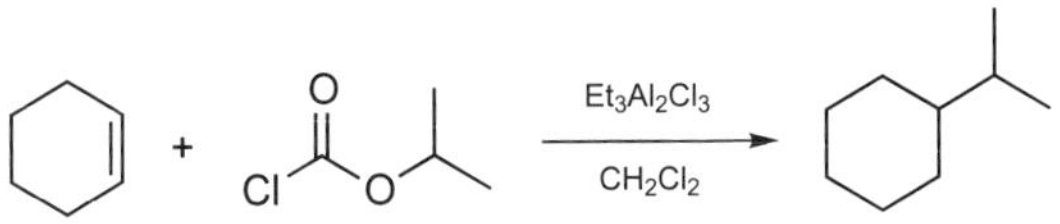

[1]Biermann, U., Metzger, J.O. *ACIEE* **38**, 3675 (1999).

N-Ethyl-2-bromo-4-methylthiazolium tetrafluoroborate.

Peptide couplings.[1] This reagent is particularly useful for coupling of hindered amino acid derivatives.

[1] Li, P., Xu, J.C. *TL* **40**, 8301 (1999).

Ethyl N-benzyloxycarbonyloxamide.

Gabriel synthesis.[1] Following *N*-alkylation of the oxamide, selective saponification removes the ethyl ester to afford RNHCOOBn. The *N*-Boc derivative is similarly transformed.

ROOC–NH–CO–COOEt —(t-BuOK / DMF, RX)→ ROOC–N(R')–CO–COOEt —(LiOH)→ ROOC–N(H)–R'

R = t-Bu, Bn

[1] Berree, F., Bazureau, J.-P., Michelot, G., Le Corre, M. *SC* **29**, 2685 (1999).

Ethyl α-chloro-α-phenylselenoacetate.

Pictet–Spengler cyclization.[1] Ethyl tetrahydroquinoline-1-carboxylates are readily prepared from a reaction (catalyzed by $SnCl_4$) of phenethylamines and the ester.

[1] Silveira, C.C., Bernardi, C.R., Braga, A.L., Kaufman, T.S. *TL* **40**, 4969 (1999).

1-Ethyl-3-(3-dimethylaminopropyl)carbodiimide.

Dehydration.[1] α-Substituted cinnamic esters are obtained stereoselectively from the benzylic alcohols [*anti*-alcohols -> (*E*)-alkenes; *syn*-alcohols -> (*Z*)-alkenes].

Ph–CH(OH)–CH(COOMe)–CH₂CH₂OTBS + Et–N=C=N–(CH₂)₃–NMe₂ —($CuCl_2$, PhMe 80°)→ Ph–CH=C(COOMe)–CH₂CH₂OTBS (99%)

[1] Sai, H., Ohmizu, H. *TL* **40**, 5019 (1999).

Ethyl diphenylphosphonoacetate.

Emmons–Wadworth reaction.[1] (Z)-2-Alkenoic esters are the overwhelmingly major products. The reaction is carried out in the presence of DBU and NaI.

[1]Ando, K., Oishi, T., Hirama, M., Ohno, H., Ibuka, T. *JOC* **65**, 4745 (2000).

Ethylenebis(triphenylphosphine)platinum(0).

Alkylation.[1] Displacement of the vinylic chlorine of 2-chloropropen-2-yl acetate by nucleophiles occurs in the presence of the Pt complex.

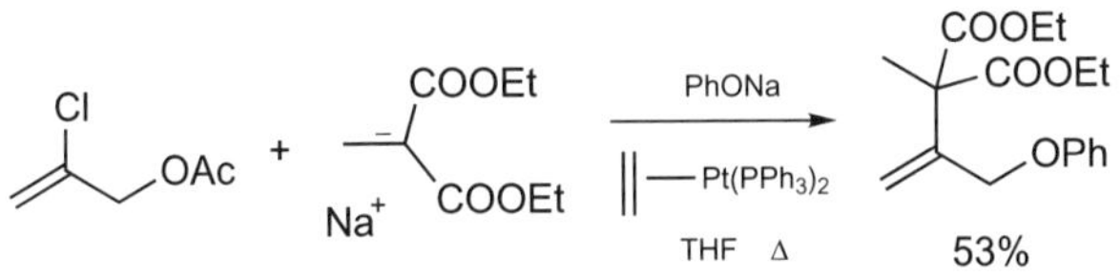

PhONa; Pt(PPh3)2; THF Δ; 53%

Hydroboration and borative coupling. *O,C*-Diboration of enones by bis(pinacolato)diboron is mediated by the Pt-complex.[2] Hydrolysis of the adducts gives β-boryl ketones (a net regioselective hydroboration of the conjugated double bond). 1,3-Dienes undergo borative coupling with *B*-silylpinacolatoborane in the presence of aldehydes.[3]

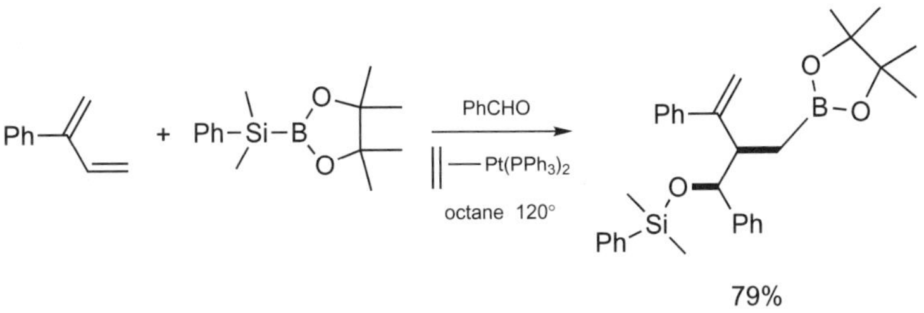

PhCHO; Pt(PPh3)2; octane 120°; 79%

Functionalization of alkynes.[4] The Pt-complex catalyzes the addition of silyl and thio groups to alkynes. Disulfides and disilanes are sources of the addends. The reaction is regioselective and stereoselective.

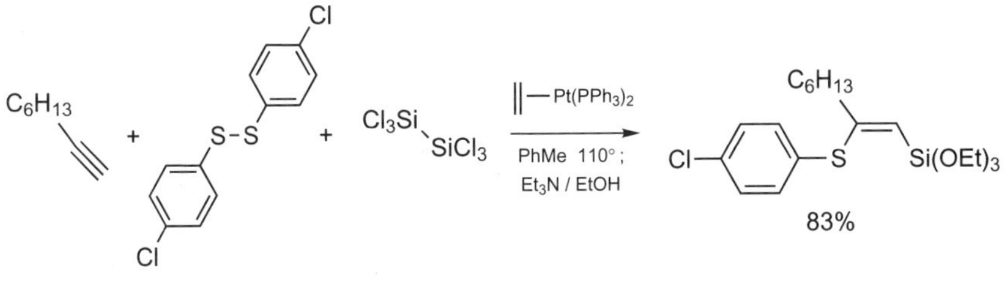

Pt(PPh3)2; PhMe 110°; Et3N / EtOH; 83%

[1]Kadota, J., Katsuragi, H., Fukumoto, Y., Murai, S. *OM* **19**, 979 (2000).
[2]Lawson, Y.G., Lesley, M.J.G., Marder, T.B., Norman, N.C., Rice, C.R. *CC* 2051 (1997).
[3]Suginome, M., Nakamura, H., Matsuda, T., Ito, Y. *JACS* **120**, 4248 (1998).
[4]Han, L.-B., Tanaka, M. *JACS* **120**, 8249 (1998).

F

Ferrocene.

Reduction of sulfoxides.[1] Together with $(CF_3CO)_2O$, ferrocene converts sulfoxides into sulfides at room temperature.

[1]Kobayashi, K., Kubota, Y., Furukawa, N. *CL* 400 (2000).

9-Fluorenylmethyl chloroformate.

9-Fluorenylmethyl esters.[1] The title reagent converts *N*-protected amino acids to 9-fluorenylmethyl esters at 0° in the presence of *i*-Pr_2NEt (1 equiv) and DMAP (10 mol%).

[1]Merette, S.A.M., Burd, A.P., Deadman, J.J. *TL* **40**, 753 (1999).

Fluorine. 13, 135; 14, 167; 15, 160; 18, 161; 19, 146; 20, 165

Heteroaromatic fluorination.[1] Fluorination of pyridine, quinoline, and quinoxaline systems occurs with fluorine–iodine mixtures. Pyridine is alkoxylated when an appropriate alcohol is present.

Fluorination at unactivated carbon sites. With fluorine gas (diluted with nitrogen) such a reaction in MeCN at the temperature of ice can be quite efficient.[2] Thus, fluorocyclohexane has been obtained in 63% yield. Note that decalin is fluorinated exclusively at an angular position whereas the secondary carbon sites are attacked by 1-chloromethyl-4- fluoro-1,4-diazoniabicyclo[2.2.2]octane bis(tetrafluoroborate).

A microreactor has been designed for use with elemental fluorine.[3]

Arylsulfur pentafluorides.[4] Diaryl disulfides undergo this transformation.

$$O_2N\text{-}C_6H_4\text{-}S\text{-}S\text{-}C_6H_4\text{-}NO_2 \xrightarrow{F_2 - N_2\ /\ MeCN} O_2N\text{-}C_6H_4\text{-}SF_5 \quad 41\%$$

[1]Chambers, R.D., Parsons, M., Sandford, G., Skinner, C.J., Atherton, M.J., Moilliet, J.S. *JCS(P1)* 803 (1999).

[2]Chambers, R.D., Parsons, M., Sandford, G., Bowden, R. *CC* 959 (2000).

[3]Chambers, R.D., Spink, R.C.H. *CC* 883 (1999).
[4]Bowden, R.D., Comina, P.J., Greenhall, M.P., Kariuki, B.M., Loveday, A., Philip, D. *T* **56**, 3399 (2000)

N-Fluorobenzenesulfonimide. 20, 165–166

N-Fluorosulfonamides.[1] Fluorine transfer from the reagent $(PhSO)_2NF$ to sulfonamides (as K salts) is facile.

[1]Taylor, D.M., Meier, G.P. *TL* **41**, 3291 (2000).

Fluoroboric acid.

β-Amino carbonyl compounds. Condensation of aldehydes, amines, and ketene silyl acetals is promoted by HBF_4 in an aqueous media (aq *i*-PrOH[1] or in the presence of a surfactant[2]).

2,3-Dihydro-γ-pyridones.[3] A facile hetero-Diels–Alder reaction takes place in water alone or aq MeOH.

Me_3SiO OMe + Ph O H_2N-Ph → HBF_4 / H_2O, MeOH → O Ph NPh 95%

[1]Akiyama, T., Takaya, J., Kagoshima, H. *SL* 1045 (1999)
[2]Akiyama, T., Takaya, J., Kagoshima, H. *SL* 1426 (1999)
[3]Akiyama, T., Takaya, J., Kagoshima, H. *TL* **40**, 7831 (1999)

1-Fluoro-3,5-dichloropyridinium triflate.

Phenol oxidation.[1] The title reagent oxidizes substituted phenols to quinones under mild conditions. When a MOM ether is present, it is concomitantly transformed (likely by fluorination at the central methylene group) and hydrolyzed on workup.

HO O O OMe N N O O HN O O → Cl Cl N+ F TfO⁻, CH_2Cl_2, 0° → O OMe HO N N HO O HN O O 69%

[1]Martinez, E.J., Corey, E.J. *OL* **1**, 75(1999).

1-Fluoro-4-hydroxy-1,4-diazoniabicyclo[2.2.2]octane bis(tetrafluoroborate).

Fluorination.[1] This new reagent, which that needs no special caution or glassware in handling, can be used for fluorination of aromatic rings, alkenes, enol ethers, and dienol acetates. In the presence of $ZnCl_2$, either mono- or difluoro derivatives of active methylene compounds can be isolated from its reactions.

Preparation of the reagent is simple. It involves passing a stream of fluorine (10% in nitrogen) to a mixture of DABCO-*N*-oxide, boron trifluoride, and tetrafluoroboric acid at 0°. On evaporation, the solid is collected and washed with DME (75% yield).

[1]Poss, A.J., Shia, G.A. *TL* **40**, 2673 (1999).

α-Fluoromalonaldehyde.

(Z)-β-Acyloxy-α-fluoroacrylaldehydes.[1] After generation from the fluorinated enol sulfonate the sodium salt of this aldehyde reacts with acid chlorides to provide the substituted acroleins.

F, F, CHOTs, F —NaOH / DMSO - H_2O→ [OHC, CHO, F, Na^+] —PhCOCl→ Ph, O, O, CHO, F

92%

[1]Funabiki, K., Fukushima, Y., Matsui, M., Shibata, K. *JOC* **65**, 606 (2000).

Fluorotris(trimethylsilyl)methane.

Fluoroallyl alcohols.[1] The reagent, obtained by silylation of lithio derivatives from tribromofluoromethane, serves as a fluorinated C_1 nucleophile. Thus, its reaction with aldehydes in the presence of KF–18-crown-6 leads to fluoroallyl alcohols.

Br, Br, Br, F —Me_3SiCl / BuLi→ $SiMe_3$, Me_3Si, $SiMe_3$, F (97%) —PhCHO / KF - 18-crown-6 / DMF→ OH, Ph, CHPh, F (72%)

[1]Shimizu, M., Hata, T., Hiyama, T. *BCSJ* **73**, 1685 (2000).

Fluorous reagents and ligands.

Improvement of organic reactions by the use of fluorous components and ionic liquids has been the subject of intense research. Imidazolium salts that carry a polyfluorinated chain (fluorous ionic liquids) are found to act as surfactants when added to conventional ionic liquids. Thus emulsification of fluoroalkanes with the ionic liquid phase is facilitated.[1]

Hydroxyl protection. Enol ether **1** has been developed.[2] A tetrahydropyranyl sulfoxide **2** is available for the formation of mixed acetals[3] upon activation with Cp_2ZrCl_2–$AgClO_4$. The fluorous bromosilane **3** form siloxanes with improved acid stability.[4] A method for selective tosylation of diol systems that is known to be catalyzed by organotin oxides is also modifiable in terms of fluorinated catalysts.[5]

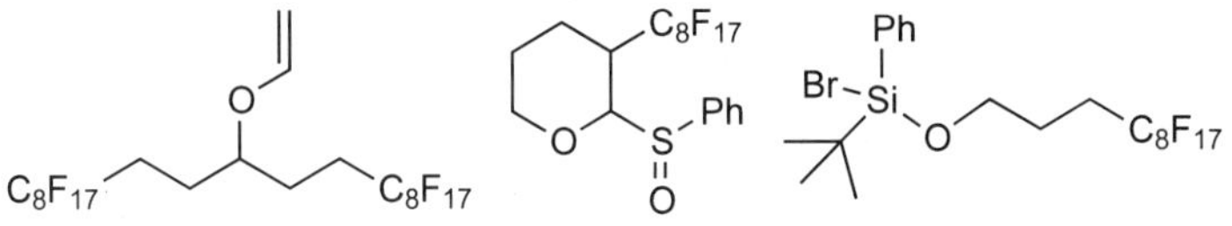

(1) (2) (3)

Silyl enol ethers.[6] The reductive silylation of enones with hydrosilanes in fluorous media has operational advantages.

Reductions. The reduction of α-alkoxyketones with SmI_2 in the presence of fluorous BINOL (chiral) derivative shows that the protonation step is enantioselective.[7] The use of fluorous tin hydrides that are highly soluble in fluorinated solvents facilitates dehalogenation.[8]

Fluorous-phase soluble hydrogenation catalysts are obtained by conversion of *N*-acryloxysuccinimide-containing fluoroacrylate polymers into phosphine ligands and hence Wilkinson catalyst analogues.[9]

Miscellaneous reactions. The Heck reaction is accomplishable in perfluorinated solvents while employing a fluorous triarylphosphine ligand.[10] A minimally fluorous diaryl diselenide used to assist tin hydride mediated cyclization can be recovered.[11]

A multicomponent coupling involving fluorous allyltin reagents has been reported,[12] and rapid Stille couplings are completed employing aryltris(polylfluoroalkyl)stannanes with microwave irradiation.[13]

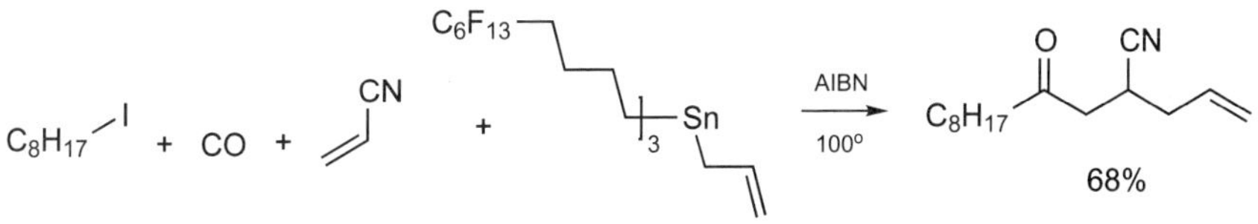

AIBN
100°
68%

When ketones are converted to enones by reaction with *p*-perfluorohexylbenzeneselenenyl chloride and subsequent oxidative elimination, the spent reagent can be recovered in the form of the diselenide by reduction and continuous fluorous extraction.[14]

Products from hydroboration with catecholborane in fluorous solvents are extractable by THF, while the catalyst $[(C_6F_{13}CH_2CH_2)_3P]_3RhCl$ (0.04 mol%) remains in the fluorous phase.[15]

Fluorous-soluble polymer ligands obtained from copolymerization of *p*-dipenylphosphinostyrene and (heptadecylfluoro)decyl acrylate are suitable for the preparation of a Rh-based hydroformylation catalyst.[16]

There is dramatic acceleration of certain Diels–Alder reactions in fluorous solvents[17] (enhancement approaching those observed in water).

Reagents involved in the oxidation of alcohols to aldehydes–ketones by oxygen under fluorous biphasic conditions are TEMPO, $CuBr·SMe_2$, and 4,4′-bis[heptadecafluoro)dodecyl]-2,2′-bipyridyl.[18] The Mn(salen) complexes that mediate epoxidation of alkenes have been modified to bear polyfluoroalkyl substituents in the aromatic rings.[19] Oxyfunctionalization of unactivated C—H sites is achieved with perfluorinated *cis*-2-butyl-3-propyloxaziridine.[20]

[1]Merrigan, T.L., Bates, E.D., Dorman, S.C., Davis, Jr., J.H. *CC* 2051 (2000).
[2]Wipf, P., Reeves, J.T. *TL* **40**, 5139 (1999).
[3]Wipf, P., Reeves, J.T. *TL* **40**, 4649 (1999).
[4]Rover, S., Wipf, P. *TL* **40**, 5667 (1999).
[5]Bucher, B., Curran, D.P. *TL* **41**, 9617 (2000).
[6]Dinh, L.V., Gladysz, J.A. *TL* **40**, 8995 (1999).
[7]Nakamura, Y., Takeuchi, S., Ohgo, Y., Curran, D.P. *T* **56**, 351 (2000).
[8]Curran, D.P., Hadida, S., Kim, S.-Y., Luo, Z. *JACS* **121**, 6607 (1999).
[9]Bergbreiter, D.E., Franchina, J.G., Case, B.L. *OL* **2**, 393 (2000).
[10]Moineau, J., Pozzi, G., Quici, S., Sinou, D. *TL* **40**, 7683 (1999).
[11]Crich, D., Hao, X., Lucas, M. *T* **55**, 14261 (1999).
[12]Ryu, I., Niguma, T., Minakata, S., Komatsu, M., Luo, Z., Curran, D.P. *TL* **40**, 2367 (1999).
[13]Olofsson, K., Kim, S.-Y., Larhed, M., Curran, D.P., Hallberg, A. *JOC* **64**, 4539 (1999).
[14]Crich, D., Barba, G.R. *OL* **2**, 989 (2000).
[15]Juliette, J.J.J., Rutherford, D., Horvath, I.T., Gladysz, J.A *JACS* **121**, 2696 (1999).
[16]Chen, W., Xu, L., Xiao, J. *CC* 839 (2000).
[17]Myers, K.E., Kumar, K. *JACS* **122**, 12025 (2000).
[18]Betzemeier, B., Cavazzini, M., Quici, S., Knochel, P. *TL* **41**, 4343 (2000).
[19]Cavazzini, M., Manfredi, A., Montanari, F., Pozzi, G. *CC* 2171 (2000).
[20]Amone, A., Foletto, S., Metrangolo, P., Pregnolato, M., Resnati, G. *OL* **1**, 281 (1999).

α-Fluorovinyltriphenylphosphonium salts.

Fluoroallyl ethers.[1] The ethers are accessible from the Wittig reaction by using these salts in the presence of NaOR. Monofluorochromenes are obtained from salicyladehyde derivatives.

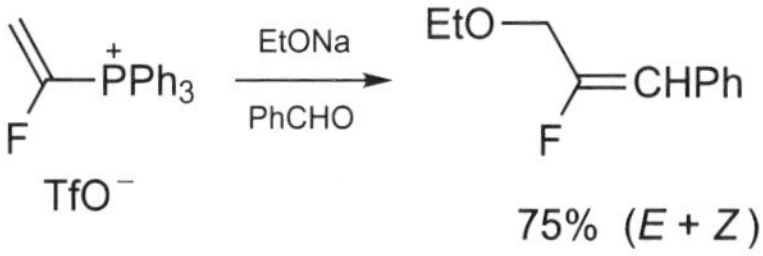

[1]Hanamoto, T., Shindo, K., Matsuoka, M., Kiguchi, Y., Kondo, M. *JCS(P1)* 103 (2000).

Formaldehyde. 20, 167

N-Methylation. By using monosodium phosphite as the reducing agent, secondary amines are methylated with HCHO. This method permits the presence of other reducible, hydrolyzable groups.[1]

[1]Davis, B.A., Durden, D.A. *SC* **30**, 3353 (2000).

Formic acid. 13, 137; 18, 163; 19, 148–149; 20, 168

N-Formylation.[1] Diphenylamine, carbazoles, 3-alkylindoles, and moderately weakly nucleophilic anilines are formylated in neat formic acid.

[1]Chakrabarty, M., Khasnobis, S., Harigaya, Y., Konda, Y. *SC* **30**, 187 (2000).

Formic pivalic anhydride.

Deoxygenation.[1] Removal of the oxygen atom from amine oxides with the mixed anhydride at the temperature of ice is quite convenient. The mixed anhydride is formed and used in situ from HCOONa and pivaloyl chloride.

[1]Rosenau, T., Potthast, A., Ebner, G., Kosma, P. *SL* 623 (1999).

Fullerenes (C_{60}/C_{70}).

[3 + 2]Cycloaddition.[1] Under photochemical conditions, fullerenes catalyze the cycloaddition of dimethyl iminodiacetate to maleic anhydride.

[1]Shi, Y., Gan, L., Wei, X., Jin, S., Zhang, S., Meng, F., Wang, Z., Yan, C. *OL* **2**, 667 (2000).

G

Gadolinium(III) chloride. 20, 169

Diels–Alder type reaction.[1] The reaction of imines with dihydropyran or dihydrofuran to give tricyclic heterocycles is catalyzed by $GdCl_3$.

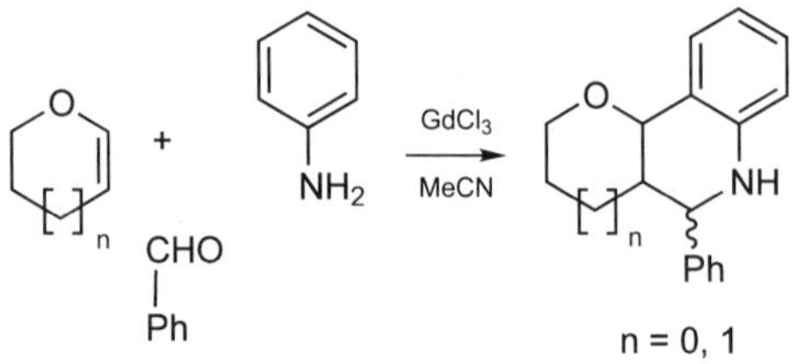

GdCl3
MeCN
NH2
CHO
Ph
NH
Ph
n = 0, 1

[1]Ma, Y., Qian, C., Xie, M., Sun, J. *JOC* **64**, 6462 (1999).

Gadolinium(III) isopropoxide.

Reductive acylation.[1] Acetates are generated from carbonyl compounds in a Meerwein–Ponndorf–Verley reduction in the presence of isopropenyl acetate. Beside gadolinium(III) isopropoxide, samarium isopropoxide can also be used to induce the reaction.

[1]Nakano, Y., Sakaguchi, S., Ishii, Y. *TL* **41**, 1565 (2000).

Gallium.

Allylation and propargylation.[1] Aldehydes are attacked by the organogallium species.

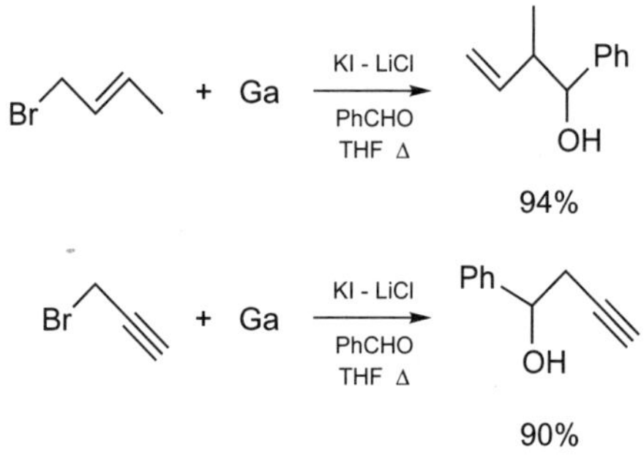

Br + Ga
KI - LiCl
PhCHO
THF Δ
Ph
OH
94%

Br + Ga
KI - LiCl
PhCHO
THF Δ
Ph
OH
90%

[1]Han, Y., Chi, Z., Huang, Y.-Z. *SC* **29**, 1287 (1999).

Gallium(III) halides. 20, 169–170

Vinylation. Introduction of vinyl groups to an *ortho*-position of substituted aromatic rings[1] and the α-position of a ketone[2] is accomplished by the catalyzed reaction with silylalkynes. If the reaction is quenched with NBS, an a-(2,2-dibromovinyl) ketone is obtained, due to bromolysis of the *gem*-digallium species.

OSiMe3 + Me3Si—≡ —GaCl3 ; H3O+→ 65%

OSiMe3 Ph + SiMe3—≡—R —GaCl3 ; H3O+→ Ph R 76–87%

Mannich-type reaction.[3] Carbinolamines formally derived from amines and CF_3CHO condense with silyl enol ethers when catalyzed by $GaCl_3$.

1,3-Diketones.[4] Dehalogenative coupling of α-bromoketones and acid chlorides with GaI_3 provides a new way to the diketones.

Directed reduction.[5] Simultaneous complexation of $GaCl_3$ to a carbonyl group and a donor (such as triple bond) activated the former toward regioselective attack by a tin hydride reagent.

CHO OHC Ph —GaCl3 / CH2Cl2 −78°; R3SnH→ OH OHC Ph

3-Alkoxy-3-pyrrolines.[6] Fischer carbene complexes and aldimines undergo cycloaddition in the presence of $GaCl_3$. The major products have a *trans*-2,5-disubstitution pattern (*trans*~*cis* ratio in the 8:2 range).

Ar NR + (OC)5Cr R′ R″O —GaCl3; ClCH2CH2Cl→ R N Ar R′ R″O

(*trans* : *cis* ~ 8 : 2)

[1]Yonehara, F., Kido, Y., Yamaguchi, M. *CC* 1189 (2000).
[2]Yamaguchi, M., Tsukagoshi, T., Arisawa, M. *JACS* **121**, 4074 (1999).
[3]Takaya, J., Kagoshima, H., Akiyama, T. *OL* **2**, 1577 (2000).
[4]Chen, R., Wu, H., Zhang, Y. *JCR(S)* 666 (1999).
[5]Asao, N., Asano, T., Ohishi, T., Yamamoto, Y. *JACS* **122**, 4817 (2000).
[6]Kagoshima, H., Akiyama, T. *JACS* **122**, 11741 (2000).

Gallium(III) nonafluorobutanesulfonate.

Friedel–Crafts acylation.[1] The effectiveness of the salt in catalytic quantities to promote acylation is indicated by the excellent yields of ketones obtained from fluorobenzene, chlorobenzene, and dichlorobenzenes.

[1]Matsuo, J., Odashima, K., Kobayashi, S. *SL* 403 (2000).

Gold(III) chloride.

Cyclization. 3-Alkynones cyclize to afford furan derivatives in the presence of gold(III) chloride.[1] Furans bearing an alkynyl chain undergo intramolecular cycloaddition further. Transformation into bicyclic phenols is noteworthy because the hydroxy group is at a *peri*-position.

R G O AuCl$_3$ MeCN 20° R O G R G OH

G = O, NTs, CH$_2$

1,2-Alkadien-4-ones behave differently toward Au(III) and Pd(II) species.[2]

R O =C= AuCl$_3$ MeCN 20° R O AuL$_n$ R O O R

(MeCN)$_2$PdCl$_2$ MeCN R O R O

[1]Hashmi, A.S.K., Frost, T.M., Bats, J.W. *JACS* **122**, 11553 (2000).
[2]Hashmi, A.S.K., Schwarz, L., Choi, J.-H., Frost, T.M. *ACIEE* **39**, 2285 (2000).

Graphite. 20, 170

Acetalization.[1] In refluxing benzene or toluene, spirodiacetals are formed when pentaerythritol serves as the condensation partner with carbonyl compounds and expansive graphite as the catalyst.

HO, OH, HO, OH + R−CHO —graphite→ R, O, O, O, O, R

[1]Jin, T.-S., Li, T.-S., Zhang, Z.-H., Yuan, Y.-J. *SC* **29**, 1601 (1999).

Grignard reagents. 13, 138–140; **14,** 171–172; **16,** 172–173; **17,** 141–142; **18,** 167–171; **19,** 151–154; **20,** 170–173

Reagent formation by exchange reactions. Pyridylmagnesium bromides and related heteroarylmagnesium reagents can be prepared by an exchange process (with *i*-PrMgBr or *i*-Pr_2Mg in THF at −25° to −40°).[1–3] The access to Grignard reagents from functionalized alkenyl and alkyl halides also benefit by this technique,[4–6] and in the preparation of $RCOOCH_2MgCl$ by this procedure, *N*-butylpyrrolidone is added as a cosolvent because it solubilizes magnesium reagents at low temperatures (NMP does not).[7] Lithium trialkylmagnesates have also been employed in the preparation of aryl- and alkenylmagnesium reagents.[8]

O, O, I —i-PrMgCl, THF, O, N, Bu, O→ O, O, MgCl

BuO, O, R, I —EtMgBr / DME→ BuO, O, R, MgBr

A method for generating rearranged reagents involves carbenoid insertion.[9]

Ph, Cl, I + ClMg, R —Et_2O, −78 ~ −20°→ Ph, MgX, R + Ph, R, MgX

(94–97 : 6–3)

Nonstabilized aziridinylmagnesium bromides are similarly prepared from the exchange of an arenesulfinyl group, with retention of configuration.[10]

Another way to prepare RMgBr from RX is to use a mixture of Mg, I_2, $BrCH_2CH_2Br$.[11] On this protocol, suitable substrates undergo Barbier reaction.[12]

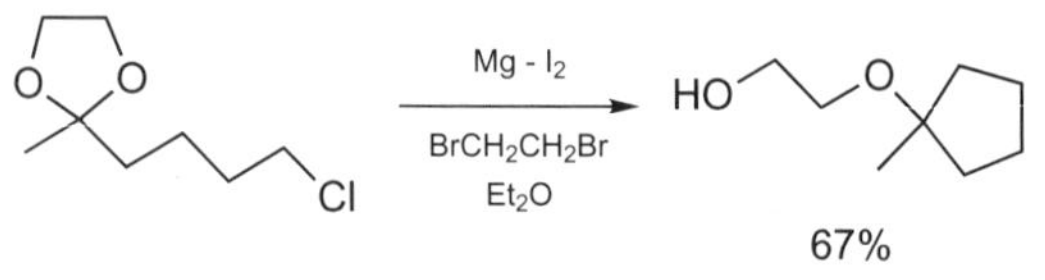

Mg - I2
BrCH2CH2Br
Et2O
HO
O
O
Cl
67%

Chiral α-chloroalkylmagnesium reagents prepared from C—S bond cleavage of chiral sulfoxides maintain their steric integrity.[13]

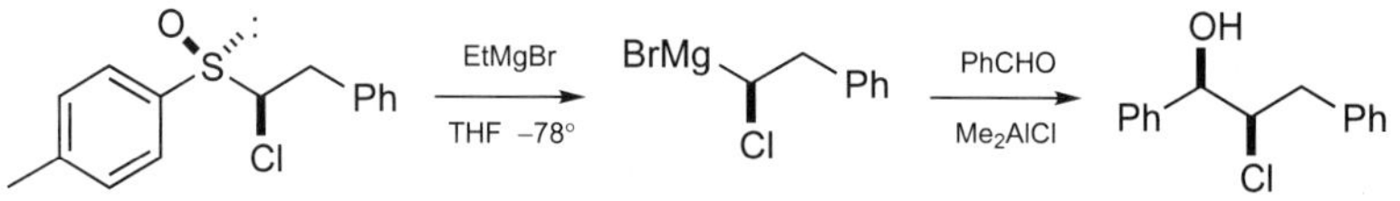

O
S
Ph
Cl
EtMgBr
THF –78°
BrMg
Ph
Cl
PhCHO
Me2AlCl
OH
Ph
Ph
Cl

Addition to the C═N bond. Several useful preparations relating to this type of addition are worthy of mention: to SAMP-hydrazones,[14] to chiral bisimines[15] and *N*-sulfinylimino esters,[16] to *N*-(diethoxyphosphoryl)aldimines,[17] and to those generated in situ from *N*-(α-benzenesulfonyl)alkyl amides.[18]

O
O
N
N
OMe
MeMgBr
PhMe 0° ;
MeOH
NH
62%

N
N
Ph
Ph
t-BuMgCl / THF ;
PhMgCl
Ph
NH
HN
Ph
Ph
89%

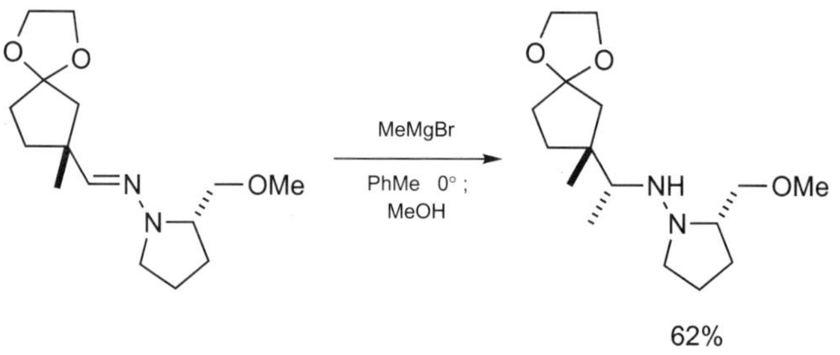

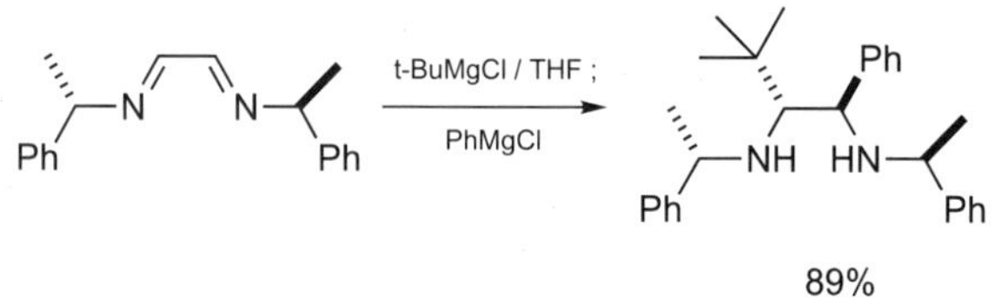

A process involving addition followed by cyclization is remarkable.[19] On the other hand, furan ring opening is featured in another reaction.[20]

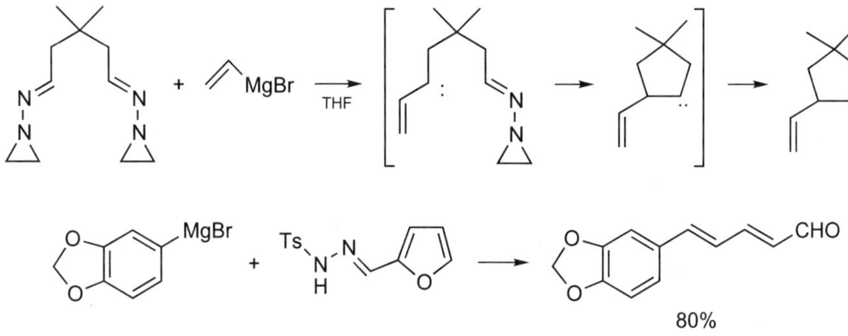

80%

Addition to multiple CC bonds. Conjugate addition of Grignard reagents to γ-hydroxy α,β-unsaturated nitriles is facilitated by chelation.[21] Thus, pretreatment of the substrates with 1 equiv of *t*-BuMgCl followed by a second RMgX completes the reaction. Propargyl alcohols are carbomagnesiated and the ensuing alkenylmagneium halides can be functionalized.[22]

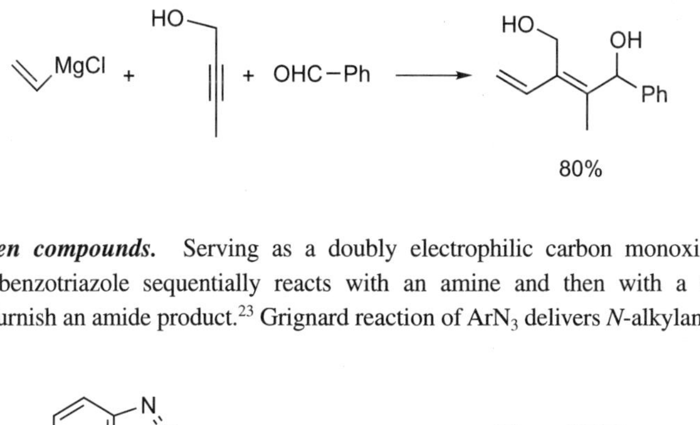

80%

Nitrogen compounds. Serving as a doubly electrophilic carbon monoxide, 1,1′-carbonylbisbenzotriazole sequentially reacts with an amine and then with a Grignard reagent to furnish an amide product.[23] Grignard reaction of ArN_3 delivers *N*-alkylanilines.[24]

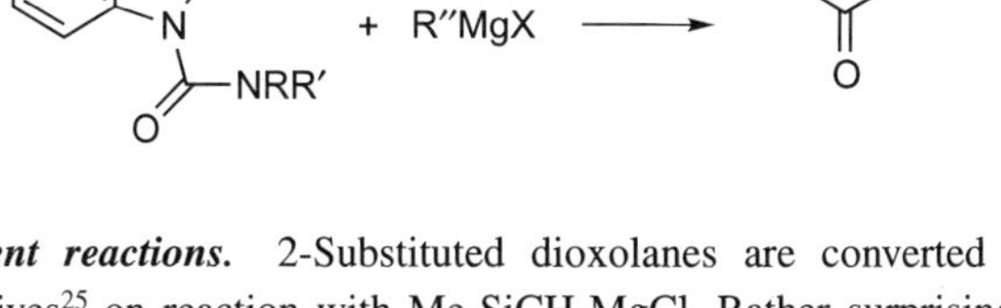

Displacement reactions. 2-Substituted dioxolanes are converted to 1,3-disilylpropane derivatives[25] on reaction with Me_3SiCH_2MgCl. Rather surprisingly, substituted aryl sulfoxides readily undergo S—C_{Ar} bond cleavage.[26]

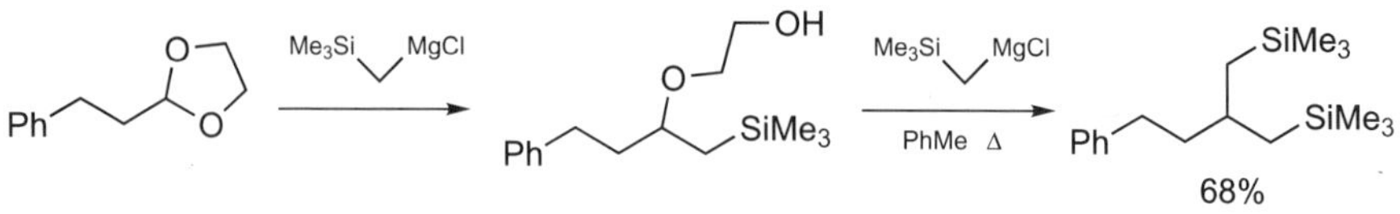

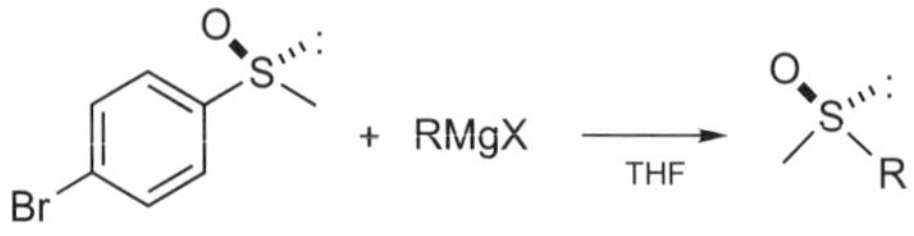

Tetrahydrofuran derivatives. With the EtMgBr–neopentyl iodide system in THF, the solvent becomes iodinated at C-2 (free radical process) and 2-aryltetrahydrofurans are obtained on addition of EtI and EtMgBr to arylmagnesium halides in THF.[27] Radical cyclization of allyl β-iodoacetals is induced by EtMgBr in DME.[6]

Elimination. 1-Bromo-1-chloroalkenes are obtained from dibromochloromethylcarbinyl acetates on exposure to EtMgBr.[28] Dehydration of β-hydroxy nitriles occurs when they are treated with MeMgCl at –78° and warmed to room temperature.[29] A method of allene synthesis also involves an analogous elimination step.[30]

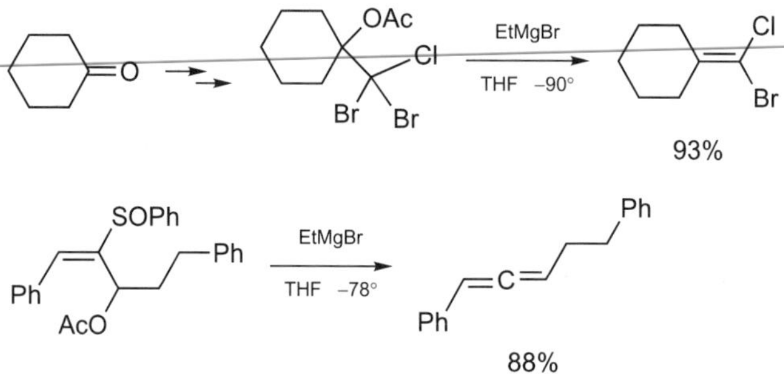

A modified Julia olefination involves reaction of sulfonylcarbanions with α-haloalkylmagnesium halides[31] instead of aldehydes. Elimination of halomagnesium phenylsulfinate immediately follows the C—C bond-forming step.

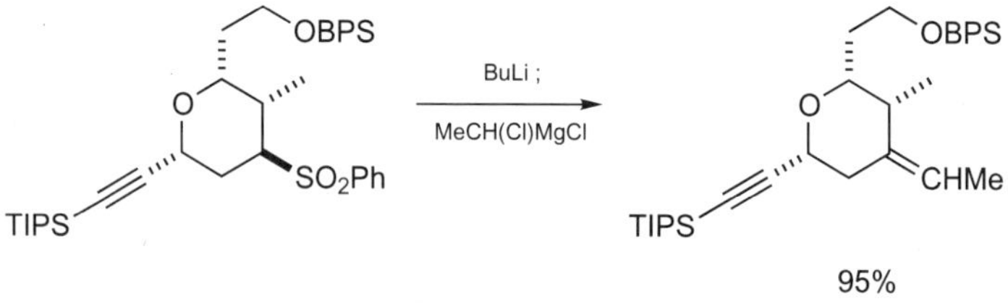

Benzylthiols.[32] Reaction of ArMgX with chlorothionoformate esters followed by reduction with $LiAlH_4$ furnishes $ArCH_2SH$.

β-Aminoketones.[33] Aqueous quench of the reaction between amides (e.g., Weinreb amides, morpholino amides) and vinylmagnesium bromide at room temperature leads to β-aminoketones.

Tetrahydrophthalides.[34] The Diels–Alder reaction of 2,4-alkadienols with acrylic esters is rendered regioselective by pretreatment with MeMgBr so that chelation directs the combination of the alkoxymagnesium species with the dienophiles.

Homologation–condensation. The reaction of cyclic β-keto esters with a vinylmagnesium halide also triggers oxy-Cope rearrangement and Claisen condensation. Bridged ring compounds can be prepared by this method.[35]

O COOMe — MgBr / $MgBr_2$ → [E OM] → O HO OMe

63%

[1] Abarbri, M., Dehmel, F., Knochel, P. *TL* **40**, 7449 (1999).
[2] Trecourt, F., Breton, G., Bonnet, V., Mongin, F., Marsais, F., Queguiner, G. *T* **56**, 1349 (2000).
[3] Felding, J., Kristensen, J., Bjerregaard, T., Sander, L., Vedso, P., Begtrup, M. *JOC* **64**, 4196 (1999).
[4] Thibonnet, J., Knochel, P. *TL* **41**, 3319 (2000).
[5] Avolio, S., Malan, C., Marek, I., Knochel, P. *SL*1820 (1999); Rottländer, M., Boymond, L., Cahiez, G., Knochel, P. *JOC* **64**, 1080 (1999).
[6] Inoue, A., Shinokubo, H., Oshima, K. *OL* **2**, 651 (2000).
[7] Avolio, S., Malan, C., Marek, I., Knochel, P. *SL* 1820 (1999).
[8] Kitagawa, K., Inoue, A., Shinokubo, H., Oshima, K. *ACIEE* **39**, 2481 (2000).
[9] Hoffmann, R.W., Knopff, O., Kusche, A. *ACIEE* **39**, 1462 (2000).
[10] Satoh, T., Matsue, R., Fujii, T., Morikawa, S. *TL* **41**, 6495 (2000).
[11] Li, J., Liao, X., Liu, H., Xie, Q., Liu, Z., He, X. *SC* **29**,1037 (1999).
[12] Huang, J.-W., Chen, C.-D., Leung, M.-K. *TL* **40**, 8647 (1999).
[13] Hoffmann, R.W., Nell, P.G. *ACIEE* **38**, 338 (1999).
[14] Enders, D., Diez, E., Fernandez, R., Martin-Zamora, E., Munoz, J.M., Papparlardo, R.R.,Lassaletta, J.M. *JOC* **64,** 6329 (1999).
[15] Roland, S., Mangeney, P. *EJOC* 511 (2000).
[16] Davis, F.A., McCoull, W. *JOC* **64,** 3396 (1999).
[17] Zwierzak, A., Napieraj, A. *S* 930 (1999).
[18] Mecozzi, T., Petrini, M. *JOC* **64,** 8970 (1999).
[19] Kim, S., Oh, D.H., Yoon, J.-Y., Cheong, J.H. *JACS* **121**, 5330 (1999).
[20] Chandrasekhar, S., Reddy, M.V., Reddy, K.S., Ramarao, C. *TL* **41**, 2667 (2000).
[21] Fleming, F.F., Wang, Q., Steward, O.W. *OL* **2**, 1477 (2000).
[22] Forgione, P., Fallis, A.G. *TL* **41**, 11 (2000).
[23] Katritzky, A.R., Monteux, D.A., Tymoshenko, D.O., Belyakov, S.A. *JCR(S)* 230 (1999).
[24] Kumar, H.M.S., Reddy, B.V.S., Anjaneyulu, S., Yadav, J.S. *TL* **40**, 8305 (1999).

[25]Hsieh, Y.-T., Luh, T.-Y. *H* **52**, 1125 (2000).
[26]Capozzi, M.A.M., Cardellicchio, C., Naso, F., Tortorella, P. *JOC* **65**, 2843 (2000).
[27]Inoue, A., Shinokubo, H., Oshima, K. *SL* 1582 (1999).
[28]Rezaei, H., Normant, J.F. *S* 109 (2000).
[29]Fleming, F.F., Shook, B.C. *TL* **41**, 8847 (2000).
[30]Satoh, T., Kuramochi, Y., Inoue, Y. *TL* **40**, 8815 (1999).
[31]Smith III, A.B., Minbiole, K.P., Verhoest, P.R., Beauchamp, T.J. *OL* **1**, 913 (1999).
[32]Nakamura, T., Matsumoto, M. *SC* **29**, 201 (1999).
[33]Gomtsyan, A. *OL* **2**, 11 (2000).
[34]Ward, D.E., Abaee, M.S. *OL* **2**, 3937 (2000).
[35]Sheehan, S.M., Lalic, G., Chen, J.S., Shair, M.D. *ACIEE* **39**, 2714 (2000).

Grignard reagents–cerium(III) chloride. **18,** 171; **19,** 154; **20,** 173

γ,γ-Disubstituted α-alkylidene-γ-butyrolactones.[1] These compounds can be synthesized in two steps from γ-keto-α,β-unsaturated esters, the second step being a Grignard reaction in the presence of $CeCl_3$.

Ph COOMe O PrMgBr - $CeCl_3$; HOAc Ph O O

53%

[1]Ballini, R., Marcantoni, F., Perella, S. *JOC* **64**, 2954 (1999).

Grignard reagents–copper salts. **18,** 171–173; **19,** 154–156; **20,** 174–175

Transpositional displacements. Based on the displacement of allylic carbamates, alkenes containing both allylic and homoallylic silyl substituents are accessible.[1]

OCONHPh $SiMe_2Ph$ BuLi / THF ; CuI - 2 LiCl ; Me_3Si MgCl $SiMe_3$ $SiMe_2Ph$

82%

Stereoselective reaction of β-alkynyl-β-lactones gives rise to 3,4-alkadienoic acids. A product of this type has been recognized as a suitable precursor of malyngolide.[2]

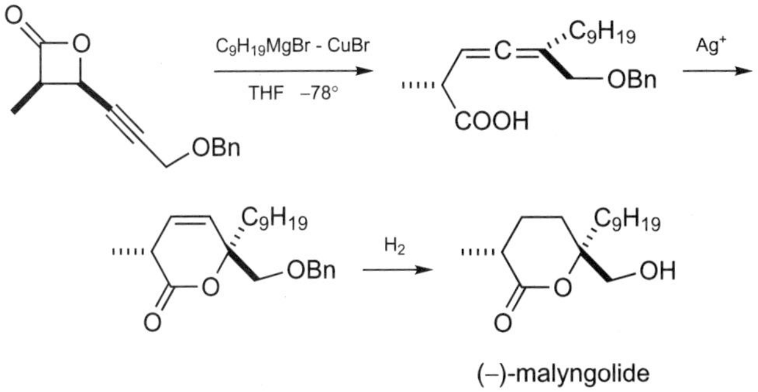

C9H19MgBr - CuBr
THF −78°
COOH
C9H19
OBn
Ag+
H2
OH
(−)-malyngolide

Primary amines.[3] A preparation of RNH_2 is via the reaction of oxime mesylate **1** with Grignard reagents (Cu-catalyzed) to give imines that are easily hydrolyzed.

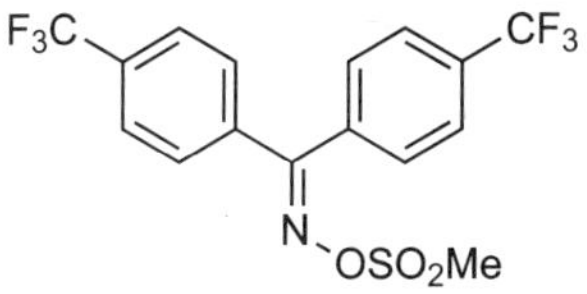

F3C CF3 N OSO2Me

(**1**)

Ketone synthesis.[4] *N*-Alkyl-2-methyleneaziridines react with RMgX (CuI) in the presence of $BF_3 \cdot OEt_2$. The metalloenamines thus generated can be hydrolyzed or trapped by electrophiles.

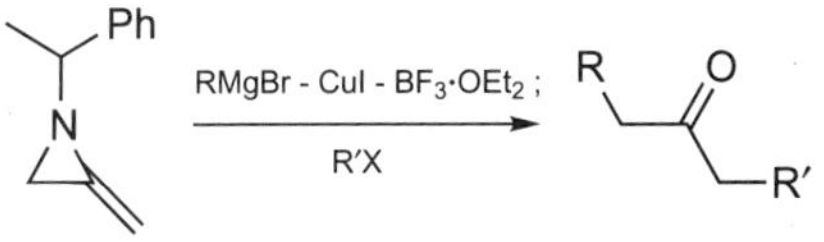

Ph N
RMgBr - CuI - BF3·OEt2 ; R′X
R O R′

[1]Smitrovich, J.H., Woerpel, K.A. *JOC* **65**, 1601 (2000).
[2]Wan, Z., Nelson, S.G. *JACS* **122**, 10470 (2000).
[3]Tsutsui, H., Ichikawa, T., Narasaka, K.*BCSJ* **72**, 1869 (1999).
[4]Hayes, J.F., Shipman, M., Twin, H. *CC* 1791 (2000).

Grignard reagents–iron(III) chloride.

Carbometallation. Grignard reagents add to strained cycloalkenes in the presence of an iron catalyst.[1] When the allylic position bears a leaving group (such as the 7-oxa-bicyclo[2.2.1]octane system a net S_N2' reaction takes place.

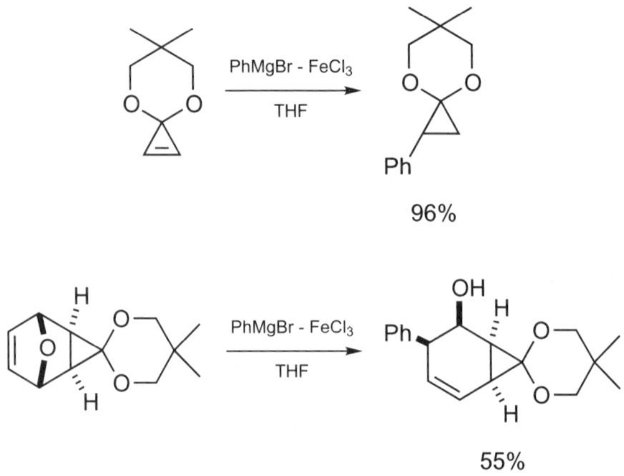

PhMgBr - $FeCl_3$, THF → 96%

PhMgBr - $FeCl_3$, THF → 55%

[1]Nakamura, M., Hirai, A., Nakamura, E. *JACS* **122**, 978 (2000).

Grignard reagents–manganese(II) chloride. 20, 175–176

Coupling reactions.[1] The coupling of RMgX with ArX (including ArF), which is activated by an electron-withdrawing substituent, proceeds readily with $MnCl_2$ as catalyst.

Allylmanganese reagents.[2] These reagents are formed from allyl ethers by a fragmentative pathway. When 2-allyloxytetrahydropyran is treated with R_3MnMgX, the ensuing fragments recombine to give 7-octene-1,5-diol.

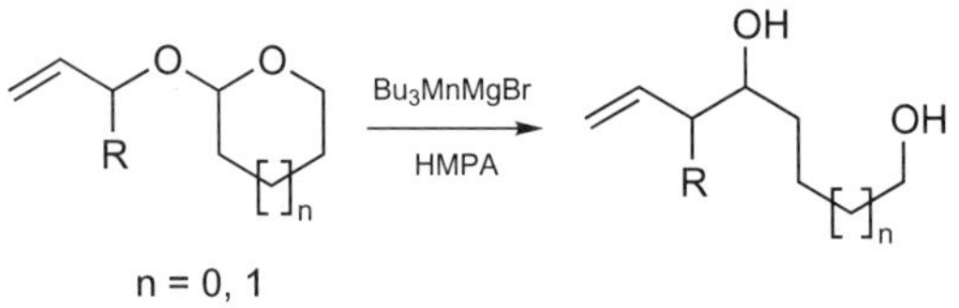

$Bu_3MnMgBr$, HMPA

n = 0, 1

[1]Cahiez, G., Lepifre, F., Ramiandrasoa, P. *S* 2138 (1999).
[2]Nishikawa, T., Nakamura, T., Kakiya, H., Yorimitsu, H., Shinokubo, H., Oshima, K. *TL* **40**, 6613 (1999).

Grignard reagents–nickel complexes. 18, 173; **19,** 156–157; **20,** 176–177

Cross couplings. Formation of unsymmetrical biphenyls is easily achieved, including those from iodophenols.[1] By adding the highly hindered 1,3-dimesitylimidazolinium salt to the reaction media, coupling with ArCl proceeds well.[2]

Alkenyl halides are transformed into alkenes while retaining their configurations.[3] 2-Substituted 1,3-dienes are generated from the readily available dienyl phosphates.[4]

Br ... Ph —MeMgBr - Ni(dppp)→ ... Ph

78%

[1]Shatayev, K.V., Ten'kovtsev, A.V., Bilibin, A.Y. *RJOC* **35**, 308 (1999).
[2]Bohm, V.P.W., Weskamp, T., Gstottmayr, C.W.K., Herrmann, W.A. *ACIEE* **39**, 1602 (2000).
[3]Uenishi, J., Kawahama, R., Izaki, Y., Yonemitsu, O. *T* **56**, 3493 (2000).
[4]Karlstrom, A.S.E., Itami, K., Backvall, J.-E. *JOC* **64**, 1745 (1999).

Grignard reagents–palladium complexes.

Cross couplings. Biphenyl synthesis by cross-coupling can be promoted by Pd complexes in the presence of 1,3-dimesitylimidazolinium chloride.[1] A chiral phosphine ligand enables the selective replacement of one of two triflyloxy groups on an arene nucleus.[2]

[1]Huang, J., Nolan, S.P. *JACS* **121**, 9889 (1999).
[2]Kamikawa, T., Hayashi, T. *T* **55**, 3455 (1999).

Grignard reagents/silver salt.

Allylamines.[1] α-Amino nitriles suffer attack by alkenylmagnesium halides.

N Ph CN —BrMg, $AgBF_4$→ N Ph

91%

[1]Agami, C., Couty, F., Evano, G. *OL* **2**, 2085 (2000).

Grignard reagents–titanium(IV) compounds. 14, 121–122; 18, 174; 19, 158–161; 20, 177–180

Functionalized cyclopropanols. Cyclopropanols in which the geminal side chain contains an α-hydroxyl,[1] a β-ester,[2] or a β-phosphono group[3] have been prepared from the precursoral esters. These are valuable synthetic intermediates. On the other hand, when the π-ligand of propenetitanium diisopropoxide [from *i*-PrMgCl and $(i\text{-PrO})_3TiCl$] is exchanged by vinyltrimethylsilane, cyclopropanols containing a nuclear silyl group are generated.[4]

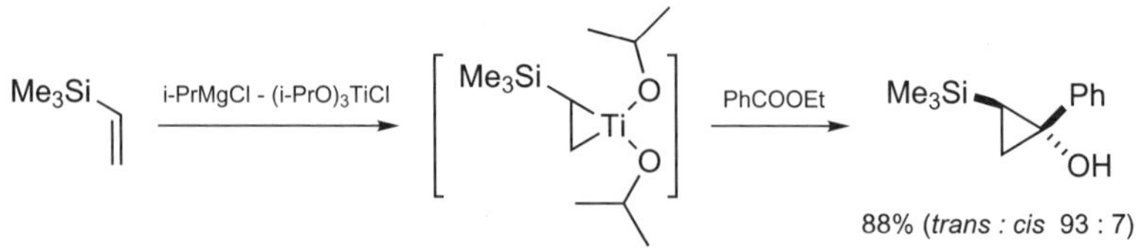

Reaction with alkenes and alkynes. In the carbosilylation of alkenes and dienes with carbon fragment that comes from a secondary or tertiary alkyl halide, promotion by $BuMgBr/Cp_2TiCl_2$ is efficient. Primary alkyl halides are less suitable contributors unless the reaction is intramolecular. The regioselectvity of this process is such that the silyl group is branched out.[5]

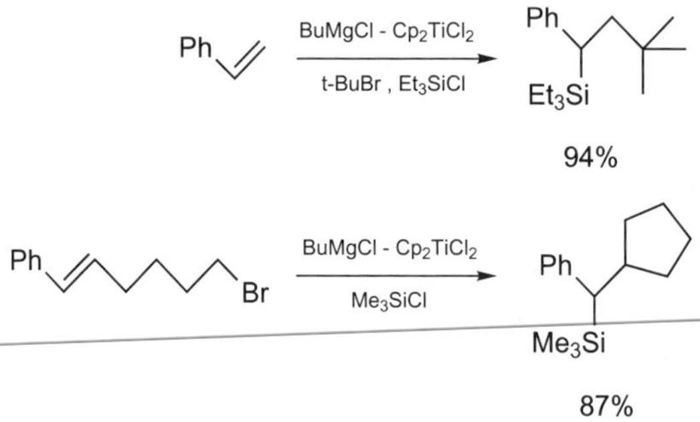

Hydrotitanation of 1-silyl- and 1-stannyl-1-alkynes leads to β-silyl (or β-stannyl)alkenyltitanates,[6] thus showing opposite regioselectivity to other related processes (hydroboration, hydroalumination, hydromagnesiation, hydrotitanation, hydrozincation, hydrozirconation).Well-defined alkenes are obtained on further reaction of the alkenyltitanates, for example, with allylic carbonates an S_N2' displacement occurs to afford 1,4-dienes.[7] If another alkyne is added before protonation of the titanacyclopropenes (to give the alkenyltitanates), then the intermediates are transformed into titanacyclopentadienes, and hence to conjugated dienes.[8]

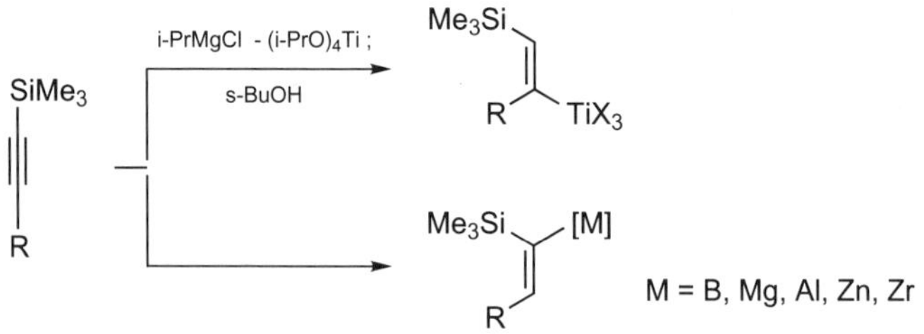

M = B, Mg, Al, Zn, Zr

A propargylic carbonate or phosphonate may be converted to a hydrazino derivative via the allenyltitanate species.[9]

Certain 2-alkylidenecycloalkanols are synthesized from alkynones by an intramolecular process initiated by cyclotitanation of the triple bond.[10]

i-PrMgCl - (i-PrO)$_4$Ti ;

X = H, I
n = 1, 2

Reductive cleavages. Diaryl disulfides[11] and ditellurides[12] are cleaved with the *i*-BuMgBr/Cp_2TiCl_2 system. Alkylation results in alkyl aryl sulfides and tellurides, respectively.

Both the removal of the ester group from a β-ketoester[13] and dechlorination of ArCl (also ArBr, ArI)[14] are accomplished at room temperature.

Allyl ethers and allylamines undergo dimerization,[15] whereas 1,6-dienes are cyclized regioselectively according to the titanium complexes used.[16]

(i-PrO)$_4$Ti

(*dl* : *meso* > 96 : 4)

51%

BuMgBr

Et$_2$O

+ Cp_2TiCl_2	92%	–
+ EBTHI-$TiCl_2$	–	49%

Cyclization. If a proper leaving group is present at an allylic position of a 1,6-diene, an opportunity for elimination exists after formation of a bicyclic titanacycle. The resulting alkyltitanium species may be functionalized.[17] A synthesis of (−)-α-kainic acid has been developed accordingly.

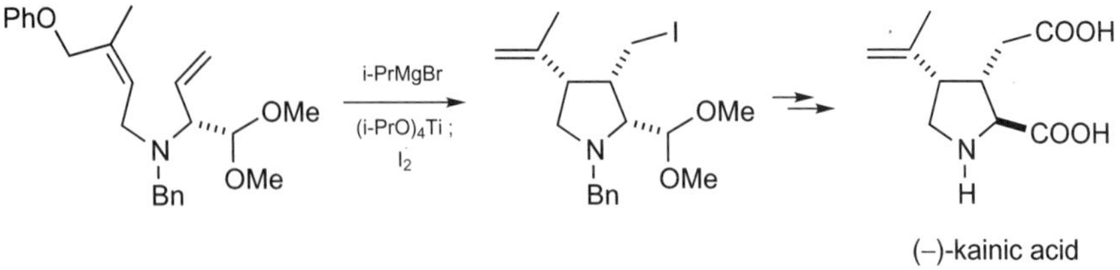

Cyclization with 1,5-chirality transfer is related to the above[18]

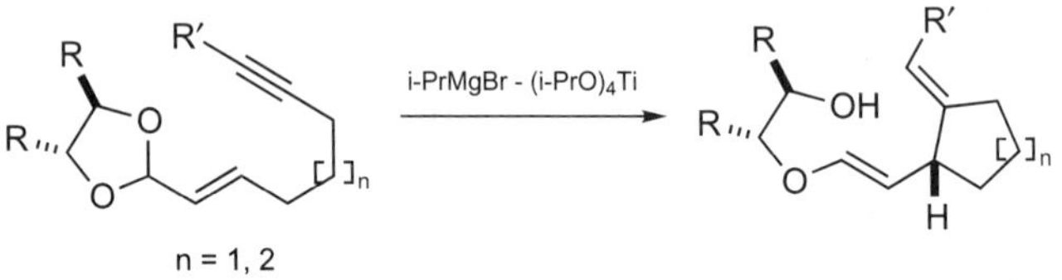

Bicyclization involving an intramolecular acylation is further extended to include an aldolization.[19] A route to bicyclic enone is based on controlled protonolysis of the titanacycle and subsequent cyclization. This synthesis is amenable to asymmetric induction.[20]

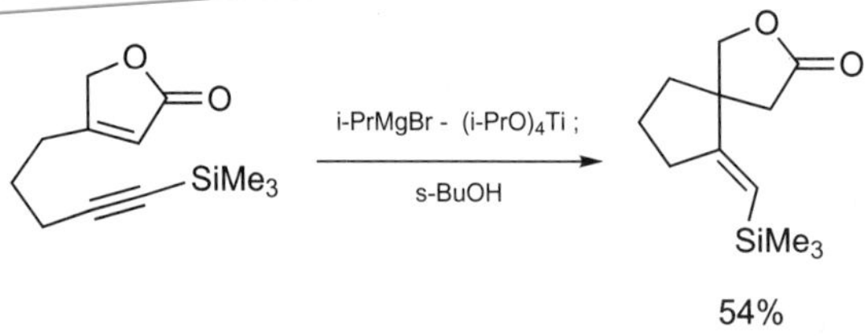

The possibility of forming a bridged tricyclic system by an intramolecular reaction of an imide[21] is of synthetic significance.

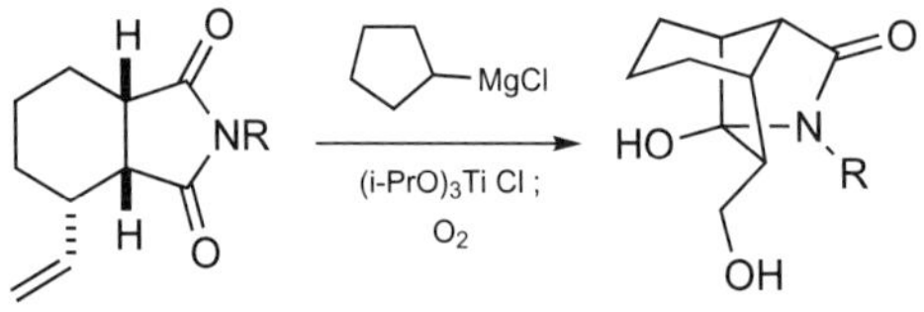

The Cp_2TiPh species generated from addition of PhMgBr to the *i*-PrMgBr–Cp_2TiCl_2 system is able to induce cyclization of δ-ketonitriles to cyclopentanolones.[22] The vinylogues afford 2-hydroxycyclopentaneacetonitriles.

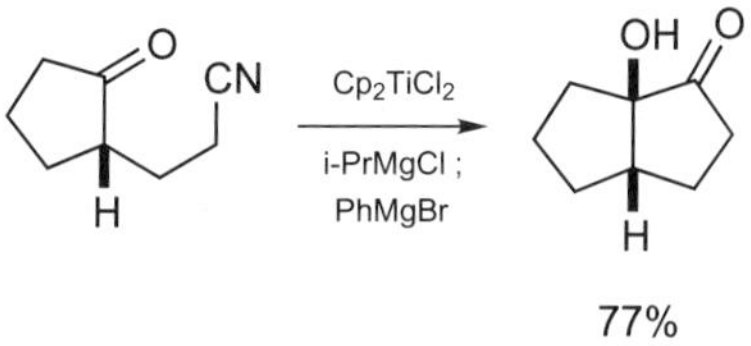

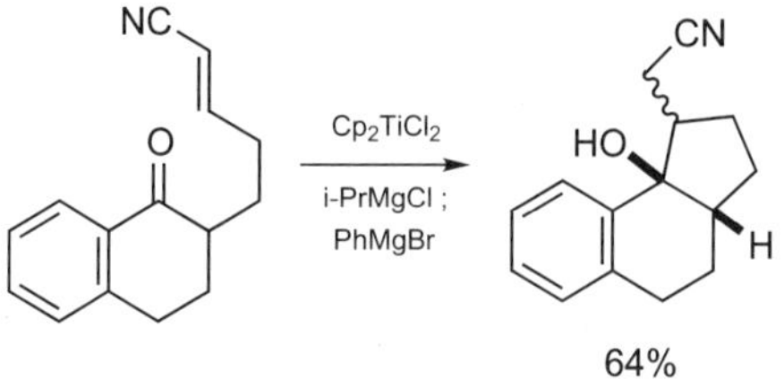

CN
Cp2TiCl2
i-PrMgCl ;
PhMgBr
OH
O
H
77%

NC
O
Cp2TiCl2
i-PrMgCl ;
PhMgBr
CN
HO
H
64%

Homoenolate and homoallenyl carbanion equivalents. Two routes to homoenolate species are based on the action of (propene)titanium diisopropoxide. Trialkoxytitanates generated from acetals of acrolein react with aldehydes and imines. Chiral cyclic acetals are similarly cleaved to afford the nucleophiles.[23] 3-Alkoxy-2-propyn-1-yl carbonates are transformed into (1-alkoxyallen)-1-yltitanates that add to carbonyl compounds with γ-selectivity.[24]

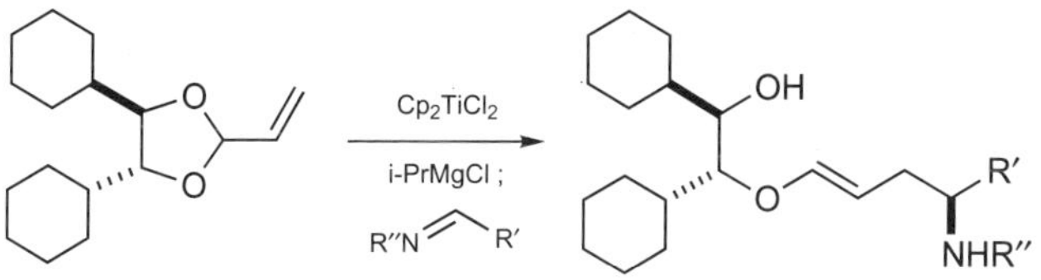

O
O
Cp2TiCl2
i-PrMgCl ;
R″N R′
OH
O
R′
NHR″

Following activation of conjugated enynes, the reaction with carbonyl compounds and imines gives products with multiple stereocenters.[25]

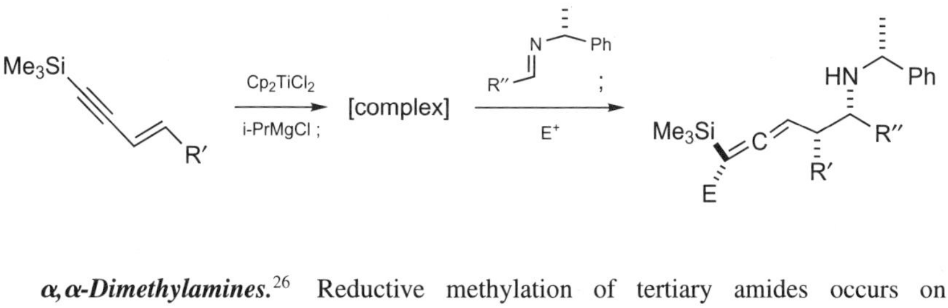

Me3Si
R′
Cp2TiCl2
i-PrMgCl ;
[complex]
N
Ph
R″
;
E+
HN
Ph
Me3Si
C
R″
R′
E

α,α-Dimethylamines.[26] Reductive methylation of tertiary amides occurs on reaction with MeMgBr in the presence of $TiCl_4$ (or $ZrCl_4$) in THF.

[1]Cho, S.Y., Cha, J.K. *OL* **2**, 1337 (2000).
[2]Raiman, M.V., Il'ina, N.A., Kulinkovich, O.G. *SL* 1053 (1999).
[3]Winsel, H., Gazizova, V., Kulinkovich, O.G., Pavlov, V., de Meijers, A. *SL* 1999 (1999).
[4]Mizojiri, R., Urabe, H., Sato, F. *TL* **40**, 2557 (1999).
[5]Nii, S., Terao, J., Kambe, N. *JOC* **65**, 5291 (2000).
[6]Urabe, H., Hamada, T., Sato, F. *JACS* **121**, 2931 (1999).
[7]Okamoto, S., Takayama, Y., Gao, Y., Sato, F. *S* 975 (2000).
[8]Hamada, T., Suzuki, D., Urabe, H., Sato, F. *JACS* **121**, 7342 (1999).
[9]An, D.K., Hirakawa, K., Okamoto, S., Sato, F. *TL* **40**, 3737 (1999).
[10]Morlender-Vais, N., Solodovnikova, N., Marek, I. *CC* 1849 (2000).
[11]Huang, X., Zheng, W.-X. *SC* **29**, 1297 (1999).
[12]Huang, X., Zheng, W.-X. *SC* **30**, 1365 (2000).
[13]Yu, Y., Zhang, Y. *SC* **29**, 243 (1999).
[14]Hara, R., Sato, K., Sun, W.-H., Takahashi, T. *CC* 845 (1999).
[15]de Meijere, A., Stecker, B., Kourdioukov, A., Williams, C.M. *S* 929 (2000).
[16]Okamoto, S., Livinghouse, T. *OM* **19**, 1449 (2000).
[17]Campbell, A.D., Raynham, T.M., Taylor, R.J.K. *CC* 245 (1999).
[18]Takayama, Y., Okamoto, S., Sato, F. *JACS* **121**, 3559 (1999).
[19]Okamoto, S., Subburaj, K., Sato, F. *JACS* **122**, 11244 (2000).
[20]Urabe, H., Hideura, D., Sato, F. *OL* **2**, 381 (2000).
[21]Sung, M.J., Lee, C.-W., Cha, J.K. *SL*561 (1999).
[22]Yamamoto, Y., Matsumi, D., Hattori, R., Itoh, K. *JOC* **64**, 3224 (1999).
[23]Teng, X., Takayama, Y., Okamoto, S., Sato, F. *JACS* **121**, 11916 (1999).
[24]Hanazawa, T., Okamoto, S., Sato, F. *OL* **2**, 2369 (2000).
[25]Hamada, T., Mizojiri, R., Urabe, H., Sato, F. *JACS* **122**, 7138 (2000).
[26]Denton, S.M., Wood, A. *SL* 55 (1999).

Grignard reagents–zinc borohydride.

Reductive alkylation. Esters are converted to secondary alcohols where the Grignard reagent contributes one alkyl group.[1]

$$\text{R–COOEt} \xrightarrow[\text{Zn(BH}_4)_2]{\text{R'MgBr}} \text{RR'CH–OH}$$

[1]Hallouis, S., Saluzzo, C., Amouroux, R. *SC* **30**, 313 (2000).

Grignard reagents–zirconium compounds. **18,** 174; **19,** 161; **20,** 180–181

Organozincation.[1] On reaction with EtMgBr–Cp_2ZrCl_2, 1-alkenes give 2-substituted dialkylzincs that can be used in coupling reactions.

Alkylsilanes.[2] Grignard reagents are isomerized by Cp_2ZrCl_2. Subsequent reaction with hydrosilanes in the presence of an alkyl bromide leads to silanes bearing a primary alkyl group.

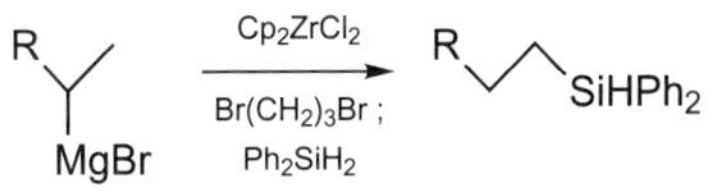

1,5-Dienes and 1,5-enynes.[3] Insertion reaction takes place when zirconacyclopentenes (derived from alkynes, ethylmagnesium bromide, and zirconocene dichloride) are treated with alkynylmetals. Protodemetallation leads to 1,5-dienes, whereas oxidation with iodine gives 1,5-enynes.

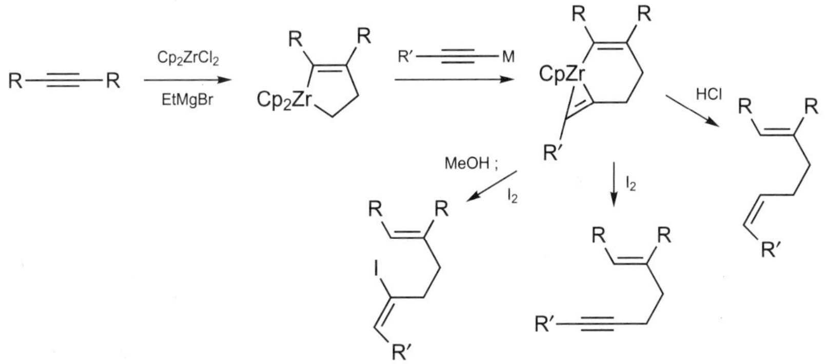

Alkenylcyclopropanes.[4] The zirconocene–ethylene complex generated from EtMgBr and Cp_2ZrCl_2 converts the carbonyl group of enones to a cyclopropane ring.

[1]Gagneur, S., Montchamp, J.-L., Negishi, E. *OM* **19**, 2417 (2000).
[2]Ura, Y., Hara, R., Takahashi, T. *CC* 875 (2000).
[3]Dumond, Y., Negishi, E. *JACS* **121**, 11223 (1999).
[4]Bertus, P., Gandon, V., Szymoniak, J. *CC* 171 (2000).

H

Hafnium(IV) chloride. 20, 182

trans-Carbosilylation of alkynes.[1] Splitting of propargyl- and allenyltrimethylsilanes and addition of the components to alkynes is effected by hafnium(IV) chloride.

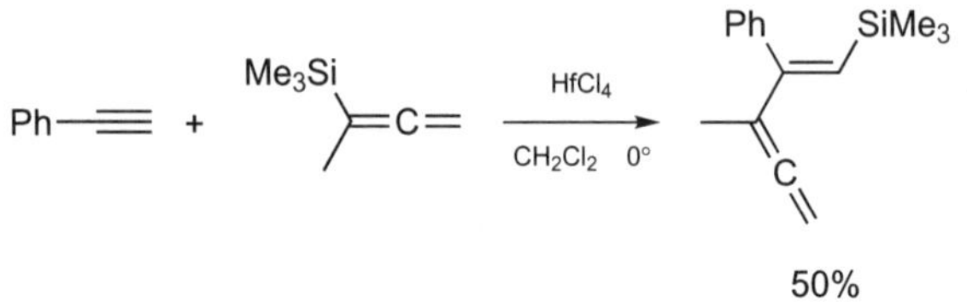

[1]Yoshikawa, E., Kasahara, M., Asao, N., Yamamoto, Y. *TL* **41**, 4499 (2000).

4-Halobenzyl bromides.

Protection of alcohols.[1] Ether formation from ROH and 4-$XC_6H_4CH_2Br$ mediated by NaH in DMF is easily achieved. Such ethers are cleaved by Pd-catalyzed processes. 4-Bromobenzyl ethers can be cleaved by treatment with $(dba)_3Pd_2$, *t*-BuONa, PhNHMe, and then $SnCl_4$. 4-Chlorobenzyl ethers are stable to these conditions but they are cleaved on changing $(dba)_3Pd_2$ to $Pd(OAc)_2$.

[1]Plante, O.J., Buchwald, S.L., Seeberger, P.H. *JACS* **122**, 7148 (2000).

Halomethyllithiums.

Homologation.[1] When alkylboranes are treated with $LiCH_2X$ before oxidative workup, insertion of a methylene group to the C—B bond is accomplished. Double homologation to some degree occurs with bromo- and iodomethyllithiums but not chloromethyllithium.

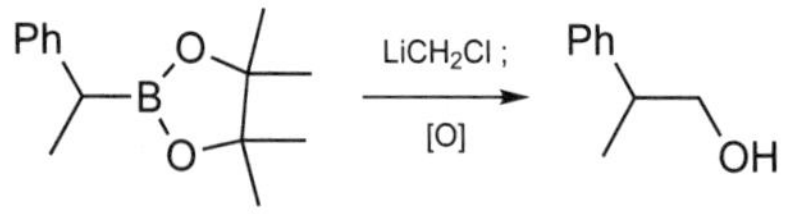

[1]Ren, L., Crudden, C.M.*CC* 721 (2000).

Hexaalkylditin. 13, 142; **14,** 173–174; **16,** 174; **17,** 143–144; **18,** 175–176; **19,** 162–163; **20**, 182–184

Aldols.[1] Treatment of a mixture of α-iodoketones and aldehydes with $(Bu_3Sn)_2$, Bu_2SnF_2, and HMPA in aq THF leads to aldols. Good diastereoselectivity is also observed.

Oxyfluoroalkylation.[2] Benzylic alcohols containing a $CH_2C_nF_{2n+1}$ chain are formed when styrenes are subjected to photooxygenation in the presence of $(Bu_3Sn)_2$.

C_6F_{13}—I + styrene (Ph) —[hν $(Bu_3Sn)_2$ / O_2 / PhH]→ $C_6F_{13}CH_2CH(OH)Ph$ 58%

Addition to propargylic alcohols.[3] Tributylmanganate reagents generated from R_4MnLi_2 and $(Bu_3Sn)_2$ effect *trans*-addition to proparglylic alcohols. The allylic alcohols thus produced bear an R group at C-3 and are stannylated at C-2.

Addition–elimination sequence. Addition of a photochemically generated free radical to an electron-deficient double bond engenders different consequence depending on the structural features of the substrates. Routes to functionalized nitroalkanes[4] and cycloalkanone oxime ethers[5] demonstrate the versatile method.

PhO(CH₂)₄I + Ph–S(=O)(=O)–CH=N⁺(O⁻)OTBS —[hν $(Me_3Sn)_2$; HCl ; KF]→ PhO(CH₂)₆NO₂ 62%

I(CH₂)₃C(=NOBz)SO₂Me + CH₂=CH–COOEt —[hν $(Me_3Sn)_2$ / PhH]→ 2-(NOBz)cyclopentyl-COOEt 83%

[1]Shibata, I., Kawasaki, M., Yasuda, M., Baba, A. *CL* 689 (1999).
[2]Yoshida, M., Ohkoshi, M., Aoki, N., Iyoda, M. *TL* **40**, 5731 (1999).
[3]Usugi, S., Tang, J., Shinokubo, H., Oshima, K. *SL* 1417 (1999).
[4]Kim, S., Yoon, J.-Y., Lim, C.J. *SL* 1151 (2000).
[5]Kim, S., Kim, N., Yoon, J.-Y., Oh, D.H. *SL* 1148 (2000).

1,1,1,3,3,3-Hexafluoro-2-propanol.

Epoxide opening. The reaction of epoxides with arylamines to afford β-amino alcohols is facilitated using the title compound as solvent.[1] Treatment of epoxides with $RSH–H_2O_2$ in the same medium result in β-hydroxy sulfoxides.[2]

[1]Das, U., Crousse, B., Kesavan, V., Bonnet-Delpon, D., Begue, J.-P. *JOC* **65**, 6749 (2000).
[2]Kesavan, V., Bonnet-Delpon, D., Begue, J.-P. *TL* **41**, 2895 (2000).

Hexamethyldisilane.

Deoxygenation. Nitroalkanes, nitrones, and heterocyclic *N*-oxides submit their oxygen atoms to hexamethyldisilane. Nitroalkanes (via nitronate anions) are converted into oximes.[1]

[1]Hwu, J.R., Tseng, W.N., Patel, H.V., Wong, F.F., Horng, D.-N., Liaw, B.R., Lin, L.C. *JOC* **64**, 2211 (1999).

Hexamethyldisilazane, HMDS. 13, 141; 18, 177–178; 19, 163–164; 20, 184–185

Amide formation. HMDS promotes condensation of acids and amines.[1]

[1]Chou, W.-C., Chou, M.-C., Lu, Y.-Y., Chen, S.-F. *TL* **40**, 3419 (1999).

Hydrazine hydrate. 13, 144; 18, 179; 20, 185

Reductive cleavage of diaryl diselenides.[1,2] Generation of arylselenide anions by hydrazine hydrate is most convenient under basic conditions.

Arylamines.[3] Reduction of $ArNO_2$ can be carried out with hydrazine hydrate over a cerium(IV)–tin(IV) oxide catalyst.

[1]Vasil'ev, A.A., Engman, L., Storm, J.P., Andersson, C.-M. *OM* **18**, 1318 (1999).
[2]Henriksen, L., Stuhr-Hansen, N. *JCS(P1)* 1915 (1999).
[3]Jyothi, T.M., Rajagopal, R., Sreekumar, K., Talawar, M.B., Sungunan, S., Rao, B.S. *JCR(S)* 674 (1999).

Hydrazoic acid.

x-Amino-z,y-Pyrandiones.[1] Introduction of an amino group to pyrandiones is by their direct exposure to NaN_3–HOAc at pH 4–5. The adducts with HN_3 eliminate dinitrogen in situ.

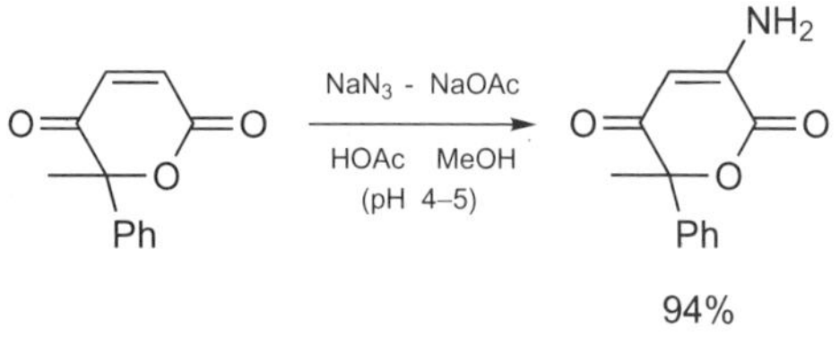

[1]Kouloccheri, S.D., Haroutounian, S.A., Apostolopoulos, C.D., Chada, R.K., Couladouros, E.A. *EJOC* 1449 (1999).

Hydrogen fluoride–amine. 16, 286–287; **18,** 181; **19,** 164–165; **20,** 185

Oxidative fluorination. γ,δ-Difluoro-α,β-unsaturated esters are produced at a Pt anode[1] when dienoic esters are subjected to electrolysis in MeCN containing 3HF–Et_3N. A chemical process employing 3HF–Et_3N and NXS converts dihydropyran into 2-fluoro-3-halotetrahydropyrans[2] and allyl alcohol into 3-bromo-2-fluoropropanol[3] (in the latter case study, only with NBS).

The combination of HF–pyridine with 1,3-dibromo-5,5-dimethylhydantoin converts dithiocarbonates into trifluoromethyl ethers.[4]

Nazarov cyclization.[5] Cross-conjugated dienones containing methylthio substituents give rise to fluorocyclopentenones.

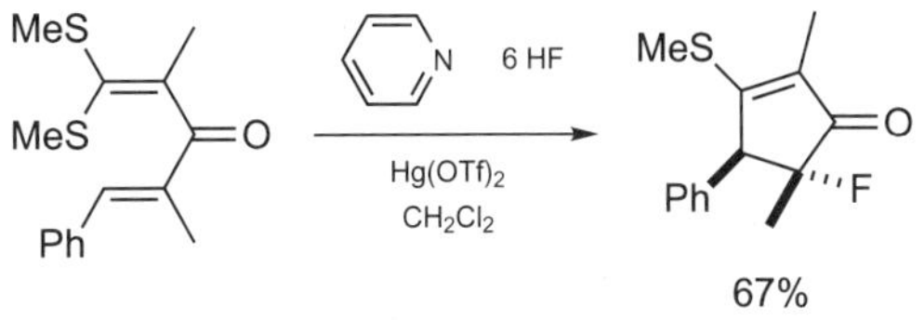

MeS, MeS, Ph, O → (pyridine 6 HF, Hg(OTf)$_2$, CH$_2$Cl$_2$) MeS, O, Ph, F — 67%

Desilylation. Desilylation promoted by 3HF–Et_3N in THF constitutes the key step in an enantioselective access of α-azido ketones.[6]

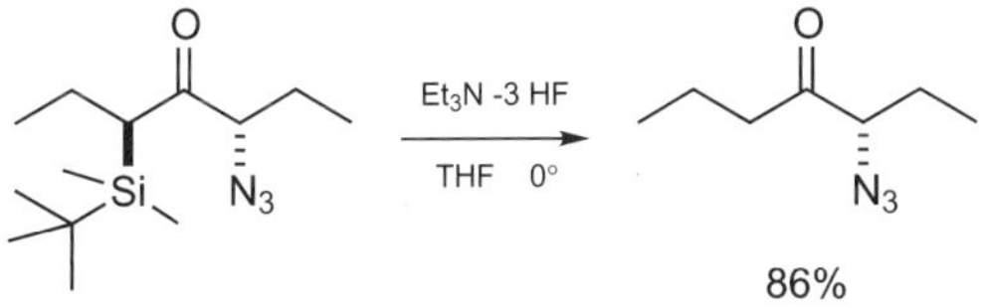

O, Si, N_3 → (Et_3N -3 HF, THF 0°) O, N_3 — 86%

1-Aryl-3-hexenopyranosiduloses.[7] 2-Acetoxyglycals serve as electrophiles toward arenes in a reaction that introduces an aryl group to C-1 of the sugar moiety.

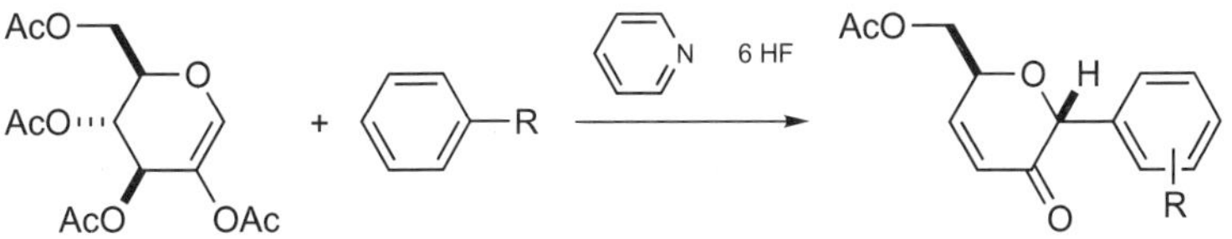

AcO, AcO, AcO, O, OAc + R → (pyridine 6 HF) AcO, O, H, O, R

[1]Dinoiu, V., Fukuhara, T., Hara, S., Yoneda, N. *JFC* **103**, 75 (2000).

[2]Shellhamer, D.F., Horney, M.J., Pettus, B.J., Pettus, T.L., Stringer, J.M., Heasley, V.L., Syvret, R.G., Dobrolsky, J.M. *JOC* **64**, 1094 (1999).

[3]Lubke, M., Skupin, R., Haufe, G. *JFC* **102**, 125 (1999).

[4]Kanie, K., Tanaka, Y., Suzuki, K., Kuroboshi, M., Hiyama, T. *BCSJ* **73**, 471 (1999).
[5]Hara, S., Okamoto, S., Narahara, M., Fukuhara, T., Yoneda, N. *SL* 411 (1999).
[6]Enders, D., Klein, D. *SL* 719 (1999).
[7]Hayashi, M., Nakayama, S., Kawabata, H. *CC* 1329 (2000).

Hydrogen iodide.

Hydrodechlorination. β-Chlorovinamidinium salts are dechlorinated by HI in dioxane.[1]

[1]Davies, I.W., Taylor, M., Hughes, D., Reider, PJ. *OL* **2**, 3385 (2000).

Hydrogen peroxide.

Activation. Catalysts for alkene epoxidation with H_2O_2 include perfluoroheptadecan-7-one,[1] butyl 2,4-bis(perfluorooctyl)phenyl selenide,[2] phenyldibutylarsine,[3] and $HReO_4$–triphenylarsine.[4] For ring contraction of cycloalkanones, an oligomeric 9,10-bis(diseleno)-anthracene effectively promotes the action of hydrogen peroxide.[5] Polyfluorinated alcohols such as hexafluoroisopropanol also activate hydrogen peroxide toward epoxidation and Baeyer–Villiger oxidation.[6]

Oxidations. Alkene formation from alkyl aryl selenides via oxidation is cleaner when the aryl group is *ortho*-substituted (vs. *para*-substitution), for example, with a nitro group.[7] In the presence of a phase-transfer catalyst, oxidation of aldehydes to carboxylic acids[8] with H_2O_2 is successfully carried out without an organic solvent, halide, or metal ion.

[1]van Vliet, M.C.A., Arends, I.W.C.E., Sheldon, R.A. *CC* 263 (1999).
[2]Betzemeier, B., Lhermitte, F., Knochel, P. *SL* 489 (1999).
[3]van Vliet, M.C.A., Arends, I.W.C.E., Sheldon, R.A. *TL* **40**, 5239 (1999).
[4]van Vliet, M.C.A., Arends, I.W.C.E., Sheldon, R.A. *JCS(P1)* 377 (2000).
[5]Giurg, M., Mlochowski, J. *SC* **29**, 2281 (1999).
[6]Neimann, K., Neumann, R. *OL* **2**, 2861 (2000).
[7]Sayama, S., Onami, T. *TL* **41**, 5557 (2000).
[8]Sato, K., Hyodo, M., Takagi, J., Aoki, M., Noyori, R. *TL* **41**, 1439 (2000).

Hydrogen peroxide, acidic. 14, 176; **15,** 167–168; **16,** 177–178; **17,** 145; **18,** 182–183; **19,** 166; **20,** 187

Degradation of acylarenes.[1] Such carbonyl compounds are readily converted to phenols with hydrogen peroxide–boric acid.

[1]Roy, A., Reddy, K.R., Mohanta, P.K., Ila, H., Junjappa, H. *SC* **29**, 3781 (1999).

Hydrogen peroxide, basic. 13, 145; **14,** 156; **15,** 167; **18,** 183–184; **20,** 187–188

Oxidations. Mildly basic conditions ($NaHCO_3$) have been used to epoxidize alkenes.[1] For epoxidation of water insoluble substrates, aq MeCN is a suitable medium. A very similar system is useful for oxidative cleavage of organostannanes containing a perfluoroalkyl group to furnish alcohols.[2]

α-Formylpyrroles yield 2-pyrrolinones.[3]

MeOOC, COOMe, CHO, N H → (H_2O_2 - $NaHCO_3$, MeOH) → MeOOC, COOMe, O, N H 85%

[1]Yao, H., Richardson, D.E. *JACS* **122**, 3220 (2000).
[2]Falck, J.R., Lai, J.-Y., Ramana, D.V., Lee, S.-G. *TL* **40**, 2715 (1999).
[3]Pichon-Santander, C., Scott, A.I. *TL* **41**, 2825 (2000).

Hydrogen peroxide, metal catalysts. 13, 145; **14,** 177; **15,** 294; **17,** 146–148; **18,** 184–185; **19,** 166–167; **20,** 188

Nitrones. *N*-Alkyl α-amino acids undergo oxidative decarboxylation to afford nitrones.[1]

Ph, N H, COOH → (H_2O_2 - Na_2WO_4, Et_4NCl - K_2CO_3, H_2O CH_2Cl_2) → Ph, N^+, O^- 70%

Oxidative esterification.[2] The H_2O_2–V_2O_5 reagent converts aromatic aldehydes to methyl esters in methanol (perchloric acid is also present).

Bromination.[3] Brominating agent for activated arenes (e.g., phenols) is formed in situ from tetrabutylammonium bromide (oxidation by H_2O_2–V_2O_5).

Oxidation.[4] In the presence of a phase-transfer catalyst (Bu_4NHSO_4), alcohols are oxidized by H_2O_2–Na_2WO_4 to afford either carboxylic acids or ketones (from primary and secondary alcohols, respectively). Because of its easiness in separation and regeneration the heterogeneous catalyst WO_3–SiO_2 is useful for cleaving cyclopentene with H_2O_2 to glutaraldehyde.[5]

Nitriles.[6] Copper powder and various copper salts catalyze the conversion of aldehydes to nitriles by ammonia and H_2O_2 in isopropanol.

[1]Ohtake, H., Imada, Y., Murahashi, S.-I. *BCSJ* **72**, 2737 (1999).
[2]Gopinath, R., Patel, B.K. *OL* **2**, 577 (2000).
[3]Bora, U., Bose, G., Chaudhuri, M.K., Dhar, S.S., Gopinath, R., Khan, A.T., Patel, B.K. *OL* **2**, 247 (2000).
[4]Bogdal, D., Lukasiewicz, M. *SL* 143 (2000).
[5]Jin, R., Xia, X., Xue, D., Deng, J.-F. *CL* 371 (1999).
[6]Erman, M.B., Snow, J.W., Williams, M.J. *TL* **41**, 6749 (2000).

Hydrogen sulfide.

Alkyl β-oxoalkanedithioates.[1] The combination of H_2S–$BF_3.OEt_2$ dealkylate one of the *S*-substituents of acyl ketene dithioacetals in refluxing dioxane.

Ph, =O, SMe, MeS — H_2S - BF_3·OEt_2 / dioxane Δ → Ph, =O, =S, MeS

67%

α-Silyl enethiols.[2] Acylsilanes react with hydrogen sulfide to provide α-silylated enethiols. A route to thiolactones passes through such intermediates.

Ph, Si, O, COOH — H_2S - HCl ; / $NaHCO_3$ → Ph, Si, SH, COOH → Ph, Si, S, O

[1]Nair, S.K., Asokan, C.V. *SC* **29**, 791 (1999).
[2]Bonini, B.F., Comes-Franchini, M., Fochi, M., Mazzanti, G., Ricci, A. *SL* 486 (1999).

Hydrosilanes. 19, 167–169; **20,** 188–192

Defunctionalization. Hydrosilanes are better reagents than hydrostannanes for deoxygenation of alcohols via their xanthates. Thus, reasonably good results are obtained by using the symmetrical tetraphenyldisilane under free radical conditions.[1] More interestingly, direct deoxygenation of primary alcohols and cleavage of ethers are possible with Et_3SiH and a catalytic amount of $(C_6F_5)_3B$.[2]

Regioselectivity is important for reductive ring opening of cyclic acetals as differentiation of the two hydroxyl groups in a diol system is often desired in synthesis. The fact that the combination of Et_3SiH with TfOH and $PhBCl_2$ operate in the opposite sense on 4,6-*O*-benzylidene sugars[3] is very valuable. Note that cleavage of benzyl esters with hydrosilane is catalyzed by $Pd(OAc)_2$.[4]

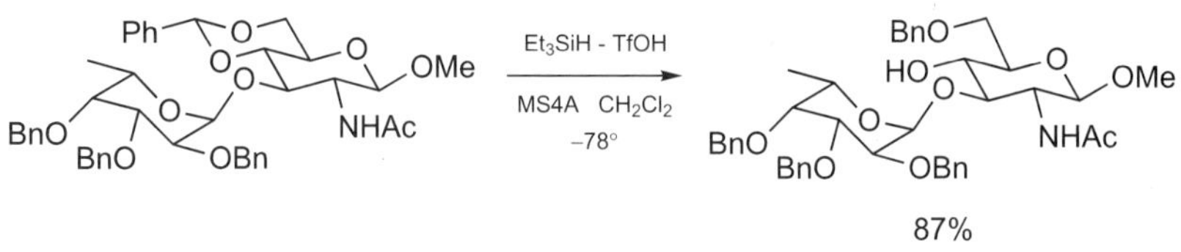

Thiono esters and lactones are converted to ethers by Ph_2SiH_2 and catalytic Ph_3SnH, Et_3B.[5] Partial dechlorination of polychloroarenes with triethylsilane and a homogeneous Rh–phosphine catalyst proceeds in high yields.[6]

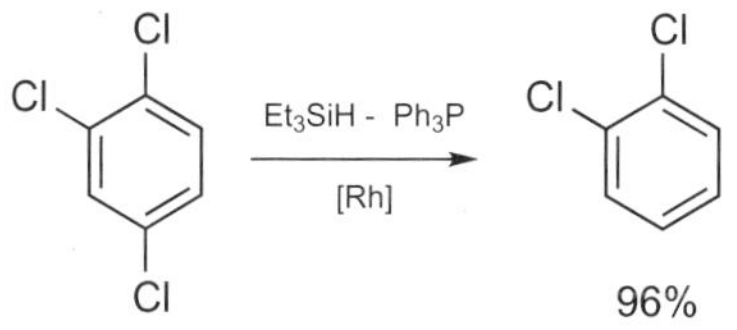

Symmetrical tetraaryldisilanes (prepared from Ar_2SiH_2 and Cp_2TiPh_2 at 120°) effectively performs hydrodehalogenation.[7] However, the radical intermediates readily form adducts with activated alkenes.

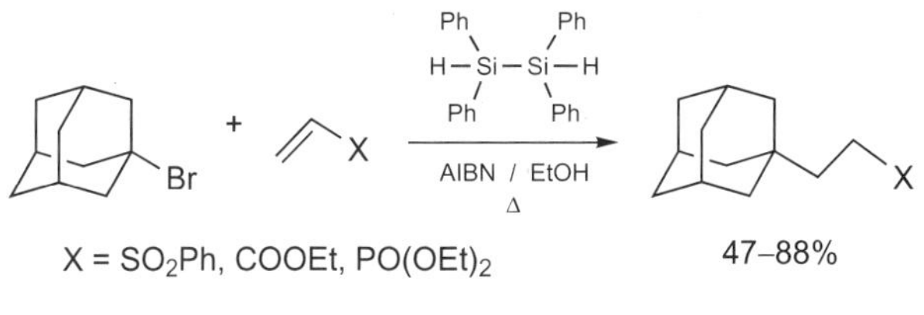

Reductions. Ketones are reduced with trialkoxysilane–histidine complexes.[8] A gold complex $(Ph_3P)AuCl$ catalyzes hydrosilylation of the C═O group.[9] Dibenzenechromium is a precatalyst in the reduction of aryl carbonyl compounds.[10] A system comprising hydrosilane and a copper salt has selective reducing power for ketones and double bonds conjugated to an aromatic ring.[11] For saturation of the enone double bond, a choice can be made of $PhMe_2SiH$ with the tris(triphenylphosphine)copper(I) fluoride–bisethanol adduct[12] or $PhSiH_3$ and $Mn(dpm)_3$.[13] Reduction of esters to alcohols can use either diphenylsilane in the presence of the $[(cod)RhCl]_2$ complex[14] or trimethoxysilane and catalytic amount of MeOLi in THF.[15]

Benzyl ethers can be prepared from THP ethers by the Me_3SiOTf-catalyzed reaction with PhCHO and Et_3SiH.[16] Analogously, amides, carbamates, and ureas are *N*-alkylated via condensation with aldehydes followed by in situ reduction with Et_3SiH.[17] The aldehyde may be replaced by a thioester that is subject to reduction in situ. Thus, a mixture of sodium triacetoxyborohydride, triethylsilane, and Pd–C catalyst is employed.[18]

NH2 Ph O O + O O NH O O SEt O Et3SiH Pd/C NaBH(OAc)3

O O NH O NH O Ph O O

93%

A regioselective conversion of 6,8-dioxabicyclo[3.2.1]octanes to oxepanes is effected by a $TiCl_4$-catalyzed reduction with Et_3SiH.[19]

Benzyl azides undergo rearrangement–reduction tandem to afford *N*-methylarylamines[20] on treatment with $Et_3SiH–SnCl_4$.

Hydrosilylation. Allyl alcohols form cyclic structures in a Rh-catalyzed hydrosilylation reaction with Ph_2SiH_2.[21] Ring opening occurs during reaction of alkylidenecyclopropanes.[22]

Ph + Et3SiH (Ph3P)3RhCl 20° Ph SiEt3

When the Pd-induced cyclization of 1,6-dienes is carried out in the presence of a hydrosilane, the products become hydrosilylated.[23] Primary alcohols can be prepared with proper hydrosilanes (e.g., pentamethyldisiloxane) to render the silylmethyl group oxidizable.[24] 5,7-Alkadienals mainly cyclize to *cis*-2-alkenylcyclopentyl silyl ethers.[25]

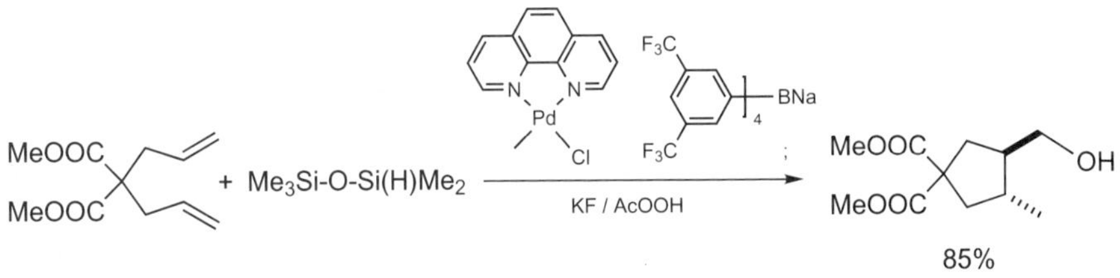

Cyclization and hydrosilylation of nitrogen-containing enynes are achieved in a Cp^*_2YMe-catalyzed process.[26]

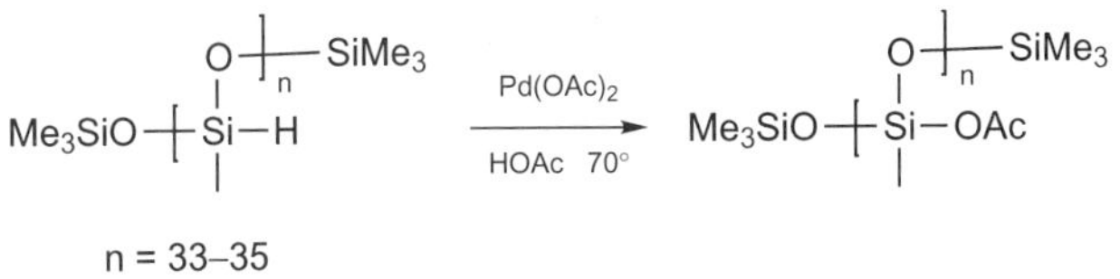

n = 33–35

Dehydrogenative silylation. 1-Alkynes and amines are converted to silyl derivatives by hydrosilanes with iridium carbonyl[27] and ytterbium–imine complexes[28] as catalyst, respectively.

Peptide synthesis.[29] An *N*-allyloxycarbonyl derivative of an α-amino acid is deblocked by a combination of $PhSiH_3$ and a Pd catalyst. When an activated ester of another amino acid is present, it will be attacked by the released amine.

Cyclocarbonylation.[30] When catalyzed by a Rh-carbonyl species and under a CO atmosphere, certain enediynes yield tricyclic ketones. Also involved in this reaction is $PhMe_2SiH$.

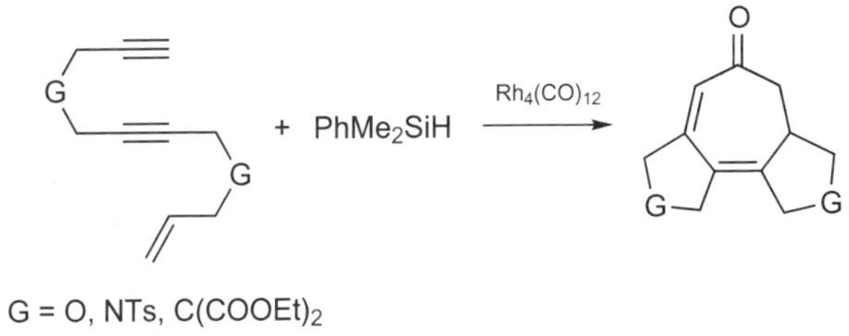

G = O, NTs, $C(COOEt)_2$

[1]Togo, H., Matsubayashi, S., Yamazaki, O., Yokoyama, M. *JOC* **65**, 2816 (2000).
[2]Gevorgyan, V., Liu, J.-X., Rubin, M., Benson, S., Yamamoto, Y. *TL* **40**, 8919 (1999).
[3]Sakagami, M., Hamana, H. *TL* **41**, 5547 (2000).
[4]Coleman, R.S., Shah, J.A. *S* 1399 (1999).
[5]Jang, D.O., Song, S.H. *SL* 811 (2000).
[6]Esteruelas, M.A., Herrero, J., Lopez, F.M., Martin, M., Oro, L.A. *OM* **18**, 1110 (1999).
[7]Yamazaki, O., Togo, H., Matsubayashi, S., Yokoyama, M. *T* **55**, 3735 (1999).
[8]LaRonde, F.J., Brook, M.A. *TL* **40**, 3507 (1999).
[9]Ito, H., Yajima, T., Tateiwa, J., Hosomi, A. *CC* 981 (2000).
[10]Le Bideau, F., Henrique, J., Samuel, E., Elschenbroich, C. *CC* 1397 (1999).
[11]Ito, H., Yamanaka, H., Ishizuka, T., Tateiwa, J., Hosomi, A. *SL* 479 (2000).
[12]Mori, A., Fujita, A., Kajiro, H., Nishihara, Y., Hiyama, T. *T* **55**, 4573 (1999).
[13]Magnus, P., Waring, M.J., Scott, D.A. *TL* **41**, 9731 (2000).
[14]Ohta, T., Kamiya, M., Kusui, K., Michibata, T., Nobutomo, M., Furukawa, I. *TL* **40**, 6963 (1999).
[15]Hojo, M., Murakami, C., Fujii, A., Hosomi, A. *TL* **40**, 911 (1999).
[16]Suzuki, T., Ohashi, K., Oriyama, T. *S* 1561 (1999).
[17]Dube, D., Scholte, A.A. *TL* **40**, 2295 (1999).
[18]Han, Y., Chorev, M. *JOC* **64**, 1972 (1999).
[19]Fujiwara, K., Amano, A., Tokiwano, T., Murai, A. *T* **56**, 1065 (2000).
[20]Lopez, F.J., Nitzan, D. *TL* **40**, 2071 (1999).
[21]Wang, X., Ellis, W.W., Bosnich, B. *CC* 2561 (1996).
[22]Bessmertnykh, A.G., Bilnov, K.A., Grishin, Y.K., Donskaya, N.A., Tveritinova, E.V., Beletskaya, I.P. *RJOC* **34**, 799 (1998).
[23]Widenhoefer, R.A., Vadehra, A. *TL* **40**, 8499 (1999).
[24]Pei, T., Widenhoefer, R.A. *OL* **2**, 1469 (2000).
[25]Sato, Y., Saito, N., Mori, M. *JACS* **122**, 2371 (2000).
[26]Molander, G.A., Corrette, C.P. *JOC* **64**, 9697 (1999).
[27]Shimizu, R., Fuchikami, T. *TL* **41**, 907 (2000).
[28]Takaki, K., Kamata, T., Miura, Y., Shishido, T., Takehira, K. *JOC* **64**, 3891 (1999).
[29]Thieriet, N., Gomez-Martinez, P., Guibe, F. *TL* **40**, 2505 (1999).
[30]Ojima, I., Lee, S.-Y. *JACS* **122**, 2385 (2000).

Hydroxylamine.

Tricyclic isoxazolidines.[1] The possibility exists for an intramolecular reaction following oximation of certain ketones that contain double bonds at proper distances. Based on this concept, a potential precursor of histrionicotoxin has been assembled.

[1]Stockman, R.A. *TL* **41**, 9163 (2000).

Hydroxy(tosyloxy)iodobenzene. 14, 179–180; **16,** 179; **17,** 150; **18,** 187; **19,** 170; **20,** 193

Cleavage of N′,N′-dialkylhydrazides.[1] Mild conditions are involved in this cleavage method. The furan ring in a substrate is not degraded by the oxidant.

Alkynyl tosylates.[2] Tosyloxy group transfer to 1-alkynes is effected in an ultrasound-assisted, AgOTf-catalyzed reaction.

[1]Wuts, P.G.M., Goble, M.P. *OL* **2**, 2139 (2000).
[2]Tuncay, A., Anaclerio, B.M., Zolodz, M., Suslick, K.S. *TL* **40**, 599 (1999).

Hypofluorous acid–acetonitrile. 18, 188; **19,** 170; **20,** 193

N-Oxides.[1] This complex is a new reagent for oxidation of tertiary amines.

[1] Dayan, S., Kol, M., Rozen, S. *S* 1427 (1999).

Hypophosphorous acid.

Radical reactions.[1] The acid mediates C—C bond formation in a similar way to Bu_3SnH but without the many problems associated with the tin reagent.

[1]Graham, S.R., Murphy, J.A., Coates, D. *TL* **40**, 2415 (1999).

Hypophosphorous acid–iodine.

Reductions.[1] Diaryl ketones are deoxygenated by this reagent system while benzaldehyde and aliphatic ketones are not affected. Chalcone undergoes saturation at the double bond only.

[1]Hicks, L.D., Han, J.K., Fry, A.J. *TL* **41**, 7817 (2000).

I

Imidazole. 20, 194

Thioethers.[1] Alkylation of thiols with alcohols under the Mitsunobu reaction conditions is catalyzed by imidazole.

[1]Falck, J.R., Lai, J.-Y., Cho, S.-D., Yu, J. *TL* **40**, 2903 (1999).

2-Imidazolylidene–palladium complexes.

Couplings. Both bis(imidazolylidene)[1] and mixed imidazolylidene–phsophine complexes[2] of palladium (**1, 2**) are good catalysts for promoting Suzuki and Stille coupling reactions.

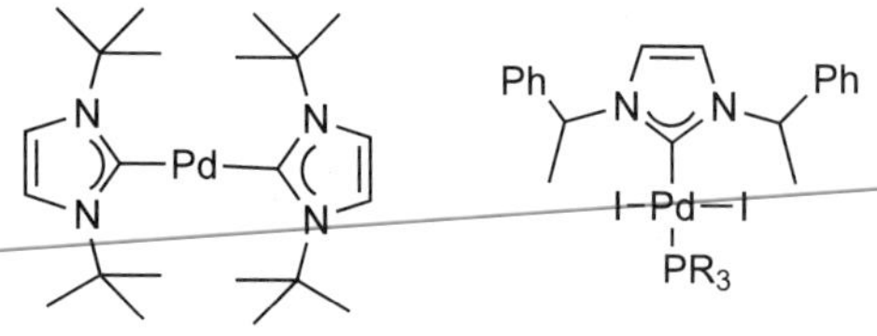

(**1**) (**2**)

[1]Bohm, V.P.W., Gstottmayr, C.W.K., Westkamp, T., Herrmann, W.A. *JOMC* **595**, 186 (2000).
[2]Westkamp, T., Bohm, V.P.W., Herrmann, W.A. *JOMC* **585**, 348 (1999).

Indium. 14, 81; **16,** 181–182; **18,** 189; **19,** 171–173; **20,** 194–197

Reductions. Reports of rather simple reductions by indium keep appearing: of α-halocarbonyl compounds in water with ultrasound assistance,[1] of *p*-nitrobenzyl ethers and esters,[2] disulfides,[3] amine *N*-oxides,[4] and organic azides[5] in alcoholic solvents in the presence of NH_4Cl. When the azide reduction is carried out in DMF the products can be directly converted to carbamates by having a chloroformate ester in the reaction vessel.[6]

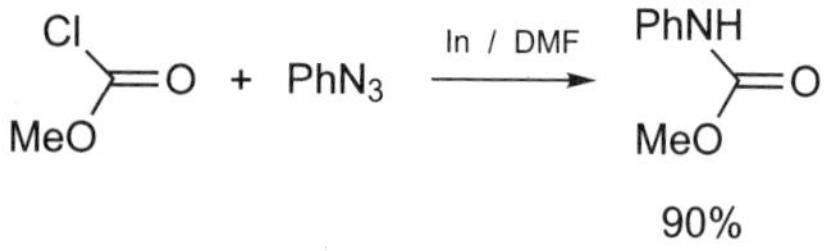

$$\text{ClCO}_2\text{Me} + \text{PhN}_3 \xrightarrow{\text{In / DMF}} \text{PhNHCO}_2\text{Me} \quad 90\%$$

In the presence of an electron acceptor (e.g., 2-bromo-2-nitropropane), 2-nitroaroyl compounds are reduced by indium to [*c*]benzoisoxazoles.[7]

Aroyl cyanides undergo reductive coupling to furnish α-diketones in moderate yields.[8]

Allylation of carbonyl compounds. The chemoselectivity of allylindium reagents toward aldehydes enables elaboration of the steroid side chain without affecting the enone system in ring A.[9] Allylation is also viable using a catalytic amount of indium and Mn–Me_3SiCl in THF.[10]

Fluorinated homoallylic alcohols are readily prepared from fluorine-containing carbonyl compounds[11] or the allylic halides.[12,13]

F F Br + H O → In / DMF,)))) → F F OH

57%

Allylation of acylsilanes[14] and α-ketoesters[15] proceeds normally. Monoallylation of α-diketones[16] is also easily realized, whereas glyoxal *N,N*-monohydrazone gives 1,7-octadiene-4,5-diol. However, sequential reactions of the glyoxal monohydrazone with RLi and then allylindium reagent lead to unsymmetrical diols.[17]

Carbonyl compounds in certain protected forms such as *gem*-diacetates[18] and hemiacetals (e.g., trifluoroacetaldehyde ethyl hemiacetal[19]) also undergo allylation, giving in these cases homoallylic acetates and homoallyl alcohols, respectively.

The employment of functionalized allylic bromides serves to construct more elaborate compounds in one step.[20,21]

O OH Br + X → In / THF ; DCC / CH_2Cl_2 → O X

X = O, NPh

The organoindium reagent derived from 5-bromo-1,3-pentadiene reacts at the central carbon atom,[22] whereas 1,4-dibromo-2-butyne behaves as a 1,3-butadien-2-ylating agent in the presence of indium in aqueous media.[23]

Br Br + O H R → In / H_2O → OH R

Different regioselectivities for the addition to azetidine-2,3-diones by organometallic reagents prepared from propargylic bromides and Zn or In are observed.[24] Allenyl derivatives are produced exclusively with In in a mixture of THF and saturated NH_4Cl.

4-Allyl-2-azetidinones are synthesized by displacement of the corresponding acetoxy derivatives.[25] A preparation of homoallylic amines can take advantage of the reactivity of allylindium reagents on iminium species that are formed in situ.[26]

π-Allylpalladium species from the Pd-catalyzed coupling of ArI with allene undergo Pd/In exchange and subsequent reaction with aldehydes, enabling a three-component assemblage of homoallylic alcohols.[27] The reaction can be further extended with an alkyne relay.[28]

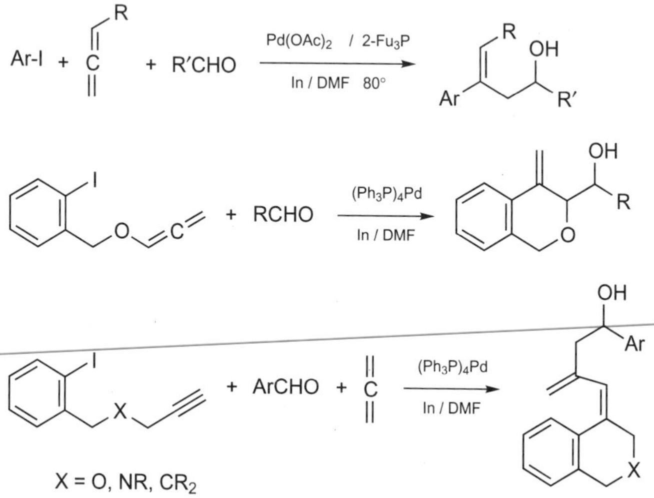

Ar-I + C + R′CHO → [$Pd(OAc)_2$ / 2-Fu_3P; In / DMF 80°]

+ RCHO → [$(Ph_3P)_4Pd$; In / DMF]

+ ArCHO + C → [$(Ph_3P)_4Pd$; In / DMF]

X = O, NR, CR_2

1,4-Dienes.[29] Allylation of functionalized hydroxylated alkynes leads to 1,4-dienes. Regioselectivity depends on the distance between the hydroxyl group and the triple bond: Propargylic alcohols give linear products due to chelation control, but branched 1,4-dienes are obtained from 4-pentynol and higher homologues. When quenched with an *N*-halosuccinimide, halogen-containing products are obtained.

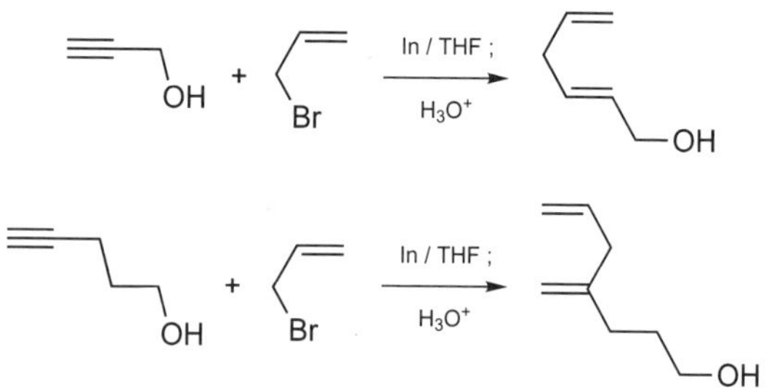

→ [In / THF ; H_3O^+]

→ [In / THF ; H_3O^+]

Allylarenes.[30] *C*-Allylation of arenes is catalyzed by indium (0.1 equiv). Calcium carbonate and molecular sieves 4Å are also present in the reaction system.

[1]Ranu, B.C., Dutta, P., Sarkar, A. *JCS(P1)* 1139 (1999).
[2]Moody, C. J., Pitts, M. R. *SL* 1575 (1999).
[3]Reddy, G.V.S., Rao, G.V., Iyengar, D.S. *SC* **30**, 859 (2000).
[4]Yadav, J.S., Reddy, B.V.S., Reddy, M.M. *TL* **41**, 2663 (2000).
[5]Reddy, G.V.S., Rao, G.V., Iyengar, D.S. *TL* **40**, 3937 (1999).
[6]Yadav, J.S., Reddy, B.V.S., Reddy, G.S.K.K. *NJC* **24**, 571 (2000).
[7]Kim, B.H., Jin, Y., Jun, Y.M., Han, R., Baik, W., Lee, B.M. *TL* **41**, 2137 (2000).
[8]Baek, H.S., Lee, S.J., Yoo, B.W., Ko, J.J., Kim, S.H., Kim, J.H. *TL* **41**, 8097 (2000).
[9]Loh, T.P., Hu, Q.-Y., Vittal, J.J. *SL* 523 (2000).
[10]Auge, J., Lubin-Germain, N., Thiaw-Woaye, A. *TL* **40**, 9245 (1999).
[11]Loh, T.-P., Zhou, J.-R., Li, X.-R. *TL* **40**, 9333 (1999).
[12]Loh, T.-P., Zhou, J.-R., Li, X.-R. *EJOC* 1893 (1999).
[13]Percy, J.M., Pintat, S. *CC* 607 (2000).
[14]Bonini, B.F., Comes-Franchini, M., Fochi, M., Laboroi, F., Mazzanti, G., Ricci, A., Varchi, G. *JOC* **64**, 8008 (1999).
[15]Loh, T.P., Huang, J.-M., Xu, K.-C., Goh, S.-H., Vittal, J.J. *TL* **41**, 6511 (2000).
[16]Nair, V., Jayan, C.N. *TL* **41**, 1091 (2000).
[17]Cere, V., Peri, F., Pollicino, S., Ricci, A. *SL* 1585 (1999).
[18]Yadav, J.S., Reddy, B.V.S., Reddy, G.S.K.K. *TL* **41**, 3376 (2000).
[19]Loh, T.-P., Li, X.-R. *T* **55**, 5611 (1999).
[20]Choudhury, P.K., Foubelo, F., Yus, M. *T* **55** 10779 (1999); *JOC* **64**, 8008 (1999).
[21]Paquette, L.A., Kern, B.E., Mendez-Andino, J. *TL* **40**, 4129 (1999).
[22]Melekhov, A., Fallis, A.G. *TL* **40**, 7867 (1999).
[23]Lu, W., Ma, J., Yang, Y., Chan, T.H. *OL* **2**, 3469 (2000).
[24]Alcaide, B., Almendros, P., Aragoncillo, C. *OL* **2**, 1411 (2000).
[25]Kang, S.K., Baik, T.-G., Jiao, X.-H., Lee, K.-K., Lee, C.H. *SL* 447 (1999).
[26]Choucair, B., Leon, H., Mire, M.-A., Lebreton, C., Mosset, P. *OL* **2**, 1851 (2000).
[27]Anwar, U., Grigg, R., Rasparini, M., Savic, V., Sridharan, V.*CC* 645 (2000).
[28]Anwar, U., Grigg, R., Sridharan, V.*CC* 933 (2000).
[29]Klaps, E., Schmid, W. *JOC* **64**, 7537 (1999).
[30]Lim, H.J., Keum, G., Kang, S.B., Kim, Y., Chung, B.Y. *TL* **40**, 1547 (1999).

Indium–Indium(III) chloride.

5-Aryl-4-penten-2-ones.[1] Reductive homoaldol reactions are effected between methyl vinyl ketone and ArCHO by In powder (200 mol%) and $InCl_3$ in aq THF.

[1]Kang, S., Jang, T.-S., Keum, G., Kang, S.B., Han, S.-Y., Kim, Y. *OL* **2**, 3615 (2000).

Indium(I) bromide.

Bromocyanomethylation.[1] The organoindium species that are formed by insertion of InBr into a C—Br bond of dibromoacetonitrile reacts with carbonyl compounds. Generation of the *anti*-bromohydrin is favored in the case of a bulky aliphatic aldehyde. Such products serve as precursors of glycidonitriles.

R, R′ C=O + Br(Br)CH–CN —InBr ; H_3O^+→ R, R′ C(OH)–CH(Br)–CN

[1]Nobrega, J.A., Goncalves, S.M.C., Peppe, C. *TL* **41**, 5779 (2000).

Indium(III) chloride. 19, 173–174; **20,** 197–198

Aldol reactions. Aldol products are obtained in good yields from reaction of ketones with glyoxylic acid monohydrate with assistance of ultrasound irradiation.[1] Substrate-control (by 1,3- + 1,5-asymmetric induction) of the aldol reaction involving γ-amino-α-ketoesters under solvent-free conditions is very effective.[2] With lithium dicyclohexylamide and $InCl_3$ the reaction of esters with aldehydes furnishes β-hydroxy esters, and that of α-bromo esters affords α,β-epoxy esters.[3] These are typical Reformatsky and Darzens reaction products, respectively.

Br–CH(CH_3)–COOMe —Cy_2NLi - $InCl_3$ / PhCHO - THF→ Ph-epoxide-COOMe 80%

Reductions. The $InCl_3$–Bu_3SnH/Ph_3P combination (catalytic in indium) reduces acid chlorides to aldehydes.[4] Aryl ketones and secondary benzyl alcohols are deoxygenated with hydrosilanes using $InCl_3$ as catalyst.[5] Deoxygenative allylation of aryl ketones occurs when allyltrimethylsilane is added.[6]

PhC(=O)Et + Me_3Si–allyl —$InCl_3$ / $Me_2Si(Cl)H$ / CH_2Cl_2→ PhCH(Et)CH_2CH=CH_2 67%

Propargylation.[7] An approach to epoxyalkynes involves indium-mediated reaction of α-chloropropargyl sulfides with carbonyl compounds. Interestingly, opposite stereoselectivity is shown by a reaction using In and $InCl_3$.

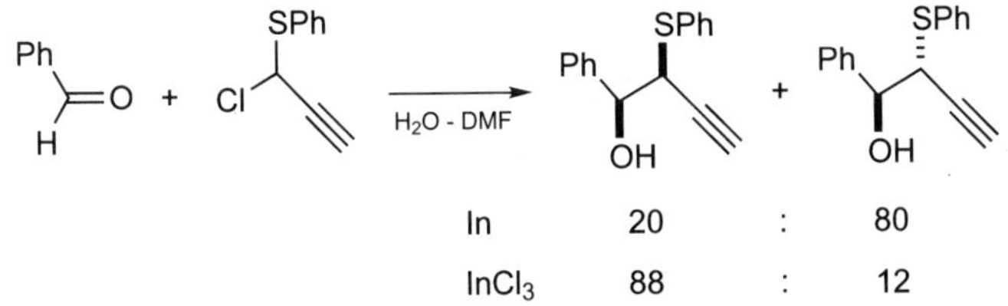

In	20	:	80
$InCl_3$	88	:	12

S_N2' Displacement. Δ^2-Glycosides of both the normal and carbon-series are produced from acetylglycals (Ferrier rearrangement) by $InCl_3$-catalyzed glycosylation with proper nucleophiles.[8,9]

Cyclization reactions. 4-Chlorotetrahydropyrans and -thiopyrans are formed from unsaturated alcohols and thiols on condensation with carbonyl compounds, which patterns after the Prins reaction.[10–12] Chlorinated oxepanes are similarly accessible.[13]

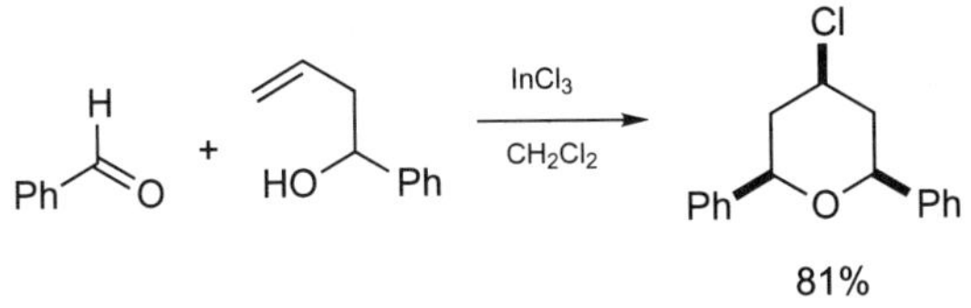

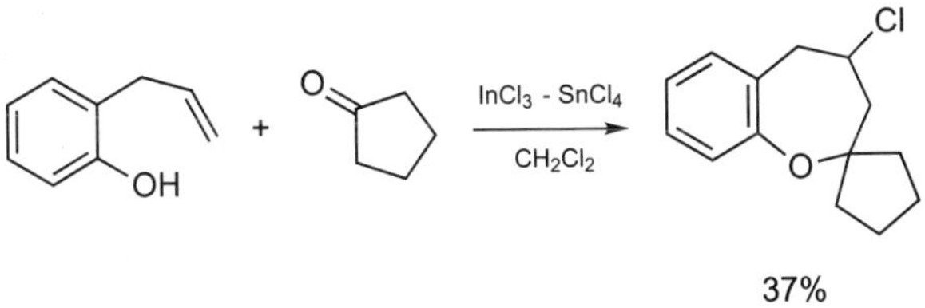

81%

37%

A one-step synthesis of α-amino-γ-lactones from arylamines, alkenes, and methyl glyoxylate is mediated by $InCl_3$ or $Sc(OTf)_3$.[14]

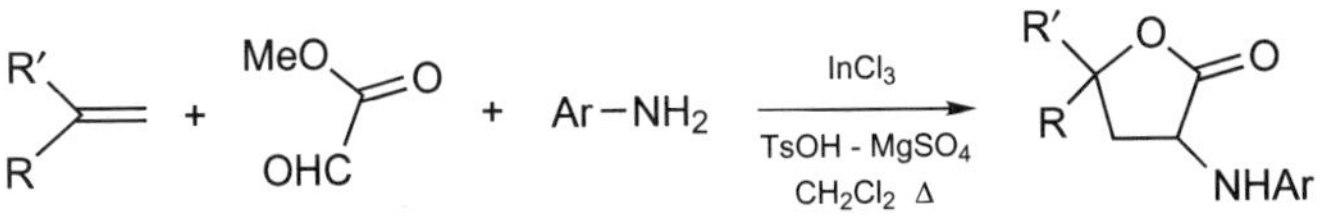

Nitrogen heterocycles. Aziridines,[15] oxazoles,[16] and quinolines[17] are formed from various $InCl_3$-catalyzed cycloadditions and cyclization reactions. The pyrroloquinoline skeleton, characteristic of several alkaloids, can be assembled in one step.[18]

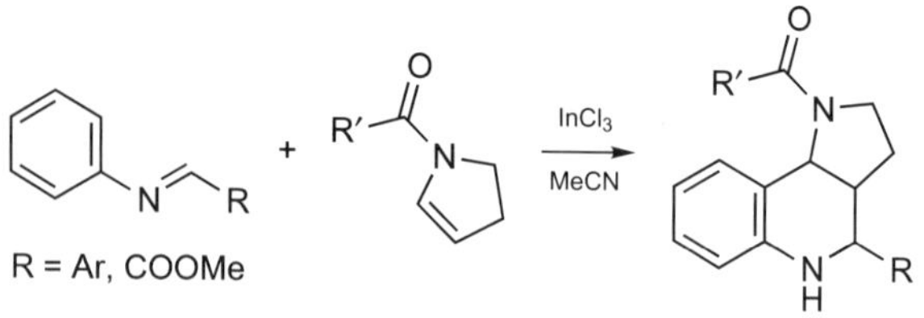

Diels–Alder reaction.[19] The dienophilic reactivity of ethyleneacetals of cycloalkenones is enhanced by $InCl_3$ through complexation.

[1]Loh, T.-P., Wei, L.-L., Feng, L.-C. *SL* 1059 (1999).
[2]Loh, T.P., Huang, J.-M., Goh, S.-H., Vittal, J.J. *OL* **2**, 1291 (2000).
[3]Hirashita, T., Kinoshita, K., Yamamura, H., Kawai, M., Araki, S. *JCS(P1)* 825 (2000).
[4]Inoue, K., Yasuda, M., Shibata, I., Baba, A. *TL* **41**, 113 (2000).
[5]Miyai, T., Ueba, M., Baba, A. *SL* 182 (1999).
[6]Yasuda, M., Onishi, Y., Ito, T., Baba, A. *TL* **41**, 2425 (2000).
[7]Engstrom, G., Morelli, M., Palomo, C., Mitzel, T. *TL* **40**, 5967 (1999).
[8]Babu, B.S., Balasubramanian, K.K. *TL* **41**, 1271 (2000).
[9]Ghosh, R., De, D., Shown, B., Maiti, S.B. *CR* **321**, 1 (1999).
[10]Yang, J., Viswanathan, G.S., Li, C.-J. *TL* **40**, 1627 (1999).
[11]Yang, J., Li, C.-J. *SL* 717 (1999).
[12]Yang, X.-F., G.S., Li, C.-J. *TL* **41**, 1321 (2000).
[13]Li, J., Li, C.-J. *H* **53**, 1691 (2000).
[14]Huang, T., Li, C.-J. *TL* **41**, 9747 (2000).
[15]Sengupta, S., Mondal, S. *TL* **41**, 6245 (2000).
[16]Sengupta, S., Mondal, S. *TL* **40**, 8685 (1999).
[17]Ranu, B.C., Hajra, A., Jana, U. *TL* **41**, 531 (2000).
[18]Hadden, M., Stevenson, P.J. *TL* **40**, 1215 (1999).
[19]Reddy, B.G., Kumareswaran, R., Vankar, Y.D. *TL* **41**, 10333 (2000).

Indium(III) hydride.

Reductions.[1] The $InH_3(PCy_3)_3$ complex reduces ketones and epoxides to alcohols. Enones and α-bromo ketones afford allylic alcohols and bromohydrins, respectively. Esters are not reduced.

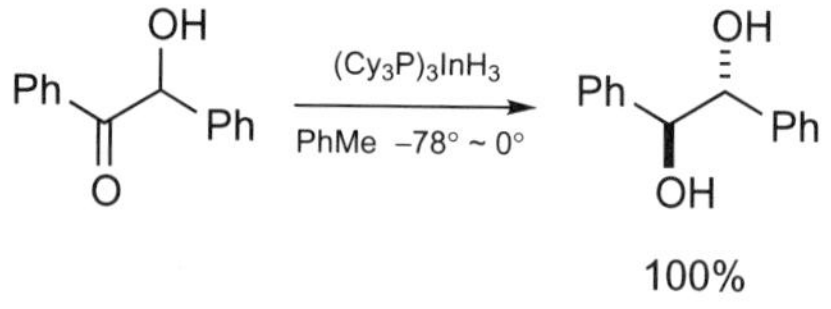

[1]Abernethy, C.D., Cole, M.L., Davies, A.J., Jones, C. *TL* **41**, 7567 (2000).

Indium(I) iodide.

Allylation and propargylation reactions. In the presence of the $(Ph_3P)_4Pd$ catalyst, indium(I) iodide form organoindium(III) reagents with allylic halides[1] and propargylic mesylates[2] that react with carbonyl compounds in various solvents.

Homoallyl alcohols.[2] Indium(III) iodide catalyzes allylation of carbonyl compounds by allylstannanes.

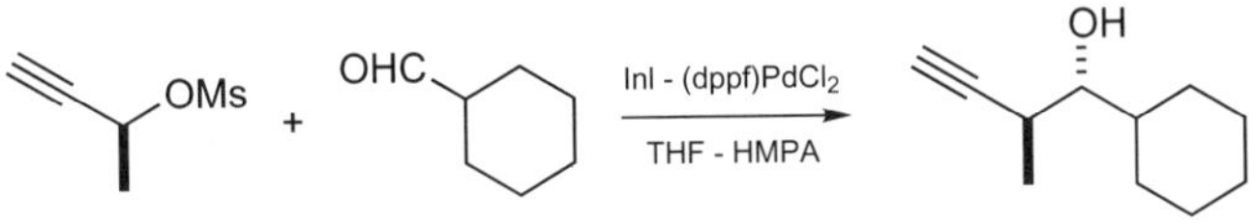

OMs + OHC — InI - (dppf)$PdCl_2$ / THF - HMPA → OH

76% (*anti* : *syn* 95 : 5)

[1]Araki, S., Kamei, T., Hirashita, T., Yamamura, H., Kawai, M. *OL* **2**, 847 (2000).
[2]Marshall, J.A., Grant, C.M. *JOC* **64**, 8215 (1999).

Indium(III) iodide.

Transacylation.[1] Alcohols and amines are acylated with an ester using InI_3 as catalyst. Selective acylation of primary alcohols (vs. secondary alcohols and phenols) and amines (vs. secondary amines and primary alcohols) is possible.

[1]Ranu, B.C., Dutta, P., Sarkar, A. *JCS(P1)* 2223 (2000).

Indium(III) triflate.

Acylation.[1] Catalyzed by this salt, alcohols and amines are acylated with acetic anhydride.

Hetero-Diels–Alder reaction.[2] The condensation of Danishefsky's diene with imines to generate *N*-substituted 2-aryl-2,3-dihydro-4-pyridones at room temperature is effected by this salt. A mixture of aldehydes, amines, and the dehydrant $MgSO_4$ can be employed instead of imines.

[1]Chauhan, K.K., Frost, C.C., Love, I., Waite, D. *SL* 1743 (1999).
[2]Ali, T., Chauhan, K.K., Frost, C.C. *TL* **40**, 5621 (1999).

Iodine. 13, 148–149; **14,** 181–182; **15,** 172–173; **16,** 182; **18,** 189–191; **19,** 174–175; **20,** 199–200

Iodohydrins. Iodohydrins are formed by opening of epoxides using either iodine–manganese(salen) complexes[1] or anhydrous HI, which is conveniently generated from iodine and AcSH.[2]

Protection and deprotection of functional groups. Alcohols and their tetrahydropyranyl ethers are interconverted with iodine as catalyst.[3,4] Etherfication is done in a nonhydroxylic solvent, whereas the ether cleavage is achieved in MeOH.

Regeneration of carbonyl compounds from oximes is accomplished by heating with iodine in MeCN,[5] and a mild workup of the products from reduction of amides with borane involves treatment with iodine.[6]

Iodinations. Arenes undergo iodination with iodine and an oxidant (e.g., potassium permanganate or manganese dioxide,[7] sodium iodate or periodate,[8] or tetrabutyl-ammonium peroxydisulfate[9]).

One method for the α-iodination of ketones calls for the assistance of selenium dioxide.[10]

Heterocyclization. A new route to α-iodo boronic acids consists of iodoetherification.[11] β-Iodobutenolides that are useful for carbon chain extension at the β-position (via Pd-catalyzed coupling) are readily accessible by iodolactonization of allenic acids[12]

R, =C=, COOH; I_2 - THF ; K_2CO_3; I, R, O, O

Unusual results are obtained from the treatment of *O*-(3-cyclohexenyl)thiocarbamidate esters with iodine.[13]

O, SMe, EtN; I_2 - THF ; Na_2SO_3; I, I, Et−N, SMe, O

69%

[1]Sharghi, H., Naeimi, H. *BCSJ* **72**, 1525 (1999); *JCR(S)* 310 (1999).
[2]Chervin, S.M., Abada, P., Koreeda, M. *OL* **2**, 369 (2000).
[3]Kumar, H.M.S., Reddy, B.V.S., Reddy, E.J., Yadav, J.S. *CL* 857 (1999).
[4]Ramasamy, K.S., Bandaru, R., Averett, D. *SC* **29**, 2881 (1999).
[5]Yadav, J.S., Sasmal, P.K., Chand, P.K. *SC* **29**, 3667 (1999).
[6]Hall, D.G., Laplante, C., Manku, S., Nagendran, J. *JOC* **64**, 698 (1999).
[7]Lulinski, P., Skulski, L. *BCSJ* **72**, 115 (1999).
[8]Lulinski, P., Skulski, L. *BCSJ* **73**, 951 (2000).
[9]Yang, S.G., Kim, Y.H. *TL* **40**, 6051 (1999).
[10]Bekaert, A., Barberan, O., Gervais, M., Brion, J.-D. *TL* **41**, 2903 (2000).
[11]Baba, S., Leroy, F., Le Goaster, C., Jegou, A., Carboni, B. *SC* **29**, 1183 (1999).
[12]Ma, S., Shi, Z., Yu, Z. *TL* **40**, 2393 (1999).
[13]Sabate, M., Llebaria, A., Molins, E., Miravitlles, C., Delgado, A. *JOC* **65**, 4826 (2000).

Iodine(I) bromide. 20, 204

Cleavage of TBS ethers.[1] Many other protecting groups are not affected when *t*-butyldimethylsilyl ethers are cleaved with IBr in MeOH.

Iodoetherification.[2] Regioselective hydration of a homopropargyl alcohol is accomplished through iodolactonization of its carbonate with subsequent deiodination and hydrolysis. This procedure is valuable for preserving epimerizable stereocenters.

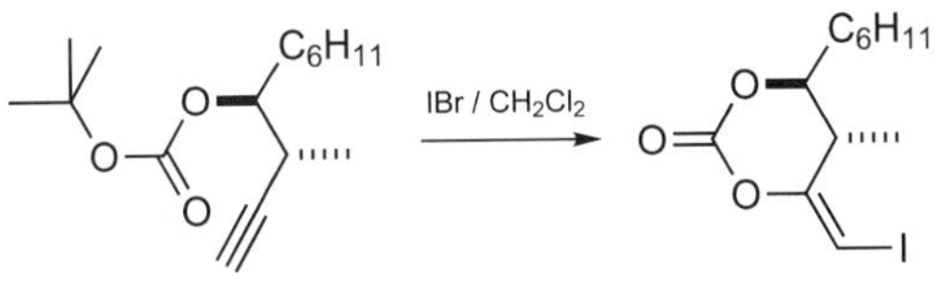

IBr / CH_2Cl_2

[1]Kartha, K. P. R, Field, R. A. *SL* 311 (1999).
[2]Marshall, J. A., Yanik, M.M. *JOC* **64**, 3798 (1999).

Iodine(I) chloride. 20, 200

Iododesilylation.[1] The silyl group of an α-silyl enone is replaced by iodine on reaction with 2 equiv of ICl and 1 equiv each of ICl and $AlCl_3$.

Iodination. The stereoselective formation of 1,2-diiodoalkenes[2] from alkynes on reaction of ICl and then NaI is instrumental to a route to highly substituted alkenes.

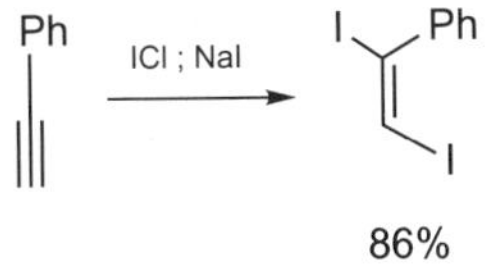

ICl ; NaI

86%

The ICl–$AgSO_4$ is a superactive iodinating agent that is capable of attacking nitrobenzene to form *m*-iodonitrobenzene (74% yield) at room temperature in 5 min.[3]

[1]Alimardanov, A., Negishi, E. *TL* **40**, 3839 (1999).
[2]Henaff, N., Whiting, A. *JCS(P1)* 395 (2000).
[3]Chaikovski, V.K., Kharlova, T.S., Filimonov, V.D., Saryucheva, T.A. *S* 748 (1999).

Iodomethyl pivalate.

Hydroxymethyl anion.[1] The title compound undergoes I–Mg exchange with *i*-PrMgCl in *N*-butylpyrrolidinone–THF to afford a useful nucleophilic reagent.

[1]Avolio, S., Malan, C., Marek, I., Knochel, P. *SL* 1820 (1999).

Iodomethylzinc 2,4,6-trichlorophenoxide.

Cyclopropanation.[1] The title compound, prepared from ArOH, Et_2Zn, and CH_2I_2, is a modified Simmons–Smith reagent, with which alkenes are transformed into cyclopropanes in excellent yields (6 examples, 90–98%). In terms of reactivity, the zinc phenoxide is comparable to bis(chloromethyl)zinc, but more reactive than bis(iodomethyl)zinc and Furukawa's reagent.

[1]Charette, A.B., Francoeur, S., Martel, J., Wilb, N. *ACIEE* **39**, 4539 (2000).

***N*-Iodosaccharin.**

Iodinations.[1] This reagent is prepared from iodine and the silver salt of saccaharin. It is effective for iodination of alkenes and activated arenes. Iodohydrins are formed when alkenes are treated with *N*-iodosaccharin in aqueous acetone.

[1]Dolenc, D. *SL* 544 (2000).

***N*-Iodosuccinimide, NIS. 16,** 185–186; **18,** 193–194; **19,** 177–178

Iodohydrins.[1] An improved synthesis of iodohydrins from alkenes involves the use of NIS in aq DME.

Cleavage of benzyl ethers. Selectivity is observed, as exemplified in the following.[2]

HO BnO BnO BnO O OMe —NIS / $MeNO_2$→ Ph O O O BnO BnO OMe 84%

Radical cyclization.[3] Medium-sized rings are formed by NIS-induced cyclization.

n =CHOMe OH —NIS / CH_2Cl_2 ; Bu_3SnH - Et_3B→ n OMe O 59%

[1]Smietana, M., Gouverneur, V., Mioskowski, C. *TL* **41**, 193 (2000).
[2]Madsen, J., Viuf, C., Bols, M. *CEJ* **6**, 1140 (2000).
[3]Nagano, H., Tada, A., Isobe, Y., Yajima, T. *SL*1193 (2000).

Iodosylbenzene. 13, 151; **16,** 186; **18,** 194; **19,** 178; **20**, 201

Oxidations. By using PhI═O (in presence of KBr) as an oxidant, alcohols are oxidized to acids and ketones in water in excellent yields.[1] When catalyzed by either poly(4-vinylpyridine)-supported sodium ruthenate[2] or a (salen)chromium complex[3] chemoselective oxidation of alcohols (e.g., allylic alcohols to alkenoic acids) occurs, which is contrary to the effect of (salen)manganese and (porphyrin)iron complexes (giving epoxy alcohols).[4]

Oxidative decarboxylation.[5] α-Amino acid derivatives undergo degradation by PhI═O/I_2.

PhI=O - I_2, CH_2Cl_2; COOH; HN; CHO; 82%

Glycosylation.[6] Oxidation of phenylthioglycosides with PhI═O in the presence of Me_3SiOTf furnishes activated glycosyl donors. Stereoselective glycosylation is achieved when such a thioglycoside is clamped to another sugar component.

PhI=O; Me_3SiOTf, MS4A CH_2Cl_2; BnO; SPh; HO; OMe; 86%

[1] Tohma, H., Takizawa, S., Maegawa, T., Kita, Y. *ACIEE* **39**, 1306 (2000).
[2] Friedrich, H.B., Singh, N. *TL* **41**, 3971 (2000).
[3] Adam, W., Gelalcha, F.G., Saha-Möller, C.R., Stegmann, V.R. *JOC* **65**, 1915 (2000).
[4] Adam, W., Stegmann, V.R., Saha-Möller, C.R. *JACS* **121**, 1879 (1999).
[5] Boto, A., Hernandez, R., Suarez, E. *TL* **40**, 5945 (1999).
[6] Wakao, M., Fukase, K., Kusumoto, S. *SL* 1912 (1999).

***p*-Iodotoluene difluoride. 19,** 178

Fluorocyclization.[1] This reagent promotes cyclization of unsaturated alcohols with HF–pyridine.

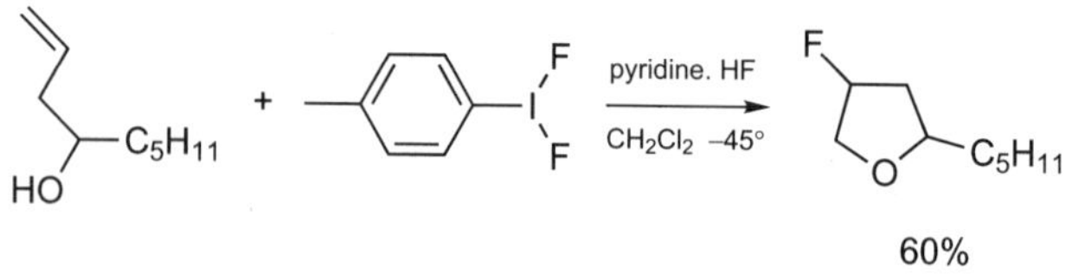

HO C5H11 + pyridine. HF / CH2Cl2 –45° → F C5H11 60%

[1]Sawaguchi, M., Hara, S., Fukuhara, T., Yoneda, N. *JFC* **104,** 277 (2000).

o-Iodoxybenzoic acid. 19, 179

Conjugated carbonyl compounds.[1] The title reagent converts saturated alcohols or carbonyl compounds to the conjugated carbonyl compounds.

OH + → THF - DMSO 74%

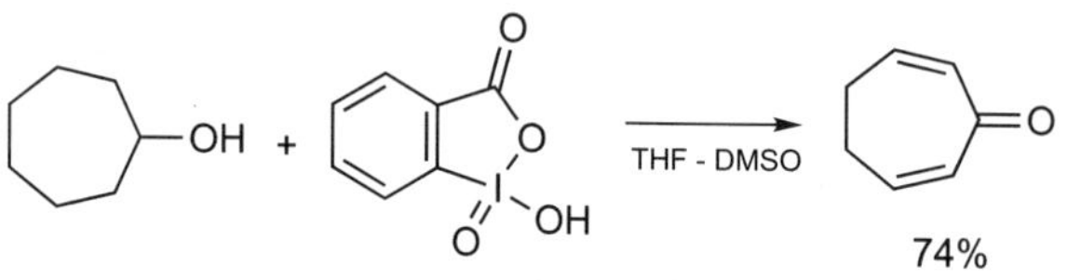

Heterocyclization.[2] Cyclization of allylic carbamates to oxazolidin-2-ones via oxidative radical cyclization. Hydrogen comes from THF.

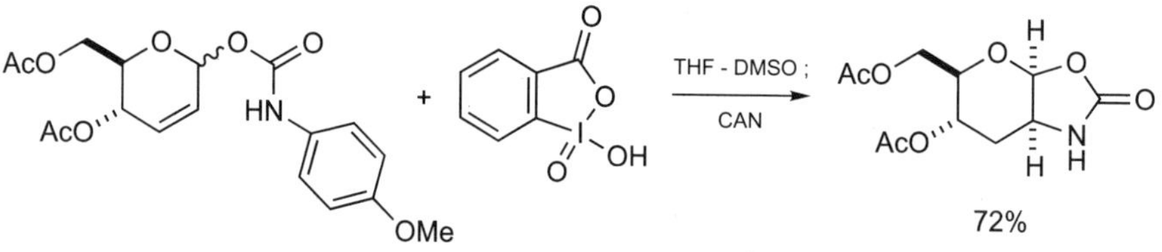

AcO AcO HN OMe + → THF - DMSO ; CAN 72%

Epoxidation.[3] The tetrabutylammonium salt, prepared on admixture with Bu_4NF, transfers its iodine-bound oxygen atom to enones.

[1]Nicolaou, K.C., Zhong, Y.-L., Baran, P.S. *JACS* **122**, 7596 (2000).
[2]Nicolaou, K.C., Baran, P.S., Zhong, Y.-L., Vega, J.A. *ACIEE* **39**, 2525 (2000).
[3]Ochiai, M., Nakanishi, A., Suefuji, T. *OL* **2**, 2923 (2000).

Ion exchange resins.

Esterification. At room temperature, saturated carboxylic acids are selectively esterified with an alcohol in the presence of Amberlyst-15 resin.[1] Alkenoic and aroic acids are left unreacted. Transesterification of alkyl formate to a dicarboxylic acid forming the monocarboxylic ester is catalyzed by Dowex-50WX2.[2]

Cyclizations. An intramolecular S_N2' process to form an oxaspirocycle as mediated by Amberlyst 15 is applicable to the synthesis of theaspirone.[3] Under the influence of Dowex-50X2-200, coumarin derivatives are obtained from phenol and propynoic acid.[4]

HO OH OH + COOH Dowex 50X2-200 80° HO O O OH 98%

Desulfinylation.[5] Sulfinamides are methanolyzed to the corresponding amines in the presence of an ion exchange resin.

Deoxygenation of epoxy ketones. An alkali metal halide on Amberlyst 15 is capable of removing the epoxide group from epoxy ketones.[6] Interestingly, LiBr or NaBr, instead of LiI, also brominates the enones at the α-position.[7]

O O Amberlyst 15 LiI O Amberlyst 15 LiBr O Br

[1] Anand, R.C., Vimal, A.M. *JCR(S)* 378 (1999).
[2] Nishiguchi, T., Ishii, Y., Fujisaki, S. *JCS(P1)* 3023 (1999).
[3] Young, J., Jung, L., Cheng, K. *TL* **41**, 3411, 3415 (2000).
[4] de la Hoz, A., Moreno, A., Vazquez, E. *SL* 608 (1999).
[5] Li, G., Kim, S.H., Wei, H.-X. *T* **56**, 719 (2000).
[6] Bovicelli, P., Righi, G., Sperandio, A. *T* **56**, 1733 (2000).
[7] Bovicelli, P., Righi, G., Sperandio, A. *TL* **40**, 5889 (1999).

Iridium complexes.

Isomerization. Cationic iridium complex effects selective isomerization of unsymmetrical diallyl ethers[1] and conjugated boronates containing an allylic ether group, including an access to γ-(siloxy)allylboronic esters.[2] The conversion of allyl homoallyl ethers to γ,δ-unsaturated carbonyl compounds[3] is promoted by $[(cod)IrCl]_2$.

[(cod)IrCl]$_2$

Cy_3P - Cs_2CO_3

[(Ph$_2$MeP)$_2$Ir(cod)]PF$_6$

H_2 / THF

78%

[(Ph$_2$MeP)$_2$Ir(cod)]PF$_6$

H_2 / THF

68%

(Tetraphenylporphyrin)iridium(III) triflate isomerizes terminal epoxides to aldehydes in refluxing dioxane.[4]

1,3-Dioxolanes.[5] Epoxides and ketones combine under the influence of [Cp*Ir-(MeCN)$_3$](PF$_6$)$_2$.

Aldol reaction.[6] A new catalyst for the Mukaiyama version of an aldol reaction is [Ir(cod)(PPh$_3$)$_2$]OTf. Actually, after activation by hydrogen, it promotes a Michael reaction of enones with silyl enol ethers and the system can be modified to continue an aldol reaction.

Reduction. Certain success has been achieved in transfer hydrogenation of ketones (from HCOOH, Et$_3$N) using [Ir(cod)Cl]$_2$ and a chiral ligand.[7]

S_N2' displacement.[8] Employment of an iridium complex instead of the more commonly used palladium catalysts successfully mediates regioselective allylic displacement.

OAc + Na^+ EtOOC$^-$COOEt

[(cod)IrCl]$_2$

(PhO)$_3$P / THF

EtOOC COOEt

94%

Friedel–Crafts alkylation.[9] A straight-chain alkyl group can be introduced into an aromatic nucleus by reaction with an alkene, using the binuclear Ir(III) complex, [Ir(μ-acac-O,O,C^3)-(acac-O,O)(acac-C^3)]$_2$, as catalyst. The result is remarkably different from the commonly known reaction pathway.

[1] Yamamoto, Y., Fujikawa, R., Miyaura, N. *SC* **30**, 2383 (2000).
[2] Yamamoto, Y., Miyairi, T., Ohmura, T., Miyaura, N. *JOC* **64**, 296 (2000).
[3] Higashino, T., Sakaguchi, S., Ishii, Y. *OL* **2**, 4193 (2000).

[4]Suda, K., Baba, K., Nakajima, S. *TL* **40**, 7243 (1999).
[5]Adams, R.D., Barnard, T.S., Brosius, K. *JOMC* **582**, 358 (1999).
[6]Matsuda, I., Hasegawa, Y., Makino, T., Itoh, K. *TL* **41**, 1405, 1409 (2000).
[7]Petra, D.G.I., Kamer, P.C.J., Spek, A.L., Schoemaker, H.E., van Leeuwen, P.W.N.M. *JOC* **65**, 3010 (2000).
[8]Takeuchi, R., Tanabe, K. *ACIEE* **39**, 1975 (2000).
[9]Matsumoto, T., Taube, D.J., Periana, R.A., Taube, H., Yoshida, H. *JACS* **122**, 7414 (2000).

Iron. 19, 179–180; **20,** 203–204

γ-Lactones.[1] In the presence of CuCl, iron mediates a reductive cyclization of halogenated esters.

Cl COOMe Cl Fe - CuCl H_2O - MeCN 140° O O + COOMe 50% 50%

[1]Somech, I., Shvo, Y. *JOMC* **601**, 153 (2000).

Iron(II) chloride. 20, 204

Chloroamination.[1] Decomposition of properly constituted unsaturated acyl azides by iron(II) chloride leads to lactams.

O O N_3 R $FeCl_2$ - Me_3SiCl EtOH 0 ~ 25° O O NH R Cl + O O NH R Cl (9 : 1)

Alkylative rearrangement.[2] Sulfonium ylide formation from allylic sulfides and diazo compounds (particularly trimethylsilyldiazomethane) and subsequent [2.3]sigmatropic rearrangement are induced by (dppe)$FeCl_2$ in 1,2-dichloroethane.

R′ R SR″ + $SiMe_3$ N_2 (dppe)$FeCl_2$ $ClCH_2CH_2Cl$ Δ R′ R SR″ $SiMe_3$

[1]Bach, T., Schlummer, B., Harms, K. *CC* 287 (2000).
[2]Carter, D.S., Van Vranken, D.L. *OL* **2**, 1303 (2000).

Iron(III) chloride. 13, 133–134; **14,** 164–165; **15,** 158–159; **16,** 167–169, 190–191; **17,** 138–139; **18,** 197; **19,** 180–181; **20,** 204–205

Acetylation. Either alcohols[1] or methoxymethyl ethers[2] are transformed into acetates in the $FeCl_3$ -catalyzed acetylation, using HOAc and Ac_2O, respectively.

Michael reactions. 2-Alkylidene derivatives of 1,3-dicarbonyl compounds behave as donors in Michael reactions. Of particular interest is the formation of biaryl precursors with quinones.[3]

$FeCl_3.6H_2O$, CH_2Cl_2; HO; 39%

Benzils.[4] Benzoins are readily oxidized by $FeCl_3 \cdot 6H_2O$ without solvent at ~80°.

Cyclobutenediones and p-benzoquinones.[5] Iron carbonyl species are formed by reaction of the $FeCl_3$–$NaBH_4$ system with carbon monoxide. When alkynes are present, they are converted to cyclobutenediones or benzoquinones in situ.

$FeCl_3$ - $NaBH_4$, CO / HOAc, MeI / THF 25°; $FeCl_3$ - $NaBH_4$, CO / HOAc, THF 75°; (65 : 35)

Cyclization. The oxyallyl intermediate derived from treatment of a cross-conjugated dienone with $FeCl_3$ is trapped by an intramolecular [4 + 3]cycloaddition if one of the α-positions of the dienone is connected to a diene. Interesting tricyclic systems can be produced in this Nazarov cyclization.[6]

FeCl$_3$, CH$_2$Cl$_2$, −30°

R = Me 65%
R = Ph 72%

Cyclopentenones are formed by treatment of dienals bearing a silylmethyl substituent at the α-position.[7] Common Lewis acids other than FeCl$_3$ are not effective for mediating this cyclization.

SiMe$_3$, CHO, FeCl$_3$, CH$_2$Cl$_2$, −25°

R = Me 78%

β-Keto esters condense with conjugated oximes at 150°–160° to afford substituted pyridines in the presence of FeCl$_3$.[8]

***t*-Butylation.**[9] FeCl$_3$ is an effective catalyst for electrophilic aromatic substitution with *t*-butyl hydroperoxide.

[1]Sharma, G.V.M., Mahalingham, A.K., Nagarajan, M., Ilangovan, A., Radhakrishnan, B. *SL*1200 (1999).
[2]Bosch, M.P., Petschen, I., Guerrero, A. *S* 300 (2000).
[3]Christoffers, J., Mann, A. *EJOC* 2511 (1999).
[4]Zhou, Y.-M., Ye, X.-R., Xin, X.-Q. *SC* **29**, 2229 (1999).
[5]Rameshkumar, C., Periasamy, M. *OM* **19**, 2400 (2000).
[6]Wang, Y., Arif, A.M., West, F.G. *JACS* **121**, 876 (1999).
[7]Kuroda, C., Koshio, H. *CL* 962 (2000).
[8]Chibiryaev, A.M., De Kimpe, N., Tkachev, A.V. *TL* **41**, 8011 (2000).
[9]Liguori, L., Bjorsvik, H.-R., Fontana, F., Bosco, D., Galimberti, L., Minisci, F. *JOC* **64**, 8812 (1999).

Iron(III) nitrate. 20, 205–206

Dehydrogenation.[1] Hydrazides (ArNHNH)$_2$CO are dehydrogenated to afford ArN=NCONHNHAr by grinding with Fe(NO$_3$)$_3$.9H$_2$O.

Oxidative alcoholysis.[2] Acylsilanes are converted to esters in alcoholic solvent in the presence of iron(III) nitrate.

Oxidation.[3] Oxidation of alcohols to carbonyl compounds in solvent-free conditions is accomplished by mixing with Fe(NO$_3$)$_3$·9H$_2$O–HZSM-5 zeolite (in equivalent weight and crushed together) and with microwave assistance.

Radical cyclization. Internal trapping of the radical generated during cleavage of a siloxycyclopropane with $Fe(NO_3)_3$ leads to a new cyclic array.[4] Various radical terminators can be used to functionalize a remote position.

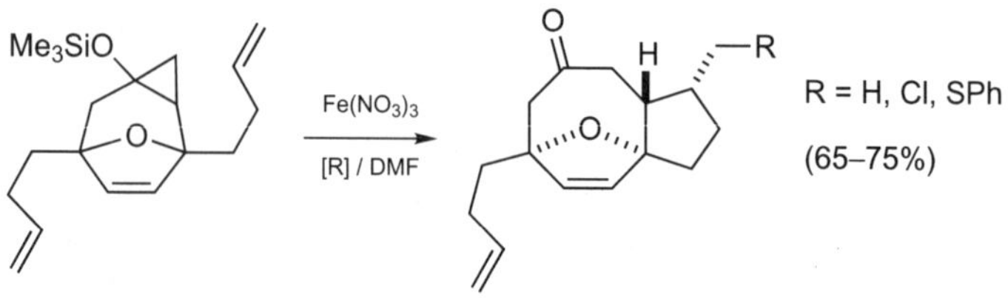

A dramatic change in the fate of the first radical intermediates is observed when $Cu(OAc)_2$ is also present in the reaction media.[5]

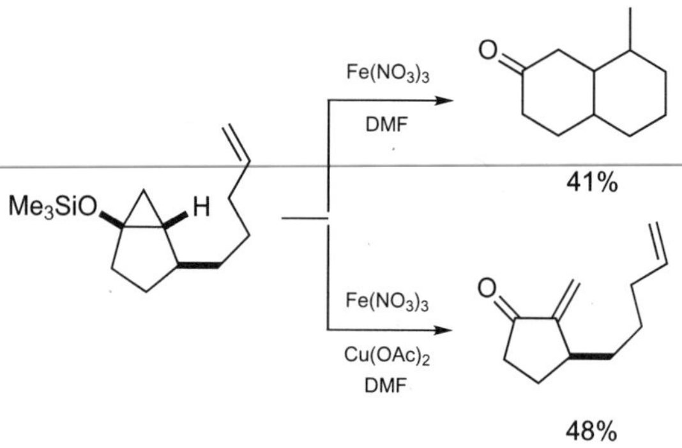

[1]Wang, H., Wang, Y.-L., Zhang, G., Li, J.-P., Wang, X.-Y. *SC* **30**, 1425 (2000).
[2]Patrocinio, A.F., Moran, P.J.S. *SC* **30**, 1419 (2000).
[3]Heravi, M.M., Ajami, D., Aghapoor, K., Ghassemzadeh, M. *CC* 833 (1999).
[4]Blake, A.J., Highton, A.J., Majid, T.N., Simpkins, N.S. *OL* **1,** 1787 (1999).
[5]Booker-Milburn, K.I., Cox, B., Grady, M., Halley, F., Marrison, S. *TL* **41**, 4651 (2000).

Iron pentacarbonyl. 13, 152; **18,** 196; **20,** 206–207

Reformatsky reaction.[1] The $Fe(CO)_5$–I_2 combination promotes Reformatsky reactions.

Cyclohexenones.[2] Carbonylation and cyclization of alkenylcyclopropanes leading to cyclohexenones with $Fe(CO)_5$ under carbon monoxide constitute an enantioselective process.

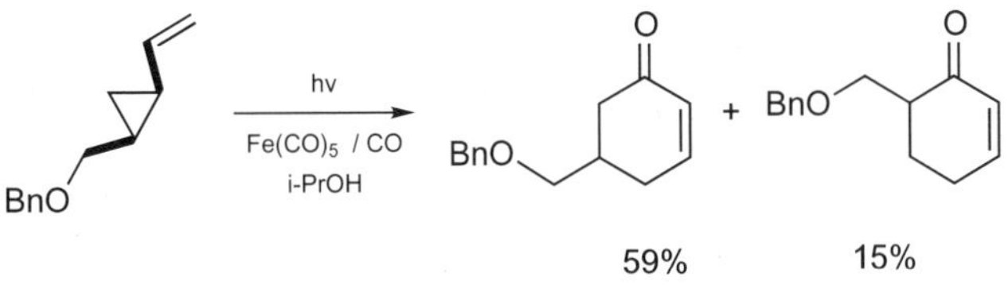

hv, $Fe(CO)_5$ / CO, i-PrOH

59% 15%

[1]Terent'ev, A.B., Vasil'eva, T.T., Kuz'mina, N.A., Mysov, E.I., Ikonnikov, N.S., Kuzne'sov, N.Yu., Belokon, Yu.N. *RCB* **48**, 1121 (1999).
[2]Taber, D.F., Kanai, K., Jiang, Q., Bui, G. *JACS* **122**, 6807 (2000).

Iron(III) perchlorate. 20, 207

Transetherification.[1] Allylic and benzylic ethers react with an alcohol to give different ethers.

Ritter reaction.[2] Iron(III) perchlorate supported on silica can be used to prepare amides from nitriles and benzylic alcohols.

[1]Salehi, P., Irandoost, M., Seddighi, B., Behbahani, F.K., Tahmasebi, D.P. *SC* **30**, 1743 (2000).
[2]Salehi, P., Motlagh, A.R. *SC* **30**, 671 (2000).

Iron(II) sulfate.

Acetalization.[1] Spirocyclic diacetals are readily prepared.

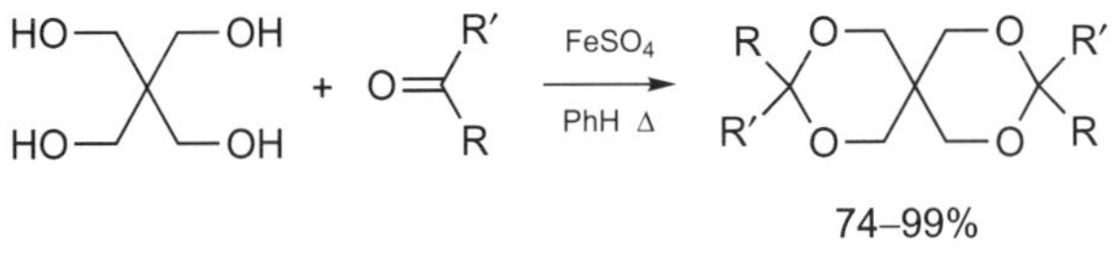

$FeSO_4$, PhH Δ

74–99%

Hydrochlorination.[2] Addition of hydrochloric acid to ynones in the presence of $FeSO_4$ is stereoselective, the products from 1-aryl-2-propyn-1-ones being predominantly (*E*)-β-chlorovinyl ketones.

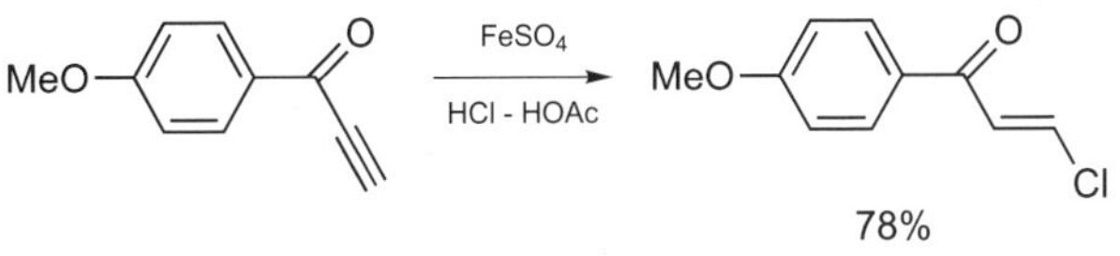

$FeSO_4$, HCl - HOAc

78%

[1]Jin, T.-S., Ma, Y.-R., Li, T.-S., Wang, J.-X. *JCR(S)* 268 (1999).
[2]Conde, J.J., Martucci, M., Olsen, M. *TL* **41**, 4709 (2000).

Iron(III) sulfate.

Nitriles.[1] Dehydration of aldoximes with iron(III) sulfate occurs in refluxing benzene.

Esterification.[2] Aroic acids and mandelic acid can be esterified in the presence of iron(III) sulfate.

[1]Desai, D.G., Swami, S.S., Mahale, G.D. *SC* **30**, 1623 (2000).
[2]Zhang, G.S., Gong, H. *SC* **29**, 1547 (1999).

N-Isocyanotriphenyliminophosphorane.

α-Diazoketones.[1] Reaction of acid chlorides with this reagent provides chlorohydrazones, which on treatment with Et_3N delivers the diazoketones.

O Cl + N CN = PPh3 → CH2Cl2 O N NH2 Cl

75%

[1]Aller, E., Molina, P., Lorenzo, A. *SL* 526 (2000).

L

Lanthanum(III) trifluoromethanesulfonate. 20, 209

Aziridines.[1] The highly *cis*-selective cycloaddition between imines and diazoacetates in protic media uses lanthanum triflate as catalyst.

Cyclization of epoxy alcohols.[2] The regioselectivity of this cyclization is greatly influenced by molecular sieves.

--	91	:	9
+ mol. sieves	12	:	88

[1]Xie, W., Fang, J., Li, J., Wang, P.G. *T* **55**, 12929 (1999).
[2]Tokiwano, T., Fujiwara, K., Murai, A. *CL* 272 (2000).

Lead(IV) acetate. 13, 155–156; **14,** 188; **16,** 193–194; **18,** 201–202; **19,** 184–185; **20,** 209–210

α-Siloxy acetates. α-Silyl alcohols are transformed into α-siloxy acetate with $Pb(OAc)_4$ via a radical Brook rearrangement.[1]

Acetoxylation.[2] β-Lactams undergo acetoxylation at the β-position.

Oxidative dearomatization. The modification of a naphthol by oxidative lactonization is critical toward a synthesis of lactonamycin. Lead(IV) acetate fulfills this need.[3]

Cleavage of catechols.[4] Lead(IV) acetate is much superior to CuCl–pyridine–methanol in the ring cleavage.

[1]Paredes, M.D., Alonso, R. *JOC* **65**, 2292 (2000).
[2]Giang, L.T., Fetter, J., Kajtar-Peredy, M., Lempert, K., Czira, G. *T* **55**, 13741 (1999).
[3]Cox, C., Danishefsky, S.J. *OL* **2**, 3493 (2000).
[4]Walsh, J.G., Furlong, P.J., Byrne, L.A., Gilheany, D.G. *T* **55**, 11519 (1999).

Lipases. 17, 133–134; **18,** 202–204; **19,** 185–188; **20,** 211–212

Resolutions. The following types of substrates have been resolved via lipase-mediated enantioselective esterification: malic and aspartic esters,[1] 3-hydroxyalken-1-yl *p*-tolyl sulfoxides,[2] β-hydroxy sulfoxides.[3] A practical method involves sequential transacetylation and sulfation, followed by extraction and treatment of the aqueous layer with methanolic HCl to recover the alcohol (the organic layer yields the acetate).[4] The use of 1-ethoxyvinyl acetate as acetyl donor in these reactions has been proposed.[5]

Coupled with in situ racemization of the unacetylated enantiomer by a ruthenium complex, the complete conversion of a secondary alcohol to the chiral acetate is efficient.[6] An alternative method involves hydrogenation of alkenyl acetates in the presence of both lipase and a Ru(II) catalyst.[7]

Simultaneous resolution of both acetate and amine with lipase has been carried out.[8]

Alcoholysis of β-lactones provides optically active β-hydroxy esters in one-half of the original quantities.[9]

Kinetic resolution.[10] Allylic alcohols are resolved with a combination of lipase and a (*p*-cymene)ruthenium complex.

Ammonolysis.[11] A preparation of primary amides catalyzed by lipase proceeds in ionic liquids at least as well as in organic media.

Partial hydrolysis and esterification. Aryl esters are hydrolyzed in the presence of alkyl esters (e.g., methyl salicylate from the *O*-benzoyl derivative), and esters of *o*-dihydroxyarenes undergo partial hydrolysis.[12,13] Transacetylation from vinyl acetate to hydroxymethylphenols occurs at the primary alcohol sites.[14]

Cleavage of N,N-dimethylhydrazones.[15] Ketones are recovered from the lipase-catalyzed reaction in good yields except the highly hindered members (e.g., menthone).

[1]Liljeblad, A., Kanerva, L.T. *TA* **10**, 4405 (1999).
[2]de la Rosa, V.G., Ordonez, M., Llera, J.M. *TA* **11**, 2991 (2000).
[3]Medio-Simon, M., Gil, J., Aleman, P., Varea, T., Asensio, G. *TA* **10**, 561 (1999).
[4]Yamano, T., Kikumoto, F., Yamamoto, S., Miwa, K., Kawada, M., Ito, T., Ikemoto, T., Tomimatsu, K., Mizuno, Y. *CL* 448 (2000).
[5]Kita, Y., Takebe, Y., Murata, K., Nata, T., Akai, S. *JOC* **65**, 83 (2000).
[6]Koh, J.H., Jung, H.M., Kim, M.-J., Pak, J. *TL* **40**, 6281 (1999).
[7]Jung, H.M., Koh, J.H., Kim, M.-J., Park, J. *OL* **2**, 2487 (2000).
[8]Garcia-Urdiales, E., Rebolledo, F., Gotor, V. *TA* **11**, 1459 (2000).
[9]Nelson, S.G., Spencer, K.L. *JOC* **65**, 1227 (2000).
[10]Lee, D., Huh, E.A., Kim, M.-J., Jung, H.M., Koh, J.H., Park, J. *OL* **2**, 2377 (2000).

[11]Lau, R.M., van Rantwijk, F., Seddon, K.R., Sheldon, R.A. *OL* **2**, 4189 (2000).
[12]Nair, R.V., Shukla, M.R., Patil, P.N., Salunkhe, M.M. *SC* **29**, 1671 (1999).
[13]Ciuffreda, P., Casati, S., Santaniello, E. *T* **56**, 317 (2000).
[14]Parmar, V.S., Prasad, A.K., Pati, H.N., Kumar, R., Azim, A., Roy, S., Errington, W. *BC* **27**, 119 (1999).
[15]Mino, T., Matsuda, T., Hiramatsu, D., Yamashita, M. *TL* **41**, 1461 (2000).

Lithium. 13, 157–158; **15,** 184; **18,** 205–206; **19,** 190–191; **20,** 212

Deamination.[1] The dimethylamino group of phenyl-substituted *N,N*-dimethylanilines suffers detachment from the aromatic ring.

α-Silylamines.[2] Reductive silylation of imines is accomplished with Li–Me_3SiCl in THF.

[1]Azzena, U., Dessanti, F., Melloni, G., Pisano, L. *TL* **40**, 8291 (1999).
[2]Bolourtchian, M., Badrian, A. *PSS* **152**, 129 (1999).

Lithium–liquid ammonia. 13, 158; **17,** 161; **18,** 206; **20,** 213

(E)-Alkenes.[1] The scope was defined for the formation from alkynes by reduction with Li/NH_3.

Reduction of heteroaromatics.[2] Indoles are reduced to cyclohexanopyrroles.

[1]Brandsma, L., Nieuwenhuizen, W.F., Zwikker, J.W., Maeorg, U. *EJOC* 775 (1999).
[2]McComas, C.C., Van Vranken, D.L. *TL* **40**, 8039 (1999).

Lithium aluminum hydride. 14, 190–191; **18,** 207; **19,** 191; **20,** 213–214

Reduction of amides. A synthesis of *N*-protected α-amino aldehydes involves reduction of morpholine amides.[1] Defluorination of secondary *o*-fluorobenzamides at room temperature takes place before carbonyl group reduction.[2] However, tertiary amides do not behave similarly.

LiAlH4
THF 25°
F O H N N O
O H N N O
77%

α-Ketols.[3] Reduction of the silyl enol ether derivatives of α-ketoesters by $LiAlH_4$ (with standard hydrolytic workup) constitutes a method for the preparation of 1-hydroxy-2-alkanones.

Hydrogenation of allylic alcohols.[4] The lithium salt of an allylic alcohol is saturated by $LiAlH_4$–$CeCl_3$.

2-Nitroalkanols.[5] Various aldehydes react with nitroalkanes to produce nitroaldols (*syn* + anti isomers) in THF with catalytic amount of $LiAlH_4$ at 0°.

[1]Douat, C., Heitz, A., Martinez, J., Fehrentz, J.-A. *TL* **41**, 37 (2000).
[2]Hendrix, J.A., Stefany, D.W. *TL* **40**, 6749 (1999).
[3]Dalla, V., Catteau, J.P. *T* **55**, 6497 (1999).
[4]Bartoli, G., Bellucci, M.C., Bosco, M., Dalpozzo, R., De Nino, A., Sambri, L., Tagarelli, A. *EJOC* 99 (2000).
[5]Youn, S.W., Kim, Y.H. *SL* 880 (2000).

Lithium amide.

β-Keto nitriles.[1] The chirality center is not disturbed in the condensation of an α-amino ester with acetonitrile when lithium amide is employed as the base.

[1]Chang, S.-J., Stuk, T.L. *SC* **30**, 955 (2000).

Lithium bromide. 18, 209–210; **19,** 192; **20,** 215

Ethylene carbonates.[1] Epoxides are transformed into cyclic carbonates with carbon dioxide (1 atm) at 100° in the presence of LiBr in NMP.

[1]Iwasaki, T., Kihara, N., Endo, T. *BCSJ* **73**, 713 (2000).

Lithium *t*-butoxide. 18, 210; **20,** 215

t-Boc amines.[1] *N*-Substituted ureas are converted to *t*-Boc amines by *t*-BuOLi in the presence of copper(II) bromide.

[1]Yamaguchi, J., Shusa, Y., Suyama, T. *TL* **40**, 8251 (1999).

Lithium alkyl(trityl)amides.

Enol silylations. With the superhindered lithium *t*-butyl(trityl)amide (and related tritylamides) ketones give more (*E*)-siloxyalkenes than lithium 2,2,6,6-tetramethylpiperidide.[1] Asymmetric enolization is possible using chiral *N*-(1-phenethyl) analogue.[2]

[1]Busch-Petersen, J., Corey, E.J. *TL* **41**, 2515 (2000).
[2]Busch-Petersen, J., Corey, E.J. *TL* **41**, 6941 (2000).

Lithium chloride. 18, 211; **20,** 218–219

Acetylation.[1] The acetylation of alcohols (including phenols), thiols, and amines by acetic anhydride is catalyzed by LiCl.

α-Methoxyacetophenones.[2] LiCl mediates the reaction of ArCN with $Ph_3P{=}CHOMe$ to give $ArCOCH_2OMe$.

[1]Sabitha, G., Reddy, B.V.S., Srividya, R., Yadav, J.S. *SC* **29**, 2311 (1999).
[2]Camuzat-Dedenis, B., Provot, O., Moskowitz, H., Mayrargue, J. *S* 1558 (1999).

Lithium chromium(I) dihydride.

Multiple reactions.[1] The reagent is prepared from chromium(III) chloride and 4 equiv of BuLi in THF. It can be isolated as a dark red-brown $LiCrH_2 \cdot THF$ complex, free from LiCl. As shown in the following diagram, it can perform hydrodehalogenation, aryl ether cleavage, coupling reactions, and polymerization.

[1]Eisch, J.J., Alila, J.R. *OM* **19**, 1211 (2000).

Lithium *N,N*-dialkylaminoborohydrides.

Benzyldialkylamine–borane complexes.[1] These reagents react with benzyl halides in THF to form the amine–borane complexes that are useful in certain synthetic purposes.

[1]Collins, C.J., Lanz, M., Goralski, G.T., Singaram, B. *JOC* **64**, 2574 (1999).

Lithium 4,4′-di-*t*-butylbiphenylide. 13, 162–163; **16,** 195–196; **17,** 164; **18,** 210–211; **19,** 192–193; **20,** 216–217

Cl–Li exchange. α-Chloroacetic and β-chloropropionic acids are converted to dilithio derivatives and their respective reactions with many substrates have been explored.[1] *N*-(Chloromethyl)carbamates[2] and *N*-(α-chloroalkoxy)carbonylpyrrolidines[3] are similarly lithiated by Li–DTTB.

3-Chloropropyl and 4-chlorobutyl phenyl ethers not only undergo Cl–Li exchange with Li–DTTB, reductive cleavage of the C—O bond also occurs. Accordingly, 1,3-dilithiopropane and 1,4-dilithiobutane are formed.[4]

The system containing nickel(II) chloride dihydrate reduces alkyl chlorides[5] and sulfonates (mesylates, triflates)[6] to hydrocarbons.

C—S bond scission. Other convenient precursors of 1,3-dilithiopropane are 1,3-bisphenylthiopropane and 1-chloro-3-phenylthiopropane.[7] As expected, β-oxyalkyllithiums are generated from 1-phenylthio-2-alkanols.[8] By virtue of its tendency of deprotonation at the α-carbon, phenyl vinyl sulfide can be considered as a 1,1-dianion equivalent of ethylene.[9] At the second stage, a C—S bond cleavage with Li—DTTB generates the required nucleophile.

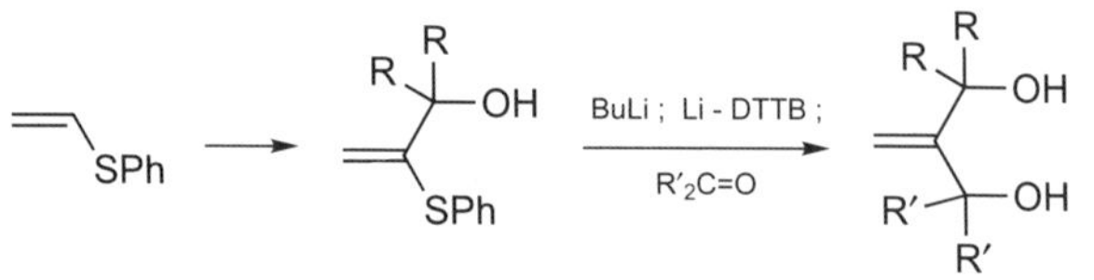

Cleavage of 2-methylene-3,3-dimethyl-4-phenyloxetane.[10] Reductive lithiation of this compound apparently leads to an *O,C*-dianion. However, only the enolate reacts with electrophiles (e.g., carbonyl compounds). Steric hindrance of the benzylic position may be the reason.

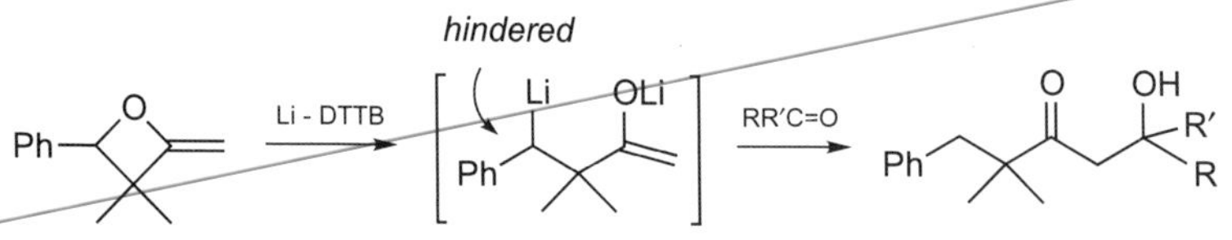

[1]Pastor, I.M., Yus, M. *TL* **41**, 5335 (2000).
[2]Ortiz, J., Guijarro, A., Yus, M. *T* **55**, 4831 (1999).
[3]Ortiz, J., Guijarro, A., Yus, M. *EJOC* 3005 (1999).
[4]Foubelo, F., Saleh, S.A., Yus, M. *JOC* **65**, 3478 (2000).
[5]Alonso, F., Radivoy, G., Yus, M. *T* **55**, 4441 (1999).
[6]Radivoy, G., Alonso, F., Yus, M. *T* **55**, 14479 (1999).
[7]Foubelo, F., Yus, M. *TL* **41**, 5047 (2000).
[8]Foubelo, F., Gutierrez, A., Yus, M. *S* 503 (1999).
[9]Foubelo, F., Gutierrez, A., Yus, M. *TL* **40**, 8173 (1999).
[10]Hashemzadeh, M., Howell, A.R. *TL* **41**, 1855 (2000).

Lithium diisopropylamide, LDA. 13, 163-164; **15,** 188–189; **16,** 196–197; **17,** 165–167; **18,** 212–214; **19,** 193–197; **20,** 218–220

Eliminations. 1-Fluoroalkenes afford 1-alkynyllithiums, which can be used to form propargylic alcohols.[1]

Deprotonation. Organolithium species are obtained from 1-alkoxy-3-phenylseleno-1-alkenes (subsequent regioselective alkylation occurs at C-3),[2] selenoamides,[3] and selenono-thioesters.[4] When those derived from α,β-unsaturated acids condense with nitriles 2-pyridones are obtained.[5]

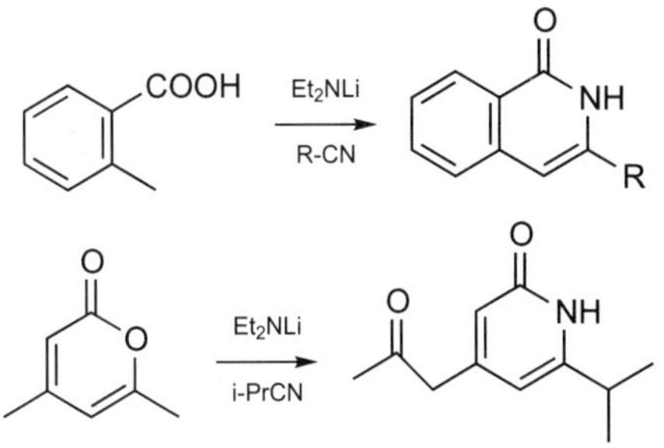

Alkylation of α-tetralone in DME using LDA as base shows rate enhancement when 1,1,4,7,10,10-hexamethyltriethylenetetramine is present. Dialkylation is also less serious.[6]

Directed lithiations. 3,4-(Methylenedioxy)halobenzenes are lithiated at C-2 with LDA.[7]

Tandem cyclization. Deprotonation of an alkyne such as that shown below initiates cyclization via an cycloalkylidene carbene intermediate.[8]

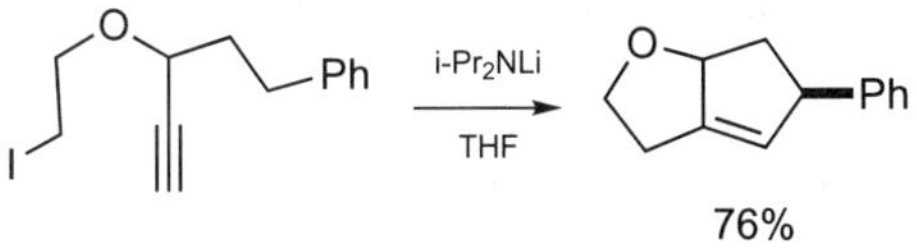

[1]Kataoka, K., Tsuboi, S. *S* 452 (2000).
[2]Nishiyama, Y., Kishimoto, Y., Itoh, K., Sonoda, N. *SL* 611 (1999).
[3]Murai, T., Suzuki, A., Ezaka, T., Kato, S. *OL* **2**, 311 (2000).
[4]Murai, T., Endo, H., Ozaki, M., Kato, S. *JOC* **64**, 2130 (1999).
[5]Brun, E. M., Gil, S., Mestres, R., Parra, M. *SL* 1088 (1999).
[6]Goto, M., Akimoto, K., Aoki, K., Shindo, M., Koga, K. *TL* **40**, 8129 (1999).
[7]Mattson, R.J., Sloan, C.P., Lockhart, C.C., Catt, J.D., Gao, Q., Huang, S. *JOC* **64**, 8004 (1999).
[8]Harada, T., Fujiwara, T., Iwazaki, K., Oku, A. *OL* **2**, 1855 (2000).

Lithium hexamethyldisilazide, LHMDS. 13, 165; **14,** 194; **18,** 215–216; **19,** 197–198; **20,** 221–222

4-Iodooxazoles.[1] After lithiation of oxazoles with LHMDS at C-2, subsequent treatment with iodine effects iodination at C-4.

Eliminations. Double elimination of β-hydroxy sulfones with LHMDS leads to alkynes,[2] while propargylic bromides are united via an alkylation–elimination process to furnish enediynes.[3]

(Me3Si)2NLi, HMPA – THF

96% (Z : E 2.2 : 1)

Alkylations. Glycine derivatives are readily alkylated at C-2 by using a combination of LHMDS and LiCl.[4]

[1]Vedejs, E., Luchetta, L.M. *JOC* **64**, 1011 (1999).
[2]Orita, A., Yoshioka, N., Struwe, P., Braier, A., Beckmann, A., Otera, J. *CEJ* **5**, 1355 (1999).
[3]Jones, G.B., Wright, J.M., Plourde, G.W., Hynd, G., Huber, R.S., Mathews, J.E. *JACS* **122**, 1937 (2000).
[4]Myers, A.G., Schnider, P., Kwon, S., Kung, D.W. *JOC* **64**, 3322 (1999).

Lithium iodide. 20, 223

Ring expansion. Cyclohexanones are formed by isomerization of methylenecyclopentane epoxides with LiI.[1] Apparently, iodohydrins are the intermediates.

LiI, CH_2Cl_2

96%

Deoxygenation.[2] Conjugated epoxy ketones are converted into enones with LiI and Amberlyst-15 resin in acetone (12 examples, 85–98%).

[1]Bouyssi, D., Cavicchioli, M., Large, S., Monteiro, N., Balme, G. *SL* 749 (2000).
[2]Righi, G., Bovicelli, C., Sperandio, A. *T* **56**, 1733 (2000).

Lithium naphthalenide, LN. 15, 190–191; **18,** 217–218; **19,** 199–200; **20,** 224–225

O—C bond cleavage. β-Ketols result from epoxy ketones on exposure to LN.[1] Various benzyllithiums are formed by reductive lithiation, for example, of 4-aryl-1,3-dioxanes,[2] *N*-substituted 2-aryloxazolidines,[3] and benzyl pivalates.[4]

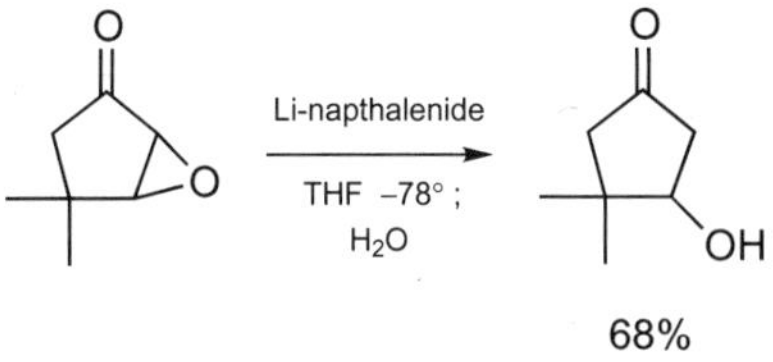
Li-napthalenide
THF –78° ;
H2O
OH
68%

Halogen–lithium exchanges. Aryllithiums are generated from fluoroarenes[5] and chlorinated nitrogen aromatic heterocycles.[6]

Partial reduction. Reduction of certain heterocycles by LN in the same pattern as the Birch reduction is observed.[7] Using this method in a large scale preparation is advantageous as liquid ammonia is not required.

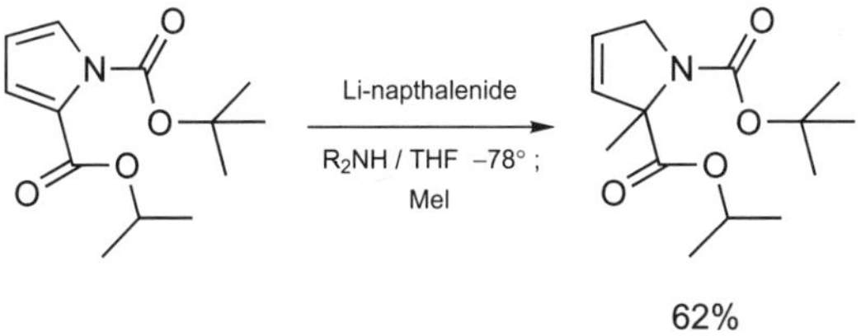
Li-napthalenide
R2NH / THF –78° ;
MeI
62%

Dialkylphosphine–borane complexes. Chiral secondary RR'PH–BH_3 complexes are accessible through reductive cleavage of thiophosphines with retention of configuration.[8]

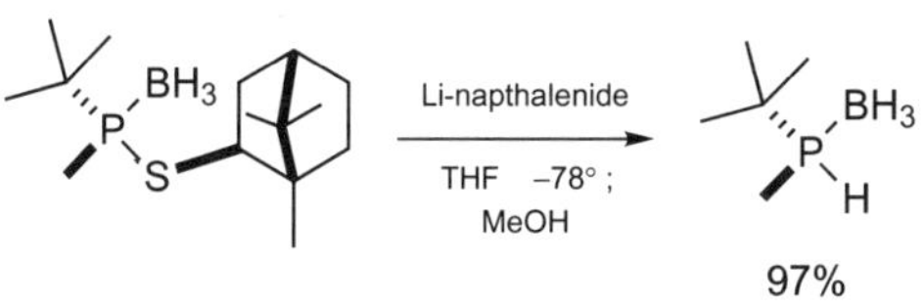
BH3
P
S
Li-napthalenide
THF –78° ;
MeOH
H
97%

Cyclization.[9] A related reagent from Li and 1-dimethylaminonaphthalene has been used to form cyclopropyl- and cyclobutylcarbinyllithiums.

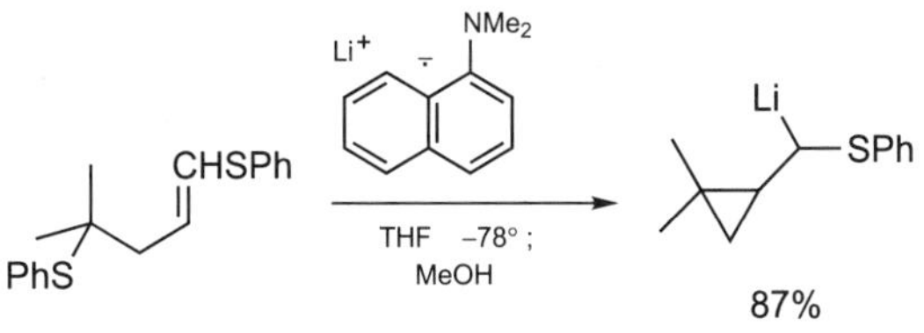
NMe2
Li+
CHSPh
PhS
THF –78° ;
MeOH
Li
SPh
87%

[1]Jankowska, R., Mhehe, G.L., Liu, H.-J. *CC* 1581 (1999).
[2]Azzena, U., Pilo, L. *S* 664 (1999).
[3]Azzena, U., Pilo, L, Piras, E. *T* **56**, 3775 (2000).
[4]Alonso, E., Guijarro, D., Martinez, P., Ramon, D.J., Yus, M. *T* **55**, 11027 (1999).
[5]Guijarro, D., Yus, M. *T* **56**, 1135 (2000).
[6]Gomez, I., Alonso, E., Ramon, D. J., Yus, M. *T* **56**, 4043 (2000).
[7]Donohoe, T.J., Harji, R.R., Cousins, R.P.C. *TL* **41**, 1331 (2000).
[8]Miura, T., Yamada, H., Kikuchi, S., Imamoto, T. *JOC* **65**, 1877 (2000).
[9]Chen, F., Mudryk, B., Cohen, T. *T* **55**, 3291 (1999).

Lithium perchlorate–diethyl ether. 18, 218–219; **19,** 200–201; **20,** 224–225

Carbonyl masking. Dithioacetalization[1] and formation of α-silylamines[2] from aldehydes exploit the Lewis acidity of $LiClO_4$—Et_2O.

CHO; Me_3SiNMe_2 - $LiClO_4$; $PhMe_2SiLi$; Ph, Si, NMe_2; 74%

Epoxide opening. The products from reaction of epoxides with diphenyl azidophosphate assisted by $LiClO_4$ are β-azido phosphates. 2-Azido-2-alkenones are obtained from epoxy ketones.[3]

In the presence of $LiClO_4$ the alkylation of the ethyl acetoacetate dianion with epibromohydrin leads to 2-ethoxycarbonylmethylene-5-hydroxymethyltetrahydrofuran.[4]

O, COOEt; NaH , BuLi - $LiClO_4$; Br, O, THF; HO—, O, COOEt; 74%

Amines.[5] In situ transformation of the Grignard reaction products of carbonyl compounds into amines is accomplished by the addition of lithium hexamethyldisilazide and $LiClO_4$.

α-Cyanohydroxylamines.[6] The three-component condensation of aldehydes *N*-alkylhydroxylamines and Me_3SiCN is conveniently accomplished at room temperature (yields in 90% range) with the aid of $LiClO_4$—Et_2O.

R−CHO + R′−NHOH + Me_3Si−CN; $LiClO_4$ Et_2O; CN, R, N, R′, OH

Baylis–Hillman reaction.[7] $LiClO_4$—Et_2O accelerates the reaction that is promoted by DABCO.

[1]Tietze, L.F., Weigand, B., Wulff, C. *S* 69 (2000).
[2]Naimi-Jamal, M.R., Mojtahedi, M.M., Ipaktschi, J., Saidi, M.R. *JCS(P1)* 3709 (1999).
[3]Masanori, M., Shioiri, T. *TL* **40**, 7105 (1999).
[4]Langer, P., Freifeld, I., Holtz, E. *SL* 501 (2000).
[5]Saidi, M.R., Javanshir, S., Mojtahedi, M.M. *JCR(S)* 330 (1999).
[6]Heydari, A., Larijani, H., Emami, J., Karami, B. *TL* **41**, 2471 (2000).
[7]Kawamura, M., Kobayashi, S. *TL* **40**, 1539 (1999).

Lithium tetrachloropalladate–copper(II) chloride.

Oxidative desilylation.[1] An allylsilane moiety is rendered electrophilic by Li_2PdCl_4–$CuCl_2$. Thus, cyclization occurs when the molecule is hydroxylated at a proper position.

R R′ OH SiMe₃ R″ → (Li_2PdCl_4, $CuCl_2$, i-PrOH) R R′ O R″

[1]Macsari, I., Szabo, K.J. *TL* **41**, 1119 (2000).

Lithium tetrafluoroborate.

S_N2' displacements.[1] Glycal acetates are converted to 2,3-dehydro glycosides and thioglycosides in the $LiBF_4$-catalyzed rreaction with alcohols and thiols, respectively.

Tetrahydropyranyl ethers.[2] Protection of alcohols by reaction with dihydropyran is promoted by $LiBF_4$ in MeCN.

[1]Babu, B.S., Balasubramanian, K.K *SC* **29**, 4299 (1999); *TL* **40**, 5777(1999).
[2]Babu, B.S., Balasubramanian, K.K. *SL* 1261 (1999).

Lithium tetrakis(pentafluorophenyl)borate.

Benzylations. Arenes undergo benzylation with $ArCH_2X$ (X = Cl, OMs) in the presence of $LiB(C_6F_5)_4$. Addition of MgO to the reaction makes it more efficient by increasing the catalyst turnover.[1]

The $LiB(C_6F_5)_4$–LiOTf/MgO combination catalyzes benzyl ether formation from alcohols (10 examples, 72–100%). The method is valuable for dealing with substrates containing base sensitive functionalities.[2]

[1]Mukaiyama, T., Nakano, M., Kikuchi, W., Matsuo, J. *CL* 1010 (2000).
[2]Nakano, M., Matsuo, J., Mukaiyama, T. *CL* 1352 (2000).

Lithium 2,2,6,6-tetramethylpiperidide, LTMP. 13, 167; **14,** 194–195; **17,** 171–172; **18,** 220–221; **19,** 202; **20,** 226–227

Conjugated dienes. Alkenyl chlorides and alkenylzirconocene chlorides are coupled with LTMP.

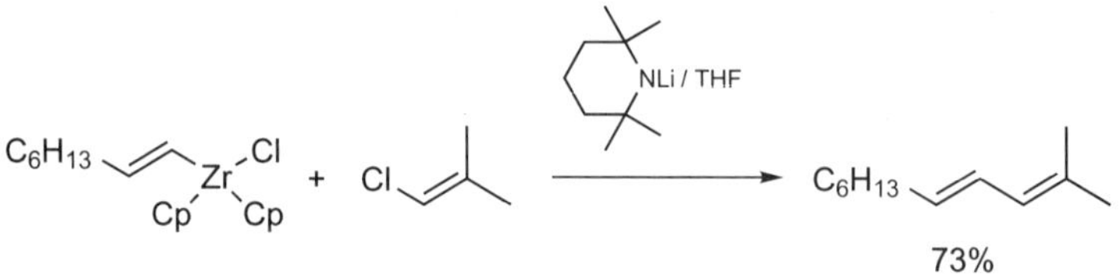

NLi / THF

C_6H_{13} Zr Cl Cp Cp + Cl → C_6H_{13}

73%

Directed lithiation. Unprotected pyridinecarboxylic acids are readily lithiated.

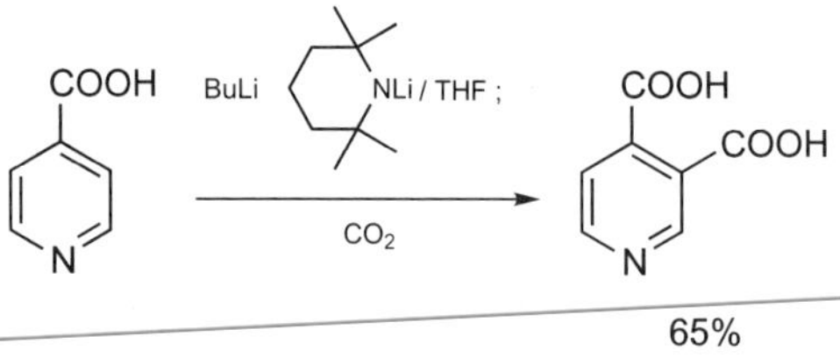

COOH BuLi NLi / THF ; CO_2 → COOH COOH N

65%

Benzyne formation.[3] Iodobenzene suffers dehydroiodination at −40° in the presence of LTMP. Trapping of the benzyne by ester enolates is followed by iodination, therefore 2-substituted iodobenzenes are generated.

[1]Kasatkin, A., Whitby, R.J. *JACS* **121**, 7039 (1999).
[2]Mongin, F., Trecourt, F., Queginer, G. *TL* **40**, 5483 (1999).
[3]Tripathy, S., LeBlanc, R., Durst, T. *OL* **1**, 1973 (1999).

Lithium tri-*t*-butoxyaluminum hydride.

Reduction.[1] One of the ester groups in a disubstitued malonic ester is reduced to a primary alcohol by this reagent.

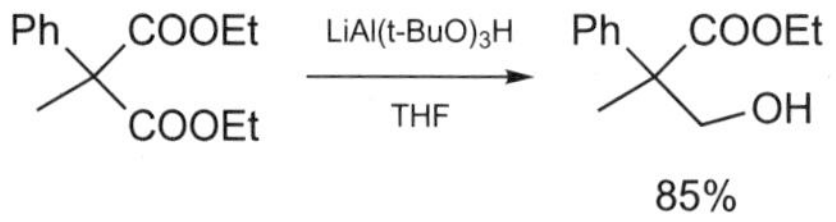

Ph COOEt COOEt → LiAl(t-BuO)$_3$H, THF → Ph COOEt OH

85%

[1]Ayers, T.A. *TL* **40**, 5467 (1999).

Lithium trifluoromethanesulfonate.

Dithioacetalization.[1] LiOTf is a highly efficient catalyst. The reaction can be performed without solvent.

Diels–Alder reactions.[2] LiOTf is a suitable substitute for $LiClO_4$ as their catalytic activities are comparable.

[1]Firouzabadi, H., Karimi, B., Eslami, S. *TL* **40**, 4055 (1999).
[2]Auge, J., Gil, R., Kalsey, S., Lubin-Germain, N. *SL* 877 (2000).

M

Magnesium. 13, 170; **15,** 194; **16,** 198–199; **18,** 224–225; **19,** 205; **20,** 229–230

Silylations. A method for silylation of tertiary alcohols employs Mg and Me_3SiCl in DMF at room temperature (12 examples, 77–99%).[1] Under similar conditions, thermodynamic silyl enol ethers are formed.[2] Interestingly, trifluoromethyl ketones provide 2,2-difluoroalkenyl silyl ethers.[3]

CF_3COPh + Me_3SiCl —(Mg / THF)→ $F_2C{=}C(Ph)OSiMe_3$ 91%

Deoxygenation. Cyclic thionocarbonates are reductively opened by Mg–MeOH.[4]

Cyclic thionocarbonate (R, COOMe) —(Mg / MeOH)→ $RCH_2CH(OH)COOMe$ R = Ph 79%

Eliminations. Unsaturated oximes are obtained from 5-bromomethylisoxazolidines that are the cycloadducts of nitrile oxides with allyl bromide. At temperature $>0°$, the double bond moves to conjugation with the oxime.[5]

5-Bromomethylisoxazoline (R) —(Mg / MeOH, −23°)→ $RC(=NOH)CH_2CH{=}CH_2$

—(Mg / MeOH, 0° ~ 25°)→ $RC(=NOH)CH{=}CHCH_3$

Cyclizations. δ,ε-Unsaturated carbonyl compounds undergo cyclization with Mg.[6] The reaction can be carried out electrochemically on a magnesium electrode and its application to a synthesis of muscone has been demonstrated.[7]

Mg / MeOH - EtOH, $HgCl_2$; COOMe; CHSPh; OH, SPh; 96%

Magnesium serves to reduce Li_2MnCl_4 to active Mn(0), with the latter species mediating a free radical cyclization of unsaturated iodo compounds.[8]

Reductions. Carbonyl group reduction is achieved with Mg–$FeCl_3 \cdot 6H_2O$ in aq DMF at room temperature.[9] For chemoselective reduction of aldehydes, the combination of Mg–$SnCl_2 \cdot 2H_2O$ in THF suffices.[10]

Barbier reactions. Allylation of aromatic aldehydes in 0.1 *N* ammonium chloride solution using a combination of Mg and an allyl bromide or iodide (but not allyl chloride) has been demonstrated.[11] Unfortunately, aliphatic aldehydes give complex mixtures. However, conjugate addition to enones[12] can be achieved with mediation by CuI.

[1]Nishiguchi, I., Kita, Y., Watanabe, M., Ishino, Y., Ohno, T., Maekawa, H. *SL* 1025 (2000).
[2]Ishino, Y., Kita, Y., Maekawa, H., Ohno, T., Yamasaki, Y., Miyata, T., Nishiguchi, I. *TL* **40**, 1349 (1999).
[3]Amii, H., Kobayashi, T., Hatamoto, Y., Uneyama, K. *CC* 1323 (1999).
[4]Rho, H.-S., Ko, B.-S. *SC* **29**, 2875 (1999).
[5]Ha, S.J., Lee, G.H., Yoon, I.K., Pak, C.S. *SC* **29**, 3165 (1999).
[6]Lee, G.H., Ha, S.J., Yoon, I.K., Pak, C.S. *TL* **40**, 2581 (1999).
[7]Kashimura, S., Murai, Y., Ishifune, M., Masuda, H., Shomomura, M., Murase, H., Shono, T. *ACS* **53**, 949 (1999).
[8]Tang, J., Shinokubo, H., Oshima, K. *T* **55**, 1893 (1999).
[9]Swami, S.S., Desai, D.G., Bhosale, D.G. *SC* **30**, 3097 (2000).
[10]Bordoloi, M., Sharmaq, R.P., Chakraborty, V. *SC* **29**, 2501 (1999).
[11]Zhang, W.-C., Li, C.-J. *JOC* **64**, 3230 (1999).
[12]Costello, D.P., Geraghty, N.W.A. *SC* **29**, 3083 (1999).

Magnesium bis(diisopropylamide).

Silyl enol ethers.[1] Kinetic enolsilylation of ketones with Me_3SiCl is observed by using this base and Et_3N, HMPA in ether at –78°.

[1]Lessene, G., Tripoli, R., Cazeau, P., Biran, C., Bordeau, M. *TL* **40**, 4037 (1999).

Magnesium bis(hexamethyldisilazide).

Aldol reactions.[1] This base, prepared in heptane from dibutylmagnesium and hexamethyldisilazane, can be isolated in crystalline form. It mediates aldol reactions in hydrocarbon solvents.

[1]Allan, J.F., Henderson, K.W., Kennedy, A.R. *CC* 1325 (1999).

Magnesium bromide. 15, 194–196; **16,** 199; **17,** 174; **18,** 226–227; **19,** 206–207; **20,** 230–232

Deprotection.[1] Trimethylsilylethoxymethyl ethers are cleaved on exposure to $MgBr_2 \cdot OEt_2$. Solvent plays a critical role, with the most effective being a mixture of ether and nitromethane.

Cyclization.[2] A remarkable formation of the skeleton of saframycin A has been reported.

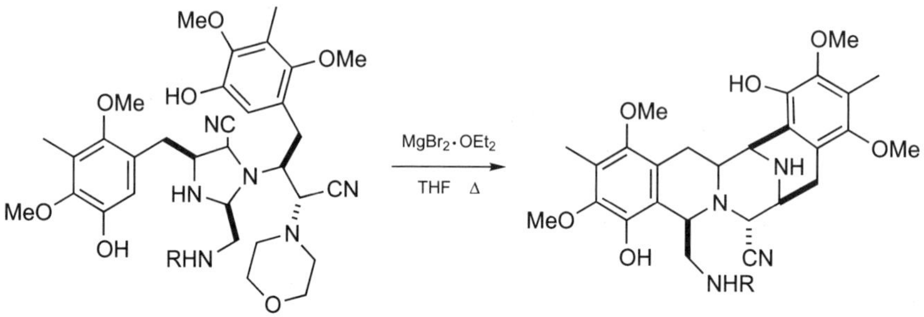

R = Fmoc

Cyanohydrination.[3] Cyanohydrination of α-alkoxy aldehydes with Me_3SiCN in dichloromethane at 0° requires an excess (5 equiv) of $MgBr_2 \cdot OEt_2$ for high diastereoselectivity (*syn*-selective). Note that $Et_4NAg(CN)_2$ is a more powerful cyanohydrinating reagent than Me_3SiCN.

[1]Vakalopoulos, A., Hoffmann, H.M.R. *OL* **2**, 1447 (2000).
[2]Myers, A.G., Kung, D.W. *OL* **2**, 3019 (2000).
[3]Ward, D.E., Hrapchak, M.J., Sales, M. *OL* **2**, 57 (2000).

Magnesium chloride–triethylamine.

Salicylaldehydes.[1] *o*-Formylation of phenols with paraformaldehyde takes place under the influence of $MgCl_2 \cdot Et_3N$.

β-Keto phosphonates. A transformation of diethylphosphonoacetic acid to β-keto phosphonates[2] is achieved with *N*-acylimidazoles in the presence of $MgCl_2 \cdot Et_3N$. However, dephosphorylation of α-fluorophosphonoacetates is also effected after *C*-acylation.[3]

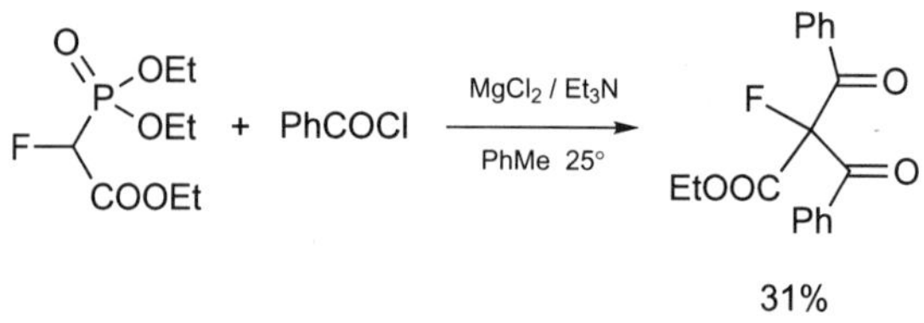

31%

[1]Hofslokken, N.U., Skattebol, L. *ACS* **53**, 258 (1999).
[2]Corbel, B., L'Hostis-Kervella, I., Haelters, J.-P. *SC* **30**, 609 (2000).
[3]Kim, D.Y., Lee, Y.M., Choi, Y.J. *T* **55**, 12983 (1999).

Magnesium iodide. 20, 232

Pyrrolidines.[1] Pyrrolidines, via ring expansion of an activated cyclopropane ring on reaction with aldimines, are catalyzed by magnesium iodide. Spiroannulated oxindoles are similarly obtained.

[1]Alper, P.B., Meyers, C., Lerchner, A., Siegel, D.R., Carreira, E.M. *ACIEE* **38**, 3186 (1999).

Manganese. 20, 233–234

Active manganese.[1] A convenient procedure for the preparation consists of reduction of Li_2MnCl_4 with Li /2-phenylpyridine in THF at room temperature. It converts organohalogen compounds into species reactive toward PhCOCl, enones and conjugated esters, and alkenyl halides.

Benzylmanganese sulfonates.[2] A preparation of these reagents is by direct reaction of tosylates and mesylates with Rieke manganese in THF at room temperature. They behave similarly to Grignard reagents.

[1]Cahiez, G., Martin, A., Delacroix, T. *TL* **40**, 6407 (1999).
[2]Kim, S.-H., Rieke, R. D. *TL* **40**, 4931 (1999).

Manganese(III) acetate. 13, 171; **14,** 197–199; **16,** 200; **17,** 175–176; **18,** 229–230; **19,** 209–210; **20,** 234–235

Perfluoroalkylation.[1] Introduction of a perfluoroalkyl group to C-3 of coumarins by reaction of sodium perfluoroalkanesulfinates is mediated by $Mn(OAc)_3{\cdot}2H_2O$.

α′-Functionalization of enones. Introduction of phenyl and acetoxy groups to the α′-position of 3-alkoxy-2-cycloalkenones (five- and six-membered) is carried out in one operation by reaction with $Mn(OAc)_3$ in benzene or a halobenzene.[2]

Radical cyclizations. When the acyl group of an anilide bears an alkylthio or alkylsulfonyl group, cyclization onto the Ar ring is initiated by $Mn(OAc)_3$ in HOAc.[3]

β-Keto amides possessing unsaturation in the side chain undergo oxidative cyclization in the presence of the $Mn(OAc)_3$–$Cu(OAc)_2$ combination.[4]

Oxidation.[5] β-Enamino carboxamides derived from cyclic β-keto amides are oxidized by $Mn(OAc)_3$–$Cu(OAc)_2$ to give enones, dienamines, or aniline derivatives, depending on the ring size.

[1]Liu, J.-T., Huang, W.-Y. *JFC* **95**, 131 (1999).
[2]Tanyeli, C., Sezen, B. *TL* **41**, 7973 (2000).
[3]Wu, Y.-L., Chuang, C.-P., Lin, P.-Y. *T* **56**, 6209 (2000).
[4]Cossy, J., Bouzide, A., Leblanc, C. *JOC* **65**, 7257 (1999).
[5]Cossy, J., Bouzide, A. *T* **55**, 6483 (1999).

Manganese dioxide. 14, 200–201; **15,** 197–198; **18,** 230–231; **19,** 210; **20,** 237–238

Nitrile oxides.[1] An alternative method for generation of these reactive species from oximes is by oxidation with MnO_2.

Oxidative chain extension.[2] Reaction conditions for oxidation of alcohols to aldehydes and the Wittig reaction with stabilized phosphonium ylides are compatible. Accordingly, three-carbon homologation of alcohols is feasible in one operation. Remarkably, the transformation is successful with unactivated alcohols such as nonanol.

OH + Ph_3P COOEt → (MnO_2 / PhMe) COOEt 80%

[1]Kiegiel, J., Poplawska, M., Jozwik, J., Kosior, M., Jurczak, J. *TL* **40**, 5605 (1999).
[2]Blackburn, L., Wei, X., Taylor, R.J.K. *CC* 1337 (1999).

Mercury(II) acetate. 15, 198–199; **17,** 176–177; **18,** 232; **19,** 211

Oxymercuration. Homoallylic alcohols undergo oxymercuration in the presence of an aldehyde, giving rise to 1,3-dioxan-4-ylmethylmercury derivatives[1] that can be functionalized.

C_8H_{17} OH + EtCHO → ($Hg(OAc)_2$ / NaCl) C_8H_{17} O O HgCl 77%

By the conventional procedure (with $NaBH_4$ work up), enol ethers are converted to alcohols.[2] It shows chemoselectivity such that a simple double bond in the same molecule is retained.

OMe → ($Hg(OAc)_2$ / H_2O - THF ; $NaBH_4$, K_2CO_3 / H_2O) OH 81%

1,2-Diketones.[3] α-Substituents of ketones such as α-sulfenyl derivatives are readily replaced by alkoxy groups in the presence of mercury(II) acetate. 1,2-Diketones are accessible therefrom.

Biaryls.[4] Homocoupling of substituted arenes to biphenyls occurs on treatment with Hg(II)–Ce(IV) salts, for example, $Hg(OAc)_2$ and $Ce(OTf)_4$.

$Hg(OAc)_2$ - $Ce(OTf)_4$, 90°

[1]Sarraf, S.T., Leighton, J.L. *OL* **2**, 403 (2000).
[2]Crouch, R.D., Mehlmann, J.F., Herb, B.R., Mitten, J.V., Dai, H.G. *S* 559 (1999).
[3]Tehrani, K.A., Boeykens, M., Tyvorskii, V.I., Kulinkovich, O., De Kimpe, N. *T* **56**, 6541 (2000).
[4]Iranpoor, N., Shekarriz, M. *JCR(S)*442 (1999).

Mercury(I) fluoride.

Decarboxylation.[1] Photodecarboxylation of arylacetic acids is induced with Hg_2F_2.

Ph, Ph, COOH → hv Hg_2F_2, MeCN → Ph, Ph, Ph, Ph 83%

[1]Habibi, M.H., Farhadi, S. *TL* **40**, 2821 (1999).

Mercury(II) tosylate.

Oxazoles.[1] Under microwave irradiation, $Hg(OTs)_2$ effects condensation of ketones with nitriles.

R, O, R′ + N, Ph → $Hg(OTs)_2$ → [R, N, Ph, +, O, R′] → R, N, R′, O, Ph

R = Ph, R′ = H 51%

[1]Lee, J.C., Song, I.-G. *TL* **41**, 5891 (2000).

Methanesulfonic acid. 20, 240

Allylation.[1] Allylation of aldehydes with allyltrimethylsilane is catalyzed by MsOH (actually Me_3SiOMs that is generated in situ).

Aziridine opening. Aziridines containing electronically biased substituents are opened with MsOH and the products may be transformed into vicinal amino alcohols.[2]

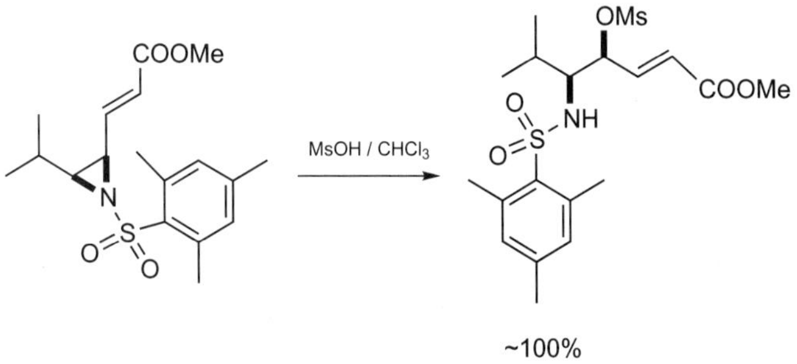

MsOH / $CHCl_3$

~100%

Fries rearrangement.[3] The MsOH–$POCl_3$ combination is useful for inducing the rearrangement.

[1] Wang, M.W., Chen, Y.J., Wang, D. *SL* 385 (2000).
[2] Tamamura, H., Yamashita, M., Nakajima, Y., Sakano, K., Otaka, A., Ohno, H., Ibuka, T., Fujii, N. *JCS(P1)* 2983 (1999).
[3] Kaboudin, B. *T* **55**, 12865 (1999).

Methoxyamine.

Amination. Conjugated dicarbonyl compounds are aminated with methoxyamine via an addition–elimination process.[1]

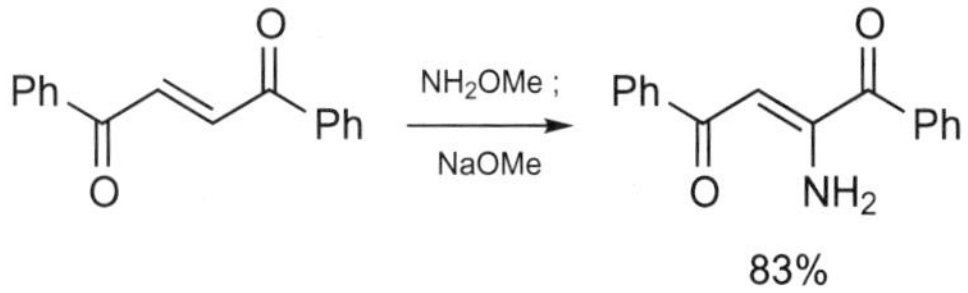

NH_2OMe ; NaOMe

83%

Nitroarenes undergo amination, but the reaction is promoted by *t*-BuOK and CuCl.[2]

[1] Seko, S., Miyake, K. *SC* **29**, 2487 (1999).
[2] Seko, S., Miyake, K., Kawamura, N. *JCS(P1)* 1437 (1999).

O-(4-Methoxybenzyl) *S*-(2-pyridyl) thiocarbonate.

4-Methoxybenzyl ethers.[1] Alcohols (primary, secondary, and tertiary) are derivatized under neutral conditions with the title reagent in the presence of AgOTf in dichloromethane.

[1]Hanessian, S., Huynh, H.K. *TL* **40**, 671 (1999).

N-Methoxycarbonylsulfamoyl triethylammonium hydroxide. (Burgess reagent).

Dehydration. Dehydration of aldoximes[1,2] and α-amido ketones[3] with Burgess reagent afford nitriles and oxazoles, respectively.

[1]Jose, B., Sulatha, M.S., Pillai, P.M., Prathapan, S. *SC* **30,** 1509 (2000).
[2]Miller, C.P., Kaufman, D.H. *SL* 1169 (2000).
[3]Brain, C.T., Paul, J.M. *SL* 1642 (1999).

***N*-Methoxy-*N*-methylformamide.**

Formylation.[1] This Weinreb amide is prepared from methyl formate and *N,O*-dimethylhydroxylamine in the presence of NaOMe. It reacts with Grignard reagents, organolithium compounds, and enolates to furnish aldehydes.

[1]Lipshutz, B.H., Pfeiffer, S.S., Chrisman, W. *TL* **40**, 7889 (1999).

2-(Methoxymethylthio)pyridine.

Methoxymethyl ethers.[1] Etherification of alcohols with this reagent is mediated by AgOTf–NaOAc. This method is suitable for derivatizing acid-sensitive alcohols. Primary, secondary, and tertiary alcohols react at comparable rates, phenols slower.

[1]Marcune, B.F., Karady, S., Dolling, U.-H., Novak, T.J. *JOC* **64**, 2446 (1999).

Methoxymethyltrimethylsilane.

Trimethylsilylmethyl ethers.[1] After radical bromination of Me_3SiCH_2OMe with NBS, reaction with an alcohol results in the mixed acetal of trimethylsilylacetaldehyde. Reduction with Dibal-H selectively removes the methoxy group.

[1]Suga, S., Miyamoto, K., Watanabe, M., Yoshida, J. *AOMC* **13**, 469 (1999).

Methylaluminoxane, MAO. 20, 242–243

Alkyne dimerization.[1] 1-Alkynes are dimerized in the head-to-tail fashion by MAO in refluxing benzene to provide 2-substituted 1-alken-3-ynes.

R—≡ + ≡—R —(MAO, PhH Δ)→ R—≡—C(=CH$_2$)R

R = i-Pr 99%

Ring formation.[2] Homoallylzirconocene chlorides couple with alkynes to form cyclopentenes under the influence of MAO.

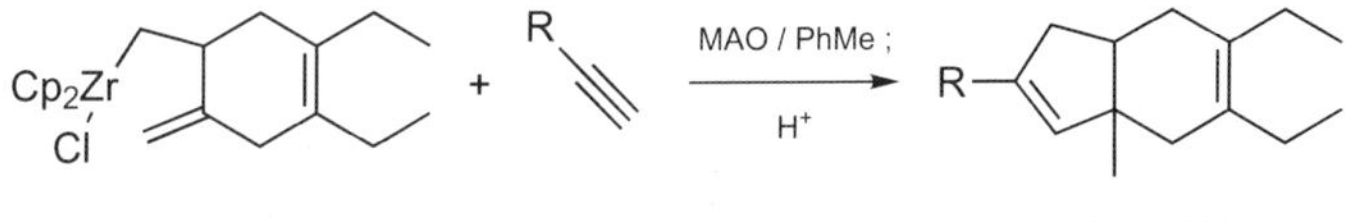

56–71%

[1]Dash, A.K., Eisen, M.S. *OL* **2**, 737 (2000).
[2]Kotora, M., Gao, G., Li, Z., Xi, Z., Takahashi, T. *TL* **41**, 7905 (2000).

Methylaluminum bis(2,6-di-*t*-butyl-4-methylphenoxide), MAD. 13, 203;**14,** 206–207; **15,** 204–205; **16,** 212; **17,** 187–188; **18,** 237; **20,** 213–214

Allylation.[1] MAD is an efficient catalyst for chemoselective allylation of aldehydes with allylstannanes.

[1]Marx, A., Yamamoto, H.*SL* 584 (1999).

Methylaluminum bis(ditriflamide).

Ethers from alcohols.[1] Allylic and benzylic alcohols form ethers in dichloromethane at room temperature under the influence of $MeAl(NTf_2)_2$. Catalytic benzylation of alcohols is possible using excess benzyl alcohol.

This reagent is prepared from 2 equiv of $AgNTf_2$ and 1 equiv of $MeAlCl_2$ in dichloromethane.

[1]Ooi, T., Ichikawa, H.,Itagaki, Y., Maruoka, K. *H* **52**, 575 (2000).

Methylammonium chloroformate.

Carbonyl regeneration.[1] With this salt adsorbed on silica, regeneration of carbonyl compounds from oximes, semicarbazones, and 2,4-dinitrophenylhydrazones under non-aqueous conditions is established.

[1]Zhang, G.-S., Chai, B. *SC* **30**, 2507 (2000).

Methyl cyanoformate.

Carbonyl group protection.[1] A C=O group is masked on conversion into the α-cyanohydrin carbonate at room temperature with NCCOOMe–*i*-Pr_2NEt. Conjugate ketones are less reactive. The adducts are reverted to ketones by treatment with NaOMe in MeOH.

[1]Berthiaume, D., Poirier, D. *T* **56**, 5995 (2000).

Methyl 2,3-dichloropropanoate.

Bisannulation.[1] Two rings are formed when the kinetic enolate of an enone reacts with this ester.

Cl, Cl, COOMe + (enone) → $(Me_3Si)_2NLi$, THF −70° → COOMe 82%

[1]Sakai, H., Hagiwara, H., Hoshi, T., Suzuki, T., Ando, M. *SC* **29**, 2035 (1999).

1-(Methyldithiocarbonyl)imidazole.

Thioureas.[1] Symmetrical and unsymmetrical thioureas can be prepared by reaction of amines with this reagent.

[1]Mohanta, P.K., Dhar, S., Samal, S.K., Ila, H., Junjappa, H. *T* **56**, 629 (2000).

3,3′-Methylenebis[1,1′-bi(2-naphthol].

Mannich-type reaction.[1] The zirconium(IV) complex **1** of this tetrol is an effective catalyst for the condensation of silyl enolates with aldimines.

(**1**)

[1]Ishitani, H., Kitazawa, T., Kobayashi, S. *TL* **40**, 2161 (1999).

2-Methylene-1,3,3-trimethylindoline.

Protection of salicylaldehydes.[1] This reagent forms spirocyclic *N,O*-acetals with salicylaldehydes. Recovery of the latter compounds is by ozonolysis.

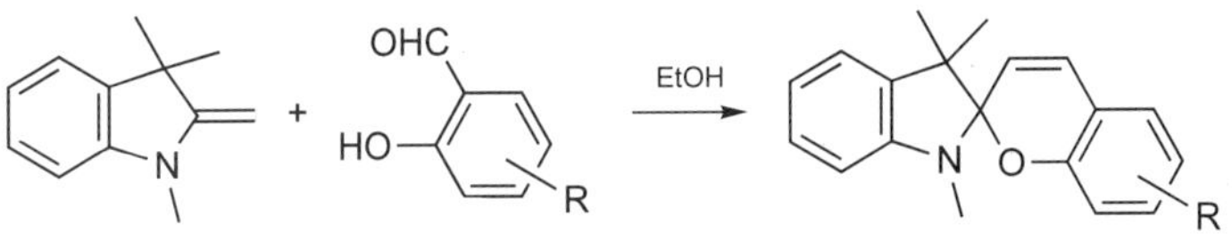

[1]Cho, Y.J., Lee, S.H., Bae, J.W., Pyun, H.-J., Yoon, C.M. *TL* **41**, 3915 (2000).

Methyl fluorosulfonyldifluoroacetate.

Trifluoromethylation.[1] The alkenyl bromine atom of an α-bromoalkenoic ester is replaced by a CF_3 group in a CuI-catalyzed reaction with the title fluorinated ester.

[1]Zhang, X., Qing, F.-L., Yang, Y., Yu, J., Fu, X.-K. *TL* **41**, 2953 (2000).

***N*-Methyl-3-(*p*-methoxyphenyl)isoindolo-1-one tetrakis(pentafluorophenyl)borate.**

Aldol reactions.[1] Compound **1** is a new catalyst for the condensation of various silyl enol ethers with aldehydes at low temperature.

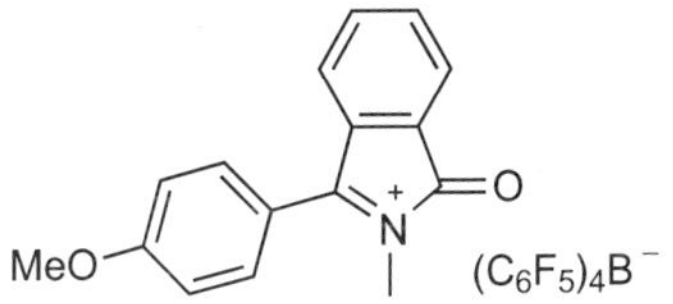

(**1**)

[1]Mukaiyama, T., Yanagisawa, M., Iida, D., Hachiya, I. *CC* 606 (2000).

***N*-Methylmorpholine *N*-oxide. 20,** 245–246

Pauson–Khand reactions. The scope of this reaction now includes as the alkene component methyl acrylate[1] and vinyl benzoate, the latter serving as an equivalent of ethylene.[2]

[1]Ahmar, M., Antras, F., Cazes, B. *TL* **40**, 5503 (1999).
[2]Kerr, W.J., McLaughlin, M., Pauson, P.L., Robertson, S.M. *CC* 2171 (1999).

2-(Methylsulfonyl-3-phenyl-1-prop-2-enyl succinimid-*N*-oxy carbonate.

Amino group protection.[1] Reagent **1** reacts with amines to form carbamates. The protection is removed by thiols and a catalytic amount of base.

(1)

[1]Carpino, L.A., Mansour, E.M.E. *JOC* **64**, 8399 (1999).

Methyltrioxorhenium–hydrogen peroxide. 19, 217; **20,** 248

Epoxidations. Epoxidation of alkenes in the NaY zeolite host have been studied.[1] Instead of hydrogen peroxide, sodium percarbonate can be used as the oxygen source.[2]

Excellent conversion and selectivity are attained when the epoxidation is performed in an ionic liquid.[3]

Oxidations. MTO and bromide ion serve as cocatalysts in the oxidation of alcohols by hydrogen peroxide.[4] Aldehydes are further transformed into methyl esters and ethers into ketones.

On the other hand, selective formation of aldehydes is observed on oxidation[5] with a system composed of MTO, TEMPO, HBr, and H_2O_2.

Desulfurization.[6] Episulfides are desulfurized with a chemical system containing triphenylphosphine, hydrogen sulfide, and catalytic amount of MTO.

[1]Adam, W., Saha-Möller, C.R., Weichold, O. *JOC* **65**, 2897 (2000).
[2]Vaino, A.R. *JOC* **65**, 4210 (2000).
[3]Owens, G.S., Abu-Omar, M.M. *CC*1165 (2000).
[4]Espenson, J. H., Zhu, Z., Zauche, T.H. *JOC* **64**, 1191 (1999).
[5]Herrmann, W. A., Zoller, J.P., Fischer, R.W. *JOMC* **579**, 404 (1999).
[6]Jacob, J., Espenson, J. H. *CC*1003 (1999).

Methyltrioxorhenium–urea–hydrogen peroxide. 19, 217–218; **20,** 248

Epoxidation.[1] Procedure modification includes the use of an ionic liquid as reaction medium.

[1]Owens, G.S., Abu-Omar, M.M. *CC* 1165 (2000).

1-Methyl-1-vinylsilacyclobutane.

Cross-coupling.[1] Pd-catalyzed coupling of silane **1** with haloarenes gives styrenes.

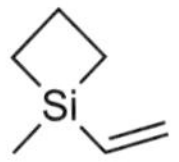

(1)

[1]Denmark, S.E., Wang, Z. *S* 999 (2000).

Mischmetal.

SmI_2 supplement.[1] This inexpensive alloy of light lanthanides (La 33%, Ce 50%, Nd 12%, Pr 4%, others 1%) can be used to replace large portion of SmI_2 in mediating reactions (e.g., Barbier reaction) that require stoichiometric quantities.

[1]Helion, F., Namy, J.-L. *JOC* **64**, 2944 (1999).

Molybdenum carbene complexes. 17, 194–195; **18,** 242–243; **19,** 219–221; **20,** 249–251

Alkene metathesis. Notable applications of the metathesis catalyzed by the Schrock catalyst (**1**) include the stereoselective synthesis of alkenylsilanes.[1]

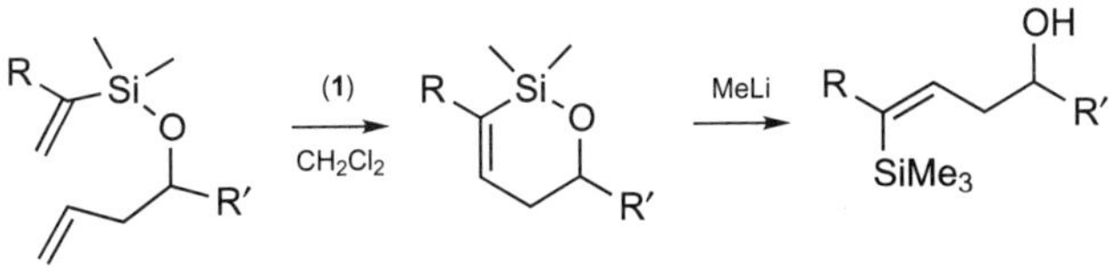

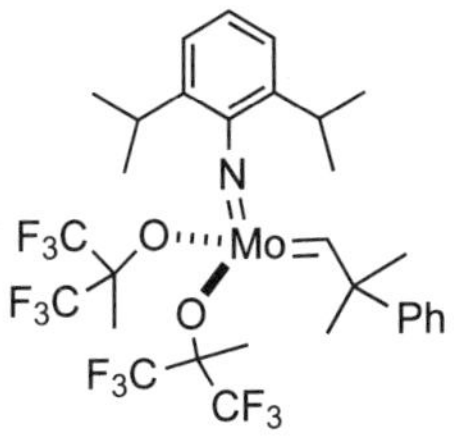

(1)
CH_2Cl_2
MeLi

(**1**)

Tandem asymmetric ring-opening metathesis and cross-metathesis has been effected, with the aid of **2**.[2]

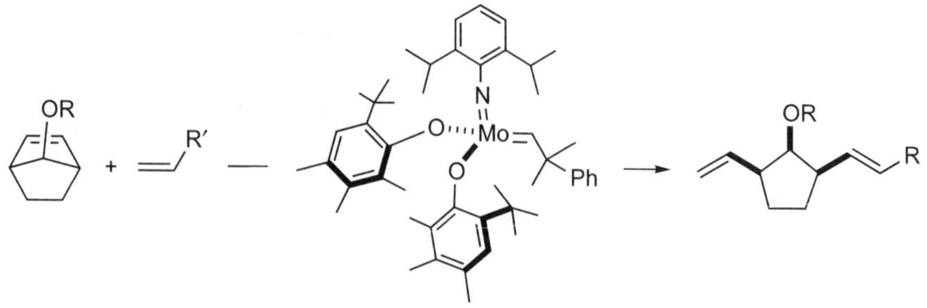

(2)

The application of [Ru] and [Mo] catalysts in consecutive steps to convert a bridged ring ether to a fused heterocycle[3] that retains all the atoms serves to illustrate the power of modern synthetic design and methodology.

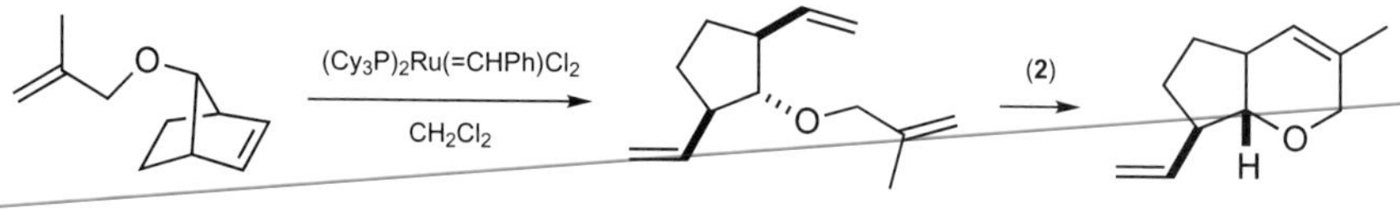

$(Cy_3P)_2Ru(=CHPh)Cl_2$

CH_2Cl_2

(2)

The tris[*N*-aryl-*N*-(*t*-butyl)amino]molybdenum(0) complexes (**3**) are good catalyst precursors readily activated in situ for metathesis of alkynes and diynes.[4]

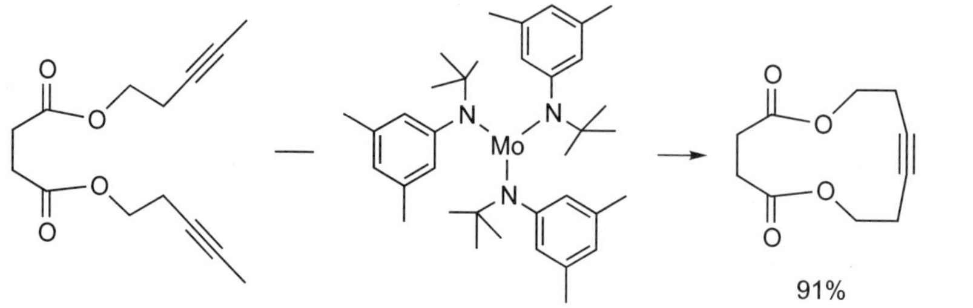

91%

(3)

Michael–aldol reaction tandem. Fischer carbene Mo complexes containing a difluoroboroxy group, that is R—C(OBF_2)═Mo$(CO)_5$, donate R to enones. The adducts may be trapped with aldehydes.[5]

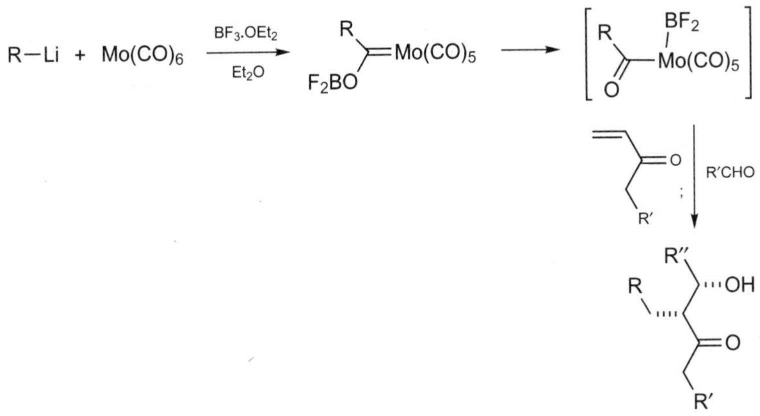

[1]Ahmed, M., Barrett, A.G.M., Beall, J.C., Braddock, D.C., Flack, K., Gibson, V.C., Procopiou, P.A., Salter, M.M. *T* **55**, 3219 (1999).
[2]La, D.S., Ford, J.G., Sattely, E.S., Bonitatebus, P.J., Schrock, R.S., Hoveyda, A.H. *JACS* **121**, 11603 (1999).
[3]Weatherhead, G.S., Ford, J.G., Alexanian, E.J., Schrock, R.S., Hoveyda, A.H. *JACS* **122**, 1828 (2000).
[4]Furstner, A., Mathes, C., Lehmann, C.W. *JACS* **121**, 9453 (1999).
[5]Barluenga, J., Rodriguez, F., Fananas, F.J., Rubio, E. *ACIEE* **38**, 3084 (1999).

Molybdenum hexacarbonyl. 13, 194–195; **15,** 225–226; **18,** 243–244; **19,** 221–222; **20,** 251–252

Metathesis. Diarylalkynes are prepared from 1-arylpropynes on heating with $Mo(CO)_6$ and *p*-chlorophenol at 140°.[1] Under the same conditions, bis[4-(prop-1-ynyl)pheny]siloxanes form cyclotrimers and cyclotetramers.[2]

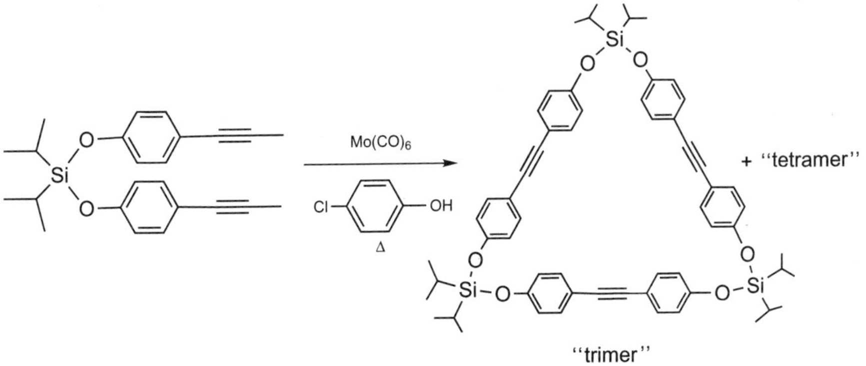

"trimer" + "tetramer"

[1]Pschirer, N.G., Bunz, U.H.F.*TL* **40**, 2481 (1999).
[2]Pschirer, N.G., Fu, W., Adams, R.D., Bunz, U.H.F.*CC* 87 (2000).

α-Morpholinyltrifluoroethyl trimethylsilyl ether.

Trifluoromethylation. This reagent is prepared from *N*-formylmorpholine and trifluoromethane. It reacts with carbonyl compounds to deliver trifluoromethylcarbinyl silyl ethers.

$$Ph_2C{=}O + Me_3SiO\text{–}CH(CF_3)\text{–}N(\text{morpholine}) \xrightarrow[\text{DME } 80^\circ]{\text{CsF}} Ph_2C(OSiMe_3)CF_3 \quad 75\%$$

[1]Billard, T., Bruns, S., Langlois, B.R.*OL* **2**, 2101 (2000).

N

Nafion-H. 14, 213; **18,** 246; **20,** 253–254

α-Hydroxy-β,γ-unsaturated esters.[1] Glycidic esters are isomerized to the unsaturated hydroxy esters.

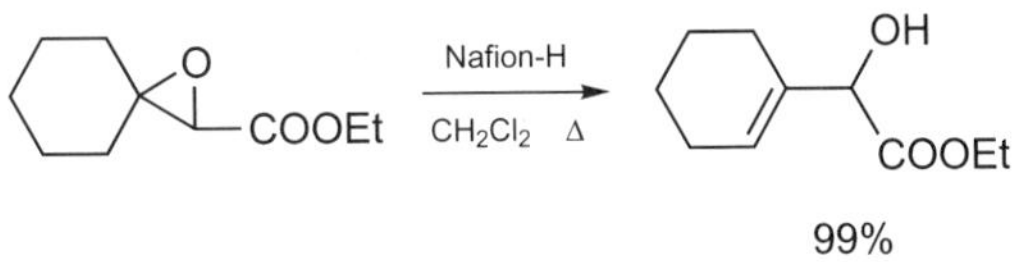

[1]Hachoumy, M., Mathew, T., Tongco, E.C., Vankar, Y.D., Prakash, G.K.S., Olah, G.A. *SL* 363 (1999).

1*H*-Naphtho[1,8-*de*]-1,2,3-triazine.

2-Amino alcohols.[1] The 2-chloromethyl derivative of the triazine undergoes Barbier reaction with carbonyl compounds. The products release 2-amino alcohols on hydrogenation. Those derived from activated aromatic ketones–aldehydes can be hydrogenolyzed also.

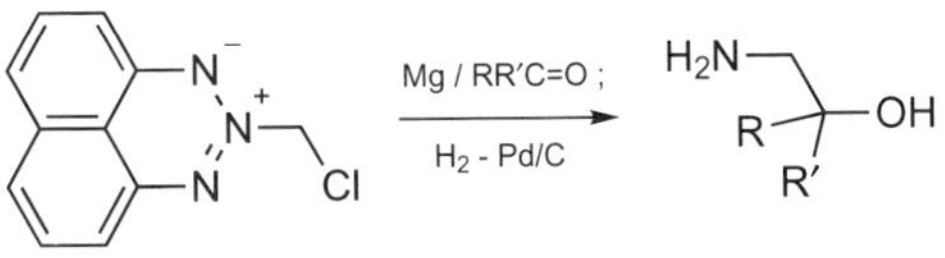

α-Amino acids.[2] Compound **1** is readily alkylated and its products are converted to esters of α-amino acids on treatment with aluminum amalgam in THF.

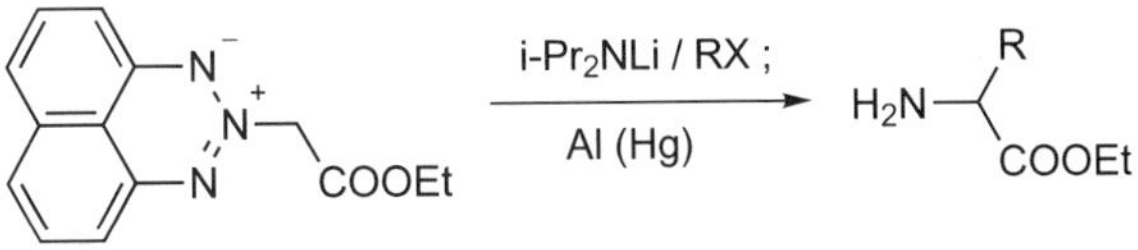

(**1**)

[1] Chandrasekhar, S., Sridhar, M.*TL* **41**, 4685 (2000).
[2] Anilkumar, R., Chandrasekhar, S., Sridhar, M.*TL* **41**, 6665 (2000).

2-Naphthylmethanol.

Functional group protection. 2-Naphthylmethyl ethers[1] and esters[2] (formation by the DCC method) are more readily hydrogenolyzed than benzyl analogues, thus permitting their differentiation. 2-Naphthylmethyl carbamates undergo selective hydrogenolysis in the presence of the 4-trifluoromethylbenzyl carbamate group.[3]

[1] Gaunt, M.J., Yu, J., Spencer, J.B. *JOC* **63**, 4172 (1998).
[2] Gaunt, M.J., Boschetti, C.E., Yu, J., Spencer, J.B. *TL* **40**, 1803 (1999).
[3] Papageorgiou, E.A., Gaunt, M.J., Yu, J., Spencer, J.B. *OL* **2**, 1049 (2000).

Nickel. 12, 355; **13,** 197; **14,** 213; **18,** 246; **19,** 224; **20,** 253–254

Reduction of nitrogen compounds. Arylamines are produced by using electrogenerated nickel (suspension from a nickel anode) to reduce nitroarenes.[1]

Deallylation.[2] Allyl aryl ethers are converted to phenols using electrogenerated nickel.

[1] Yasuhara, A., Kasano, A., Sakamoto, T. *JOC* **64**, 2301 (1999).
[2] Yasuhara, A., Kasano, A., Sakamoto, T. *JOC* **64**, 4211 (1999).

Nickel, Raney. 13, 265–266; **14,** 270; **15,** 278; **17,** 296; **18,** 246; **20,** 254

Reductions. Chemoselective reduction of aldehydes in the presence of ketones is quite general, except for very highly hindered aldehydes.[1]

O
CHO
Raney Ni
THF 25°
O
OH
98%

Selective hydrogenation of enones,[2] particularly assisted by ultrasonic irradiation,[3] is a synthetically valuable process. Ultrasonically activated nickel reduces nitroarenes to the azoxyarene stage.[4] On the other hand, with formic acid or ammonium formate as the hydrogen source, the reduction results in arylamines.[5]

The aromatic ring of a phenol is saturated under alkaline conditions (1% KOH, no organic solvent).[6] Interestingly, 4-methoxybiphenyl is reduced to phenylcyclohexane (69% yield).

Raney Ni-Al, KOH - H_2O: 4-methoxybiphenyl → cyclohexylbenzene, 69%

Raney Ni-Al, KOH - H_2O: phenol → cyclohexanol, 93%

Nascent Raney nickel from the Ni—Al alloy reduces BINOL and the analogous diamine to the 5,6,7,8,5′,6′,7′,8′-octahydro derivatives in good yields.[7] Chiral substrates afford products with 97.5–99% ee.

[1]Barrero, A.F., Alvarez-Manzaneda, E.J., Chahboun, R., Meneses, R. *SL* 197 (2000).
[2]Barrero, A.F., Alvarez-Manzaneda, E.J., Chahboun, R., Meneses, R. *SL* 1663 (1999).
[3]Wang, X., Xu, M., Lian, H., Pan, Y., Shi, Y. *SC* **29**, 129 (1999).
[4]Wang, X., Xu, M., Lian, H., Pan, Y., Shi, Y. *SC* **29**, 3031 (1999).
[5]Gowda, D.C., Gowda, A.S.P., Baba, A.R., Gowda, S. *SC* **30**, 2889 (2000).
[6]Tsukinoki, T., Kanda, T., Liu, G.-B., Tsuzuki, H., Tashiro, M. *TL* **41**, 5865 (2000).
[7]Guo, H., Ding, K. *TL* **41**, 10061 (2000).

Nickel–carbon.

Cross-coupling. Ni–C is an inexpensive reagent for coupling of ArCl with organometallic reagents (e.g., RZnI)[1] and arylation of secondary amines.[2]

[1]Lipshutz, B.H., Blomgren, P.A. *JACS* **121,** 5819 (1999).
[2]Lipshutz, B.H., Ueda, H. *ACIEE* **39,** 4492 (2000).

Nickel acetate–2,2′-bipyridine. 20, 255

Biaryls.[1] The bipyridine complex of $Ni(OAc)_2$ has catalytic activity in coupling of aryl halides with LiH–*t*-BuOLi.

[1]Massicot, F., Schneider, R., Fort, Y. *JCR(S)* 664 (1999).

Nickel(II) acetylacetonate. 17, 201; **18,** 247–248; **19,** 225–226; **20,** 255–256

Cross-couplings. The Ni-catalyzed coupling of organostannanes with hypervalent iodonium reagents is also extendable to a carbonylative process, yielding ketones.[1] The coupling between functionalized benzylzinc reagents and primary haloalkanes[2] is performed in the presence of Bu_4NI in addition to $Ni(acac)_2$.

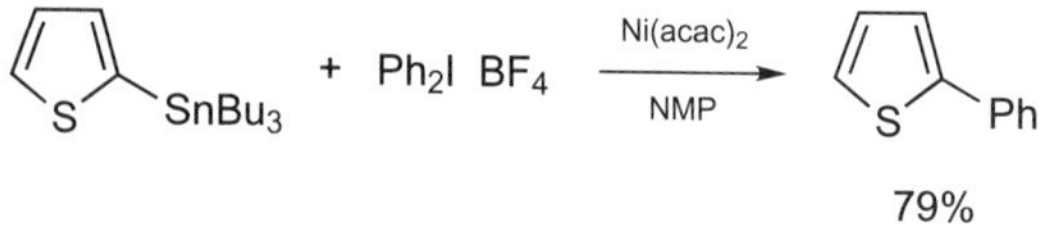

The coupling between two tetrahedral carbon centers from functionalized diorganozinc reagents and alkyl iodides is efficiently catalyzed by $Ni(acac)_2$ while adding *p*-trifluoromethylstyrene as promoter.[3]

***N*-arylation.**[4] Secondary amines undergo arylation with ArCl in the presence of NaH, bipyridine, and $Ni(acac)_2$.

Homoallylation. Carbonyl compounds afford 4-alkenols from a Ni-catalyzed condensation with 1,3-dienes that is promoted by dialkylzinc (as alternative to the previously reported triethylborane). With diethylzinc there is an overall reductive alkylation but dimethylzinc also donates a methyl group to the product.[5,6]

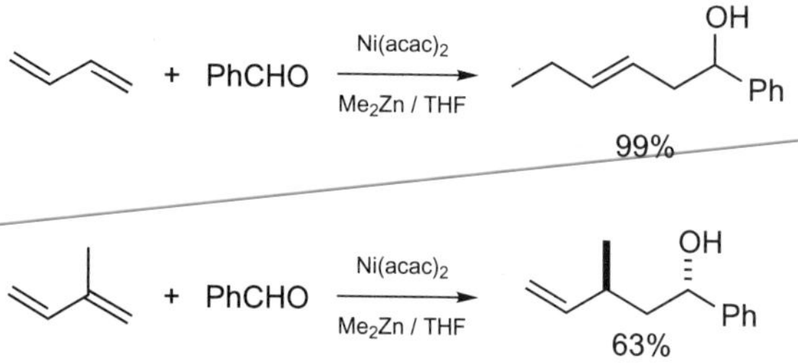

Cyclization of aldehydes bearing a distant diene unit[7] is synthetically valuable.

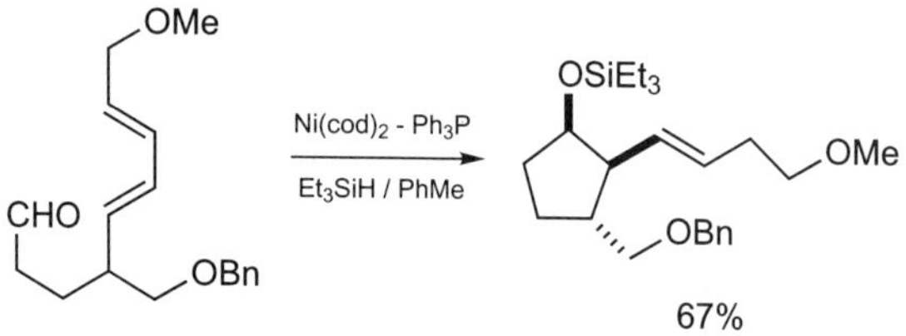

[2 + 2 + 2]Cycloaddition. Enones and alkynes combine to furnish indanones. The catalytic system consists of $Ni(acac)_2$–PPh_3, Me_3Al, and phenol.[8] There is a change in regioselectivity when the reaction is promoted by $Ni(acac)_2$ and an oxazole ligand.[9] (Note a different reaction pathway that proceeds by carbozincation of alkynes and conjugate addition.[10])

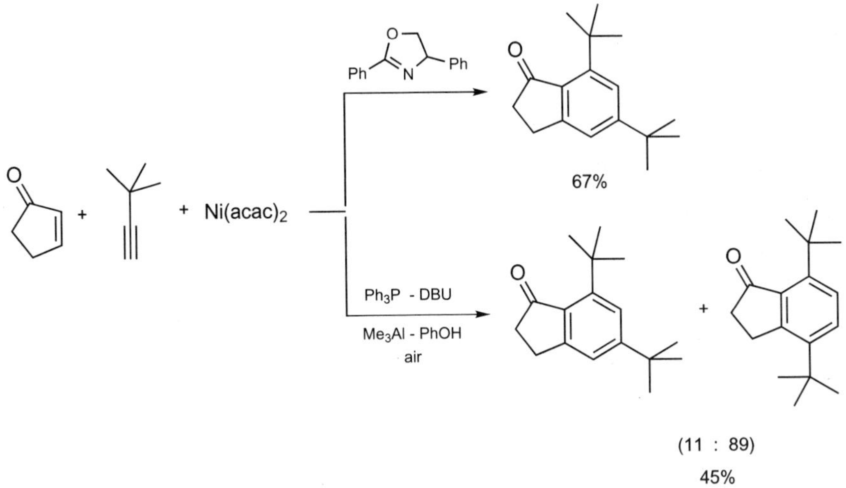

Ph, Ph (oxazoline)

+ Ni(acac)$_2$

67%

Ph_3P - DBU

Me_3Al - PhOH

air

(11 : 89)

45%

Pentasubstituted benzene derivatives are expediently formed. The observed regioselectivity indicates the possible development of a synthetic route for taiwanin C based on this process.[11]

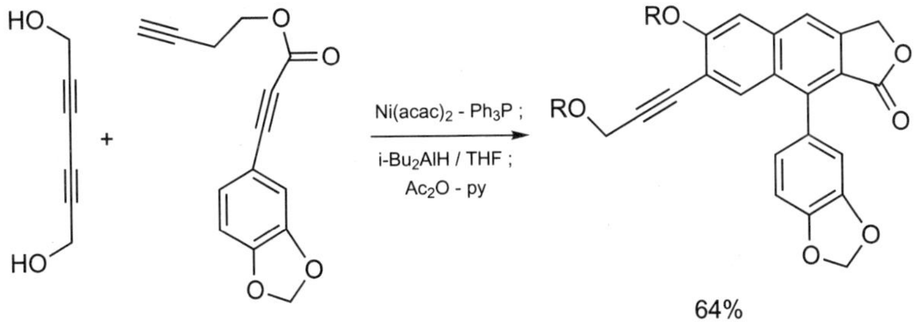

HO

HO

+

Ni(acac)$_2$ - Ph_3P ;

i-Bu_2AlH / THF ;

Ac_2O - py

RO

RO

64%

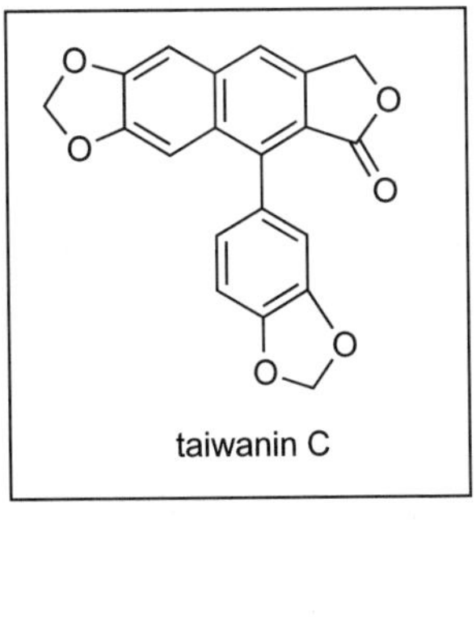

taiwanin C

1,4-Silaboration.[12] *B*-Silylpinacolatoboranes are split and add to 1,3-dienes by the $Ni(acac)_2$–*i*-Bu_2AlH combination. The major products possess a (*Z*)-configuration, and cyclic dienes furnish adducts with *cis*-B,Si substituents.

[1]Kang, S.-K., Ryu, H.-C., Lee, S.-W. *JCS(P1)* 2661 (1999).
[2]Piber, M., Jensen, A.E., Rottlander, M., Knochel, P. *OL* **1**, 1323 (1999).
[3]Giovannini, R., Stüdemann, T., Devasagayaraj, A., Dussin, G., Knochel, P. *JOC* **64**, 3544 (1999).
[4]Brenner, E., Schneider, R., Fort, Y. *T* **55**, 12829 (1999).
[5]Kimura, M., Fujimatsu, H., Ezoe, A., Shibata, K., Shimizu, M., Matsumoto, S., Tamaru, Y.*ACIEE* **38**, 397 (1999).
[6]Kimura, M., Matsuo, S., Shibata, K., Tamaru, Y. *ACIEE* **38**, 3386 (1999).
[7]Sato, Y., Takimoto, M., Mori, N. *JACS* **122**, 1624 (2000).
[8]Mori, N., Ikeda, S., Sato, Y. *JACS* **121**, 2722 (1999).
[9]Ikeda, S., Kondo, H., Mori, N. *CC* 815 (2000).
[10]Ikeda, S., Cui, D.-M., Sato, Y. *JACS* **121**, 4712 (1999).
[11]Sato, Y., Ohashi, K., Mori, M. *TL* **40**, 5231 (1999).
[12]Suginome, M., Matsuda, T., Yoshimoto, T., Ito, Y. *OL* **1**, 1567 (1999).

Nickel boride.

Reductive amination.[1] Secondary benzylamines are rapidly formed when aromatic aldehydes and ketones are treated with RNH_2 and nickel boride.

[1]Saxena, I., Borah, R., Sarma, J.C. *JCS(P1)* 503 (2000).

Nickel bromide–amine complexes.

Coupling reactions. Electrochemical coupling reactions with $(bpy)NiBr_2$ catalysis include those between alkenyl bromides and α-halo esters,[1] aryl and pyridyl halides,[2] as well as aryl and chlorophosphines.[3]

Cyclization–carboxylation.[4] Cyclization of *o*-haloarylalkenes is induced by $(cyclam)NiBr_2$. Under a carbon dioxide atmosphere, carboxylation occurs. However, yields in most cases are not impressive.

Cl O — e CO_2 / $Ni(cyclam)Br_2$ / Bu_4NBF_4 / DMF → COOH O

70%

Allylation.[5] Formation of homoallylic alcohols by reaction of aldehydes with allyl bromide under the influence of $NiBr_2$–$CrCl_3$ also requires an electron source. Tetrakis-(dimethylamino)ethylene is a suitable candidate.

[1]Cannes, C., Condon, S., Durandetti, M., Perichon, J., Nedelec, J.-Y. *JOC* **65**, 4575 (2000).
[2]Gosmini, C., Nedelec, J.-Y., Perichon, J. *TL* **41**, 5039 (2000).

[3]Budnikova, Y., Kargin, Y., Nedelec, J.-Y., Perichon, J. *JOMC* **575**, 63 (1999).
[4]Olivero, S., Dunach, E. *EJOC* 1885 (1999).
[5]Kuroboshi, M., Goto, K., Mochizuki, M. *SL* 1930 (1999).

Nickel bromide–dppe.

Diarylmethanols.[1] Zinc in combination with the (dppe)$NiBr_2$ complex mediates the reaction between aryl bromides and aromatic aldehydes. This process tolerates functional groups such as ketone, ester, amide, and nitrile.

[1]Majumdar, K.K., Cheng, C.-H. *OL* **2,** 2295 (2000).

Nickel chloride.

Claisen rearrangement. Allyl esters of *N*-acylglycine form chelates on treatment with LDA and $NiCl_2$. Such chelates undergo stereoselective rearrangement.[1]

i-Pr_2NLi - $NiCl_2$, THF

3-Hexenedioic esters. 3-Halopropenoic esters are reductively dimerized by Zn with $NiCl_2$ as catalyst.[2]

EtOOC–CH=CH–Cl → ($NiCl_2$.6H_2O, Zn / H_2O, pyridine) → [EtOOC, Cl, Ni, Cl, COOEt] → EtOOC, COOEt

66% (*E* : *Z* 3 : 1)

Cross-coupling. Suzuki coupling tolerable to water is achieved using the phosphine-free complex $(Et_3N)_2NiCl_2$.[3] A soluble nickel species prepared from treatment of Ph_3P and $NiCl_2$ with BuLi effects the coupling of RZnI with ArCl.[4]

[1]Kazmaier, U., Maier, S. *JOC* **64**, 4574 (1999).
[2]Kotora, M., Matsumura, H., Takahashi, T. *CL* 236 (2000).
[3]Leadbeater, N.E., Resouly, S.M. *T* **55**, 11889 (1999).
[4]Lipshutz, B.H., Blomgren, P.A., Kim, S.K. *TL* **40**, 197 (1999).

Nickel chloride–phosphine complexes. 14, 125; **15,** 122; **16,** 124; **18,** 250; **19,** 227–228; **20,** 258–259

Couplings. A nickel catalyst prepared from reduction of (dppf)$NiCl_2$ with BuLi has been employed to synthesize arylalkenes by coupling arylboronic acids with enolphosphates.[1]

A water-soluble Ni catalyst is obtained[2] when (dppe)$NiCl_2$ is treated with Zn and sodium triphenylphosphinotrimetasulfonate.

Alkyldesulfurization.[3] Thioglycolic is the leaving group in the Ni-catalyzed reaction with organozinc reagents.

Me_2Zn, $(MePh_2P)_2NiCl_2$, THF Δ — 80%

[1]Nan, Y., Yang, Z. *TL* **40**, 3321 (1999).
[2]Galland, J.-C., Savignac, M., Genet, J.-P. *TL* **40**, 2323 (1999).
[3]Srogl, J., Liu, W., Marshall, D., Liebeskind, L.S. *JACS* **121**, 9449 (1999).

Nickel iodide.

[2 + 2 + 2]Cycloaddition. Conjugated carbonyl compounds are benzannulated with alkynes.[1] Different reaction patterns may emerge on varying the substrates.

$(Ph_3P)_2NiI_2$, Zn - ZnI_2, THF Δ — 80%

$(Ph_3P)_2NiCl_2$, Zn / CH_2Cl_2

R = Pr 62%
R = Bu 71%

[1]Sambaiah, T., Li, L.-P., Huang, D.-J., Lin, C.-H., Rayabarapu, D.K., Cheng, C.-H. *JOC* **64**, 3663 (1999).

Nickel perchlorate.

Hetero-Diels–Alder reaction. When catalyzed by nickel perchlorate the condensation of *N*-acrylyloxazolidinone with thiabutadienes gives dihydrothiapyrans.[1]

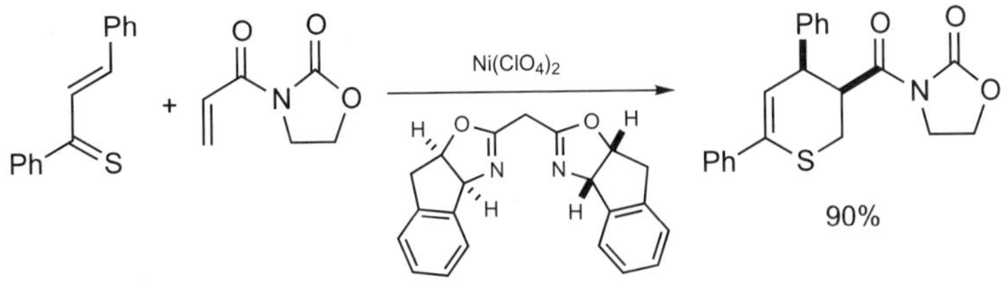

Reduction.[2] For transfer hydrogenation of aryl ketones, nickel perchlorate, in which the cation is complexed to a macrocyclic ligand, may serve as catalyst.

[1]Saito, T., Takekawa, K., Takahashi, T. *CC* 1001 (1999).
[2]Phukan, P., Sudalai, A. *SC* **30**, 2401 (2000).

Nickel tetrafluoroborate–bipyridine complex.

Deallylation and depropargylation. Allyl and proparyl ethers of phenols and propargyl aroates are cleaved electrolytically with the Ni-catalyst.[1] Amines are released from allyl carbamates by this procedure.[2]
When the propargyl ether moiety is ortho to an aldehyde, an intramolecular propargyl group transfer occurs to a large extent.[3]

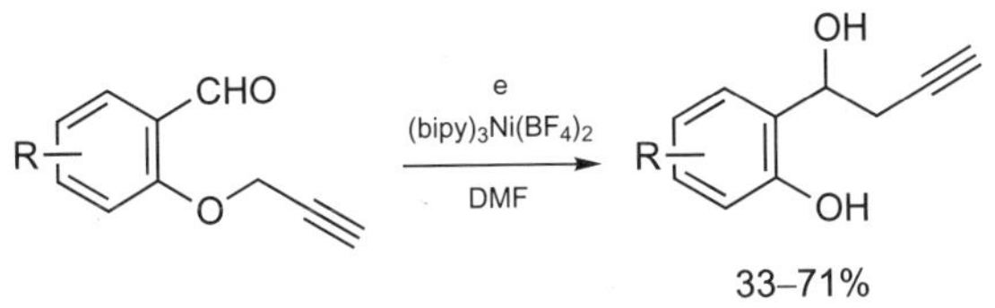

Carboxylation. The electrochemical carboxylation is followed by lactonization when the molecule is appended with a neighboring epoxide.[4]

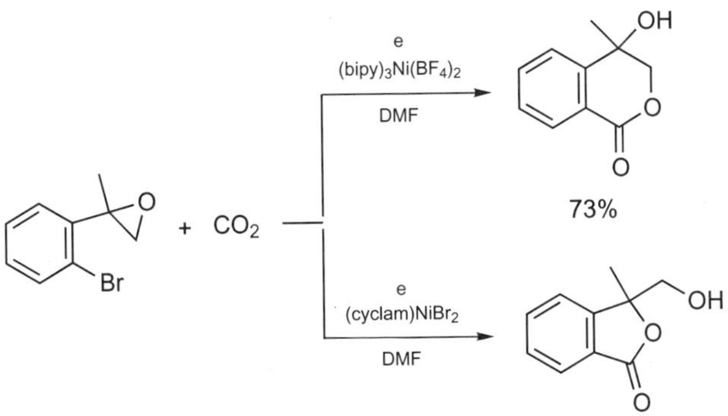

[1]Olivero, S., Rolland, J.-P., Dunach, E., Labbe, E. *OM* **19**, 2798 (2000); Olivero, S., Dunach, E. *TL* **38**, 6193 (1997).
[2]Franco, D., Dunach, E. *TL* **41**, 7333 (2000).
[3]Franco, D., Dunach, E. *TL* **40**, 2951 (1999).
[4]Tascedda, P., Dunach, E. *SL* 245 (2000);

Nitric oxide. 19, 229; **20,** 259–260

Alkyl nitrites.[1] Alcohols react with NO in organic solvent (air is present) to give RONO directly.

Oxidation.[2] Under the influence of *N*-hydroxyphthalimide, NO is used as an oxidizing agent for benzyl ethers (to aldehydes).

NO, *N*-hydroxyphthalimide: phthalan → phthalaldehyde (CHO, CHO) 80% + phthalide 12%

NO, *N*-hydroxyphthalimide: 2-methyl-6-(*tert*-butoxymethyl)naphthalene → 6-methyl-2-naphthaldehyde (CHO) 85%

N-(1-Adamantyl) amides.[2] Adamantyl radical is formed by hydrogen abstraction. In situ interception with a nitrile results in the generation of the amide.

[1]Grossi, L., Strazzari, S. *JOC* **64**, 8076 (1999).
[2]Eikawa, M., Sakaguchi, S., Ishii, Y. *JOC* **64**, 4676 (1999).

2-Nitrobenzenesulfonamide.

Primary amines.[1] After *N*-alkoxycarbonylation (Boc, Cbz, etc.) of the sulfonamide, alkylation at the nitrogen atom and treatment with a thiolate anion complete the preparation of a protected primary amine.

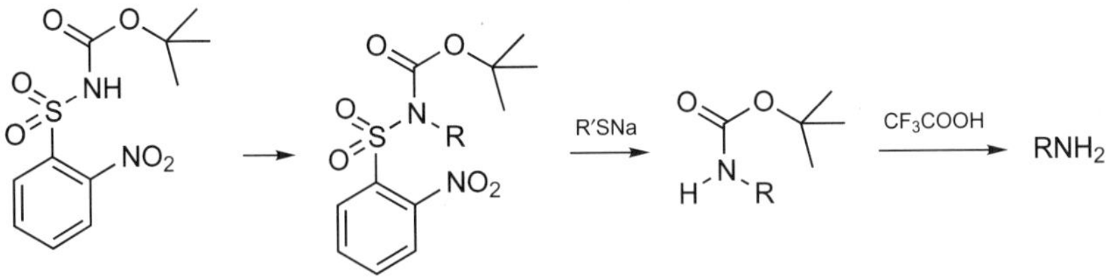

[1]Fukuyama, T., Cheung, M., Kan, T. *SL* 1301 (1999).

2-Nitrobenzenesulfonyl chloride.

Sulfonamides.[1] The reagent sulfonylates the primary amino group(s) of a polyamine.

[1]Favre-Reguillon, A., Segat-Dioury, F., Nait-Bouda, L., Cosma, C., Siaugue, J.-M., Foos, J., Guy, A. *SL* 868 (2000).

Nitrogen dioxide. 15, 219; **18,** 252–253

Decarbamoylation.[1] Primary amines are recovered from monoalkylureas on treatment with NO_2.

[1]Collet, H., Boiteau, L., Taillades, J., Commeyras, A. *TL* **40**, 3355 (1999).

Nitrolic acids.

Nitrile oxides.[1] Nitrolic acids are prepared from nitroalkanes or alkyl bromides in DMSO containing HOAc and $NaNO_2$. They decompose under neutral conditions to afford nitrile oxides.

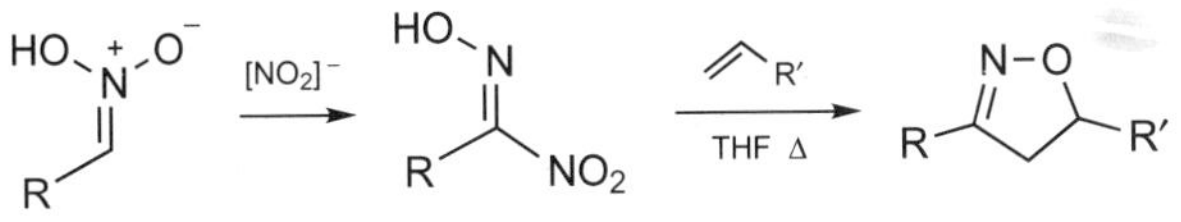

[1]Matt, C., Gissot, A., Wagner, A., Mioskowski, C. *TL* **41**, 1191 (2000).

4-Nitrophenyl formate.

N-Formylation.[1] Protection of amino group takes place by transacylation. Primary amines are selectively formylated.

[1]Orelli, L.R., Garcia, M.B., Niemevz, F., Perillo, I.A. *SC* **29**, 1819 (1999).

O

Organoaluminum reagents.

Allyl alcohols. Stereoselective addition of alkenylaluminums to chiral aldehydes enables creation of quaternary carbon centers at the allylic position after esterification and S_N2' displacement with cuprates.[1]

CHO + Al Ph $\xrightarrow{CH_2Cl_2}$ OH Ph

76%

1,1,1-Trifluoro-3-alkyn-2-ols. Substitution of optically active 1-benzyloxy-2,2,2-trifluoroethyl tosylate with lithium alkynylaluminates is an excellent method for access to the precursors of these unique alcohols.[2]

CF_3 BnO OTs + Li R $\xrightarrow{Et_3Al / Et_2O}$ CF_3 BnO R

77%

[1]Spino, C., Beaulieu, C. *ACIEE* **39**, 1930 (2000).
[2]Matsutani, H., Kusumoto, T., Hiyama, T. *CL* 529 (1999).

Organocopper reagents. 13, 207–209; **14,** 218–219; **15,** 221–227; **16,** 232–238; **17,** 207–218; **18,** 257–262; **19,** 232–235; **20,** 264–267

Silylcuprations. Alkynes and allenes are readily converted to useful reagents for further applications. Thus, a synthesis of β-silyl alanine derivatives is readily achived.[1]

O N O O $\xrightarrow[THF\ -78°]{(Me_3Si)_2Cu(CN)Li_2}$ $SiMe_3$ O N O O

79%

To direct a ring formation[2] following the silylcupration is also synthetically significant.

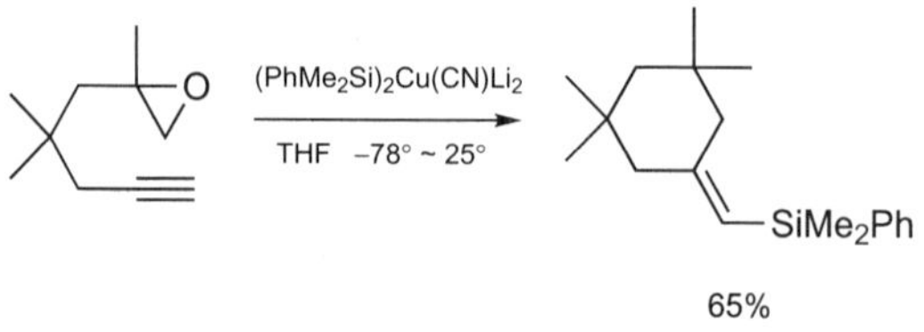

$(PhMe_2Si)_2Cu(CN)Li_2$

THF −78° ~ 25°

$SiMe_2Ph$

65%

A route to 3-silylmethyl-1,4-alkadienes[3] involves silylcupration of allenes and trapping of the ensuing organocopper species with allyl phosphate.

The 3-silyl-propen-2-ylcopper reagent derived from allene is used to react with enones and the products can be cyclized to give 3-methylenecyclopentanols.[4]

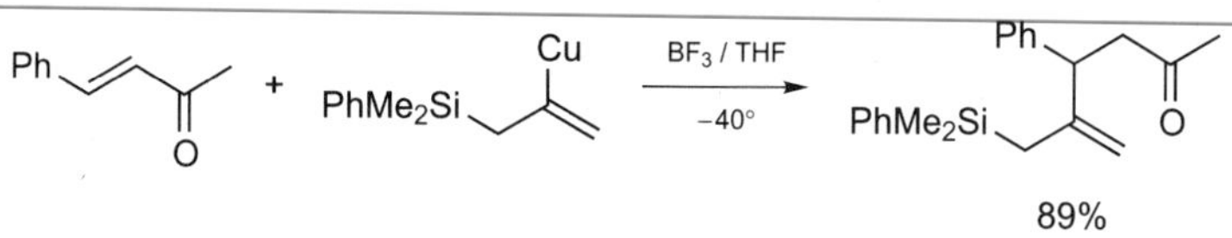

Ph O + $PhMe_2Si$ Cu

BF_3 / THF

−40°

Ph $PhMe_2Si$ O

89%

Stannylcupration is analogous to silylcupration. Such a reaction on 1,4-diyn-3-ols is regioselective and stereoselective.[5]

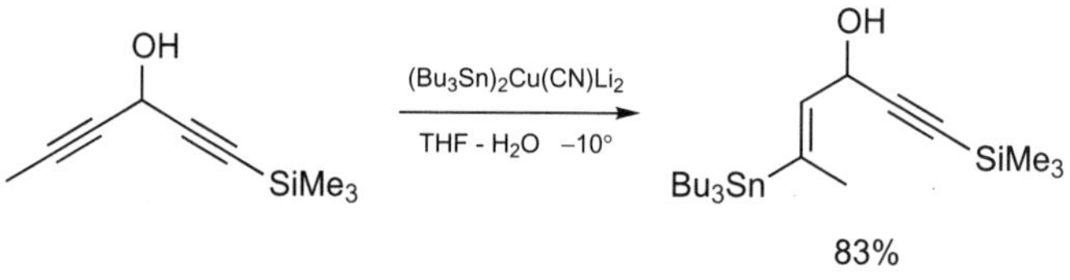

OH $SiMe_3$

$(Bu_3Sn)_2Cu(CN)Li_2$

THF - H_2O −10°

OH Bu_3Sn $SiMe_3$

83%

Conjugate additions. Methoxyallylcopper reagents are formed from the allyl ethers via lithiation (with *s*-BuLi) and they give monoenol ethers of 1,6-dicarbonyl compounds on conjugate addition to enones.

N-(*t*-Butoxycarbonyl)dimethylamine is lithiated at one of the *N*-methyl groups. A route to β-alkylidenepyrrolidinones is developed on further conversion to the cuprate reagents and subsequent addition to 2,3-alkadienoic esters.[6]

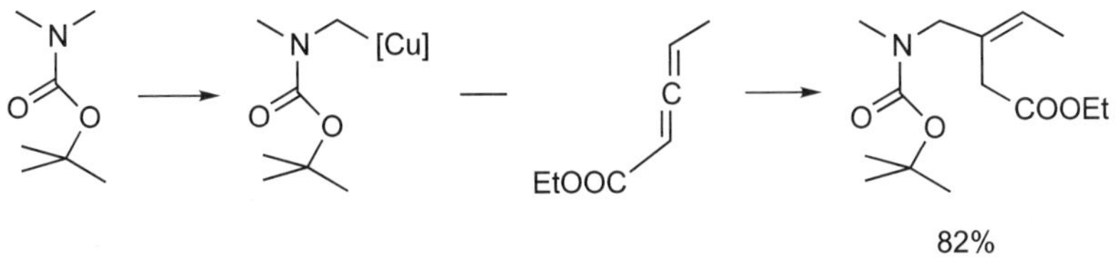

82%

Addition of organocuprates to cross-conjugated enynones leads to the dienones.[7] Dimethyl sulfide stabilizes the cuprate adduct intermediates when it is employed as a reaction medium.[8]

The conjugate addition–trapping protocol has been applied to preparation of phosphonates. Under CO carbonylation occurs prior to the addition and γ-ketophosphonates are formed.[9]

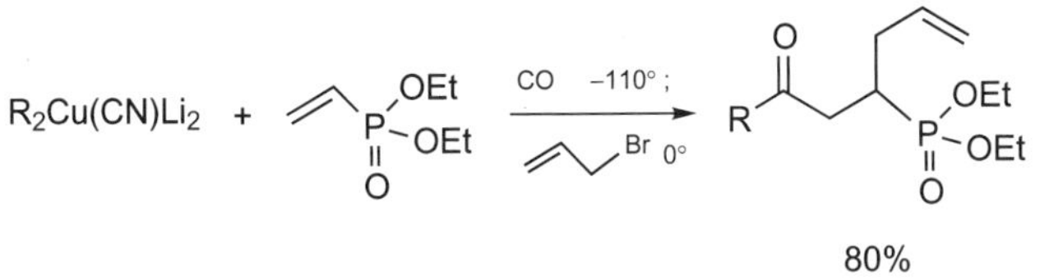

$R_2Cu(CN)Li_2$ + ... CO −110° ; ... Br 0° ...

80%

Many methods have been developed for the synthesis of chiral 3-substituted cycloalkanones. A clever design incorporates a neomenthol moiety at the α-position prior to treatment with organocopper reagents.[10]

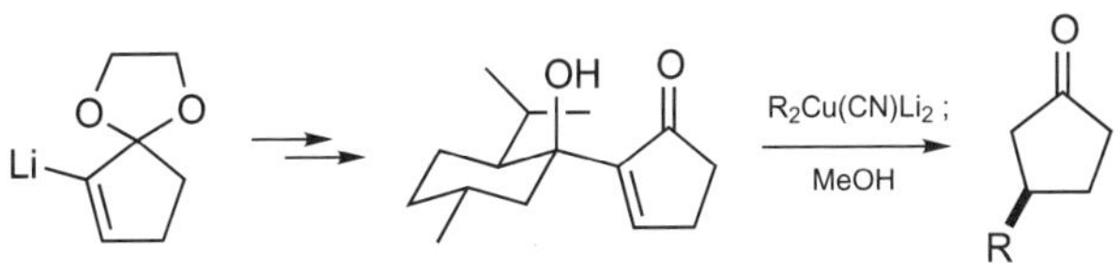

$R_2Cu(CN)Li_2$; MeOH

S_N2' displacements. 1-Bromo-1,2-alkadienes undergo transpositional substitution to give 1-alkynes. The chirality of the allene moiety determines the configuration of the entrant group.[11] (Note the Zn-based cuprate reagents can be used[12]).

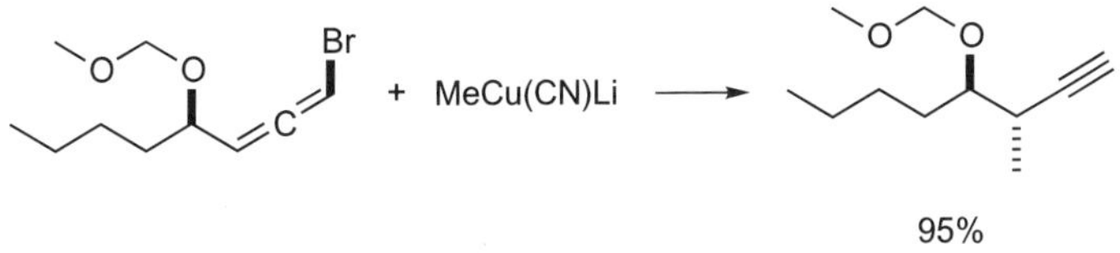

+ MeCu(CN)Li

95%

Olefination.[13] Under the influence of $BF_3 \cdot OEt_2$, acetals are converted to 2-substituted 1-alkenes on reaction with $Me_3SiCH_2Cu(PBu_3) \cdot LiI$.

Furans and pyrroles.[14] Of 2-alkynyl-1,3-dithiolanes, the consecutive reactions with cuprate reagents and aldehydes or aldimines lead to the five-membered heterocycles.

Bu$_2$CuLi / THF ; PhCHO ; CF$_3$COOH

60%

[1]Reginato, G., Mordini, A., Valacchi, M., Grandini, E. *JOC* **64**, 9211 (1999).
[2]Fleming, I., de Marigorta, M.E. *JCS(P1)* 889 (1999).
[3]Liepins, V., Karlstrom, A.S.E., Backvall, J.E. *OL* **2**, 1237 (2000).
[4]Barbero, A., Garcia, C., Pulido, F.J. *TL* **40**, 6649 (1999).
[5]Betzer, J.-F., Pancrazi, A. *S* 629 (1999).
[6]Dieter, R.K., Lu, K. *TL* **40**, 4011 (1999).
[7]Lee, P.H., Park, J., Lee, K., Kim, H.-C. *TL* **40**, 7109 (1999).
[8]Kingsbury, C.L., Sharp. K.S., Smith, R.A.J. *T* **55**, 14693 (1999).
[9]Li, N.-S., Yu, S., Kabalka, G.W. *OM* **18**, 1811 (1999).
[10]Funk, R.L., Yang, G. *TL* **40**, 10739 (1999).
[11]Bernard, N., Chemla, F., Normant, J.F. *TL* **40**, 1649 (1999).
[12]Caporusso, A.M., Filippi, S., Barontini, F., Salvadori, P. *TL* **41**, 1227 (2000).
[13]Suzuki, T., Oriyama, T. *SL* 859 (2000).
[14]Lee, C.-F., Yang, L.-M., Hwu, T.-Y., Feng, A.-S., Tseng, J.-C., Luh, T.-Y. *JACS* **122**, 4992 (2000).

Organogallium reagents.

Conjugate additions. Both triallylgallium[1] and lithium tetraorganogallates[2] are suitable addends for electron-deficient alkenes.

[1]Araki, S., Horie, T., Kato, M., Hirashita, T., Yamamura, H., Kawai, M. *TL* **40**, 2331 (1999).
[2]Han, Y., Huang, Y.-Z., Fang, L., Tao, W.T. *SC* **29**, 867 (1999).

Organolithium reagents. 20, 268–269

Ketones. At low temperatures, 3,4-dihydropyranones react with RLi. The adducts can be properly manipulated to afford the 1,5-diketones.[1]

THF, –75° ; Me$_3$SiCl ; H$_3$O$^+$

77%

Ketones, including chiral α-amino ketone derivatives,[2] are also produced from the organolithium reaction with morpholine amides.

Addition reactions. A route to *C*-arylglycosides involves reaction of lithioglycals with quinones derivatives.[3]

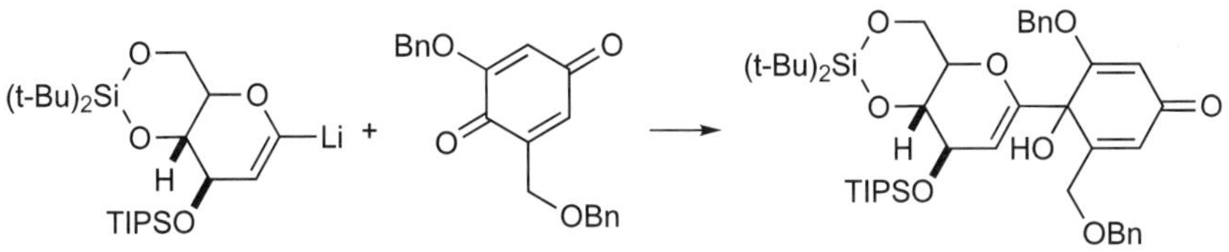

Addition of RLi to cinnamaldehyde oxime ethers gives allylic amines. Those containing a chirality center at the carbinyl site are subject to 1,4-asymmetric induction, therefore optically active α-amino acid derivatives are readily synthesized (upon cleavage of the double bond).[4]

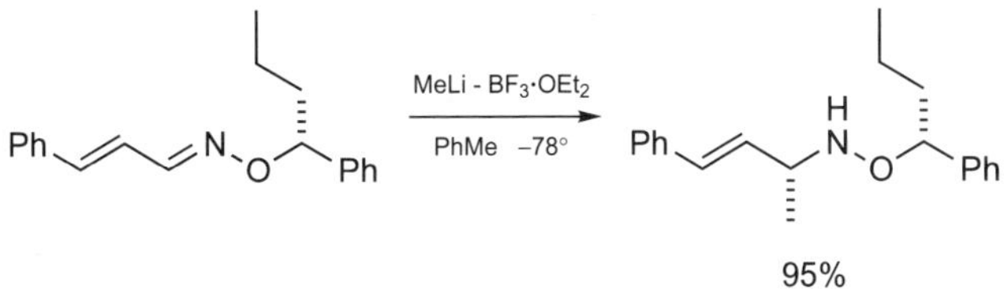

The adducts derived from a bicyclic oxazolidine bearing an α-cyano group undergo ring expansion on reduction with $LiAlH_4$.[5]

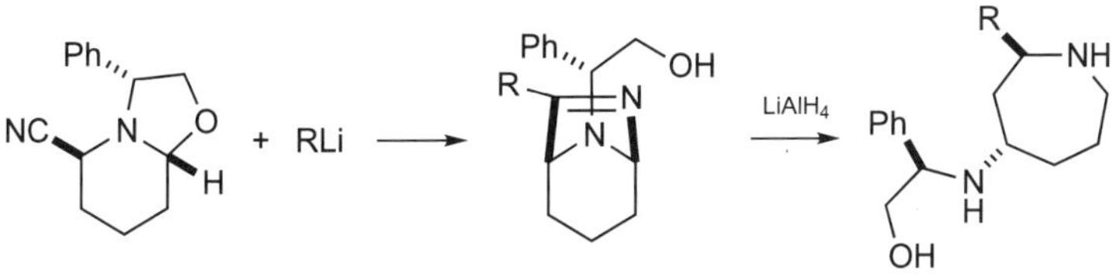

Sulfines provide dithioacetal monoxides in the addition reaction.[6] A controlled reaction can avoid attack on the carbonyl group.

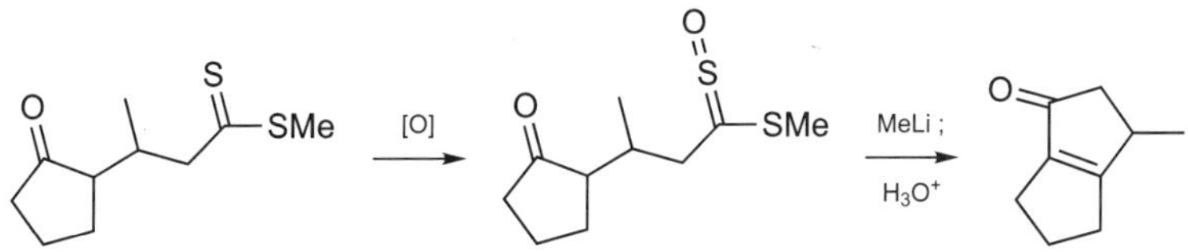

Butenolides and 1-isoindolinones. Reaction of enals containing an iron substitutent with RLi results in the formation of γ-substituted (with R) butenolides.[7] Thus, the addition also induces cyclocarbonylation.

CHO
Fe(CO)$_2$Cp
MeLi / THF
50%

3-Substituted 1-isoindolinones are synthesized from *N,N*-dimethylaminophthalimide via reaction with organolithiums (or Grignard reagents), deoxygenation with Et_3SiH-CF_3COOH, and N—N bond cleavage with Zn in HOAc.[8]

N–NMe$_2$
RLi
N–NMe$_2$
OH
R
85–92%
Et_3SiH - CF_3COOH ;
Zn - HOAc
N–H
R

Eliminations. Thionocarbonates derived from 1,2-diols are transformed into iodo thiocarbonates with RI, which decompose into alkenes on treatment with RLi.[9] The three-step process represents a reductive deoxygenation.

Alkynes are formed when trichloromethylcarbinyl tosylates react with MeLi. This reaction is suitable for the synthesis of ethynylcyclopropane.[10]

Organoiodine compounds.[11] Iodoarenes, iodoalkenes, and iodoalkynes are prepared from organolithiums by reaction with 2,2,2-trifluoroethyl iodide.

Homopropargylic alcohols. In the opening of hetero-substituted epoxides with alkynyllithiums, trimethylgallium shows good catalytic reactivity.[12]

α-Amino acids.[13] Chiral amino acids (L-alanine, L-methionine, L-leucine, L-homocysteine) have been prepared via carboxylation of the proper *N*-(α-lithioalkyl)-4-phenyloxazolidin-2-ones. The configuration of the chelated organolithium species is such that the side chain would not sterically interact with the phenyl group. Regardless of the original configuration in the precursorial stannane, rapid equilibration to the more stable isomer after Sn–Li exchange ensures the generation of the predicted (desired) product in each instance.

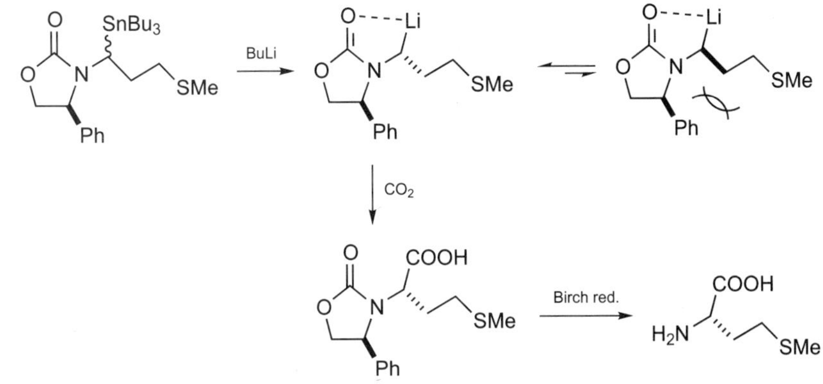

Zirconacycles → carbocycles.[14] Replacement of the zirconium atom by carbon in zirconacycles is achieved by reaction with α,α-dihaloalkyllithiums.

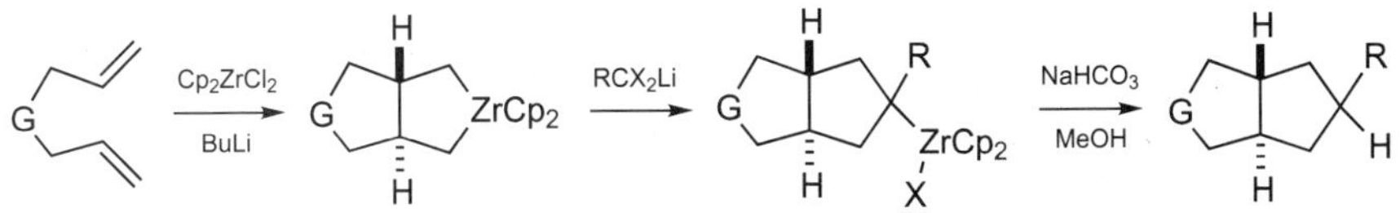

[1]Harrowven, D.C., Hannam, J.C. *T* **55**, 9333 (1999).
[2]Sengupta, S., MondaL, S., Das, D. *TL* **40**, 4107 (1999).
[3]Parker, K.A., Georges, A.T. *OL* **2**, 497 (2000).
[4]Moody, C.J., Gallagher, P.T., Lightfoot, A.P., Slawin, A.M.Z. *JOC* **64**, 4419 (1999).
[5]Cutri, S., Bonin, M., Micouin, L., Froelich, O., Quirion, J.-C., Husson, H.-P. *TL* **41**, 1179 (2000).
[6]Corbin, F., Alayrac, C., Metzner, P. *EJOC* 2859 (1999).
[7]Moller, C., Mikulas, M., Wierschem, F., Rück-Braun, K. *SL* 182 (2000).
[8]Deniau, E., Enders, D. *TL* **41**, 2347 (2000).
[9]Adiyaman, M., Jung, Y.-J., Kim, S., Saha, G., Powell, W.S., FitzGerald, G.A., Rokach, J. *TL* **40**, 4019 (1999).
[10]Wang, Z., Campagna, S., Yang, K., Xu, G., Pierce, M.E., Fortunak, J.M., Confalone, P.N. *JOC* **65**, 1889 (2000).
[11]Blackmore, I.J., Boa, A.N., Murray, E.J., Dennis, M., Woodward, S. *TL* **40**, 6671 (1999).
[12]Ooi, T., Morikawa, J., Ichikawa, H., Maruoka, K. *TL* **40**, 5881 (1999).
[13]Jeanjean, F., Fournet, G., Le Bars, D., Gore, J. *EJOC* 1297 (2000).
[14]Vicart, N., Whitby, R.J. *CC* 1241 (1999).

Organotellurium reagents. 19, 237–239

Organocopper reagents.[1] Dialkenyl tellurides are readily transformed into copper reagents for conjugate addition, on reaction with CuCN–RLi. When the tellurides contain both alkyl and alkenyl groups they are also transformed into alkenylcuprates.

Enynes.[2] (*Z*)-Alkenyl butyl tellurides couple with 1-alkynes at room temperature in the presence of $PdCl_2$–CuI.

[1]Araujo, M.A., Barrientos-Astigarraga, R.E., Ellensohn, R.M., Comasseto, J.V. *TL* **40**, 5115 (1999).
[2]Zeni, G., Comasseto, J.V. *TL* **40**, 4619 (1999).

Organotin reagents. 19, 239–240

Preparation. Aryltributylstannanes are available from coupling of ArI with Bu_3SnH using a catalyst system containing $PdCl_2/(PMePh_2)_2$ and KOAc.[1] Reaction of potassium triphenylstannate (from Ph_3SnH and *t*-BuOK) with haloarenes (and haloheteroarenes) provides Ph_3SnAr.[2]

When alkenyl selenides are available, they can be converted to the alkenylstannanes by reaction with Bu_3SnH in the presence of AIBN.[3]

SePh / R / MeOOC / NHBz —(Bu_3SnH - AIBN, PhMe Δ)→ $SnBu_3$ / R / MeOOC / NHBz —(H^+)→ R / MeOOC / NHBz

Metal exchange.[4] Allenyltributylstannane is transformed into propargylmetal halides by reaction at low temperatures with MX_n, where M = P, As, Sb, Ge, B. The allenyl isomers are favored at higher temperatures.

Conjugate addition.[5] Contrasting diastereoselectivities are observed for the addition of R_3SnLi and $R_3Sn(Et_2)ZnLi$ to γ-alkoxy-α,β-unsaturated esters.

COOMe —(THF, −78°)→ Bu_3Sn / COOMe + Bu_3Sn / COOMe

Bu_3SnLi	58%	100	:	0
$Bu_3SnZnEt_2Li$	84%	0	:	100

[1]Murata, M., Watanabe, S., Masuda, Y. *SL* 1043 (2000).
[2]Lockhart, M.T., Chopa, A.B., Rossi, R.A. *JOMC* **582**, 229 (1999).
[3]Berkowitz, D.B., McFadden, J.M., Chisowa, E., Semerad, C.L. *JACS* **122**, 11031 (2000).
[4]Guillemin, J.-C., Malagu, K. *OM* **18**, 5259 (1999).
[5]Krief, A., Provins, L., Dumont, W. *ACIEE* **38**, 1946 (1999).

Organotitanium reagents.

3-Hydroxy aldehydes.[1] On treatment with MeLi and then $(BuO)_4Ti$, silyl enol ethers of aldehydes form enoxytitanates that condense with carbonyl compounds. It should be emphasized that in conventional aldol reactions aldehydes behave as acceptors. In this method, the donor role is established.

β-Alkenyl ketones.[2] 1-Silyl-1-alken-2-yltitanium reagents generated from 1-silylalkynes with *i*-PrMgCl/$(i\text{-PrO})_4Ti$ are used as Michael donors in the presence of $Li_2Cu(CN)Cl_2$.

[1]Yachi, K., Shinokubo, H., Oshima, K. *JACS* **121**, 9465 (1999).
[2]Urabe, H., Hamada, T., Sato, F. *JACS* **121**, 2931 (1999).

Organozinc reagents. 13, 220–222; **14,** 233–235; **15,** 238–240; **16,** 246–248; **17,** 228–234; **18,** 264–265; **19,** 240–241; **20,** 270–275

Preparations. Functionalized organozinc reagents have been prepared from alkenes via stereoselective and regioselective hydroboration and B–Zn exchange.[1] An interesting synthesis of carboxylic acids by chain extension from (ω-1)-unsaturated acids involves prior attachment to a resin (in the form of polymeric esters).[2]

OBn; Ph; Et_2BH - Me_2S ; i-Pr_2Zn; H; OBn; Ph; Zn; CuCN.2LiCl; Br; H; OBn; Ph; H; 65%

An access to unsymmetrical 1,6-diketones by conjugate addition of γ-keto zincates to enones relies on successful enolization and a Sn–Zn exchange reaction of β-stannyl ketones.[3]

Lithiation and Li–Zn exchange make 3-chloroallenylzinc bromides available from propargyl chlorides. These reagents react with aldimines to form 2-alkynylaziridines.[4] Allenylzinc reagents prepared from chiral propargylic mesylates have been used to synthesize *anti*-homopropargylic alcohols.[5]

AcO; OMs; +; OHC; $Pd(OAc)_2$ - Ph_3P; Et_2Zn; AcO; OH; 75% (*anti* : *syn* 95 : 5)

Electrogenerated zinc (Zn anode) reacts rapidly with functionalized alkyl iodides and cross-coupling reactions can be effected in situ.[6]

Mixed diorganozinc reagents containing a nontransferable neopentyl or neophyl group are obtained by a metathetic reaction between dineopentylzinc and another R_2Zn, or by reaction of neopentyllithium with RZnX. Neophyl analogues are similarly prepared. These reagents have been used in asymmetric alkylation of aldehydes.[7]

Addition to C=X bond. Allylzinc reagents add to the carbonyl group of β-keto phosphonates.[8] With respect to a catalytic enantioselective synthesis of chiral pyrimidyl carbinols, an automultiplication phenomenon is elucidated.[9]

CHO + OH (catalyst) i-Pr_2Zn ; HCl - H_2O OH

Addition of organozincs to iminium salts[10] and oxime ethers[11] furnish amines. Diethylzinc used in the latter process is a chain-transfer agent.

α-Seleno zinc reagents such as $ArSeCH_2ZnCl$ have been developed.[12] Their addition to aldehydes is mediated by CuCN–LiCl and promoted by $BF_3 \cdot OEt_2$.

Conjugate additions. Allylzinc halides add to β-nitrostyrenes in DMF at room temperature (12 examples, 76–90%).[13] This result is contrary to the report on substitution of the nitro group under microwave irradiation,[14] therefore the situation needs clarification.

An enyne group that is introduced to the β-position of an enone in the nickel-catalyzed addition is generated from alkynylzincation.[15]

O + $[R-\equiv]_2Zn$ + $R-\equiv$ Ni(acac)$_2$ / Me_3SiCl / THF → O R R

R = Bu 67%

Air facilitates the addition of diethylzinc to 6-alkoxy- and 6-acyloxypyran-3-ones, due to formation of EtZnOOEt, which is a catalyst.[16]

Addition to alkynes and dienes. Organozincation of alkynes and dienes generates new organozinc species that are exploited synthetically. For example, alkenylzincs derived from propargyltrimethylsilane are useful for the synthesis of 1,3-disubstituted 1,3-dienes.[17]

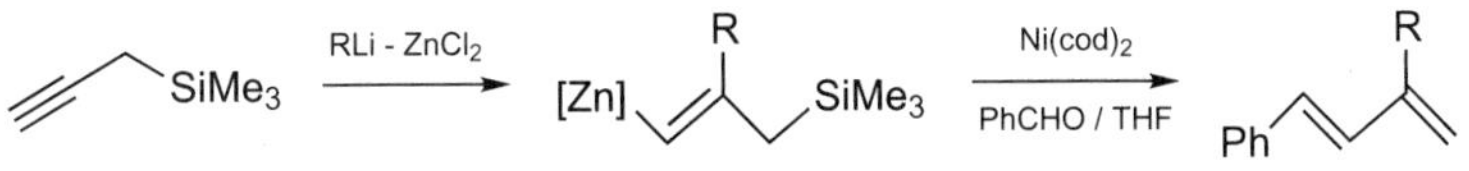

The addition of allylzinc bromide (allylmagnesium bromide + $ZnBr_2$) to alkynyllithiums gives rise to *gem*-bimetalloalkenes. Chlorination of the latter species with $PhSO_2Cl$ sets up a Fritsch–Buttenberg–Wiechell rearrangement.[18]

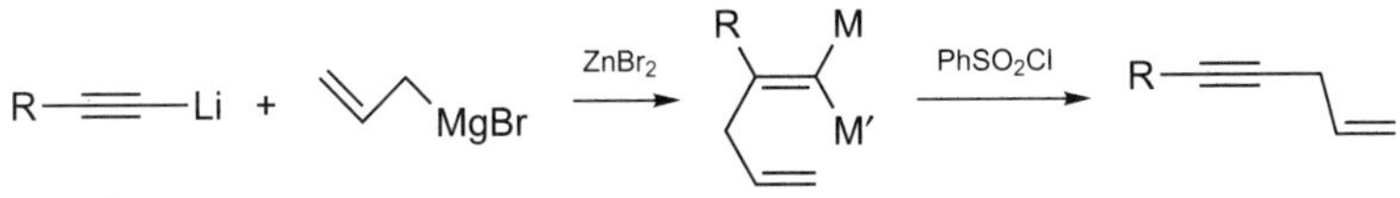

Coupling reactions. Cross-coupling reactions of organozincs with activated triflates need the presence of cuprates.[19] Those involving tellurium compounds and CuI are catalyzed by Pd(0) complexes.[20] Regioselectivity in the Pd(0)-catalyzed coupling of electron-deficient alkenyl halides with allenic/propargylic zinc reagents depends on the substitution patterns of the latter species therefore either alkenynoic or alkatrienoic esters result.[21]

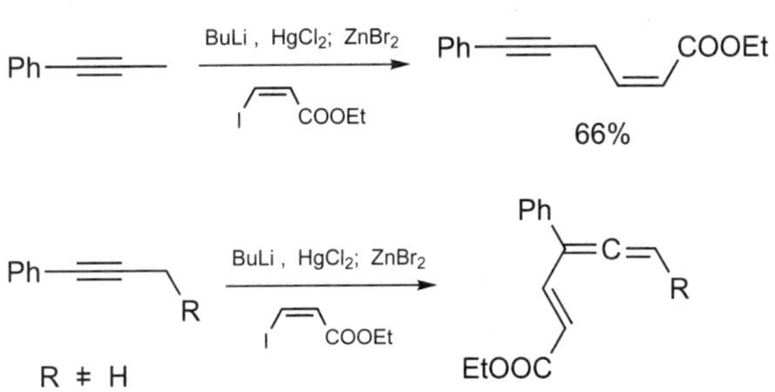

Other coupling partners to organozinc reagents include heterocyles such as 2-methylthiobenzothiazole,[22] alkenyl aryl iodonium triflates (alkenyl group transfer for synthesis of trisubstituted alkenes),[23] and aryl heteroaryl ethers.[24] Improved nickel-catalyzed cross-coupling conditions between *ortho*-substituted aryl iodides–nonaflates and alkylzinc iodides in solution and in the solid phase have been defined.[25]

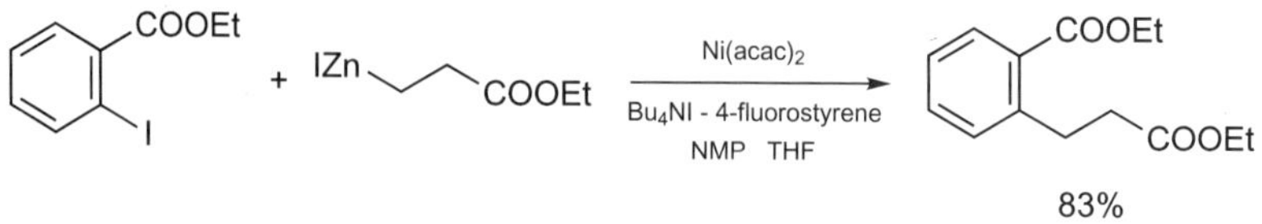

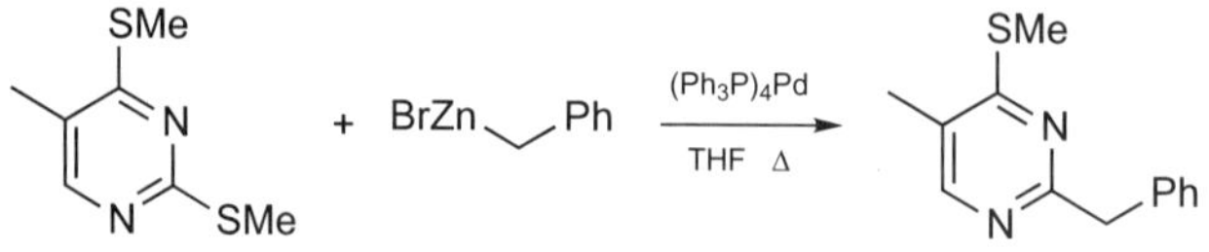

Alkylations. With Pd-catalyst to open all allylic oxabridge diorganozincs intercept the π-allylpalladium species by donating their σ-ligands.[26]

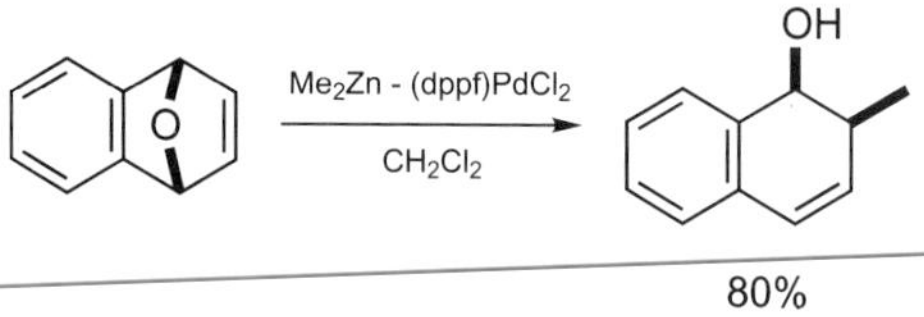

Elimination. Chiral terminal allenes have been prepared from 2-bromo-2-alken-1-yl mesylates bearing a protected amino group by treatment with diethylzinc.[27] An alternative way to set up the elimination is by alkylation of alkenylcopper species containing an α-sulfonylalkenyl (or sulfinylalkenyl) group with α-iodoalkylzinc reagents.[28] The latter protocol allows preparation of trisubstituted allenes.

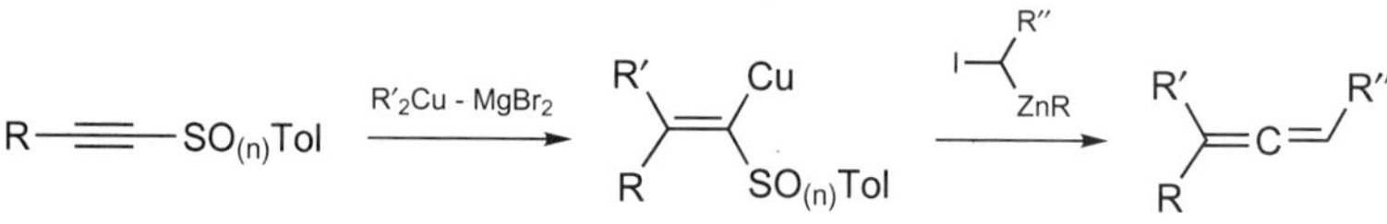

[1]Boudier, A., Hupe, E., Knochel, P. *ACIEE* **39**, 2294 (2000).
[2]Jackson, R.F.W., Oates, L.J., Block, M.H. *CC* 1401 (2000).
[3]Ryu, I., Ikebe, M., Sonoda, N., Yamamoto, S., Yamamura, G., Komatsu, M. *TL* **41**, 5689 (2000).
[4]Chemla, F., Hebbe, V., Normant, J.F. *TL* **40**, 8093 (1999).
[5]Marshall, J.A., Adams, N.D. *JOC* **64**, 5201 (1999).
[6]Kurono, N., Sugita, K., Takasugi, S., Tokuda, M. *T* **55**, 6097 (1999).
[7]Lutz, C., Jones, P., Knochel, P. *S* 312 (1999).
[8]Lentsch, L.M., Wiemer, D.F. *JOC* **64**, 5205 (1999).
[9]Shibata, T., Morioka, H., Hayase, T., Choji, K., Soai, K. *JACS* **118**, 471 (1996).
[10]Millot, N., Piazza, C., Avolio, S., Knochel, P. *S* 941 (2000).

[11]Bertrand, M.P., Feray, L., Nouguier, R., Perfetti, P. *JOC* **64**, 9189 (1999).
[12]Duan, D.-H., Huang, X. *JCR(S)* 26 (1999).
[13]Kumar, H.M.S., Reddy, B.V.S., Reddy, P.T., Yadav, J.S. *TL* **40**, 5387 (1999).
[14]Hu, Y., Yu, J., Yang, S., Wang, J.-X., Yin, Y. *SC* **29**, 1157 (1999).
[15]Ikeda, S., Kondo, K., Sato, Y. *CL* 1227 (1999).
[16]van der Deen, H., Kellogg, R.M., Feringa, B.L. *OL* **2**, 1593 (2000).
[17]Qi, X., Montgomery, J. *JOC* **64**, 9310 (1999).
[18]Rezael, H., Yamanoi, S., Chemla, F., Normant, J.F. *OL* **2**, 419 (2000).
[19]Lipshutz, B.H., Vivian, R.V. *TL* **40**, 2871 (1999).
[20]Dabdoub, M.J., Dabdoub, V.B., Marino, J.P. *TL* **41**, 433, 437 (2000).
[21]Ma, S., Zhang, A., Yu, Y., Xia, W. *JOC* **65**, 2287 (2000).
[22]Angiolelli, M.E., Casalnuovo, A.L., Selby, T.P. *SL* 905 (2000).
[23]Hinkle, R.J., Leri, A.C., David, G.A., Erwin, W.M. *OL* **2**, 1521 (2000).
[24]Brigas, A.F., Johnstone, R.A.W. *JCS(P1)* 1735 (2000).
[25]Jensen, A.E., Dohle, W., Knochel, P. *T* **56**, 4197 (2000).
[26]Lautens, M., Renaud, J.-L., Hiebert, S. *JACS* **122**, 1804 (2000).
[27]Ohno, H., Toda, A., Oishi, S., Tanaka, T., Takemoto, Y., Fujii, N., Ibuka, T. *TL* **41**, 5131 (2000).
[28]Varghese, J.P., Knochel, P., Marek, I. *OL* **2**, 2849 (2000).

Organozirconium reagents.

Alkylations. Acylzirconocenes react with allyl and propargyl halides to give ketones.[1] In the latter case, allenyl ketones are obtained, as a result of the S_N2' mechanism.

Cl
ZrCp2
Ph
O
Cu+
DMF
Br
Ph
O
91%
Br
Ph
C
O
61%

Under the influence of $(Ph_3P)_4Ni$, α,α-difluoro-β,γ-unsaturated esters are formed from the reaction of alkenylzirconium reagents with bromodifluoroacetic esters.[2]

Acylations. Alkenylzirconocene chlorides that are generated from hydrozirconation of alkynes readily undergo copper-catalyzed acylations. Thus, enones containing tin and selenium substituents are available from alkynylstannanes[3] and alkynylselenides,[4] respectively. Alkenyl alkynyl ketones are obtained when the reaction is carried out under carbon monoxide with alkynyliodonium salts.[5]

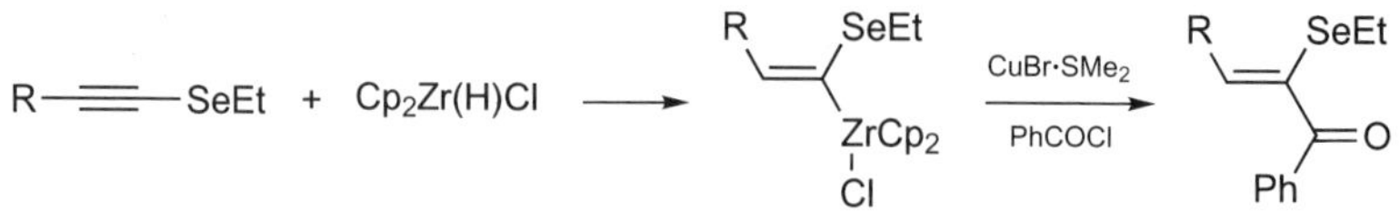

After hydrozirconation with Cp_2ZrEt_2 and reaction with an chloroformic ester, alkynes directly (requiring no catalyst) afford conjugated esters.[6]

Heterofunctionalizations. Practically all the substrates are alkenylzirconocene chlorides. They behave well in halogenation,[7] phosphorylation,[8] sulfenylation,[9] sulfinylation,[10] selenenylation[11,12] and selenoacylation,[13] as well as tellurylation,[11,14] which gives rise to the substituted alkenes.

α-Cyanoalkenylation.[15] Lithiated epoxynitriles give 2-cyano-1,3-dienes on reaction with alkenylzirconocene chlorides.

R = Hx 64%

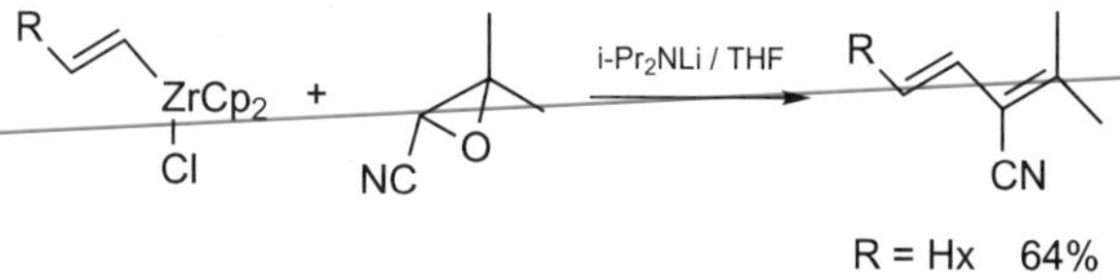

Homoallylic alcohols and 1,3-dienes. Insertion of carbenoids to the C—Zr bond of alkenylzirconocene chlorides by *gem*-chloroalkyllithium generates reactive reagents that can be exploited in a carbon chain extension. Thus, consecutive reactions of the zirconocene species with $Me_3SiCH(Cl)Li$ and aldehydes lead to homoallylic alcohols.[16] Similar insertion with 1-chloro-1-lithioalkenes gives rise to conjugated dienes.[17] This method is adaptable to the synthesis of more extended conjugated systems.

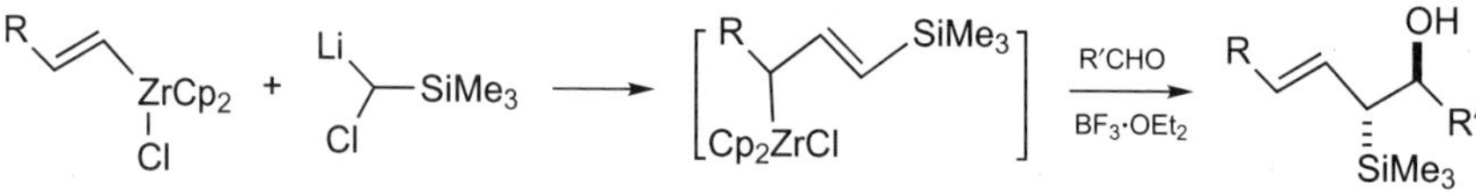

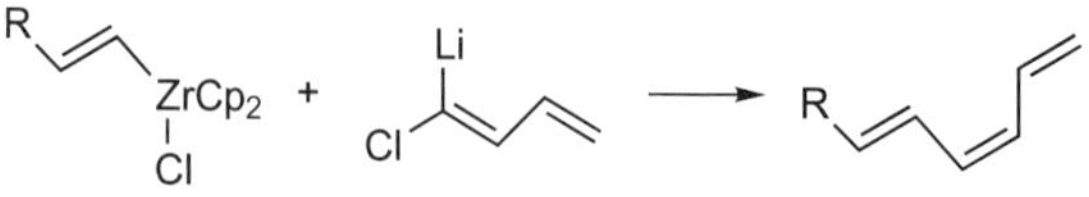

Carbocycles.[18] Either five- or six-membered carbocycles are obtained from zirconacyclopentanes and alkynes in the presence of CuCl.

[1]Hanzawa, Y., Narita, K., Taguchi, T. *TL* **41**, 109 (2000).
[2]Schwaebe, M.K., McCarthy, J.R., Whitten, J.P. *TL* **41**, 791 (2000).
[3]Zhong, P., Xiong, Z.-X., Huang, X. *SC* **30**, 3245 (2000).
[4]Sun, A., Huang, X. *S* 775 (2000).
[5]Sun, A., Huang, X. *T* **55**, 13201 (1999).
[6]Takahashi, T., Xi, C., Ura, Y., Nakajima, K. *JACS* **122**, 3228 (2000).
[7]Huang, X., Zhong, P. *SC* **29**, 3425 (1999).
[8]Zhong, P., Huang, X., Xiong, Z.-X. *SL* 721 (1999).
[9]Huang, X., Xu, X.-H., Zheng, W.-X. *SC* **29**, 2399 (1999).
[10]Huang, X., Zhong, P., Guo, M.-P. *JOMC* **603**, 249 (2000).
[11]Park, C.P., Sung, J.W., Oh, D.Y. *SL* 1055 (1999).
[12]Ma, Y., Huang, X. *SC* **29**, 429 (1999).
[13]Zhong, P., Xiong, Z.-X., Huang, X. *SC* **30**, 887 (2000).
[14]Huang, X., Liang, C.-G. *SC* **30**, 1737 (2000).
[15]Kasatkin, A.N., Whitby, R.J. *TL* **41**, 6201 (2000).
[16]Kasatkin, A.N., Whitby, R.J. *TL* **40**, 9353 (1999).
[17]Kasatkin, A.N., Whitby, R.J. *JACS* **121**, 7039 (1999).
[18]Liu, Y., Shen, B., Kotora, M., Takahashi, T. *ACIEE* **38**, 949 (1999).

Osmium tetroxide. 13, 222–225; **14,** 233–239; **15,** 240–241; **16,** 249–253; **17,** 236–240; **18,** 265–267; **19,** 241–242; **20,** 275–276

Modifications. A recoverable and reusable catalyst has been prepared from an acrylonitrile–butadiene–styrene polymer and OsO_4.[1]

Dihydroxylations. By using molecular oxygen to sustain the oxidation, high atom-efficiency is attained.[2] A new cocatalyst duet is *N*-methylmorpholine and the flavin analogue 1,3-dimethyl-5-ethyl-5,10-dihydrobenzopteridine-2,4-dione.[3]

An allylic trichloroacetamino group in cycloalkenes directs dihydroxylation with OsO_4 and quinuclidine *N*-oxide as the oxidizing system.[4] An interesting change in the diastereoselectivity by variation of the oxidant composition has been observed.[5]

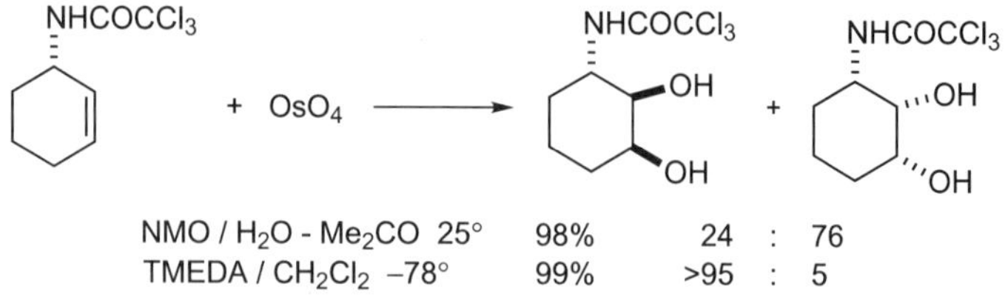

NHCOCCl3
+ OsO4
NHCOCCl3
OH
OH
+
NHCOCCl3
OH
OH
NMO / H2O - Me2CO 25° 98% 24 : 76
TMEDA / CH2Cl2 −78° 99% >95 : 5

Stereoselectivity in the catalytic dihydroxylation of acyclic allylic alcohols can be enhanced.[6]

Functional alkanes bearing a 2,3-dihydroxylated pattern are readily obtained, for example, aldehydes from 1-acetoxy-2-alkenyl phenyl sulfones[7] and esters from ketene acetals.[8]

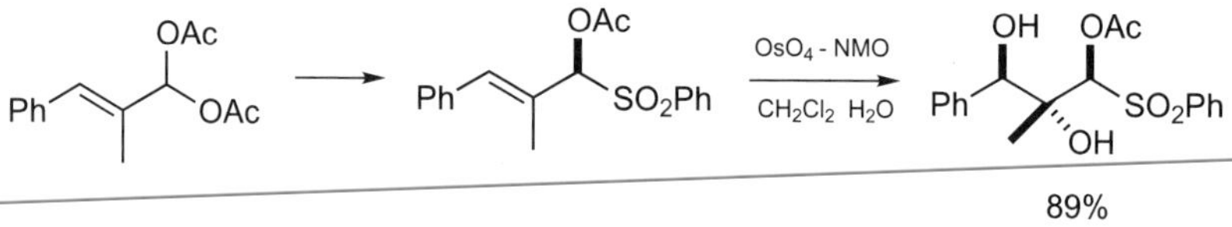

OAc
Ph
OAc
OAc
Ph
SO2Ph
OsO4 - NMO
CH2Cl2 H2O
OH OAc
Ph
SO2Ph
OH
89%

In situ oxidation of the diols derived from terminal alkenes results in α-hydroxy carboxylic acids.[9]

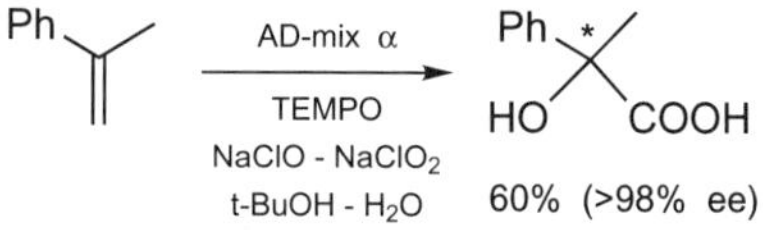

Ph
AD-mix α
TEMPO
NaClO - NaClO2
t-BuOH - H2O
Ph
*
HO
COOH
60% (>98% ee)

Aminohydroxylations. Baylis–Hillman alkenes give predominantly *syn*-diols,[10] whereas the reaction with α,β-unsaturated phsophonates gives rise to α-hydroxy-β-aminophosphonates.[11]

OH
COOMe
K2OsO2(OH)4
TsN(Cl)Na
MeCN - H2O
OH
COOMe
OH
NHTs
+
OH
COOMe
HO
NHTs

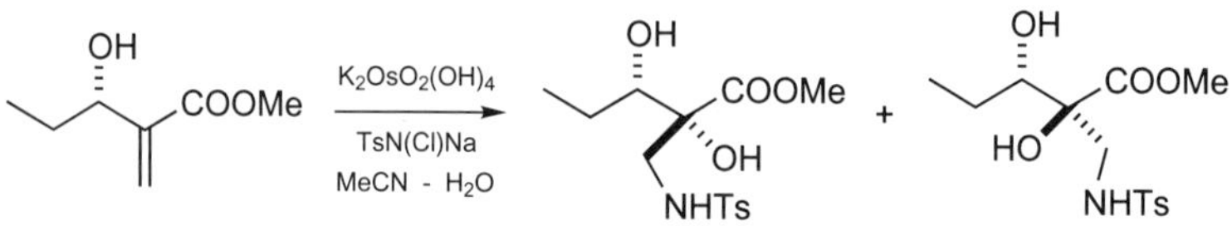

Amino-substituted heterocycles (e.g., 2-aminopyrimidine),[12] sodium *N*-chloro-*t*-butylsulfonamide[13], and primary amides have been developed as a nitrogen source for these reactions.[14] The untenable situation in using *t*-butyl hypochlorite (3 equiv, freshly prepared) as cooxidant in industrial settings is amended by replacing it with 1,3-dichloro-5,5-dimethylhydantoin.[15]

Aryl 2-alkenoates show a reversed regioselectvity in the asymmetric aminohydroxylation.[16]

$K_2OsO_2(OH)_4$
$(DHQ)_2$-AQN
Z-NH_2 / t-BuOCl
n-PrOH / H_2O

NHZ OH O Br + OH NHZ O Br

(7 : 1)

[1]Kobayashi, S., Endo, M., Nagayama, S. *JACS* **121**, 11229 (1999).
[2]Dobler, C., Mehltretter, G., Beller, M. *ACIEE* **38**, 3026 (1999).
[3]Bergstad, K., Jonsson, S.Y., Bäckvall, J.-E.. *JACS* **121**, 10424 (1999).
[4]Blades, K., Donohoe, T.J., Winter, J.J.G., Stemp, G. *TL* **41**, 4701 (2000).
[5]Donohoe, T.J., Blades, K., Helliwell, M., Moore, P.R., Winter, J.J.G., Stemp, G. *JOC* **64**, 2980 (1999).
[6]Donohoe, T.J., Waring, M.J., Newcombe, N.J. *SL* 149 (2000).
[7]Trost, B.M., Crawley, M.L., Lee, C.B. *JACS* **122**, 6120 (2000).
[8]Monenschein, H., Drager, G., Jung, A., Kirschning, A. *CEJ* **5**, 2270 (1999).
[9]Aladro, F.J., Guerra, F.M., Moreno-Dorado, F.J., Bustamante, J.M., Jorge, Z.D., Massanet, G.M. *TL* **41**, 3209 (2000).
[10]Pringle, W., Sharpless, K.B. *TL* **40**, 5151 (1999).
[11]Thomas, A.A., Sharpless, K.B. *JOC* **64**, 8379 (1999).
[12]Goossen, L.J., Liu, H., Dress, K.R., Sharpless, K.B. *ACIEE* **38**, 1080 (1999).
[13]Gontcharov, A.V., Liu, H., Sharpless, K.B. *OL* **1**, 783 (1999).
[14]Demko, Z.P., Bartsch, M., Sharpless, K.B. *OL* **2**, 2221 (2000).
[15]Barta, N.S., Sidler, D.R., Somerville, K.B., Weissman, S.A., Larsen, R.D., Reider, P.J. *OL* **2**, 2821 (2000).
[16]Morgan, A.J., Masse, C.E., Panek, J.S. *OL* **1**, 1949 (1999).

Osmium trichloride–potassium ferricyanide.

Dihydroxylations. An efficient dihydroxylation protocol employs these oxidant couples and quinuclidine methanesulfonamide in aqueous *t*-butanol.[1]

[1]Eames, J., Mitchell, H.J., Nelson, A., O'Brien, P., Warren, S., Wyatt, P. *JCS(P1)* 1095 (1999).

Oxalyl chloride. 17, 241–242; **18,** 267–268; **19,** 243; **20,** 277

Diaryl ketones.[1] A Friedel–Crafts acylation method is shown in the following equation:

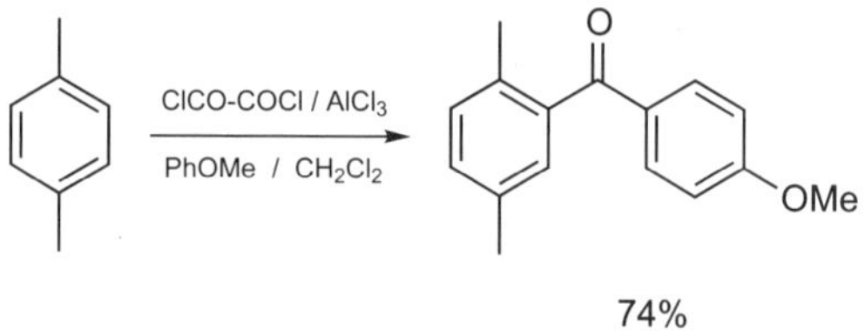

ClCO-COCl / AlCl3
PhOMe / CH2Cl2
O
OMe
74%

[1]Taber, D.F., Sethuraman, M.R. *JOC* **65**, 254 (2000).

***S*-(1-Oxido-2-pyridyl)-1,1,3,3-tetramethyluronium salts.**

Amides.[1] Carboxylic acids are converted to amides at room temperature by a combination of NH_4Cl and *i*-Pr_2NEt in DMF using the substituted uronium tetrafluoroborate or hexafluorophospahate (**1**) as condensing agent.

+NMe2
X−
N
S
NMe2
O−

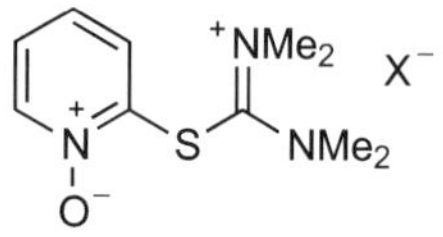

(**1**)

[1]Bailen, M.A., Chinchilla, R., Dodsworth, D.J., Najera, C. *TL* **41**, 9809 (2000).

1-Oxo-2,2,6,6-tetramethylpiperidine chloride.

β-Aminoxy chlorides.[1] The title compound behaves as an electrophilic agent toward activated alkenes. Allied oxoammonium salts can be used to form the chlorohydrin derivatives.

N
O
Cl−
+
R
CHCl3
R = Ph, OEt,...
N
O
R
Cl

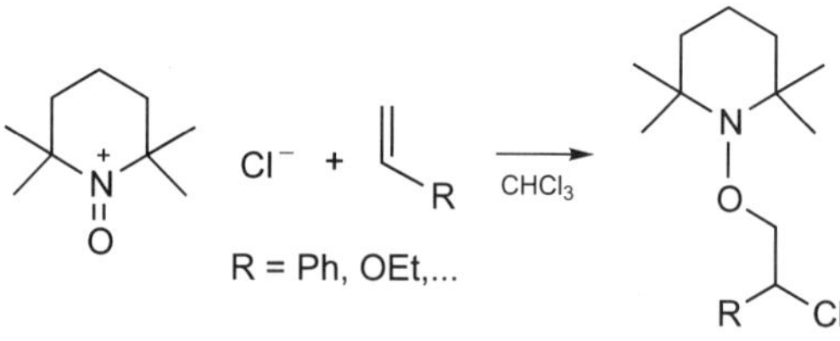

[1]Takata, T., Tsujino, Y., Nakanishi, S., Nakamura, K., Yoshida, E. *CL* 937 (1999).

Oxygen. 18, 268–269; **19,** 243–244; **20,** 277–279

Epoxidations. In the presence of perfluoroacetone and *N*-hydroxyphthalimide, hydrogen peroxide generated in situ from oxygen and 1-phenylethanol epoxidizes alkenes without a metal catalyst,[1] although there is also an alternative[2] in using *N*-hydroxyphthalimide, $Mo(CO)_6$, and $Co(OAc)_2$.

Oxidations. Benzylic alcohols undergo aerial oxidation to aromatic aldehydes in the presence of many different catalysts: OsO_4/CuCl–pyridine (for benzylic alcohols only OsO_4 is needed),[3] Ru-on-hydroxyapatite,[4] hydrotalcite-supported Pd(II),[5] and $CuBr{\cdot}SMe_2$ under fluorous biphasic conditions.[6]

A remarkable effect of quaternary ammonium bromides in the *N*-hydroxyphthalimide-catalyzed benzylic oxidation has been noted.[7]

Aromatic aldehydes themselves are oxidized to acid [8] by molecular oxygen in ionic liquid under the influence of $Ni(acac)_2$. Excellent yields and mild conditions are characteristic of the $VOCl_3$-catalyzed oxidation of α-hydroxy carbonyl compounds (to the dicarbonyl compounds).[9]

β-Ketoesters are oxidized at the α-position with oxygen in the presence of either manganese(II) acetate[10] or cobalt(II) chloride.[11] Cyclic allyl phosphonates give γ-acetoxy α,β-unsaturated phosphonates when they are exposed to oxygen in HOAc containing $PdCl_2$ and isopentyl nitrite.[12] Tertiary amine oxides are formed under cooxidation conditions, that is, O_2, Fe_2O_3, and isovaleraldehyde.[13]

Oxidation of organomercury compounds via formation of TEMPO derivatives and cleavage with Zn–HOAc completes the functionalization of alkenes.[14] Without TEMPO the oxidative capture of a primary radical generated from organomercurial is inefficient, and the reductive pathway (loss of functionality) becomes competitive.

HgCl

O_2 - $NaBH_4$ TEMPO / DMF ;

Zn / HOAc - H_2O

100°

OH

50%

Alkyl halides and tosylates are oxidized to carbonyl compounds by oxygen using CuCl–Kieseelguhr as catalyst.[15]

Sulfoxidation of saturated hydrocarbons[16] with O_2–SO_2 is catalyzed by $VO(acac)_2$.

$$\text{1-adamantanecarboxylic acid} \xrightarrow[\text{VO(acac)}_2\ /\ \text{HOAc}]{O_2 - SO_2} \text{3-sulfo-1-adamantanecarboxylic acid (COOH, } SO_3H)$$

Oxidation of unsaturated compounds. Methyl ketones are produced from 1-alkenes using molecular oxygen as oxidant [catalyst: $Pd(OAc)_2$–pyridine].[17] On the other hand, methyl 3,3-dimethoxypropanoate is formed when ethyl acrylate is oxidized on activated carbon-supported molybdovanadophosphate and $Pd(OAc)_2$ in acidic ethanol.[18] Treatment of enones with $LiAlH_4$ under dry oxygen gives 1,3-diols.[19]

$$\text{PhCH=CHCOPh} \xrightarrow[\text{THF}]{O_2 - LiAlH_4} \text{PhCH(OH)CH}_2\text{CH(OH)Ph}$$

90% (*syn : anti* 1 ; 1)

Cyclization of diarylamines are also effected although an analogous process for the less reactive diphenyl ether requires $Pd(OCOCF_3)_2$ and $Sn(OAc)_2$.[20]

Tyrosinase initiates conversion of phenol into *o*-benzoquinone, which forms Diels-Alder adducts.[21] Employment of oxygen instead of other cooxidants in the oxidative cyclization of compounds containing a silyl enol ether and alkenyl side chain is desirable.[22]

$$\text{PhOH} + \text{CH}_2\text{=CHOEt} \xrightarrow[\text{CHCl}_3]{O_2\ /\ \text{Tyrosinase}} \text{bicyclic dione (OEt)}$$

70%

$$\text{OSiMe}_2\text{(t-Bu) diene (MOMO, allyl)} \xrightarrow[\text{Pd(OAc)}_2]{O_2\ /\ \text{Me}_2\text{SO}} \text{MOMO bicyclic enone} \rightarrow\rightarrow GA_{111} \text{ (HOOC, H, HOOC, H)}$$

81%

A catalyst prepared by encapsulation of $CuCl_2$ in zeolite X is useful for oxidative cleavage of enamines to afford amides.[23]

A formal hydration of enones and dienones is achieved by reaction with $PhSiH_3$–$Mn(dpm)_3$ in isopropanol under oxygen followed by work up with $(EtO)_3P$.[24]

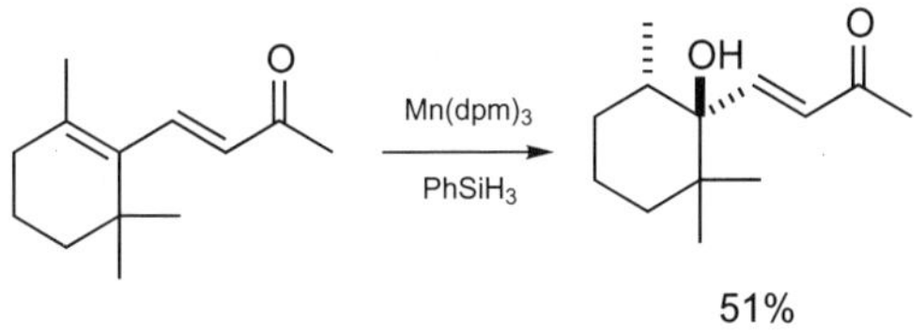

Mn(dpm)3
PhSiH3
OH
51%

Oxidation of phenols. Phenols including 2-naphthol are oxidized to 2,2′-dihydroxybiaryls in the presence of $VO(acac)_2$.[25] Hydroquinones give quinones.[26]

α-Hydroxy-γ-butyrolactones.[27] Secondary alcohols and methyl acrylate are combined in an oxidative fashion when they are subject to oxidation with *N*-hydroxyphthalimide, $Co(acac)_3$, and $Co(OAc)_2$ under oxygen.

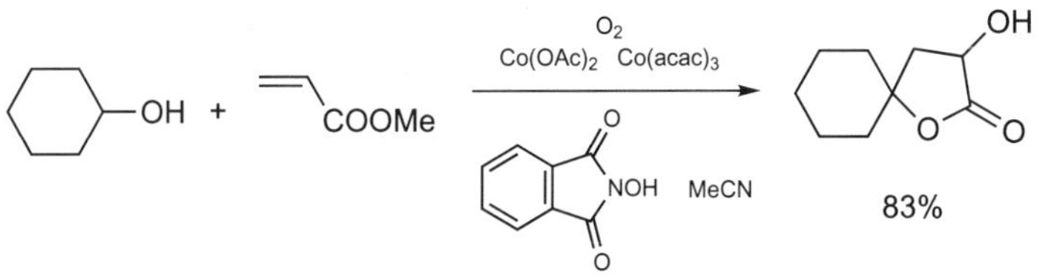

OH + COOMe
O2
Co(OAc)2 Co(acac)3
NOH MeCN
OH
83%

Acetyladamantanes.[28] Acetyl groups are introduced into the bridgehead positions of adamantane when the hydrocarbon and biacetyl are submitted to oxidation conditions [O_2–$Co(OAc)_2$ in HOAc, 60°].

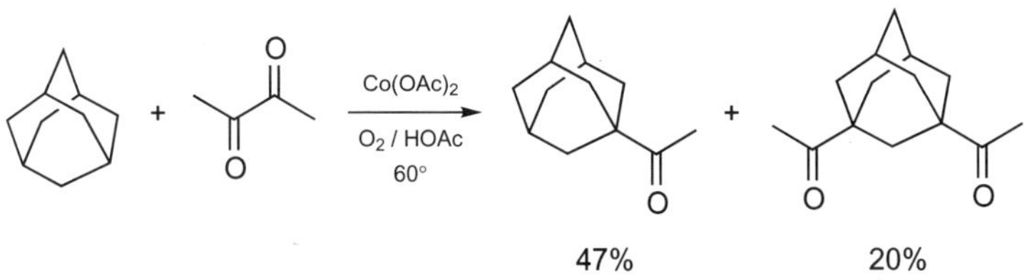

Co(OAc)2
O2 / HOAc
60°
47% 20%

Alkylation of carbonyl compounds and derivatives. The $O_2/Co(OAc)_2–Mn(OAc)_2$ system is useful to accomplish α-alkylation of ketones with 1-alkenes.[29] Acetals also add to acrylic esters under O_2 in the presence of catalytic amounts of $Co(OAc)_2$ and *N*-hydroxyphthalimide to afford α-hydroxy-γ-oxo ester acetals.[30] The adducts of methyl vinyl ketone suffer oxidative degradation in situ.

O O R H + COOMe —(O_2 / $Co(OAc)_2$; N-hydroxyphthalimide)→ O O R OH COOMe

Cleavage of 1,3-oxathiolanes.[31] Carbonyl compounds are regenerated on heating 2-substituted oxathiolanes with $VOCl_3$ in CF_3CH_2OH under O_2.

[1]Iwahama, T., Sakaguchi, S., Ishii, Y. *H* **52**, 693 (2000).
[2]Iwahama, T., Hatta, G., Sakaguchi, S., Ishii, Y. *CC* 163 (2000).
[3]Coleman, K.S., Coppe, M., Thomas, C., Osborn, J.A. *TL* **40**, 3723 (1999).
[4]Yamaguchi, K., Mori, K., Mizugaki, T., Ebitani, K., Kaneda, K. *JACS* **122**, 7144 (2000).
[5]Nishimura, T., Kakiuchi, N., Inoue, M., Uemura, S. *CC* 1245 (2000).
[6]Betzmeier, B., Cavazzini, M., Quici, S., Knochel, P. *TL* **41**, 4343 (2000).
[7]Matsunaka, K., Iwahama, T., Sakaguchi, S., Ishii, Y. *TL* **40**, 2165 (1999).
[8]Howarth, J. *TL* **41**, 6627 (2000).
[9]Kirihara, M., Ochiai, Y., Takizawa, S., Takahata, H., Nemoto, H. *CC* 1387 (1999).
[10]Christoffers, J. *JOC* **64**, 7668 (1999).
[11]Baucherel, X., Levoirier, E., Uziel, J., Juge, S. *TL* **41**, 1385 (2000).
[12]Attolini, M., Peiffer, G., Maffei, M. *T* **56**, 2693 (2000).
[13]Wang, F., Zhang, H., Song, G., Lu, X. *SC* **29**, 11 (1999).
[14]Hayes, P., Suthers, B.D., Kitching, W. *TL* **41**, 6175 (2000).
[15]Hashemi, M.M., Beni, Y.A. *JCR(S)* 434 (1999).
[16]Ishii, Y., Matsunaka, K., Sakaguchi, S. *JACS* **122**, 7390 (2000).
[17]Nishimura, T., Kakiuchi, N., Onoue, T., Ohe, K., Uemura, S. *JCS(P1)* 1915 (2000).
[18]Kishi, A., Sakaguchi, S., Ishii, Y. *OL* **2**, 523 (2000).
[19]Csaky, A.G., Maximo, N., Plumet, J., Ramila, A. *TL* **40**, 6485 (1999).
[20]Hagelin, H., Oslob, J.D., Akermark, B. *CEJ* **5**, 2413 (1999).
[21]Muller, G.H., Lang, A., Seithel, D.R., Waldmann, H. *CEJ* **4**, 2513 (1998).
[22]Toyota, M., Odashima, T., Wada, T., Ihara, M. *JACS* **122**, 9036 (2000).
[23]Ebitani, K., Nagashima, K., Mizugaki, T., Kaneda, K. *CC* 869 (2000).
[24]Magnus, P., Payne, A.H., Waring, M.J., Scott, D.A., Lynch, V. *TL* **41**, 9725 (2000).
[25]Hwang, D.-R., Chen, C.-P., Uang, B.-J. *CC* 1207 (1999).
[26]Hwang, D.-R., Chen, C.-P., Wang, S.-K., Uang, B.-J. *SL* 77 (1999).
[27]Iwahama, T., Sakaguchi, S., Ishii, Y. *CC* 613 (2000).
[28]Kishi, A., Kato, S., Sakaguchi, S., Ishii, Y. *CC* 1421 (1999).
[29]Iwahama, T., Sakaguchi, S., Ishii, Y. *CC* 2317 (2000).
[30]Hirano, K., Iwahama, T., Sakaguchi, S., Ishii, Y. *CC* 2457 (2000).
[31]Kirihara, M., Ochiai, Y., Arai, N., Takizawa, S., Momose, T., Nemoto, H. *TL* **40**, 9055 (1999).

Oxygen, singlet. 13, 228–229; **14,** 247; **15,** 243; **16,** 257–258; **17,** 251–253; **18,** 269–270; **19,** 244

Allylic alcohols. Access to *o*-(2-hydroxy-3-methylbut-3-enyl)phenols is through photooxygenation with reductive workup.[1]

Alkoxydioxines.[2] Dienol ethers form such cycloadducts with singlet oxygen.

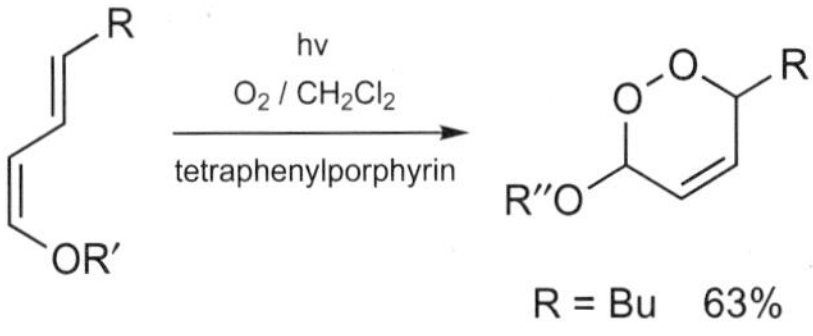

hv, O_2 / CH_2Cl_2, tetraphenylporphyrin

R = Bu 63%

Oxidative cleavage.[3] The *N*-arylamino derivatives of piperidine and pyrrolidine undergo ring fission to afford (aryl)diazenylalkanals or their acetals. When water is excluded from the reaction media, the α-carbon can be functionalized (e.g., cyanation in the presence of Me_3SiCN).

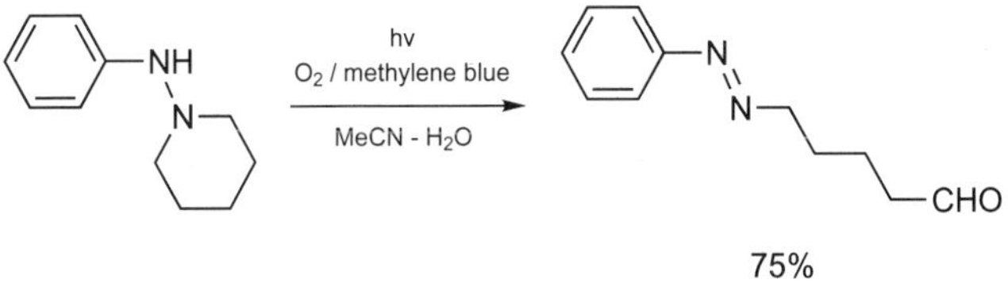

hv, O_2 / methylene blue, MeCN - H_2O

75%

[1]Helesbeux, J.-J., Guilet, D., Seraphin, D., Duval, O., Richomme, P., Bruneton, J. *TL* **41**, 4559 (2000).
[2]Dussault, P.H., Han, Q., Sloss, D.G., Symonsbergen, D.J. *T* **55**, 11437 (1999).
[3]Cocquet, G., Ferroud, C., Guy, A. *T* **56**, 2975 (2000).

P

Palladacycles.

Many palladacycles, besides those derived from bidenate phosphines, are available for catalyzing organic reactions. These varieties include *O,N-*, *N,N-*, *N,C-*, *P,N-*, *P,C*-ligated species. In the following, a selection of palladacycle catalysts bearing such as well as more exotic bidentate diphosphines are presented.

Reductions. Acenaphthoquinonediimine ligated palladium complexes and a polymer-bound palladaisoindoline have found use as catalysts for semihydrogenation of alkynes[1] and reduction of several common unsaturated compounds (including the nitro group),[2] respectively.

Coupling reactions. Various (Heck, Stille, Suzuki, Sonogashira, and Ullmann) coupling reactions are mediated by a stable palladacycle **1**.[3]

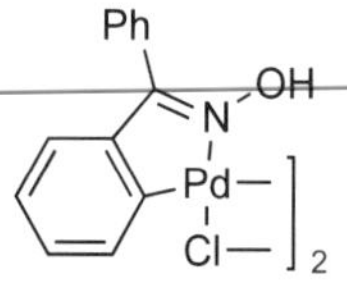

(**1**)

Acceleration of the intramolecular coupling of phenol and aryl halide moieties that is effected with a *P,C*-palladacycle by a base is realized.[4] *o*-Aminophenyldiphenylphosphine is a ligand that forms effective Pd complexes for the Heck reaction.[5]

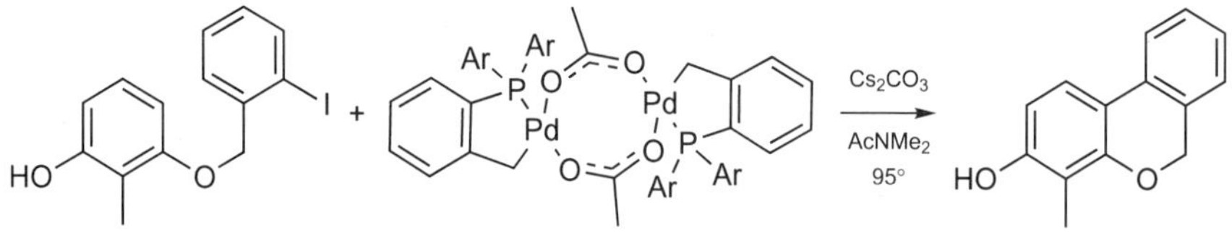

Cs_2CO_3, $AcNMe_2$, 95° — 80%

Note that palladacycles typified by **2** are poisoned by 1,4-dienes,[6] in contrast to other Pd(0) catalyst systems. Accordingly, the Heck reaction cannot employ these catalysts. On the other hand, **3** has a high thermal stability and broad scope of application (aryl couplings).[7]Complex **4** is also a highly active catalyst for the Heck reaction (3 examples, 95–100%).[8]

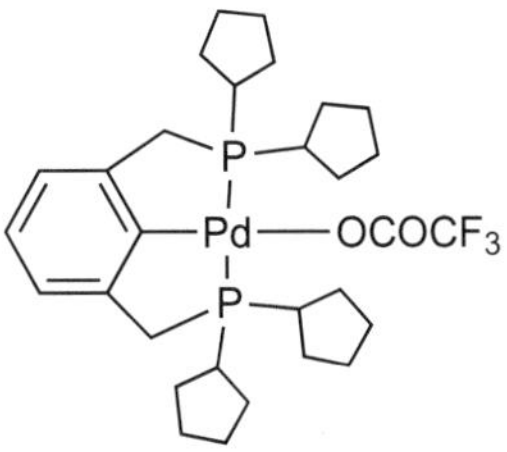

(**2**) 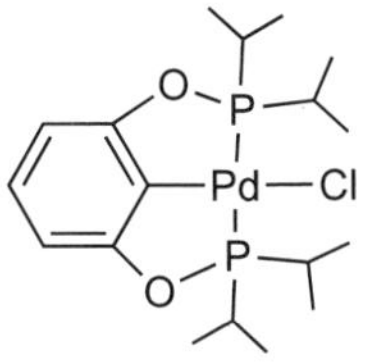 (**3**) 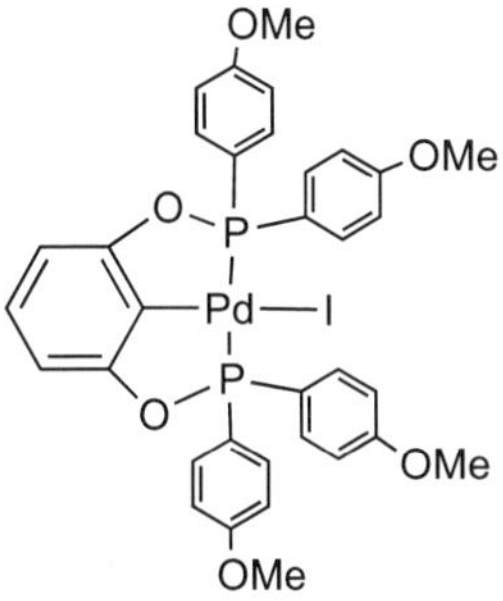 (**4**)

Other stable and efficient catalysts for the Heck reaction are pyrazole- and benzothiazole-based palladacycles.[9] They are capable of promoting Barbier-type reactions.[10] Note that palladacycles derived from aromatic imines decompose to form metallic Pd as the active species.[11]

Sulfur-containing palladacycles such as **5,** which are stable to air, water, and heat, show excellent utility in the phosphine-free Heck reaction (turnover number up to 1,850,000)[12] and Suzuki couplings at room temperature.[13]

Aromatic ketones are also accessible by coupling of $ArB(OH)_2$ with thioesters $RCOS(CH_2)_4X$. Palladacycle **5** and NaI are present in the reaction milieu.[14]

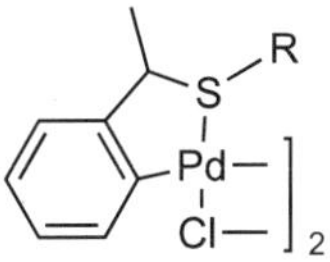

(**5**)

Displacements. A rather unusual displacement of alkenyl chlorides with carbon nucleophiles has been reported.[15]

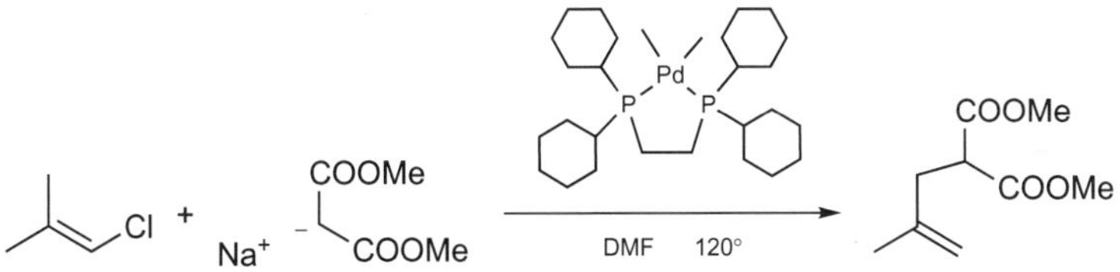

DMF 120°

76%

Addition to multiple bonds. A water-soluble Pd catalyst (picolinic acid as one of the ligands) is applicable to the synthesis of 2-arylpropanoic acids from styrenes by carbonylation.[16]

[1]van Laren, M.W., Elsevier, C.J. *ACIEE* **38**, 3715 (1999).
[2]Islam, M., Bose, A., Mal, D., Saha, C.R. *JCR(S)* 44 (1998).
[3]Alonso, D.A., Najera, C., Pacheco, M.C. *OL* **2**, 1823 (2000).
[4]Hennings, D.D., Iwasa, S., Rawal, V.H. *JOC* **62**, 2 (1997).
[5]Reddy, K.R., Surekha, K., Lee, G.-H., Peng, S.-M., Liu, S.-T. *OM* **19**, 2637 (2000).
[6]Kiewel, K., Liu, Y., Bergbreiter, D.E., Sulikowski, G.A. *TL* **40**, 8945 (1999).
[7]Morales-Morales, D., Redon, R., Yung, C., Jensen, C.M. *CC* 1619 (2000).
[8]Miyazaki, F., Yamaguchi, K., Shibasaki, M. *TL* **40**, 7379 (1999).
[9]Gai, X., Grigg, R., Ramzan, M.I., Sridharan, V., Collard, S., Muir, J.E. *CC* 2053 (2000).
[10]Gai, X., Grigg, R., Collard, S., Muir, J.E. *CC* 1765 (2000).
[11]Nowotny, M., Hanefeld, U., van Koningsveld, H., Maschmeyer, T. *CC* 1877 (2000).
[12]Gruber, A.S., Zim, D., Eberling, G., Monteiro, A.L., Dupont, J. *OL* **2**, 1287 (2000).
[13]Zim, D., Gruber, A.S., Eberling, G., Dupont, J., Monteiro, A.L. *OL* **2**, 2881 (2000).
[14]Savarin, C., Srogl, J., Liebeskind, L.S. *OL* **2**, 3229 (2000).
[15]Reetz, M.T., Wanninger, K., Hermes, M. *CC* 535 (1997).
[16]Jayasree, S., Seayad, A., Chaudhari, R.V. *CC* 1239 (2000).

Palladium–carbon. 13, 230–232; 15, 245; 18, 273; 19, 247; 20, 280–281

Hydrogenolysis. 2-Naphthylmethyl esters are more readily hydrogenolyzed in the presence of Pd–C than benzyl analogues, therefore two such esters present in a molecule are easily differentiated.[1,2] Quinolin-4-ylmethyl esters including carbamates and mixed carbonates have the same selectivities in transfer hydrogenolysis.[3]

A practical synthesis of polysubstituted pyrrole-2-carboxylic esters is via the corresponding benzyl esters, the other functional groups are not affected under hydrogenolysis conditions.[4]

β-Aryl-β-lactams undergo ring fission in a transfer hydrogenolysis that is also assisted by microwave irradiation.[5] Regioselective transacetylation after hydrogenolysis of certain *C*-arylglycosides has been observed.[6]

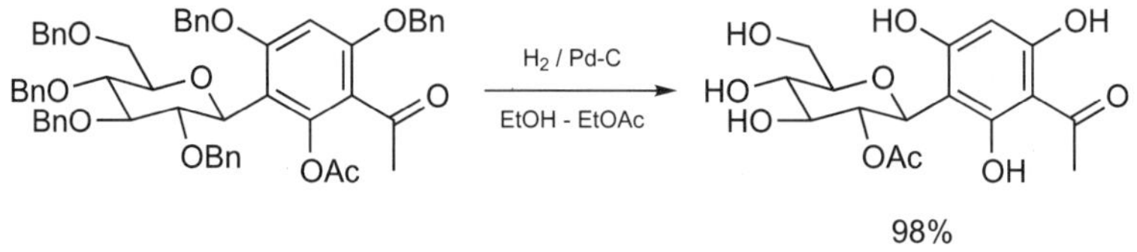

Hydrogenolysis of terminal epoxides to afford secondary alcohols is best accomplished with the Pd–C(en) catalyst in methanol.[7]

Hydrogenation. The Pd–C(en) catalyst shows selectivity in hydrogenation such that an epoxide group is retained.[8] The same system provides a means for the hydrogenation of alkenes that also contain an acid-labile TBS ether.[9]

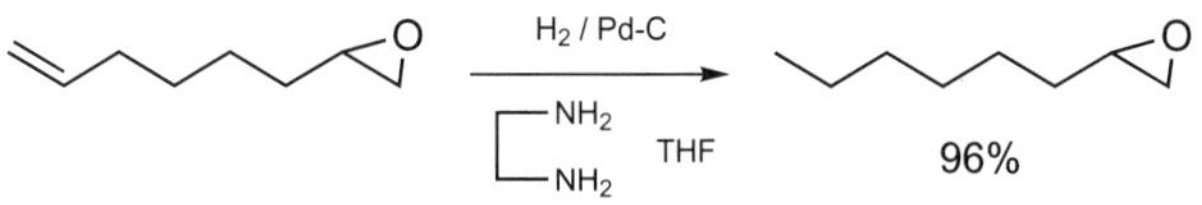

It is possible to retain a benzyloxy group during hydrogenation of γ-amino-α,β-unsaturated esters.[10]

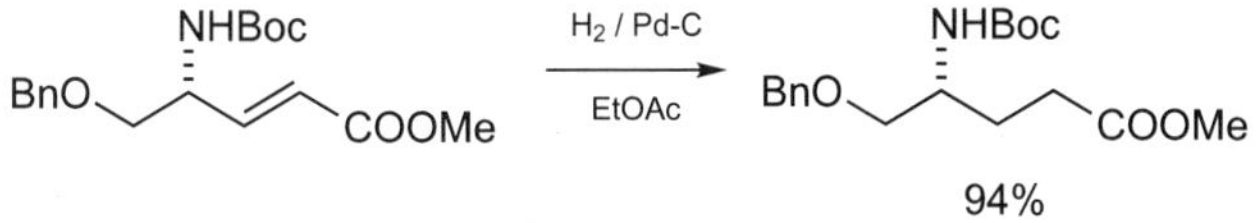

Trimethylsilyl ethers of Baylis–Hillmann adducts undergo *syn*-selective hydrogenation.[11]

Catalytic reduction in methanol using decaborane as the hydrogen source successfully saturates alkenes and alkynes[12] and converts nitroarenes to arylamines.[13] Concurrent *N*-alkylation to afford ArNHCHRR′ occurs when a carbonyl compound (RR′C=O) is present.[14]

Dehydrogenation. Under an ethylene atmosphere, allylic and benzylic alcohols undergo hydrogen transfer and are thereby transformed into ketones.[15]

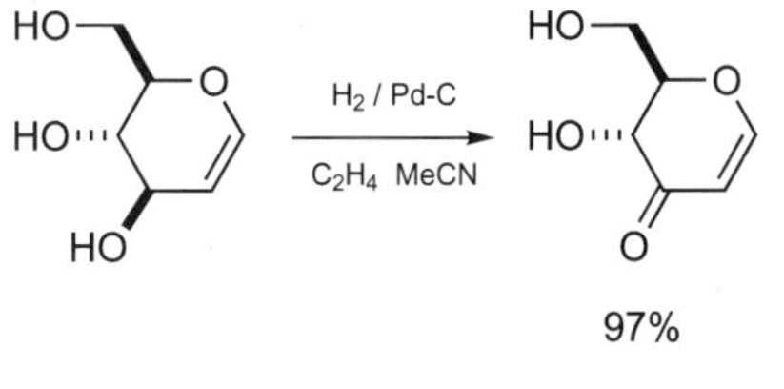

Biaryls. Homocoupling of aryl halides has been effected by Pd–C in an aqueous media (PEG-400, NaOH, H_2O) under hydrogen[16] or in air in the presence of Zn and 18-crown-6.[17] The considerable improvement in yield observed with added crown ether is intriguing.

Amidocarbonylation. A useful procedure for the Pd-catalyzed reaction is available.[18]

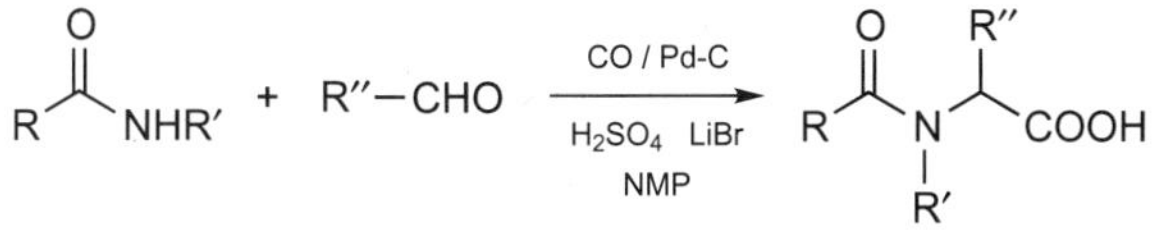

Enolsilylation. The conversion of carbonyl compounds to silyl enol ethers with triethylsilane in toluene at 100° is performed by Pd–C while adding an amine and halide as cocatalysts.[19]

[1]Gaunt, M.J., Boschetti, C.E., Yu, J., Spencer, J.B. *TL* **40**, 1803 (1999).
[2]Papageorgiou, E.A., Gaunt, M.J., Yu, J.-Q., Spencer, J.B. *OL* **2**, 1049 (2000).
[3]Boutros, A., Legros, J.-Y., Fiaud, J.-C *TL* **40**, 7329 (1999).
[4]Narkunan, K., Ciufolini, M.A. *S* 673 (2000).
[5]Banik, B.K., Barakat, K.J., Wagle, P.R., Manhas, M.S., Bose, A.K. *JOC* **64**, 5746 (1999).
[6]Kumazawa, T., Akutsu, Y., Matsuba, S., Sato, S., Onodera, J. *CR* **320**, 129 (1999).
[7]Sajiki, H., Hattori, K., Hirota, K. *CC* 1041 (1999).
[8]Sajiki, H., Hattori, K., Hirota, K. *CEJ* **6**, 2200 (2000).
[9]Hattori, K., Sajiki, H., Hirota, K. *TL* **41**, 5711 (2000).
[10]Misiti, D., Zappia, G., Delle Monache, G. *S* 873 (1999).
[11]Mateus, C.R., Almeida, W.P., Coelho, F. *TL* **41**, 2533 (2000).
[12]Bae, J.W., Cho, Y.J., Lee, S.H., Yoon, C.M. *TL* **41**, 175 (2000).
[13]Lee, S.H., Park, Y.J., Yoon, C.M. *TL* **41**, 887 (2000).
[14]Bae, J.W., Cho, Y.J., Lee, S.H., Yoon, C.-O.M., Yoon, C.M. *CC* 1857 (2000).
[15]Hayashi, M., Yamada, K., Nakayama, S. *S* 1869 (1999); *JCS(P1)* 1501 (2000).
[16]Mukhopadhyay, S., Rothenberg, G., Wiener, H., Sasson, Y. *T* **55**, 14763 (1999).
[17]Venkatraman, S., Li, C.-J. *TL* **41**, 4831 (2000).
[18]Beller, M., Moradi, W.A., Eckert, M., Neumann, H. *TL* **40**, 4523 (1999).
[19]Igarashi, M., Sugihara, Y., Fuchikami, T. *TL* **40**, 711 (1999).

Palladium–various supports.

Semihydrogenation. Alkynes are hydrogenated to (*Z*)-alkenes with Pd on pumice.[1]

C—C Bond couplings. Different supported Pd catalysts have been tried. For example, Suzuki coupling based on Pd–thiourea resin[2] and nanosize Pd on poly(*N*-vinylpyrrolidone),[3] Heck reactions using Pd–$CaCO_3$,[4] or Pd–SiO_2.[5] Colloidal Pd stabilized by tetraoctylammonium formate that is generated in situ from simple palldium salts is effective in catalyzing Suzuki and Heck reactions (phosphine-free conditions).[6]

Hydrogenolysis. Palladium nanoparticles show excellent activity for hydrogenolysis of benzylated carbohydrates that are attached to solid supports.[7]

[1]Gruttadauria, M., Noto, R., Deganello, G., Liotta, L.F. *TL* **40**, 2857 (1999).
[2]Zhang, T.Y., Allen, M.J. *TL* **40**, 5813 (1999).
[3]Li, Y., Hong, X.M., Collard, D.M., El-Sayed, M.A. *OL* **2**, 2385 (2000).
[4]Brunner, H., Le Cousturier de Courcy, N., Genet, J.-P. *TL* **40**, 4815 (1999).
[5]Zhao, F., Bhanage, B.M., Shirai, M., Arai, M. *CEJ* **6**, 843 (2000).
[6]Reetz, M.T., Westermann, E. *ACIEE* **39**, 165 (2000).
[7]Kanie, O., Grotenbreg, G., Wong, C.-H. *ACIEE* **39**, 4545 (2000).

Palladium(II) acetate. 13, 232–233; **14,** 248; **15,** 245–247; **16,** 259–263; **17,** 255–259; **18,** 274–277; **19,** 248–251; **20,** 281–283

Coupling reactions. Microwave assists the Suzuki coupling in aqueous media.[1] A route to biaryl-2-carboxylic acids is via coupling of 1-hydroxy-1,2-benziodoxo-3(1*H*)-one.[2]

Complexation of Pd with a bisimidazol-2-ylidene unit is favorable to catalyzing the Suzuki coupling.[3]

Aromatic alkylation and Suzuki or Heck coupling are combined in one operation by proper design of a catalytic system.[4,5] Norbornene acts as a platform for regioselective *o*-alkylation.

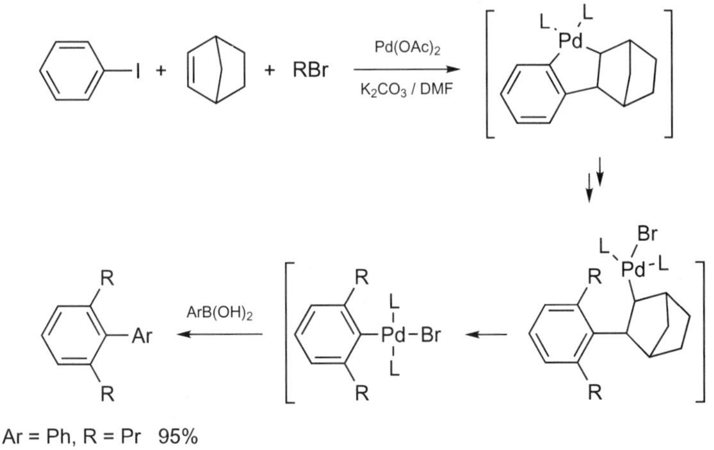

Pd(OAc)$_2$ K$_2$CO$_3$ / DMF

ArB(OH)$_2$

Ar = Ph, R = Pr 95%

Other variations of the biaryl synthesis involve K[ArBF$_3$] and diaryliodonium salts[6] or arenediazonium salts.[7]

Heck reactions with arenediazonium trifluoroacetates[8] are more efficient because these salts are more soluble in organic solvents than the tetrafluoroborates. α-Benzylidene-γ-butyrolactones are successfully prepared from α-methylene-γ-butyrolactone and arenediazonium salts[9] or ArI.[10] The coupling of cycloalkenes with ArI in water at 225° (no other solvents) can be catalyzed by Pd(OAc)$_2$-NaOAc alone, but yields are low to moderate.[11]

Arylsilanols,[12] aryltributylstannanes,[13] and arylantimony chlorides[14] are suitable coupling partners. When a hypervalent iodine heterocycle undergoes ring opening upon coupling, an aryl iodide is left behind that can be further manipulated.[15]

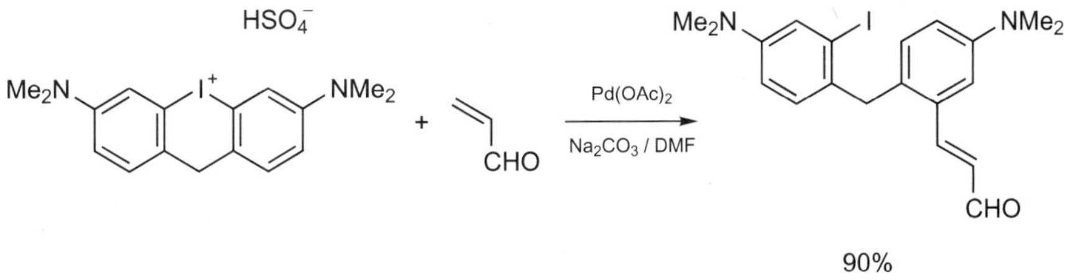

Pd(OAc)$_2$ Na$_2$CO$_3$ / DMF

90%

Ring formation is observed from reaction of aryl halides with alkynes.[16–18]

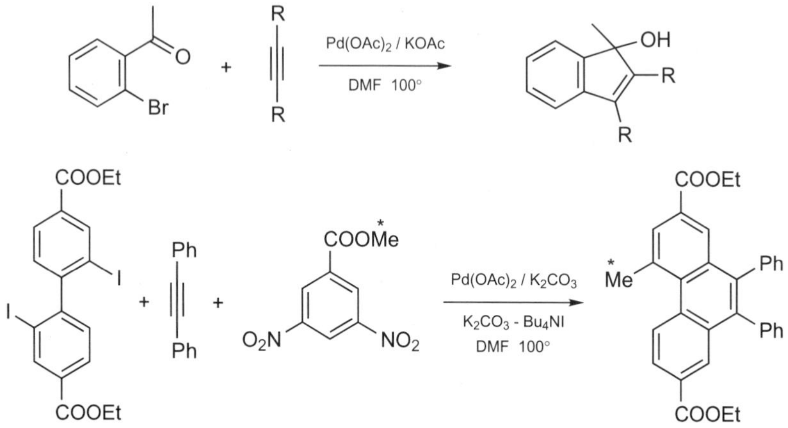

Triarylbismuthines are decomposed to give biaryls in the presence of $Pd(OAc)_2$ at room temperature in the air.[19]

Reactions of alkynes. Hydroarylation,[20] carbonylation,[21] and the formation of α-substituted acrylamides[22] are some of the reactions mediated by $Pd(OAc)_2$. While an intramolecular version of the former completes a synthesis of coumarins and quinolinones, the latter two incorporate CO group(s) from carbon monoxide.

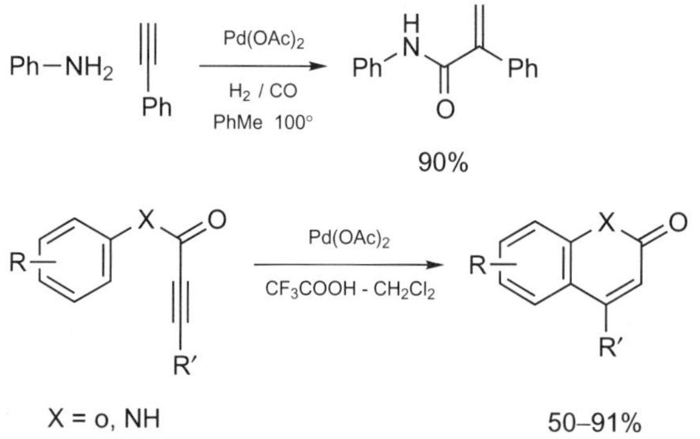

A method for the preparation of isoindolo[2,1-α]indoles prescribes treatment of *N*-(*o*-iodoaryl)aldimines and arylalkynes with $Pd(OAc)_2$ and a base. The coupling is followed by an annulation event.[23]

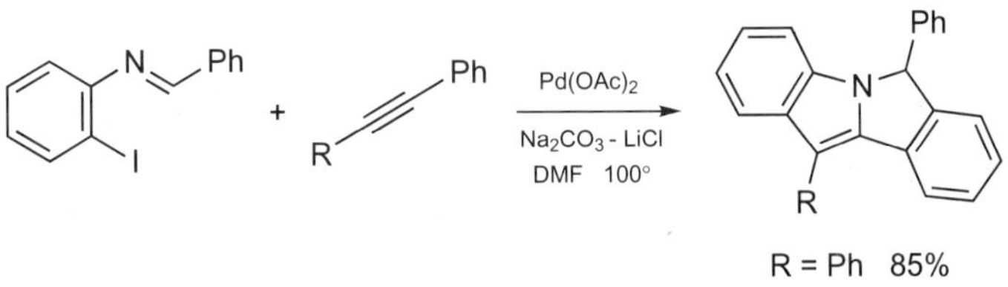

Carbonyl compounds. Allylic alcohols are oxidized under ethylene with $Pd(OAc)_2$ as catalyst.[24] A more conventional hydrogen acceptor is molecular oxygen in such oxidation.[25] Under such conditions, cyclobutanols that are *gem*-vinylated or arylated are cleaved (further transformations are also possible).[26]

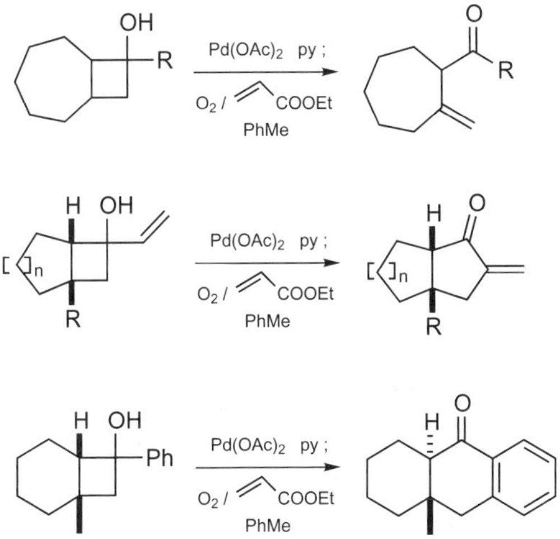

Suzuki coupling under carbon monoxide leads to diaryl ketones.[27] Ligandless $Pd(OAc)_2$ also catalyzes the reaction of ArCOCl with $NaBAr_4$ to afford ketones.[28]

Carbonyl compounds are regenerated from hydrazones by Pd-catalyzed reaction with tin(II) chloride.[29]

Miscellaneous reactions. Cyclization accompanies oxidation of an allene moiety as a result of the participation of an internal nitrogen nucleophile.[30]

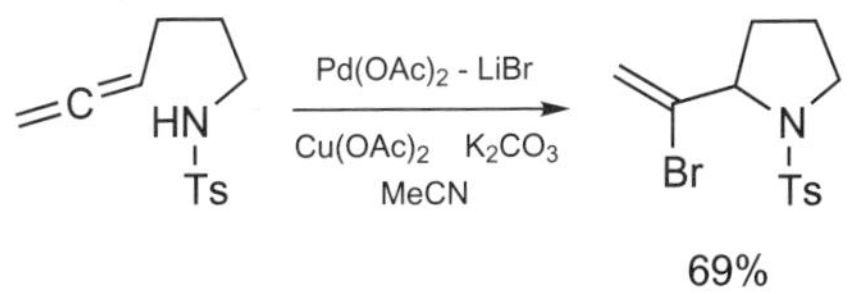

Chromenes are formed by dehalogenative coupling of *o*-haloaryl ethers that bear a haloalkene side chain.[31]

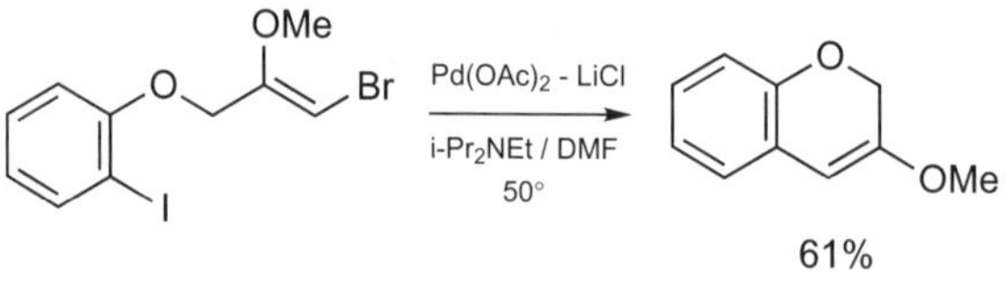

Allylic *N*-tosylcarbamates (from allylic alcohols and TsN=C=O) afford 1,3-transposed allylic *N*-tosylamines via ionization with $Pd(OAc)_2$–LiBr in DMF, decarboxylation, and substitution.[32]

Silyl esters are formed by oxidative functionalization of hydrosilanes with $Pd(OAc)_2$ and carboxylic acids.[33]

Bisphosphines are oxidized to the monooxides,[34] thus bidentate phosphorus ligands of mixed oxidation states are readily accessible.

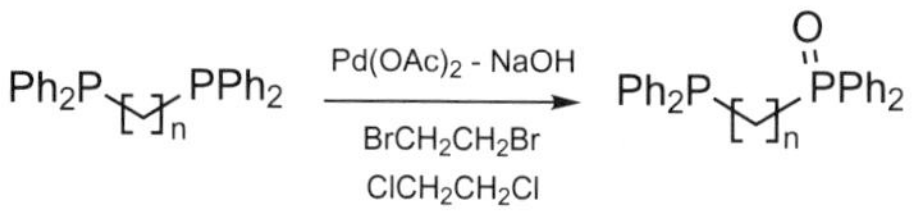

Hydroarylation. Catalytic Ar—H bond activation is observed with a complexed Pd species and the delivery of the aryl group to alkenes shows some asymmetric induction.[35] For heterocycles such as methylfuran, pyrroles, and indoles, Ar—H activation by $Pd(OAc)_2$ is adequate in their addition to 2-alkynoic esters in a *trans*-fashion.[36]

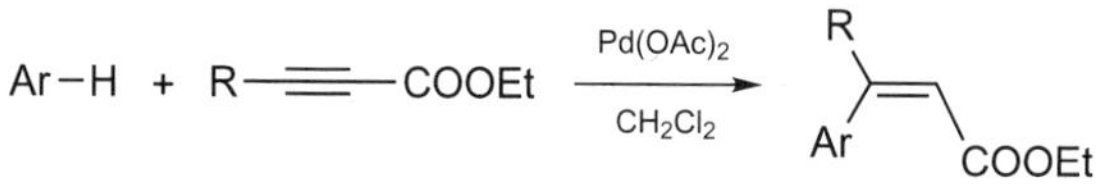

Carboxylation.[37] Many arenes are converted to carboxylic acids in good yields under atmospheric pressure of CO under the influence of $Pd(OAc)_2$ but without phosphine ligands. Potassium peroxysulfate and trifluoroacetic acid are present in the reaction media.

Rearrangement-coupling.[38] This effective synthetic process involving an alkene and allylic alcohol in juxtaposition is applicable to the synthesis of (+)-equilenin. Solvent determines the diastereoselectivity.

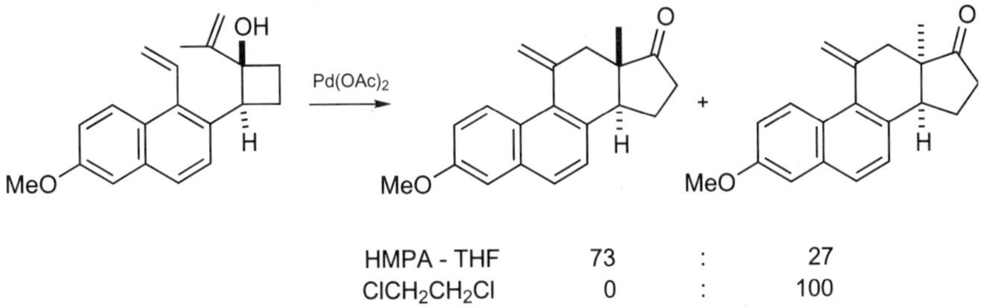

HMPA - THF 73 : 27
$ClCH_2CH_2Cl$ 0 : 100

Conjugate addition. The addition of Me_3SiCN to enones derivatives of glycals[39] is catalyzed by $Pd(OAc)_2$. Five-membered azacycles (lactams, imidazolinones, oxazolidinones) that bear a 4-oxoalkylidene group adjacent to the nuclear nitrogen atom are readily available by a Pd-catalyzed intramolecular addition and conjugate addition tandem.[40]

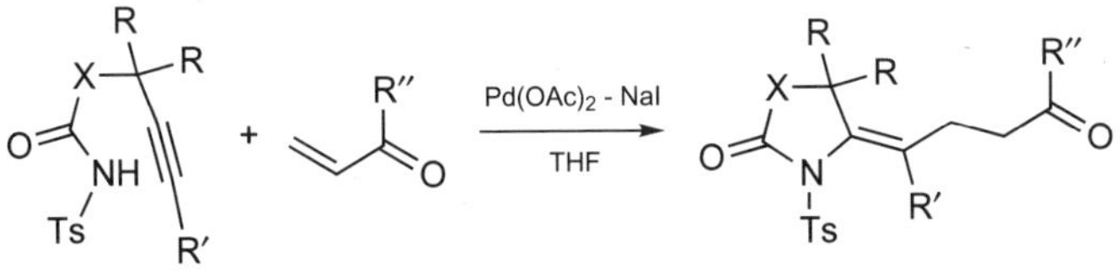

$Pd(OAc)_2$ - NaI
THF

[1]Blettner, C.G., Konig, W.A., Stenzel, W., Schotten, T. *JOC* **64**, 3885 (1999).
[2]Xia, M., Chen, Z. *SC* **30**, 63 (2000).
[3]Zhang, C., Trudell, M.L. *TL* **41**, 595 (2000).
[4]Catellani, M., Motti, E., Minari, M. *CC* 157 (2000).
[5]Catellani, M., Cugini, F. *T* **55**, 6595 (1999).
[6]Xia, M., Chen, Z. *SC* **29**, 2457 (1999).
[7]Darses, S., Michaud, G., Genet, J.-P. *EJOC* 1875 (1999).
[8]Colas, C., Goeldner, M.*E JOC* 1357 (1999).
[9]Brunner, H., Le Cousturier de Courcy, N., Genet, J.-P. *TL* **40**, 4815 (1999).
[10]Arcadi, A., Chiarini, M., Marinelli, F., Berente, Z., Kollar, L. *OL* **2**, 69 (2000).
[11]Gron, L.U., Tinsley, A.S. *TL* **40**, 227 (1999).
[12]Hirabayashi, K., Ando, J., Kawashima, J., Nishihara, Y., Mori, A., Hiyama, T. *BCSJ* **73**, 1409 (2000).
[13]Hirabayashi, K., Ando, J., Nishihara, Y., Mori, A., Hiyama, T. *SL* 99 (1999).
[14]Matoba, K., Motofusa, S., Cho, C.S., Ohe, K., Uemura, S. *JOMC* **574**, 3 (1999).
[15]Liang, Y., Luo, S., Liu, C., Wu, X., Ma, Y. *T* **56**, 2961 (2000).
[16]Quan, L.G., Gevorgyan, V., Yamamoto, Y. *JACS* **121**, 3545 (1999).
[17]Mandal, A.B., Lee, G.-H., Liu, Y.-H., Peng, S.-M., Leung, M.-K. *JOC* **65**, 332 (2000).
[18]Larock, R.C., Doty, M.J., Han, X. *JOC* **64**, 8770 (1999).
[19]Ohe, T., Tanaka, T., Kuroda, M., Cho, C.S., Ohe, K., Uemura, S. *BCSJ* **72**, 1851 (1999).
[20]Jia, C., Lu, W., Oyamada, J., Kitamura, T., Matsuda, K., Irie, M., Fujiwara, Y. *JACS* **122**, 7252 (2000); Jia, C., Piao, D., Kitamura, T., Fujiwara, Y. *JOC* **65**, 7516 (2000);
[21]Sakurai, Y., Sakaguchi, S., Ishii, Y. *TL* **40**, 1701 (1999).
[22]Ali, B.E., El-Ghanam, A., Fettouhi, M., Tijani, J. *TL* **41**, 5761 (2000).
[23]Roesch, K.R., Larock, R.C. *OL* **1**, 1551 (1999).
[24]Hayashi, M., Yamada, K., Arikita, O. *T* **55**, 8331 (1999).
[25]Nishimura, T., Onoue, T., Ohe, K., Uemura, S. *JOC* **64**, 6750 (1999).
[26]Nishimura, T., Ohe, K., Uemura, S. *JACS* **121**, 2645 (1999).

[27] Xia, M., Chen, Z. *JCR(S)* 400 (1999).
[28] Bumagin, N.A., Korolev, D.N. *TL* **40**, 3057 (1999).
[29] Mino, T., Hirota, T., Fujita, N., Yamashita, M.*S* 2024 (1999).
[30] Jonasson, C., Karstens, W.F.J., Hiemstra, H., Bäckvall, J.E. *TL* **41**, 1619 (2000).
[31] Edvardsen, K.R., Benneche, T., Tius, M.A. *JOC* **65**, 3085 (2000).
[32] Lei, A., Lu, X. *OL* **2**, 2357 (2000).
[33] Chauhan, M., Chauhan, B.P.S., Boudjouk, P. *OL* **2**, 1027 (2000).
[34] Grushin, V.V. *JACS* **121**, 5831 (1999).
[35] Mikami, K., Hatano, M., Terada, M. *CL* 55 (1999).
[36] Lu, W., Jia, C., Kitamura, T., Fujiwara, Y. *OL* **2**, 2927 (2000).
[37] Lu, W., Yamaoka, Y., Taniguchi, Y., Kitamura, T., Takaki, K., Fujiwara, Y. *JOMC* **580**, 290 (1999).
[38] Nemoto, H., Yoshida, M., Fukumoto, K., Ihara, M. *TL* **40**, 907 (1999).
[39] Hayashi, M., Kawabata, H., Shimono, S., Kakehi, A. *TL* **41**, 2591 (2000).
[40] Lei, A., Lu, X.*OL* **2**, 2699 (2000).

Palladium(II) acetate–phase-transfer catalyst. 20, 284–286

Heck reaction. Heck reaction of 2,5-dimethoxy-2,5-dihydrofuran leads to β-arylbutenolides. It only requires a reduction step to furnish 3-arylfurans.[1]

+ MeO OMe

$Pd(OAc)_2$ - R_4NCl

i-Pr_2NEt / DMF 80°; H_3O^+

66%

An aryl group can be introduced to a substituted β-posittion of an α,β-unsaturated ketone[2] or other electron-deficient alkenes[3] by the Heck reaction in the presence of a phase-transfer catalyst.

Deallylation.[4] Water-insoluble allyl substrates are cleaved by $Pd(OAc)_2$ using water-soluble phosphine ligand and per(2,6-di-*O*-methyl)-β-cyclodextrin.

Unsymmetrical biaryls. Coupling of ArI with Ar′Br proceeds with good selectivity employing electron-deficient Ar′Br under specified conditions: A catalytic system composed of $Pd(OAc)_2$, Bu_4NBr, *i*-Pr_2NEt in refluxing *p*-xylene.[5]

A simple procedure for Suzuki coupling involves heating the substrates with $Pd(OAc)_2$ and K_3PO_4 in DMF. A beneficial additive is Bu_4NBr.[6] When $(EtS)_2PdCl_2$ is used, the quaternary ammonium salt is not required.

[1] Taniguchi, T., Nagata, H., Kanada, R.M., Kadota, K., Takeuchi, M., Ogasawara, K. *H* **52**, 67 (2000).
[2] Xia, M., Chen, Z. *SC* **30**, 1281 (2000).
[3] Gurtler, C., Buchwald, S.L. *CEJ* **5**, 3107 (1999).
[4] Widehem, R., Lacroix, T., Bricout, H., Monflier, E. *SL* 722 (2000).
[5] Hassan, J., Hathroubi, C., Gozzi, C., Lemaire, M. *TL* **41**, 8791 (2000).
[6] Zim, D., Monteiro, A.L., Dupont, J. *TL* **41**, 8199 (2000).

Palladium(II) acetate–tertiary phosphine. 13, 91, 233–234; **14,** 249, 250–253; **15,** 247–248; **16,** 264–268; **17,** 259–269; **18,** 277–281; **19,** 252–256; **20,** 286–289

Aryl coupling reactions. A highly active catalytic system for Suzuki coupling contains $Pd(OAc)_2$, KF, and di-*t*-butyl-2-biphenylphosphine,[1] but the most efficient catalyst to date is the combination of $Pd(OAc)_2$ and bis(1-adamantyl)butylphosphine, which is effective to couple deactivated ArCl on very low Pd loading (down to 0.001 mol%) yet showing turnover number in the 10,000–20,000 range.[2] Products are obtained in good yields.

Allylations. Allylation of enamines with 1-allylbenzotriazoles[3] is accomplished in the presence of $Pd(OAc)_2$–Ph_3P and $ZnBr_2$. The stepwise twofold allylation of cycloalkanones leading to bridged ring systems[4] uses 2-methylene-1,3-propylene diacetate. Allylation of active methylene compounds with allylic alcohols is promoted by triethylborane.[5]

OAc; OAc; $Pd(OAc)_2$ - Ph_3P; MeCN 40°; 77%

Arylamines undergo *N*-allylation with allylic alcohols.[6] The reaction system also contains titanium tetraisopropoxide and molecular sieves. Some dialkylation occurs when the amino group is unhindered.

Homoallyl alcohols. Benzaldehyde reacts with various allylic alcohols to give homoallylic alcohols in the Pd-catalyzed reaction that is promoted by triethylborane.[7] A three-component condensation that produces arylated homoallyl alcohols derives the allyl moiety from allene and aryl iodides.[8] This Pd-catalyzed reaction is mediated by metallic indium. A relayed process that forms an isochromane system[9] illustrates the power and intricacy of such a reaction.

CHO + =C= + I–Ph; In - $Pd(OAc)_2$; P; 3; OH; Ph; 64%

I; O; + =C= + H–C(=O)Ph; In - $Pd(OAc)_2$; P; 3; Ph; OH; O; 52%

Cyclizations. In a catalytic process 2-aryl-3-alkenylindoles are generated from aldimines derived from *o*-alkynylanilines.[10] Cyclization reactions of enynes and dienynes under reductive conditions (HCOOH)[11] or in the presence of aryl iodides[12] to participate in a coupling process are useful.

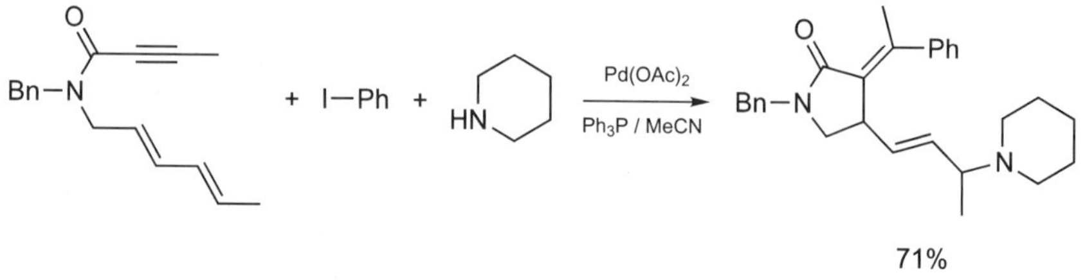

Bn—N + I—Ph + HN Pd(OAc)$_2$ / Ph$_3$P / MeCN → Bn—N … Ph … N

71%

Cyclization that accompanies coupling efficiently delivers α-benzylidene-γ-lactones from homopropargylic chloroformates.[13]

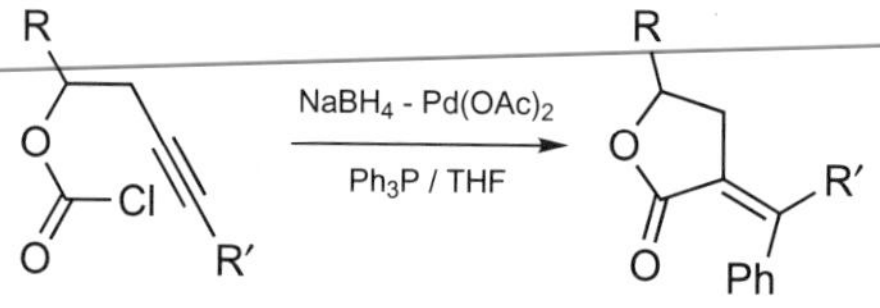

R … O … Cl … R′ NaBH$_4$ - Pd(OAc)$_2$ / Ph$_3$P / THF → R … O … R′ … Ph

An alkyne/5-hydroxy-2-pentynoic ester coupling leads to dihydropyran derivatives. Formation of the homologous heterocycle from ethyl 6-hydroxy-2-hexynoate proceeds in two stages, the second stage is promoted by Pd(OCOCF$_3$)$_2$.[14]

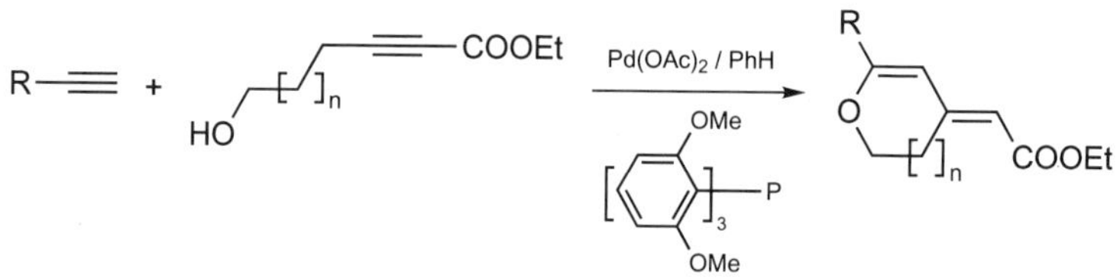

R—≡ + HO …]$_n$ … COOEt Pd(OAc)$_2$ / PhH, [OMe … OMe]$_3$P → R … O …]$_n$ … COOEt

Annulation. Benzyne is probably generated from *o*-trimethylsilylphenyl tosylate by CsF. A union of two benzyne molecules and the alkyne to afford a phenanthrene is observed in the presence of Pd(OAc)$_2$, tri-*o*-tolylphosphine, and an alkyne.[15]

A synthesis of 9-alkylidene-9*H*-fluorenones from ArI and alkynylarenes involves a rearrangement step.[16]

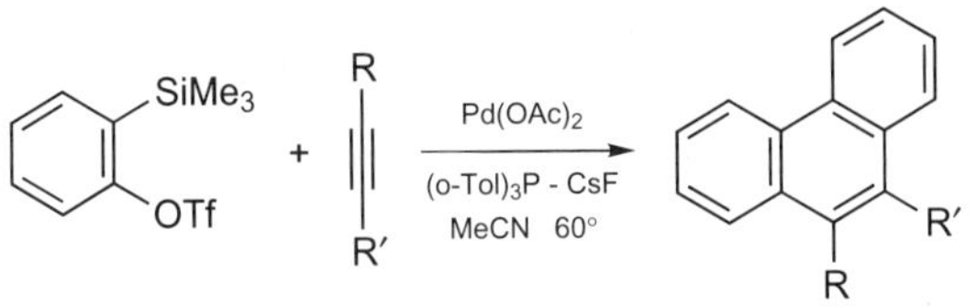

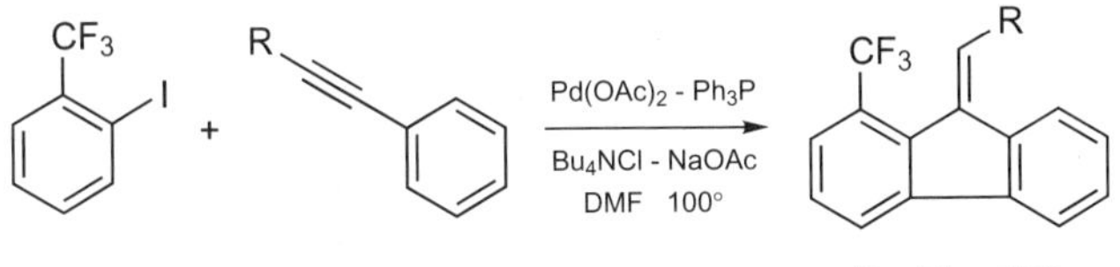

Triarylphosphines. $ArPPh_2$ are synthesized by a Pd-catalyzed reaction of aryl bromides or triflates with Ph_3P in DMF at 110°. Thus, phosphines containing functional groups[17] such as a ketone are obtained by this method. Useful *P,N*-ligands are also readily accessible.[18]

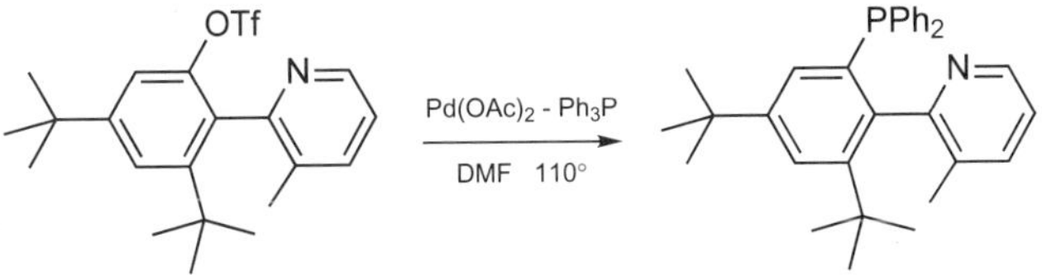

Allylic displacements.[19] The $Pd(OAc)_2$-Ph_3P-K_2CO_3 catalyst system in an ionic liquid has been used for effecting allylic displacements.

[1]Wolfe, J.P., Singer, R.A., Yang, B.H., Buchwald, S.L. *JACS* **121**, 9550 (1999).
[2]Zapf, A., Ehrentraut, A., Beller, M. *ACIEE* **39**, 4153 (2000).
[3]Katritzky, A.R., Huang, Z., Fang, Y. *JOC* **64**, 7625 (1999).
[4]Buono, F., Tenaglia, A. *JOC* **65**, 3869 (2000).
[5]Tamaru, Y., Horino, Y., Araki, M., Tanaka, S., Kimura, M. *TL* **41**, 5705 (2000).
[6]Yang, S.-C., Hung, C.-W. *S* 1747 (1999).
[7]Kimura, M., Tomizawa, T., Horino, Y., Tanaka, S., Tamaru, Y. *TL* **41**, 3627 (2000).
[8]Anwar, U., Grigg, R., Rasparini, M., Savic, V., Sridharan, V. *CC* 645 (2000).
[9]Anwar, U., Grigg, R., Sridharan, V. *CC* 933 (2000).
[10]Takeda, A., Kamijo, S., Yamamoto, Y. *JACS* **122**, 5662 (2000).
[11]Oh, C.H., Jung, H.H., Kim, J.S., Cho, S.W. *ACIEE* **39**, 752 (2000).
[12]Xie, X., Lu, X. *TL* **40**, 8415 (1999).
[13]Grigg, R., Savic, V. *CC* 2381 (2000).
[14]Trost, B.M., Frontier, A.J. *JACS* **122**, 11727 (2000).
[15]Saito,S., Tsuboya, N., Chounan, Y., Nogami, T., Yamamoto, Y., Radhakrishnan, K.V., Yoshikawa, E. *TL* **40**, 7533 (1999).

[16]Tian, C., Larock, R.C. *OL* **2**, 3329 (2000).
[17]Kwong, F.Y., Chan, K.S. *CC* 1069 (2000).
[18]Kwong, F.Y., Chan, K.S. *OM* **19**, 2058 (2000).
[19]Chen, W., Xu, L., Chatterton, C., Xiao, J. *CC* 1247 (1999).

Palladium(II) acetate–tertiary phosphine–base. 20, 289–292

Aryl coupling reactions. A new catalyst for Heck and Sonogashira couplings and allylic displacements is made from glass beads coated with triarylphosphine (**1**) that carry a dimethylguanidinyl group in each benzene ring.[1]

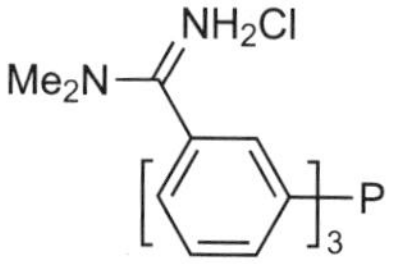

(**1**)

By taking advantage of *o*-alkylation to precede a Heck reaction, α-alkylidene cycloaromatic compounds are synthesized in one step.[2]

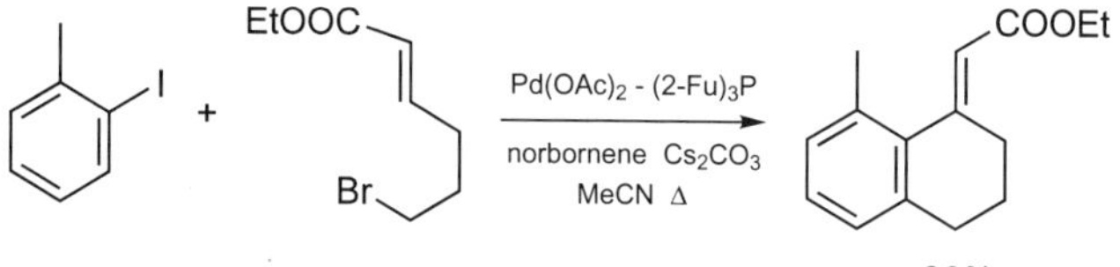

Pd(OAc)$_2$ - (2-Fu)$_3$P; norbornene Cs_2CO_3, MeCN Δ — 90%

Alkynylsilanes can be used to couple with aryl iodides, and the process is useful for the preparation of arylalkynylamides.[3]

In promoting the Suzuki coupling of aryl chlorides and bromides with arylboronic acids, the Pd(OAc)$_2$–(*i*-PrO)$_3$P system is adequate.[4]

Arylation. For *N*-arylation of lactams with ArBr, a catalytic system composed of Pd(OAc)$_2$, 1,1′-bis(diphenylphosphino)ferrocene, and *t*-BuONa has been employed.[5] For arylation of amides,[6] and amines,[7,8] Xantphos (**1**) appears to be the ligand of choice.

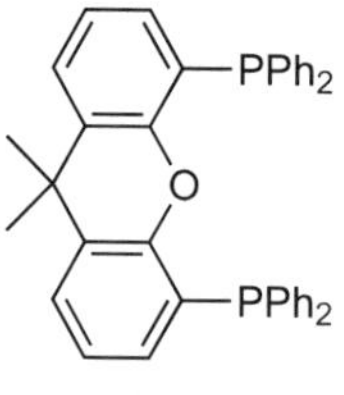

(**1**)

A route to 2-aryl-2*H*-indazoles involves intramolecular *N'*-arylation of *N*-aryl-*N*-(*o*-halobenzyl)hydrazines.[9]

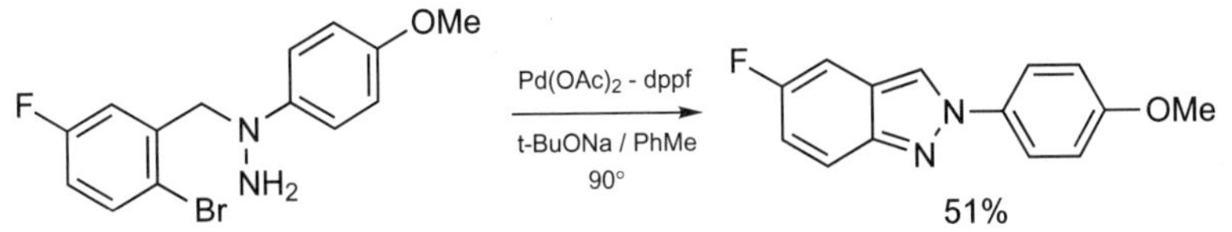

A new route to biaryl-2-carboxylic acid derivatives is found through regioselective *o*-arylation (and *o,o'*-diarylation)[10] of benzanilides with ArBr or ArOTf. Reactive benzanilides must contain N—H.

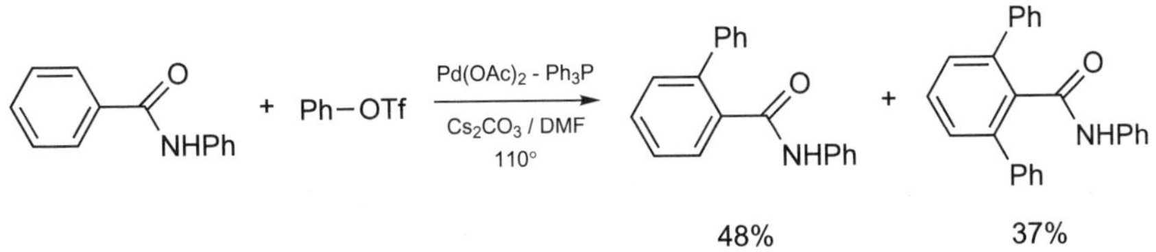

A synthesis of arylamidines[11] from ArBr, *t*-butyl isonitrile, and amines is promoted by $Pd(OAc)_2$ or $PdCl_2$ in combination with dppf and Cs_2CO_3.

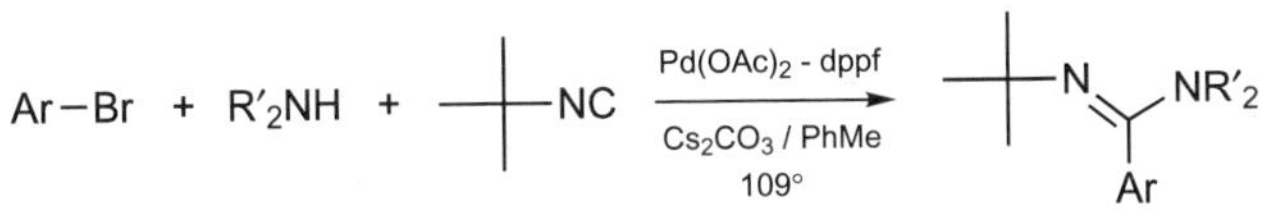

P—C bond formation. Alkenyldiarylphosphines are formed when Ar_2PH are alkylated with enol triflates.[12] Cocatalysis by Pd(0) and Cu(I) species is effective for accomplishing *P*-arylation of ArP(H)Me–borane complexes.[13]

Cyclizations. A great number of substrates designed for cyclization have been studied, including those leading to 1-sulfinylmethylene-2-methylenecycloalkanes[14] and to pyrroles.[15] Intramolecular Heck reaction involving enol triflate and allylsilane moieties is useful for the synthesis of cyclic compounds.[16]

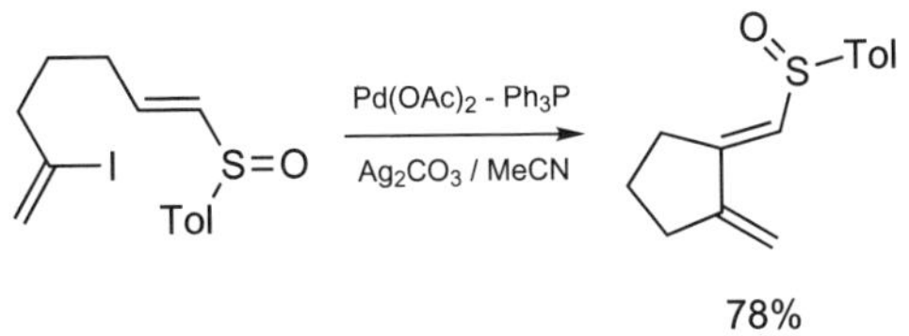

N-Allylbenzotriazoles in which a remote halogen atom is also present in the α-position of the allyl group undergo stepwise substitution reactions with primary amines.[17] Intramolecular allylic displacement of the benzotriazole occurring in the second stage is induced by a Pd-catalyst.

Formation of a bridged ring skeleton by an intramolecular Heck reaction is the key feature in a synthesis of cytisine.[18]

TfO; N; O; $Pd(OAc)_2$; O; N; PPh_2; i-Pr_2NEt / PhMe; N; O; 52%

Intramolecular addition to a carbonyl group (to afford 1-indanols and 1-tetralols) through activation of an *o*-bromoarene by $Pd(OAc)_2$–phosphine[19] is chemoselectively superior to that relying on Br–Li exchange. However, different reaction patterns are revealed in the following,[20,21] in which α-arylation of ketones[22] must be involved.

Br; Br; $Pd(OAc)_2$ - Ph_3P; Cs_2CO_3 160°; Ph; O; Ph; Ph; Ph; O; 100%; CHO; CHO; 95%

CHO; Br; +; Ph; O; Ph; $Pd(OAc)_2$ - Ph_3P; K_2CO_3 / xylene; Ph; OH; Ph; 72%

Sequential Stille coupling and cyclization to give α-pyrones is readily achieved.[23]

R; I; COOH; +; $SnBu_3$; R′; C; R″; $Pd(OAc)_2$ - Ph_3P; Na_2CO_3 - Bu_4NBr; DMF 25°; R; O; O; R′; R″

Alkenylation. Alkenylation of enol ethers at the α-position furnishes 2-alkoxy-1,3-dienes (or enones).[24]

2-Alkylidenecyclopentanones.[25] Coupling-induced ring expansion of 1-(1-alkynyl)cyclobutanols proceeds in moderate yields. Some of these products are relatively difficult to prepare by other methods.

OH + I–Ph → $Pd(OAc)_2$ - Ph_3P / i-Pr_2NEt - Bu_4NCl / DMF 80° → O, Ph 60%

[1]Leese, M.P., Williams, J.M.J. *SL* 1645 (1999).
[2]Lautens, M., Piguel, S. *ACIEE* **39**, 1045 (2000).
[3]Koseki, Y., Omino, K., Anzai, S., Nagasaka, T. *TL* **41**, 2377 (2000).
[4]Zapf, A., Beller, M. *CEJ* **6**, 1830 (2000).
[5]Shakespeare, W.C. *TL* **40**, 2035 (1999).
[6]Yin, J., Buchwald, S.L. *OL* **2**, 1101 (2000).
[7]Harris, M.C., Geis, O., Buchwald, S.L. *JOC* **64**, 6019 (1999).
[8]Guari, Y., van Es, D.S., Reek, J.N.H., Kamer, P.C.J., van Leeuwen, P.W.N.M. *TL* **40**, 3789 (1999).
[9]Song, J.J., Yee, N.K. *OL* **2**, 519 (2000).
[10]Kametani, Y., Satoh, T., Miura, M., Nomura, M. *TL* **41**, 2655 (2000).
[11]Saluste, C.G., Whitby, R.J., Furber, M. *ACIEE* **39**, 4156 (2000).
[12]Gilbertson, S.R., Fu, Z., Starkey, G.W. *TL* **40**, 8509 (1999).
[13]Al-Masum, M., Livinghouse, T. *TL* **40**, 7731 (1999).
[14]Segorbe, M.M., Adrio, J., Carretero, J.C. *TL* **41**, 1983 (2000).
[15]Grigg, R., Savic, V. *CC* 873 (2000).
[16]Tietze, L.F., Modi, A. *EJOC* 1959 (2000).
[17]Katritzky, A.R., Yao, J., Yang, B. *JOC* **64**, 6066 (1999).
[18]Coe, J.W. *OL* **2**, 4205 (2000).
[19]Quan, L.G., Lamrani, M., Yamamoto, Y. *JACS* **122**, 4827 (2000).
[20]Terao, Y., Satoh, T., Miura, M., Nomura, M. *T* **56**, 1315 (2000).
[21]Terao, Y., Satoh, T., Miura, M., Nomura, M. *BCSJ* **72**, 2345 (1999).
[22]Fox, J.M., Huang, X., Chieffi, A., Buchwald, S.L. *JACS* **122**, 1360 (2000).
[23]Rousset, S., Abarbri, M., Thibonnet, J., Duchene, A., Parrain, J.-L. *CC* 1987 (2000).
[24]Valin, K.S.A., Larhed, M., Johansson, K., Hallberg, A. *JOC* **65**, 4537 (2000).
[25]Larock, R.C., Reddy, C.K. *OL* **2**, 3325 (2000).

Palladium(II) acetate–tertiary phosphine–carbon monoxide. 20, 292

Thiocarbonylation. Extension of the method of carbonylation in the presence of a thiol to 1,3-dienes produces β,γ-unsaturated thioesters.[1] Enynes furnish branched products[2] and there is further deviation of reaction pattern by using *o*-iodothiophenol in the reaction.[3]

I, SH + C → $Pd(OAc)_2$ - dppf / i-Pr_2NEt - CO / PhH 100° → O, S 87%

Heterocycles. 5-Vinyloxazolidin-2-ones expel carbon dioxide and then incorporate CO to form dihydropyridones.[4] 3.4-Cycloindoles are accessible from *peri*-nitro 1-methylenetetralin and analogues by a reductive cyclization process.[5] Interestingly, the simplest way to prepare the substrates is by an intramolecular Heck reaction.

Pd(OAc)$_2$ - Ph$_3$P
CO / EtOH
87%

Pd(OAc)$_2$ - (o-Tol)$_3$P
Et$_3$N
Pd(OAc)$_2$ - dppp
CO / DMF 120°
X = O, NAc

[1]Xiao, W.-J., Vasapollo, G., Alper, H. *JOC* **65**, 4138 (2000).
[2]Xiao, W.-J., Vasapollo, G., Alper, H. *JOC* **65**, 2080 (2000).
[3]Xiao, W.-J., Alper, H. *JOC* **64**, 9646 (1999).
[4]Knight, J.G., Ainge, S.W., Harm, A.M., Harwood, S.J., Maughan, H.I., Armour, D.R., Hollinshead, D.M., Jaxa-Chamiec, A.A. *JACS* **122**, 2944 (2000).
[5]Söderberg, B., Rector, S.R., O'Neil, S.N. *TL* **40**, 3657 (1999).

Palladium(II) acetylacetonate–2,6-xylyl isocyanide.

Silaboration.[1] Regioselective silaboration of monosubstituted allenes at the internal double bond is observed with this catalyst system. However, perfluoroalkylallenes show a different reaction profile.

[1]Suginome, M., Ohmori, Y., Ito, Y. *SL* 1567 (1999).

Palladium(II) chloride. 13, 234–235; **15,** 248–249; **16,** 268–269; **18,** 282; **19,** 257–258; **20,** 293–394

Coupling reactions. A new catalyst for the Heck reaction consists of $PdCl_2$ and Ph_4PBr intercalated in montmorillonite K10 clay.[1] Stille coupling of (*Z*)-1,2-bis(trimethylstannyl)alkenes with hypervalent iodonium salts takes place initially at C-1 with retention of configuration.[2] Replacement of both stannyl groups is complete with 2 equiv of the coupling reagents.

Ph, $SnMe_3$ / $SnMe_3$ —($PdCl_2$; Ph_2IBF_4, DMF)→ Ph, $SnMe_3$ / Ph → Ph, Ph / Ph

Coupling of organostannanes and organotellurium chlorides under a CO atmosphere affords ketones.[3] Either $PdCl_2$ or CuI is used as catalyst.

Arylation of salicylaldehydes.[4] *o*-Hydroxybenzophenones are obtained when salicylaldehydes and diaryliodonium salts are brought together with $PdCl_2$–LiCl. This same reaction had been realized before using ArI instead of the iodonium salts.

OH, CHO + $[Cl-C_6H_4]_2I^+$ Br^- —($PdCl_2$ - LiCl; DMF)→ OH, O, Cl 89%

Deallylation.[5] Allyl aryl ethers suffer cleavage electrochemically. $PdCl_2$–bipyridine is the catalyst. Under such conditions $O \rightarrow C$ allyl transfer within 2-allyloxybenzaldehydes takes place.

Halosilanes.[6] Hydrosilanes are converted to halosilanes in halocarbon solvents such as CCl_4 and CH_2Br_2 in the presence of catalytic amounts of $PdCl_2$.

Hydrogenation.[7] The double bond of cinnamic acid derivatives is saturated by subjecting it to $PdCl_2$ in HCOOH in an alkaline medium. This transfer hydrogenation also reduces benzoylformic acid to mandelic acid in 80% yield.

[1]Varma, R.S., Naicker, K.P., Liesen, P.J. *TL* **40**, 2075 (1999).
[2]Kang, S.-K., Lee, Y.-T., Lee, S.-H. *TL* **40**, 3573 (1999).
[3]Kang, S.-K., Lee, S.-W., Ryu, H.-C. *CC* 2117 (1999).
[4]Xia, M., Chen, Z.-C. *SC* **30**, 531 (2000).
[5]Franco, D., Panyella, D., Rocamora, M., Gomez, M., Clinet, J.C., Muller, G., Dunach, E. *TL* **40**, 5685 (1999).
[6]Ferreri, C., Constantino, C., Romeo, R., Chatgilialoglu, C. *TL* **40**, 1197 (1999).
[7]Arterburn, J.B., Pannala, M., Gonzalez, A.M., Chamberlin, R.M. *TL* **41**, 7847 (2000).

Palladium(II) chloride–copper(II) chloride–carbon monoxide. 20, 294

Carbonylation. By changing reaction conditions, 1-alkynes are converted to either β-chloro-α,β-unsaturated esters[1] or 2-substituted dialkyl maleates.[2]

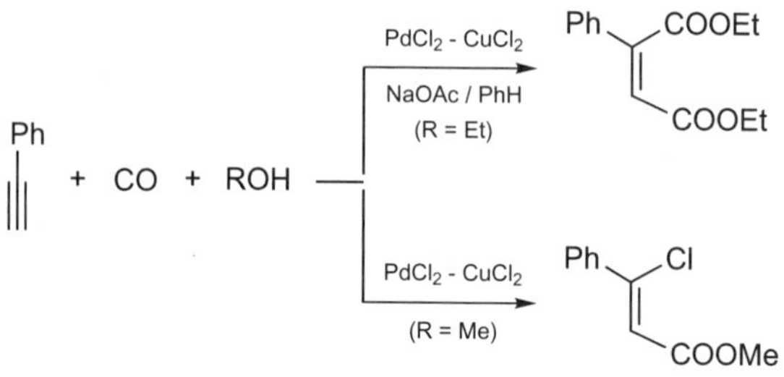

Propargylic acetates form β-alkoxy-α,β-unsaturated esters[3] when acetals are added to the carbonylation media.

Cleavage of cycloalkanones. Diesters are the major products. The cleavage preferentially occurs at the less highly substituted C—C bond.

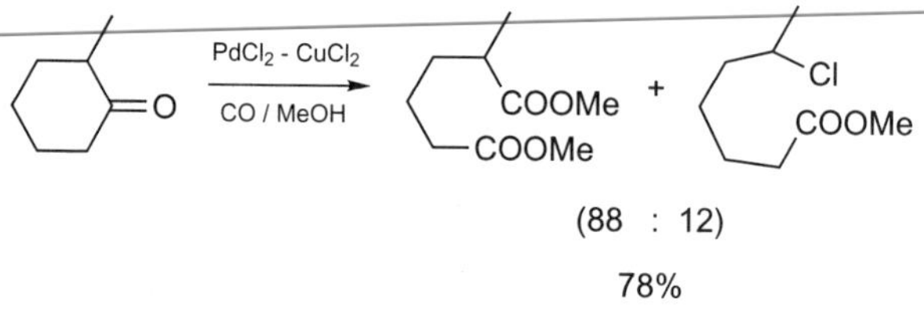

(88 : 12)

78%

[1]Li, J., Jiang, H., Feng, A., Jia, L. *JOC* **64**, 5984 (1999).
[2]Li, J., Jiang, H., Jia, L. *SC* **29**, 3733 (1999).
[3]Okumoto, H., Nishihara, S., Nakagawa, H., Suzuki, A. *SL* 217 (2000).
[4]Hamed, O., El-Qisairi, A., Henry, P.M. *TL* **41**, 3021 (2000).

Palladium(II) chloride–copper(II) chloride–oxygen. 18, 283; **19,** 261–262; **20,** 294

Oxidation. As an extension of the lactone synthesis from 1-trimethylsilyl-4-hydroxy-1-alkynes the corresponding 4-alkoxycarbonylamino derivatives give the *N*-protected γ-lactams.[1] A stereoselective synthesis of substituted tetrahydropyrans is based on carbonylation of 6-hydroxy-1-alkenes.[2]

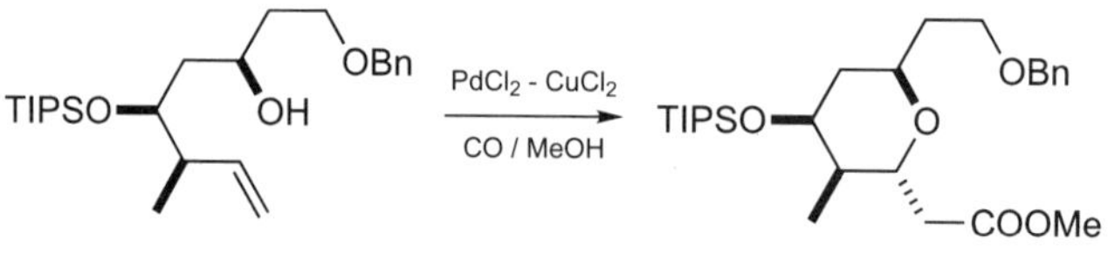

3,3-Dimethoxypropanoic esters are generated when acrylic esters are subjected to oxidation with oxygen–methanol. A reaction medium composing supercritical carbon dioxide is useful.[3]

[1]Duan, H.D., Gore, J., Vatele, J.-M. *TL* **40**, 6765 (1999).
[2]White, J.D., Hong, J., Robarge, L.A. *TL* **40**, 1463 (1999).
[3]Jia, L., Jiang, H., Li, J. *CC* 985 (1999).

Palladium(II) chloride–copper(I) iodide–triphenylphosphine.

Coupling reactions. The variety of enynes and alkynylarenes available from alkyne–iodoalkene/arene coupling is unlimited. It is worth indicating that propargylic alcohols arising therefrom can undergo isomerization to enones.[1] The latter species that can further react with guanidines provides pyrimidines in one operation.[2]

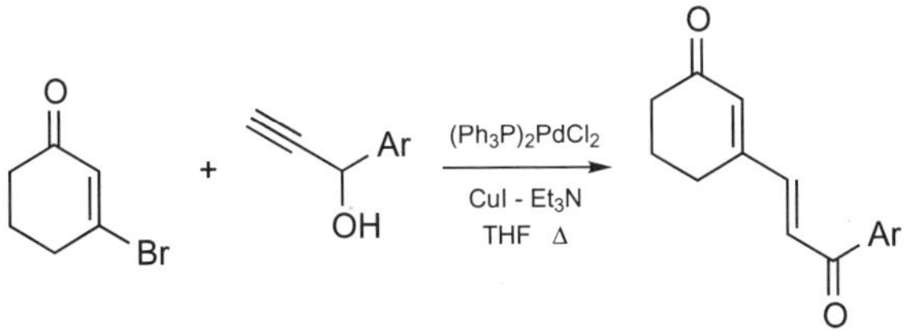

Br, O, Ar, OH, $(Ph_3P)_2PdCl_2$, CuI - Et_3N, THF Δ, O, Ar, O

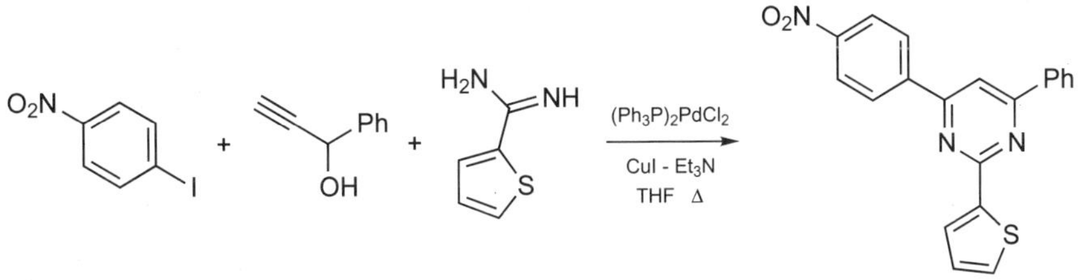

O_2N, I, Ph, OH, H_2N, NH, S, $(Ph_3P)_2PdCl_2$, CuI - Et_3N, THF Δ, O_2N, Ph, N, N, S

57%

The 1-alkyne-acetone adducts can be used instead of the alkynes in the coupling, when reaction conditions (e.g., presence of NaOH) are conducive to their decomposition in situ.[3]

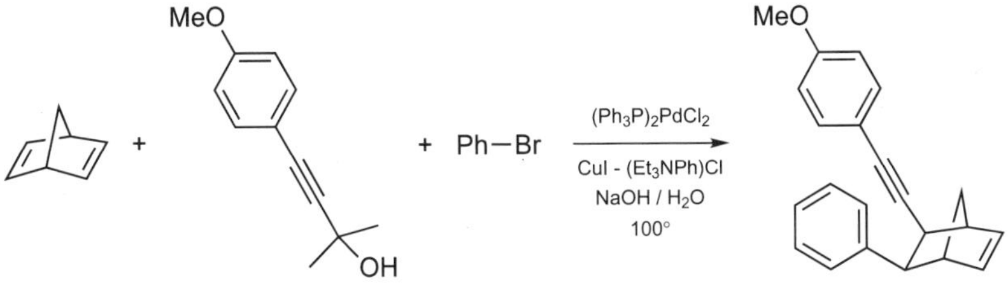

MeO, OH, Ph–Br, $(Ph_3P)_2PdCl_2$, CuI - $(Et_3NPh)Cl$, NaOH / H_2O, 100°, MeO

94%

[1]Müller, T.J.J., Ansorge, M., Aktah, D. *ACIEE* **39**, 1253 (2000).
[2]Müller, T.J.J., Braun, R., Ansorge, M. *OL* **2**, 1967 (2000).
[3]Choi, .-K., Tomita, I., Endo, T. *CL* 1253 (1999).

Palladium(II) chloride–tertiary phosphine. 19, 261; **20,** 295–298

Coupling reactions. With a ligand based on gluconamide in which a 4-(diphenylphosphino)benzyl group is attached to the nitrogen atom, the Suzuki coupling can be carried out in water.[1] In the synthesis of alkenylarenes by this coupling method, it is not necessary to employ alkenyl triflates as the corresponding mesylates, tosylates, and phosphates appear to be adequate.[2]

The Heck reaction of allenyl carbinols under carbon dioxide leads to cyclic carbonates.[3]

HO + Ph—I → $(Ph_3P)_2PdCl_2$; CO_2 - K_2CO_3; $MeCONMe_2$ 100° → Ph

B-(2-Alkenyl)pinacolatoboranes are now readily prepared from bis(pinacolato)diboron and 1-alkenyl halides or triflates.[4] Closure of the unsaturated macrocycle representing access to a potential precursor of phomactin D by an intramolecular Suzuki coupling[5] is remarkable, despite low yielding of this step at the present stage of development.

CN H OTBS, I, BBN → (dppf)$PdCl_2$ - Ph_3As; Cs_2CO_3 / H_2O - DMF → CN H OTBS

Telomerization. The Pd-catalyzed head-to-head telomerization of isoprene with amines provides 3,6-dimethyl-2,7-octadienylamines.[6]

+ N–H → $PdCl_2$ - Ar_3P; MeOH → NEt_2

78%

Ketones. 2-Alkynones can be prepared from Pd-catalyzed reaction between 1-alkynylstibines and acyl chlorides.[7]

Generation of the cobalt complexes of 4,5-didehydro-2,3-benzotropones[8] is rather surprising.

dppm; $(Ph_3P)_2PdCl_2$; Et_2NH / DMF, 80°; $Co(CO)_6$; $Co(CO)_4dppm$

1,4-Diol monoacetates undergo Pd-catalyzed elimination of HOAc. This method has been used in a synthesis of (−)-shikimic acid.[9]

OAc; OH; $(Ph_3P)_2PdCl_2$; $HCOONH_4$; MeCN Δ; O; 80%

N-Arylamines.[10] Arylation of secondary amines in good regioselectivity is observed when the catalyst is supported on NaY zeolite. Such catalysts are easily separated after the reaction by filtration.

Hydrostannylation.[11] Regiochemical *syn*-addition of Bu_3SnH to arylalkynes affords α-stannylated styrenes when the aromatic ring contains an electron-withdrawing group in the *p*-position or *o*-substituent of any electronic nature.

$(Ph_3P)_2PdCl_2$; Bu_3SnH / THF; R; $SnBu_3$; +; R; $SnBu_3$; R

R = H	38	62
R = Br	100	0

[1]Ueda, M., Nishimura, M., Miyaura, N. *SL* 856 (2000).
[2]Huffman, M.A., Yasuda, N. *SL* 471 (1999).
[3]Uemura, K., Shiraishi, D., Noziri, M., Inoue, Y. *BCSJ* **72**, 1063 (1999).
[4]Takahashi, K., Takagi, J., Ishiyama, T., Miyaura, N. *CL* 126 (2000).
[5]Kallan, N.C., Halcomb, R.L. *OL* **2**, 2687 (2000).
[6]Maddock, S.M., Finn, M.G. *OM* **19**, 2684 (2000).

[7]Kakusawa, N., Yamaguchi, K., Kurita, J., Tsuchiya, T. *TL* **41**, 4143 (2000).
[8]Iwasawa, N., Satoh, H. *JACS* **121**, 7951 (1999).
[9]Yoshida, N., Ogasawara, K. *OL* **2**, 1461 (2000).
[10]Djakovitch, L., Wagner, M., Kohler, K. *JOMC* **592**, 225 (1999).
[11]Liron, F., Le Garrec, P., Alami, M. *SL* 246 (1999).

Palladium(II) chloride–triphenylphosphine–carbon monoxide. 20, 298–299

Lactams. Depending on the nature of the imines and their reaction partners, lactams arising from the Pd-catalyzed reaction under carbon monoxide vary.[1,2]

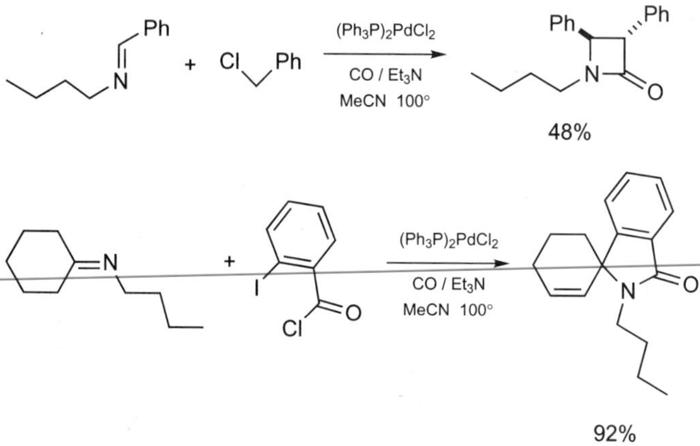

$(Ph_3P)_2PdCl_2$; CO / Et_3N; MeCN 100°

48%

$(Ph_3P)_2PdCl_2$; CO / Et_3N; MeCN 100°

92%

Heterocycles. Many different heterocycles are synthesized via a carbonylation process: hydantoins from aldehydes and urea,[3] flavones from ethynylarenes and *o*-iodophenol acetates,[4] and α-substituted α,β-unsaturated lactones from iodoalkenols.[5]

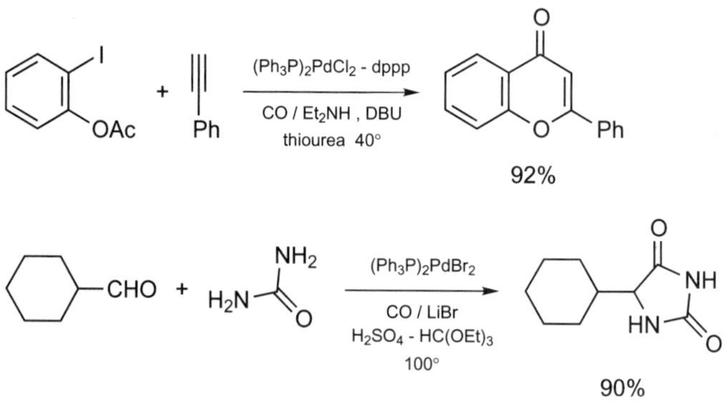

$(Ph_3P)_2PdCl_2$ - dppp; CO / Et_2NH, DBU; thiourea 40°

92%

$(Ph_3P)_2PdBr_2$; CO / LiBr; H_2SO_4 - $HC(OEt)_3$; 100°

90%

Thiocarbamates.[6] The carbonylative condensation of amines and thiols by Pd-catalysis avoids using phosgene.

Aromatic aldehydes.[7] Electrochemical deiodoformylation of ArI is accomplished with this catalytic system under carbon monoxide and in the presence of formic acid.

3-Alkenamides.[8] These amides are prepared from allyl carbonates and amines under the carbonylation reaction conditions.

(dppp)PdCl$_2$ - CO; i-Pr$_2$NEt / DMF; 52%

[1]Cho, C.S., Jiang, L.H., Shim, S.C. *SC* **29**, 2695 (1999).
[2]Cho, C.S., Wu, X., Jiang, L.H., Shim, S.C., Kim, H.R *JHC* **36**, 297 (1999).
[3]Beller, M., Eckert, M., Moradi, W.A., Neumann, H *ACIEE* **38**, 1455 (1999).
[4]Miao, H., Yang, Z. *OL* **2**, 1765 (2000).
[5]Liao, B., Negishi, E.-I. *H* **52**, 1241 (2000).
[6]Jones, W.D., Reynolds, K.A., Sperry, C.K., Lachicotte, R.J. *OM* **19**, 1661 (2000).
[7]Carelli, I., Chiarotto, I., Cacchi, S., Pace, P., Amatore, C., Jutand, A., Meyer, G *EJOC* 1471 (1999).
[8]Loh, T.-P., Cao, G.-Q., Yin, Z. *TL* **40**, 2649 (1999).

Palladium(II) chloride–tertiary phosphine–silver(I) oxide.

Cyclopropanes. Coupling of cyclopropylboronic acids with various activated halides, for example, benzylic,[1] allylic,[2] and acyl halides,[3] leads to substituted cyclopropanes.

(dppf)PdCl$_2$; Ag$_2$O - K$_2$CO$_3$, PhMe 80°; 78%

[1]Chen, H., Deng, M.-Z. *JCS(P1)* 1609 (2000).
[2]Chen, H., Deng, M.-Z. *JOC* **65**, 4444 (2000).
[3]Chen, H., Deng, M.-Z. *OL* **2**, 1649 (2000).

Palladium(II) hydroxide–carbon. 19, 262; **20,** 299–300

Hydrogenolysis.[1] *N′,N′*-Dialkylformamidines serve as protected primary amines. Removal of the =CHNR$_2$ group is readily achieved by hydrogenolysis.

H$_2$; Pd(OH)$_2$ - C, t-BuOH - H$_2$O; 98%

N-Methylation.[2] Monoalkyl amino acids are further methylated on hydrogenation using formaldehyde as the carbon donor.

[1] Vincent, S., Mioskowski, C., Lebeau, L. *JOC* **64**, 991 (1999).
[2] Song, Y., Sercel, A.D., Johnson, D.R., Colbry, N.L., Sun, K.L., Roth, B.D. *TL* **41**, 8225 (2000).

Palladium(II) iodide–potassium iodide.

Heterocyclizations. Cyclization of 5-hydroxylated 3-en-1-ynes leads to furans.[1] The analogous thiols afford thiophenes.[2] In methanol and under a carbon monoxide atmosphere (CO/air = 9:1) 5-hydroxy-1-alkynes are transformed into 2(*E*)-(methoxycarbonyl)methylenetetrahydrofurans.[3]

OH + CO —(PdI_2 - KI, MeOH)→ COOMe

Oxidative cyclocarbonylation of 2-amino alcohols is achieved.[4] A convenient synthesis of oxaolidin-2-ones is thereby developed.

Substituted maleic anhydrides.[5] The catalyzed formation of butenolides from alkynes and CO is changed to maleic anhydrides by adding carbon dioxide in the system.

R ≡ + CO —(PdI_2 - KI, H_2O - dioxane)→ butenolide (R, O, O)
R ≡ + CO —(PdI_2 - KI, H_2O - dioxane, CO_2)→ maleic anhydride (R, O, O, O)

[1] Gabriele, B., Salerno, G., Lauria, E. *JOC* **64**, 7687 (1999).
[2] Gabriele, B., Salerno, G., Fazio, A. *OL* **2**, 351 (2000).
[3] Gabriele, B., Salerno, G., de Pascali, F., Costa, M., Chiusoli, G.P. *JOMC* **593/594**, 409 (2000).
[4] Gabriele, B., Salerno, G., Brindisi, D., Costa, M., Chiusoli, G.P. *OL* **2**, 625 (2000).
[5] Gabriele, B., Salerno, G., Costa, M., Chiusoli, G.P. *CC* 1381 (1999).

Palladium(II) iodide–thiourea.

Benzo[b]furan-3-carboxylic esters.[1] Cyclization and alkoxycarbonylation is observed when 2-alkynylphenols are exposed to PdI_2–thiourea–CBr_4 under CO and in the presence of Cs_2CO_3. Carbon tetrabromide serves as the reoxidant of Pd(0) species.

$$\text{R-C}_6\text{H}_3\text{(OH)-C}\equiv\text{C-R'} + \text{CO} \xrightarrow[\text{CBr}_4\text{ - Cs}_2\text{CO}_3,\ \text{MeOH}]{\text{PdI}_2\text{ - CS(NH}_2)_2} \text{2-R'-3-(COOMe)benzofuran (R)}$$

79–85%

[1]Nan, Y., Miao, H., Yang, Z. *OL* **2**, 297 (2000).

Pentamethylcyclopentadienylbis(triphenylphosphine)ruthenium complexes.

Aldol reaction.[1] The Rh-complexes that also contain the ethyl cyanoacetate anion are active in promoting both aldol and Michael reactions, for example, involving cyanoacetic esters.

[1]Murahashi, S.-I., Take, K., Naota, T., Takaya, H. *SL* 1016 (2000).

3,3-Pentamethylenediaziridine.

Aziridination.[1] Both *cis*- and *trans*-α,β-unsaturated amides are transformed into *cis*-aziridine derivatives.

HN–NH + BuLi → HN; N; O; O; N

82%

[1]Hori, K., Sugihara, H., Ito, Y.N., Katsuki, T. *TL* **40**, 5207 (1999).

1,2,2,6,6-Pentamethylpiperidine.

Base.[1] Hindered secondary amines that undergo *N*-alkylation by using this tertiary amine as base can avoid racemization of chirality centers.

COOMe; NH; COOMe + TsO–CH₂CH₂–OTs; N–; PhMe Δ → COOMe; N; COOMe; COOMe; N; COOMe

93%

[1]Insaf, S.S., Witiak, D.T. *S* 435 (1999).

Perfluoroarenes.

N-Arylation.[1] Amines are perfluoroarylated with perfluoroarenes in THF. The amino group is attached to the *p*-position in the reaction involving perfluorotoluene or pentafluoropyridine. Generally, primary amines are more reactive but secondary amines also undergo perfluoroarylation at a higher temperature and with added triethylamine.

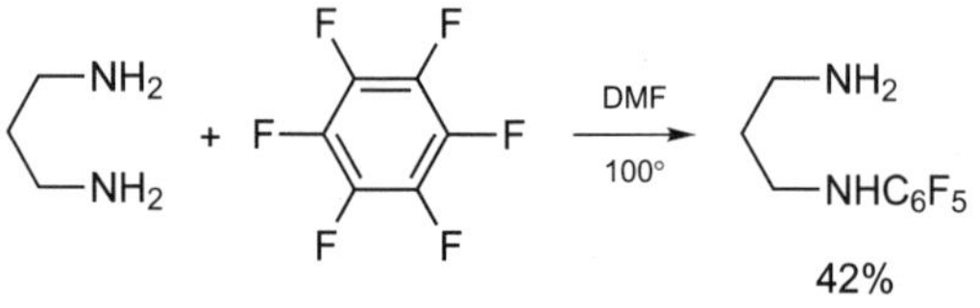

NH_2, NH_2 + F, F, F, F, F, F; DMF, 100° → NH_2, NHC_6F_5; 42%

[1]Beletskaya, I.P., Artamkina, G.A., Ivushkin, V.A., Guilard, R. *TL* **41**, 313 (2000).

Phase-transfer catalysts. 13, 239–240; **15,** 252–253; **18,** 286–289; **19,** 264–267; **20,** 302–303

Alkylations. For alkylation and Michael reactions involving enolates a new phase-transfer catalyst is the Ru-complex **1.**[1]

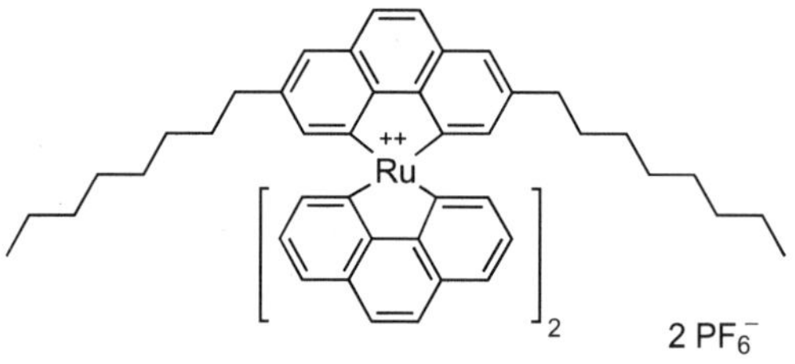

++ Ru; 2; 2 PF_6^-

(1)

Phenols undergo *O*-alkylation in the presence of PEG-400 without solvent.[2] 2-Methallyloxyphenol is prepared from catechol by a microwave-assisted monoalkylation.[3]

Alkylation of *N*-sulfonyl derivatives of α-amino acid derivatives proceeds well under solid–liquid phase-transfer catalytic conditions,[4] whereas imines from α-amino acid *t*-butyl esters undergo asymmetric alkylation[5] with activated halides in the presence of the chiral quaternary ammonium salt (**2**).

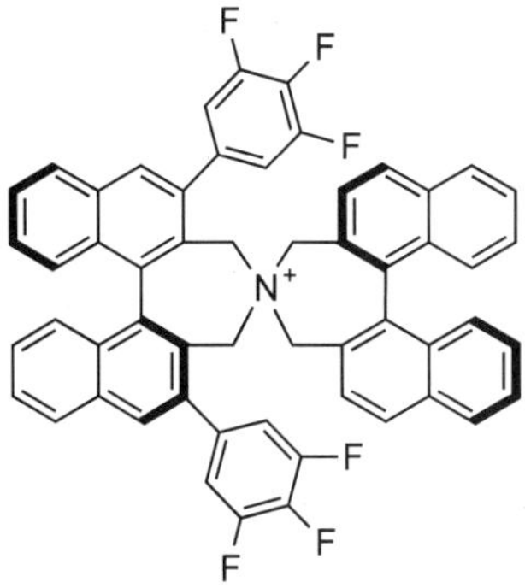

(2)

Wittig reaction. The poly(ethylene glycol)-linked bis(triphenylphosphine) (**3**) is a precursor of Wittig reagents.[6] Upon *P*-alkylation, the phosphonium salts are used.

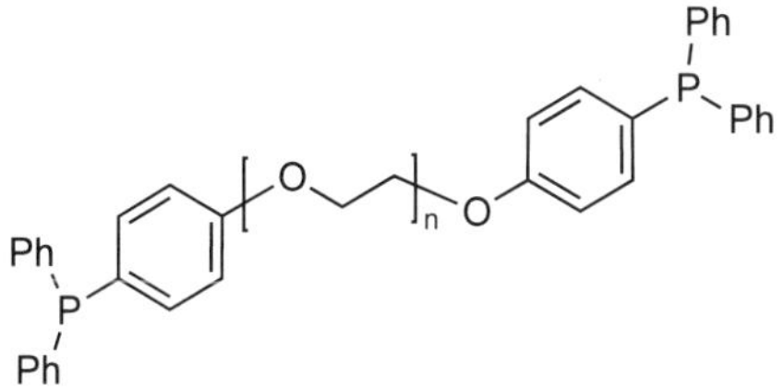

(3)

Cyclocondensations. The cyclopropanation of conjugated sulfones with dimethylsulfoxonium methylide generated under PTC conditions is higher yielding than employing the *t*-BuOK–DMSO system.[7] Related to the the well-known dichlorocyclopropanation is the formation of 2,2-dichloroepisulfides from thioketones.[8]

The asymmetric Darzens reaction of α-chlorocycloalkanones[9] using quaternized cinchona alkaloids as phase-transfer catalyst and LiOH as base in Bu_2O at room temperature is usually high yielding. However, the moderate ee is disappointing.

Oxidations. Oxidation of benzyl ethers with $KMnO_4$ in the presence of benzyltriethylammonium chloride gives alkyl benzoates in refluxing dichloromethane,[10] and under similar conditions, *N*-phenyl azacycles afford lactams.[11]

Degradation of aldehydes is accomplished via their silyl enol ethers by oxidative cleavage with hydrogen peroxide in a two-phase system with added PTC.[12]

Epoxidation of enones with H_2O_2–LiOH in water and a solid catalyst prepared from DABCO and Merrifield resin is essentially operated in a triphase situation.[13]

Isomerizations. Under PTC conditions (presence of tetrahexylammonium bromide) 1,3-diarylpropynes can be isomerized to diarylallenes by KOH at room temperature.[14] A Brook rearrangement of acylsilanes occurs upon addition of cyanide ion.[15]

KCN / Bu_4NBr, H_2O / CH_2Cl_2 — 95%

Hydrolysis. Hydrolytic cleavage of silyl ethers under PTC conditions shows selectivity. Phenolic TBS ethers are preferentially removed.[16]

OTBS, OTBS — NaOH (solid) / dioxane, Bu_4NHSO_4 → OH, OTBS — 76%

Heck reaction. The sodium salt of a disulfonated triphenylphosphine serves as ligand for the Pd-catalyzed Heck reaction when the latter is conducted in a mixture of toluene and ethylene glycol.[17]

Hydroboration.[18] Controlled generation of borane in the organic phase is critical to the success of hydroboration under PTC conditions that consist of stirring mixtures of the alkenes, bromobutane, an onium salt, and aq $NaBH_4$.

Dichloromethylation. Insertion of dichlorocarbene into tertiary C—H bonds geminal to alkoxy and siloxy groups proceeds with retention of configuration.[19] This reaction is synthetically valuable.

NaOH - $CHCl_3$, $C_{16}H_{33}N(Me_3)Cl$, 80° → Cl_2HC, Ph — 86%

[1]Tzalis, D., Knochel, P. *TL* **40**, 3685 (1999).
[2]Cao, Y.-Q., Pei, B.-G. *SC* **30**, 1759 (2000).
[3]Li, J., Pang, J., Cao, G., Xi, Z. *SC* **30**, 1337 (2000).
[4]Albanese, D., Landini, D., Lupi, V., Penso, M. *EJOC* 1443 (2000).
[5]Ooi, T., Takeuchi, M., Kameda, M., Maruoka, K. *JACS* **122**, 5228 (2000).
[6]Sieber, F., Wentworth, P., Toker, J.D., Wentworth, A.D., Metz, W.A., Reed, N.N., Janda, K.D. *JOC* **64**, 5188 (1999).
[7]Padmaja, A., Reddy, K.R., Reddy, N.S., Reddy, D.B. *PSS* **152**, 91 (1999).
[8]Mloston, G., Romanski, J., Swiatek, A., Heimgartner, H. *HCA* **82**, 946 (1999).
[9]Arai, S., Shirai, Y., Ishida, T., Shioiri, T. *CC* 49 (1999).

[10]Markgraf, J.H., Choi, B.Y. *SC* **29**, 2405 (1999).
[11]Markgraf, J.H., Stickney, C.A. *JHC* **37**, 109 (2000).
[12]Sakaguchi, S., Yamamoto, Y., Sugimoto, T., Yamamoto, H., Ishii, Y. *JOC* **64**, 5954 (1999).
[13]Anand, R.V., Singh, V.K. *SL* 807 (2000).
[14]Oku, M., Arai, S., Katayama, K., Shioiri, T. *SL* 493 (2000).
[15]Takeda, K., Ohnishi, Y. *TL* **41**, 4169 (2000).
[16]Crouch, R.D., Steiff, M., Frie, J.L., Cadwaller, A.B., Beris, D.C. *TL* **40**, 3133 (1999).
[17]Thorpe, T., Brown, S.M., Crosby, J., Fitzjohn, S., Muxworthy, J.P., Williams, J.M.J. *TL* **41**, 4503 (2000).
[18]Albanese, D., Landini, D., Maia, A., Penso, M. *SL* 997 (2000).
[19]Masaki, Y., Arasaki, H., Shiro, M. *CL* 1180 (2000).

1-(1-Phenoxyalkyl)benzotriazoles.

Diketones.[1] These reagents are acyl anion equivalents. Thus, alkylation with 1, ω-dihaloalkanes followed by acid hydrolysis leads to diketones.

BuLi / Br(CH₂)₄Br — THF −78° ; aq. HCl , MeOH

[1]Katritzky, A.R., Huang, Z., Fang, Y., Prakash, I. *JOC* **64**, 2124 (1999).

O-(Phenylcarbamoyl)hydroxylamine.

Nitriles.[1] Direct conversion of aldehydes to nitriles is completed on heating with $PhNHCOONH_2$.TsOH in THF (11 examples, 87–97%).

[1]Coskun, N., Arikan, N. *T* **55**, 11943 (1999).

Phenyl chloro(thionoformate).

Dealkylation.[1] Unhindered aliphatic tertiary amines afford thionocarbamates of secondary amines on reaction with the title compound at room temperature. Subsequent *S*-methylation (with Me_2SO_4) and hydrolysis accomplishes *N*-dealkylation. The thionoformate reagent is superior to ClCOOPh in terms of its efficiency.

AcO + Cl(C=S)OPh — CH_2Cl_2, 20° → AcO

96%

Dehydration.[2] Alkanamides are dehydrated to nitriles by this reagent, but formamides give isonitriles.

[1]Millan, D.S., Prager, R.H. *AJC* **52**, 841 (1999).
[2]Bose, D.S., Goud, P.R. *TL* **40**, 747 (1999).

Phenyl(fluoro)iodine triflate. 20, 304–305

Oxidative rearrangement.[1] (PhIF)OTf induces rearrangement of nonterminal alkynes in an alcohol to give α-branched esters.

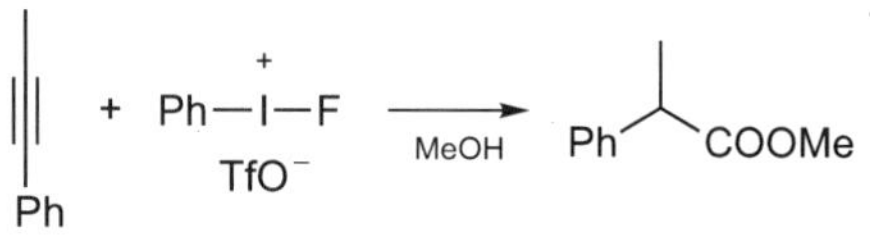

[1]Pirguliyev, N. Sh., Brel, V. K., Zefirov, N. S., Stang, P. J. *MC* 189 (1999).

Phenyliodine(III) bis(trifluoroacetate). 13, 241–242; **14,** 257; **15,** 257–258; **16,** 274–275; **18,** 289–290; **19,** 267–268; **20,** 305

Cyclizations. Annulation accompanies oxidation of *N*-sulfonyl-4-methoxyanilines with $PhI(OCOCF_3)_2$ in the presence of 1-alkenes.[1] 5-Methoxyindoles or indolines are produced in this one-pot reaction. Cyclization of 3-arylpropyl azides where the aromatic ring is activated (e.g., 3-methoxylated) to fused quinone imines is observed.[2]

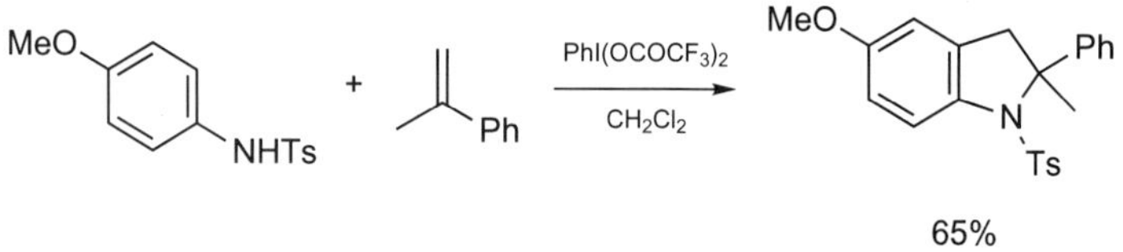

65%

2-(*m*-Methoxyaryl)ethyl benzyl sulfides are similarly cyclized to *S*-benzyl-2,3-dihydrobenzothiophenes.[3]

Dehydrogenation.[4] 1, 4-Dihydropyridines are rapidly aromatized on exposure to $PhI(OCOCF_3)_2$ although addition of sulfur and microwave irradiation of the mixtures has the same effect.

Dearomatization.[5] 2-Substituted 1-naphthols are allylated at C-2 when they are exposed to $PhI(OCOCF_3)_2$ and an allylsilane. 1-Trimethylsiloxy-1,3-butadiene also react with the naphthoxyiodonium species.

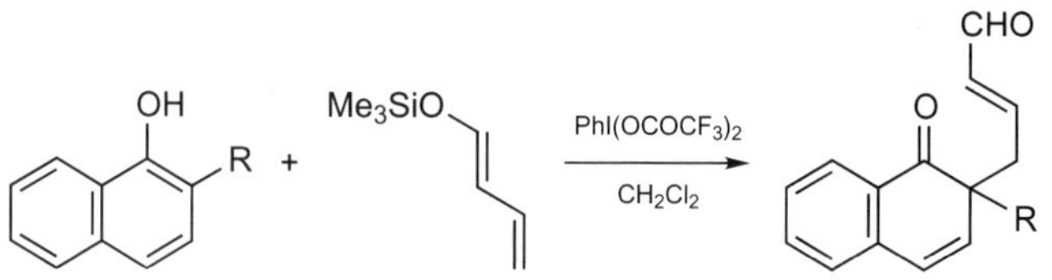

[1]Tohma, H., Watanabe, H., Takisawa, S., Maegawa, T., Kita, Y. *H* **51**, 1785 (1999).
[2]Kita, Y., Egi, M., Ohtsubo, M., Saiki, T., Okajima, A., Takada, T., Tohma, H. *CPB* **47**, 241 (1999).
[3]Kita, Y., Egi, M., Tohma, H. *CC* 143 (1999).
[4]Varma, R.S., Kumar, D. *JCS(P1)* 1755 (1999).
[5]Quideau, S., Looney, M.A., Pouysegu, L. *OL* **1**, 1651 (1999).

Phenyliodine(III) diacetate. 13, 242–243; **14,** 258–259; **15,** 258; **16,** 275–276; **17,** 280–281; **18,** 290–291; **19,** 268–270; **20,** 305–307

Diaryliodonium sulfonates. These salts are prepared from $ArB(OH)_2$ upon admixture with $PhI(OAc)_2 \cdot 2TfOH$ in dichloromethane.[1]

Oxidations. 4-Hydoxy-2-cyclobutenones are oxidized to γ-acetoxybutenolides[2] with $PhI(OAc)_2$. 2-Methoxyphenols give *o*-quinones that can be trapped as Diels–Alder adducts.[3,4]

Acylnitroso compounds generated by oxidation of hydroxamic acids react with alkenes in situ. *O*-Acetylation and reduction afford *N*-allyl amides.[5]

2-(β-Indolylethyl)oxazolines are transformed into a tetracyclic system.[6]

With a (salen)CrCl complex as catalyst, the oxidation of allylic alcohols to enones with $PhI(OAc)_2$ is chemoselective.[7]

Ring contraction. A facile synthesis of 2-alkylidenecycloalkanecarboxylic esters[8] is by oxidation of 2-alkylidenecycloalkanones with $PhI(OAc)_2$.

Benzyne precursors. Cyclic siloxanes such as **1** are readily prepared from *o*-dibromoarenes via Grignard reaction with dimethylchlorosilane. *o*-Silylaryliodonium triflates that are obtained from oxidation decompose to benzynes on exposure to fluoride ion.[9]

(**1**) — $PhI(OAc)_2$, TfOH / CH_2Cl_2 → Ph–I(TfO)–aryl–Si(Me)$_2$OH 87% — Bu_4NF, furan →

Only the *m*-silyl group of methyl 3,4-bis(trimethylsilyl)benzoate is replaced by the [PhI] residue on reaction with $PhI(OAc)_2$.[10]

N-Sulfonylamine activation. Iminoiodine(III) compounds (which can serve as nitrene sources) are produced when sulfonamides are combined with $PhI(OAc)_2$.[11] In the presence of sulfides, *N*-sulfonylsulfilimines ($R_2S{=}NSO_2R'$) are formed.[12]

Amination.[13] A combination of $PhI(OAc)_2$ and RNH_2 (R = Ts, Ns, Ms, CF_3CO) in the presence of Mn(porphyrin) or Ru(porphyrin) are effective in aminating a variety of hydrocarbons.

[1] Carroll, M.A., Pike, V.W., Widdowson, D.A. *TL* **41**, 5393 (2000).
[2] Ohno, M., Oguri, I., Eguchi, S. *JOC* **64**, 8995 (1999).
[3] Rao, P.D., Chen, C.-H., Liao, C.-C. *CC* 713 (1999).
[4] Kurti, L., Szilagyi, L., Antus, S., Nogradi, M. *EJOC* 2579 (1999).
[5] Adam, W., Bottke, N., Krebs, O., Saha-Möller, C.R. *EJOC* 1963 (1999).
[6] Braun, N.A., Bray, J.D., Ciufolini, M.A. *TL* **40**, 4985 (1999).
[7] Adam, W., Hajra, S., Herderich, M., Saha-Möller, C. *OL* **2**, 2773 (2000).
[8] Varma, R.S., Kumar, D. *S* 1288 (1999).
[9] Kitamura, T., Meng, Z., Fujiwara, Y. *TL* **41**, 6611 (2000).
[10] Kitamura, T., Wasai, K., Todaka, M., Fujiwara, Y. *SL* 731 (1999).
[11] Dauben, P., Dodd, R.H. *OL* **2**, 2327 (2000).
[12] Ou, W., Chen, Z.-C. *SC* **29**, 4443 (1999).
[13] Yu, X.-Q., Huang, J.-S., Zhou, X.-G., Che, C.-M. *OL* **2**, 2233 (2000).

Phenyliodine(III) diacetate–bromine.

Hunsdiecker reaction.[1] Decarboxylative bromination is achieved photochemically by this reagent combination.

o-nitrobenzoic acid (COOH, NO_2) — $PhI(OAc)_2$, Br_2 / CH_2Br_2, hν Δ → o-bromonitrobenzene (Br, NO_2) 51%

[1]Camps, P., Lukach, A.E., Pujol, X., Vazquez, S. *T* **56**, 2703 (2000).

Phenyliodine(III) diacetate–iodine. 20, 307

Iodination.[1] An acid-catalyzed iodination of activated arenes uses the oxidant combination. An oxidative decarboxylation of α-amino acid derivatives results in the introduction of an iodine atom at the β-position.[2]

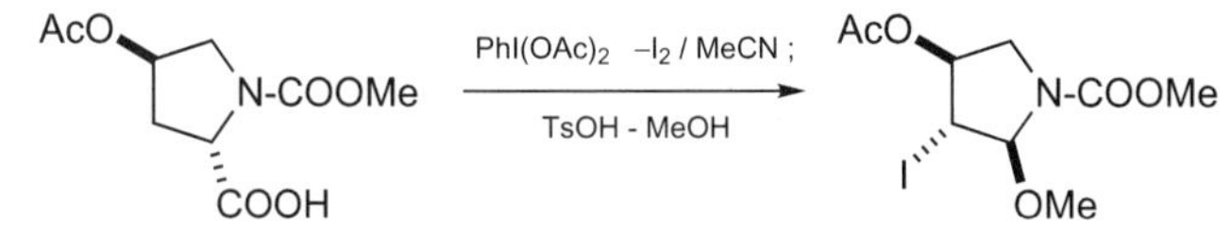

Phthalides. *o*-Alkylbenzoic acids are transformed into phthalides.[3] 3-substituted phthalides with a hydroxyl group in the side chain give spiroacetals.[4]

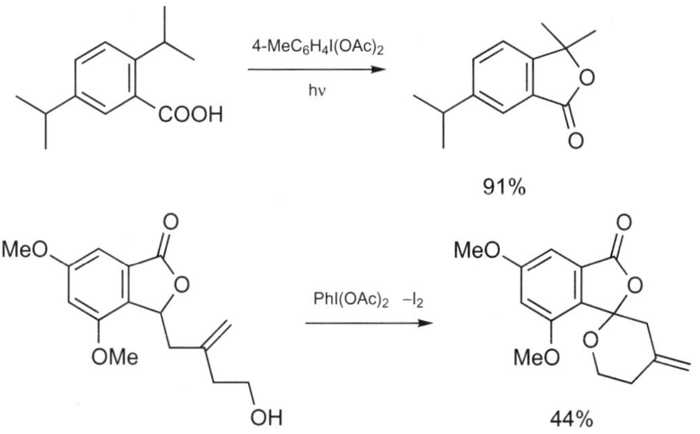

Remote functionalization.[5] The combined reagent changes an alcohol into a hypoiodite for subsequent iodination of an unactivated C—H bond. Ultrasound is helpful.

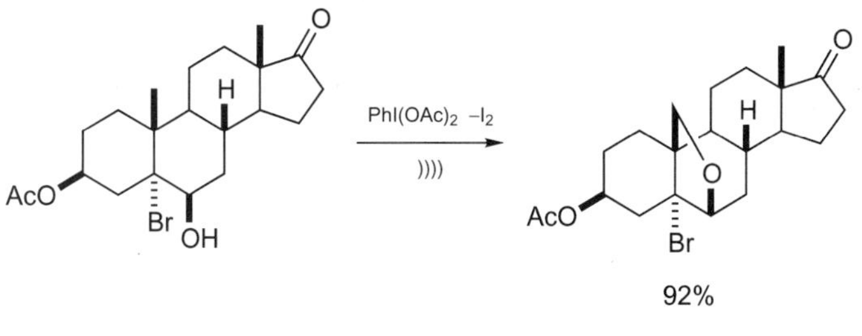

Dealkylation. The alkyl group of an *N*-alkylsulfonamide is detached by the reagent couple with ultrasound assistance.[6] However, arylethylsulfonamides undergo iodinative heterocyclization.[7]

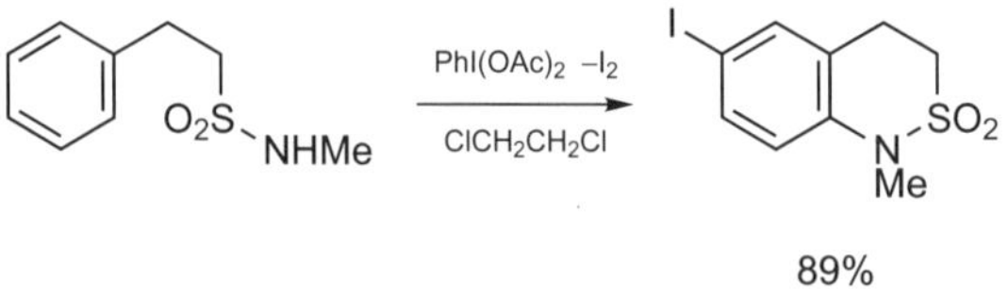

$PhI(OAc)_2$–I_2, $ClCH_2CH_2Cl$; 89%

Cycloaddition. The following reaction is a most unusual oxidative 1,3-dipolar cycloaddition involving a benzene ring as the dipolarophile.[8]

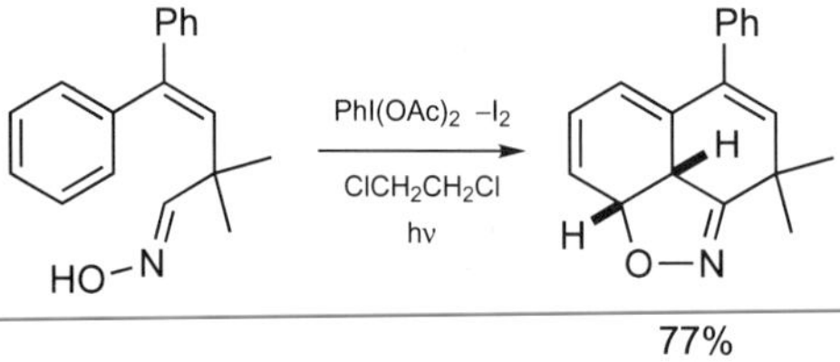

$PhI(OAc)_2$–I_2, $ClCH_2CH_2Cl$, hν; 77%

Fragmentation. Angular alcohols suffer fragmentation on treatment with $PhI(OAc)_2$–I_2, thereby creating medium-sized rings.[9] However, the regioselectivity can be subverted by placing an azido group in a subangular carbon atom.[10]

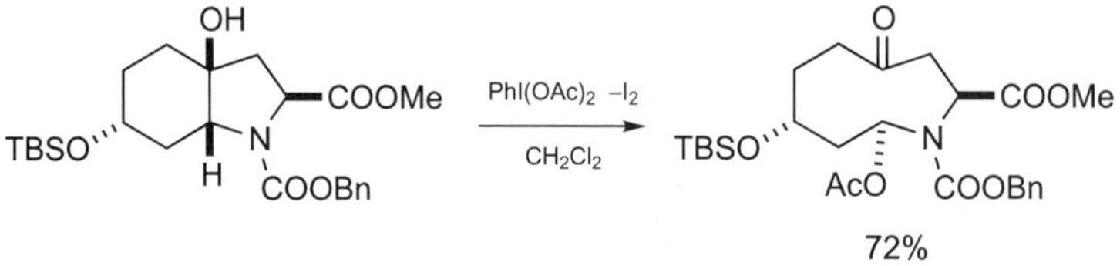

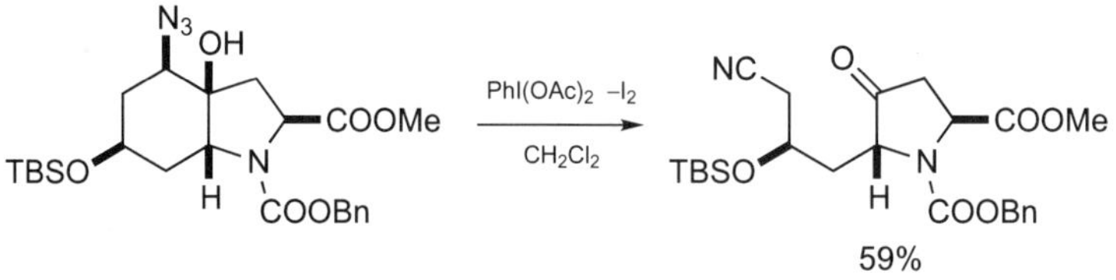

$PhI(OAc)_2$–I_2, CH_2Cl_2; 72%

$PhI(OAc)_2$–I_2, CH_2Cl_2; 59%

[1]Kryska, A., Skulski, L. *JCR(S)* 590 (1999).
[2]Boto, A., Hernandez, R., Suarez, E. *TL* **41**, 2494 (2000).
[3]Muraki, T., Togo, H., Yokoyama, M. *JCS(P1)* 1713 (1999).
[4]Brimble, M.A., Caprio, V.E., Johnston, A.D., Sidford, M.H. *TL* **41**, 3955 (2000).

[5]Costa, S.C.P., Moreno, M.J.S.M., Sa e Melo, M.L., Neves, A.S.C. *TL* **40**, 8711 (1999).
[6]Katohgi, M., Yokoyama, M., Togo, H. *SL* 1055 (2000).
[7]Togo, H. , Harada, Y., Yokoyama, M. *JOC* **65**, 926 (2000).
[8]Barnes, J.C., Horspool, W.H., Hynd, G. *CC* 425 (1999).
[9]Wipf, P., Li, W. *JOC* **64**, 4576 (1999).
[10]Wipf, P., Mareska, D.A. *TL* **41**, 4723 (2000).

Phenyliodine(III) diacetate–trimethylsilyl azide.

Oxidation. 3-*O*-Silylated glycals are oxidized to afford the enones.[1] Other silyl ethers are not affected.

Et3SiO, OTBS, Et3SiO, O → (PhI(OAc)2 - Me3SiN3, PhMe 25°) Et3SiO, OTBS, O=, O

95%

Note that the $PhI(OAc)_2$–NaN_3 combination converts ArCHO into aroyl azides.[2]

β-Bromoalkyl azides.[3] Bromide ion is oxidized in situ by $PhI(OAc)_2$ to induce bromoazidation of alkenes.

[1]Kirschning, A., Hary, U., Plumeier, C., Ries, M., Rose, L. *JCS(P1)* 519 (1999).
[2]Chen, D.-J., Chen, Z.-C. *TL* **41**, 7361 (2000).
[3]Kirschning, A., Abdul Hashem, M., Monenschein, H., Rose, L., Schoning, K.-U. *JOC* **64**, 6522 (1999)

Phenyliodine(III) dicyanide.

Alkene functionalizations.[1] This reagent can be used for oxidative activation of reagents (e.g., PhSeSePh) to initiate a reaction with alkenes. With participation of proper nucleophiles present in the reaction media addition reactions completed.

R → (Phi(CN)2 - KSCN, PhSeSePh / MeCN) PhSe, SCN, R

R = Hx 80%

[1]Margarita, R., Mercanti, C., Parlanti, L., Piancatelli, G. *EJOC* 1865 (2000)

α-(Phenylseleno)acrylic esters.

Pyrrolidine-3-carboxylic esters.[1] The selenoacrylic esters are obtained from adducts of PhSeCl and acrylic esters by base-promoted dehydrochlorination. After Michael addition of allylic amines to the selenoacrylates, radical generation via selenium abstraction starts the free radical cyclization.

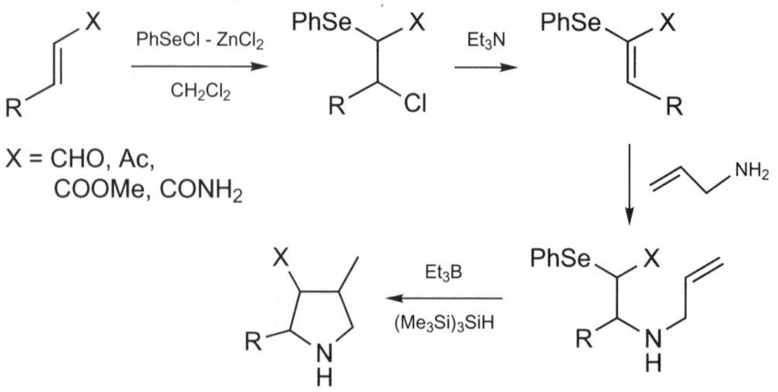

X = CHO, Ac, COOMe, $CONH_2$

[1]Berlin, S., Engman, L. *TL* **41**, 3701 (2000).

3-Phenylthio-2-(*N*-cyanoimino)thiazolidine.

Sulfenylation.[1] This reagent (**1**) readily transfers its PhS group to amines and enolates.

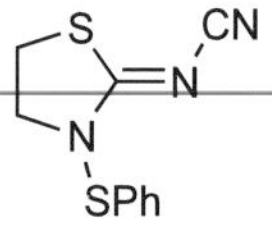

(**1**)

[1]Tanaka, T., Azuma, T., Fang, X., Uchida, S., Iwata, C., Ishida, T., In, Y., Maezaki, N. *SL* 33 (2000).

Phenyl trifluoroacethioate.

Trifluoromethylation.[1] Photochemical reaction of $CF_3C{=}O(SPh)$ (also CF_3SO_2SPh) with alkenes gives alkanes bearing a trifluoromethyl group, although in rather low yields. The trifluoromethyl radical generated on decomposition of the reagent is the addend.

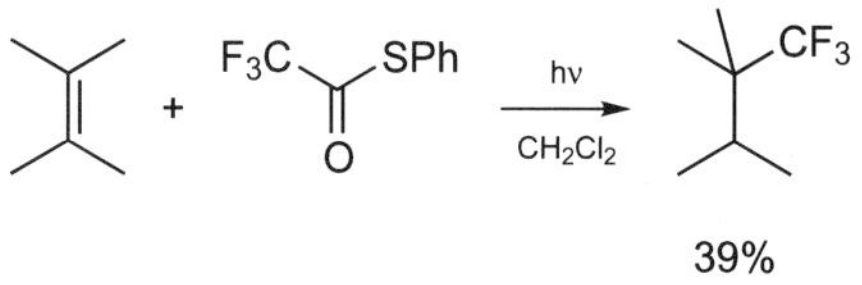

39%

[1]Billard, T., Roques, N., Langlois, B.R. *TL* **41**, 3069 (2000).

Phenyl trifluoromethanesulfonethiolate.

Alkenyl triflones.[1] This reagent, CF_3SO_2SPh, as well as CF_3SO_2SePh, react with alkenes at 80°, and the adducts undergo elimination on oxidation.

F_3C-SO_2SPh + cyclohexene $\xrightarrow{80°}$ PhS / SO_2CF_3 $\xrightarrow{[O]}$ SO_2CF_3 63%

[1]Billard, T., Langlois, B.R. *T* **55**, 8665 (1999).

Phenyltrifluorosilane.

Azomethine ylides.[1] These reactive dipoles are generated from *N*-(α-silylalkyl)imino ethers by treatment with $PhSiF_3$. It is necessary to provide proper dipolarophiles as trapping agents for their use in ring formation.

Ph, $SiMe_3$, N, OMe, Ph + NPh (O, O) $\xrightarrow[CH_2Cl_2]{PhSiF_3}$ Ph, N, Ph, NPh (O, O) 99%

[1]Washizuka, K.-I., Minakata, S., Ryu, I., Komatsu, M. *T* **55**, 12969 (1999).

2-Phenyl-2-(trimethylsilyl)ethanol.

Carboxyl protection.[1] DCC-mediated esterification transforms carboxylic acids into esters. Rapid regeneration of the acid (as salts) is achieved by treatment with Bu_4NF.

[1]Wagner, M., Kunz, H. *SL* 400 (2000).

Phosgene.

Phosphonothioformic esters.[1] Consecutive reaction of phosgene with a thiol (in the presence of triethylamine) at low temperature and trialkyl phosphite furnishes $RSC({=}O)PO(OR')_2$.

[1]Salomon, C.J., Breuer, E. *SL* 815 (2000).

Phosphazene bases.

Diaryl sulfides.[1] The phosphazene base P_2—Et (**1**) is useful in the CuBr-promoted reaction between ArSH and Ar′I in toluene. Bases of this class are very strong and hindered, they generate highly nucleophilic "naked" anions from various acids (including carbon acids).

(1)

[1]Palombo, C., Oiarbide, M., Lopez, R., Gomez-Bengoa, E. *TL* **41**, 1283 (2000).

Phosphinic acid.

Radical cyclization.[1] Phosphinic acid has been used to induce reductive cyclization of haloalkenes.

H_3PO_2 - $NaHCO_3$

AIBN / EtOH Δ

87%

[1]Yorimitsu, H., Shinokubo, H., Oshima, K. *CL* 104 (2000).

Phosphines.

Preparation. 2-Hydroxyethylphosphines are conveniently prepared from red phosphorus, via reduction with Na in *t*-BuOH and reaction of the resulting $NaPH_2$ with epoxides in liquid ammonia.[1]

Carbamates[2] Reaction of organic azides with Me_3P, treatment of the ensuing iminophosphoranes with chloroformic esters, and hydrolytic cleavage of the aminophosphonium salts complete the conversion.

Transesterification.[3] Activation of 2,2,2-trihaloethyl esters by phosphines enables exchange of the alkoxy residues.

Furfurylidenetrialkylphosphoranes.[4] These Wittig reagents are generated from conjugated 2-alken-4-ynones by the addition of R_3P. In situ reaction with carbonyl compounds leads to 2-alkenylfurans.

Bu_3P / R″CHO

CH_2Cl_2

Baylis–Hillman reactions.[5] BINOL and a teriary phosphine (e.g., Bu_3P) constitute a cooperative catalytic system to promote the reaction (higher rates and yields). Application to catalytic asymmetric synthesis is indicated.

Mukaiyama aldol reactions.[6] Silyl keteneacetals undergo condensation with aldehydes under the influence of tris(2,4,6-trimethoxyphenyl)phosphine. Triphenylphosphine is slightly less efficient and aliphatic phosphines such as *t*-Bu_3P are much inferrior.

Copper(I) hydride complexes. Stabilization of CuH with various phosphines enables the use of such complexes to reduce carbonyl groups (in the presence of a double bond).[7]

Mitsunobu reactions.[8] An improved reagent pair consists of diphenyl(2-pyridyl)-phosphine and di-*t*-butyl azodicarboxylate.

Hydroboration.[9] The nature of the phosphine additive plays a critical role in the hydroboration of allenes with pinacolatoborane under the catalysis of $(dba)_2Pt$. Thus, different regioisomeric adducts can be prepared by variation of such additive.

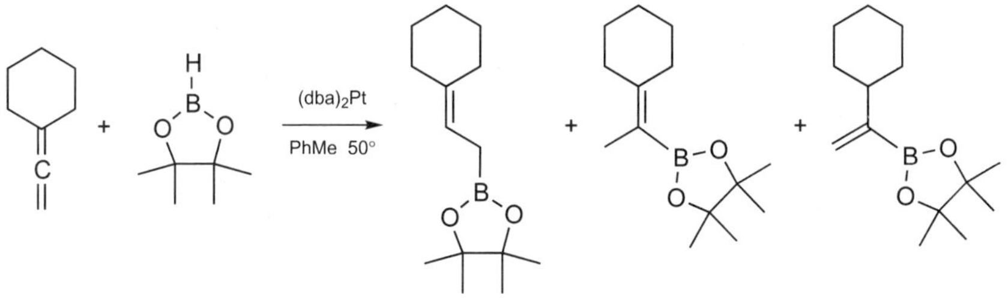

$(dba)_2Pt$, PhMe 50°

Additive			
+ Cy_3P	56	0	44
+ t-Bu_3P	1	0	99
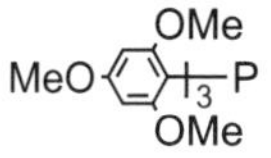 + (2,4,6-$(MeO)_3C_6H_2)_3P$	0	100	0

[1]Arbuzova, S.N., Brandsma, L., Gusarova, N.K., van der Kerk, A.H.T.M., van Hooijdonk, M.C.J.M., Trofimov, B.A. *S* 65 (2000).
[2]Ariza, X., Urpi, F., Vilarrasa, J. *TL* **40**, 7515 (1999).
[3]Hans, J.J., Driver, R.W., Burke, S.D. *JOC* **64**, 1430 (1999).
[4]Kuroda, H., Hanaki, E., Kawakami, M. *TL* **40**, 3753 (1999).
[5]Yamada, Y.M.A., Ikegami, S. *TL* **41**, 2165 (2000).
[6]Matsukawa, S., Okano, N., Imamoto, T. *TL* **41**, 103 (2000).
[7]Chen, J.-X., Daeuble, J.F., Brestensky, D.M., Stryker, J.M. *T* **56**, 2153 (2000).
[8]Kiankarimi, M., Lowe, R., McCarthy, J.R., Whitten, J.P. *TL* **40**, 4497 (1999).
[9]Yamamoto, Y., Fujikawa, R., Yamada, A., Miyaura, N. *CL* 1069 (1999).

Phthalic anhydride.

Hydrolysis of amides.[1] Primary amides are 'hydrolyzed' to carboxylic acids by heating in the dry state with phthalic anhydride at 240–250°. Thus, pivalamide affords pivalic acid in 91% yield (the side product is phthalimide).

[1]Chemat, F. *TL* **40**, 3855 (1999).

Picoline hydrochloride.

Chlorination.[1] Aliphatic aldehydes are converted to α,α-dichloro derivatives with chlorine using picoline hydrochloride as catalyst. On further reaction, the dichloro-carboxylic acids are formed.

[1]Bellesia, F., de Buyck, L., Ghelfi, F. *SL* 146 (2000).

Platinum/montmorillonite.

Deoxygenation.[1] A method for the reductive removal of the oxygen atom from a carbonyl group consists of high-temperature hydrogenation with a Pt-on-K10 montmorillonite clay catalyst in diglyme.

[1]Torok, B., London, G., Bartok, M. *SL* 631 (2000).

Platinum(II) chloride. 19, 272

Cyclization. Novel cyclization patterns of 1,6-enynes induced by $PtCl_2$ are revealed.[1,2]

COOMe — $PtCl_2$, PhMe 80° → MeOOC — 70%

Ts—N — SiMe$_3$ — $PtCl_2$, PhMe 80° → Ts—N — SiMe$_3$ — 73%

COOMe — $PtCl_2$, PhMe 80° → MeOOC — 73%

X, X, R — $PtCl_2$, R′OH → X, X, H, OR′, R

X = COOMe, SO_2Ph...

E, E — $PtCl_2$, MeOH → E, E, OMe

E = COOMe

When the ene portion of the substrates is allylically silylated or stannylated, the products are still 1,4-dienes.[3]

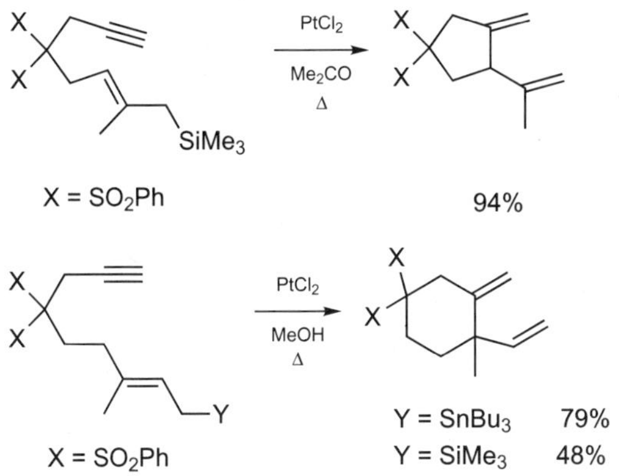
PtCl2
Me2CO
Δ
SiMe3
X = SO2Ph
94%
PtCl2
MeOH
Δ
X = SO2Ph
Y = SnBu3 79%
Y = SiMe3 48%

Allylation.[4] Allylic chlorosilanes are activated by $PtCl_2$ and submit their allyl residues to aldehydes. Such reactions are *syn*-selective.

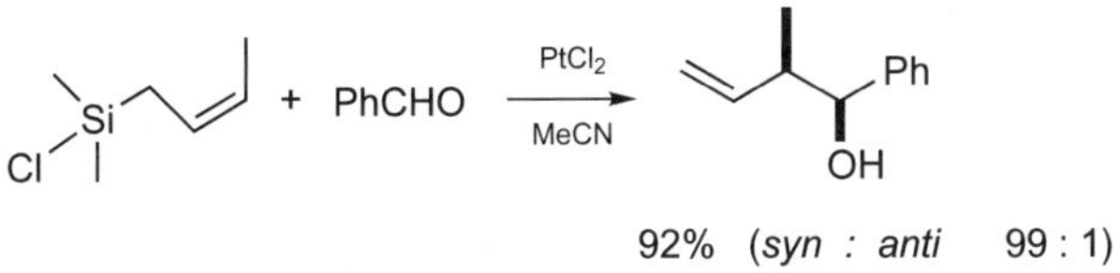
+ PhCHO
PtCl2
MeCN
Ph
OH
92% (syn : anti 99 : 1)

[1]Fürstner, A., Szillat, H., Stelzer, F. *JACS* **122**, 6785 (2000).
[2]Mendez, M., Munoz, M.P., Echavarren, A.M. *JACS* **122**, 11549 (2000).
[3]Fernandez-Rivas, C., Mendez, M., Echavarren, A.M. *JACS* **122**, 1221 (2000).
[4]Fürstner, A., Voigtlander, D. *S* 959 (2000).

Platinum(IV) oxide.

Hydrogenation. Partial saturation of a 2,3-bipyridyl after selective quaternization constitutes a key step in a synthesis of cytisine. PtO_2 serves well as the hydrogenation catalyst.[1]

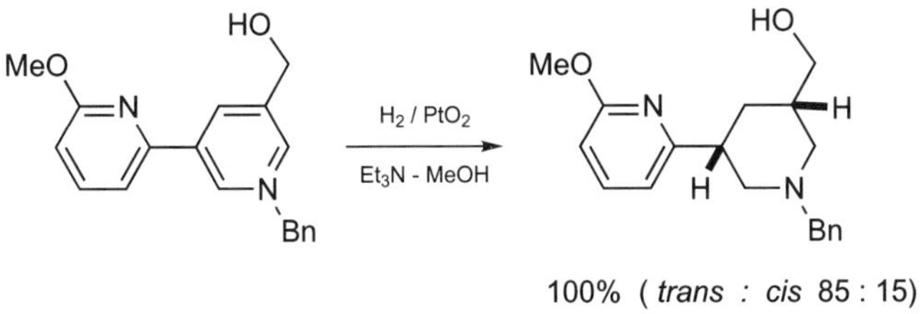
HO
MeO
N
N
Bn
H2 / PtO2
Et3N - MeOH
HO
MeO
N
H
H
N
Bn
100% (trans : cis 85 : 15)

[1]O'Neill, B.T., Yohannes, D., Bundesmann, M.W., Arnold, E.P. *OL* **2**, 4201 (2000).

Poly(methylhydrosiloxane), PMHS. 20, 311

Reductions. The reduction of carbonyl compounds with PMHS requires promoters. These include 2,8,9-trimethyl-1-phospha-2,5,8,9-tetraazabicyclo[3.3.3]undecane[1] and zinc compounds.[2,3] Conjugated aldehydes are converted to allylic alcohols.[2]

Imines are similarly saturated in the presence of zinc chloride[4] or a chiral titanocene difluoride,[5] the latter reaction on ketimines proceeds with excellent enantioselectivity (up to 99% ee).

Hydrogenolysis. PMHS is a useful hydrogen source in the Pd–C-catalyzed hydrogenolysis of aziridines[6] and organic azides.[7] The amine products can be converted to *t*-butyl carbamates in situ by $(Boc)_2O$.

Benzylidene acetals are cleaved reductively with PMHS in the presence of $AlCl_3$. A less hyghly substituted hydroxy group is exposed.[8]

PMHS; $AlCl_3$; Et_2O - CH_2Cl_2

Hydrostannylation and Stille coupling. In situ reduction of triorganochlorostannanes with PMHS during Pd-catalyzed hydrostannylation[9] is convenient. Modification of reaction conditions enables an efficient synthesis of dienes (Stille coupling).[10]

Me_3SnCl - $(2\text{-Fu})_3P$; $(Ph_3P)_2PdCl_2$ - $(dba)_3Pd_2$; PMHS / $NaHCO_3$; H_2O - Et_2O

[1]Wang, Z., Wroblewski, A.E., Verkade, J.G. *JOC* **64**, 8021 (1999).
[2]Mimoun, H. *JOC* **64**, 2582 (1999).
[3]Mimoun, H., de Saint Laumier, J.Y., Giannini, L., Scopelliti, R., Floriani, C. *JACS* **121**, 6158 (1999).
[4]Chandrasekhar, S., Reddy, M.V., Chandraiah, L. *SC* **29**, 3981 (1999).
[5]Hansen M.C., Buchwald, S.L. *OL* **2**, 713 (2000).
[6]Chandrasekhar, S., Ahmed, M. *TL* **40**, 9325 (1999).
[7]Chandrasekhar, S., Chandraiah, L., Reddy, C.R., Reddy, M.V. *CL* 780 (2000).
[8]Chandrasekhar, S., Reddy, Y.R., Reddy, C.R. *CL* 1273 (1999).

[9]Maleczka, R.E., Terrell, L.R., Clark, D.H., Whitehead, S.L., Gallagher, W.P., Terstiege, I. *JOC* **64**, 5958 (1999).
[10]Maleczka, R.E., Gallagher, W.P., Terstiege, I. *JACS* **122**, 384 (2000).

Potassium.

Deoxygenation.[1] Ultrasonically dispersed potassium reduces azoxyarenes to azoarenes at room temperature.

[1]Wang, X., Xu, M., Chen, J., Pan. Y., Shi, Y. *SC* **30**, 2253 (2000).

Potassium *t*-butoxide. 13, 252–254; **15,** 271–272; **17,** 289–290; **18,** 296–297; **19,** 273–275; **20,** 11–313

Eliminations. A new preparation of α-substituted acrylonitriles involves a twofold elimination of quaternary ammonium salts.[1]

NMe2 N NMe2 OSiMe3 MeI / Me2CO ; t-BuOK / THF CN OSiMe3 57%

1,3-Elimination of phosphine oxide with stereocontrol enables synthesis of optically active cyclopropyl ketones.[2]

O Ph P Ph Bu Ph O OSiMe3 t-BuOK / t-BuOH Ph O Bu 83%

Double elimination of [$PhSO_2H$] and benzoic acid, which results in substituted arenes, is the last step in an alkylative aromatization of cyclohexone.[3]

O + Ph–SO2 BuLi ; PhCOCl / THF ; t-BuOK 75%

Wittig and Emmons–Wadsworth reactions. An alkyne synthesis by chain extension of aldehydes uses dihalomethyltriphenylphosphonium salts as Wittig reagent precursors.[4] Note that temperature is important in determining whether the products are 1,1-diiodoalkenes, 1-iodoalkynes, or 1-alkynes.

Ph_3P - CHI_3, t-BuOK, –78° → 80%; t-BuOK: –50° R = I 97%; +25° R = H 81%

For polyvinylogation, 4-diethoxyphosphonyl-1,3-dienolates and their *O*-acylated and *O*-silylated derivatives are efficient reagents.[5]

+ PhCHO; t-BuOK / THF, –70° → 75%

2-Alkylideneglutaronitriles are readily formed by consecutive Michael and Wittig reactions.[6] The process involves addition of the cyanomethylphosphonate anion to acrylonitrile and quenching the homologated anion with aldehydes.

t-BuOK / THF - HMPA ; PhCHO ; HCl - H_2O → 72%

Cyclizations. Synthesis of substituted indoles from *o*-haloanilines by condensation with ketones[7] or nitriles[8] is promoted by *t*-BuOK in DMSO.

Furans are formed from α-propargyl ketones.[9] Alkynyl benzyl sulfides cyclize to give 2-aryl-2,3-dihydrothiophenes.[10] Deprotonation at the benzylic position initiates the cyclization.

t-BuOK / MeCN, 0° → 75%

Aromatic substitutions. 2-Nitroaryl-1,3-dithianes are assembled by the vicarious nucleophilic substitution on nitroarenes.[11]

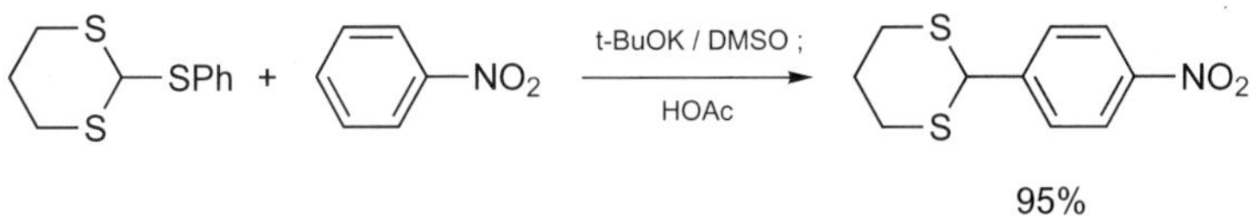

t-BuOK / DMSO ; HOAc

95%

Amides. Conversion of esters to amides in a solvent-free process uses microwave heating with solid *t*-BuOK.[12] Oxidative decyanation of α-aminonitriles derived from aldehydes with *t*-BuOK in DMSO also afford amides.[13]

Hydrolysis. Amino esters (α- and β-) are cleaved under nonaqueous conditions (*t*-BuOK–THF) at 0° or below. This method proves valuable for kinetic resolution of aziridine esters.[14]

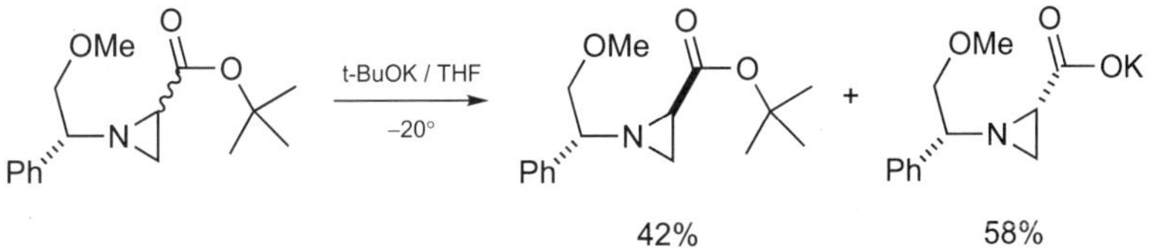

t-BuOK / THF, −20°

42% 58%

Claisen rearrangement. Allyl allenyl ethers resulting from isomerization of allyl propargyl ethers with *t*-BuOK are ready to undergo Claisen rearrangement.[15] A route to α-substituted acroleins is established.

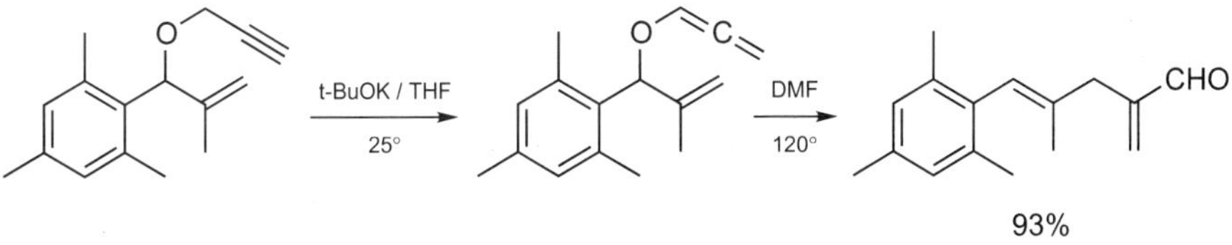

t-BuOK / THF, 25°; DMF, 120°

93%

Aldol reactions. A reagent containing *t*-BuOK and $(BuO)_4Ti$ is highly effective.[16]

Addition to styrenes. A catalytic amount of *t*-BuOK in DMSO promotes the addition of ketones and imines to styrenes at 40°. Clean addition of nitriles is conducted at room temperature in either DMSO or NMP.[17]

[1]Tsvetkov, N.P., Vakhmistrov, V.E., Koldobsky, A.B., Kalinin, V.N. *RCB* **48**, 1685 (1999).
[2]Nelson, A., Warren, S. *JCS(P1)* 3425 (1999).
[3]Orita, A., Yaruva, J., Otera, J. *ACIEE* **38**, 2267 (1999).
[4]Michel, P., Gennet, D., Rassat, A. *TL* **40**, 8575, 8579 (1999).
[5]Mohamed-Hachi, A., About-Jaudet, E., Combret, J.-C., Collignon, N.*S* 1188 (1999).
[6]Shen, Y., Zhang, Z. *SC* **30**, 445 (2000).
[7]Baumgartner, M.T., Nazareno, M.A., Murguia, M.C., Pierini, A.B., Rossi, R.A. *S* 2053 (1999).
[8]Moskalev, N., Makosza, M. *H* **52**, 533 (2000).
[9]MaGee, D.I., Leach, J.D., Setiadji, S. *T* **55**, 2847 (1999).
[10]McConachie, L.K., Schwan, A.L. *TL* **41**, 5637 (2000).
[11]Kim, W.-K., Paik, S.-C., Lee, H., Cho, C.-G. *TL* **41**, 5111 (2000).
[12]Varma, R.S., Naicker, K.P. *TL* **40**, 6177 (1999).
[13]Enders, D., Amaya, A.S., Pierre, F. *NJC* **23**, 261 (1999).
[14]Alezza, V., Bouchet, C., Micouin, L., Bonin, M., Husson, H.-P. *TL* **41**, 655 (2000).
[15]Parsons, P.J., Thomson, P., Taylor, A., Sparks, T. *OL* **2**, 571 (2000).
[16]Han, Z., Yorimitsu, H., Shinokubo. H., Oshima, K. *TL* **41**, 4415 (2000).
[17]Rodriguez, A.L., Bunlaksananusorn, T., Knochel, P. *OL* **2**, 3285 (2000).

Potassium bromide–oxidant.

N-Bromosaccharin.[1] *N*-Bromination of saccharin by the KBr–$KBrO_3$ combination and sulfuric acid in aq HOAc at room temperature yields 74% of the product that is a source of electrophilic bromine.

Bromination of anilines.[2] Good selectivity (*p*-bromination) is observed with KBr and sodium perborate in HOAc at room temperature.

[1]Zajc, B. *SC* **29**, 1779 (1999).
[2]Roche, D., Prasad, K., Repic, O., Blacklock, T.J. *TL* **41**, 2083 (2000).

Potassium chlorate–hydrochloric acid.

Aryliodine(III) dichloride.[1] When used in a two-phase system (H_2O–CCl_4), chlorine generated from the reagent combination converts ArI to $ArICl_2$.

[1] Krassowska-Swiebocka, B., Prokopienko, G., Skulski, L. *SL* 1409 (1999).

Potassium *trans*-dichlorotetracyanoplatinate(IV).

Disulfides.[1] Rapid and quantitative intramolecular disulfide bond formation from dithiols is effected by this reagent.

[1]Shi, T., Rabenstein, D.L. *JOC* **64**, 4590 (1999).

Potassium fluoride. 13, 256–257; **15,** 272; **18,** 297–298; **19,** 275–276; **20,** 313

Trifluoromethylthiolation.[1] In combination with thiophosgene, KF (or other fluoride salts) serves to introduce a SCF_3 group to aromatic substrates.

$Cl_2C{=}S$ - KF, –15°

85%

Allylation.[2] Desilylative allylation is promoted by KF and CuI.

KF - CuI, THF - DMSO

67% (*E* : *Z* 90 : 10)

[1]Tavener, S.J., Adams, D.J., Clark, J.H. *JFC* **95**, 171 (1999).
[2]Takeda, T., Uruga, T., Gohroku, K., Fujiwara, T. *CL* 821 (1999).

Potassium fluoride–alumina. 20, 313

Alkylations. KF–Al_2O_3 is a convenient base for *N*-alkylation of lactam-type compounds in DMF.[1,2]

Azlactones.[3] Formation of these products from *N*-acylglycine and aldehydes in acetic anhydride in the presence of KF–Al_2O_3 apparently requires a lower temperature.

Desulfonylation.[4] Sulfonates and sulfonamides suffer X—S bond cleavage in the dry state on microwave irradiation.

[1]Blass, B.E., Drowns, M., Harris, C.L., Liu, S., Portlock, D.E. *TL* **40**, 6545 (1999).
[2]Blass, B.E., Burt, T.M., Liu, S., Portlock, D.E., Swing, E.M. *TL* **41**, 2063 (2000).
[3]Wang, Y., Shi, D., Lu, Z., Dai, G. *SC* **30**, 707 (2000).
[4]Sabitha, G., Abraham, S., Reddy, B.V.S., Yadav, J.S. *SL*1745 (1999).

Potassium hexamethyldisilazide.

Cyclopentenes.[1] On dehydrobromination of 2-substituted 1,2-dibromoalkanes with KHMDS carbenes of the RR′C═C: type are generated. Intramolecular hydrogen abstraction and cyclization follow.

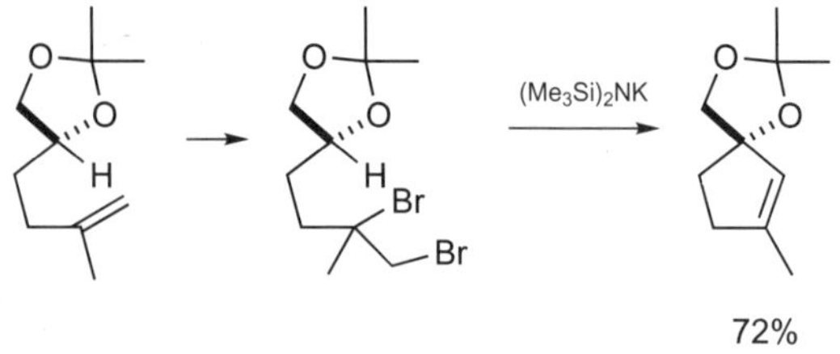

(Me$_3$Si)$_2$NK

72%

Benzyl nitriles.[2] Tertiary benzyl nitriles are prepared by substitution of aryl fluorides.

[1]Taber, D.F., Christos, T.E., Neubert, T.D., Batra, D. *JOC* **64**, 9673 (1999).
[2]Caron, S., Vazquez, E., Wojcik, J.M. *JACS* **122**, 712 (2000).

Potassium hydride.

Indoles.[1] *o*-Alkynylanilines undergo cyclization (5-*endo-dig* process) on treatment with KH in NMP. Sodium hydride is largely ineffective (<5% yield) in bringing about the reaction.

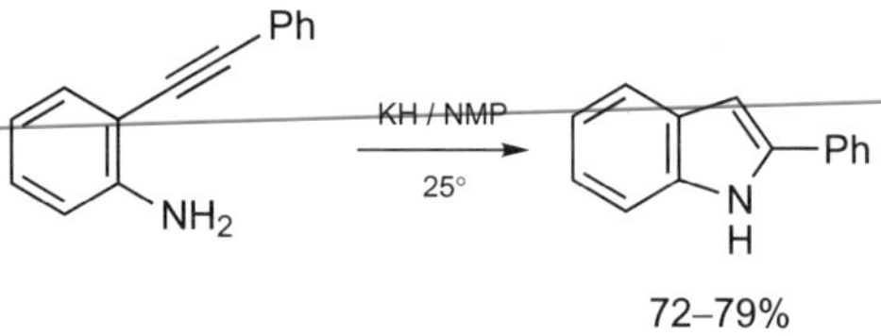

KH / NMP, 25°

72–79%

Ring expansion.[2] 2-Cyano-1-vinylcycloalkanols are induced to expand by two carbon units on heating with KH and 18-crown-6 in THF.

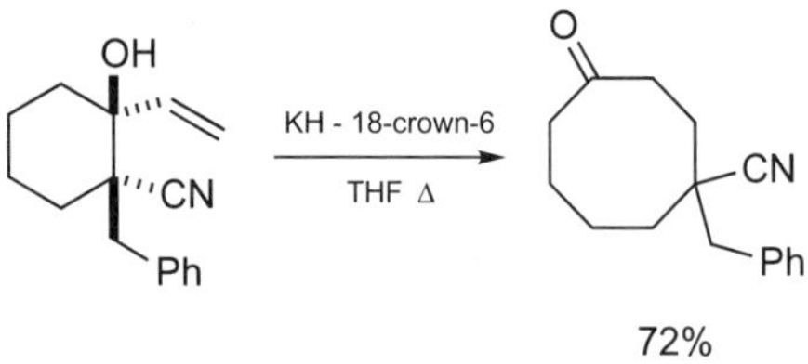

KH - 18-crown-6, THF Δ

72%

[1]Rodriguez, A.L., Koradin, C., Dohle, W., Knochel, P. *ACIEE* **39**, 2488 (2000).
[2]Shia, K.-S., Jan, N.-W., Zhu, J.-L., Ly, T.W., Liu, H.-J. *TL* **40**, 6753 (1999).

Potassium hydroxide.

Ketone cleavage.[1] Enolizable ketones are cleaved to furnish carboxylic acids with KOH in DMF.

Nitro aldols.[2] Condensation of nitroalkanes with aldehydes can be performed in dry media (powdered KOH as condensation agent).

$RNH_2 \rightarrow ROH$.[3] Primary aliphatic amines are converted to alcohols in good yields by KOH in diethylene glycol at 210°.

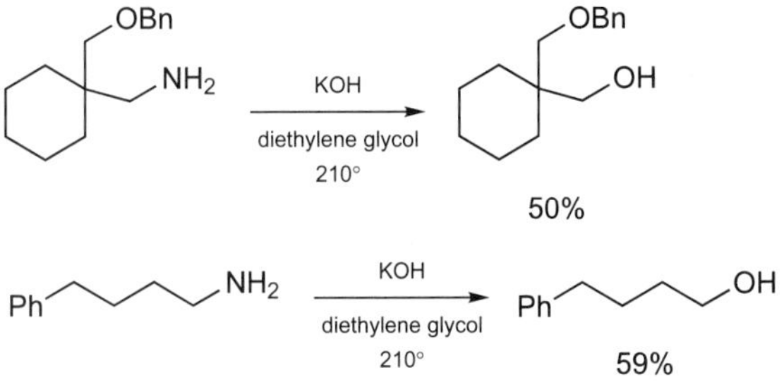

[1]Zabjek, A., Petric, A. *TL* **40**, 6077 (1999).
[2]Ballini, R., Bosica, G., Parrini, M. *CL* 1105 (1999).
[3]Rahman, S.M.A., Ohno, H., Maezaki, N., Iwata, C., Tanaka, T. *OL* **2**, 2893 (2000).

Potassium monoperoxysulfate, Oxone®. 13, 259; **14,** 267; **15,** 274–275; **16,** 285; **18,** 300; **19,** 277; **20,** 313–315

Epoxidations. Combination of Oxone and the iminium salt derived from pyrrolidine and *o*-trifluoromethylbenzaldehyde is effective for epoxidation of alkenes.[1] In the case of an active alkene (e.g., trisubstituted alkene), pyrrolidine is an adequate catalyst.[2] Other types of mediators include α-functionalized ketones (e.g., α-acetaminoacetone)[3] and the *N,N′*-dialkylalloxans **1.**[4]

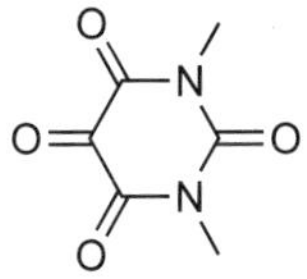

(**1**)

Oxidations. Oxidation of alcohols to aldehydes with Oxone–TEMPO under mild conditions has been reported.[5] Acetals are oxidized to esters, with release of one mole of the constituent alcohols.[6] Accordingly, this oxidation can be used to cleave THP ethers.

The double bond of an enol is readily cleaved, therefore a route to α-keto esters from arylacetic esters via hydroxymethylenation is viable because of the high yields in both steps.[7]

Based on the facile oxidation of sulfides to sulfones with Oxone, phenyl vinyl sulfone (a multipurpose synthetic reagent) can be obtained in three steps from 2-bromoethanol, that is, displacement with PhSH, oxidation, and dehydration (via mesylation).[8]

Cyclopropyl aryl ethers are conveniently prepared[9] from phenols in a manner that the last two steps involve sulfide oxidation and reductive removal of a benzenesulfonyl group.

Oxone(R)

$CHCl_3$

100%

Oxone in water oxidizes *o*-iodobenzoic acid to 2-iodoxybenzoic acid at 70° (yield 79–81%).[10] It also can be used in the workup of hydroboration.[11]

Chlorination. Amides and carbamates undergo *N*-chlorination[12] by NaCl–Oxone. Under the same conditions, oximes afford α-chloro nitroalkanes.[13] When the chlorine source is absent, regeneration of carbonyl compounds from oximes and hydrazones occurs.[14]

Desilylation.[15] A solution of Oxone in 50% aq MeOH is effective for cleaving primary and aromatic *t*-butyldimethylsilyl ethers at room temperature. In a molecule containing both types of ethers, the aliphatic silyl group can be selectively removed by virtue of the faster rates of hydrolysis.

[1]Armstrong, A., Ahmed, G., Garnett, I., Goacolou, K., Wailes, J.S. *T* **55**, 2341 (1999).
[2]Adamo, M.F.A., Aggarwal, V.K., Sage, M.A. *JACS* **122**, 8317 (2000).
[3]Armstrong, A., Hayter, B.R. *T* **55**, 11119 (1999).
[4]Carnell, A.J., Johnstone, R.A.W., Parsy, C.C., Sanderson, W.R. *TL* **40**, 8029 (1999).
[5]Bolm, C., Magnus, A.S., Hildebrand, J.P. *OL* **2**, 1173 (2000).
[6]Curini, M., Epifano, F., Marcotullio, M.C., Rosati, O. *SL* 777 (1999).
[7]Mahmood, S.J., McLaughlin, M., Hossain, M.M. *SC* **29**, 2957 (1999).
[8]Lee, J.W., Lee, C.-W., Jung, J.H., Oh, D.Y. *SC* **30**, 2897 (2000).
[9]Hollingsworth, G.J., Dinnell, K., Dickinson, L.C., Elliott, J.M., Kulagowski, J.J., Swain, C.J., Thomson, C.G. *TL* **40**, 2633 (1999).
[10]Frigerio, M., Santagostino, M., Sputore, S. *JOC* **64**, 4532 (1999).
[11]Ripin, D.H.B., Cai, W., Brenek, S.J. *TL* **41**, 5817 (2000).
[12]Curini, M., Epifano, F., Marcotullio, M.C., Rosati, O., Tsadjout, A. *SL* 813 (2000).
[13]Curini, M., Epifano, F., Marcotullio, M.C., Rosati, O., Rossi, M. *T* **55**, 6211 (2000).
[14]Bose, D.S., Narsaiah, A.V., Lakshminarayana, V. *SC* **30**, 3121 (2000).
[15]Sabitha, G., Syamala, M., Yadav, J.S. *OL* **1**, 1701 (1999).

Potassium organotrifluoroborates.

Alkylations. Allylation of aldehydes with the stable allyltrifluoroborate reagent proceeds at low temperatures by using $BF_3{\cdot}OEt_2$ as catalyst. Crotylation is performed analogously.[1] In the presence of $(acac)Rh(CO)_2$ and a phosphine the alkenyl- and aryltrifluoroborates add to aldehydes and enones, the latter in a conjugate fashion.[2]

[1]Batey, R.A., Thadani, A.N., Smil, D.V., Lough, A.J. *S* 990 (2000).
[2]Batey, R.A., Thadani, A.N., Smil, D.V. *OL* **1**, 1683 (2000).

Potassium permanganate. **13,** 258–259; **14,** 267; **15,** 273–274; **18,** 301; **19,** 277–278; **20,** 315–316

Oxidations. The oxidation of alcohols to carbonyl compounds using $KMnO_4$–Al_2O_3 (no solvent)[1] is very rapid.

Ketones are also produced from amines[2] by oxidation with $KMnO_4$ supported on copper(II) sulfate pentahydrate in dichloromethane. Bridged alkenes such as norbornene are cleaved with the same reagent. This method is an alternative to ozonolysis.[3]

Aromatization.[4] 1,4-Cyclohexadienes are dehydrogenated with $KMnO_4$–Al_2O_3 in acetone at 0°.

[1]Hajipour, A.R., Mallakpour, S.E., Imanzadeh, G. *CL* 99 (1999).
[2]Noureldin, N.A., Bellegarde, J.W. *S* 939 (1999).
[3]Goksu, S., Altundas, R., Sutbeyaz, Y. *SC* **30**, 1615 (2000).
[4]McBride, C.M., Chrisman, W., Harris, C.E., Singaram, B. *TL* **40**, 45 (1999).

Potassium selenocyanate.

Dialkyl diselenides.[1] A convenient access to RSe-SeR from RBr is by reaction with KSeCN in aq DMF in the presence of K_2CO_3. Alkyl selenocyanates are hydrolyzed in situ and the resulting alkyl selenide ions undergo aerial oxidation.

[1]Krief, A., Dumont, W., Delmotte, C. *ACIEE* **39**, 1669 (2000).

Potassium superoxide.

Hydrosilylation and the Tishchenko reaction. Aromatic aldehydes give both $ArCH_2OSiEt_3$ and $ArCOOCH_2Ar$ on treatment with KO_2–18-crown-6 in benzene or THF at 40°.[1]

Sulfoxides. Sulfides are oxidized to sulfoxides with KO_2 and Me_3SiCl. This reaction is mediated by the trimethylsilylperoxy radical.[2]

[1]Le Bideau, F., Coradin, T., Gourier, D., Henique, J., Samuel, E. *TL* **41**, 5215 (2000).
[2]Chen, Y.-J., Huang, Y.-P. *TL* **41**, 5233 (2000).

Potassium trimethylsilanolate.

Nitrile hydrolysis.[1] Refluxing with Me_3SiOK in THF converts RCN to primary amides. At least in the case of PhCN, the counterion of different alkaline salts has a dramatic effect on the product yields: with Me_3SiOLi, 5%, with Me_3SiONa, 40%, and with Me_3SiOK, 83%.

[1]Merchant, K.J. *TL* **41**, 3747 (2000).

Pyridinium fluorochromate.

Oxidation.[1] Trimethylsilyl ethers are directly oxidized to carbonyl products.

[1]Ho, T.-L., Jana, G.H. *JCCS(T)* **46**, 639 (1999).

(2-Pyridyldimethylsilyl)methyllithium.

Hydroxymethylation.[1] The organolithium reagent is obtained by lithiation of 2-trimethylsilylpyridine. Its alkylation with RX and oxidative desilylation to provide RCH_2OH are quite strightforward.

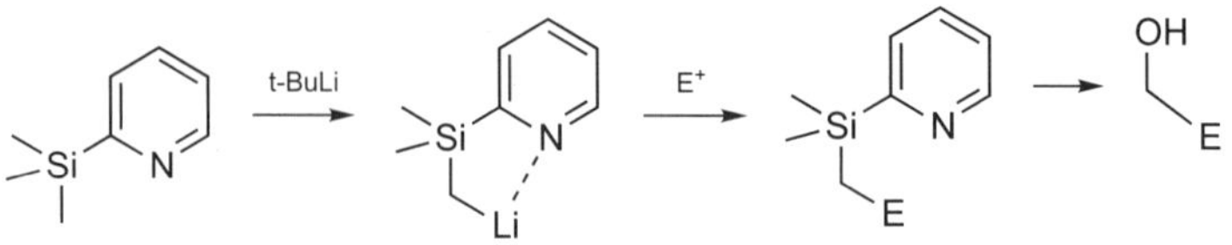

[1]Itami, K., Mitsudu, K., Yoshida, J. *TL* **40**, 5537 (1999).

2-Pyridyl *N*-methoxy-*N*-methylcarbamate.

Ketone synthesis.[1] It is not necessary to prepare individual Weinreb reagents for the preparation of ketones. Sequential exposure of the title pyridyl carbamate to two different organometallic compounds is the most expedient procedure.

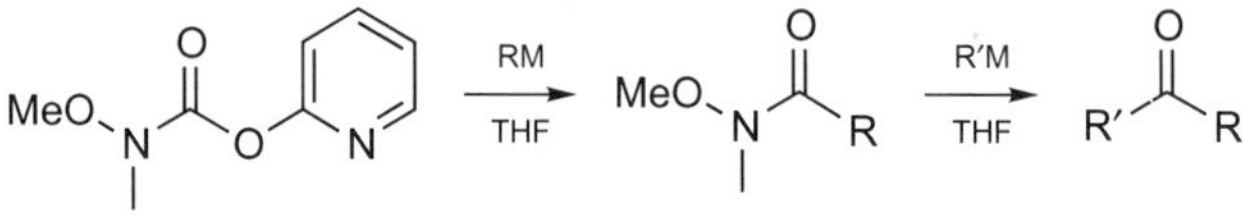

[1]Lee, N.R., Lee, J.I. *SC* **29**, 1249 (1999).

2-Pyridyl methoxymethyl sulfide.

O-Methoxymethylation.[1] The MOM ethers are formed by reaction of the title reagent with alcohols in the presence of AgOTf.

[1]Marcune, B.F., Karady, S., Dolling, U.-H., Novak, T.J. *JOC* **64**, 2446 (1999).

Q

Quinoline-2-thiol.

Alkanethiols.[1] The reagent is converted to mixed sulfides by alkylation. Thiols are released from the mixed sulfides on reduction with $NaBH_3CN$. Accordingly, this reagent can be considered as a source of hydrosulfide.

$NaBH_3CN$ / HOAc, H_2O

95%

[1]Zhang, J., Matteucci, M.D. *TL* **40**, 1467 (1999).

Quinuclidine *N*-Oxide.

Alkylations.[1] Quinuclidine *N*-oxide offers an alternative choice to HMPA in alkylation reactions that typically require such an additive.

[1]O'Neil, I.A., Lai, J.Y.Q., Wynn, D. *CC* 59 (1999).

R

Rhodium.

Hydrogenation.[1] Rhodium in nanosize (from reduction of rhodium trichloride with $NaBH_4$ and stabilized with water-soluble alkylammonium salts) is effective for hydrogenation of arenes at atmospheric pressure and room temperature under biphasic conditions.

[1]Schulz, J., Roucoux, A., Patin, H. *CEJ* **6**, 618 (2000).

Rhodium carbonyl clusters. 13, 288; **15,** 334; **18,** 305–306; **19,** 280–281; **20,** 317–318

Hydroformylation. A phosphine plays a crucial role to suppress hydrogenation during hydxoformylation of 4-vinylpyridine.[1] The *N*-vinyl group of a divinylpyrrole can be retained to achieve selective hydroformylation of a carbon-linked vinyl residue.[2]

1-Indanone-3-carboxylic esters.[3] Carbonylation of ethynylarenes in the presence of alcohols leads to the indanone esters.

R–C₆H₄–C≡CH + CO + R'OH →($Rh_6(CO)_{16}$, dioxane 130°) R-indanone-3-COOR'

[1]Caiazzo, A., Settambolo, R., Pontorno, L., Lazzaroni, R. *JOMC* **599**, 298 (2000).
[2]Lazzaroni, R., Settambolo, R., Mariani, M., Caiazzo, A. *JOMC* **592**, 69 (1999).
[3]Yoneda, E., Kaneko, T., Zhang, S.-W., Onitsuka, K., Takahashi, S. *TL* **40**, 7811 (1999).

Rhodium carbonyl clusters–hydrosilanes.

Silylfunctionalizations. The reaction of alkynes with this reagent combination can give β-silylacrylamides,[1] α-silylmethyl α,β-unsaturated aldehydes,[2] or cyclization products,[3] depending on reaction conditions (additives) and substrate structures.

1-ethynylcyclohexanol + H-SiMe₂Ph →(CO / $Rh_4(CO)_{12}$, PhH 100°) cyclohexylidene-CH(CHO)CH₂SiMe₂Ph

81%

[1]Matsuda, I., Takeuchi, K., Itoh, K. *TL* **40**, 2553 (1999).
[2]Matsuda, I., Niikawa, N., Kuwabara, R., Inoue, H., Nagashima, H., Itoh, K. *JOMC* **574**, 133 (1999).
[3]Fukuta, Y., Matsuda, I., Itoh, K. *TL* **40**, 4703 (1999).

Rhodium carboxylates. 13, 266; **15,** 278–286; **16,** 289–292; **17,** 298–302; **18,** 306–307; **19,** 281–285; **20,** 318–320

Carbenoid insertions. Advantage has been taken of the insertion reaction in the formation of dehydropeptides,[1] α-silyl α-amino acid derivatives,[2] functionalized tetrahydrofuranones,[3] 1,4-dioxenes,[4] and spirane systems.[5]

Ph, O, MeOOC, N2 + HO, HO; $Rh_2(OAc)_4$; TsOH / PhH; Ph, O, MeOOC, O; 42%

N2, N2, O, O; $Rh_2(OAc)_4$; CH_2Cl_2; O, O; 27%

The insertion at C-2 and C-5 of *N*-Boc-pyrrolidine in the presence of $Rh_2[(S)\text{-DOSP}]_4$ gives rise to C_2-symmetric compounds.[6] On the other hand, by using $Rh_2[(R)\text{-DOSP}]_4$ as the catalyst, the reaction leads to mixtures of diastereomers and regioisomers.

N–Boc + N2, Ar, COOMe; $Rh_2(S\text{-DOSP})_4$; Ar, H, COOMe, N–Boc, H, COOMe, Ar

A key building block for synthesis of eburnamonine was assembled by the reaction taking advantage of the preferred insertion into a tertiary C—H bond by Rh-carbenoids.[7]

OTBDPS, O, O, MeOOC, N2; $Rh_2(OAc)_4$; OR, O, O, MeOOC; H, N, N, O

(–)-eburnamonine

An exceptionally reactive and selective chiral dirhodium(II) carboxamides, $Rh_2[(4S)\text{-}MEAZ]_4$, has the potential to significantly broaden the applicability of asymmetric synthesis using diazocarbonyl compounds.[8]

Rearrangements. A variety of reactions catalyzed by Rh(II) carboxylates, including Wolff rearrangement of 3-diazo-2,4-dioxo-1,2,3,4-tetrahydroquinolines leading to oxindoles due to in situ decarboxylation,[9] insertion followed by [2,3]sigmatropic rearrangement[10] and Claisen rearrangement,[11] serve to affirm their synthetic potential.

SiMe3 + N2 — Rh2(OAc)4 → SiMe3 SPh

86%

Four-membered rings.[12] Rearrangement of α-diazo thiol esters to substituted ketenes and cycloaddition of the latter to alkenes lead to cyclobutanones. A similar addition to imines gives β-lactams.

PhS N2 + — Rh2(OAc)4, ClCH2CH2Cl, Δ → PhS

73–78%

Diheterobicycloalkanes.[13] Hydroformylation of 1-alkenes that contain in-chain heteroatoms is terminated by cyclization.

NH NH2 — H2 / CO, Rh2(OAc)4, phosphine → N NH

80%

[3+2]Cycloadditions.[14] Carbonyl ylides generated from δ- and ε-carbonyl-α-diazoketones undergo cycloaddition with quinones. Chemoselectivity (C=C vs. C=O) of the reaction is dependent on the solvent and catalyst.

$Rh_2(OCOAr)_4$

61%

$Rh_2(OCOAr)_4$

64%

Wolff rearrangement. Silyl ketenes are readily formed from α-silyl-α-diazoketones on exposure to rhodium(II) octanoate.[15]

$Rh_2(OCOC_7H_{15})_4$, PhH 20°; R′NH_2, CH_2Cl_2

[1]Buck, R.T., Clarke, P.A., Coe, D.M., Drysdale, M.J., Ferris, L., Haigh, D., Moody, C.J., Pearson, N.D., Swann, E. *CEJ* **6**, 2160 (2000).
[2]Bolm, C., Kasyan, A., Drauz, K., Gunther, K., Raabe, G. *ACIEE* **39**, 2288 (2000).
[3]Lacrampe, F., Leost, F., Doutheau, A. *TL* **41**, 4773 (2000).
[4]Hilgenkamp, R., Brogan, J.B., Zercher, C.K. *H* **51**, 1073 (1999).
[5]Aburel, P.S., Undheim, K. *JCS(P1)* 1891 (2000).
[6]Davies, H.M.L., Hansen, T., Hopper, D.W., Panaro, S.A. *JACS* **121**, 6509 (1999).
[7]Wee, A.G.H., Yu, Q. *TL* **41**, 587 (2000).
[8]Doyle, M.P., Davies, S.B., Hu, W. *OL* **2**, 1145 (2000).
[9]Lee, Y.R., Suk, J.Y., Kim, B.S. *TL* **40**, 8219 (1999).
[10]Carter, D.S., Van Vranken, D.L. *TL* **40**, 1617 (1999).
[11]Wood, J.L., Moniz, G.A., Pflum, D.A., Stoltz, B.M., Holubec, A.A., Dietrich, H.-J. *JACS* **121**, 1748 (1999).
[12]Lawlor, M.D., Lee, T.W., Danheiser, R.L. *JOC* **65**, 4375 (2000).
[13]Bergmann, D.J., Campi, E.M., Jackson, W.R., Patti, A.F. *CC* 1279 (1999).
[14]Pirrung, M.C., Kaliappan, K.P. *OL* **2**, 353 (2000).
[15]Marsden, S.P., Pang, W.-K. *CC* 1199 (1999).

Rhodium perfluorocarboxylates.

Alkoxycarbonylmethyl enol ethers.[1] A convenient preparation of these enol ethers from carbonyl compounds is by a $Rh_2(OCOCF_3)_4$-catalyzed reaction of a diazoacetic ester. The enol etherification is applicable to α-pyridone.

Rh$_2$(OCOCF$_3$)$_4$

ClCH$_2$CH$_2$Cl

Δ

COOEt

N_2

69%

[1]Busch-Petersen, J., Corey, E.J. *OL* **2**, 1641 (2000).

Rhodium(III) chloride.

Aryl ketones.[1] Benzyl alcohols and alkenes are united by a catalytic reaction using polymeric phosphine ligated $RhCl_3$ hydrate.

OH

$RhCl_3 \times H_2O$

Ph_3P

PhMe Δ

N NH$_2$

72%

[1]Jun, C.-H., Hong, H.-S., Huh, C.-W. *TL* **40**, 8897 (1999).

Ruthenium–carbene complexes. 18, 308; **19,** 285–289; **20,** 320–323

The prevalent catalysts for metathetic reactions of alkenes or alkynes are **1**, **2**, **3**, **4**, and **5**. Those containing imidazol-2-ylidene ligands are readily prepared from **1** by ligand exchange.[1] Complex **6** is a precatalyst for various ring-closing metathesis (RCM) reactions.[2] Its counterion affects reactivity and selectivity. The immobilization of the Ru complexes to polymer yields a series of RCM catalysts that are recyclable and possess comparable or better reactivity than their homogeneous counterparts, particularly in dealing with highly hindered substrates.[3–5]

The four-coordinate complexes **7**, which possess trigonal pyramidal geometries, are moderately effective RCM catalysts, but reactions involving them are greatly accelerated by the addition of HCl.[6] Also prepared from **1** and allyl 2-bromo-2-methylpropanaote is a multifunctional catalyst **8** that is capable of mediating three mechanistically distinct reactions[7]: ring-opening metathetic polymerization, atom-transfer radical polymerization, and hydrogenation.

The use of tris(hydroxymethyl)phosphine is recommended for removal of the ruthenium residue from the products.[8] Addition of a modest amount of $Pb(OAc)_4$ (1.5 equiv relative to the Grubbs catalyst) at the end of the reaction rids all the colored impurities.[9]

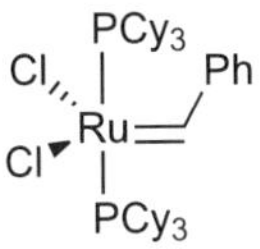

(**1**) 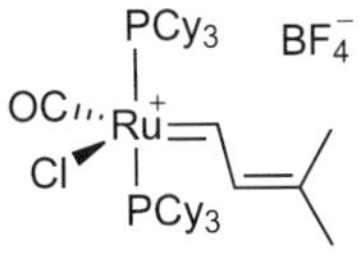 (**2**) 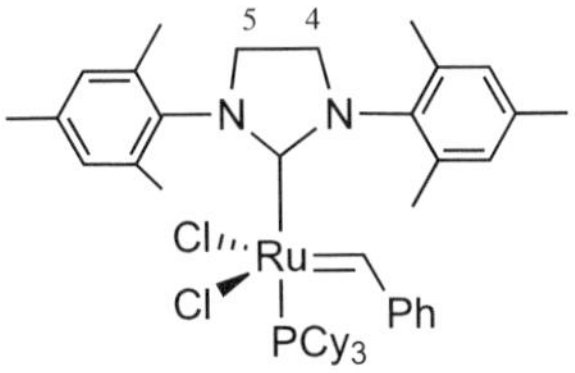 (**3s**) C_4-C_5 saturated
(**3u**) C_4-C_5 unsaturated

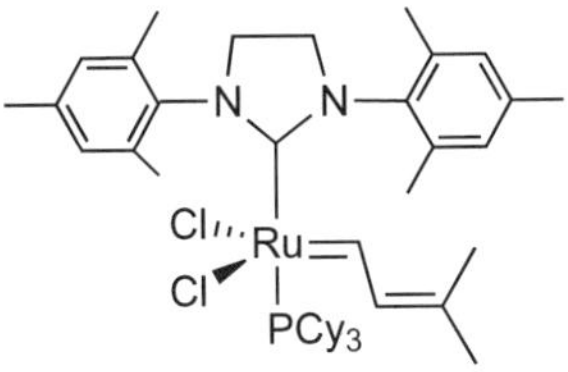

(**4**) 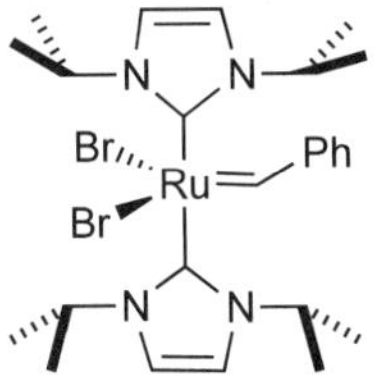 (**5**) 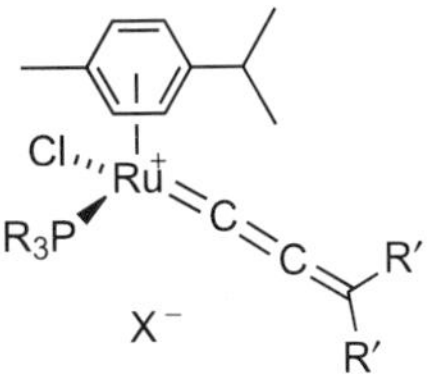 (**6**)

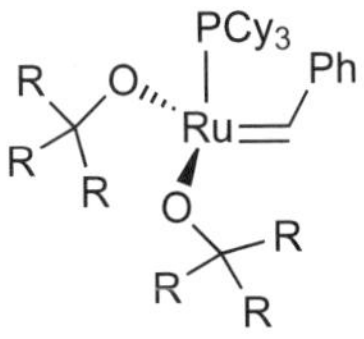

R = Me, CF_3

(**7**) (**8**) 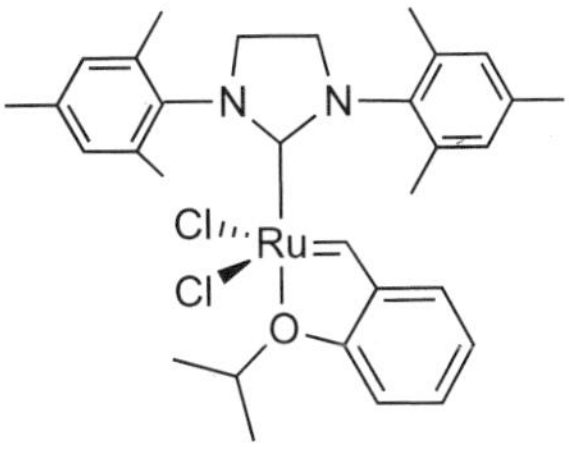 (**9**)

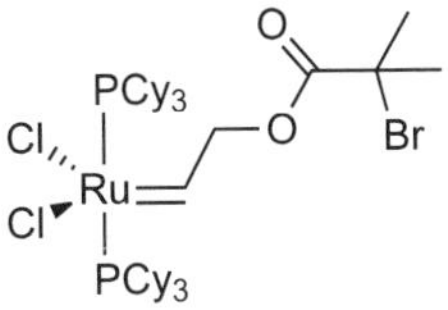

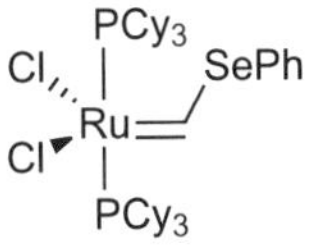

(**10**)

Metathetic ring closure. Catalyst **1** can be regarded as the standard workhorse for RCM and the scope of its applications continues to expand. Thus, its use in the elaboration of cyclic structures including azaspirocycles,[10] 3-pyrrolines,[11,12] and those containing phosphine oxides,[13] phosphinates,[14] dioxasilanes,[15] sulfonamides.[16] 1-(Dialkoxyboryl)vinylcycloalkenes are obtained from *x*-alken-1-ynyl boronates.[17]

Formation of a *trans*-cyclooctene system reveals a profound stereochemical effect of the metathesis (using either **1** or the Schrock catalyst). The other diastereoisomer does not undergo cyclization.[18]

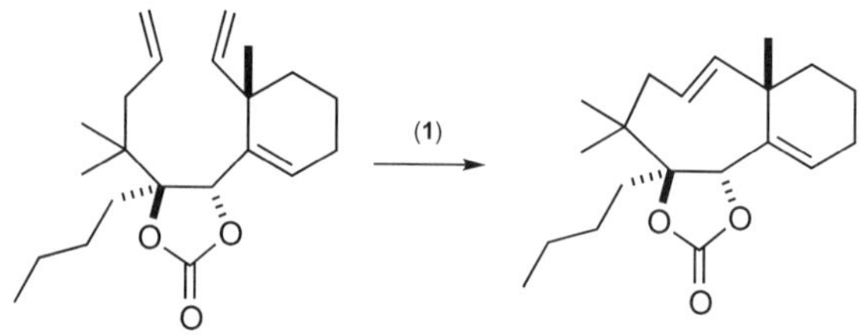

(1)

On elaboration of 1,5-hexadiene-3,4-diol through a 1,3-dioxolane derivative, the RCM delivers a precursor of either *exo*-brevicomin or *endo*-brevicomin.[19] A route to (+)-malyngolide also exploits the RCM process.[20] A successful cyclization–fragmentation approach to medium-sized rings is based on the RCM reaction of 2-hydroxycycloalkanones that are substituted with proper alkenyl groups at both α- and α′-positions.[21]

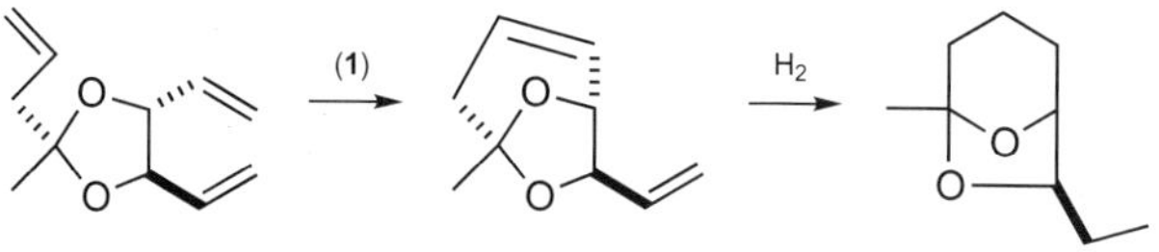

(1) H_2

exo-brevicomin

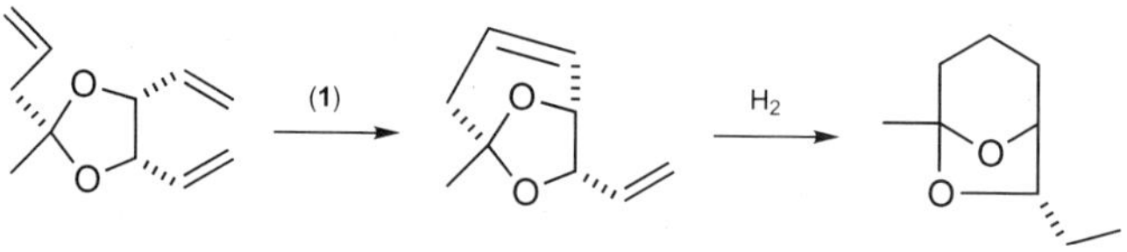

(1) H_2

endo-brevicomin

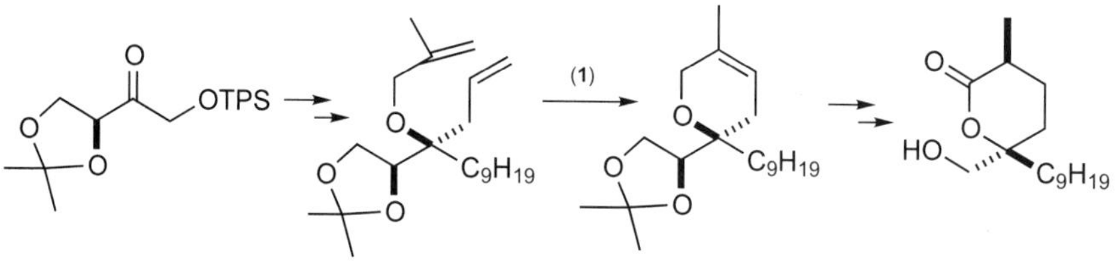

OTPS (1) C_9H_{19} C_9H_{19} HO C_9H_{19}

92%

(+)-malyngolide

$OSiMe_3$ HO n (1) ; $Pb(OAc)_4$ COOMe n

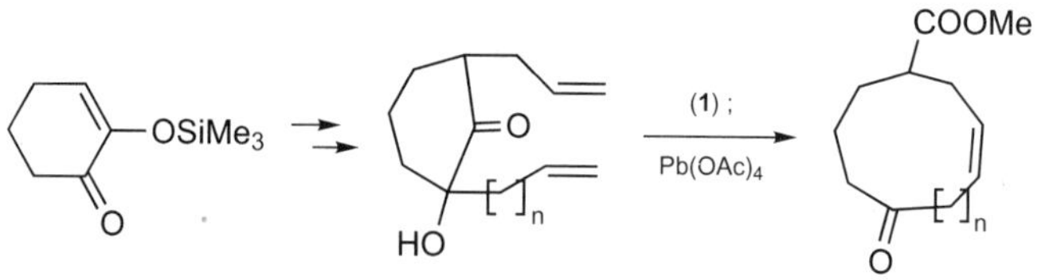

n = 0, 1

Structural modifications such as those shown below[22,23] are pleasing because of their high efficiency.

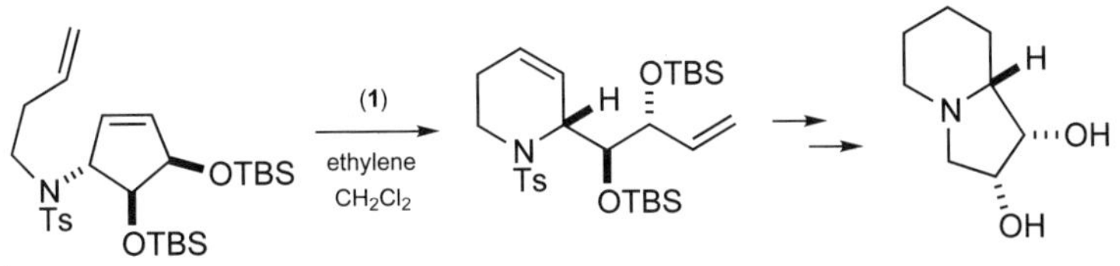

Higher activity of catalysts embodying imidazolin-2-ylidene ligands (e.g., **3u**) for RCM has been observed.[24] Such complexes as **3u** and **5** allow formation of tetrasubstituted cycloalkenes,[25,26] and various functionalized analogues (e.g., 1-cyanoalkenes[27]). The highly active catalyst **9** is recoverable and recyclable.[28]

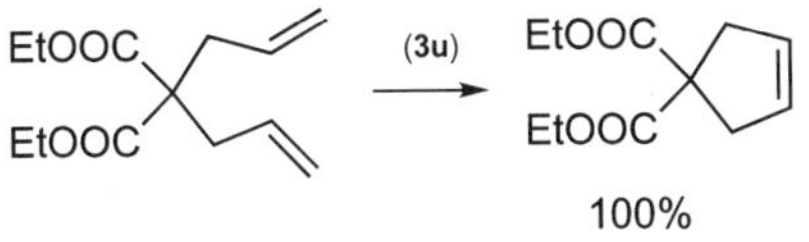

100%

Cross-metathesis. Functionalization of terminal alkenes by the metathetic method using catalyst **1** has been well established. The reaction between styrene and vinylsilanes gives ω-silylstyrenes,[29] between allylarenes and acrylonitrile leads to 4-aryl-2-butenonitriles.[30] Alternatively, homo-metathesis of two allylarene molecules to give 1,4-diaryl-2-butene is first carried out and the cross-metathesis follows.[31] Also of interest is the homo-metathesis of monosubstituted allenes to symmetrical allenes.[32]

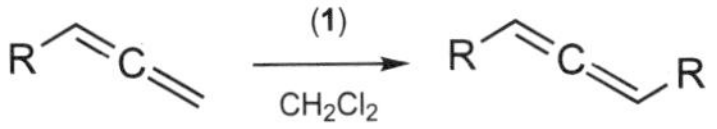

Both catalysts **1** and **2** are effective in promoting cross-metathesis leading to various conjugated dienes from alkenes and alkynes.[33,34] Chiral 2-(α-acetoxybenzyl)-1,3-butadiene is obtained from (*R*)-3-acetoxy-3-phenylpropyne via cross-metathesis with ethylene.[35] Furthermore, the reaction of 1,6-diynes with alkenes is even more intriguing:[36]

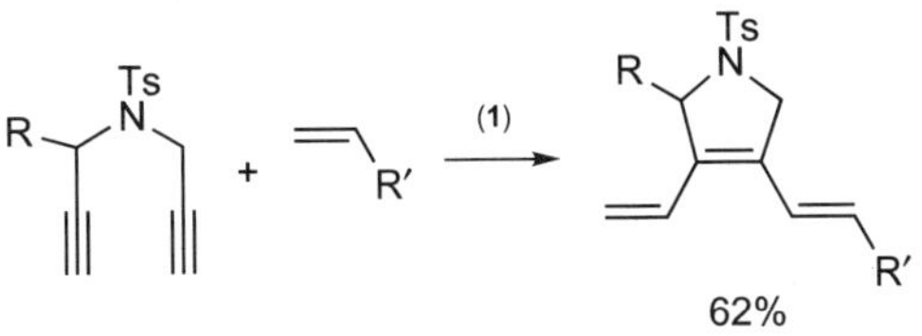

62%

The scope is further expanded by using catalysts **3s**[37,38] and **4**.[39] For example, it allows the preparation of trisubstituted alkenes by an intermolecular reaction for the first time,[40] and 1,5-cyclooctadienes (e.g., a precursor of aristeriscanolide[41]). The intramolecular version is a useful preparation of some other interesting molecules.[42]

(3)
PhH Δ
74%
aristeriscanolide

Technically significant is the finding that **4** can be generated in situ from **1** and 1,3-dimesityl-4,5-dihydroimidazolium tetrafluoroborate (treatment of the salt with *t*-BuOK in THF at room temperature for <1 min). The RCM is performed in the presence of ethereal HCl.[43]

Complex **10** is a very active catalyst for ring-opening cross-metathesis of norbornene derivatives.[44]

[2 + 2 + 2]Cycloaddition. Formation of benzene derivatives from three alkyne units is catalyzed by several transition metal reagents. 1,2,4-Trisubstituted benzenes are the major products from reactions using the Grubbs catalyst containing Ph_3P ligands.[45] Differences in regioselectivty for reactions promoted by the Grubbs and Wilkinson catalysts have been noted.[46]

RO
(1)
CH_2Cl_2
major
minor

TsN + R →

(1)	5–6	:	1
$(Ph_3P)_3RhCl$	1	:	1.5–10

Radical addition.[47] Carbon tetrachloride adds to alkenes in the presence of $(Ph_3P)_2Ru(=CHPh)Cl_2$.

[1]Scholl, M., Ding, S., Lee, C.W., Grubbs, R.H. *OL* **1**, 953 (1999).
[2]Fürstner, A., Liebl, M., Lehmann, C.W., Picquet, M., Kunz, R., Bruneau, C., Touchard, D., Dixneuf, P.H. *CEJ* 1847 (2000).
[3]Jafarpour, L., Nolan, S.P. *OL* **2**, 4075 (2000).
[4]Yao, Q. *ACIEE* **39**, 3896 (2000).
[5]Schürer, S.C., Gessler, S., Buschmann, N., Blechert, S. *ACIEE* **39**, 3898 (2000).
[6]Sanford, M.S., Henling, L.M., Day, M.W., Grubbs, R.H. *ACIEE* **39**, 3451 (2000).
[7]Bielawski, C.W., Louie, J., Grubbs, R.H. *JACS* **122**, 12872 (2000).
[8]Maynard, H.D., Grubbs, R.H. *TL* **40**, 4137 (1999).
[9]Paquette, L.A., Schloss, J.D., Efremov, I., Fabris, F., Gallou, F., Mendez-Andino, J., Yang, Y. *OL* **2**, 1259 (2000).
[10]Wright, D.L., Schulte, J.P., Page, M.A. *OL* **2**, 1847 (2000).
[11]Evans, P., Grigg, R., Monteith, M. *TL* **40**, 5247 (1999).
[12]Bujard, M., Briot, A., Gouverneur, V., Mioskowski, C. *TL* **40**, 8785 (1999).
[13]Trevitt, M., Gouverneur, V., Mioskowski, C. *TL* **40**, 7333 (1999).
[14]Bujard, M., Gouverneur, V., Mioskowski, C. *JOC* **64**, 2119 (1999).
[15]Hoye, T.R., Promo, M.A. *TL* **40**, 1429 (1999).
[16]Hanson, P.R., Probst, D.A., Robinson, R.E., Yau, M. *TL* **40**, 4761 (1999).
[17]Renaud, J., Graf, C.-D., Oberer, L. *ACIEE* **39**, 3101 (2000).
[18]Bourgeois, D., Pancrazi, A., Ricard, L., Prunet, J. *ACIEE* **39**, 726 (2000).
[19]Burke, S.D., Müller, N., Beaudry, C.M. *OL* **1**, 1827 (1999).
[20]Carda, M., Castillo, E., Rodriguez, S., Marco, J.A. *TL* **41**, 5511 (2000).
[21]Rodriguez, J.R., Castedo, L., Mascarenas, J.L. *OL* **2**, 3209 (2000).
[22]Ovaa, H., Stragies, R., van der Marel, G.A., van Boom, J., Blechert, S. *CC*1501 (2000).
[23]Voigtmann, U., Blechert, S. *OL* **2**, 3971 (2000).
[24]Scholl, M., Trnka, T.M., Morgan, J.P., Grubbs, R.H. *TL* **40**, 2247 (1999).
[25]Fürstner, A., Thiel, O.R., Ackermann, L., Schanz, H.-J., Nolan, S.P. *JOC* **65**, 2204 (2000).
[26]Ackermann, L., Fürstner, A., Westkamp, T., Kohl, F.J., Herrmann, W.A. *TL* **40**, 4787 (1999).
[27]Gessler, S., Randl, S., Blechert, S. *TL* **41**, 9973 (2000).
[28]Garber, S.B., Kingsbury, J.S., Gray, B.L., Hoveyda, A.H. *JACS* **122**, 8168 (2000).
[29]Pietraszuk, C., Marciniec, B., Fischer, H. *OM* **19**, 913 (2000).
[30]Blanco, O.M., Castedo, L. *SL* 557 (1999).
[31]Blackwell, H.E., O'Leary, D.J., Chatterjee, A.K., Washenfelder, R.A., Bussmann, D.A., Grubbs, R.H. *JACS* **122**, 58 (2000).
[32]Ahmed, M., Arnauld, T., Barrett, A.G.M., Braddock, D.C., Flack, K., Procopiou, P.A. *OL* **2**, 551 (2000).
[33]Schürer, S.C., Blechert, S. *TL* **40**, 1877 (1999).
[34]Yi, C.S., Lee, D.W., Chen, Y. *OM* **18**, 2043 (1999).
[35]Smulik, J.A., Diver, S.T. *JOC* **65**, 1788 (2000).
[36]Stragies, R., Schuster, M., Blechert, S. *CC* 237 (1999).
[37]Smulik, J.A., Diver, S.T. *OL* **2**, 2271 (2000).
[38]Stragies, R., Voightmann, U., Blechert, S. *TL* **41**, 5465 (2000).
[39]Chatterjee, A.K., Morgan, J.P., Scholl, M., Grubbs, R.H. *JACS* **122**, 3783 (2000).
[40]Chatterjee, A.K., Grubbs, R.H. *OL* **1**, 1751 (1999).
[41]Limanto, J., Snapper, M.L. *JACS* **122**, 8071 (2000).
[42]Mori, M., Kitamura, T., Sakakibara, N., Sato, Y. *OL* **2**, 543 (2000).
[43]Morgan, J.P., Grubbs, R.H. *OL* **2**, 3153 (2000).
[44]Katayama, H., Urushima, H., Nishioka, T., Wada, C., Nagao, M., Ozawa, F. *ACIEE* **39**, 4513 (2000).
[45]Das, S.K., Roy, R. *TL* **40**, 4015 (1999).

[46]Witulski, B., Stengel, T., Fernandez-Hernandez, J.M. *CC* 1965 (2000).
[47]Simal, F., Demonceau, A., Noels, A.F. *TL* **40**, 5689 (1999).

Ruthenium(III) chloride. 13, 268; **14,** 271–272; **19,** 289–290; **20,** 324

Quinolines. Quinolines are obtained by following the procedure of an indole synthesis from arylamines by changing the reaction partners to 3-aminopropanol.[1] Yields are moderate (11 examples, 29–46%).

[1]Cho, C. S., Oh, B.H., Shim, S. C. *JHC* **36,** 1175 (1997).

Ruthenium(III) chloride–sodium periodate. 18, 310; **19,** 290; **20,** 324

α-Acetoxy-N-acetyl amides.[1] The availability of chiral pyrimidylalkanols is the impetus for the oxidative cleavage reaction. No racemization occurs.

$RuCl_3$ - $NaIO_4$, MeCN CCl_4 H_2O; AcNH, OAc, 80%

Oxidation. Primary alcohols are converted to carboxylic acids using EtOAc to replace carbon tetrachloride as solvent.[2] Bicyclo[2.2.1]heptane-2,3-diones are obtained from the Diels–Alder adducts of a 5,5-dimethoxy-1,2,3,4-tetrahalocyclopentadiene.[3]

$RuCl_3 . 3H_2O$, $NaIO_4$, MeCN - H_2O

X = Cl, Br

A $RuCl_2$ complex **(1)** in combination with $NaIO_4$ has been used to oxidize trifluoromethylcarbinols.[4]

(1)

Dihydroxylation. Alkenes afford *cis*-1,2-diols instead of suffering oxidative cleavage by replacement of the solvent EtOAc with acetone.[5] However, 1,6-dienes afford tetrahydropyran derivatives in a mixture of EtOAc, MeCN, and H_2O.[6]

$RuCl_3 \cdot 3H_2O$, $NaIO_4$, MeCN - H_2O, EtOAc

63%

[1]Tanji, S., Kodaka, Y., Shibata, T., Soai, K. *H* **52**, 151 (2000).
[2]Prashad, M., Lu, Y., Kim, H.-Y., Hu, B., Repic, O., Blacklock, T.J. *SC* **29**, 2937 (1999).
[3]Khan, F.A., Prabhudas, B., Dash, J., Sahu, N. *JACS* **122**, 9558 (2000).
[4]Kesavan, V., Bonnet-Delpon, D., Begue, J.-P., Srikanth, A., Chandrasekaran, S. *TL* **41**, 3327 (1999).
[5]Shing, T.K.M., Tam, E.K.W. *TL* **40**, 2179 (1999).
[6]Piccialli, V. *TL* **41**, 3731 (2000).

S

Samarium. 14, 275; **17,** 305–307; **18,** 311; **19,** 291; **20,** 325–326

Reductive dimerization. Alkenes, which are geminally substituted with two activators, undergo dimerization at the β-position. Cyclization follows.[1]

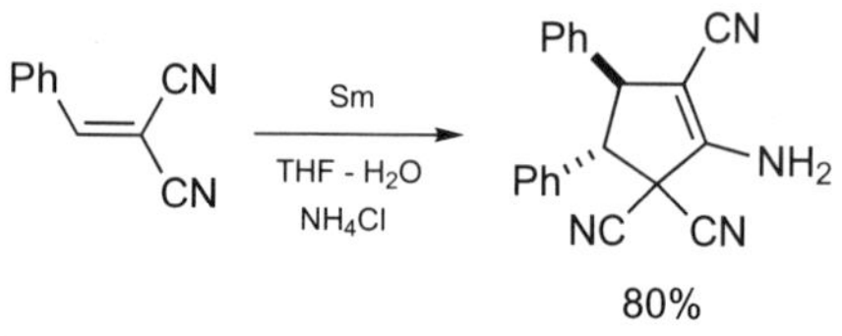

Cleavage of diaryl diselenides. With Sm–Me_3SiCl the cleavage is accelerated by water.[2]

[1]Wang, L., Zhang, Y. *T* **55**, 10695 (1999).
[2]Wang, L., Zhang, Y. *SC* **29**, 3107 (1999).

Samarium–metal halides.

Samarium selenides. Reductive cleavage of ArSeSeAr by samarium is catalyzed by a great number of metal halides: bismuth(III) chloride,[1] cadmium chloride,[2] chromium(III) chloride,[3] cobalt(II) chloride,[4] potassium iodide,[5] and titanium(IV) chloride.[6] The resulting samarium arylselenides readily react with various organic halides.

Reductive desulfonylation.[7] β-Keto sulfones gives ketones on treatment with Sm–$HgCl_2$ in THF containing water.

Cleavage of diaryl ditellurides. Michael donors are generated from ArTeTeAr on reaction with Sm–$ZrCl_4$.[8]

[1]Zhan, Z., Lu, G., Zhang, Y. *JCR(S)* 280 (1999).
[2]Zheng, Y., Bao, W., Zhang, Y. *SC* **30**, 1731 (2000).
[3]Liu, Y., Zhang, Y. *SC* **29**, 4043 (1999).
[4]Chen, R., Zhang, Y. *SC* **30**, 1331 (2000).
[5]Lu, G., Zhang, Y. *SC* **29**, 219 (1999).
[6]Zhou, L.-H., Zhang, Y. *SC* **29**, 533 (1999); *JCR(S)* 28 (1999).
[7]Guo, H., Zhang, Y. *SC* **30**, 2559 (2000).
[8]Zhang, S., Zhang, Y. *SC* **30**, 285 (2000).

Samarium–iodine. 19, 292

Reduction and reductive dimerizations. ArN=CH_2 are reduced to ArNHMe in methanol.[1] *N*-Alkyl aldimines undergo reductive dimerization.[2]

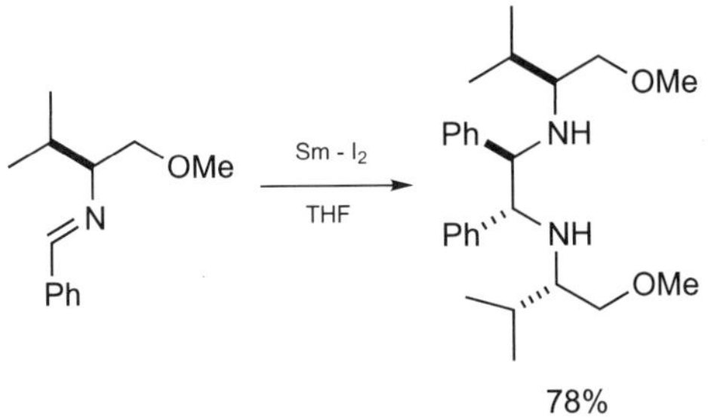

78%

[1]Banik, B.K., Zegrocka, O., Banik, I., Hackfeld, L., Becker, F.F. *TL* **40**, 5731 (1999).
[2]Yamada, R., Negoro, N., Okaniwa, M., Miwa, Y., Taga, T., Yanada, K., Fujita, T. *SL* 537 (1999).

Samarium(II) iodide. 13, 270–272; **14,** 276–281; **15,** 282–284; **16,** 294–300; **17,** 307–311; **18,** 312–316; **19,** 292–296; **20,** 327–335

Reductions. β-Ketols[1] and 3-alkoxy ketones[2] are reduced to *anti*-1,3-diols and the monoethers, respectively. For the directed reduction, the alkoxy group must complex effectively with Sm, however, this is not observed with TBS and benzyl ethers.

The N—O bond in hydroxylamines and hydroxamic acids is rapidly cleaved by SmI_2.[3] However, a bridged oxazine has been found to undergo ring contraction at or above room temperature.[4]

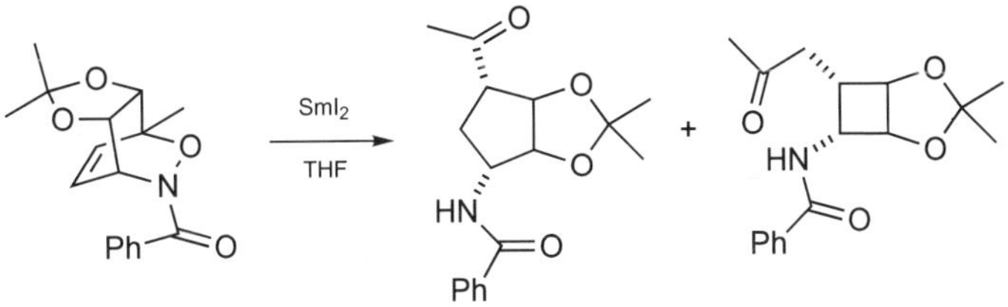

Desulfonylation of *N*-sulfonylamides[5] with SmI_2 is found to be superior to the methods using either tin hydride or Zn–$TiCl_4$, because of lower reaction temperature and higher yields. The carbon-bound sulfonyl group at the α-position of cinnamonitriles is readily removed.[6] Under slightly different conditions, the double bond is also saturated.[7]

Resolution of *trans*-2,5-disbstituted pyrrolidines[8] via the carbamates that constitute a mandelate ester is enabled by the facile reductive cleavage of *O*-acylmandelic esters.

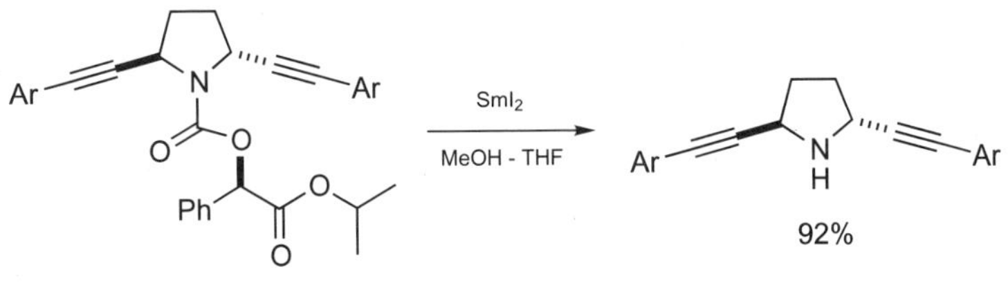

92%

Deamination of α-amino carbonyl compounds[9] occurs when they are treated with SmI_2 in the presence of HMPA and a sufficient amount of MeOH as a proton source. By this reaction, proline esters are converted to piperidones [and thence application to a synthesis of (−)-adalinine].[10] Pyridinemethanols are deoxygenated.[11]

Enolates are generated from α-cyanoketones. Such species undergo regioselective alkylation.[12]

Reductive couplings. A synthesis of tartaric acid derivatives containing quaternary carbon centers is by reductive coupling of α-ketoamides.[13]

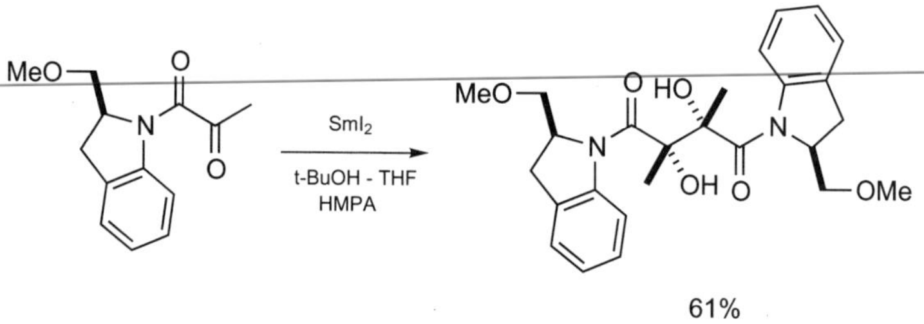

61%

The diastereoselectivity for the reductive coupling of PhCHO in THF is increased (favoring the *anti–syn* isomer to 6:1) by adding tetraglyme.[14]

2-Substituted cyclopropane-1,1-dicarboxylic esters undergoes ring opening and dimerization.[15] Efficient coupling of two aldimine molecules or aldimines with carbonyl compounds (imine slowly added to the ketone and reagents) has been carried out with two metal iodides: SmI_2–NiI_2.[16]

Aldol and Reformatsky reactions. The cross-aldol reaction exemplified below[17] is readily promoted by SmI_2.

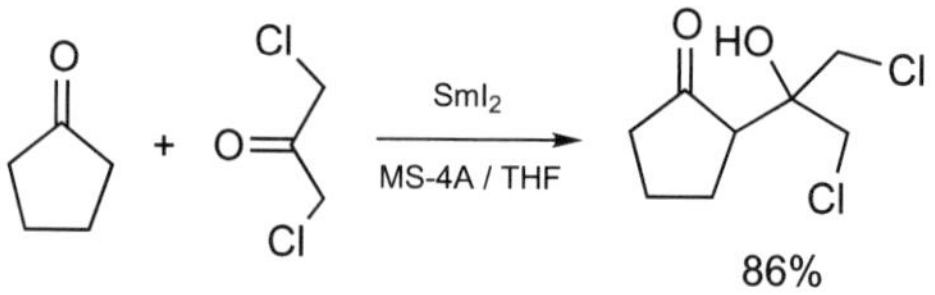

86%

Condensation reactions with donor species generated by reductive cleavage of α-substituted carbonyl compounds are regioselective. Thus, the synthesis of α-(1-hydroxyalkyl)-β-ketols is quite straightforward from α,β-epoxy ketones.[18] Samarium enolates derived from α-(pyridylthio)glycyl peptides are also applicable to the synthesis of unnatural peptides by reaction with various carbonyl compounds.[19]

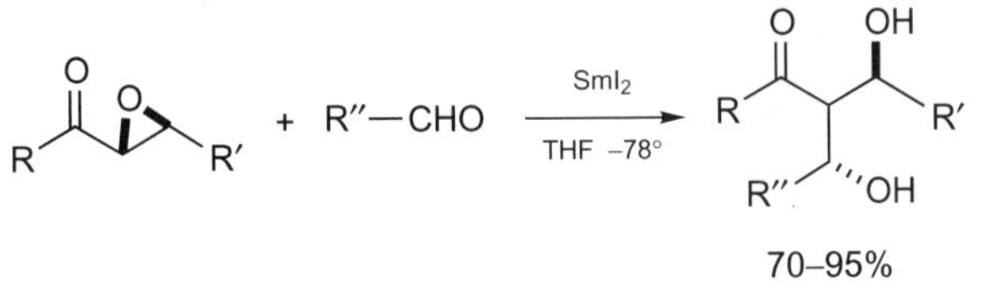

+ R″—CHO $\xrightarrow[\text{THF } -78°]{SmI_2}$

70–95%

An access to optically active β-hydroxy acids uses the bromoacetyl derivatives of chiral oxazolidin-2-ones to react with aldehydes.[20] 1,3-Diketones are obtained from reaction of α-haloketones with acyl cyanides.[21]

Dimerization of enones is terminated by an intramolecular aldol reaction.[22] A bridged hydroxy ketone is formed by a reductive coupling between an enone and an aldehyde side chain. The net result is that of an aldol reaction involving a homoenolate.[23,24]

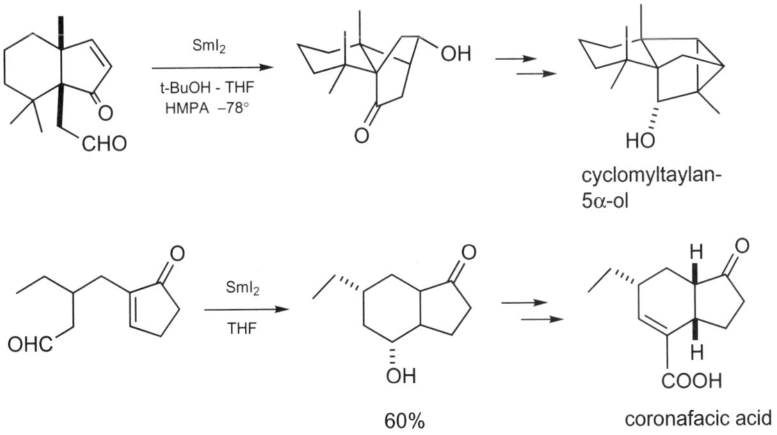

SmI_2, t-BuOH - THF, HMPA −78°

cyclomyltaylan-5α-ol

SmI_2, THF

60%

coronafacic acid

As expected, bromoacetonitrile reacts with carbonyl compounds to give β-hydroxy nitriles.[25]

δ-Lactones.[26] The SmI_2–*i*-PrSH or SmI_2–MeSSMe combination induces redox cyclization of δ-keto aldehydes to furnish lactone products. The thiol facilitates operation of the catalytic cycle.

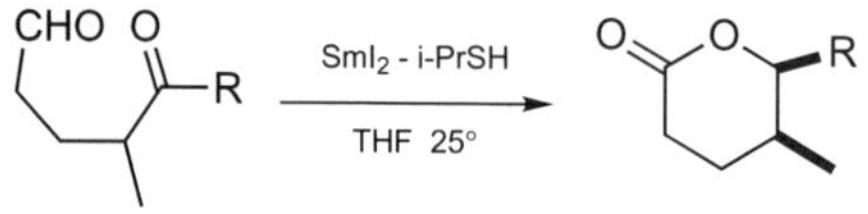

Barbier reactions. Allylic alcohols are produced from alkenyl halides and carbonyl compounds[27] on mediation by SmI_2. This process includes the preparation of the Baylis–Hillman adducts from α-bromoacrylamides and carbonyl compounds.[28] Barbier reaction involving 1-chloromethylbenzotriazole[29] affords alcohols that are valuable synthetic intermediates. *N*-(ω-Iodoalkyl)-*O,O'*-bis(*t*-butyldimethylsilyl)tartrimide undergoes cyclization quite efficiently and the reaction has been applied to a synthesis of (+)-lentiginosine.[30]

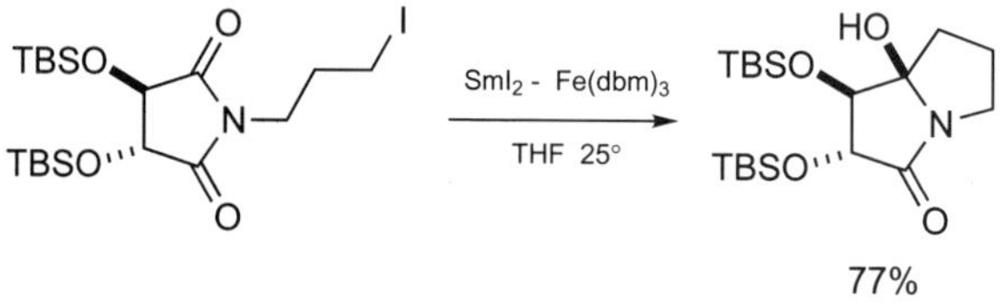

A route that leads carbohydrates to carbocycles involves an intramolecular alkylation.[31]

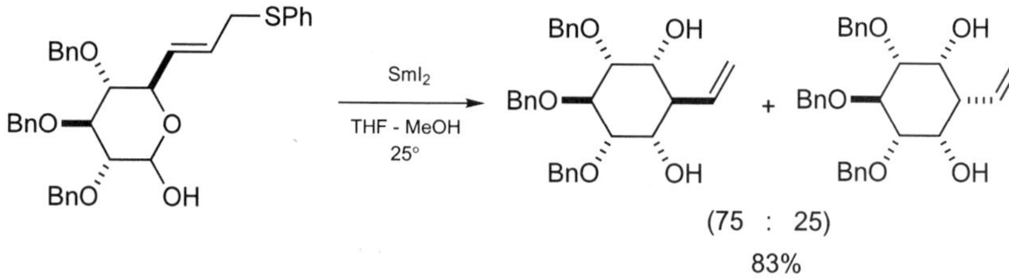

The condensation of alkyl halides with nitriles, tertiary amides, and esters afford ketone products with SmI_2–NiI_2.[32,33]

Reaction of ketyls. The coupling between ketones and nitriles to afford α-ketols[34] likely involves ketyl intermediates. Conjugate addition of the ketyl derived from α-ketols shows stereoselectivity owing to chelation effects.[35]

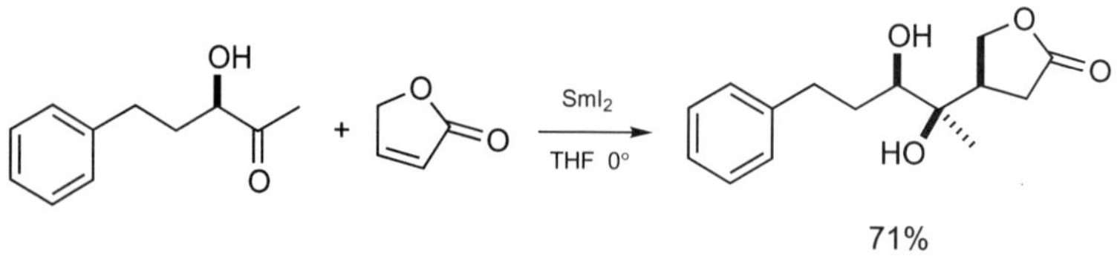

Ketones bearing a remote propargyl ester group afford 2-alkynylcycloalkanols[36] on exposure to SmI_2 and $(Ph_3P)_4Pd$.

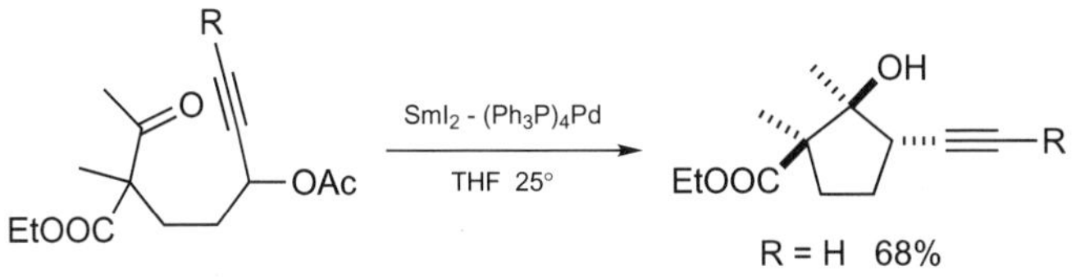

A three-step protocol for alkylation of 1,10-phenanthroline(at C-2 and C-2–C-9) involves two (first and third) reactions mediated by SmI_2. Thus, condensation of the parent heterocycle with ketones is followed by *O*-methylation and reduction.[37]

Cyclizations. Reductive cyclization that leads to cyclopropanes[38] and cyclobutanes[39] are remarkably efficient.

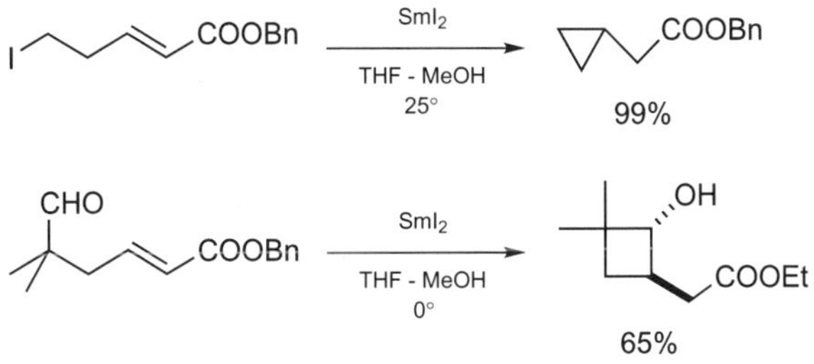

2,3-Disubstituted pyrrolidines[40] and piperidines[41] are readily obtained from tertiary amines that bear an alkenyl chain and a benzotriazolyl group at the α-position of another chain. The far terminus of the double bond becomes nucleophilic upon cyclization.

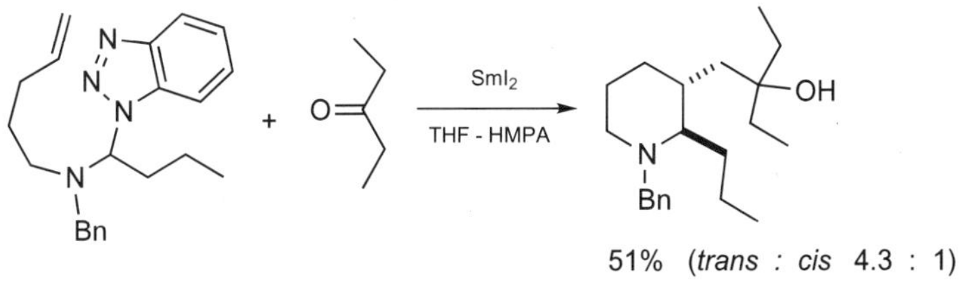

An activated *o*-methoxy group is replaced via ketyl attack and elimination.[42]

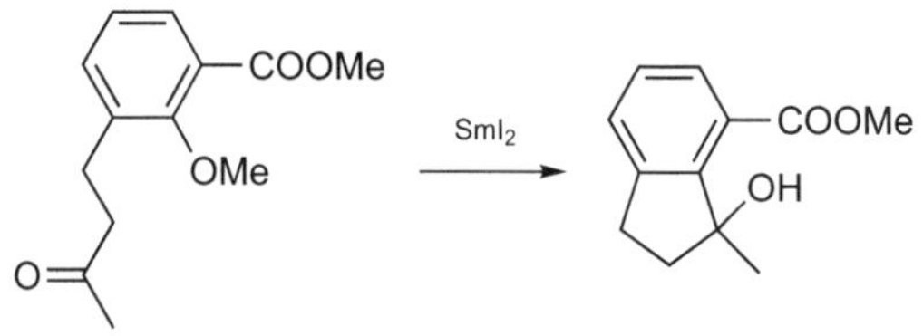

A very interesting and highly useful synthetic process is that which dearomatizes a benzene residue through addition of a ketyl in the side chain and thereby a fused ring system is created.[43] If an ethynyl group is present at an *o*-position cyclization gives an eight-membered ring.

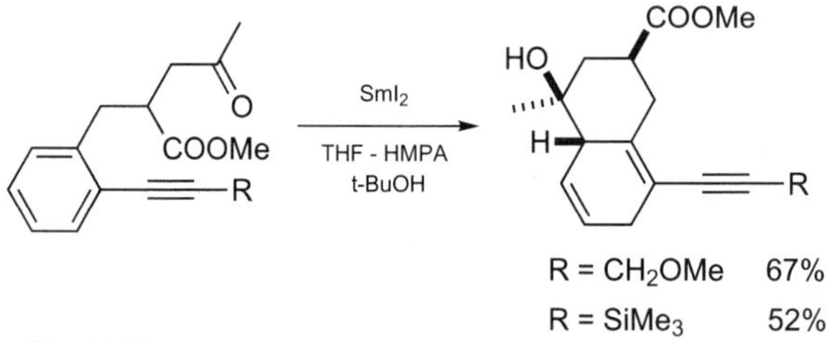

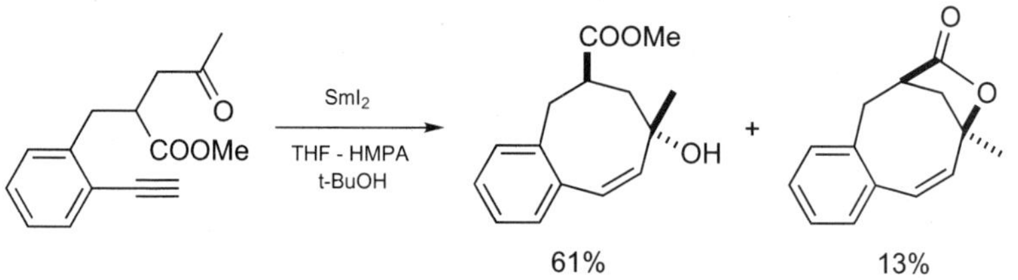

R = CH_2OMe 67%
R = $SiMe_3$ 52%

61% 13%

Fully substituted *sym*-triazines are formed from cyclotrimerization of nitriles,[44] which reaction is catalyzed by SmI_2 and an amine.

Eliminations and cycloadditions. 1,1-Dihalo-2-alkyl acetates are transformed into (*Z*)-alkenyl halides[45] on treatment with SmI_2. Azomethine ylides are generated from bis(tosylmethyl)amines $RN(CH_2Ts)_2$. Trapping of these 1,3-dipolar species with alkenes or alkynes furnishes pyrrolidines and 3-pyrrolines, respectively.[46]

By a cycloaddition process, oxazolidines are formed from epoxides and imines.[47] Contact of epoxides with SmI_2 apparently results in their activation (ring opening).

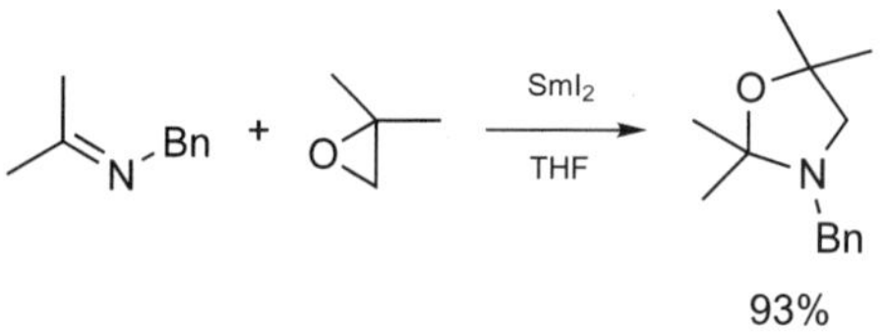

93%

[1]Keck, G.E., Wager, C.A., Sell, T., Wager, T.T. *JOC* **64**, 2172 (1999).
[2]Keck, G.E., Wager, C.A. *OL* **2**, 2307 (2000).
[3]Keck, G.E., Wager, C.A., McHardy, S.F. *T* **55**, 11755 (1999).
[4]McAuley, B.J., Nieuwenhuyzen, M., Sheldrake, G.N. *OL* **2**, 1457 (2000).
[5]Knowles, H.S., Parsons, A.F., Pettifer, R.M., Rickling, S. *T* **56**, 979 (2000).
[6]Guo, H., Zhang, Y. *JCR(S)* 342 (1999).
[7]Guo, H., Zhang, Y. *SC* **30**, 1879 (2000).
[8]Hanamoto, T., Shimomoto, N., Kikukawa, T., Inanaga, J. *TA* **10**, 2951 (1999).
[9]Honda, T., Ishikawa, F. *CC* 1065 (1999).
[10]Honda, T., Kimura, M. *OL* **2**, 3925 (2000).
[11]Kato, Y., Mase, T. *TL* **40**, 8823 (1999).
[12]Zhu, J.-L., Shia, K.S., Liu, H.-J. *TL* **40**, 7055 (1999).
[13]Kim, S.M., Byun, I.S., Kim, Y.H. *ACIEE* **39**, 728 (2000).
[14]Pedersen, H.L., Christensen, T.B., Enemaerke, R.J., Daasbjerg, K., Skrypdstrup, T. *EJOC* 565 (1999).
[15]Yamashita, M., Okiyama, K., Ohhara, T., Kawasaki, I., Michihiro, Y., Sakamaki, K., Ito, S., Ohta, S. *CPB* **47**, 1439 (1999).
[16]Machrouhi, F., Namy, J.-L. *TL* **40**, 1315 (1999).
[17]Lu, L., Chang, H.-Y., Fang, J.-M. *JOC* **64**, 843 (1999).
[18]Mukaiyama, T., Arai, H., Shiina, I. *CL* 580 (2000).
[19]Ricci, M., Madariaga, L., Skrydstrup, T. *ACIEE* **39**, 243 (2000).
[20]Fukuzawa, S., Matsuzawa, H., Yoshimitsu, S. *JOC* **65**, 1702 (2000).
[21]Baek, H.S., Yoo, B.W., Keum, S.R., Yoon, C.M., Kim, S.H., Kim, J.H. *SC* **30**, 31 (2000).
[22]Zhou, L., Zhang, Y. *SC* **30**, 597 (2000).
[23]Sakai, H., Hagiwara, H., Ito, Y., Hoshi, T., Suzuki, T., Ando, M. *TL* **40**, 2965 (1999).
[24]Sono, M., Hashimoto, A., Nakashima, K., Tori, M. *TL* **41**, 5115 (2000).
[25]Caracoti, A., Flowers II, R.A. *TL* **41**, 3039 (2000).
[26]Hsu, J.-L., Chen, C.-T., Fang, J.-M. *OL* **2**, 1989 (2000).
[27]Kunishima, M., Yoshimura, K., Nakata, D., Hioki, K., Tani, S. *CPB* **47**, 1196 (1999).
[28]Youn, S.W., Park, H.S., Kim, Y.H. *CC* 2005 (2000).
[29]Huang, Z.-Z., Jin, H.-W., Duan, D.-H., Huang, X. *JCR(S)* 564 (1999).
[30]Ha, D.-C., Yun, C.-S., Lee, Y. *JOC* **65**, 621 (2000).
[31]Kan, T., Nara, S., Ozawa, T., Shirahama, H., Matsuda, F. *ACIEE* **39**, 355 (2000).
[32]Kang, H.Y., Song, S.E. *TL* **41**, 937 (2000).
[33]Molander, G.A., Machrouhi, F. *JOC* **64**, 4119 (1999).
[34]Zhou, L., Zhang, Y., Shi, D. *S* 91 (2000).
[35]Matsuda, F., Kawatsura, M., Hosaka, K., Shirahama, H. *CEJ* **5**, 3252 (1999).
[36]Aurrecoechea, J.M., Fananas, R., Arrate, M., Gorgojo, J.M., Aurrekoetxea, N. *JOC* **64**, 1893 (1999).
[37]O'Neill, D.J., Helquist, P. *OL* **1**, 1659 (1999).
[38]David, H., Alfonso, C., Bonin, M., Doisneau, G., Guillerez, M.-G., Guibe, F. *TL* **40**, 8557 (1999).
[39]Johnston, D., McCusker, C.F., Muir, K., Procter, D.J. *JCS(P1)* 681 (2000).
[40]Aurrecoechea, J.M., Fernandez, A., Gorgojo, J.M., Saornil, C. *T* **55**, 7345 (1999).
[41]Katritzky, A.R., Luo, Z., Fang, Y., Feng, D., Ghiviriga, I. *JCS(P2)* 1375 (2000).
[42]Tanaka, T., Wakayama, R., Maeda, S., Mikamiyama, H., Maezaki, N., Ohno, H. *CC* 1287 (2000).
[43]Nandanan, E., Dinesh, C.U., Reissig, H.-U. *T* **56**, 4267 (2000).
[44]Xu, F., Sun, J.-H., Yan, H.-B., Shen, Q. *SC* **30**, 1017 (2000).
[45]Concellon, J.M., Bernad, P.L., Perez-Andres, J.A. *ACIEE* **38**, 2384 (1999).
[46]Katritzky, A.R., Feng, D., Fang, Y. *SL* 590 (1999).
[47]Nishitani, T., Shiraishi, H., Sakaguchi, S., Ishii, Y. *TL* **41**, 3389 (2000).

Samarium(III) iodide.

β-Amino esters.[1] Promoted by SmI_3 the condensation of silyl ketene acetals with aldimines is highly *anti*-selective.

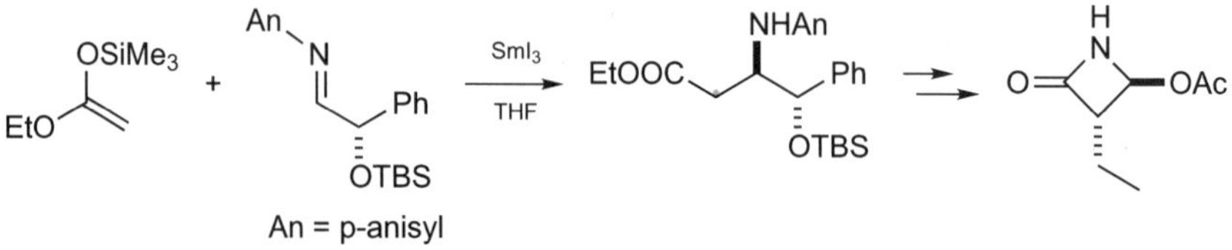

[1]Hayakawa, R., Shimizu, M. *CL* 591 (1999).

Samarium(II) iodide–samarium(III) triflate.

Reduction.[1] With this combination of reagents, methanol and base (KOH), reduction of carboxylic acids to primary alcohols is observed. Aldehydes are hardly affected.

[1]Kamochi, Y., Kudo, T. *TL* **41**, 341 (2000).

Samarium(III) isopropoxide.

Pyrroles.[1] Imines and nitroalkenes undergo condensation to afford 1,3,4-trisubstituted pyrroles. For the promotion of this reaction, samarium(III) isopropoxide is better than several other Sm compounds and isopropoxides of lanthanum and ytterbium.

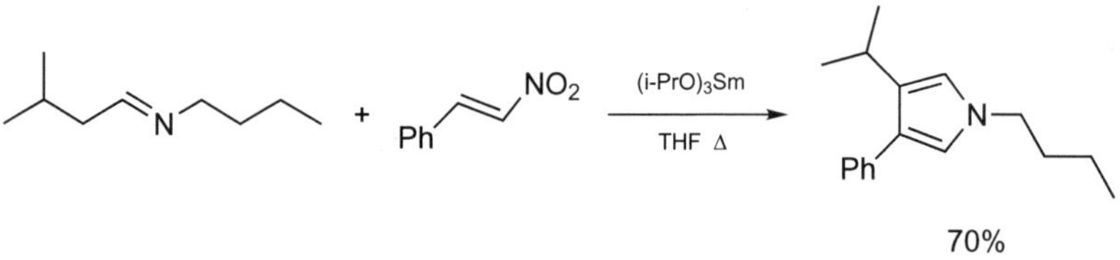

[1]Shiraishi, H., Nishitani, T., Nishihara, T., Sakaguchi, S., Ishii, Y. *T* **55**, 13957 (1999).

Samarium(III) triflate.

Glycosylation.[1] Samarium(III) triflate catalyzes the reaction of glycosyl 2-pyridyl sulfones with alcohols. This method is applicable to the preparation of di- and trisaccharides containing both furanose and pyranose residues. The difference in reactivity from thioglycosides that are not activated by the hard Lewis acid $Sm(OTf)_3$ is a synthetically valuable feature.

[1]Chang, G.X., Lowary, T.L. *OL* **2**, 1505 (2000).

Samarium(II) triflate–nickel(II) iodide.

Alkylations.[1] Samarium(II) triflate is prepared by reduction with Sm in DME containing catalytic amounts of Hg. The solvated product is freed of solvent and combined with NiI_2 for condensation of ketones and acrylic esters to afford γ-lactones.

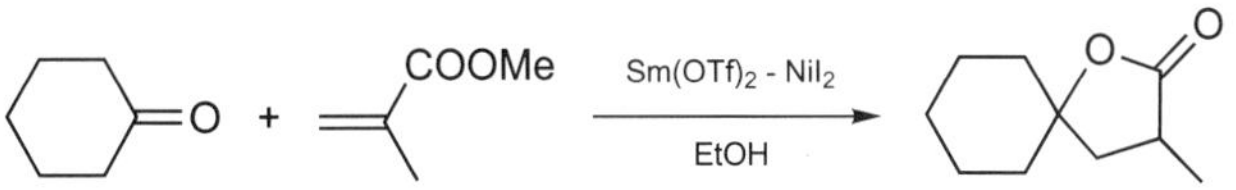

Various reactions.[1] The binary salt is a very useful reagent for promoting Barbier reaction, Mukaiyama-aldol reaction, Michael reaction, Mannich reaction, Diels–Alder reaction, as well as the reductive coupling of carbonyl compounds and of imines.

[1]Collin, J., Giuseppone, N., Machrouhi, F., Namy, J.-L., Nief, F. *TL* **40**, 3161 (1999).

Scandium(III) triflate. 18, 317–318; 19, 300–302; 20, 335–337

Allylation. When tetrallylgermane is used as the allylating agent for carbonyl compounds, the presence of water in the reaction medium is indispensable.[1]

Aldol reactions. To conduct the $Sc(OTf)_3$-catalyzed Mukaiyama aldol reaction in water, the presence of a surfactant is very advantageous.[2] Calixarenesulfonate salts can be employed as surfactants.[3] A hydrophobic microenvironment is created to protect silyl enol ethers, thereby increasing the yields of the products.

Vinylogous aldol reactions leading to δ-hydroxy-α,β-unsaturated aldehydes[4] from alkenyloxiranes and aldehydes are promoted by $Sc(OTf)_3$. Such oxiranes possess amphoteric characters in that they behave as acceptors toward allylborate reagents.

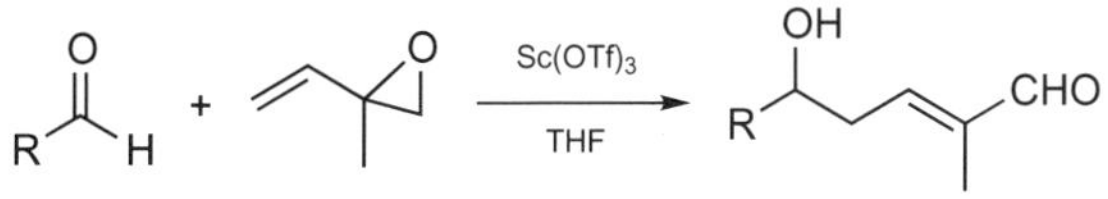

Mannich reactions. β-Amino ketones and esters are readily obtained from a $Sc(OTf)_3$-catalyzed condensation of silyl enolates, aldehydes, and amines in water containing a surfactant.[5] Under the influence of $Sc(OTf)_3$, *O*-trimethylsilylnitronates add to imines in MeCN to provide β-amino nitroalkanes.[6]

Mechanistically related to the Mannich reaction is the formation of hydroxyaryl-glycine derivatives in a three-component reaction.[7]

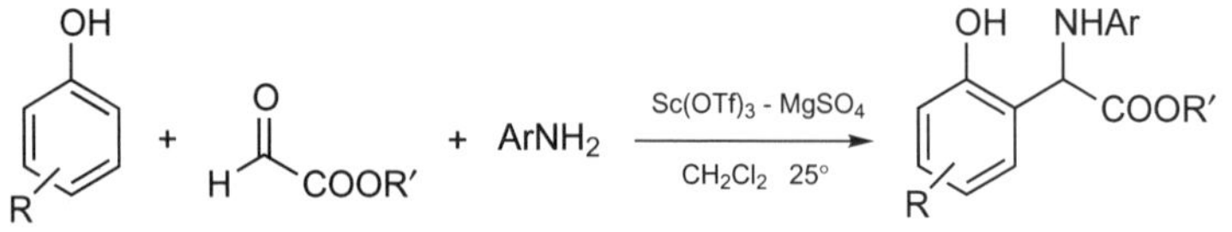

Tishchenko reaction. Reduction of β-hydroxy ketones with isobutyraldehyde while catalyzed by $Sc(OTf)_3$ is stereoselective, *anti*-1,3-diol monoisobutyrates are the major products.[8]

Cyclocondensation. A Prins-type reaction between aldehydes and 3-butenol leads to 4-tetrahydropyranols and ethers.[9] $Sc(OTf)_3$ assists epoxide opening by an intramolecular attack of an enolate, resulting in the formation of three-, four-, and five-membered rings.[10]

Friedel–Crafts alkylation. Secondary alkyl mesylates are adequate alkyl donors in this reaction.[11] Both $Sc(OTf)_3$ and TfOH can be used as the catalyst. It has also been reported that $Sc(OTf)_3$ immobilized in ionic liquid forms a recyclable system for arene alkylation with alkenes.[12]

1,1-Diarylalkenes are formed in the reaction of arenes with 1-phenylalkynes. Triflates of Sc, In, and Zr are suitable catalysts.[13]

Hydrolysis.[14] Esters bearing a coordinative group at a proximal position are hydrolyzed selectively under mild conditions, in the presence of $Sc(OTf)_3$.

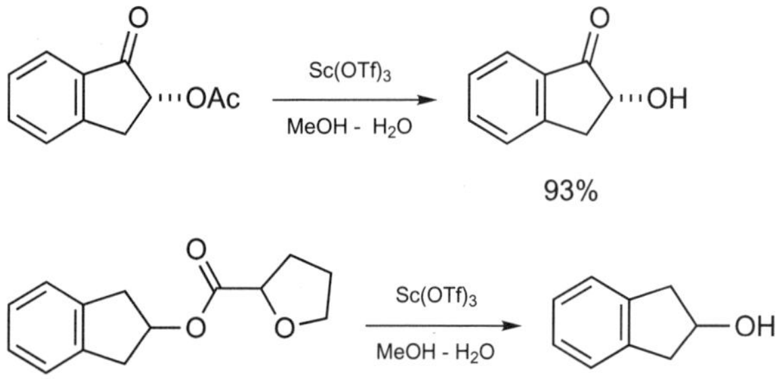

Silyl ethers.[15] Silylation of alcohols at room temperature using methallyl(*t*-butyl)dimethylsilane as TBS group donor is catalyzed by $Sc(OTf)_3$.

[1]Akiyama, T., Iwai, J., Sugano, M. *T* **55**, 7499 (1999).
[2]Manabe, K., Kobayashi, S. *SL* 547 (1999).
[3]Tian, H.-Y., Chen, Y.-J., Wang, D., Zeng, C.-C., Li, C.-J. *TL* **41**, 2529 (2000).
[4]Lautens, M., Ouellet, S.G., Raeppel, S. *ACIEE* **39**, 4079 (2000).
[5]Kobayashi, S., Busujima, T., Nagayama, S. *SL* 545 (1999).
[6]Anderson, J.C., Peace, S., Pih, S. *SL* 850 (2000).
[7]Huang, T., Li, C.-J. *TL* **41**, 6715 (2000).
[8]Gillespie, K.M., Munslow, I.J., Scott, P. *TL* **40**, 9371 (1999).
[9]Zhang, W.-C., Li, C.-J. *T* **56**, 2403 (2000).
[10]Crotti, P., Di Bussolo, V., Favero, L., Macchia, F., Pineschi, M., Napolitano, E. *T* **55**, 5853 (1999).
[11]Kotsuki, H., Ohishi, T., Inoue, M., Kojima, T. *S* 603 (1999).
[12]Song, C.E., Shim, W.H., Roh, E.J., Choi, J.H. *CC* 1695 (2000).
[13]Tsuchimoto, T., Maeda, T., Shirakawa, E., Kawakami, Y. *CC* 1573 (2000).
[14]Kajiro, H., Mitamura, S., Mori, A., Hiyama, T. *BCSJ* **72**, 1553 (1999).
[15]Suzuki, T., Watahiki, T., Oriyama, T. *TL* **41**, 8903 (2000).

Scandium(III) tris(perfluoroalkanesulfonyl)methides.

Debenzylation.[1] Benzyl ethers, *N*-benzylamides, and benzyl ester(s) are efficiently cleaved by catalysis of $Sc(CTf_3)_3$.

Friedel-Crafts acylation.[2] Scandium tris(perfluorobutanesulfonyl)methide is a multipurpose catalyst. Besides promoting Friedel–Crafts reactions, it is also useful in the Diels–Alder reaction and Meerwein–Ponndorf–Verley reduction.

[1]Ishihara, K., Hiraiwa, Y., Yamamoto, H. *SL* 80 (2000).
[2]Nishikido, J., Yamamoto, F., Nakajima, H., Mikami, Y., Matsumoto, Y., Mikami, K. *SL* 1990 (1999).

Selenium. **18,** 318; **20,** 337

Alkenylselenium compounds. Selenium is easily inserted into the C—Zr bond of alkenylzirconocene derivatives. The products can be oxidized to dialkenyl diselenides[1] or acetylated.[2]

Selenides and diselenides.[3] By manipulation of reaction conditions it is possible to prepare either RSeR′ or RSeSeR′.

$$\text{BuLi} + 2\,\text{Se} \longrightarrow \text{BuSe-SeLi} \xrightarrow{\text{RX}} \text{BuSe-SeR}$$

$$\text{BuSe-SeLi} \xrightarrow{\text{BuLi}} 2\,\text{BuSeLi} \xrightarrow{\text{R'X}} \text{BuSeR'}$$

[1]Huang, X., Wang, J.-H. *SC* **30**, 301 (2000).
[2]Huang, X., Wang, J.-H. *SL* 560 (2000).
[3]Krief, A., Van Wemmel, T., Redon, M., Dumont, W., Delmotte, C. *ACIEE* **38**, 2245 (1999).

Selenium–carbon monoxide.

Ureas.[1] Nitroarenes undergo reductive carbonylation and the in situ trapping with unhindered secondary amines leads to unsymmetrical ureas.

Indoles.[2] 2-Nitrostyrenes afford indoles in the Se-catalyzed cyclization. The substrates include 2-nitrostyrene itself and various α- and β-substituted homologues.

R′ R NO2 Se / CO Et3N - DMF 100° R′ R N H

R = H, Ar, ...

R′ = H, Me, ...

[1]Yang, Y., Lu, S. *TL* **40**, 4845 (1999).
[2]Nishiyama, Y., Maema, R., Ohno, K., Hirose, M., Sonoda, N. *TL* **40**, 5717 (1999).

Silica gel. 15, 282; **18,** 319; **19,** 303–304; **20,** 338–339

Selective reactions. Desilylation of triethylsilyl ethers in the presence of *t*-butyldimethylsilyl ethers is accomplished with a mesoporous silica in methanol at room temperature.[1] Selective esterification of nonaromatic carboxylic acids using $NaHSO_4$–silica in methanol has also been described.[2]

Oxidations. Photochemical degradation of α-hydroxy acids and phenylacetic acid derivatives (oxidative decarboxylation)[3] takes place in the presence of a mesoporous silica.

With acidic potassium dichromate adsorbed on silica-zirconia, regioselective allylic oxidation is achieved.[4] This reagent is better than CrO_3–3,5-dimethylpyrazole for oxidation of 1-menthene derivatives because the 3-keto products are largely absent.

OH $K_2Cr_2O_7$/ ZrO_2-SiO_2 O OH

Diels–Alder reaction. Adsorption on silica gel of certain alkyne–$Co_2(CO)_6$ complexes that possess diene and dienophile units separated by a suitable distance serves to shift the equilibrium toward their cycloadducts, as compared with that in solution.[5]

[1]Itoh, A., Kodama, T., Masaki, Y. *SL* 357 (1999).
[2]Das, B., Venkataiah, B., Madhusudhan, P. *SL* 59 (2000).
[3]Itoh, A., Kodama, T., Inagaki, S., Masaki, Y. *OL* **2**, 331 (2000).
[4]Baptistella, L.H.B., Sousa, I.M.O., Gushikem, Y., Aleixo, A.M. *TL* **40**, 2695 (1999).
[5]Iwasawa, N., Sakurada, F., Iwamoto, M. *OL* **2**, 871 (2000).

Silicon tetrafluoride.

3-Fluoroalkanols.[1] Oxetanes in which one of the α-positions is benzylic, allylic, or propargylic are opened by SiF_4 regioselectively. Also, the ring opening is influenced by additives. For example, good results are obtained in the presence of Bu_4NF, whereas the reaction seems to be suppressed by diisopropylethylamine–water, which is effective in promoting epoxide opening.

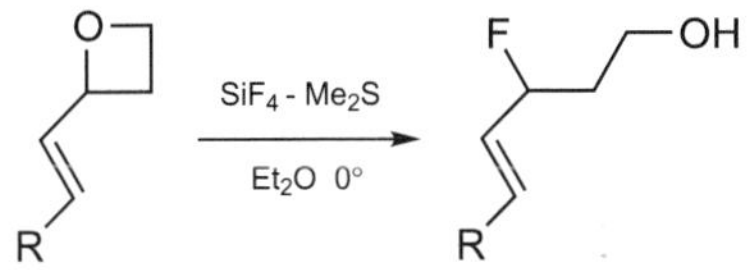

Glycosyl fluorides.[2] Glycals are transformed into glycosyl fluorides with a combination of SiF_4 and an oxidant such as 1,3-dibromo-5,5-dimethylhydantoin, phenyliodine(III) acetate (H_2O). In such cases, a bromine atom or a hydroxyl group is also introduced.

Fluorination.[3] Fluorination of arylalkenes with xenon difluoride is enhanced by SiF_4. The reagent combination also transforms aromatic aldehydes into difluoromethoxyarenes.

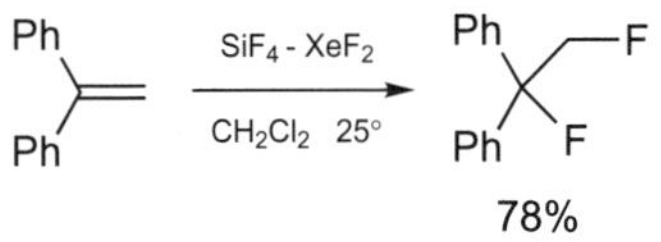

78%

[1]Shimizu, M., Kanemoto, S., Nakahara, Y. *H* **52**, 117 (2000).
[2]Shimizu, M., Nakahara, Y., Yoshioka, H. *JFC* **97**, 57 (1999).
[3]Tamura, M., Takagi, T., Quan, H.-D., Sekiya, A. *JFC* **98**, 163 (1999).

Silver.

Claisen rearrangement.[1] Together with KI in acetic acid, silver effects Claisen rearrangement of allyloxyanthraquinones.

[1]Sharghi, H., Aghapour, G. *JOC* **65**, 2813 (2000).

Silver acetate.

Cycloadditions.[1] 1,3-Dipolar cycloadditions of isocyanoacetic esters are catalyzed by AgOAc. In the absence of dipolarophiles, the esters dimerize to give imidazole-4-carboxylic esters.

CN⌒COOMe + MeOOC–CH=CH$_2$ —AgOAc, MeCN→ (pyrroline product with MeOOC, COOMe, N–H)

[1]Grigg, R., Lansdell, M.I., Thornton-Pett, M. *T* **55**, 2025 (1999).

Silver carbonate.

2-Methylenetetrahydrofurans.[1] 4-Alkynols cyclize on exposure to silver carbonate. An oxygen functionality at the propargylic position has a remarkable accelerating effect.

OH —Ag_2CO_3, PhH 80°→ 99%

1-Amino sugars.[2] Displacement of an anomeric bromine atom by an acetamino group is achieved when glycosyl bromides are treated with silver carbonate in MeCN in the dark.

AcO, OAc, O, AcO, AcO, $CONH_2$, Br —Ag_2CO_3, MeCN→ AcO, OAc, O, AcO, AcO, CN, NHAc 76%

Oxidation.[3] Oxidation of a naphthol can go beyond the quinone stage.

[1]Pale, P., Chuche, J. *EJOC* 1019 (2000).
[2]Gyollai, V., Somsak, L., Szilagyi, L. *TL* **40**, 3969 (1999).
[3]Hauser, F.M., Yin, H. *OL* **2**, 1045 (2000).

Silver nitrate. 18, 320; **19,** 305–306; **20,** 340

Cyclization.[1] 2-Alkynylbenzoic acids afford lactones by the action of $AgNO_3$. 3-Substituted isocoumarins are the major products.

Carbonyl compounds from a-amino nitriles. A synthesis of α-hydroxy enones from the α-amino nitrile derivatives of enals starts from alkylation with aldehydes and the hydrolysis of the products.[2]

Barbier reaction.[3] A catalytic amount of $AgNO_3$ is important in the Zn-mediated reaction of benzylic halides with ArCHO in buffer solutions (pH ~ 12) as less bibenzyls are formed.

[1]Bellina, F., Ciucci, D., Vergamini, P., Rossi, R. *T* **56**, 2533 (2000).
[2]Pierre, F., Enders, D. *TL* **40**, 5301 (1999).
[3]Bieber, L.W., Storch, E.C., Malvestiti, I., da Silva, M.F. *TL* **39**, 9393 (1998).

Silver(I) oxide. 18, 321; **20,** 341

Coupling of 1-alkynes.[1] The Pd(0)-catalyzed reaction of terminal alkynes with alkenyl and aryl halides in THF at 60° proceeds in good yields when Ag_2O is added as activator. The salts. Bu_4NX (X = OH, F) have similar effects.

[1]Mori, A., Kawashima, J., Shimada, T., Suguro, M., Hirabayashi, K., Nishihara, Y. *OL* **2**, 2935 (2000).

Silver tetrafluoroborate. 13, 273–274; **18,** 322

Cleavage of S-(2-trimethylsilyl)ethyl group.[1] Facile Ag(I)-mediated S–C bond cleavage is the basis for the use of the TSE group in thiol protection. The TSE-substituted thioglycosides are stable toward most reagents for carbohydrate transformations except the very strongly hard and soft Lewis acids and desulfonylating conditions, therefore their many applications can be envisaged.

[1]Grundberg, H., Andergran, M., Nilsson, U.J. *TL* **40**, 1811 (1999).

Silver tosylate–urea.

Allylation.[1] Silver tosylate is used in combination with urea, a Lewis base catalyst, to promote allylation of aldehydes with allyltrichlorosilane.

[1]Chataigner, I., Piarulli, U., Gennari, C. *TL* **40**, 3633 (1999).

Silver trifluoromethanesulfonate. 13, 274–275; **14,** 282–283; **16,** 302; **17,** 314; **18,** 322–323; **19,** 306; **20,** 342

N-Alkylation.[1] Silver triflate is essential for an intramolecular alkylation of oxazole that contains an aziridine moiety. A 1,3-dipolar cycloaddition is triggered henceforth.

OAc OTBS O N NMe I AgOTf / MeCN OAc OTBS N NMe O 40%

[1]Vedejs, E., Klapars, A., Naidu, B.N., Piotrowski, D.W., Tucci, F.C. *JACS* **122**, 5401 (2000).

Sodamide. 20, 342

Alkylation.[1] Allyl phenyl sulfide undergoes alkylation using DME-activated sodamide. The process is adaptable to a synthesis of phenyl 1-vinylcycloalkyl sulfides by a twofold alkylation with 1,ω-dihaloalkanes.

PhS + I I $NaNH_2$ DME 0° PhS 75%

Desilylation.[2] The C—Si bond between an aromatic nucleus and the silicon atom of an aryltrialkylsilane suffers reductive severance on treatment with sodamide in liquid ammonia.

[1]Choppin, S., Gros, P., Fort, Y. *SC* **30**, 795 (2000).
[2]Sun, G.-R., He, J.-B., Jie, H.-J., Pittman, C.U. *SL* 619 (2000).

Sodium 13, 277; **18,** 323–324; **20,** 342–343

Aromatic acylation.[1] An acyl group is introduced to the skeleton of a polycyclic aromatic hydrocarbon such as naphthalene when it is treated with sodium and a carboxylic ester in THF at room temperature. Aldehydes are formed when formic esters or DMF are used.

Desulfonylation.[2] A general method for cleaving *N*-tosylaziridines uses sodium naphthalenide (Na + naphthalene) in DME at –78°.

Me3Si ... N Ts —Na(naphthalene) / DME –78°→ Me3Si ... N H

69%

[1]Periasamy, M., Reddy, M.R., Bharathi, P. *SC* **29**, 677 (1999).
[2]Bergmeier, S.C., Seth, P.P. *TL* **40**, 6181 (1999).

Sodium alkanethiolates.

Arenethiols.[1] Sodium *t*-butanethiolate is useful for demethylation of aryl methyl sulfides.

Decomplexation.[2] Dicobalt hexacarbonyl complexes of alkynes are efficiently decomposed at room temperature with NaSMe in DMF.

[1]Pincharti, A., Dallaire, C., van Bierbeek, A., Gingras, M. *TL* **40**, 5479 (1999).
[2]Davis, D.S., Shadinger, S.C. *TL* **40**, 7749 (1999).

Sodium alkoxide–aryloxide clusters.

Ester interchange.[1] These clusters are mild but highly effective catalysts (e.g., in the preparation of *t*-butyl esters by reaction of methyl esters with *t*-butyl acetate, while removing methyl acetate continuously). They are kinetically less basic and therefore useful in dealing with enolizable esters.

[1]Kissling, R.M., Gagne, M.R. *OL* **2**, 4209 (2000).

Sodium amalgam. 18, 324; **19,** 306; **20,** 343

Desulfonylation.[1] Cleavage of the C-S bond of alkyl aryl sulfones shows rate dependence on the aryl portion. *p*-Fluorophenylsulfonyl and 2-naphthalenesulfonyl groups are more readily removed than the corresponding phenylsulfonyl analogues.

[1]Clive, D.L.J., Yeh, V.S.C. *SC* **30**, 3267 (2000).

Sodium azide. 18, 325–326; **19,** 307; **20,** 343

1,2-Azido alcohols.[1] The regioselectivity for unsymmetrical epoxide opening with aqueous sodium azide is influenced by pH.

+ NaN_3 → N_3 OH + OH N_3

pH 9.5	35	:	65
pH 4.2	80	:	20

Azides. A direct conversion of alcohols to azides involves activation with bis(2,4-dichlorophenyl) chlorophosphate.[2]

[1]Fringuelli, F., Piermatti, O., Pizzo, F., Vaccaro, L. *JOC* **64**, 6094 (1999).
[2]Yu, C., Liu, B., Hu, L. *OL* **2**, 1959 (2000).

Sodium bis(2-methoxyethoxy)aluminum hydride. 15, 290.

3,5-Alkadienols.[1] This reagent (commercially known as Red-Al®) reduces homopropargylic alcohols to the (*E*)-alkenols. 3,5-Alkadienols, and hence conjugated trienes, are readily acquired from the corresponding enynols.

Cl OH Ph → (Red-Al, Et_2O) Cl OH Ph 79%

[1]Crousse, B., Mladenova, M., Ducept, P., Alami, M., Linstrumelle, G. *T* **55**, 4353 (1999).

Sodium bisulfate.

Esterification.[1] Heating a mixture of carboxylic acids and alcohols with $NaHSO_4.H_2O$ results in ester formation.

[1]Li, Y.-Q. *SC* **29**, 3901 (1999).

Sodium borohydride. 13, 278–279; **15,** 290; **16,** 304; **18,** 326–327; **19,** 307–309; **20,** 344–345

Reaction media. While alcoholic solvents are used in conventional reduction of carbonyl compounds with sodium borohydride, they can be replaced with hexane if alumina[1] or silica gel[2] is added (heterogeneous conditions). In a refluxing mixture of DME and methanol, $NaBH_4$ reduces methyl methoxybenzoates to the corresponding benzylic alcohols.[3] Stereoselectivity changes are observed in the reduction of 2-substituted cyclohexanones when conducted in DMSO.[4]

Transesterification to the nascent alcohol (alkoxide ion) during reduction, such as that observed in the case of 2,5-bis(4′-methoxybenzoyloxy)benzaldehyde, is largely prevented by adding some acetic acid.[5]

Of interest is the reaction of aryl isocyanates with $NaBH_4$ and trifluoroacetic acid. *N,N*-Bis(2,2,2-trifluoroethyl)anilines are generated in good yields.[6]

Reduction of carboxylic acids and lactones. Acids are activated by cyanuric chloride to facilitate their reduction with $NaBH_4$ to primary alcohols.[7] By controlling the amounts of $NaBH_4$, oxazolidinones can be converted to either the lactols or alcohols.[8]

Reduction of conjugated imines. On forming the tricarbonyliron complexes, conjugated imines (i.e., 1-azadienes) are susceptible to reduction to give saturated amines.[9]

Primary amines are obtained by reductive amination of carbonyl compounds in two steps.[10] The iminating agent is made up from $(i\text{-PrO})_4Ti$, NH_4Cl, and Et_3N, and the reducing agent is $NaBH_4$.

Nonchelate control. Diastereoselectivity for reduction is important in a synthetic context. Accessibility to defined stereoisomers by reduction of α-amino-β-hydroxy ketones is desirable. Different profiles from reduction with zinc borohydride and sodium borohydride (with slight modification of the substrates) are observed. The results are accountable in terms of chelate and nonchelate transition states.[11]

$NaBH_4$/ MeOH	71%	1 : 19
$Zn(BH_4)_2$/ THF	84%	49 : 1

5-Acetoxy-1-naphthol.[12] This compound is obtained (59% yield) from the diacetate by reaction with $NaBH_4$ in a mixture of ethanol and toluene at room temperature.

[1]Yakabe, S., Hirano, M., Morimoto, T. *CJC* **76**, 1916 (1998).
[2]Yakabe, S., Hirano, M., Morimoto, T. *SC* **29**, 295 (1999).
[3]Zanka, A., Ohmori, H., Okamoto, T. *SL* 1636 (1999).
[4]Barros, M.T., Maycock, C.D., Ventura, M.R. *TL* **40**, 557 (1999).
[5]Pugh, C. *OL* **2**, 1329 (2000).
[6]Turnbull, K., Krein, D.M. *S* 391 (1999).
[7]Falomi, M., Porcheddu, A., Taddei, M. *TL* **40**, 4395 (1999).
[8]Reddy, G.V., Rao, G.V., Iyengar, D.S. *TL* **40**, 2653 (1999).
[9]Akisanya, J., Danks, T.N., Garman, R.N. *JOMC* **603**, 240 (2000).
[10]Bhattacharyya, S., Neidigh, K.A., Avery, M.A., Williamson, J.S. *SL* 1781 (1999).
[11]Chung, S.-K., Lee, J.-M. *TA* **10**, 1441 (1999).
[12]Becher, J., Matthews, O.A., Nielsen, M.B., Raymo, F.M., Stoddart, J.F. *SL* 330 (1999).

Sodium borohydride–iodine. 17, 316; 18, 328; 19, 309; 20, 346

Reduction of carboxylic acids.[1] After derivatizing into the pentachlorophenyl esters, acids are reduced to alcohols by this reagent system. This method is applicable to *N*-protected amino acids and peptides.

Imides to amines.[2] The last step of a convenient route to C_2-symmetric 3,4-disubstituted pyrrolidines involves reduction of the corresponding succinimides.

Ph N–Bn O O → $NaBH_4$ - I_2, THF Δ → Ph Ph N–Bn 76%

[1]Naqvi, T., Bhattacharya, M., Haq, W. *JCR(S)* 424 (1999).
[2]Rao, V.D., Periasamy, M. *S* 703 (2000).

Sodium borohydride–metal salt.

Reduction of enones. Dependence of reduction pathways on reaction parameters is very apparent. In a typical example, 1,2-reduction changes to 1,4-reduction on adding cobalt(II) chloride. Micellar conditions are conducive to further reduction in the latter case, giving rise to saturated alcohols.[1]

Cross-conjugated dienones are converted by $NaBH_4$–$CeCl_3$(heptahydrate) in methanol to ethers with the conjugated diene system.[2] Transpositional methanolysis of the initial products accounts for the result.

Ph–CH=CH–CO–CH=CH–Ph $\xrightarrow[\text{MeOH}]{NaBH_4 - CeCl_3\ 7H_2O}$ Ph–CH=CH–CH=CH–CH(OMe)–Ph 86%

Reduction of nitrogen functionalities. Addition of $CoCl_2$ to the reduction milieu causes reduction of azides to primary amines.[3] Nitriles are also saturated using $NaBH_4$–$NiCl_2$ in methanol at 0°. In order to avoid the predominant formation of $(RCH_2)_2NH$ from RCN direct conversion of the products to acetamides (or *N*-Boc-amines) is advisable.[4] The complete reduction of oximes with $NaBH_4$–$TiCl_3$ has found a useful application to a synthesis of α-amino esters.[5]

NOH, COOEt, OEt $\xrightarrow[\text{L-tartaric acid (aq. buffer)}]{NaBH_4 - TiCl_3}$ NH_2, COOEt, OEt 82%

Tin(IV) bis(1,2-benzenedithiolate) has been used to mediate the reduction of azides to amines.[6]

[1]Aramini, A., Brinchi, L., Germani, R., Savelli, G. *EJOC* 1793 (2000).
[2]Barluenga, J., Fananas, F.J., Sanz, R., Garcia, F., Garcia, N. *TL* **40**, 4735 (1999).
[3]Fringuelli, F., Pizzo, F., Vaccaro, L. *S* 646 (2000).
[4]Caddick, S., Haynes, A.K.de K., Judd, D.B., Williams, M.R.V. *TL* **41**, 3513 (2000).
[5]Boukhris, S., Souizi, A. *TL* **40**, 1669 (1999).
[6]Bosch, I., Costa, A.M., Martin, M., Urpi, F., Vilarrasa, J. *OL* **2**, 397 (2000).

Sodium borohydride–chalcogenide.

Elimination. Under phase-transfer conditions, the $NaBH_4$–Te combination is effective for the removal of all the oxygen function of 5-tosyloxyoxazolidin-2-ones to afford allyl amines.[1] Probably the displacement by a telluride ion initiates the process (with subsequent loss of elemental Te in the elimination step).

Bn, N, O, O, Ph, OTs $\xrightarrow[\text{adogen 464, } H_2O - PhMe]{NaBH_4 - Te}$ Bn–NH, Ph 93%

Another elimination process that transforms 1,2-diol dimesylates (e.g., ribonucleoside derivatives) to alkenes uses $NaBH_4$ in conjunction with catalytic quantity of bis(4-perfluorohexylpheny) diselenide.[2] The diselenide is readily recovered.

[1]Xu, Q., Dittmer, D.C. *TL* **40**, 2255 (1999).
[2]Crich, D., Neelamkavil, S., Sartillo-Piscil, F. *OL* **2**, 4029 (2000).

Sodium borohydride–zirconium(IV) chloride.

N—O bond scission. Heterocyclic *N*-oxides[1] and nitro compounds (both aliphatic and aromatic)[2] are reduced to the amine stage by this reagent pair.

C—O bond cleavage. Alcohols are recovered after treatment of allyl ethers $NaBH_4$–$ZrCl_4$.[3] The Lewis acidity of zirconium chloride also facilitates the cleavage of the C—O bond of acetals to allow reduction to give ether products.[4]

[1]Chary, K.P., Mohan, G.H., Iyengar, D.S. *CL* 1339 (1999).
[2]Chary, K.P., Ram, S.R., Iyengar, D.S. *SL* 683 (2000).
[3]Chary, K.P., Mohan, G.H., Iyengar, D.S. *CL* 1223 (1999).
[4]Chary, K.P., Laxmi, Y.R.S., Iyengar, D.S. *SC* **29**, 1257 (1999).

Sodium bromate. 18, 330; **20,** 347

Oxidations. Ethers are susceptible to oxidation by $NaBrO_3$. Thus, THF furnishes γ-butyrolactone (80% yield)[1] and tetrahydropyranyl benzyl ether gives benzaldehyde (95% yield).[2]

Debenzylation.[3] Benzyl ether and benzylidene derivatives of carbohydrates are cleaved with $NaBrO_3$–$Na_2S_2O_4$ in a biphasic reaction system. Esters are stable under such conditions.

OAc, AcO, OMe, BnO, OAc → ($NaBrO_3$ - $Na_2S_2O_4$; H_2O / EtOAc; 25°) → OAc, AcO, OMe, HO, OAc; 96%

[1]Metsger, L., Bittner, S. *T* **56**, 1905 (2000).
[2]Mohammadpoor-Baltork, I., Nourozi, A.R. *S* 487 (1999).
[3]Adinolfi, M., Barone, G., Guariniello, L., Iadonisi, A. *TL* **40**, 8439 (1999).

Sodium chlorite. 20, 348

Oxidations. The oxidation of primary alcohols to acids[1] with sodium chlorite is catalyzed by TEMPO and bleach. On the other hand, when the oxidation is carried out at 0° with a silica-supported TEMPO in the presence of KBr and in a biphasic solvent system (aqueous phase pH 9.1) primary alcohols afford aldehydes.[2]

[1]Zhao, M., Li, J., Manoa, E., Song, Z., Tschaen, D.M., Grabowski, E.J.J., Reider, P.J. *JOC* **64**, 2564 (1999).
[2]Bolm, C., Fey, T. CC 1795 (1999).

Sodium cyanoborohydride.

α-Hydroxy esters.[1] α-Chloroglycidic esters that are accessible from Darzens-type condensation are effectively converted to the hydroxy esters by $NaBH_3CN$.

Tributyltin hydride.[2] It is advantageous to use only a catalytic amount of Bu_3SnCl to generate Bu_3SnH for synthetic use. The combination with $NaBH_3CN$ is one such option. Introduction of a carbon substituent to C-2 of indoles is accomplished by subjecting the 2-iodoindole derivatives to these reagents in the presence of AIBN and alkene addends.

Homoallyic alcohols.[3] Alkenyl epoxides are reduced regioselectively by $NaBH_3CN$ (with zeolite H-ZSM 5) in 1,2-dichloroethane.

Ph / O → ($NaBH_3CN$, H-ZSM 5, $ClCH_2CH_2Cl$) → Ph / OH

[1]Grison, C., Coutrot, F., Comoy, C., Lemilbeau, C., Coutrot, P. *TL* **41**, 6571 (2000).
[2]Fiumana, A., Jones, K. *CC* 1761 (1999).
[3]Gupta, A., Vankar, Y.D. *TL* **40**, 1369 (1999).

Sodium formate.

Reductions.[1] Aldehydes are reduced to alcohols with HCOONa in supercritical water. The conditions can be controlled such that ketones are not affected.

Deiodination.[2] Iodopyrroles undergo Pd-catalyzed dehalogenation when HCOONa is used as a hydrogen source.

[1]Bryson, T.A., Jennings, J.M., Gibson, J.M. *TL* **41**, 3523 (2000).
[2]Leung, S.H., Edington, D.G., Griffith, T.E., James, J.J. *TL* **40**, 7189 (1999).

Sodium hexamethyldisilazide. 18, 332; **20,** 349

4-Iminopyrimidines.[1] A dramatic change in chemoselectivity is observed in the condensation of amidines to alkynyl cyanides.

H H N N + Ph—≡—CN → Ph NH N N + HN Ph N N

+ $(Me_3Si)_2NNa$	3	:	97
+ (no base)	97	:	3

Twofold alkylation.[2] Alkylation of 3,3′-bis(oxindole) derivatives with 2,3-*O,O*-propylidene-threose-1,4-ditosylate in the presence of MHMDS gives different diastereomers in accordance to the alkali metal ion M, whether M = Na or M = Li.

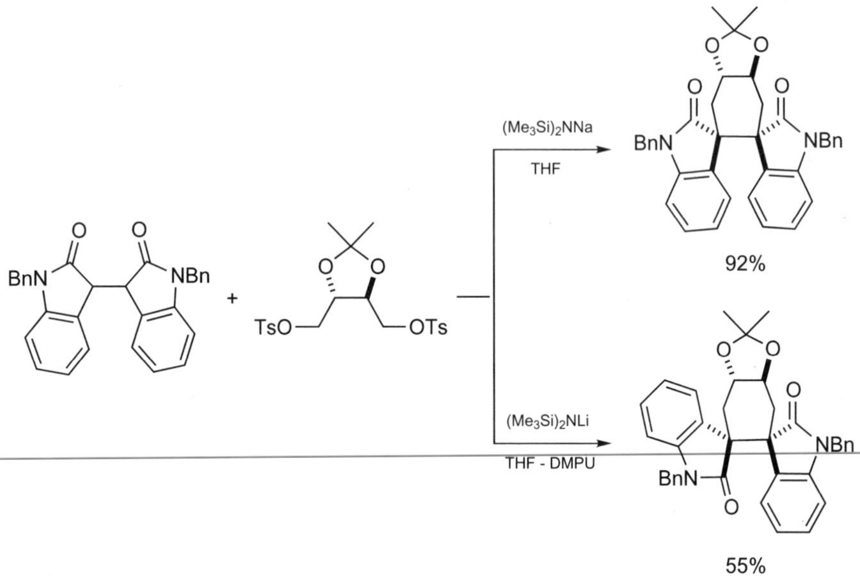
BnN
NBn
+
TsO
OTs
(Me3Si)2NNa
THF
92%
(Me3Si)2NLi
THF - DMPU
55%

[1]McCauley, J.A., Theberge, C.R., Liverton, N.J. *OL* **2**, 3389 (2000).
[2]Overman, L.E., Larrow, J.F., Stearns, B.A., Vance, J.M. *ACIEE* **39**, 213 (2000).

Sodium hydride. 14, 288; **16,** 307–308; **18,** 333; **19,** 312–313; **20,** 349–350

Cyclizations. Alkynones in which the two functional groups are separated by two and three methylene units are converted to furans and pyrans,[1] respectively, on treatment with sodium hydride. *N*-Sulfonylated allylamines that bear a leaving group at the distal allylic position give aziridine derivatives by an intramolecular S_N2' displacement.[2]

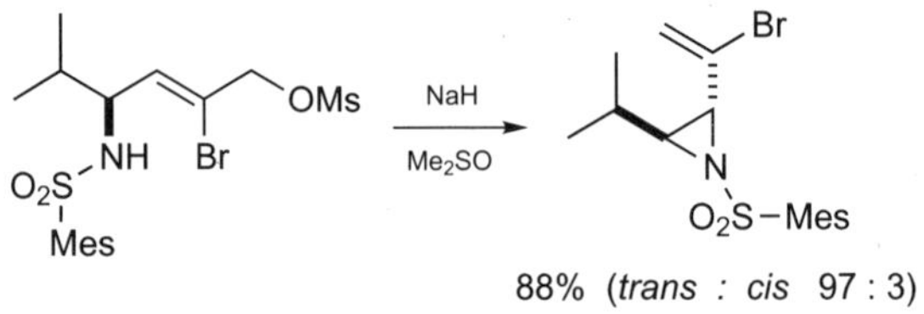
OMs
NH
Br
O2S
Mes
NaH
Me2SO
Br
N
O2S−Mes
88% (trans : cis 97 : 3)

Carbanion formation as a result of an O → C silyl group migration initiated by the generation of a proximal alkoxide ion is exploitable in carbocyclization.[3]

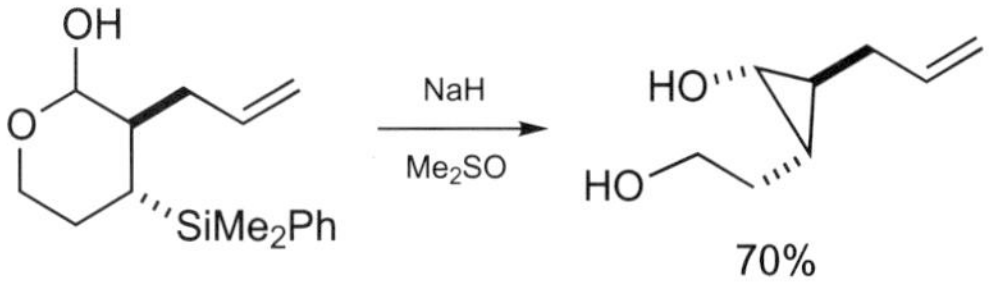

Double annulation is observed when propylenedimalononitrile (and analogues) and 2-alkynones react.[4] The size of the alkyl groups in the alkynones has an enormous effect on the nature of the second ring.

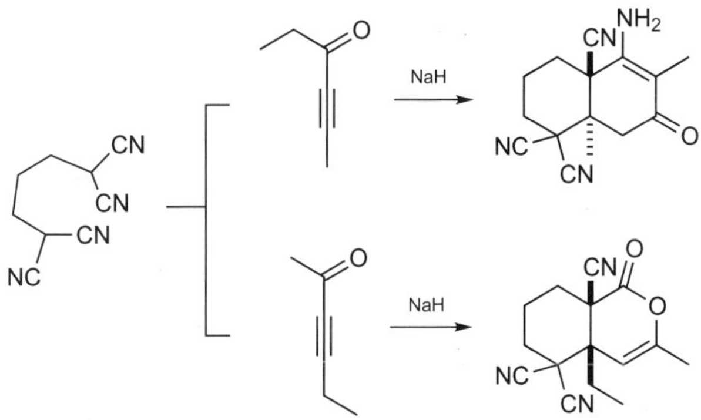

Claisen rearrangement.[5] Base-promoted reaction of allylic malonates represents an alternative method for the ester Claisen rearrangement. For acid sensitive compounds, this protocol is clearly advantageous.

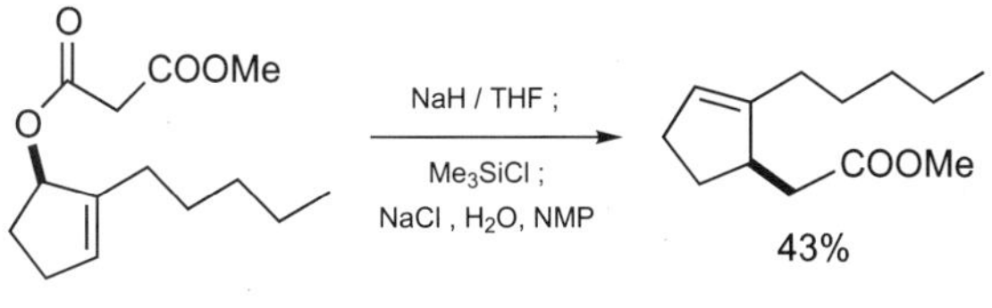

Dehydrohalogenation.[6] A combination of NaH and 2,8,9-trimethyl-1-phospha-2,5,8,9-tetraazabicyclo[3.3.3]undecane hydrochloride generates a powerful base for dehydrohalogenation.

[1]Nicola, T., Vieser, R., Eberbach, W. *EJOC* 527 (2000).
[2]Ohno, H., Toda, A., Takemoto, Y., Fujii, N., Ibuka, T. *JCS(P1)*2949 (1999).

[3]Fleming, I., Mandal, A.K. *CC* 923 (1999).
[4]Grossman, R.B., Skaggs, A.J., Kray, A.E., Patrick, B.O. *OL* **1**, 1583 (1999).
[5]Fehr, C., Galindo, J. *ACIEE* **39**, 569 (2000).
[6]Liu, X., Verkade, J.G. *JOC* **64**, 4840 (1999).

Sodium hypochlorite. 15, 293; **16,** 308; **17,** 316; **18,** 334–335; **19,** 313; **20,** 350

Oxidation of nitrogen compounds. The following reactions have been reported: hydroxylamines to nitrones,[1] *N*-alkyl α-amino nitriles to *N*-alkylformimidoyl cyanides,[2] arylamines to quinones.[3]

Benzylic oxidations.[4] With a nickel salt as catalyst, substituted toluenes are oxidized by NaClO to carboxylic acids.

Isoxazolines.[5] Applying the conventional method that converts aldoximes to nitrile oxides to 2,2-di-(3-butenyl)malonoaldoxime leads to spirocyclic products. Development of such compounds into chiral ligands is expected.

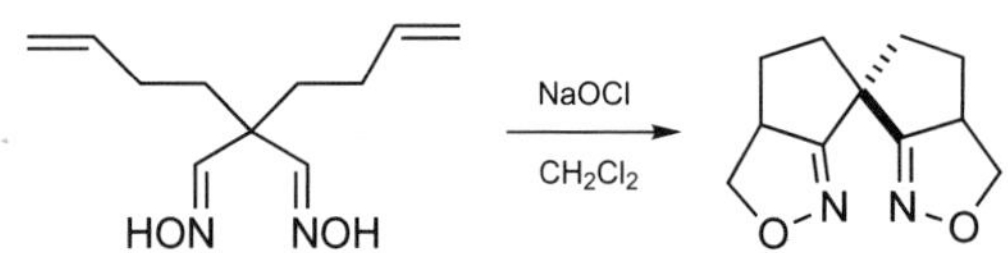

[1]Cicchi, S., Corsi, M., Goti, A. *JOC* **64**, 7243 (1999).
[2]Perosa, A., Selva, M., Tundo, P. *TL* **40**, 7573 (1999).
[3]Hashemi, M.M., Beni, Y.A. *JCR(S)* 672 (1999).
[4]Yamazaki, S. *SC* **29**, 2211 (1999).
[5]Arai, M.A., Arai, T., Sasai, H. *OL* **1**, 1795 (1999).

Sodium iodide.

1-Iodoalkynes.[1] 1-Alkynes are iodinated by anodic oxidation in MeOH with NaI as the supporting electrolyte. Note that replacing NaI with LiCl leads to 1,1-dichloro-2,2-dimethoxyalkanes.

[1]Nishiguchi, I., Kanbe, O., Itoh, K., Maekawa, H. *SL* 89 (2000).

Sodium naphthalenide.

Reduction of pyridines.[1] Sodium naphthalenide or Na/ammonia can be used to reduce electron-deficient pyridines to the 1,2-dihydro derivatives.

Desulfonylation.[2] Aziridines are released from their *N*-tosyl derivatives with sodium naphthalenide in THF.

[1]Donohoe, T.J., McRiner, A.J., Sheldrake, P. *OL* **2**, 3861 (2000).
[2]Bergmeier, S.C., Seth, P.P. *TL* **40**, 6181 (1999).

Sodium nitrite.

Nitrosation. Nitrosation of secondary amines[1] and thiols[2] under heterogeneous conditions uses $NaNO_2$–oxalic acid in dichloromethane and *t*-butanol, respectively.

The more conventional method ($NaNO_2$–HCl) suffices to convert β-hydroxyalkyl ureas to oxazolidin-2-ones.[3] *N*-Nitrosation of the products can be prevented by introducing an organic cosolvent (e.g., EtOAc).

Nitration.[4] Nitration of phenols stopping at the mononitro or dinitro stage can be controlled with the $NaNO_2$–oxalic acid reagent system.

[1]Zolfigol, M.A. *SC* **29**, 905 (1999).
[2]Zolfigol, M.A., Nematollahi, D., Mallakpour, S.E. *SC* **29**, 2277 (1999).
[3]Suzuki, M., Yamazaki, T., Ohta, H., Shima, K., Ohi, K., Nishiyama, S., Sugai, T. *SL* 189 (2000).
[4]Zolfigol, M.A., Ghaemi, E., Madrakian, E. *SC* **30**, 1689 (2000).

Sodium percarbonate. **19,** 314

Oxidations.[1] A chromium-catalyzed oxidation of alcohols with sodium percarbonate is accomplished in benzotrifluoride. $PhCF_3$ is a useful replacement for 1,2-dichloroethane as solvent.

[1]Delaval, N., Bouquillon, S., Henin, F., Muzart, J. *JCR(S)* 286 (1999).

Sodium periodate. **15,** 294; **18,** 338–339; **19,** 315

Carbodiimides.[1] Formation of carbodiimides from selenoureas is observed on reaction with $NaIO_4$ in DMF. The selenoureas are obtained from nitrile oxides and selenonoamides.

$$\text{Ar-NH-C(=Se)-NH-R} \xrightarrow[\text{DMF }\Delta]{\text{NaIO}_4} \text{Ar-N=C=N-R}$$

Oxidations.[2] Epoxidation of alkenes and hydroxylation (and partial oxidation to ketones) of alkanes have been carried out with $NaIO_4$ and manganese(III) porphyrin complex on an ion-exchange resin.

[1]Koketsu, M., Suzuki, N., Ishihara, H. *JOC* **64**, 6473 (1999).
[2]Mirkhani, V., Tangestaninejad, S., Moghadam, M. *JCR(S)* 722 (1999).

Sodium tetracarbonylcobaltate.

β-Lactams.[1] The solvent-dependent activities of $NaCo(CO)_4$, namely, as a Lewis acid in nonpolar solvent for acylation using anhydrides, and as nucleophilic catalyst in polar aprotic solvents, are known. It can be used in cycloadditions by activating ketenes.

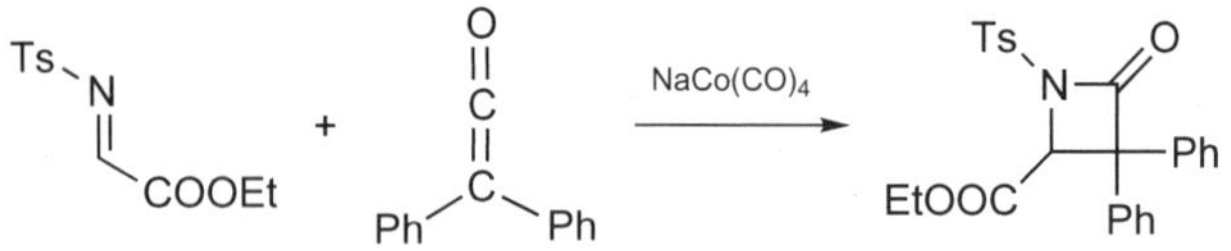

[1] Wack, H., Drury III, W.J., Taggi, A.E., Ferraris, D., Lectka, T. *OL* **1**, 1985 (1999).

Sodium triacetoxyborohydride. 13, 283; **16,** 309–310; **18,** 340; **19,** 315–316; **20,** 352

Reduction of indoles.[1] The stereoselective reduction of tricarbonylchromium complexes of indole derivatives by $NaBH(OAc)_3$–CF_3COOH proceeds after protonation. The hydride attack may involve mediation of the metal atom

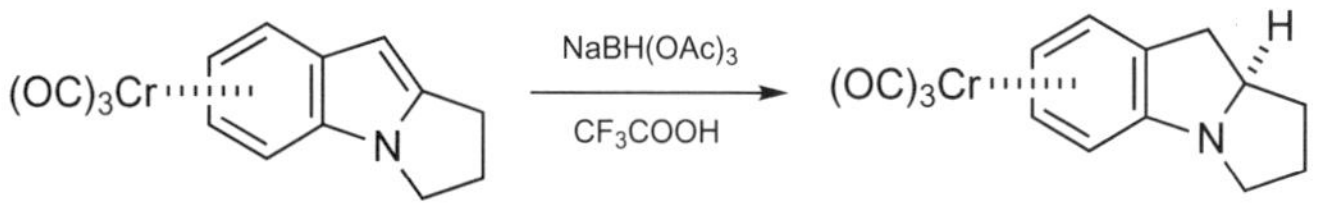

[1] Jones, G.B., Guzel, M., Mathews, J.E. *TL* **41**, 1123 (2000)

Sulfur dioxide. 20, 354

Thiophene 1-oxides. Treatment of zirconacyclopentadienes directly affords the thiophene 1-oxides.[1]

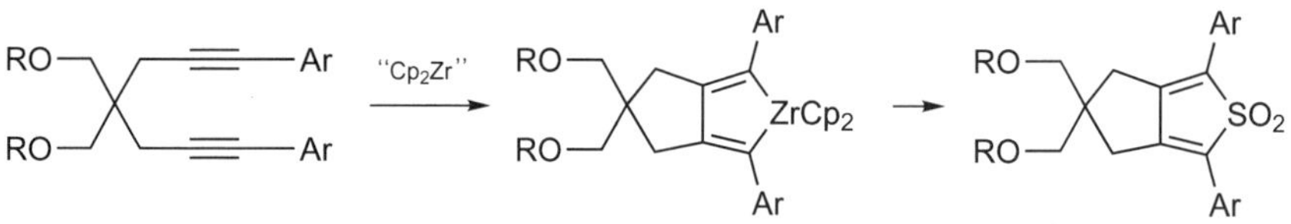

[1] Jiang, B., Tilley, T.D. *JACS* **121**, 9744 (1999).

T

Tantalum(V) chloride.

Allylation.[1] $TaCl_5$ is useful for promoting allylation of aldehydes with allyltrimethylsilane as well as acetylation of alcohols, therefore these two reactions can be achieved without isolation of the homoallylic alcohols.

[1]Chandrasekhar, S., Mohanty, P.K., Raza, A. *SC* **29**, 257 (1999).

Tetrabutylammonium borohydride. 18, 344; 20, 356

Reduction of bromo compounds. Hydrodebromination is achieved with Bu_4NBH_4 in THF. Since under the same conditions a double bond is hydroborated, bromoalkenes and bromoalkynes are converted to saturated alcohols.[1]

[1]Narasimhan, S., Swarnalakshmi, S., Balakumar, R., Velmathi, S. *SC* **29**, 685 (1999).

Tetrabutylammonium bromide. 20, 356

δ-Ketoesters. A catalytic amount of Bu_4NBr promotes the Michael reaction between tin enolates with α,β-unsaturated esters.

[1]Yasuda, M., Ohigashi, N., Shibata, I., Baba, A. *JOC* **64**, 2180 (1999).

Tetrabutylammonium fluoride, TBAF. 13, 286–287; 14, 293–294; 15, 298, 304; 17, 324–326; 18, 344–345; 19, 319–321; 20, 357–359

Desilylation. The powerful desilylation ability of Bu_4NF makes it useful for generating various carbanion equivalents. Synthetic exploitations of this reactivity include access to homoallylamines from imines and an allylsilane,[1] to β-hydroxy-α-diazoalkanoic esters from α-silyl-α-diazoacetic esters and aldehydes,[2] and α-fluorovinylation of carbonyl compounds by 1-fluoro-1-methyldiphenylsilylsilylethylene.[3] Benzyne is formed by treatment of phenyl 2-trimethylsilylphenyliodonium triflate[4] with Bu_4NF.

SiPh2Me, F + H–C(=O)–Ph → (Bu_4NF, THF) → OH, Ph, F

61%

A bicyclo[6.2.2]dodecatriene instead of the condensed 8:6-fused ring system is formed by a fluoride ion-triggered intramolecular alkylation.[5] Complexation with fluoride ion to render bis(allylsilanes) dramatically reactive is indicated.[6]

Protection of amines as triisopropylsilyl carbamates has been proposed, with the unmasking simply by exposure to Bu_4NF.[7]

A synthesis of unsymmetrical azines is based on the formation of trisiopropylsilylhydrazones and their reaction with different carbonyl compounds in the presence of Bu_4NF.[8]

A combination of Bu_4NF and HOAc selectively cleaves *t*-butyldiphenylsilyl ethers without affecting *t*-butyldimethylsilyl ethers.[9]

Formation of transient bicyclo[2.2.1]hept-2-en-5-yne from phenyl-(3-trimethylsilylbornadien-2-yl)iodonium triflate is a reality shown by trapping experiments.[10] An unusual elimination reaction is involved in the generation of perfluoroalkylethylenes from 1-iodo-1-trimethylsilyl-2-perfluoroalkylethanes.[11] The substrates are available from radical addition of perfluoroalkyl iodides to vinyltrimethylsilanes.

Intramolecular alkylation accompanying desilylation of a silyl enol ether is expected in the case that a *cis*-decalin is formed.[12] For acquiring the *trans*-isomer the cyclization must rely on activation by the nitrile group alone.

Cl OTBS NC R R $\xrightarrow[\text{THF}]{Bu_4NF}$ NC O R R

Aldol reactions. α-Isocyanoalkanoic esters condense with aldehydes in the presence of Bu_4NF to afford oxazolines.[13] The reaction of hydrosilyl enol ethers with aldehydes afford 1,3-diols as a result of aldol reaction and subsequent reduction.[14]

$OSiHMe_2$ + H(C=O)Ph $\xrightarrow[\text{THF, −78°}]{Bu_4NF}$ OH OH Ph 65%

Indole derivatives. A method for hydrolysis of indole-3-acetonitriles[15] is by exposing them to air in the presence of Bu_4NF. 2-Substituted indoles are conveniently prepared from 2-alkynylanilines in refluxing THF with Bu_4NF.[16]

Reactions of sulfur compounds. *N,N*-Bis(sulfonyl) arylamines lose one of the sulfonyl groups on heating with Bu_4NF in THF.[17] Cyclic disulfides are obtained from 1, ω-dithiocyanotoalkanes.[18]

Coupling reactions. Two types of Pd-catalyzed aryl–aryl coupling: [ArX + Ar′$Si(OMe)_3$][19] and [ArX + Ar′$_4$Sn][20] are assisted by Bu_4NF.

[1]Wang, D.-K., Zhou, Y.-G., Tang, Y., Hou, X.-L., Dai, L.-X. *JOC* **64**, 4233 (1999).
[2]Kanemasa, S., Araki, T., Kanai, T., Wada, E. *TL* **40**, 5059 (1999).
[3]Hanamoto, T., Harada, S., Shindo, K., Kondo, M. *CC* 2397 (1999).
[4]Kitamura, T., Yamane, M., Inoue, K., Todaka, M., Fukatsu, N., Meng, Z., Fujiwara, Y. *JACS* **121**, 11674 (1999).
[5]Fujishima, H., Takeshita, H., Toyota, M., Ihara, M. *CC* 893 (1999).
[6]Shibato, A., Itagaki, Y., Tayama, E., Hokke, Y., Asao, N., Maruoka, K. *T* **56**, 5373 (2000).
[7]Lipshutz, B.H., Papa, P., Keith, J.M. *JOC* **64**, 3792 (1999).
[8]de Pomar, J.C.J., Soderquist, J.A. *TL* **41**, 3285 (2000).
[9]Higashibayashi, S., Shinko, K., Ishizu, T., Hashimoto, K., Shirahama, H., Nakata, M. *SL* 1306 (2000).
[10]Kitamura, T., Kotani, M., Yokoyama, T., Fujiwara, Y. *JOC* **64**, 680 (1999).
[11]Szlavik, Z., Tarkanyi, G., Gomory, A., Rabai, J. *OL* **2**, 2347 (2000).
[12]Fleming, F.F., Shook, B.C., Jiang, T., Steward, O.W. *OL* **1**, 1547 (1999).
[13]Ito, Y., Higuchi, N., Murakami, M. *H* **52**, 91 (2000).
[14]Miura, K., Nakagawa, T., Suda, S., Hosomi, A. *CL* 150 (2000).
[15]Laronze, M., Laronze, J.-Y., Nemes, C., Sapi, J. *EJOC* 2285 (1999).
[16]Yasuhara, A., Kanamori, K., Kaneko, M., Numata, A., Kondo, Y., Sakamoto, T. *JCS(P1)* 529 (1999).
[17]Yasuhara, A., Kameda, M., Sakamoto, T. *CPB* **47**, 809 (1999).
[18]Burns, C.J., Field, L.D., Morgan, J., Ridley, D.D., Vignevich, V. *TL* **40**, 6489 (1999).
[19]Mowery, M.E., DeShong, P. *OL* **1**, 2137 (1999).
[20]Fugami, K., Ohnuma, S., Kameyama, M., Saotome, T., Kosugi, M. *SL* 63 (1999).

Tetrabutylammonium hydroxide. 20, 359

Partial hydrolysis.[1] The partial hydrolysis of dimethyl esters can be achieved by using dry Bu_4NOH in THF or DME.

[1]Hasegawa, T., Yamamoto, H. *SL* 84 (1999).

Tetrabutylammonium nitrate–trifluoroacetic anhydride.

Nitration.[1] This reagent system nitrates 1-deazapurine nucleosides. The regioselectivity depends on existing substitution patterns.

[1]Deghati, P.Y.F., Bieraugel, H., Wanner, M.J., Koomen, G.-J. *TL* **41**, 569 (2000).

Tetrabutylammonium peroxydisulfate. 19, 322

C═O regeneration. Oximes[1] and semicarbazones[2] are cleaved.

Oxidations. Benzyl ethers are removed by oxidation with $(Bu_4N)_2S_2O_8$ and alcoholysis[3]. Primary amines are oxidized to nitriles with Ni–Cu formates as catalyst.[4]

Oxidative cycloaddition.[5] The oxidation of 1,3-dicarbonyl compounds in the presence of cyclic enol ethers leads to fused acetals.

[1]Chen, F., Liu, A., Yan, Q., Liu, M., Zhang, D., Shao, L. *SC* **29**, 1049 (1999).
[2]Chen, F., Liu, J.-D., Fu, H., Peng, Z.-Z., Shao, L.-Y. *SC* **30**, 2295 (2000).
[3]Chen, F., Peng, Z.-Z., Fu, H., Meng, G., Cheng, Y., Lu, Y.-X. *SL* 627 (2000).
[4]Chen, F., Peng, Z.-Z., Fu, H., Liu, J.-D., Shao, L.-Y. *JCR(S)* 726 (1999).
[5]Chen, F., Fu, H., Meng, G., Cheng, Y., Hu, Y.-L. *S* 1091 (2000).

Tetrabutylammonium tribromide.

Ether cleavage.[1] Bu_4NBr_3 in methanol cleaves several kinds of ethers but TBS ethers are the most susceptible. Thus, desilylation can be achieved in the presence of acetonides and THP ethers.

[1]Gopinath, R., Patel, B.K. *OL* **2**, 4177 (2000).

Tetrabutylammonium triorganodifluorostannates.

Diarylmethanes.[1] With $(Ph_3P)_4Pd$ as catalyst, unsymmetrical $ArCH_2Ar'$ are obtained from a cross-coupling reaction between aryl triflates and $Bu_4N[Bn_3SnF_2]$ in DMF.

Disulfides.[2] $Bu_4N[R_3SnF_2]$ act as nucleophiles toward sulfur. Oxidative dimerization of the thiols initially formed, results in disulfides.

[1]Martinez, A.G., Barcina, J.O., del R.C. Heras, M., de F. Cerezo, A. *OL* **2**, 1377 (2000).
[2]Kerverdo, S., Fernandez, X., Poulain, S., Gingras, M. *TL* **41**, 5841 (2000).

Tetrachlorophthalimide.

Primary amines.[1] When used in the Mitsunobu reaction, the litle compound converts primary and secondary alcohols into the corresponding protected amines.

[1]Jia, Z.J., Kelberlau, S., Olsson, L., Anilkumar, G., Fraser-Reid, B. *SL* 565 (1999).

Tetracobalt dodecacarbonyl.

Pauson–Khand reaction.[1] With cyclohexylamine as activator, catalytic amount of $Co_4(CO)_{12}$ mediates the Pauson–Khand reaction in DME under CO.

[1]Krafft, M.E., Bonaga, L.V.R. *ACIEE* **39**, 3676 (2000).

Tetraethylammonium hydrogen carbonate. 20, 360

Carbonates.[1] Treatment of 1,2-diols with Et_4NHCO_3 in MeCN at room temperature results in the formation of cyclic carbonates.

Sulfides.[2] Et_4NHCO_3 or $(Et_4N)_2CO_3$ can serve as base in the alkylation of thiols.

[1]Casadei, M.A., Cesa, S., Feroci, M., Inesi, A. *NJC* **23**, 433 (1999).
[2]Feroci, M., Inesi, A., Rossi, L. *SC* **29**, 2611 (1999).

Tetraethylammonium peroxydicarbonate.

Oxazolidin-2-ones.[1] This reagent is available from electrochemical reaction of Et_4NClO_4, CO_2, and O_2. It carboxylates 2-amino alcohols and the products undergo cyclization to oxazolidin-2-ones on further treatment with TsCl.

[1]Feroci, M., Inesi, A., Muccianti, V., Rossi, L. *TL* **40**, 6059 (1999).

Tetrakis(acetonitrile)copper(I) hexafluorophosphate.

Cyclic imines.[1] Alkynylamines cyclize under the influence of $(MeCN)_4CuPF_6$.

Diaryl ethers.[2] The phenol–aryl halide coupling is promoted by Cs_2CO_3 and $(MeCN)_4CuPF_6$.

Epoxidation.[3] Conversion of alkenes to epoxides with MCPBA, using $(MeCN)_4CuPF_6$ as catalyst proceeds at low temperature (e.g., $-20°$).

[1]Muller, T.E., Grosche, M., Herdtweck, E., Pleier, A.-K., Walter, E., Yan, Y.-K. *OM* **19**, 170 (2000).
[2]Kalinin, A.V., Bower, J.F., Riebel, P., Snieckus, V. *JOC* **64**, 2986 (1999).
[3]Andrus, M.B., Poehlein, B.W. *TL* **41**, 1013 (2000).

Tetrakis(triphenylphosphine)palladium(0). **13,** 289–294; **14,** 295–299; **15,** 300–304; **16,** 317–323; **17,** 327–331; **18,** 347–349; **19,** 324–331; **20,** 362–368

Allylic displacements. New types of substrates for this $(Ph_3P)_4Pd$-catalyzed substitution are β-allyloxyacrylic esters and β-allyloxyvinyl sulfones.[1] These vinylogous carbonates and sulfonates contain better leaving groups.

Regioselective displacement of allylic acetates that contain a homoallylic silyl group gives allylsilanes.[2] Allenyonitriles are obtained by reacting propargyl carbonates with Me_3SiCN.[3]

R′ R″ R OCOOR + Me_3Si-CN $\xrightarrow[\text{THF } \Delta]{(Ph_3P)_4Pd}$ R NC =C= R′ R″

Displacement that is followed by cyclization enables the synthesis of chromenes[4] and γ-hydroxyalkyl-γ-lactones.[5]

Tandem coupling and cyclization. Functionalized allenes are converted to α-arylvinylated cyclopropanes[6] and heterocycles including epoxides,[7] tetrahydrofurans,[8] and oxazolidinones.[9] Furan derivatives are formed from allenyl ketones.[10]

O C $C_{12}H_{25}$ + Ph-I $\xrightarrow[\text{PhMe } 80°]{(Ph_3P)_4Pd,\ Ag_2CO_3 - Et_3N}$ Ph $C_{12}H_{25}$ O

75%

In the presence of sodium alkoxide (for generating alkoxyimino nucleophiles for the cyclization), 2-alkynylbenzonitriles afford isoindole derivatives.[11]

Cyclizations and cycloadditions. Treatment of *N,N′*-diacylhydrazines[12] and γ,δ-unsaturated ketone *O*-pentafluorobenzoyloximes[13] with $(Ph_3P)_4Pd$ generates 1,3,4-oxadiazoles and substituted pyrroles, respectively.

$(Ph_3P)_4Pd$, Et_3N / DMF

A [4 + 2]cycloaddition between enynes and 1,3-diynes, with the latter serving as the two-carbon component, results in substituted benzene derivatives. The adducts derived from 1,4-di-*t*-butoxy-1,3-butadiyne are readily converted to coumaranones.[14]

$(Ph_3P)_4Pd$, THF 0°

91%

Alkylidenemalononitriles undergo cycloaddition with Me_3SiCN and then *N*-allylation to give 3,5-disubstituted tetrazoles in one step.[15]

$(Ph_3P)_4Pd$, THF 60°

93%

Addition to alkynes. Nucleophiles add to alkynes in different addition modes. Thus, α-substituted vinylphosphonium salts[16] and 1,2-diphosphonylalkanes[17] are prepared from 1-alkynes, but apparently an isomerization–hydroalumination pathway prevails when 2-alkynes react with amines under similar conditions. The products are allylamines.[18] (Note the formation of *N*-benzyl-2-styrylpiperidine in an intramolecular reaction, but the corresponding *N*-tosyl derivative fails to cyclize.)

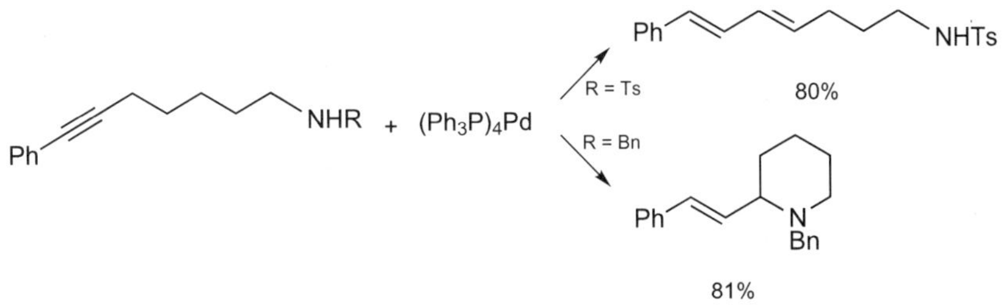

Acylations. 2-Alkynoic esters are made from 1-alkynes and a chloroformic ester in the presence of $(Ph_3P)_4Pd$ and base (DMAP–1,2,2,6,6-pentamethylpiperidine).[19]

Stille coupling. The usefulness of tosylates in the Stille coupling facilitates preparation of arylcoumarin inhibitors of gyrase B.[20] Benzylic and allylic boronates are readily obtained from $RSnBu_3$ and bromomethylboronates.[21]

The Stille coupling is accelerated by CuCl. An effective system applicable to sterically congested substrates has been developed.[22] Preparation of functionalized dienes by homocoupling of alkenylstannanes using slightly different conditions enables completion of a synthesis of (−)-wodeshiol.[23]

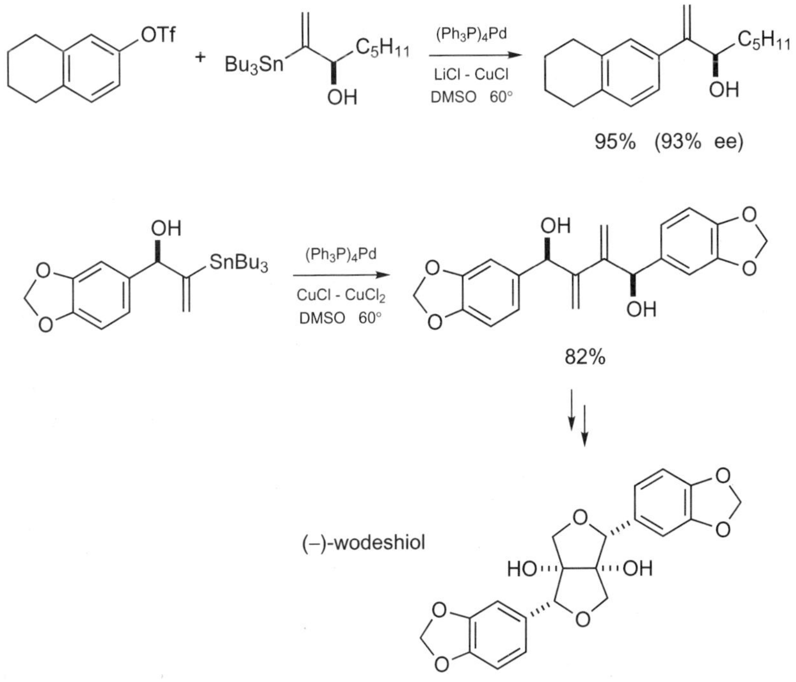

1-Alkynes are converted to (*E*)-alkenes in a microwave-assisted hydrosilylation-Stille coupling process,[24] and both reactions are catalyzed by $(Ph_3P)_4Pd$. Interestingly, arylation of tributylstannylacetylene is achieved via a Stille coupling and then *C*-stannylation.[25]

$$Bu_3Sn{-}C{\equiv}CH + Ph{-}I \xrightarrow[\text{THF}\ 60^\circ]{(Ph_3P)_4Pd} Ph{-}C{\equiv}CH + Bu_3Sn{-}I \xrightarrow{\text{LDA}} Ph{-}C{\equiv}C{-}SnBu_3\quad 83\%$$

Stille coupling between $RSeSnBu_3$ and R′X is applicable to the synthesis of diorganyl selenides RSeR′.[26]

Suzuki coupling. A practical transformation of ArX to ArMe is by the Suzuki coupling with $[MeBO]_3$.[27] Diarylmethanes are obtained from $ArCH_2Br$ and arylboronic acids.[28]

The coupling of 3-pyrrolin-3-yl triflates with arylboronic acids leads to 3-arylpyrroles because of concurrent dehydrogenation.[29] A caveat of the coupling involving haloanilines is that deamination[30] also occurs to some extent.

A route to alkynylarenes and enynes from 1-alkynes involves formation of alkynylboronic esters [lithioalkynes + $(i\text{-PrO})_3B$] and Suzuki coupling in situ.[31] Suzuki cross-coupling using thallium(I) ethoxide[32] as promoter is superior to TlOH because of its stability, commercial availability, and ease of use. A convenient method for assembling conjugated polyenes is assured.

In a two-stage coupling of *N*-allyl-*N*-2-bromoallylamine, the *N*-sulfonyl derivatives are most suitable as β-elimination of alkylpalladium intermediates after the initial intramolecular Heck reaction is suppressed (likely by coordination).[33]

The coupling of arylboronic acids with acid chlorides is the basis of a ketone synthesis.[34] It is found that the Suzuki coupling in an ionic liquid has several advantages: reduced catalyst concentration, no homocoupling, and reaction in the air.[35]

Other coupling reactions. A model study has demonstrated the utility of intramolecular Heck reaction in the construction of the morphine skeleton.[36] It is surprising that only the desirable regioisomer is formed.

The 2-azabicyclo[3.3.1]nonane framework, a portion of the strychnos alkaloids, is accessible from an intramolecular coupling of 4-*N*-(2-haloallyl)aminocyclohexanones.[37]

Organobismuth dialkoxides couple with electron-deficient aryl and alkenyl triflates.[38]

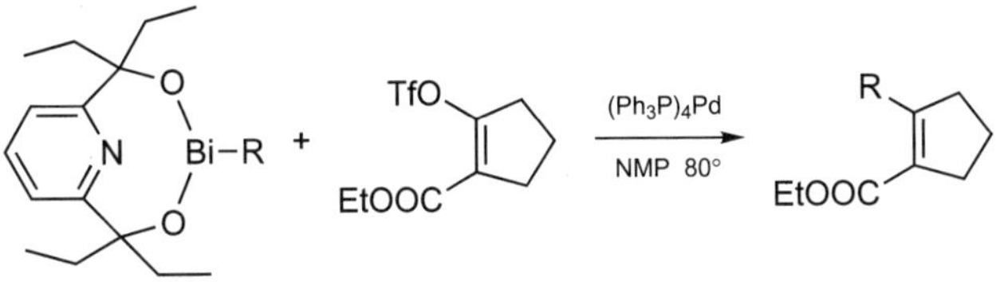

A testimony to the efficiency of constructing highly unsaturated carbon skeletons by Pd-catalyzed reactions is delineated in a synthesis of xerulin.[39]

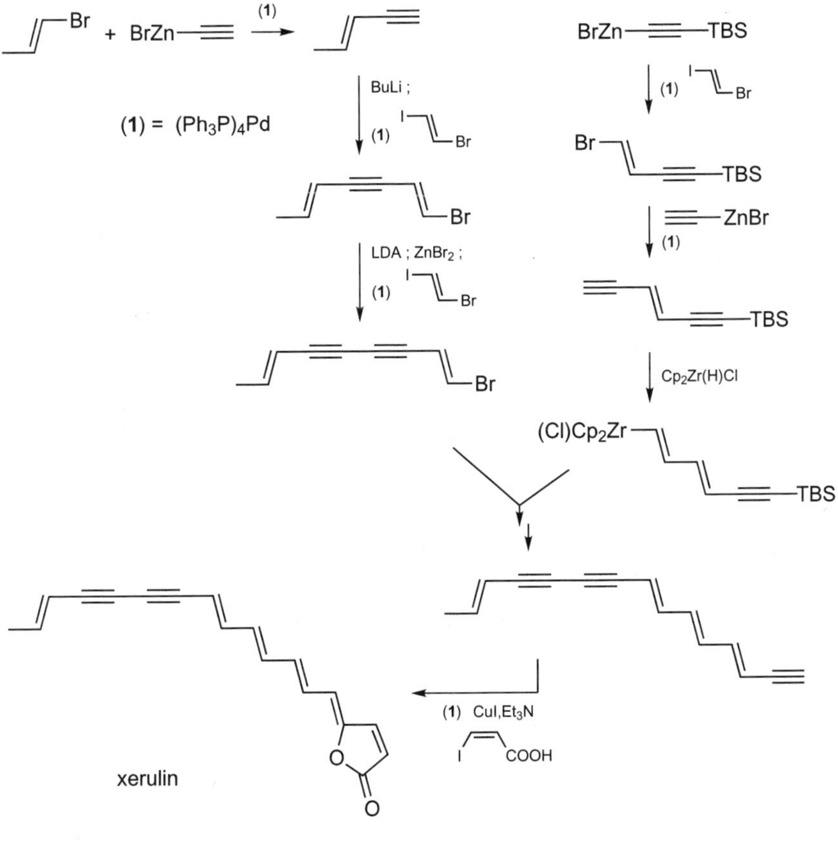

Rearrangements. Allyl esters are directly converted to isocyanates[40] when the modified Curtius rearrangement is carried out in the presence of $(Ph_3P)_4Pd$. A versatile construction of the core structure of antibiotic CP-263114 is highlighted in a reaction sequence consisting of carbonylation, lactonization, and siloxy–Cope rearrangement.[41]

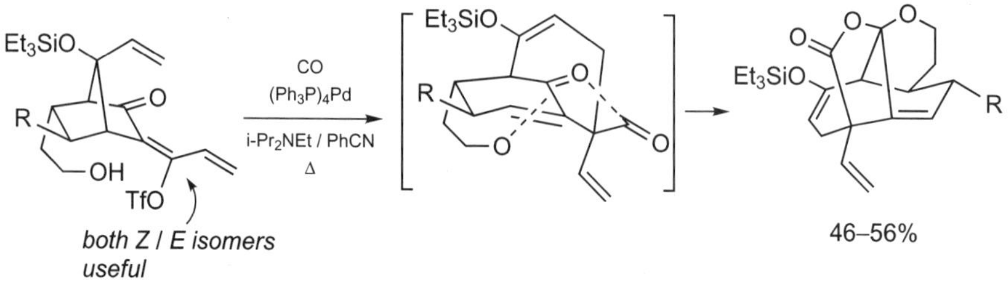

[1]Evans, P.A., Brandt, T.A., Robinson, J.E. *TL* **40**, 3105 (1999).
[2]Macsari, I., Hupe, E., Szabo, K.J. *JOC* **64**, 9547 (1999).
[3]Tsuji, Y., Taniguchi, M., Yasuda, T., Kawamura, T., Obora, Y. *OL* **2**, 2635 (2000).
[4]Nay, B., Peyrat, J.-F., Vercauteren, J. *EJOC* 2231 (1999).
[5]Rudler, H., Parlier, A., Cantagrel, F., Harris, P., Bellassoued, M. *CC* 771 (2000).
[6]Ma, S., Zhao, S. *OL* **2**, 2495 (2000).
[7]Ma, S., Zhao, S. *JACS* **121**, 7943 (1999).
[8]Kang, S.-K., Baik, T.-G., Kulak, A.N. *SL* 324 (1999).
[9]Kang, S.-K., Baik, T.-G., Hur, Y. *T* **55**, 6863 (1999).
[10]Ma, S., Zhang, J. *CC* 117 (2000).
[11]Wei, L.-M., Lin, C.-F., Wu, M.-J. *TL* **41**, 1215 (2000).
[12]Lutun, S., Hasiak, B., Couturier, D. *SC* **29**, 111 (1999).
[13]Tsutsui, T., Narasaka, K. *CL* 45 (1999).
[14]Gevorgyan, V., Quan, L.G., Yamamoto, Y. *JOC* **65**, 568 (2000).
[15]Gyoung, Y.S., Shim, J.-G., Yamamoto, Y *TL* **41**, 4193 (2000).
[16]Arisawa, M., Yamaguchi, M. *JACS* **122**, 2387 (2000).
[17]Allen, Jr., A., Manke, D.R., Lin, W. *TL* **41**, 151 (2000).
[18]Kadota, I., Shibuya, A., Lutete, L.M., Yamamoto, Y. *JOC* **64**, 4570 (1999).
[19]Bottcher, A., Becker, H., Brunner, M., Preiss, T., Henkelmann, J., DeBakker, C., Gleiter, R. *JCS(P1)* 3555 (1999).
[20]Schio, L., Chatreaux, F., Klich, M. *TL* **41**, 1543 (2000).
[21]Falck, J.R., Bondlela, M., Ye, J., Cho, S.-D. *TL* **40**, 5647 (1999).
[22]Han, X., Corey, E.J. *JACS* **121**, 7600 (1999).
[23]Han, X., Corey, E.J. *OL* **1**, 1871 (1999).
[24]Maleczka, Jr., R.E., Lavis, J.M., Clark, D.H., Gallagher, W.P. *OL* **2**, 3655 (2000).
[25]Antonelli, E., Rosi, P., Sterzo, C.L., Viola, E. *JOMC* **578**, 210 (1999).
[26]Nishiyama, Y., Tokunaga, K., Sonoda, N. *OL* **1**, 1725 (1999).
[27]Gray, M., Andrews, I.P., Hook, D.F., Kitteringham, J., Voyle, M. *TL* **41**, 6237 (2000).
[28]Chowdhury, S., Georghiou, P.E. *TL* **40**, 7599 (1999).
[29]Lee, C.-W., Chung, Y.J. *TL* **41**, 3423 (2000).

[30]Hird, M., Seed, A.J., Toyne, K.J. *SL* 438 (1999).
[31]Castanet, A.-S., Colobert, F., Schlmam, T. *OL* **2**, 3559 (2000).
[32]Frank, S.A., Chen, H., Kunz, R.K., Schnaderbeck, M.J., Roush, W.R. *OL* **2**, 2691 (2000).
[33]Lee, C.-W., Oh, K.S., Kim, K.S., Ahn, K.H. *OL* **2**, 1213 (2000).
[34]Haddach, M., McCarthy, J.R. *TL* **40**, 3109 (1999).
[35]Mathews, C.J., Smith, P.J., Welton, T. *CC* 1249 (2000).
[36]Frey, D.A., Duan, C., Hudlicky, T. *OL* **1**, 2085 (1999).
[37]Sole, D., Peidro, E., Bonjoch, J. *OL* **2**, 2225 (2000).
[38]Rao, M.L.N., Shimada, S., Tanaka, M. *OL* **1**, 1271 (1999).
[39]Negishi, E., Alimardanov, A., Xu, C. *OL* **2**, 65 (2000).
[40]Okumoto, H., Nishihara, S., Yamamoto, S., Hino, H., Nozawa, A., Suzuki, A. *SL* 991 (2000).
[41]Bio, M.M., Leighton, J.L. *OL* **2**, 2905 (2000).

Tetrakis(triphenylphosphine)palladium(0)–copper(I) iodide. 18, 349–350; 20, 369

Coupling reactions. α-Fluorovinylstannanes are a source of α-fluorostyrenes by virtue of their capacity of partaking in the Stille coupling.[1] A synthesis of (*Z*)-alkenes is based on the Pd(0)-catalyzed exchange of alkenyl tellurides or selenides with zinc followed by the Negishi coupling.[2]

1,2-Alkadien-4-ynes are formed by the regioselective coupling of propargyl substrates with 1-alkynes.[3]

Me_3Si + C_5H_{11} Cl ——$(Ph_3P)_4Pd$ - CuI / i-Pr_2NH——> C_5H_{11} =C= Me_3Si

84%

[1]Chen, C., Wilcoxen, K., Zhu, Y.-F., Kim, K., McCarthy, J.R. *JOC* **64**, 3476 (1999).
[2]Dabdoub, M.J., Dabdoub, V.B., Marino, J.P. *TL* **41**, 433, 437 (2000).
[3]Condon-Gueugnot, S., Linstrumelle, G. *T* **56**, 1851 (2000).

Tetrakis(triphenylphosphine)platinum(0). 20, 369–370

Hydrosilylation. This reaction has been studied using either transition metal complexes such as $(Ph_3P)_4Pt$ or radical initiators.[1]

Diboration of methylenecyclopropanes. Regioselective functionalization with ring opening of the substrates occurs on reaction with bis(pinacolato)diboron.[2] The adducts can be transformed in various ways to useful substances.

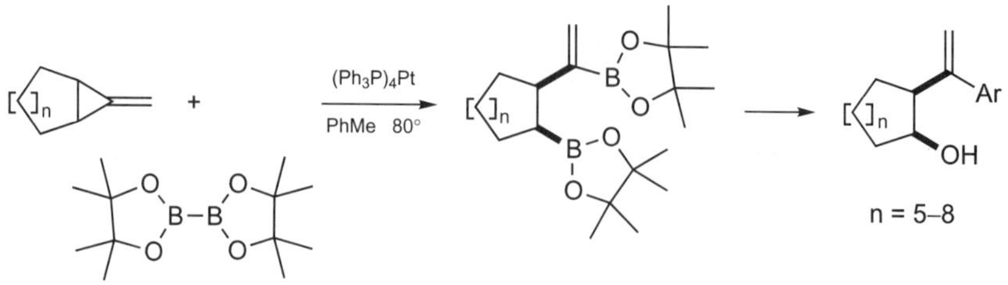

Allylic displacement.[3] The Pt(0) complex is effective in mediating the reaction.

[1]Itoh, M., Iwata, K., Kobayashi, M. *JOMC* **574**, 241 (1999).
[2]Ishiyama, T., Momota, S., Miyaura, N. *SL* 1790 (1999).
[3]Blacker, A.J., Clarke, M.L., Loft, M.S., Mahon, M.F., Humphries, M.E., Williams, J.M.J. *CEJ* **6**, 353 (2000).

α-Tetralol.

Carboxyl protection. Tetralyl esters can be selectively cleaved by $Me_3SiCl-NaI$ in MeCN at room temperature without affecting benzhydryl esters and *p*-methoxybenzyl esters.

[1]Slade, C.J., Pringle, C.A., Summer, I.G. *TL* **40**, 5601 (1999).

N,N,N′,N′-Tetramethyl-*S*-(1-oxidopyridin-2-yl)isothiouronium salts.

Peptide coupling.[1] The tetrafluoroborate and hexafluorophosphate salts are developed as peptide coupling reagents, illustrated by the preparation of dipeptides and tripeptides.

[1]Bailen, M.A., Chinchilla, R., Dodsworth, D.J., Najera, C. *JOC* **64**, 8936 (1999).

2,2,6,6-Tetramethylpiperidinoxyl, TEMPO.

N-Alkoxy-2,2,6,6-tetramethylpiperidines.[1] TEMPO replaces the catecholboryl group from its *B*-alkyl derivatives. Thus, alcohols in a protected form are obtained from hydroboration if so desired, using this oxidant in the workup.

Oxidation.[2] The presence of TEMPO is essential to the smooth autoxidation of primary alcohols to aldehydes catalyzed by $(Ph_3P)_3RuCl_2$. Without TEMPO, further conversion of the aldehydes to carboxylic acids is observed.

[1]Ollivier, C., Chuard, R., Renaud, P. *SL* 807 (1999).
[2]Dijksman, A., Arends, I.W.C.E., Sheldon, R.A. *CC* 1591 (1999).

Tetrapropylammonium perruthenate, TPAP. 20, 370

Isomerization.[1] TPAP is a catalyst that converts allylic alcohols to ketones (8 examples, 41–92%).

[1]Marko, I.E., Gautier, A., Tsukazaki, M., Llobet, A., Plantalech-Mir, E., Urch, C.J., Brown, S.M. *ACIEE* **38**, 1960 (1999).

Thallium(III) acetate.

Oxidation of β,γ-unsaturated acids.[1] Either butenolides or degraded allylic acetates are formed.

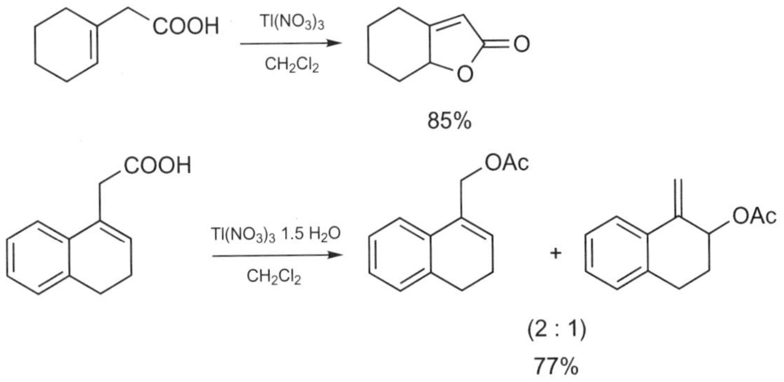

COOH $\xrightarrow[CH_2Cl_2]{Tl(NO_3)_3}$ 85%

COOH $\xrightarrow[CH_2Cl_2]{Tl(NO_3)_3\ 1.5\ H_2O}$ OAc + OAc (2 : 1) 77%

[1]Ferraz, H.M.C., Grazini, M.V.A., Silva, Jr., L.F., Longo, Jr., L.S. *SC* **29**, 1953 (1999).

Thallium(III) nitrate, TTN. 16, 326; **18,** 351; **19,** 334; **20,** 371

Cyclopentyl 2-hydroxyalkyl ketones.[1] 2-(1-Cyclohexenyl)ethanols undergo ring contraction on treatment with TTN in aq HOAc.

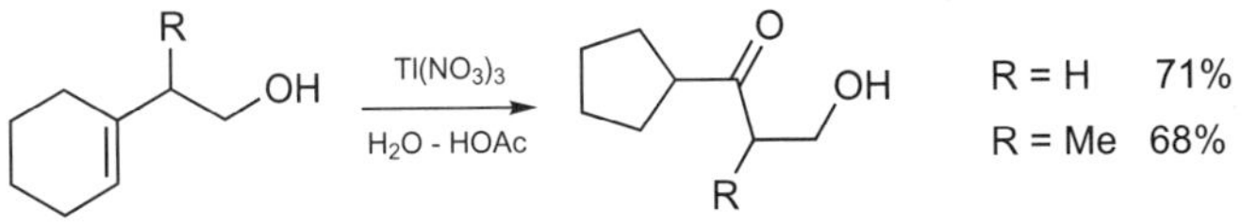

R, OH $\xrightarrow[H_2O\ -\ HOAc]{Tl(NO_3)_3}$ O, OH, R

R = H 71%
R = Me 68%

Ring expansion.[2] The ring expansion of unsymmetrical 1-vinylcyclobutanols shows different regioselectivities as effected by $Tl(NO_3)_3$ and by $(PhCN)_2PdCl_2$.

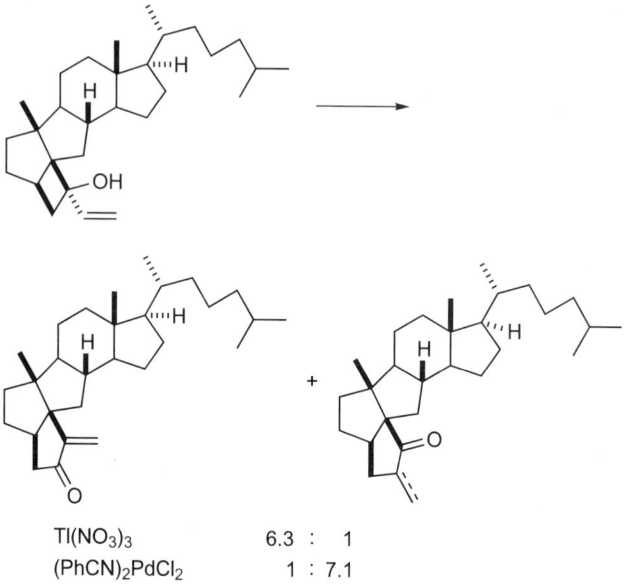

$Tl(NO_3)_3$ 6.3 : 1

$(PhCN)_2PdCl_2$ 1 : 7.1

[1]Ferraz, H.M.C., Santos, A.P., Silva, Jr., L.F., de O.Viera, T. *SC* **30**, 751 (2000).
[2]Kocovsky, P., Dunn, V., Gogoll, A., Langer, V. *JOC* **64**, 101 (1999).

1-Thionoacyl 6-nitrobenzotriazoles.

Thiono esters.[1] At room temperature and in the presence of imidazole, the benzotriazole derivatives transfer the RC=S group to alcohols.

[1]Shalaby, M.A., Rapoport, H. *JOC* **64**, 1065 (1999).

Thionyl chloride–benzotriazole.

Chlorides and nitriles. The reagent combinant transforms alcohols to alkyl chlorides and acids to acid chlorides,[1] while aldoximes are dehydrated.[2]

[1]Chaudhari, S.S., Akamanchi, K.G. *SL* 1763 (1999).
[2]Chaudhari, S.S., Akamanchi, K.G. *SC* **29**, 1741 (1999).

Thionyl chloride–triflic acid.

Diaryl sulfoxides.[1] A Friedel–Crafts-type reaction between arenes and thionyl chloride is catalyzed by TfOH.

[1]Olah, G.A., Marinez, E.R., Prakash, G.K.S. *SL* 1397 (1999).

Thiourea. 19, 336; **20,** 371–372

Episulfides.[1] In a conversion of epoxides to episulfides that is catalyzed by a tin(IV)–porphyrin complex, thiourea furnishes the sulfur atom.

1,3-Dioxolane cleavage.[2] Hydrolysis of cyclic acetals is accomplished with thiourea in aqueous ethanol at reflux temperature. Some selectivity is shown in substrates such as 1,2;5,6-di-*O*-isopropylidenefuranoses in which the terminal acetonide is cleaved.

[1]Tangestaninejad, S., Mirkhani, V. *SC* **29**, 2079 (1999).
[2]Majumdar, S., Bhattacharjya, A. *JOC* **64**, 5682 (1999).

Thulium(II) iodide.

Alkylation.[1] $TmI_2(dme)_2$ is more powerful than $SmI_2(thf)_X$–HMPA in mediating the reaction of RX with ketones. It is particularly valuable in a situation where HMPA must be avoided.

[1]Evans, W.J., Allen, N.T. *JACS* **122**, 2118 (2000).

Tin. 13, 298; **17,** 333–334; **18,** 352; **20,** 372–373

Alkylation. The allylating agent derived from tin and allyl bromide in water consists of two species, the initially formed allyltin(II) bromide is more reactive.[1]

Tin in combination with Me_3SiCl mediates the reaction of bromoacetonitrile and bromomethyl ketones with aldehydes.[2]

[1]Chan, T.H., Yang, Y., Li, C.J. *JOC* **64**, 4452 (1999).
[2]Sun, P., Shi, B. *JCR(S)* 318 (1999).

Tin(II) bromide. 14, 303–304; **18,** 352; **19,** 336–337

Allylation.[1] With $SnBr_2$ as promoter, the carbonyl allylation shows interesting regioselectivity and diastereoselectivity in the presence or absence of Bu_4NBr.

Br + PhCHO —($SnBr_2$; H_2O - CH_2Cl_2)→ OH Ph + OH Ph

----	94	6
+ Bu_4NBr	5	95

[1]Ito, A., Kishida, M., Kurusu, Y., Masuyama, Y. *JOC* **65**, 494 (2000).

Tin(II) chloride. 13, 298–299; **15,** 309–310; **16,** 329; **18,** 353–354; **19,** 337–338; **20,** 373

Addition to enones.[1] The addition of tributylstannylacetic esters to enones proceeds in different manners according to the catalyst:1,2-addition is promoted by $SnCl_2$ and 1,4-addition by Me_3SiCl.

R, R', SnBu3, MeOOC, O, OH, COOMe

$SnCl_2$ / MeCN < 1 : > 99

Me_3SiCl / $MeNO_2$ > 99 : < 1

Hydrostannylation.[2] Allenes undergo hydrostannylation on exposure to $(Ph_3P)_2PdCl_2$, $SnCl_2$ and HCl in DMF. Allylation of aldehydes by the ensuing allyltin trichloride reagents is completed at room temperature.

$SnCl_2$ - $(Ph_3P)_2PdCl_2$ / HCl - DMF

[1]Yasuda, M., Matsukawa, Y., Okamoto, K., Sako, T., Kitahara, N., Baba, A. *CC* 2149 (2000).
[2]Chang, H.-M., Cheng, C.-H. *OL* **2**, 3439 (2000).

Tin(IV) chloride. 13, 300–301; **14,** 304–306; **15,** 311–313; **17,** 335–340; **18,** 354–356; **19,** 338–339; **20,** 373–375

Cyclizations. Activation of an allenyl group with $SnCl_4$ and linkage to a *C*-nucleophile constitute a ring formation process.[1] The reaction of *N*-arylaldimines with allylsilanes (or allylgermanes) in the presence of $SnCl_4$ leads to 2,4-disubstituted 1,2,3,4-tetrahydroquinolines[2] due to rapid interception of the carbocationic intermediates.

1,3-Butadien-2-ylation.[3] Treatment of 2-tributylstannyl-1,3-butadiene with $SnCl_4$ renders it reactive as a nucleophile toward carbonyl compounds.

Mukaiyama aldol reaction.[4] Chemoselective ionization of a mixed-*O,S*-acetal by Lewis acid is noted. Thus, it is possible to synthesize β-alkoxyketones or β-organothioketones at will.

SnCl4
OMe
62%
OSiMe3
+
MeO SPh
TiCl4
SPh
86%

Cleavage of p-methoxybenzyl ethers.[5] By using $SnCl_4$–PhSH to perform this ether cleavage, functional groups sensitive to DDQ or CAN can be retained. Thus, organothio, allyloxy, siloxy, acyloxy groups, and acetonides survive the reaction conditions.

[1]Kitagawa, O., Suzuki, T., Fujiwara, H., Taguchi, T. *TL* **40**, 2549 (1999).
[2]Akiyama, T., Suzuki, M., Kagoshima, H. *H* **52**, 529 (2000).
[3]Luo, M., Iwabuchi, Y., Hatakeyama, S. *SL* 1109 (1999).
[4]Braga, A.L., Dornelles, L., Silveira, C.C., Wessjohann, L.A. *S* 562 (1999).
[5]Yu, W., Su, M., Gao, X., Yang, Z., Jin, Z. *TL* **41**, 4015 (2000).

Tin(II) iodide.

Allylation and proparylation/allenylation. The SnI_2–Bu_4NI–NaI system activates allylic alcohols to act as nucleophiles toward aldehydes.[1] Slow reactions are observed when SnI_2 is replaced with $SnBr_2$. No reaction occurs with a corresponding reagent system containing $SnCl_2$.

The analogous propargylation or allenylation uses propargyl mesylates.[2] 1-Alkyn-3-yl mesylates and 2-alkyn-1-yl mesylates give different products.

R
OMs
SnI2 - NaI - Bu4NI / DMI ;
R′CHO
OH
R
R′
C

R
OMs
SnI2 - NaI - Bu4NI / DMI ;
R′CHO
R
OH
R′

[1]Masuyama, Y., Ito, T., Tachi, K., Ito, A., Kurusu, Y. *CC* 1261 (1999).
[2]Masuyama, Y., Watanabe, A., Ito, A., Kurusu, Y. *CC* 2009 (2000).

Tin(IV) oxide.

Meerwein–Ponndorf–Verley reduction.[1] Hydrous SnO_2 catalyzes the vapor-phase reduction of carbonyl compounds with isopropanol.

Transacylation.[1] Passing a mixture of an ester and an alcohol or amine through hydrous SnO_2 completes the acyl exchange.

[1]Waghoo, G., Jayaram, R.V., Joshi, M.V. *SC* **29**, 513 (1999).

Tin(II) triflate. 13, 301–302; **14,** 306–307; **15,** 313–314; **17,** 341–344; **18,** 357–358; **19,** 340; **20,** 376

Ethers from ethyleneacetals.[1] *C,O*-Dialkylation of ethyleneacetals is effected with silanes and silyl ethers using $Sn(OTf)_2$ as catalyst.

Ph + Me_3Si + BnO–$SiMe_3$ → ($Sn(OTf)_2$, MeCN, –20°) → OBn / Ph 93%

Aziridine opening.[2] Hydroxylated compounds (alcohols, water) open *N*-tosylaziridines in the presence of a Lewis acid [$Sn(OTf)_2$ or $BF_3.OEt_2$].

[1]Suzuki, T., Oriyama, T. *SC* **29**, 1263 (1999).
[2]Prasad, B.A.B., Sekar, G., Singh, V.K. *TL* **41**, 4677 (2000).

Titanium(II) chloride–*N,N,N′,N′*-tetramethylethylenediamine.

Alkylidenation. Methylenation of esters with $CH_2(ZnI)_2$ is promoted by (tmeda)$TiCl_2$.[1] Borylalkenes[2] and germanyalkenes[3] are similarly prepared from $[M]CH(ZnX)_2$.

ZnBr / ZnBr + O H Ph → ($TiCl_4$, THF) → Ph 42% (*E* / *Z* 62 : 38)

Reductive couplings.[4] Benzaldehyde is converted to *syn*-hydrobenzoin with a $TiCl_2$–amine system. Moderate asymmetric induction is observed when the reaction is carried out in the presence of a chiral *trans-N,N,N′,N′*-tetramethyl-1,2-diaminocyclohexane.

[1]Matsubara, S., Ukai, K., Mizuno, T., Utimoto, K. *CL* 825 (1999).
[2]Matsubara, S., Otake, Y., Hashimoto, Y., Utimoto, K. *CL* 747 (1999).
[3]Matsubara, S., Yoshino, H., Utimoto, K., Oshima, K. *SL* 495 (2000).
[4]Matsubara, S., Hashimoto, Y., Okano, T., Utimoto, K. *SL* 1411 (1999).

Titanium(III) chloride.

4-Oxo-2-alkenylphosphonates.[1] A synthesis of this class of apparently useful compounds involves nitrile oxide cycloaddition with allylic phosphonates, conversion of the resulting isoxazolines to the conjugated oximes, and hence to the enones. The last step is readily achieved by the use of $TiCl_3$.

TiCl3 / DMF, HCl - H2O

64%

[1]Lee, S.Y., Lee, B.S., Lee, C.-W., Oh, D.Y. *JOC* **65**, 256 (2000).

Titanium(III) chloride–lithium. 19, 340; 20, 377

Depropargylation.[1] *N*-Propargylamines undergo cleavage on treatment with $TiCl_3$–Li in THF at room temperature.

[1]Rele, S., Talukdar, S., Banerji, A. *TL* **40**, 767 (1999).

Titanium(III) chloride–zinc. 19, 341; 20, 377

McMurry reaction.[1] Substituted phenanthrenes have been synthesized from 2,2′-diformylbiaryls by the McMurry reaction.

TiCl3 (DME) - Zn/Cu

45%

[1]Gies, A.-E., Pfeffer, M. *JOC* **64**, 3650 (1999).

Titanium(IV) chloride. 13, 304–309; 14, 309–311; 15, 317–320; 16, 332–337; 17, 344–347; 18, 359–361; 19, 341–344; 20, 377–379

Functional group exchanges. $TiCl_4$ serves as a catalyst in the conversion of THP ethers,[1] silyl ethers,[2] and propargyl esters[3] to the corresponding esters and ethers. A direct synthesis of α-(benzotriazol-1-yl)alkyl ethers involves treatment of dialkyl ethers with 1-chlorobenzotriazole and $TiCl_4$.[4]

Aldol and Mannich reactions. Formation of cyclopentenones is readily achieved by a twofold Mukaiyama aldol reaction between 1,3-bis(trimethylsiloxy)-1,3-dienes and 1,2-diketones.[5]

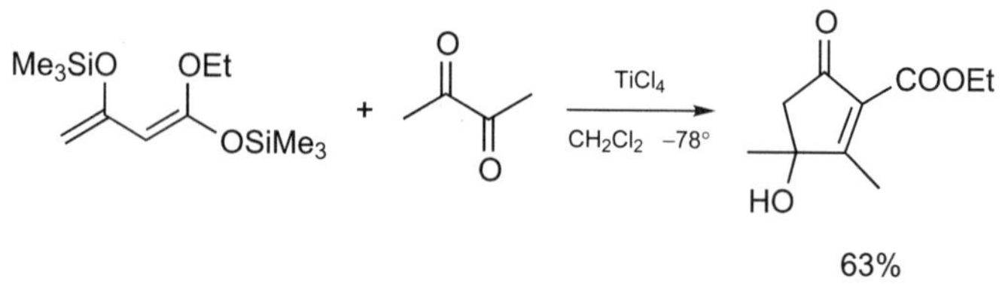

Activated imines and aminal derivatives react with malonic esters to give precursors of β-amino acids,[6] while conjugated aldimines undergo 1,4- and 1,2-additions, both induced by $TiCl_4$.[7]

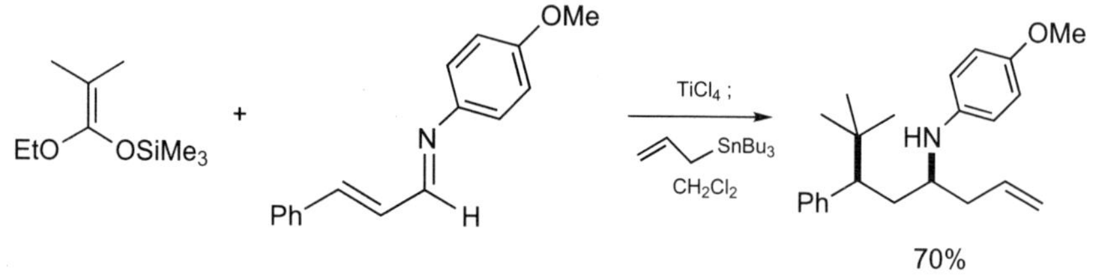

Baylis–Hillman reaction. When $TiCl_4$ is used to promote the condensation, a chlorine atom is introduced into the adducts.[8–11]

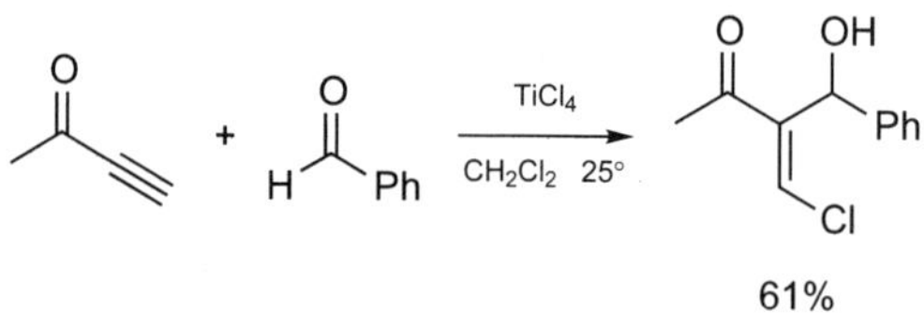

Biaryls. On treatment with $TiCl_4$ in nitromethane, 2-naphthol and derivatives afford binaphthol derivatives.[12] The oxidative coupling proceeds particularly well with substrates containing electron-donating groups.

N,N-Dialkylarylamines also afford substituted biaryls, via nuclear titanation,[13] but *N*-dealkylation is observed with hindered amines and those containing a *p*-substituent.

Cyclizations. Tetrahydrofurans and pyrrolidines bearing a 2-silylmethyl group are formed when alkenylsilanes and 5-substituted 1-pentenylsilanes are exposed to $TiCl_4$ at room temperature.[14,15]

A tetracyclic framework containing six contiguous stereocenters emerges from treatment of an aryltrienone with $TiCl_4$.[16] The Nazarov cyclization triggers additional C—C bond formation.

Schmidt reaction.[17] Lactams are formed when cycloalkanones react with RCH_2N_3 in the presence of $TiCl_4$. Interestingly, TfOH induces a Mannich reaction, indicating the more rapid formation of the $[RNH{=}CH_2]^+$ species.

[1]Chandrasekhar, S., Ramachandar, T., Reddy, M.V., Takhi, M. *JOC* **65**, 4729 (2000).
[2]Iranpoor, N., Zeynizadeh, B. *SC* **29**, 2123 (1999).
[3]Bartels, A., Mahrwald, R., Quint, S. *TL* **40**, 5989 (1999).
[4]Katritzky, A.R., Voronkov, M.V., Pastor, A., Tatham, D. *H* **51**, 1877 (1999).
[5]Langer, P., Kohler, V. *OL* **2**, 1597 (2000).
[6]Milenkovic, A., Fache, F., Faure, R., Lemaire, M. *SC* **29**, 1535 (1999).
[7]Shimizu, M., Morita, A., Kaga, T. *TL* **40**, 8401 (1999).
[8]Wei, H.-X., Kim, S.H., Caputo, T.D., Purkiss, D.W., Li, G. *T* **56**, 2397 (2000).
[9]Kataoka, T., Kinoshita, H., Kinoshita, S., Iwamura, T., Watanabe, S. *ACIEE* **39**, 2358 (2000).
[10]Shi, M., Jiang, J.-K., Feng, Y.-S. *OL* **2**, 2397 (2000).
[11]Li, G., Gao, J., Wei, H.-X., Enright, M. *OL* **2**, 617 (2000).
[12]Doussot, J., Guy, A., Ferroud, C. *TL* **41**, 2545 (2000).
[13]Periasamy, M., Jayakumar, K.N., Bharathi, P. *JOC* **65**, 3548 (2000).
[14]Miura, K., Hondo., T., Takahashi, T., Hosomi, A. *TL* **41**, 2129 (2000).
[15]Miura, K., Hondo., T., Nakagawa, T., Takahashi, T., Hosomi, A. *OL* **2**, 385 (2000).
[16]Bender, J.A., Arif, A.M., West, F.G. *JACS* **121**, 7443 (1999).
[17]Desai, P., Schildknegt, K., Agrios, K.A., Mossman, C., Milligan, G.L., Aube, J. *JACS* **122**, 7226 (2000).

Titanium(IV) chloride–amines. 20, 380

Reductive cyanation.[1] Aromatic ketones condense with aminoacetonitrile in the presence of $TiCl_4$–Et_3N. α-Substituted arylacetonitriles are obtained upon treatment of the resulting imines with K_2CO_3 in refluxing DMF.

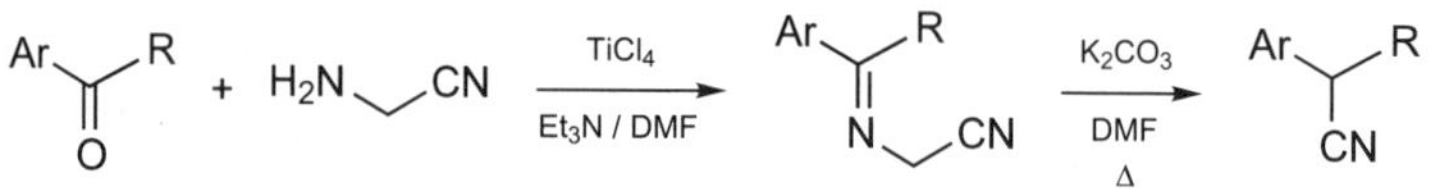

Reductive couplings.[2] The low-valent titanium species formed by interaction of $TiCl_4$ with Et_3N is capable of transforming ArCHO and ArCH═NAr′ to hydrobenzoins and 1,2-diarylethylenediamines, respectively.

Aldol and Dieckmann reactions.[3] These condensation reactions are readily achieved using $TiCl_4$–Bu_3N and a catalytic amount of Me_3SiCl. The effectiveness of the reagent system can be gauged by the reaction conditions (e.g., cyclization of dimethyl adipate in dichloromethane at −78° affords methyl 2-oxocyclpentanecarboxylate in 95% yield).

Michael reaction.[4] The conjugate addition of *N*-acyloxazolidin-2-ones to nitroalkenes is best promoted by a combination of $TiCl_4$ and *i*-Pr_2NEt. Excellent stereoselectivity is observed from reactions of substrates in which the heterocycle is substituted at C-4 with an isopropyl group (stereocontroller) and *gem*-dimethylated at C-5.

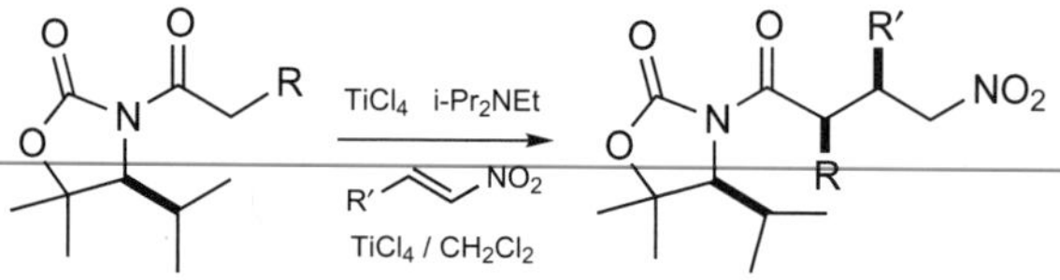

syn-2,3-Disubstituted 4-pentenamides.[5] The reaction of allylamines with acid chlorides proceeds via *N*-acylation and enolate–Claisen rearrangement.

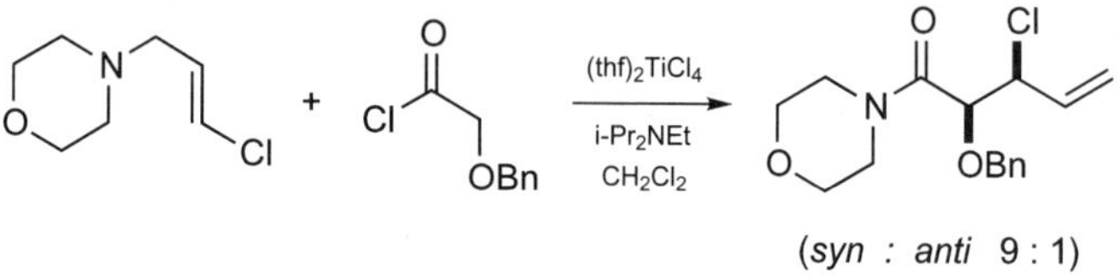

(*syn* : *anti* 9 : 1)

[1]Selva, M., Bomben, A., Tundo, P. *SC* **29**, 1561 (1999).
[2]Periasamy, M., Srinivas, G., Karunakar, G.V., Bharathi, P. *TL* **40**, 7577 (1999).
[3]Yoshida, Y., Matsumoto, N., Hamasaki, R., Tanabe, Y. *TL* **40**, 4227 (1999).
[4]Brenner, M., Seebach, D. *HCA* **82**, 2365 (1999).
[5]Yoon, T.P., Dong, V.M., MacMillan, D.W.C. *JACS* **121**, 9726 (1999).

Titanium(IV) chloride–lithium–trimethylsilyl chloride–nitrogen. 20, 380–381

Tetrahydroindoles.[1] The [Ti—N] complex induces the formation of 2-substituted 4,5,6,7-tetrahydroindoles from 2-(2-alkynyl)cyclohexanones. To obtain reasonable yields of the products, the alkynyl moiety must carry an electron-withdrawing group.

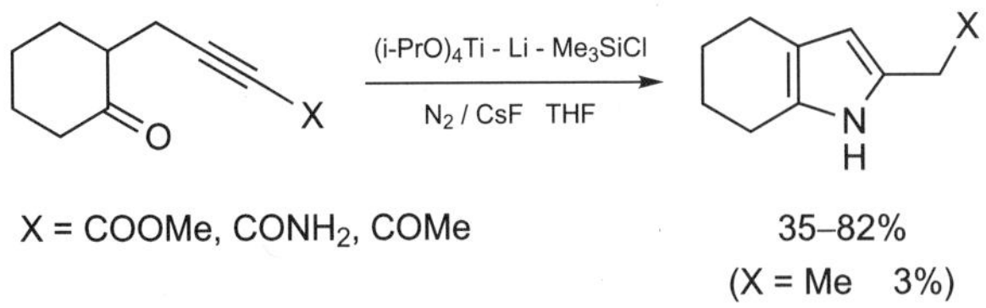
(i-PrO)4Ti - Li - Me3SiCl
N2 / CsF THF
X
O
N
H
X = COOMe, CONH2, COMe
35–82%
(X = Me 3%)

[1]Akashi, M., Nishida, M., Mori, M. *CL* 465 (1999).

Titanium(IV) chloride–samarium. 20, 381

Ketones.[1] Reductive acylation of ketones by nitriles is accomplished using $TiCl_4$–Sm in THF.

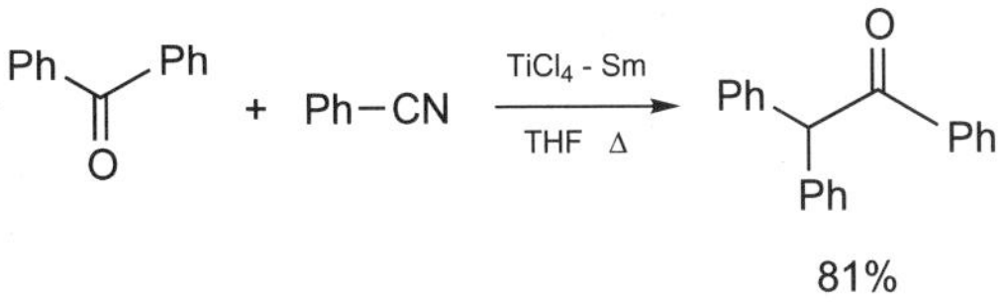
Ph
Ph
O
+ Ph−CN
TiCl4 - Sm
THF Δ
O
Ph
Ph
Ph
81%

[1]Zhou, L., Zhang, Y. *T* **56**, 2953 (2000).

Titanium(IV) chloride–tetrabutylammonium iodide.

Prenyl ether cleavage.[1] The cleavage shows preference to ethers possessing a proximal coordinative site to allow chelation control.

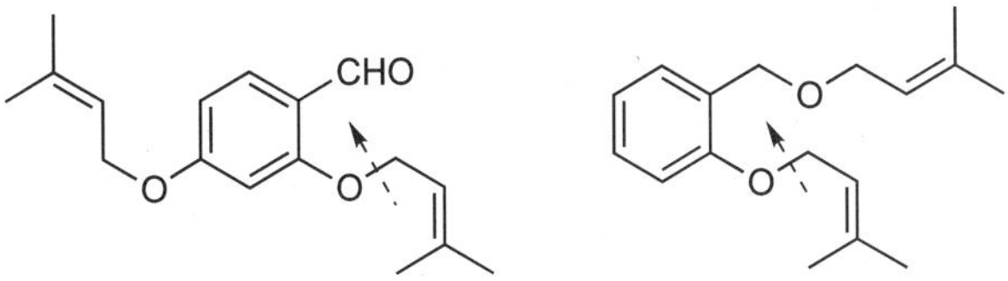
CHO
O
O
O
O

Reductive couplings.[2] $TiCl_4$–Bu_4NI appears to have similar reactivity as $TiCl_4$–Et_3N in effecting reductive coupling of ArCHO. Enolates generated from α-haloketones by this reagent undergo aldol reaction with aldehydes, providing predominantly the *syn*-isomer.

[1]Tsuritani, T., Shinokubo, H., Oshima, K. *TL* **40**, 8121 (1999).
[2]Tsuritani, T., Ito, S., Shinokubo, H., Oshima, K. *JOC* **65**, 5066 (2000).

Titanium(IV) chloride–zinc. 20, 381

Reductive couplings. Formation of 4,5-diarylimidazolidines from imines is readily effected.[1] The reagent system for coupling ArCHO to afford *syn*-hydrobenzoins contains TMEDA.[2,3] (The complex is prepared from $TiCl_4$, Zn, TMEDA, and $PbCl_2$. Note that an alternative complex for the same purpose is derived from $TiCl_4$, Mn, Me_3SiCl, and a Schiff base, but diastereoselectivity depends on the Schiff base.[4])

[1]Li, J., Wang, S., Hu, J., Chen, W. *TL* **40**, 1961 (1999).
[2]Li, T., Cui, W., Liu, J., Zhao, J., Wang, Z. *CC* 139 (2000).
[3]Oshiki, T., Kiriyama, T., Tsuchida, K., Takai, K. *CL* 334 (2000).
[4]Bandini, M., Cozzi, P.G., Morganti, S., Umani-Ronchi, A. *TL* **40**, 1997 (1999).

Titanium(II) halide–copper.

Reductive couplings.[1] Aliphatic and aromatic aldehydes undergo reductive dimerization on exposure to $TiBr_2$–Cu.

β-Hydroxy carbonyl compounds.[2] Aldol and Reformatsky reactions involving debrominative enolization of α-bromoketones and α-bromo thioesters with $TiCl_2$–Cu–*t*-BuCN are readily achieved. Acceptors are limited to aliphatic aldehydes because pinacol formation by aromatic aldehydes predominates under the conditions.

[1]Mukaiyama, T., Kagayama, A., Igarashi, K. *CL* 336 (2000).
[2]Mukaiyama, T., Kagayama, A., Igarashi, K., Shiina, I. *CL* 1157 (1999).

Titanium(IV) iodide.

Reductions. Chemoselective reduction of sulfoxides to sulfides[1] and α-diketones to α-ketols[2] employs the title reagent in MeCN at 0°. Aldehydes are dimerized to give 1,2-diols (*dl-* >> *meso-*) on exposure to TiI_4, with or without adding Cu.[3,4]

Aldol reactions.[5] Methoxyallene oxide forms a titanium enolate on treatment with TiI_4. Addition of aldehydes or acetals completes the aldol reactions.

[1]Shimizu, M., Shibuya, K., Hayakawa, R. *SL* 1437 (2000).
[2]Hayakawa, R., Sahara, T., Shimizu, M. *TL* **41**, 7939 (2000).
[3]Mukaiyama, T., Yoshimura, N., Igarashi, K. *CL* 838 (2000).
[4]Hayakawa, R., Shimizu, M. *CL* 724 (2000).
[5]Hayakawa, R., Shimizu, M. *OL* **2**, 4079 (2000).

Titanium tetraisopropoxide. **13,** 311–313; **14,** 311–312; **15,** 322; **16,** 339; **17,** 347–348; **18,** 363–364; **19,** 346–347; **20,** 381–382

β-Amino acid derivatives. (i-PrO)$_4$Ti plays a critical role in the Mannich-type condensation.[1]

	syn	:	anti
no additive	30	:	70
+ (i-PrO)$_4$Ti	92	:	8

Allylic displacements. In Pd(0)-catalyzed reaction, titanates generated from the pronucleophiles on treatment with (i-PrO)$_4$Ti serve adequately.[2]

Aldol reactions. A high degree of *syn*-diastereoselectivity is exhibited in the aldol reaction mediated by titanium(IV) alkoxides in the presence of α-hydroxy acids. With chiral α-hydroxy acids, asymmetric induction is observed.[3]

In conjunction with Ph$_3$P, (i-PrO)$_4$Ti promotes condensation of bromomethyl perfluoroalkyl ketones with aldehydes. Allylic alcohols are obtained owing to the reducibility of the ketone by (i-PrO)$_4$Ti in the Meerwein–Ponndorf fashion.[4]

[1]Kise, N., Ueda, N. *JOC* **64**, 7511 (1999).
[2]Poli, G., Giambastiani, G., Mordini, A. *JOC* **64**, 2962 (1999).
[3]Mahrwald, R. *OL* **2**, 4011 (2000).
[4]Shen, Y., Zhang, Y., Zhou, Y. *JCS(P1)* 1759 (1999).

Titanocene bis(triethyl phosphite). 20, 383

Desulfurative alkylation and acylation. Titanocene bis(triethyl phosphite) promotes desulfurative alkylation of allylic sulfides (dithioacetals) with *t*-alkyl halides.[1]

$Cp_2Ti[P(OEt)_3]_2$ 58%

$Cp_2Ti[P(OEt)_3]_2$ 73%

Dithioacetals $RCH(SPh)_2$ and vinylogues [i.e., RCH(SPh)CH=CHSPh] become carbanion equivalents $[RCH_2]^-$ when they are treated with $Cp_2Ti[P(OEt)_3]_2$ and Mg. Thus, a subsequent reaction with nitriles furnishes ketones.[2] Conjugated dienes and alkenylcyclopropanes are also obtained.[3]

$Cp_2Ti[P(OEt)_3]_2$ / Mg , mol. sieves ; 60%

$Cp_2Ti[P(OEt)_3]_2$ 53%

$Cp_2Ti[P(OEt)_3]_2$ R = H.... 80–85%

Cyclizations. Cyclization on elimination of a dithioacetal unit simultaneous with a thioester carbonyl by $Cp_2Ti[P(OEt)_3]_2$, leads to 2,3-dihydrothiophenes.[4] The analogous reaction of ω-alkanoyloxyalkanal bis(phenylthioacetals) gives ω-hydroxy ketones due to hydrolysis of the cyclic enol ether products.[5]

The cycloelimination is also applicable to dithioacetals with a remote C═CH$_2$ group,[6] apparently involving the release of [Cp$_2$Ti═CH$_2$] (a retro-Tebbe reaction).

[1]Takeda, T., Nozaki, N., Saeki, N., Fujiwara, T. *TL* **40**, 5353 (1999).
[2]Takeda, T., Taguchi, H., Fujiwara, T. *TL* **41**, 65 (2000).
[3]Takeda, T., Takagi, Y., Saeki, N., Fujiwara, T. *TL* **41**, 8377 (2000).
[4]Rahim, M.A., Fujiwara, T., Takeda, T. *SL* 1029 (1999).
[5]Rahim, M.A., Fujiwara, T., Takeda, T. *T* **56**, 763 (2000).
[6]Fujiwara, T., Kato, Y., Takeda, T. *H* **52**, 147 (2000).

Titanocene dicarbonyl. 19, 347–348; 20, 384

Intramolecular ene reaction.[1] Enynes and dienynes undergo this cycloisomerization in the presence of Cp$_2$Ti(CO)$_2$.

[1]Sturla, S.J., Kablaoui, N.M., Buchwald, S. L. *JACS* **121**, 1976 (1999).

Titanocene dichloride–diisobutylaluminum hydride.

Reductive cyclization.[1] On treatment with the combination of Cp$_2$TiCl$_2$ and *i*-Bu$_2$AlH, certain alkadienal acetals undergo cyclization as a result of attack by crotyltitanium species formed in situ.

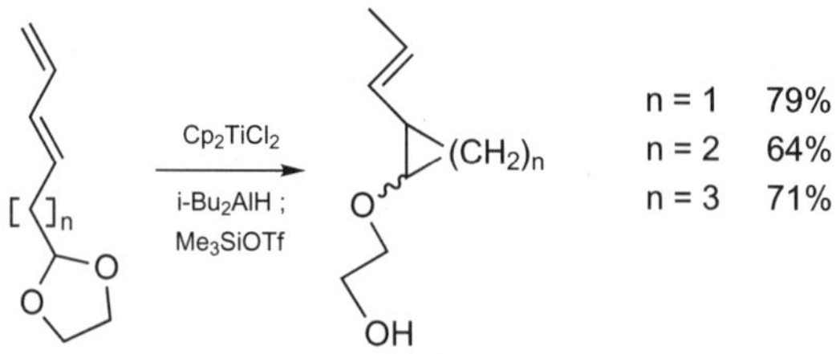

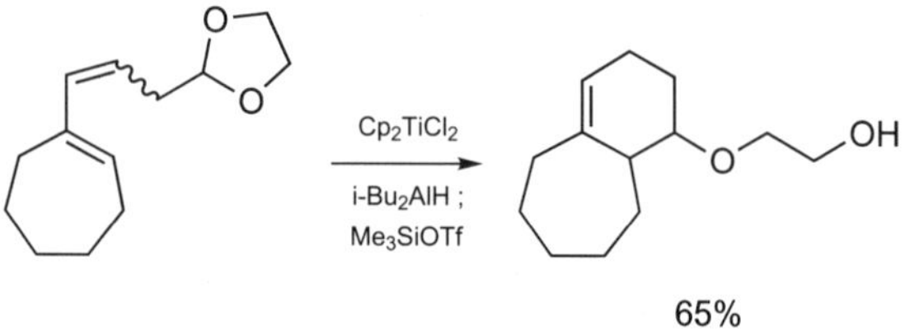

[1]Thery, N., Szymoniak, J., Moise, C. *EJOC* 1483 (2000).

Titanocene dichloride–manganese. 20, 384

Glycals.[1] This combination of reagents, which generates Cp_2TiCl, is effective for eliminations such as formation of glycals from per-*O*-acetylglycosyl bromides.

Reductive couplings.[2] Aldehydes are converted to trimethylsilyl ethers of 1,2-diols with Cp_2TiCl_2–Mn and Me_3SiCl.

Reductive cyclization.[3] Epoxides bearing a 5-pentenyl chain undergo cyclization.

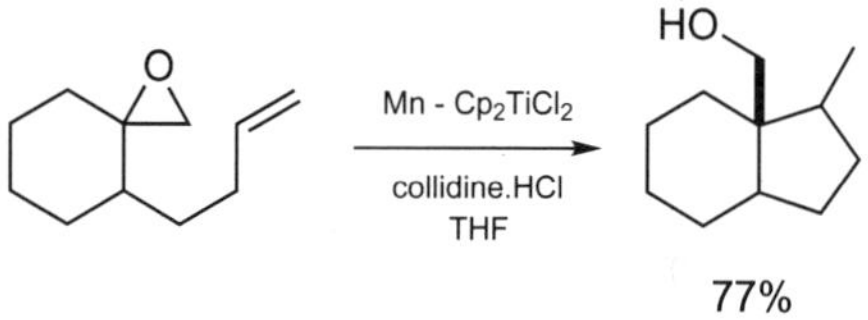

[1]Hanien, T., Krintel, S.L., Daasbjerg, K., Skrydstrup, T. *TL* **40**, 6087 (1999).
[2]Dunlap, M.S., Nicholas, K.M. *SC* **29**, 1097 (1999).
[3]Gansauer, A., Pierobon, M. *SL* 1357 (2000).

Titanocene dichloride–samarium.

Diaryl disulfides.[1] Arenesulfonyl derivatives (e.g., chlorides) undergo reductive coupling.

[1]Huang, Y., Guo, H., Zhang, Y., Wang, Y. *JCR(S)* 214 (1999).

Titanocene dichloride–zinc. 20, 384–385

Bicyclotetrahydrofurans.[1] A synthetic method for elaborating neolignans such as sesamin is by a formal 1,3-dipolar cycloaddition between an epoxide and a styrenic double bond.

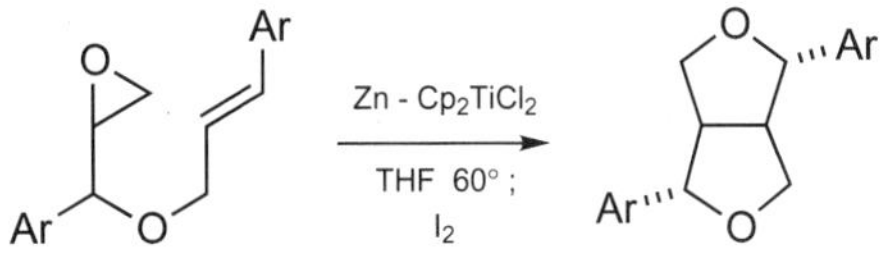

Zn - Cp_2TiCl_2 / THF 60°; I_2

[1]Rana, K.K., Guin, C., Roy, S.C. *TL* **41**, 9337 (2000).

Titanocene methylidenes.

Deoxygenation[1]. The Tebbe reagent deoxygenates amine oxides, sulfoxides, and selenoxides.

[1]Nicolaou, K.C., Koumbis, A.E., Snyder, S.A., Simonsen, K.B. *ACIEE* **39**, 2529 (2000).

***p*-Toluenesulfonylacetylene.**

Thiol protection.[1] As a Michael reaction donor, this reagent forms adducts with thiols. Regeneration of the thiols is by treatment with pyrrolidine in MeCN.

[1]Arjona, O., Iradier, F., Medel, R., Plumet, J. *JOC* **64**, 6090 (1999).

***p*-Toluenesulfonyl chloride.**

α-Chloroketones.[1] TsCl can serve as a chlorinating agent for ketones. Treatment of a ketone with LDA and then TsCl in THF at 0° gives rise to the α-chloroketone. A polymer-bound TsCl is also useful.

Alkenyl tolyl sulfones.[2] A convenient synthesis involves reaction of alkenylzirconocene chlorides (derived from 1-alkynes) with TsCl.

O-Tosylation. The use of TMEDA as base makes tosylation of alcohols with TsCl faster and prevents formation of alkyl chlorides.[3] Alternatively, TMEDA or Me_3NHCl[4] in catalytic amount (Et_3N as base) also shows an advantage.

[1]Brummond, K.M., Gesenberg, K.D. *TL* **40**, 2231 (1999).
[2]Duan, D.-H., Huang, X. *SL* 317 (1999).
[3]Yoshida, Y., Shimonishi, K., Sakakura, Y., Okada, S., Aso, N., Tanabe, Y. *S* 1633 (1999).
[4]Yoshida, Y., Sakakura, Y., Aso, N., Okada, S., Tanabe, Y. *T* **55**, 2183 (1999).

***p*-Toluenesulfonyl cyanide.**

β-Chloro sulfones.[1] TsCN adds to alkenes in the presence of $TiCl_4$ to afford β-chloro sulfones.

[1]Morgan, P.E., McCague, R., Whiting, A. *TL* **40**, 4857 (1999).

β-(*p*-Toluenesulfonylhydrazono) phosphonates.

Pyrazoles.[1] A synthesis of polysubstituted pyrazoles is completed in one step by the Emmons–Wadsworth reaction of these reagents with aldehydes.

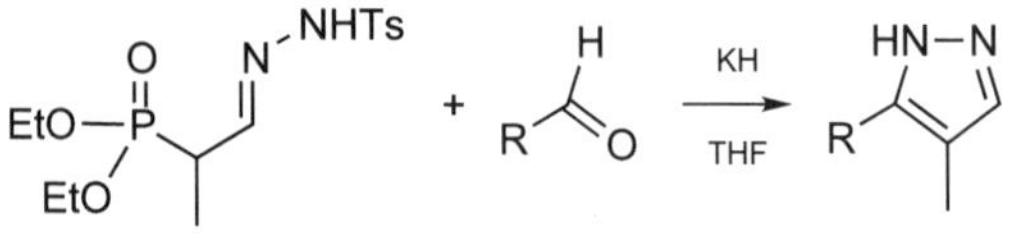

[1]Almirante, N., Benicchio, A., Cerri, A., Fedrizzi, G., Marazzi, G., Santagostino, M. *SL* 299 (1999).

2,8,9-Trialkyl-1-phospha-2,5,8,9-tetraazabicyclo[3.3.3]undecanes. 19, 370

trans-Stilbene epoxides.[1] Stereoselective deoxygenative dimerization of aromatic aldehydes are observed with the trimethyl base (**1**, R = Me) at room temperature.

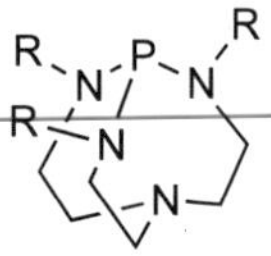

(**1**)

Condensation reactions. The triisopropyl base is an efficient promoter for the nitroaldol (Henry) reaction,[2] and condensation between alkanonitriles and carbonyl compounds.[3]

Alcohol–ester interchange. The trimethyl base can be used as a catalyst for acylation of alcohols (esters, and particularly enol esters, as acyl donors) as well as deacylation of esters.[4] (Desilylation of TBS ethers is also reported.[5])

[1]Liu, X., Verkade, J.G. *JOC* **65**, 4560 (2000).
[2]Kisanga, P.B., Verkade, J.G. *JOC* **64**, 4298 (1999).
[3]Kisanga, P., McLeod, D., D'Sa, B., Verkade, J.G. *JOC* **64**, 3090 (1999).
[4]Ilankumaran, P., Verkade, J.G. *JOC* **64**, 3086 (1999).
[5]Yu, Z., Verkade, J.G. *JOC* **65**, 2065 (2000).

Triallylborane. 20, 386–387

Reductive allylation.[1] *N*-Allylanilines are formed when nitroarenes are heated with triallylborane in toluene. The diallylated amines are minor products.

Homoallylamines. A synthesis of homoallylamines from nitriles consists of consecutive treatment with diisobutylaluminum hydride and triallylamine.[2] Lactams undergo deoxygenative diallylation.[3]

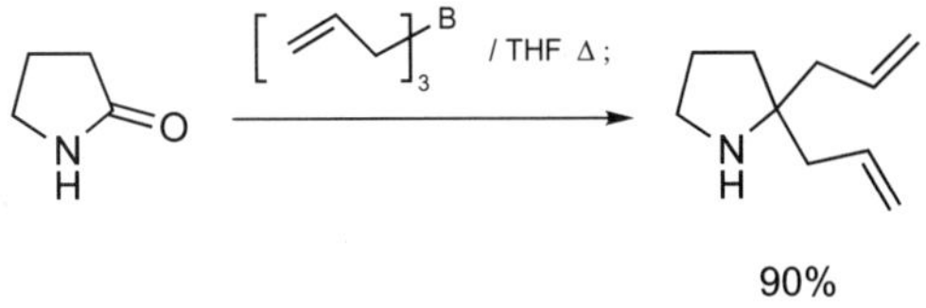

90%

[1]Bubnov, Yu.N., Pershin, D.G., Ignatenko, A.V., Gurskii, M.E. *MC* 108 (2000).
[2]Watanabe, K., Kuroda, S., Yokoi, A., Ito, K., Itsuno, S. *JOMC* **581**, 103 (1999).
[3]Bubnov, Yu.N., Pastukhov, F.V., Yampolsky, I.V., Ignatenko, A.V. *EJOC* 1503 (2000).

Triarylbismuthine dichlorides.

Arylation.[1] Enolates derived from α,β-unsaturated carbonyl compounds and related systems are arylated at room temperature with Ar_3BiCl_2 to provide the deconjugated products.

[1]Arnauld, T., Barton, D.H.R., Normant, J.-F., Doris, E. *JOC* **64**, 6915 (1999).

1,5,7-Triazabicyclo[4.4.0]dec-5-ene.

Epoxidation.[1] This bicyclic guanidine (**1**) promotes epoxidation of electron-deficient alkenes with *t*-BuOOH at room temperature.

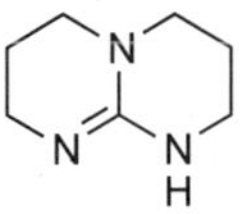

(**1**)

[1]Genski, T., Macdonald, G., Wei, X., Lewis, N., Taylor, R.J.K. *SL* 795 (1999).

Tributoxysilyl hydrosulfide.

Defunctionalization.[1] Methoxymethyl ethers are hydrogenolyzed on heating with $(t\text{-BuO})_3SiSH$.

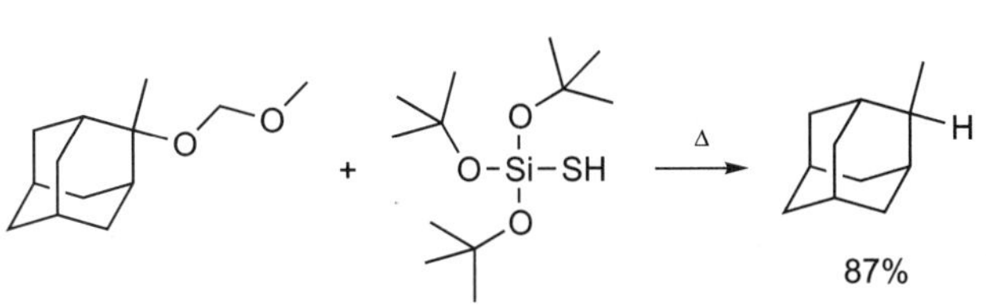

87%

[1]Dang, H.-S., Franchi, P., Roberts, B.P. *CC* 498 (2000).

Tributylphosphine. 20, 387–388

Transacylation.[1] 2,2,2-Trihaloethyl esters are converted to other esters via reductive fragmentation, formation of phosphonium carboxylates, and alcoholysis of the activated esters. For the analogous amidation, hexamethylphosphoramide is used.

Reduction of disulfides. Optimum conditions are established for the reduction of disulfides to thiols[2] using Bu_3P–H_2O in THF. The transformation of alcohols to *S*-xanthates in one step is accomplished with a mixture of Bu_3P and $(i\text{-PrOCSS})_2$.[3]

Dimerization.[4] Activated alkenes are dimerized in the presence of Bu_3P under pressure.

CN —(Bu_3P, 50°)→ CN, CN

1,3-Dipoles.[5] 2-Alkynoic esters are presented as zwitterionic conjugated esters in which nucleophilic and electrophilic sites are located at the α- and γ-carbon atoms, respectively, when they are treated with Bu_3P and dipolarophiles. For example, trapping with aldimines leads to 2,5-disubstituted 2,5-dihydropyrrole-3-carboxylic esters.

R, COOEt + TsN, R′ —(Bu_3P, PhH)→ Ts, N, R, R′, COOEt

Nitriles.[6] A modified Mitsunobu reaction that transforms alcohols to nitriles employs Bu_3P and *N,N,N′,N′*-tetramethylazodicarboxamide as activator and acetone cyanohydrin as pronucleophile.

[1]Hans, J.J., Driver, R.W., Burke, S.D. *JOC* **65**, 2114 (2000).
[2]Ayers, J.T., Anderson, S.R. *SC* **29**, 351 (1999).
[3]Gueyrard, D., Tatibouet, A., Gareau, Y., Rollin, P. *OL* **1**, 521 (1999).
[4]Jenner, G. *TL* **41**, 3091 (2000).
[5]Xu, Z., Lu, X. *TL* **40**, 549 (1999).
[6]Tsunoda, T., Uemoto, K., Nagino, C., Kawamura, M., Kaku, H., Ito, S. *TL* **40**, 7355 (1999).

Tributyl[2-(trimethylsilyl)ethoxymethoxymethyl]stannane.

Hydroxymethyl anion equivalent.[1] After undergoing Sn–Li exchange (with BuLi), reagent **1** becomes reactive toward various electrophiles. The (trimethylsilyl)-ethoxymethyl group of the products can be removed any time afterward.

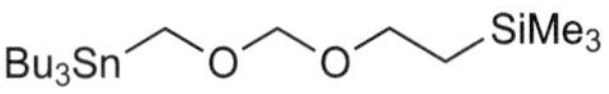

Bu_3Sn–O–O–$SiMe_3$

(1)

[1]Fernandez-Megia, E., Ley, S.V. *SL* 455 (2000).

Tributyltin hydride. 13, 316–319; **14,** 312–318; **15,** 325–333; **16,** 343–350; **17,** 351–361; **18,** 368–371; **19,** 352–353; **20,** 389–391

Reduction. Besides the reduction of carbonyl compounds with Bu_3SnH to alcohols in hydroxylic solvents (e.g., MeOH),[1] that of hydroxy ketones and diketones in dichloromethane[2] in the presence of $BF_3{\cdot}OEt_2$ is particularly interesting due to high stereoselectivity.

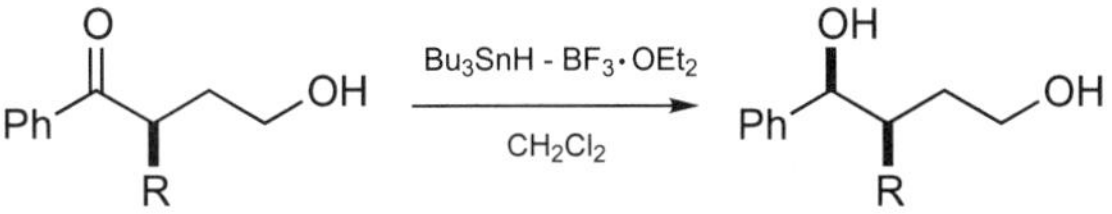

Bu_3SnH - $BF_3{\cdot}OEt_2$ / CH_2Cl_2

Aromatic aldehydes bearing an *o*-alkenyl substituent is reduced selectively when Me_3Al is added.[3] This selectivity is a manifest of chelative activation.

Reductive amination.[4] A method for the conversion of carbonyl compounds to amines is by reaction with Bu_3SnH and ammonium salts (derived from primary or secondary amines) in DMF at room temperature.

Hydrostannylation.[5] Regioselective hydrostannylation of several types of alkynes (to yield mainly α-stannyl derivatives) is catalyzed by a molybdenum complex.

[1]Kamimura, K., Wada, M. *TL* **40**, 9059 (1999).
[2]Ooi, T., Uraguchi, D., Morikawa, J., Maruoka, K. *OL* **2**, 2015 (2000).
[3]Asao, N., Shimada, T., Yamamoto, Y. *TL* **41**, 9533 (2000).
[4]Suwa, T., Sugiyama, E., Shibata, I., Baba, A. *SL* 556 (2000).
[5]Kazmaier, U., Schauss, D., Pohlmann, M. *OL* **1**, 1017 (1999).

Tributyltin hydride–2,2′-azobis(isobutyronitrile). 19, 353–357; **20,** 391–394

Defunctionalizations. Heteroarenesulfonyl groups are reductively severed without affecting a geminal C—F bond.[1] A clean deoxygenation of alcohols via reduction of the corresponding thionocarbonates calls for binding the tin hydride to a polymer and using trimethoxysilane to recycle the spent reagent.[2]

A method for synthesis of chiral cyclohexene derivatives from 4-substituted cyclohexyl halides involves derivatization to the Grignard reagent, reaction with menthyl (*S*)-2-bromophenylsulfinate, and then treatment with Bu_3SnH, AIBN, and methylaluminum diphenoxide with irradiation by a sun lamp.[3]

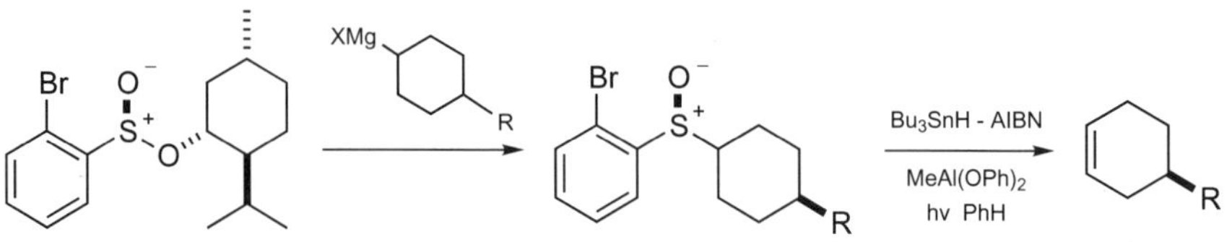

Group-transfer reactions. A *Si*- or *P*-linked aryl group separated by five bonds to a carbon radical has the tendency to migrate to the carbon center. This reaction pattern can be exploited in a synthesis of 2-hydroxyalkylbiaryls from 2-bromobenzyl alcohols.[4,5]

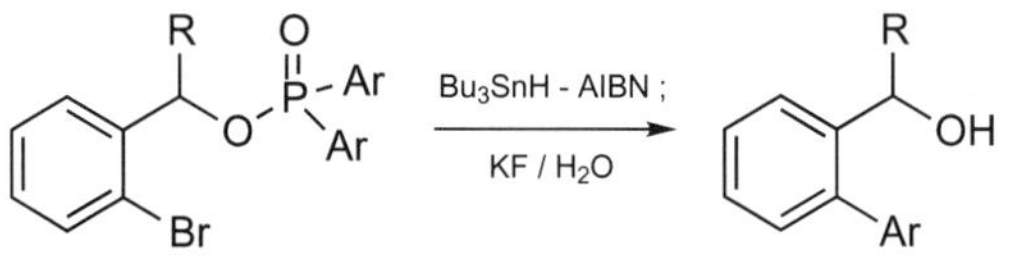

This process is also applicable to arylation of secondary aliphatic radicals,[6] and a carbon radical generated from O—S bond homolysis followed by radical transfer can be trapped.[7]

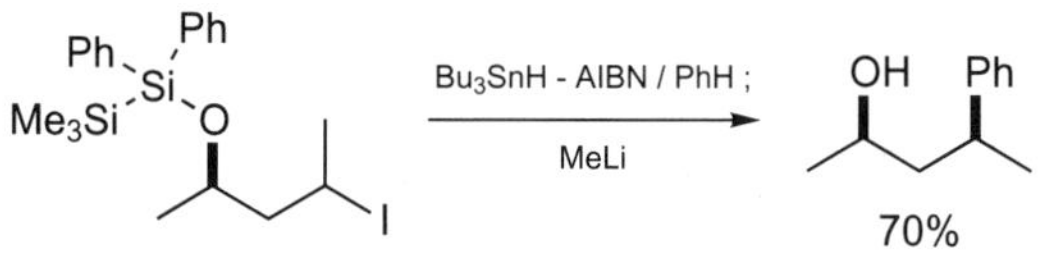

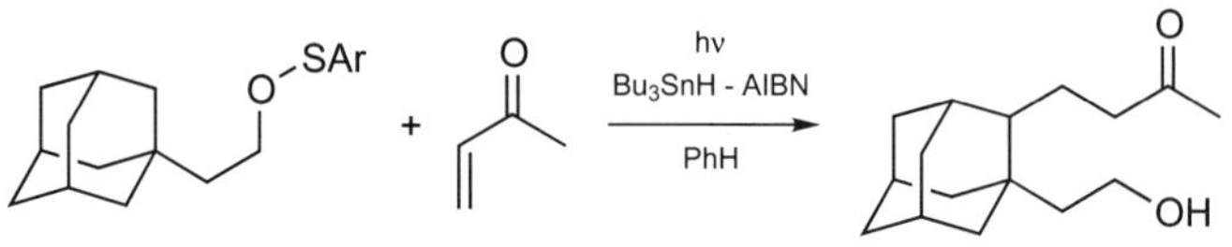

1,2-Migration of an acyl group to a nitrogen radical generated from α-azido-β-keto esters to form amides(lactams)[8] is a pathway not observed in ionic reactions.

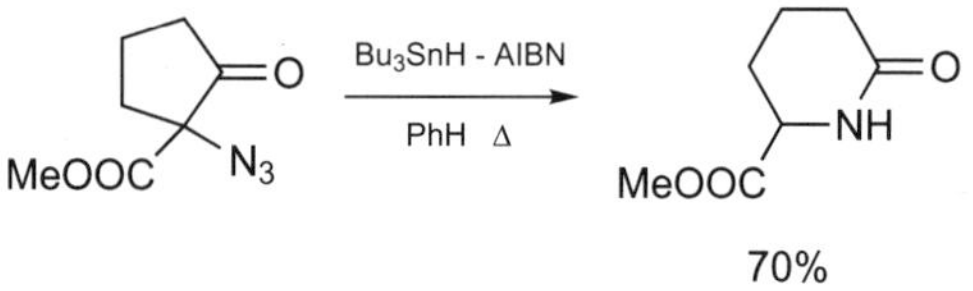

Amidoyl radicals are generated from amidoyl selenides. In the presence of an electron-rich alkene such as enol derivative, addition occurs.[9]

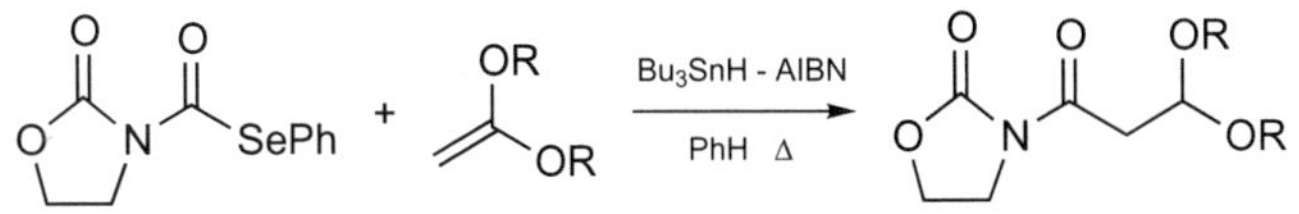

Reductive cyclization. The Bu_3Sn radical adds to *O*-alkylaldoximes and the ensuing carbon radicals are liable to cyclization with a carbonyl group[10] or conjugated system.[11]

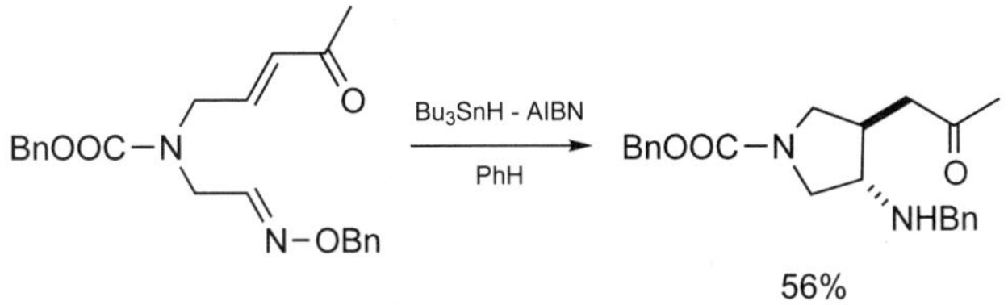

Cyclizations. Bromoarenes form radicals that can be exploited in synthesis, including cyclization routes to aporphines–indolo[2,1-*a*]isoquinolines,[12] and protoberberine–pavine alkaloids.[13]

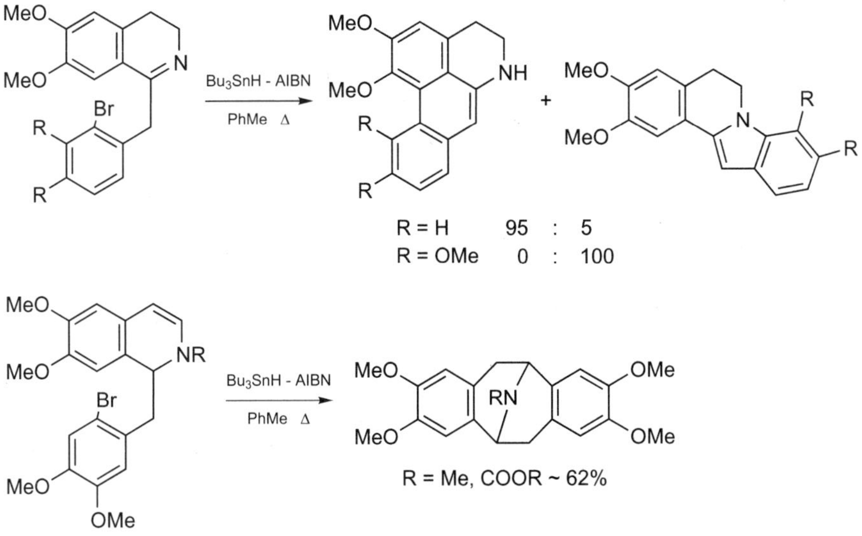

Ring closure is expected for a radical precursor set with an unsaturation four bonds away when such a compound is treated with Bu_3SnH–AIBN. The versatility of such cyclizations is derived from allowance of many varieties of substitution patterns and heteroatoms between the reactive centers. Furthermore, as shown in a synthesis of 2,4-disubstituted pyrrolidines,[14] diastereoselectivity may be controlled in certain cases.

PhSe, Ph, N, R — Bu_3SnH - AIBN, PhH hv → Ph, N, R + Ph, N, R

R = H 38 : 1

R = P(O)Ph_2 1 : 24

Based on this method, expedient access to conjugated exocyclic dienes,[15] γ-sultams,[16] indoles,[17] and α-oximino-γ-lactones[18] has been devised.

Br, $OCPh_3$, N, O, O — Bu_3SnH - AIBN, PhMe Δ → OH, N, O, O

80%

The regioselectivity of cyclization through bond formation at either the α- or β-position of a pyrrole depends on the electronic nature of the *N*-substituent.[19]

MeO, Br, R, N, N, O, SEM — Bu_3SnH - AIBN, PhMe Δ → MeO, RN, N, O, SEM + MeO, R, N, N, O, SEM

R = Me 43%

R = COOMe 15% 32%

Tandem cyclization is valued for synthetic efficiency. As illustrated, approaches to the BCD-ring segment of steroids[20] and a precursor of (+)-paniculatine[21] are worth mentioning.

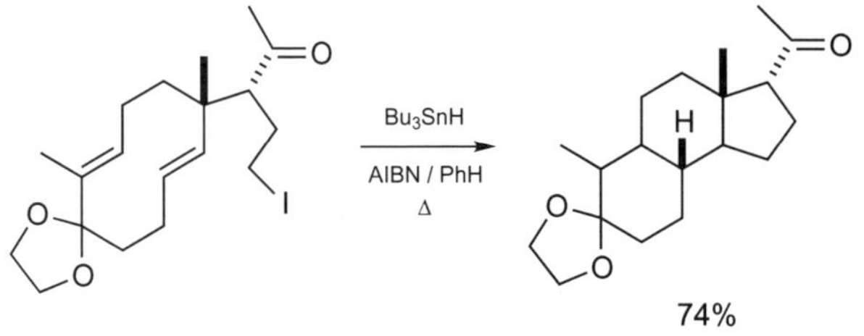

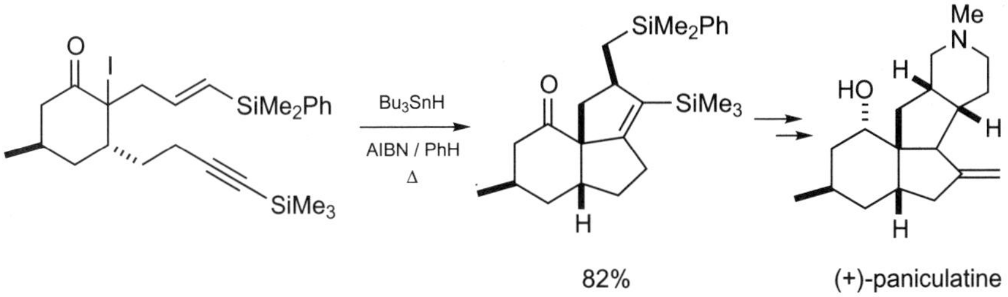

Bu3SnH
AIBN / PhH
Δ

74%

Bu3SnH
AIBN / PhH
Δ

82%

(+)-paniculatine

Also notable is the formation of a bicyclo[3.1.1]heptane skeleton from an acyclic 1,7-diyne.[22]

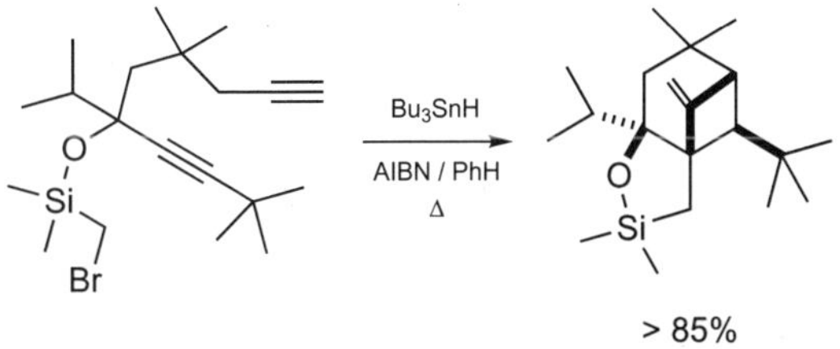

Bu3SnH
AIBN / PhH
Δ

> 85%

A process involving cyclization–fragmentation of 2-(ω-bromoalkyl)-1,1-dicyanocyclopropanes[23] is a radical alternative to the Thorpe–Ziegler reaction of dinitriles. Thus, 2-aminocyclooctenenitrile is formed in 56% yield.

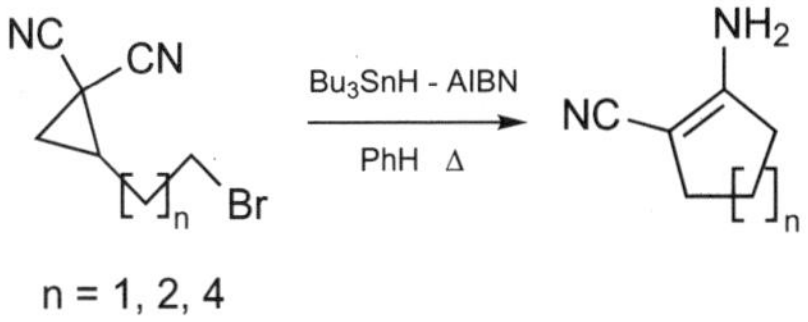

Bu3SnH - AIBN
PhH Δ

n = 1, 2, 4

[1]Wnuk, S.F., Rios, J.M., Khan, J., Hsu, Y.-L. *JOC* **65**, 4169 (2000).
[2]Boussagnet, P., Delmond, B., Dumartin, G., Pereyre, M. *TL* **41**, 3377 (2000).
[3]Imboden, C., Villar, F., Renaud, P. *OL* **1**, 873 (1999).
[4]Clive, D.L.J., Kang, S. *TL* **41**, 1315 (2000).
[5]Studer, A., Bossart, M., Vasella, T. *OL* **2**, 985 (2000).

[6]Amrein, S., Bossart, M., Vasella, T., Studer, A. *JOC* **65**, 4281 (2000).
[7]Petrovic, G., Cekovic, Z. *OL* **2**, 3769 (2000).
[8]Benati, L., Nanni, D., Sangiorgi, C., Spagnolo, P. *JOC* **64**, 7836 (1999).
[9]Keck, G.E., Grier, M.C. *SL* 1657 (1999).
[10]Naito, T., Nakagawa, K., Nakamura, T., Kasei, A., Ninomiya, I., Kiguchi, T. *JOC* **64**, 2003 (1999).
[11]Naito, T., Fukumoto, D., Takebayashi, K., Kiguchi, T. *H* **51**, 489 (1999).
[12]Orito, K., Uchiito, S., Satoh, Y., Tatsuzawa, T., Harada, R., Tokuda, M. *OL* **2**, 307 (2000).
[13]Orito, K., Satoh, Y., Nishizawa, H., Harada, R., Tokuda, M. *OL* **2**, 2535 (2000).
[14]Besev, M., Engman, L. *OL* **2**, 1589 (2000).
[15]Sha, C.-K., Zhan, Z.-P., Wang, F.-S. *OL* **2**, 2011 (2000).
[16]Leit, S.M., Paquette, L.A. *JOC* **64**, 9225 (1999).
[17]Tokuyama, H., Yamashita, T., Reding, M.T., Kaburagi, Y., Fukuyama, T. *JACS* **121**, 3791 (1999).
[18]Clive, D.L.J., Subedi, R. *CC* 237 (2000).
[19]Escolano, C., Jones, K. *TL* **41**, 8951 (2000).
[20]Tomida, S., Doi, T., Takahashi, T. *TL* **40**, 2363 (1999).
[21]Sha, C.-K., Lee, F.-K., Chang, C.-J. *JACS* **121**, 9875 (1999).
[22]Bogen, S., Fensterbank, L., Malacria, M. *JOC* **64**, 819 (1999).
[23]Curran, D.P., Liu, W. *SL* 117 (1999).

Tributyltin hydride–triethylborane. 15, 333; **16,** 350; **17,** 363–364; **18,** 372; **20,** 394

Cyclization.[1] Addition of Bu_3SnH to β-allenyl-*O*-benzoyl oximes also causes cyclization.

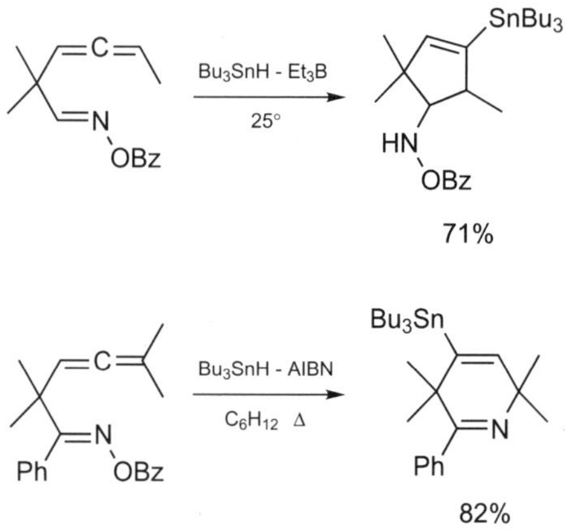

Bu_3SnH - Et_3B, 25° → 71%

Bu_3SnH - AIBN, C_6H_{12} Δ → 82%

Radical addition. Conjugate addition of primary radical to acrylic acid is realizable.[2]

[1]Departure, M., Diwok, J., Grimaldi, J., Hatem, J. *EJOC* 275 (2000).
[2]Wu, B., Avery, B.A., Avery, M.A. *TL* **41**, 3797 (2000).

Tricarbonyl(pentamethylcyclopentadienyl)rhenium.

Borylation.[1] Under photochemical conditions, alkanes are borylated at the terminal carbon on reaction with bis(pinacolato)diboron, with $Cp^*Re(CO)_3$ as a catalyst and under a CO atmosphere.

[1]Chen, H., Hartwig, J.F. *ACIEE* **38**, 3391 (1999).

Trichloroacetonitrile.

Hydroxyl protection.[1] Alcohols combine with Cl_3CCN in a DBU-catalyzed reaction. Three sets of conditions are available for the cleavage of trichloroacetimidates, using DBU in MeOH, $TsOH \cdot H_2O$ in a mixture of dichoromethane and MeOH, or $Zn–NH_4Cl$ in refluxing ethanol.

Acid chlorides.[2] Sensitive acid chlorides can be prepared from carboxylic acids by treatment with $Ph_3P–Cl_3CCN$ at room temperature.

[1]Yu, B., Yu, H., Hui, Y., Han, X. *SL* 753 (1999).
[2]Jang, D.O., Park, D.J., Kim, J. *TL* **40**, 5323 (1999).

Trichloronitromethane.

Dissulfides.[1] Treatment of thiols with Cl_3CNO_2–NaOEt in MeCN leads to disulfides. *S*-Nitrosothiols are the intermediates.

[1]Demir, A.S., Igdir, A.C., Mahasneh, A.S. *T* **55**, 12399 (1999).

Trichlorosilane. 18, 373; 19, 360

Reduction. Carbonyl compounds are reduced by Cl_3SiH that is activated by *N*-formylpyrrolidine.[1] Previously, DMF had been used.

Radical reduction (photochemical conditions) with this reagent is subject to alkoxy-direction[2] when the proper functionality is present. Thus, α-ketols give predominantly *trans*-1,2-diols, with a *trans/cis* ratio up to 134 : 1 in one case has been observed.

[1]Iwasaki, F., Onomura, O., Mishima, K., Maki, T., Matsumura, Y. *TL* **40**, 7507 (1999).
[2]Enholm, E.J., Schulte, J.P. *JOC* **64**, 2610 (1999).

Triethylborane. 20, 395

Imine adducts.[1] Stable adducts $RCH{=}N(BEt_3)H$ are formed from $RCH{=}NSiMe_3$ by treatment with triethylborane.

N-Ethylation.[2] Primary amines are converted to RNHEt in two steps: *N*-Benzoyloxylation with dibenzoyl peroxide and reaction with triethylborane.

Alkylative amination.[3] Aldimines RCH=NOBn are susceptible to free radical addition. Thus, a mixture of RCHO, $BnONH_2$, and R′I react in the presence of $BF_3 \cdot OEt_2$ and Et_3B to afford RR′CHNHOBn.

Radical reactions.[3] In the presence of air (or oxygen), Et_3B promotes radical formation. A tin-free cyclization of ω-iodo carbonyl compounds is effected.[4] Regioselective arylation can take advantage of 1,4-aryl migration from tin to a carbon radical center.[5]

α-Hydroxyalkylation at C-2 of THF is conveniently accomplished with RCHO.[6]

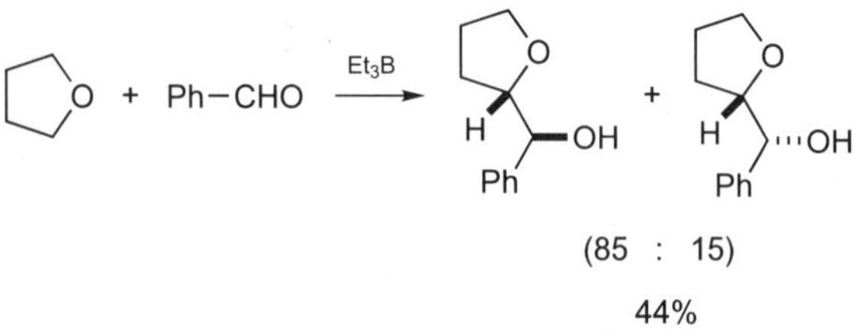

Substitution of the nitro group of nitroarenes with an alkyl residue on reaction with a trialkylborane/*t*-BuOK combination can proceed in good yields.[7]

[1]Chen, G.-M., Brown, H.C. *JACS* **122**, 4217 (2000).
[2]Phanstiel, O., Wang, Q.X., Powell, D.H., Ospina, M.P., Leeson, B.A. *JOC* **64**, 803 (1999).
[3]Miyabe, H., Yamakawa, K., Yoshioka, N., Naito, T. *T* **55**, 11209 (1999).
[4]Devin, P., Fensterbank, L., Malacria, M. *TL* **40**, 5511 (1999).
[5]Wakabayashi, K., Yorimitsu, H., Shinokubo, H., Oshima, K. *OL* **2**, 1899 (2000).
[6]Yoshimitsu, T., Tsunoda, M., Nagaoka, H. *CC* 1745 (1999).
[7]Shifman, A., Palani, N., Hoz, S. *ACIEE* **39**, 944 (2000).

Trifluoroacetic acid, TFA. 14, 322–323; **15,** 338–339; **18,** 375–376; **20,** 395–396

Cleavage of p-methoxybenzyl ethers.[1] Simple exposure to CF_3COOH effects the cleavage of the ether group at C-5 of furanose derivatives while retaining the primary ether at C-6.

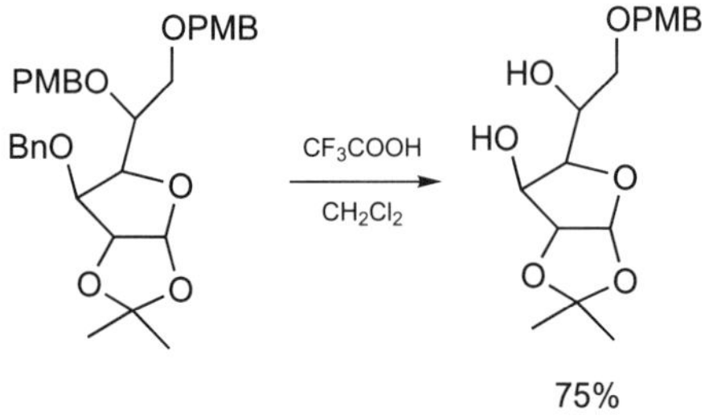

Isomerization. The Baylis–Hillman adducts from aromatic aldehydes and acrylic esters undergo stereoselective isomerization to (*E*)-2-hydoxymethylcinnamic esters.[2] Adducts derived from *o*-nitroaraldehydes behave differently due to intervention of an intramolecular redox reaction, leading eventually to *N*-oxides of 4-hydroxyquinoline-3-carboxylic esters.[3]

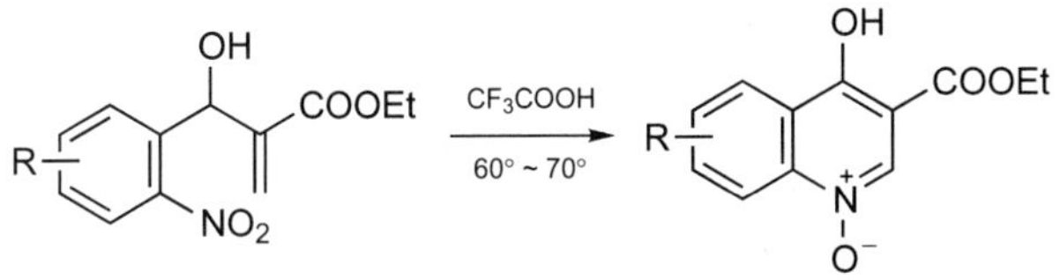

Simmons–Smith reaction.[4] A dramatic acceleration of the cyclopropanation by CF_3COOH is probably due to formation of a more reactive species $CF_3COOZnCH_2I$.

[1]Bouzide, A., Sauve, G. *TL* **40**, 2883 (1999).
[2]Kim, H.S., Kim, T.Y., Lee, K.Y., Chung, Y.M., Lee, H.J., Kim, J.N. *TL* **41**, 2613 (2000).
[3]Kim, J.N., Kim, T.Y., Lee, K.Y., Kim, H.S., Kim, T.Y. *OL* **2**, 343 (2000).
[4]Yang, Z., Lorenz, J.C., Shi, Y. *TL* **39**, 8621 (1998).

Trifluoroacetic anhydride, TFAA. 18, 376–377; **19,** 361; **20,** 396–397

Dehydration. Endocyclic enecarbamates are prepared from the corresponding lactams via reduction and subsequent dehydration with TFAA–2,6-lutidine.[1]

Regiochemically divergent lactonization processes of a hydroxynaphthoquinone derivative induced by TFAA and Me_3SiOTf are observed.[2]

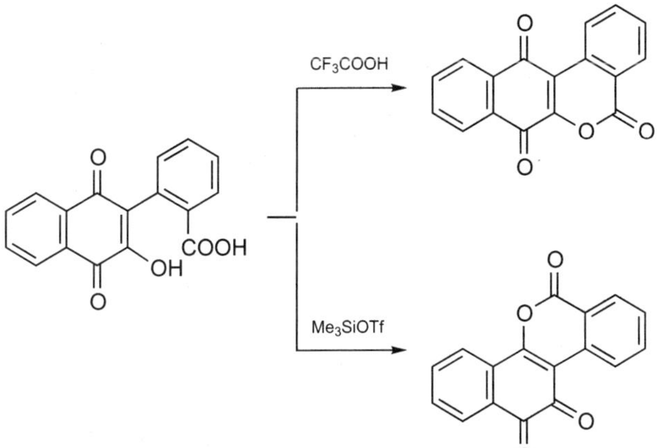

Trifluoromethyl α-aminoalkyl ketones.[3] The Dakin–West reaction using TFAA–pyridine on *N*-substituted α-amino acids affords the fluorinated ketones.

Rearrangement. 2-Pyrrolidinemethanol derivatives undergo stereoselective ring expansion to give piperidin-3-ols.[4] Carboxylic acids are converted to nitriles[5] with one less carbon by TFA–TFAA–$NaNO_2$.

Nitrodeboration.[6] *ipso*-Substitution of arylboronic acids to give nitroarenes is accomplished with TFAA and NH_4NO_3.

Reductive functionalization of (arylseleninyl)acetates.[7] When the Pummerer-type rearrangement of $PhSe(O)CH_2COOR$ is carried out in the presence of a reactive silane (e.g., allylsilanes, Me_3SiN_3) carbon chain homologation or functionalization of the esters results.

[1]Oliveira, D.F., Miranda, P.C.M.L., Correia, C.R.D. *JOC* **64**, 6646 (1999).
[2]Qabaja, G., Perchellet, E.M., Perchellet, J.-P. *TL* **41**, 3007 (2000).
[3]Kawase, M., Hirabayashi, M., Kumakura, H., Saito, S., Yamamoto, K. *CPB* **48**, 114 (2000).
[4]Cossy, J., Dumas, C., Pardo, D.G. *EJOC* 1693 (1999).
[5]Smushkevich, Y.I., Smushkevich, V.Y., Usorov, M.I. *JCR(S)* 1727 (1999).
[6]Salzbrunn, S., Simon, J., Prakash, G.K.S., Petasis, N.A., Olah, G.A. *SL* 1485 (2000).
[7]Shimada, K., Kikuta, Y., Koganebuchi, H., Yonezawa, F., Aoyagi, S., Takikawa, Y. *TL* **41**, 4637 (2000).

N-(Trifluoroacetyl)succinimide.

Trifluoroacetylation.[1] Alcohols, phenols, and amines are acylated by the title reagent (14 examples, 83–99%).

[1]Katritzky, A.R., Jang, B., Qiu, G., Zhang, Z. *S* 55 (1999).

2,2,2-Trifluoroethyl carbamates.

Ureas.[1] The carbamates $RNHCOOCH_2CF_3$ are obtained by an electrochemical reaction of primary amides ($RCONH_2$) in CF_3CH_2OH. They undergo aminolysis to afford unsymmetrical ureas.

[1]Matsumura, Y., Satoh, Y., Onomura, O., Maki, T. *JOC* **65**, 1549 (2000).

Trifluoromethanesulfonic acid (triflic acid). **14,** 323–324; **15,** 339; **18,** 377; **19,** 362–363; **20,** 398–399

Allylation.[1] Triflic acid is capable of catalyzing the reaction of allyltributylstannane with aldehydes (not ketones) in water.

Michael reaction.[2] For conducting a Michael reaction of β-ketoesters with conjugated esters and ketones at room temperature under solvent-free conditions, triflic acid is useful.

Cyclizations. A stereocontrolled synthesis of trisubstituted tetrahydropyrans by condensation of homoallylic alcohols with aldehydes is developed.[3] Treatment of THP ethers derived from unsaturated alcohols with triflic acid leads to oxygen heterocycles.[4]

CF_3SO_3H

HO

65%

Dealkylation.[5] Protodemethylation is the key to functionalization of an unactivated ethyl group in the following reaction sequence.

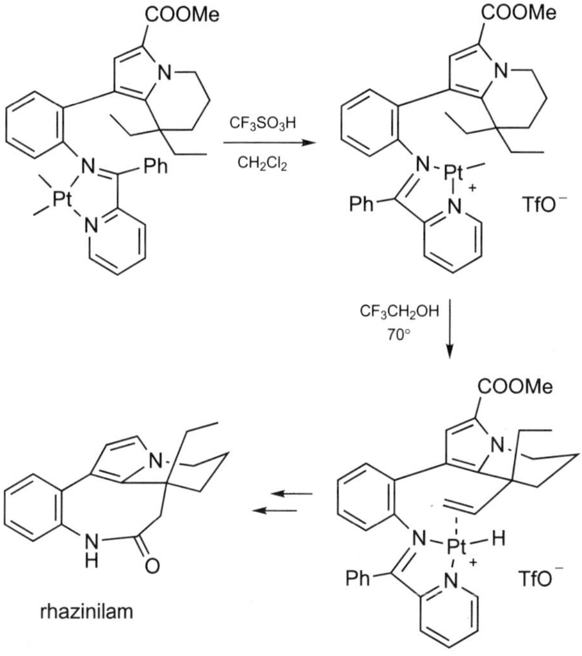

[1]Loh, T.-P., Xu, J. *TL* **40**, 2431 (1999).
[2]Kotsuki, H., Arimura, K., Ohishi, T., Maruzasa, R. *JOC* **64**, 3770 (1999).
[3]Cloninger, M.J., Oveman, L.E. *JACS* **121**, 1092 (1999).
[4]Dixon, D.J., Ley, S.V., Tate, E.W. *JCS(P1)* 1829 (2000).
[5]Johnson, J.A., Sames, D. *JACS* **122**, 6321 (2000).

Trifluoromethanesulfonic anhydride (triflic anhydride). 13, 324–325; **14,** 324–326; **15,** 339–340; **16,** 357–358; **18,** 377–378; **19,** 363–365; **20,** 399

Nitrile oxide generation.[1] *O*-Silylated hydroxamic acids are identified as stable precursors of nitrile oxides, by virtue of their susceptibility to transformation by Tf_2O–Et_3N.

1,3,4-Oxadiazoles.[2] Tf_2O–pyridine is a mild reagent for the cyclodehydration of diacylhydrazines.

[1]Muri, D., Bode, J.W., Carreira, E.M. *OL* **2**, 539 (2000).
[2]Liras, S., Allen, M.P., Segelstein, B.E. *SC* **30**, 437 (2000).

Trifluoromethanesulfonyl azide.

Diazocarbonyl compounds.[1] The title reagent is particularly useful for the introduction of an α-diazo group to α-nitro carbonyl compounds.

[1]Charette, A.B., Wurz, R.P., Ollevier, T. *JOC* **65**, 9252 (2000).

3-(Trifluoromethanesulfonyloxy)-3-trifluoromethylpropeniminium triflate.

2-Trifluoromethylquinolines.[1] Reagent **1** transforms arylamines into 2-trifluoromethylquinolines in one step.

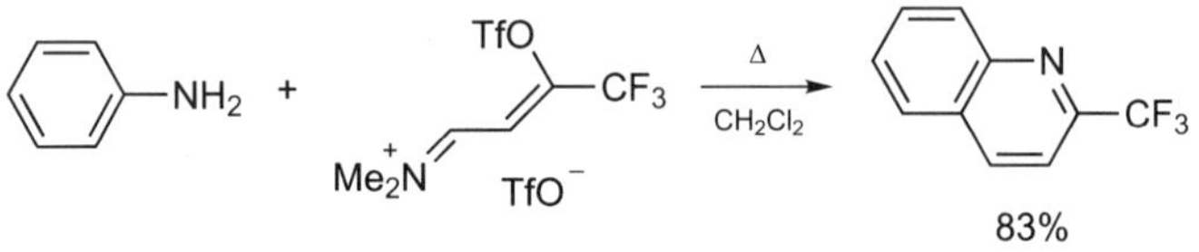

(**1**)

83%

[1]Baraznenok, I.L., Nenajdenko, V.G., Balenkova, E.S. *EJOC* 937 (1999).

2-(Trifluoromethanesulfonyloxy)vinyl aryliodonium triflates.

Aryliodonium salts.[1] Unsymmetrical diaryliodonium triflates and aryl(alkynyl)-iodonium triflates are readily synthesized from the title compounds by reaction with ArLi and alkynyllithium reagents, respectively. Ethylene is one of the byproducts.

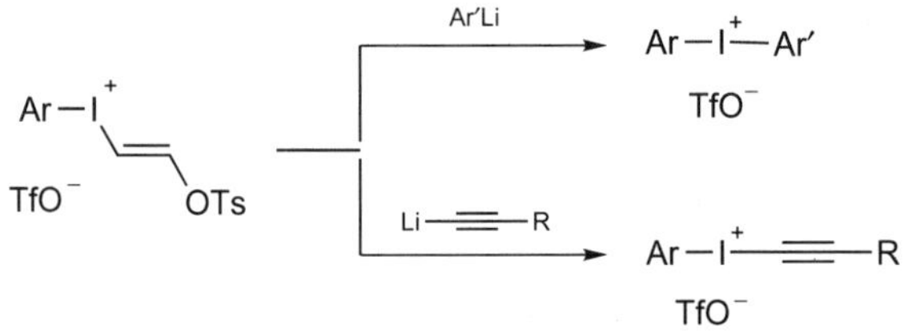

[1]Pirguliyev, N.Sh., Brel, V.K., Akhmedov, N.G., Zefirov, N.S. *S* 81 (2000).

***S*-Trifluoromethyldiarylsulfonium triflates.**

Trifluoromethylation. These reagents are prepared from $ArS(O)CF_3$. They donate the trifluoromethyl group to suitable nucleophiles (e.g., arenes).

[1]Yang, J.-J., Kirchmeier, R.L., Shreeve, J.M. *JOC* **63**, 2656 (1998).

(Trifluoromethyl)trimethylsilane. 15, 341; **18,** 378–379; **19,** 366–367; **20,** 400

Trifluoromethylation. Imines show similar reactivity as carbonyl compounds in accepting the Me_3Si and CF_3 groups from Me_3SiCF_3.[1] In the presence of CsF of KF, the title reagent converts 1-alkynes to alkynyltrimethylsilanes.[2]

[1]Blazejewski, J.-C., Anselmi, E., Wilmshurst, M.P. *TL* **40**, 5475 (1999).
[2]Ishizaki, M., Hoshino, O. *T* **56**, 8813 (2000).

Tri-2-furanylgermane–triethylborane.

Defunctionalization.[1] Organobromides, iodides, and xanthates are defunctionalized with this reagent via radical intermediates. Only catalytic amount of the germane is required if $NaBH_4$ is used in conjunction.

[1]Nakamura, T., Yorimitsu, H., Shinokubo, H., Oshima, K. *SL* 1415 (1999).

Triisobutylaluminum. 19, 367–368

Reductive rearrangement. Extension of the previously discovered reaction to enol ethers establishes a homologation route for alcohols, for examples., $ArCH_2OC({=}CH_2)Me$ to $ArCH_2CH_2CH(OH)Me$.[1]

Carbocycles are formed on subjecting unsaturated *S*-, *Se*-, and *C*-glycosides to *i*-Bu_3Al.[2] Claisen rearrangement followed by reduction of the resulting ketones is accomplished in one step.[3]

[1]du Roizel, B., Sollogoub, M., Pearce, A.J., Sinay, P. *CC* 1507 (2000).
[2]Sollogoub, M., Mallet, J.-M., Sinay, P. *ACIEE* **39**, 362 (2000).
[3]Wang, W., Sollogoub, M., Sinay, P. *ACIEE* **39**, 2466 (2000).

1-Triisopropylsiloxy-1,2-propadiene.

Acrylic acid α-anion equivalent.[1] O → C Silyl migration occurs on treatment of the title compound with *t*-BuLi in THF at −78°. The ensuing lithium enolate can be alkylated. Quenching with aldehydes generates Baylis–Hillman adducts.

[1]Stergiades, I.A., Tius, M.A. *JOC* **64**, 7457 (1999).

Triisopropylsilyl triflate. 20, 401

2-(Triisopropylsiloxy)acrolein.[1] This reagent is a valuable component of [4 + 3] cycloadditions. It is conveniently prepared from 2-methoxy-2-methyl-1,3-dioxan-5-one by reaction with *i*-$Pr_3SiOTf–Et_3N$.

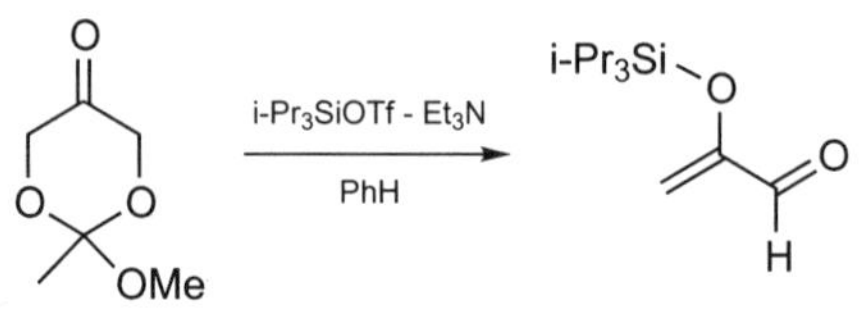

Silyl carbamates.[2] Primary and secondary amines are protected as triisopropoxy-carbonyl derivatives on consecutive treatment with carbon dioxide ($Et_3N–CH_2Cl_2$, −78°) and *i*-Pr_3SiOTf. The silyl carbamates are decomposed by Bu_4NF at ice temperature.

[1]Harmata, M., Sharma, U. *OL* **2**, 2703 (2000).
[2]Lipshutz, B.H., Papa, P., Keith, J.M. *JOC* **64**, 3792 (1999).

Trimethylaluminum. 15, 341–342; **17,** 372–375; **18,** 365–367; **19,** 369–370

Peptide synthesis.[1] Mediated by Me_3Al, thiol esters condense with amines to form amides. Note the reactivity differentiation of two thiol esters in the amino acid components in the following equation.

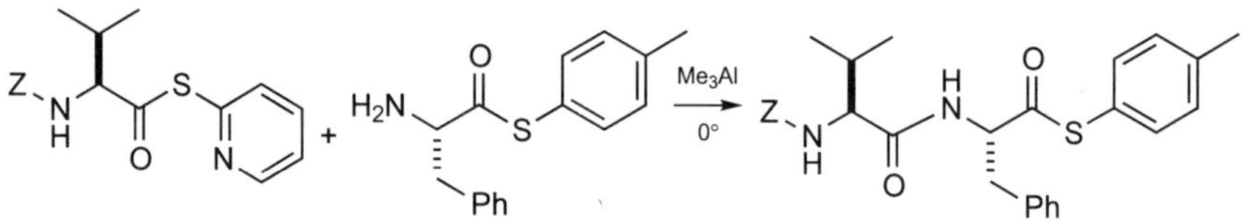

Epoxide opening. Regioselective opening of epoxides substituted with different activating groups at the two carbon atoms on reaction with $Me_3Al–H_2O$ is observed.[2] [A report is on epoxide opening with Et_3Al using Lewis base, e.g., $(Me_2N)_3P$, as catalyst.[3]]

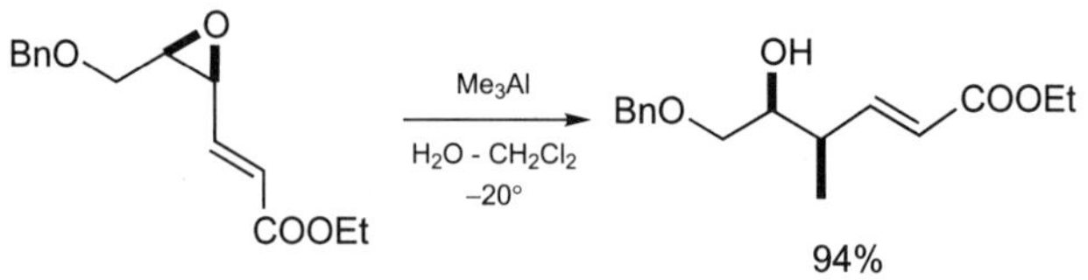

94%

Diels–Alder reaction.[4] The Diels–Alder reaction is facilitated and rendered stereoselective on using Me_3Al to tether both diene and dienophile components when each of which contains an allylic hydroxy group.

[1]Kurosu, M. *TL* **41**, 591 (2000).
[2]Abe, N., Hanawa, H., Maruoka, K., Sasaki, M., Miyashita, M. *TL* **40**, 5369 (1999).
[3]Schneider, C., Brauner, J. *TL* **41**, 3043 (2000).
[4]Bertozzi, F., Olsson, R., Frejd, T. *OL* **2**, 1283 (2000).

Trimethyl orthoformate.

Methyl 7,7-dimethoxyalkanoates.[1] Admixture of 2-acylcyclohexanones with $HC(OMe)_3$–MeOH in the presence of TsOH at room temperature causes ring cleavage to afford the ketoester acetals in good yields (3 examples, 80–93%).

[1]Martins, M.A.P., Bastos, G.P., Sinhorin, A.P., Flores, A.F.C., Bonacorso, H.G., Zanatta, N. *SL* 789 (1999).

Trimethylsiloxytrioxorhenium.

Isomerization of allylic alcohols.[1] Primary and secondary allylic alcohols and their silyl ethers are equilibrated in the presence of $(Me_3SiO)ReO_3$.

[1]Bellemin-Laponnaz, S., Le Ny, J.P., Osborn, J.A. *TL* **41**, 1549 (2000).

Trimethylsilyl azide. 13, 24–25; **14,** 25; **15,** 342–343; **16,** 17; **18,** 379–380; **19,** 371–372; **20,** 403

β-Azido alcohols.[1] Opening of epoxides with the title reagent at room temperature is catalyzed by Bu_4NF under solvent-free conditions.

[1]Schneider, C. *SL* 1840 (2000).

Trimethylsilyl chloride. 15, 89; **16,** 85–86; **18,** 381; **19,** 374–375; **20,** 404–405

Functional group manipulations. Aliphatic TBS ethers are hydrolyzed with a catalytic amount of Me_3SiCl and 1 equiv of H_2O in MeCN.[1] ArOTBS survives such treatment. Me_3SiCl also catalyzes the selective acetylation of aliphatic alcohols with trimethyl orthoacetate.[2]

Temporary protection of α-amino acids (and *N*-derivatives) as trimethylsilyl esters facilitates certain transformations, including conversion of the free amino group to an isocyanate.[3] Thus, Fmoc-derivatives of α-amino esters undergo elimination to give α-isocyanatocarboxylic esters on reaction with Me_3SiCl–Et_3N.[4]

When de-*N*-(*t*-butoxycarbonylation) is carried out in MeOH, esterification also occurs.[5]

Selective reaction of the primary amine in the presence of a secondary amine after trimethylsilylation of both has been demonstrated.[6] The monosilylated primary amine is still derivatizable.

Reductive acylation.[7] Azides are converted to amides with a mixture of an anhydride and Me_3SiCl.

Hydrohalogenation of alkynyl ethers.[8] α-Haloalkenyl ethers are formed. Such ethers are useful as acyl anion equivalents.

Bis(indol)-3-ylalkanes.[9] Nitrones condense with indoles under acidic conditions. When Me_3SiCl is used as promoter, the adducts serve as alkylating agents.

Me_3SiCl, CH_2Cl_2, 25° — 85%

HCl / MeOH, 0° — 94%

[1]Grieco, P.A., Markworth, C.J. *TL* **40**, 665 (1999).
[2]Sabitha, G., Reddy, B.V.S., Reddy, G.S.K., Yadav, J.S. *NJC* **24**, 63 (2000).
[3]Weiberth, F.J. *TL* **40**, 2895 (1999).
[4]Chong, P.Y., Petillo, P.A. *TL* **40**, 4501 (1999).
[5]Chen, B.-C., Skoumbourdis, A.P., Guo, P., Bednarz, M.S., Kocy, O.R., Sundeen, J.E., Vite, G.D. *JOC* **64**, 9294 (1999).
[6]Wang, T., Zhang, Z., Meanwell, N.A. *TL* **40**, 6745 (1999).
[7]Barua, A., Bez, G., Barua, N.C.*SL* 553 (1999).
[8]Yu, W., Jin, Z. *JACS* **122**, 9840 (2000).
[9]Chalaye-Mauger, H., Denis, J.-N., Averbuch-Pouchot, M.-T., Vallee, Y. *T* **56**, 791 (2000).

Trimethylsilyl chlorosulfate.

Sultones.[1] Upon admixture with iodosylbenzene, trimethylsilyl chlorosulfate provides nascent sulfur trioxide, which can be trapped by alkenes.

Me_3Si + $Me_3Si-OSO_2Cl$ —(PhI=O, CH_2Cl_2)→ 57%

[1]Bassindale, A.R., Katampe, I., Maesano, M.G., Patel, P., Taylor, P.G. *TL* **40**, 7417 (1999).

Trimethylsilyl cyanide. 13, 87–88; **14,** 107; **15,** 102–104; **17,** 89; **18,** 381–382; **19,** 375; **20,** 405

α-Cyanohydrin derivatives. Acetals[1] and acylals[2] exchange one of their oxy groups in catalyzed reactions with Me_3SiCN. The exchange using KCN–DMSO works well with acylals derived from aliphatic aldehydes but not for $ArCH(OCOR')_2$, therefore this alternative method is an improvement.

Regioselective 1,2- or 1,4-addition to β-alkoxyvinyl ketones is realized by proper modification of the reagent.[3]

EtO–CH=CH–CO–CHF_2 + Me_3Si-CN —(Et_3N, 0°)→ EtO–CH=CH–C(CN)(CHF_2)($OSiMe_3$) 80%

EtO–CH=CH–CO–C_2F_5 + Me_3Si-CN —(I_2, 50° ~ 60°)→ EtO–CH(CN)–CH=C($OSiMe_3$)–C_2F_5 89%

Isonitriles.[4] In the presence of a silver(I) salt, Me_3SiCN delivers the CN group to alkenes to furnish isonitriles.

—(Me_3SiCN, $AgClO_4$)→ NC 91%

Nitriles by substitution. Glycosyl cyanides are formed from glycal esters (Pd-catalyzed reaction),[5] whereas S_N2 reaction of alkyl halides requires fluoride ion to activate Me_3SiCN (hypervalent silicate).[6]

Glycosylation.[7] A glycoside with one free hydroxyl group and another protected sugar (except the OH at the anomeric center) condense on treatment with the aid of an acid anhydride and Me_3SiClO_4, which is prepared in situ from Me_3SiCl and $AgClO_4$.

[1]Tanaka, N., Masaki, Y. *SL* 1277 (1999).
[2]Sandberg, M., Sydnes, L.K. *OL* **2**, 6870 (2000).
[3]Kruchok, I.S., Gerus, I.I., Kukhar, V.P. *T* **56**, 6533 (2000).
[4]Kitano, Y., Chiba, K., Tada, M. *SL* 288 (1999).
[5]Hayashi, M., Kawabata, H., Arikita, O. *TL* **40**, 1729 (1999).

[6]Soli, E.D., Manoso, A.S., Patterson, M.C., DeShong, P., Favor, D.A., Hirschmann, R., Smith III, A.B. *JOC* **64**, 3171 (1999).
[7]Wakao, M., Nakai, Y., Fukase, K., Kusumoto, S. *CL* 27 (1999).

Trimethylsilyldiazomethane. 20, 405–406

Homologation.[1] One-carbon extension of alkenes via hydroboration (with catecholborane) is by further treatment with Me_3SiCHN_2 before oxidative workup.

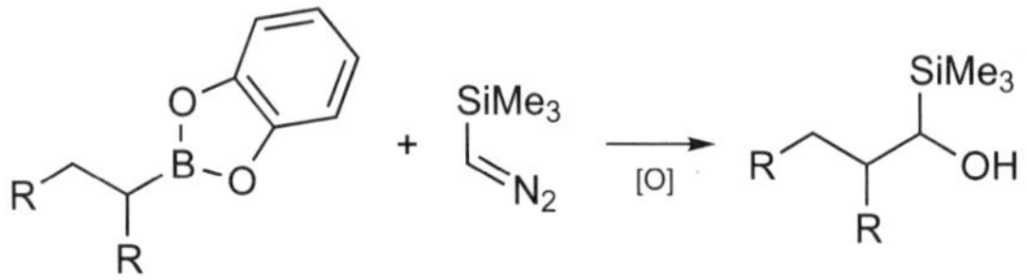

[2,3]-Sigmatropic rearrangement.[2] Transpositional rearrangement with one-carbon insertion of allylic sulfides occurs in the reaction with Me_3SiCHN_2 that is catalyzed by $FeCl_2$(dppe).

Cycloadditions. Direct cycloaddition of Me_3SiCHN_2 to *N*-sulfonylimines afford aziridines with high *cis*-selectivity.[3] Due to stereoselective ring opening and replacement of the silyl group by a carbon chain offered by the products, synthetic potentials are indicated.

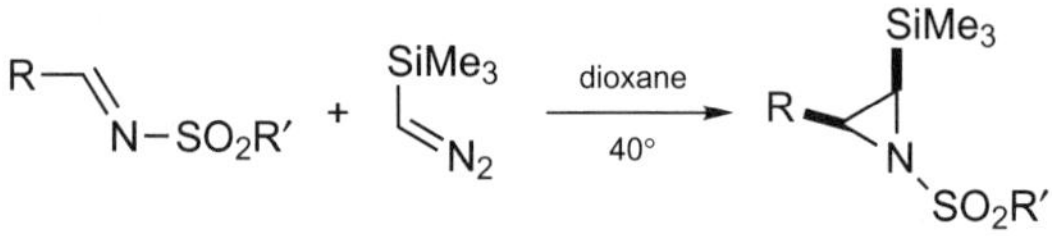

Alkylidenecyclopropanes are formed by reaction of the lithiated trimethylsilyldiazomethane with carbonyl compounds in the presence of alkenes.[4]

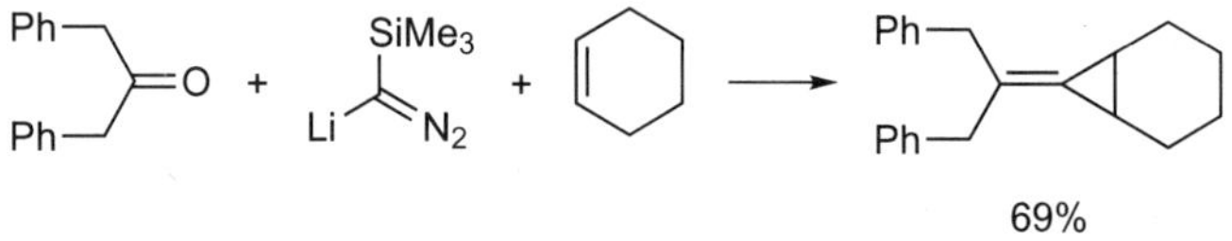

69%

[1]Goddard, J.-P., Le Gall, T., Mioskowski, C. *OL* **2**, 1455 (2000).
[2]Carter, D.S., van Vranken, D.L. *OL* **2**, 1303 (2000).
[3]Aggarwal, V.K., Ferrara, M. *OL* **2**, 4107 (2000).
[4]Sakai, A., Aoyama, T., Shioiri, T. *T* **55**, 3687 (1999).

Trimethylsilyldi(ethyl)amine. 18, 382; **19,** 376; **20,** 407

Alkylation. Cyclopropane formation is observed[1] in the reaction of aldehydes with dimethyl 2,3-dihalopropanoate in the presence of Et_2NSiMe_3. This process is a modification of the Michael reaction between aldehydes and methyl vinyl ketone that leads to δ-ketoaldehydes.[2]

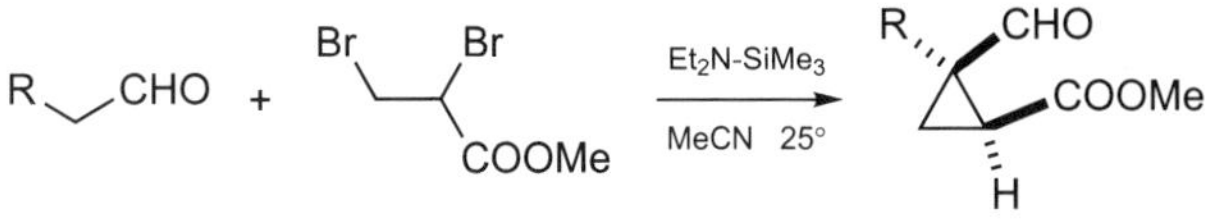

Aldol reactions.[3] Reaction between two aldehydes proceeds under solvent-free conditions. It is perhaps useful for the preparation of α-substituted cinnamaldehydes.

Ether cleavage.[4] A combination of Et_2NSiMe_3 and MeX (X = Br, I) shows the same reactivity as Me_3SiX in effecting ether cleavage.

[1]Hagiwara, H., Komatsubara, N., Hoshi, T., Suzuki, T., Ando, M. *TL* **40**, 1523 (1999).
[2]Hagiwara, H., Kato, M. *TL* **37**, 5139 (1996).
[3]Hagiwara, H., Ono, H., Komatsubara, N., Hoshi, T., Suzuki, T., Ando, M. *TL* **40**, 6627 (1999).
[4]Oshita, J., Iwata, A., Kanetani, F., Kunai, A. *JOC* **64**, 8024 (1999).

Trimethylsilyl *N,N*-dialkylcarbamates.

Enamines.[1] Reaction of the title compounds with ketones without solvent leads to enamines. More than 2 equiv of the reagents are employed because they serve as dehydrating agent besides donating the amino group. This procedure is particularly valuable for the preparation of *N,N*-dimethyl enamines and the like because conventional methods are not suitable due to the volatile nature of the anhydrous amines.

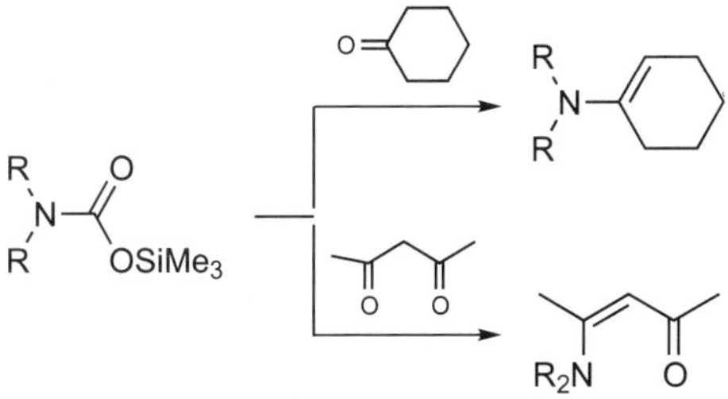

[1] Kardon, F., Mörtl, M., Knausz, D. *TL* **41**, 8937 (2000).

4-(Trimethylsilylethoxymethoxy)benzyl bromide.

Substituted benzyl ethers.[1] Protection of alcohols in the form of substituted benzyl ethers is accomplished by alkylation of RONa in DMF. These ethers are cleaved with Bu_4NF.

[1]Jobron, L., Hindsgaul, O. *JACS* **121**, 5835 (1999).

Trimethylsilyl fluorosulfonyldifluoroacetate.

gem-Difluorocyclopropanation.[1] Formation of *gem*-difluorocyclopropanes from electron-deficient alkenes such as acrylic esters is surprisingly efficient using this reagent and catalytic amount of NaF.

[1]Tian, F., Kruger, V., Bautista, O., Duan, J.-X., Li, A.-R., Dolbier, Jr., W.R., Chen, Q.-Y. *OL* **2**, 563 (2000).

Trimethylsilyl iodide. 16, 188–189; **18,** 383; **19,** 376–377; **20,** 407

Ketones from alkenyl sulfoxides.[1] The functional group transformation does not occur with the corresponding sulfides or sulfones.

Me$_3$Si-I

50%

[1]Aversa, M.C., Barattucci, A., Bonaccorsi, P., Bruno, G., Giannetto, P., Policicchio, M. *TL* **41**, 4441 (2000).

Trimethylsilylmethyllithium.

β-Keto silanes.[1] 1-Bora-1-bromoalkenes are homologated and refunctionalized on reaction with Me_3SiCH_2Li followed by oxidative workup. 1-Trimethylsilyl-2-alkanones are obtained in good yields (6 examples, 63–75%).

Li⌄SiMe$_3$

H_2O_2 - NaOAc

70%

[1]Bhat, N.G., Martinez, C., De Los Santos, J. *TL* **41**, 6541 (2000).

(Trimethylsilyl)methylenetriphenylphosphorane.

4H-Chromen-4-ones.[1] This Wittig reagent reacts with a silyl carboxylate to give the α-acylphosphorane by elimination through siloxane elimination. Opportunity for an intramolecular Wittig reaction presents itself in the case of a silyl 2-acyloxybenzoate and the formation of a 4*H*-chromen-4-one is observed.

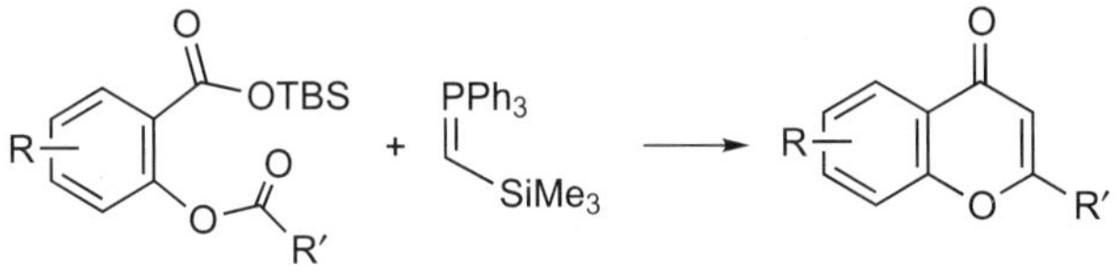

[1]Kumar, P., Bodas, M.S. *OL* **2**, 3821 (2000).

2-Trimethylsilyl nitrate.

α-Nitro ketones.[1] Functionalization of alkenes by Me_3SiONO_2–DMSO or Me_3SiONO_2–CrO_3 in MeCN is characterized by the generation of products in a high oxidation state. Under the same conditions, cyclic ethers are oxidized to lactones.

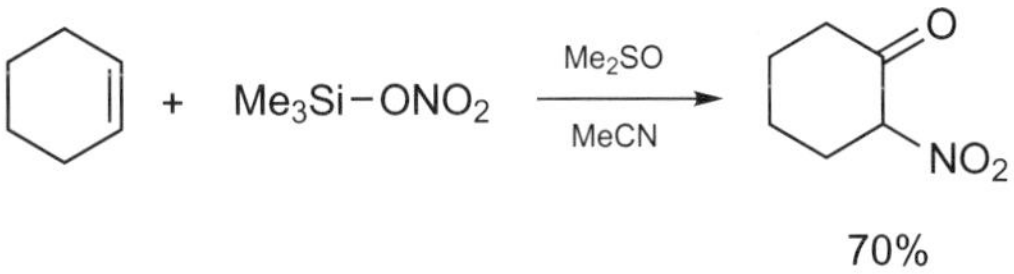

[1]Shahi, S.P., Gupta, A., Pitre, S.V., Reddy, M.V.R., Kumareswaran, R., Vankar, Y.D. *JOC* **64**, 4509 (1999).

Trimethylsilyl phenyl selenide. 20, 408

β-Nitroalkenyl selenides.[1] β-Nitroalkenyl sulfoxides undergo group exchange reaction with $Me_3SiSePh$.

[1]Abe, H., Fujii, H., Yamasaki, A., Kinome, Y., Takeuchi, Y., Harayama, T. *SC* **30**, 543 (2000).

Trimethylsilyl phenyl telluride.

Reductive silylation.[1] Quinones are converted to trimethylsilyl ethers of hydroquinones with $Me_3SiSePh$ in THF at room temperature in excellent yield. The reaction is initiated by a single electron transfer from the reagent to the quinones.

[1]Yamago, S., Miyazoe, H., Iida, K., Yoshida, J. *OL* **38**, 3671 (2000).

Trimethylsilyl(tributylstannane).

Acylsilanes.[1] The reaction of acyl chlorides with $Me_3SiSnBu_3$ is Pd-catalyzed.

β-Aryl allylsilanes.[2] The Pd(0)-catalyzed reaction of allenes with $Me_3SiSnBu_3$ in the presence of ArI proceeds by addition and Stille coupling.

C + Ph-I + $Bu_3Sn-SiMe_3$ —$(dba)_2Pd$→ Ph, $SiMe_3$ 85%

Ring expansion.[3] A Barbier-type reaction occurs when 2-(2-iodoallyl)cyclohexanones are treated with a mixture of $Me_3SiSnBu_3$ and CsF. Fragmentation follows the intramolecular addition if a leaving group is installed at C-3 of the substrate cyclohexanones and, as a result, 4-cyclooctenones are formed.

MsO, I, O —$Bu_3Sn-SiMe_3$, CsF→ O 70% (in THF) → $SnBu_3$, O 86% (in DMF)

[1]Geng, F., Maleczka, R.E. *TL* **40**, 3113 (1999).
[2]Wu, M.-Y., Yang, F.-Y., Cheng, C.-H. *JOC* **64**, 2471 (1999).
[3]Imai, A.E., Sato, Y., Nishida, M., Mori, M. *JACS* **121**, 1217 (1999).

Trimethylsilyl trifluoromethanesulfonate. 13, 329–331; **14,** 333–335; **15,** 346–350; **16,** 363–364; **17,** 379–386; **18,** 383–384; **19,** 379–381; **20,** 408–410

Desilylation.[1] On treatment with Me_3SiOTf and then MeOH *t*-butyldimethylsilyl ethers are hydrolyzed chemoselectively. *t*-Butyldiphenylsilyl ethers are not affected.

Elimination.[2] β-Siloxyalkyltri-2-furanylgermanes that are adducts of silyl enol ethers with $(Fu)_3GeH$ defunctionalize on reaction with Me_3SiOTf. Thus, hydrodesilyloxylation is achieved by combining the two reactions.

Iminium salts.[3] The condensation of acetals with Me_3SiNR_2 is catalyzed by Me_3SiOTf.

Aldol reactions. The Mukaiyama–aldol reaction has been extended to using peroxyacetals[4] and trimethyl orthoformate[5] as acceptors. The latter process serves to introduce the lactonic carbonyl group across the bicyclic core of (+)-CP263114.

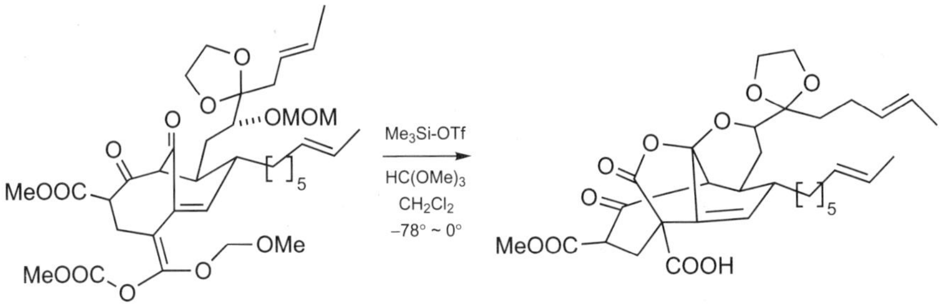

Me3Si-OTf
HC(OMe)3
CH2Cl2
−78° ~ 0°

1,ω-Bisacylsilanes undergo cyclization (five-, six-membered rings)[6] with Me_3SiOTf.

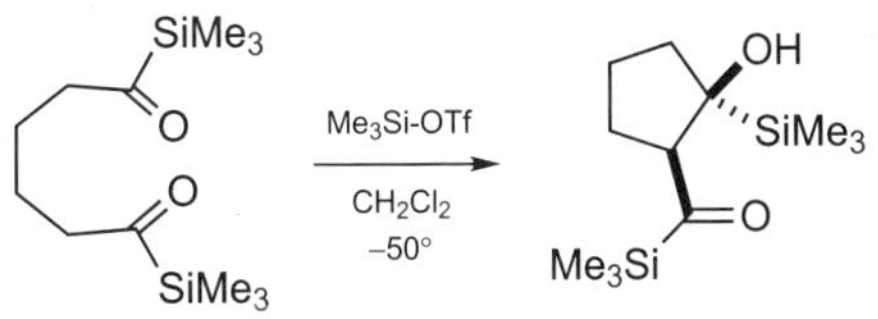

Me3Si-OTf
CH2Cl2
−50°
78%

Ring expansion. Activated fused cyclopropanes undergo cleavage of an internal C—C bond attendant the attack by silyl enol ethers which leads to products containing a larger ring.[7,8]

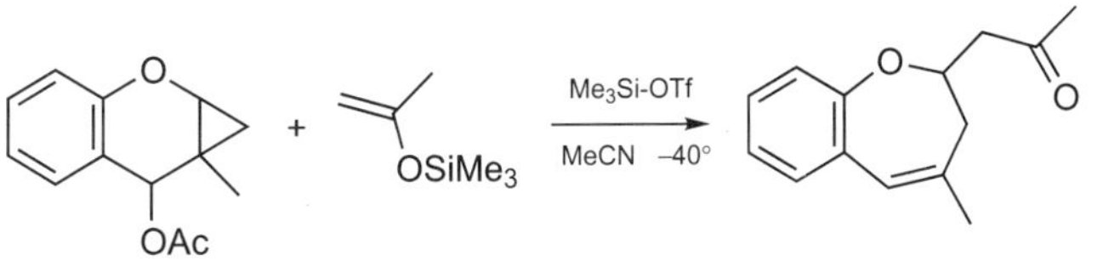

Me3Si-OTf
MeCN −40°
81%

Cyclopropanation. β-Stannyl carbocations collapse by formation of a three-membered ring. Methods for generating such active intermediates, and hence compounds containing a cyclopropane unit, are available.[9,10]

[1]Hunter, R., Hinz, W., Richards, P. *TL* **40**, 3643 (1999).
[2]Tanaka, S., Nakamura, T., Yorimitsu, H., Shinokubo, H., Oshima, K. *OL* **2**, 1911 (2000).
[3]Grotjahn, D.B., Albers, R.A., Beckman, J. *SL* 633 (2000).
[4]Dussault, P.H., Lee, R.J., Schultz, J.A., Suh, Y.S. *TL* **41**, 5457 (2000).
[5]Chen, C., Layton, M.E., Sheehan, S.M., Shair, M.D. *JACS* **122**, 7424 (2000).
[6]Bouillon, J.-P., Portella, C. *EJOC* 1571 (1999).
[7]Sugita, Y., Kawai, K., Hosoya, H., Yokoe, I. *H* **51**, 2029 (1999).
[8]Sugita, Y., Hosoya, H., Yokoe, I. *H* **53**, 1251 (2000).
[9]Sugawara, M., Yoshida, J. *CC* 505 (1999).
[10]Sugawara, M., Yoshida, J. *TL* **40**, 1717 (1999).

Trimethylsulfonium methylsulfate–dimethyl sulfide.

Methylide precursor.[1] Carbonyl compounds are transformed into epoxides in the presence of KOH when they react with the mixture of these two reagents pretreated with sulfuric acid at 40°.

[1]Forrester, J., Jones, R.V.H., Preston, P.N., Simpson, E.S.C. *JCS(P1)* 3333 (1999).

Triphenylbismuth–copper(II) acetate.

N-Phenylation.[1] A method for the preparation of 1-aryl-1-phenylhydrazines is via phenylation of *N*-arylaminophthalimides.

[1]Aoki, Y., Saito, Y., Sakamoto, T., Kikugawa, Y. *SC* **30**, 131 (2000).

Triphenylphosphine. 18, 385–386; 19, 382–383; 20, 411–412

Alkyl sulfinates.[1] Reductive esterification of $ArSO_2Cl$ with Ph_3P, and in situ sulfinylation, is accomplished in one flask. By using a chiral alcohol in this process, two diastereomers are obtained.

[8 + 2]Cycloaddition.[2] Zwitterionic adducts from Ph_3P and allenyl ketones or 2,3-butadienoic esters are dipolarophiles toward tropone.

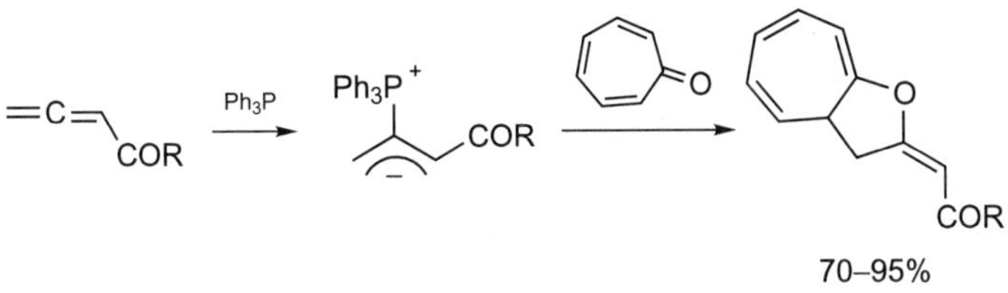

=C=\ COR —Ph₃P→ Ph₃P⁺ COR → (cycloheptatrienone) → COR

70–95%

[1]Watanabe, Y., Mase, N., Tateyama, M., Toru, T. *TA* **10**, 737 (1999).
[2]Kumar, K., Kapur, A., Ishar, M.P.S. *OL* **2**, 787 (2000).

Triphenylphosphine–diethyl azodicarboxylate. 13, 332; **14,** 336–337; **17,** 389–390; **18,** 387; **19,** 384–385; **20,** 413

Dehydration. The Mitsunobu reaction conditions induce conversion of 1,2-diols to carbonyl compounds.[1]

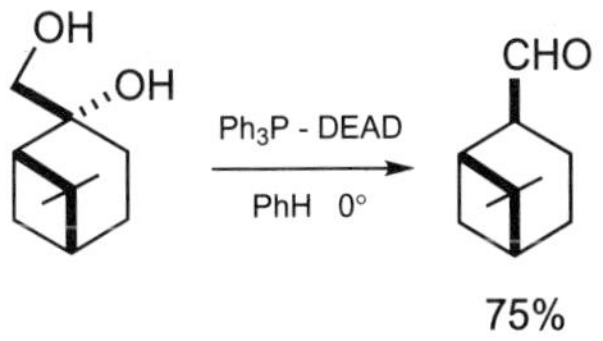

OH, OH —Ph_3P - DEAD, PhH 0°→ CHO

75%

The homobenzopyran system is readily elaborated from 4-(2-hydroxyaryl)-2-butenols[2] at room temperature as shown in the following equation. Dihydrobenzofurans are not formed.

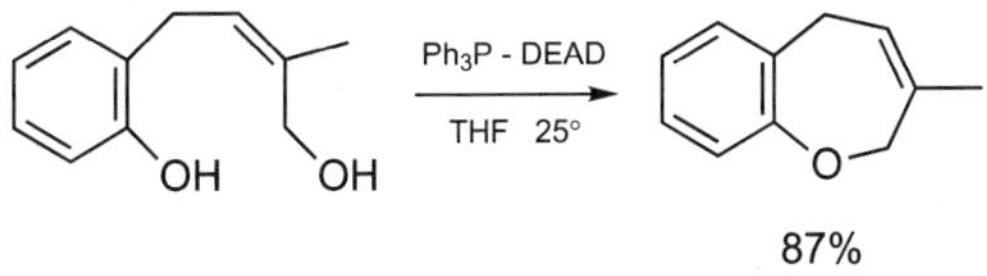

OH OH —Ph_3P - DEAD, THF 25°→ O

87%

Inverted sugars.[3] A synthesis of L-hexoses from D-glucono-1,5-lactones involves ring opening and reclosure via *O*-benzylhydroxamic acid intermediates. The latter step is accomplished by a Mitsunobu reaction.

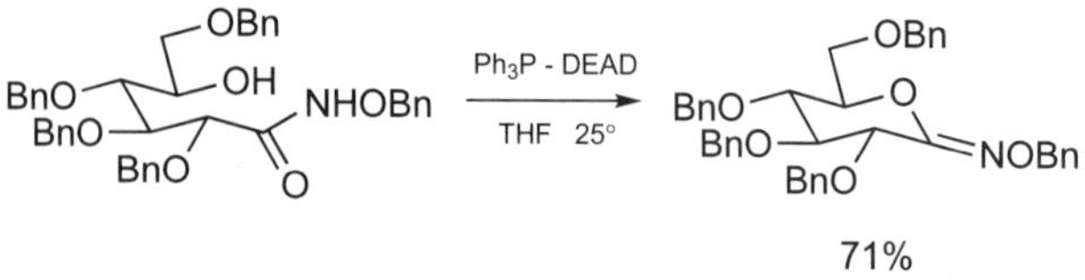

N-Alkylations. The Mitsunobu protocol is applicable to *N*-alkylation of protected *S*-methylisothioureas, and hence the preparation of unsymmetrically substituted guanidines.[4] 1,1-Dialkylhydrazines are similarly accessible.[5]

[1] Barrero, A.F., Alvarez-Manzaneda, E.J., Chahboun, R. *TL* **41**, 1959 (2000).
[2] Yamaguchi, S., Furihara, K., Miyazawa, M., Yokoyama, H., Hirai, Y. *TL* **41**, 4787 (2000).
[3] Takahashi, H., Hitomi, Y., Iwai, Y., Ikegami, S. *JACS* **122**, 2995 (2000).
[4] Kim, H.-O., Mathew, F., Ogbu, C. *SL* 193 (1999).
[5] Brosse, N., Pinto, M.-F., Jamart-Gregoire, B. *JOC* **65**, 4370 (2000).

Triphenylphosphine–diisopropyl azodicarboxylate. 15, 352–353; **17,** 390; **18,** 387–388; **19,** 385; **20,** 413–414

Isocyanates.[1] *N*-Carboxylation of amines with carbon dioxide followed by treatment with Ph_3P–DIAD delivers RN=C=O.

β-Lactams.[2] Hydroxamic acids obtained from aminolysis of β-lactones with *O*-benzylhydroxylamine undergo cyclization again with Ph_3P–DIAD.

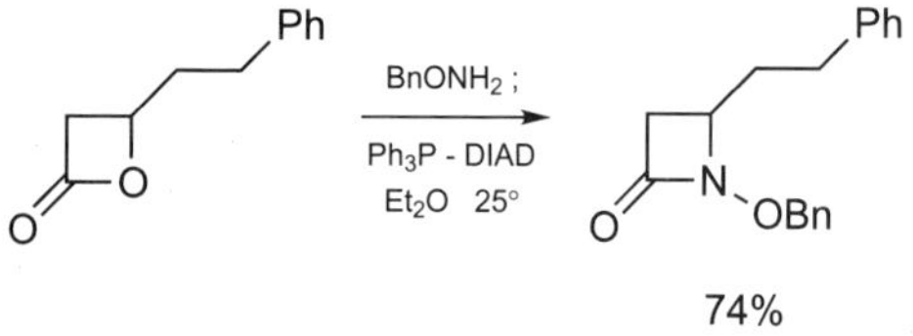

[1] Horvath, M.J., Saylik, D., Jackson, W.R., Elmes, P.S., Lovel, C.G., Moody, K. *TL* **40**, 363 (1999).
[2] Yang, H.W., Romo, D. *JOC* **64**, 7657 (1999).

Triphenylphosphonium bromide.

Bromohydrins.[1] As a source of hydrogen bromide, $[Ph_3PH]Br$ is useful in opening of epoxides.

[1] Afonso, C.A.M., Vieira, N.M.L., Motherwell, W.B. *SL* 382 (2000).

Triphosgene. 18, 388; **19,** 386; **20,** 415–416

β-Lactams. Carboxylic acids are converted to ketenes by a mixture of triphosgene and triethylamine. The nascent ketenes can be trapped with imines.[1]

[1]Krishnaswamy, D., Bhawal, B.M., Deshmukh, A.R.A.S. *TL* **41**, 417 (2000).

Triruthenium dodecacarbonyl. 18, 308; **19,** 386–387; **20,** 416–417

Aminations. Allylic amination with nitroarenes[1] is catalyzed by the Ru complexes made up of $Ru_3(CO)_{12}$ and a bis(arylimino)acenaphthene under CO.

Aromatic ketimines are formed by hydroamination of 1-alkynes with arylamines.[2,3] Interestingly, styrenes afford adducts from *o*-aminoarylation.[3]

Ph–NH(Me) + Ph–C≡CH —[$Ru_3(CO)_{12}$, 70°]→ Ph–N(Me)–C(=CH$_2$)–Ph 85%

Ph–NH(Me) + PhCH=CH$_2$ —[$Ru_3(CO)_{12}$, 70°]→ o-(MeNH)C$_6$H$_4$–CH(CH$_3$)Ph 83%

Cyclocarbonylations. Formation of γ- and δ-lactones from the proper allenyl alcohols via carbonylation and cyclization by heating with $Ru_3(CO)_{12}$–CO is observed.[4] γ-Lactones with a different substitution pattern are formed by a [2 + 2 + 1]cycloaddition from ketones (including α-diketones), ethylene, and carbon monoxide.[5]

PhC(=O)COOMe —[$Ru_3(CO)_{12}$, (4-F_3CC$_6$H$_4$)$_3$P, CO - ethylene, PhMe 160°]→ γ-lactone (Ph, COOMe) 94%

Under a CO atmosphere and in the presence of $Ru_3(CO)_{12}$, 1-aza-1,3-dienes are converted to unsaturated γ-lactams.[6] Bicyclic lactams are formed in the reaction of 5-alkynaldimines.[7]

When ethylene is also present in the reaction involving conjugated azadienes the products are deconjugated and ethylated at the α-position.[8]

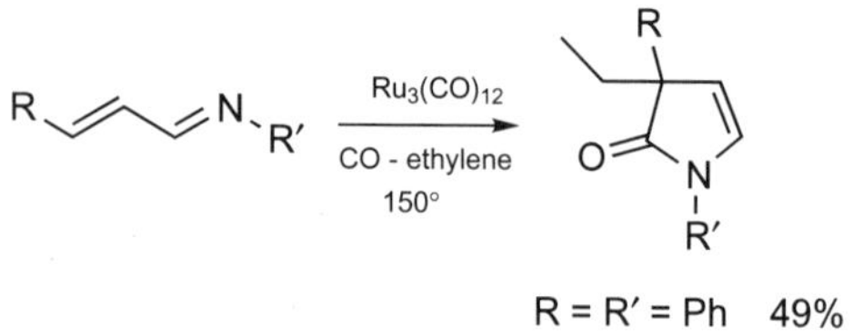

$Ru_3(CO)_{12}$, CO - ethylene, 150°

R = R′ = Ph 49%

3-Alkoxy-2-cyclopetenones.[9] In a cycloaddition process with alkenes, cyclobutenediones participate after losing [CO]. This method is complementary to the Pauson–Khand reaction

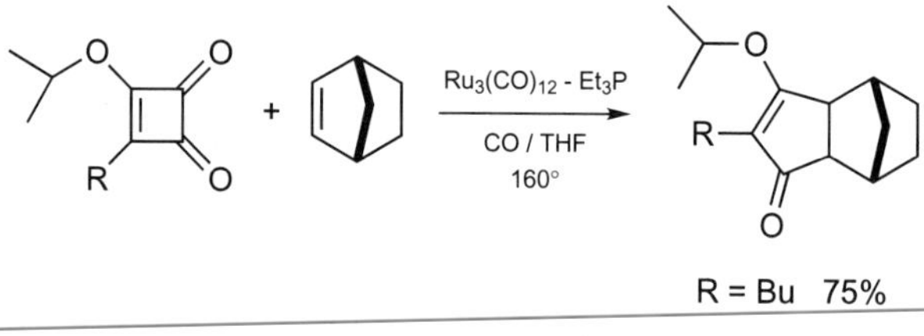

$Ru_3(CO)_{12}$ - Et_3P, CO / THF, 160°

R = Bu 75%

Deacylation.[10] Removal of an entire acyl side chain from an aromatic ketone occurs when the acyl group is ortho to a potential ligand for ruthenium. Accordingly, this process exhibits excellent regioselectivity in the cases of certain diacylarenes. The importance of the coordination is shown by the fate of a β-(2-pyridyl)-α,β-unsaturated ketone.

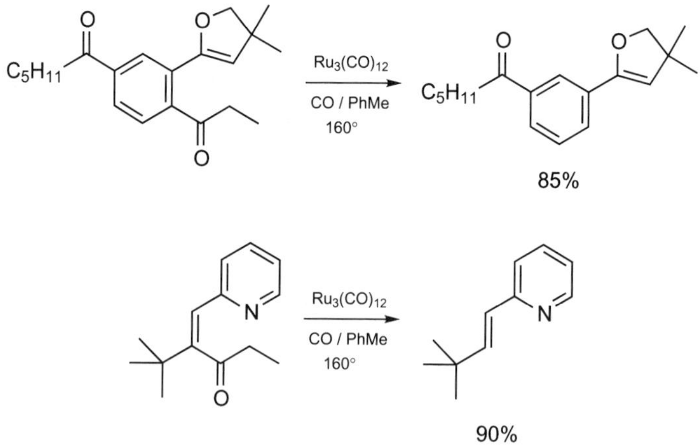

$Ru_3(CO)_{12}$, CO / PhMe, 160°

85%

$Ru_3(CO)_{12}$, CO / PhMe, 160°

90%

Ketone synthesis.[11] A method for ketone synthesis is based on *N*-allylation of 3-methyl-2-aminopyridine and exploiting the coordination ability of the pyridine moiety to stabilize cyclic rhodia–ketimine intermediates that are active in insertion to 1-alkenes. In situ hydrolysis of the demetalated ketimine products affords ketones.

$Ru_3(CO)_{12}$, PhMe 130°; H_3O^+

[1]Ragaini, F., Cenini, S., Tollari, S., Tummolillo, G., Beltrami, R. *OM* **18**, 928 (1999).
[2]Tokunaga, M., Eckert, M., Wakatsuki, Y. *ACIEE* **38**, 3222 (1999).
[3]Uchimaru, Y. *CC* 1133 (1999).
[4]Yoneda, E., Kaneko, T., Zhang, S.-W., Onitsuka, K., Takahashi, S. *OL* **2**, 441 (2000).
[5]Chatani, N., Tobisu, M., Asaumi, T., Fukumoto, Y., Murai, S. *JACS* **121**, 7160 (1999).
[6]Morimoto, T., Chatani, N., Murai, S. *JACS* **121**, 1758 (1999).
[7]Chatani, N., Morimoto, T., Kamitani, A., Fukumoto, Y., Murai, S. *JOMC* **579**, 177 (1999).
[8]Berger, D., Imhof, W. *CC* 1457 (1999).
[9]Kondo, T., Nakamura, A., Okada, T., Suzuki, N., Wada, K., Mitsudo, T. *JACS* **122**, 6319 (2000).
[10]Chatani, N., Ie, Y., Kakiuchi, F., Murai, S. *JACS* **121**, 8645 (1999).
[11]Jun, C.-H., Lee, H., Park, J.-B., Lee, D.-Y. *OL* **1**, 2161 (1999).

Tris(acetonitrile)cyclopentadienylruthenium(I) hexafluorophosphate.

Addition to alkynes and allenes. An ene-type reaction[1] between alkynes and alkenes proceeds regioselectively at room temperature with $[CpRu(MeCN)_3]PF_6$. In the presence of bromide salt(s) an atom-economical coupling of 1-alkynes and enones proceeds.[2]

C_6H_{13} + ; $[CpRu(MeCN)_3]PF_6$, LiBr - $SnBr_4$, Me_3CO; Br, O, C_6H_{13}; 88% (*E* : *Z* 1 : 6.6)

Couplings involving allenyl alcohols and amines (without halide incorporation) are accompanied by the formation of heterocycles.[3,4] Many functional groups including basic amines are tolerated in these reactions.

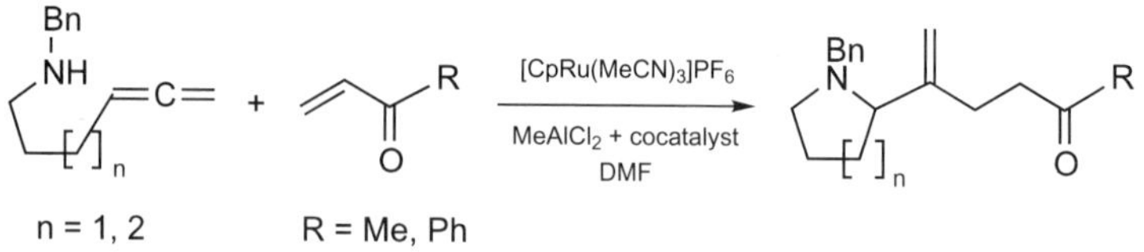

Cycloisomerizations. 1,6-Enynes and 1,7- analogues cyclize to furnish 2-vinyl-1-methylenecycloalkanes[5] in the presence of $[CpRu(MeCN)_3]PF_6$. The regioselectivity is contrary to that catalyzed by Pd-catalysts.

Substitution pattern of the substrates have profound effects on the products of the Ru-catalyzed reaction. For example, cycloheptene derivatives may be generated.[6]

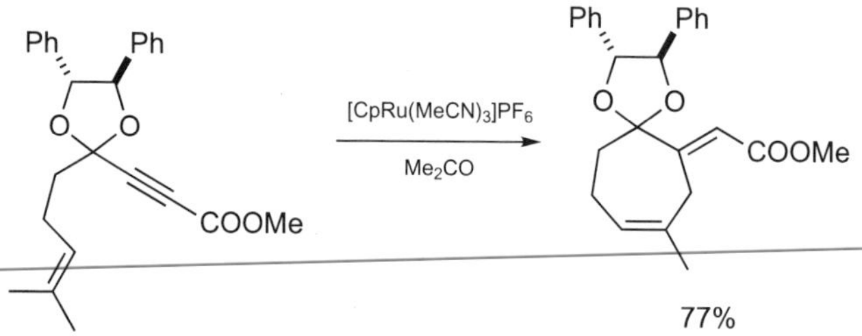

Cycloadditions. When the double bond of 1,6- and 1,7-enynes is connected to a ketone[7] or cyclopropane unit,[8,9] the cycloaddition reaction pathways prevail.

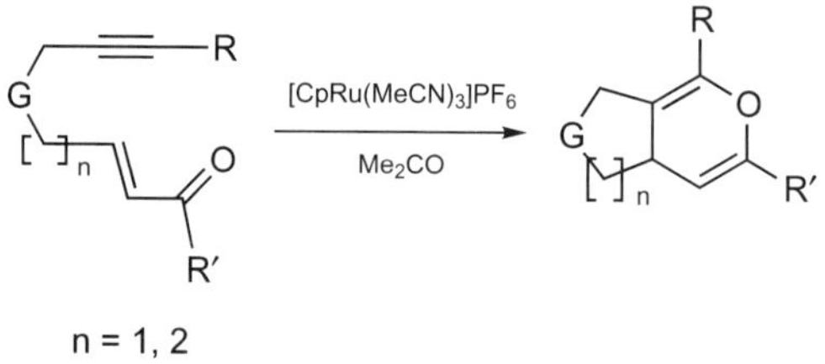

[1]Trost, B.M., Machacek, M., Schnaderbeck, M.J. *OL* **2**, 1761 (2000).
[2]Trost, B.M., Pinkerton, A.B. *ACIEE* **39**, 360 (2000).
[3]Trost, B.M., Pinkerton, A.B., Kremzow, D. *JACS* **122**, 12007 (2000).
[4]Trost, B.M., Pinkerton, A.B. *JACS* **121**, 10842 (1999).
[5]Trost, B.M., Toste, F.D. *JACS* **122**, 714 (2000).
[6]Trost, B.M., Toste, F.D. *JACS* **121**, 9728 (1999).
[7]Trost, B.M., Brown, R.E., Toste, F.D. *JACS* **122**, 5877 (2000).
[8]Trost, B.M., Toste, F.D., Shen, H. *JACS* **122**, 2379 (2000).
[9]Trost, B.M., Shen, H. *OL* **2**, 2523 (2000).

Tris(dibenzylideneacetone)dipalladium. 14, 339; **15,** 353–355; **16,** 372; **17,** 394; **18,** 389–393; **19,** 388–390; **20,** 417–420

Substitution reactions of allylic systems. An access to allylic thiols[1] is based on substitution by MeCOSK with $(dba)_3Pd_2$-dppb as catalyst. 2-Alkylidene-3-alkyl-1,4-benzodioxanes are readily formed in a reaction between catechols and propargylic carbonates.[2]

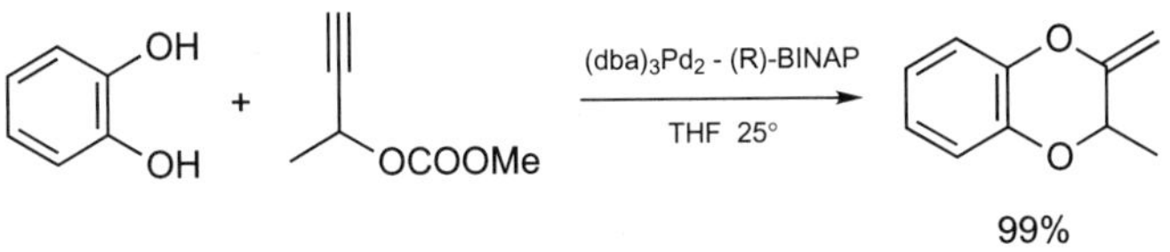

(dba)3Pd2 - (R)-BINAP; THF 25°; OH; OH; OCOOMe; O; O; 99%

β-Siloxy-ε-acetoxy-γ,δ-unsaturated carbonyl compounds undergo reductive transposition on exposure to $HCOONH_4$, $(dba)_3Pd_2$, and Bu_3P in DMF to afford the δ,ε-unsaturated products with (all) *syn*-isomers predominating.[3] In conjunction with the enantioselective aldol access to the substrates, this process expands the methodology of existing polyketide synthesis.

N-Arylation. The scope and limitations of the Pd–BINAP combination for *N*-arylation with ArBr has been defined.[4] 2,4-Disubstituted azetidines are arylated without any problem.[5] Sodium alkoxides containing a β-hydrogen atom are shown to be suitable bases in the *N*-arylation reaction.[6]

Besides BINAP, biphenylphosphine ligands have found use in these reactions, particularly the *N*-arylation of indoles[7] and vinylogous amides.[8] Moreover, the efficiency of the reaction enables protection of alcohols as 4-halobenzyl ethers,[9] because such esters release the alcohol moiety on conversion to ROC_6H_4NHBn and treatment with $SnCl_4$.

Usually, only one halogenated site of a polyhaloarene is substituted.[10] The reaction also exhibits chemoselectivity as generally expected.

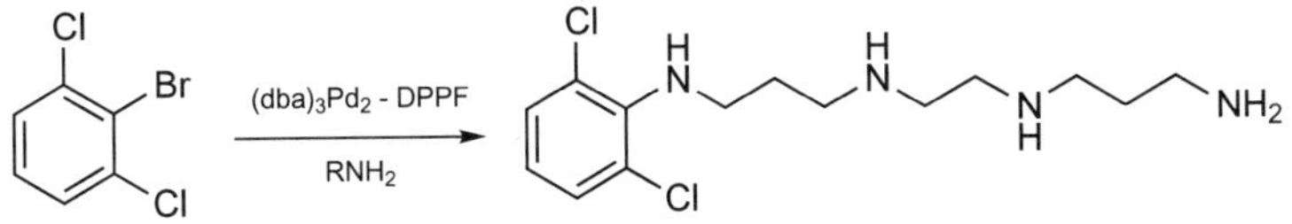

Cl; Br; Cl; (dba)3Pd2 - DPPF; RNH2; Cl; H N; H N; N H; NH2; Cl

Aryl hydrazides such as $ArN(Boc)NH_2$ are available by this method (from ArBr and $BocNHNH_2$).[11] A phosphine-free catalyst system for for *N*-arylation specifies addition of 1,3-bis(2,6-diisopropylphenyl)imidazolium chloride.[12] Apparently, carbene derived from this additive becomes ligated to the Pd atoms.

Other aromatic substitutions. Aryl cyanides can be prepared from aryl chlorides in the Pd(0)-catalyzed reaction with Zn–$Zn(CN)_2$,[13] and from aryl bromides and iodides, with CuCN.[14]

Benzannulated oxacycles are formed by ring closure of *o*-bromoarylalkanols.[15]

Suzuki and Stille couplings. The coupling of sterically hindered ArBr with phenylboronic acid in the presence of $(dba)_3Pd_2$, $(MeO)_3P$, and K_3PO_4 is not problematic.[16] By using a $(dba)_3Pd_2$–*t*-Bu_3P–KF catalytic system, the Suzuki coupling operates on a wide range of aryl and alkenyl halides, typically at room temperature, with a reactivity profile showing ArCl > ArOTf.[17] Very similar reaction conditions are effective to achieve Stille cross-couplings.[18]

1,1-Dibromoalkenes afford (*E*)-bromostilbenes by the Suzuki[19] or the Stille protocol.[20] Diarylacetylenes are obtained when the Stille coupling is conducted in the presence of a tertiary amine.

MeOOC–C6H4–CH=CBr2 (Br, Br) + Bu3Sn–(2-furyl)

→ $(dba)_3Pd_2$, $(2\text{-Fu})_3P$ / PhMe, 100° → MeOOC–C6H4–CH=C(Br)–(2-furyl) 80%

→ $(dba)_3Pd_2$, $(4\text{-An})_3P$ / DMF, i-Pr_2NEt 80° → MeOOC–C6H4–C≡C–(2-furyl) 91%

Facile Suzuki coupling is also effected in the presence of the 1,3-dimesitylimidazol-2-ylidene ligand.[21]

Related to the original Suzuki coupling is the ketone synthesis from acid chloride and trialkyl boranes[22] and thiol esters with arylboronic acids.[23] The latter procedure is Pd-catalyzed and Cu-mediated.

Heck reaction. The possibility of applying the Heck reaction to achieve multibond formation is once again substantiated in the elaboration of the xestoquinone skeleton.[24]

OMe, OTf, OMe, O, O — $(dba)_3Pd_2$, (S)-BINAP, PMP / PhMe → OMe, OMe, O, O → O, O, O, O, X

X = O halenaquinone

X = H,H xestoquinone

Synthetically useful methods evolve from coupling after a regioselective ring opening of cyclobutanols[25] and capture of *p*-allylpalladium species that are generated from an intramolecular Heck reaction.[26]

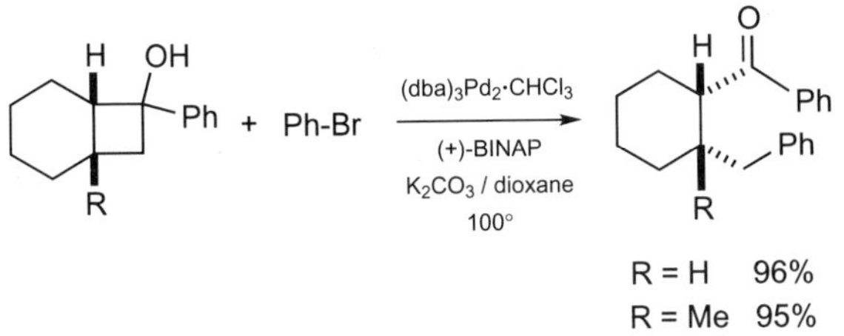

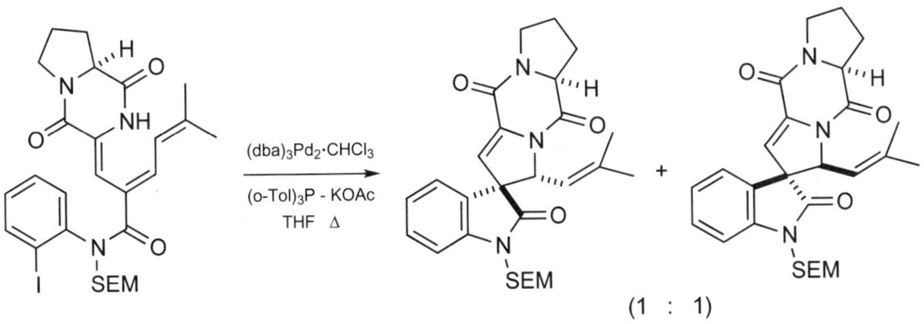

Other coupling reactions. For the Kumada coupling (ArMgX + Ar′X), the effectiveness of a carbene ligand for Pd formed in situ from 1,3-bis(2,6-diisopropylphenyl)-imidazolium chloride has been noted.[27] In a low-temperature Sonogashira coupling leading to diarylacetylenes, the ligand of preference is tris(mesityl)phosphine.[28]

Symmetrical and unsymmetrical 1,3-diynes have been prepared from 1,1-dibromo-1-alkenes.[29] Alkynylpalladium intermediates undergo homocoupling when CuI is present in the reaction medium. On the other hand, the persistent intermediates can await addition of other 1-alkynes for their ultimate transformation.

Zircona-arenes are readily formed from aryllithiums. The coupling products derived from such species and ArX afford further opportunity for regioselective functionalization (e.g., iodination).[30]

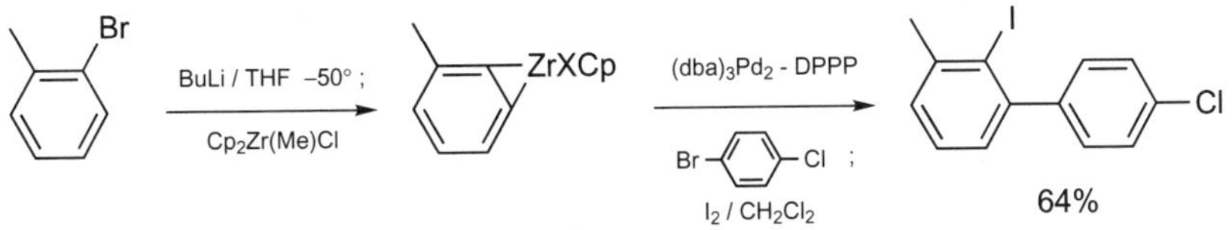

The Negishi coupling is applicable to the synthesis of β- and γ-amino acid derivatives.[31]

I–CH2–CH(NHBoc)–CH2–COOMe → IZn–CH2–CH(NHBoc)–CH2–COOMe —(dba)3Pd2, (o-Tol)3P, Ph—I→ Ph–CH2–CH(NHBoc)–CH2–COOMe 73%

[2 + 2 + 2]Cycloadditions. Several versions of this reaction have been developed, leading to hexabenzotriphenylene (trimerization of benzyne)[32] and heavily substituted dihydroisobenzofurans.[33]

SiMe3 / OTf —CsF - (dba)3Pd2, MeCN, 25°→ 39%

Isomerization of cyclopropanols. Ring opening of cyclopropanols gives enones on Pd-catalysis. Formation of α-methylene ketones is favored by the use of $(dba)_3Pd_2$, molecular sieve 4Å, and benzoquinone (reoxidant) in toluene.[34]

R, R′, HO —(dba)3Pd2, p-benzoquinone, MS 4A / PhMe→ R, R′, O

Reductive cyclization.[35] 1,6-Enynes cyclize to give 2-methyl-1-methylenecyclopentanes on treatment with $(dba)_3Pd_2$, Ph_3P, and Et_3SiH. Interestingly, a different isomer predominates when the same substrates are exposed to $Pd(OAc)_2$, Ph_3P, and HCOOH.

$(dba)_3Pd_2$ - Ph_3P
Et_3SiH - HOAc
PhMe 50°

47% + 5%

$Pd(OAc)_2$ - Ph_3P
HCOOH
PhMe 60°

69%

[1]Divekar, S., Safi, M., Soufiaoui, M., Sinou, D. *T* **55**, 4369 (1999).
[2]Labrosse, J.-R., Lhoste, P., Sinou, D. *TL* **40**, 9025 (1999); *OL* **2**, 527 (2000).
[3]Hughes, G., Lautens, M., Wen, C. *OL* **2**, 107 (2000).
[4]Wolfe, J. P., Buchwald, S. L., Tomori, H., Sadighi, J.P., Yin, J. *JOC* **65**, 1144, 1158 (2000).
[5]Marinetti, A., Hubert, P., Genet, J.-P. *EJOC* 1815 (2000).
[6]Prashad, M., Hu, B., Lu, Y., Draper, R., Har, D., Repic, O., Blacklock, T.J. *JOC* **65**, 2612 (2000).
[7]Old, D.W., Harris, M.C., Buchwald, S.L. *OL* **2**, 1403 (2000).
[8]Edmondson, S.D., Mastracchio, A., Parmee, E.R. *OL* **2**, 1109 (2000).
[9]Plante, O.J., Buchwald, S. L., Seeberger, P.H. *JACS* **122**, 7148 (2000).
[10]Beletskaya, I.P., Bessmertnykh, A.G., Guilard, R. *SL* 1459 (1999).
[11]Wang, Z., Skerlj, R.T., Bridger, G.J. *TL* **40**, 3543 (1999).
[12]Huang, J., Grasa, G., Nolan, S.P. *OL* **1**, 1307 (1999).
[13]Jin, F., Confalone, P.N. *TL* **41**, 3271 (2000).
[14]Sakamoto, T., Ohsawa, K. *JCS(P1)* 2323 (1999).
[15]Torraea, K.E., Kuwabe, S.-I., Buchwald, S.L. *JACS* **122**, 12907 (2000).
[16]Griffiths, C., Leadbeater, N.E. *TL* **41**, 2487 (2000).
[17]Littke, A.F., Dai, C., Fu, G.C. *JACS* **122**, 4020 (2000).
[18]Littke, A.F., Fu, G.C. *ACIEE* **38**, 2411 (1999).
[19]Shen, W. *SL* 737 (2000).
[20]Shen, W., Wang, L. *JOC* **64**, 8873 (1999).
[21]Zhang, C., Huang, J., Trudell, M.L., Nolan, S.P. *JOC* **64**, 3804 (1999).
[22]Kabalka, G.W., Malladi, R.R., Tejedor, D., Kelley, S. *TL* **41**, 999 (2000).
[23]Liebeskind, L.S., Srogl, J. *JACS* **122**, 11260 (2000).
[24]Lau, S.Y.W., Keay, B.A. *SL* 605 (1999).
[25]Nishimura, T., Uemura, S. *JACS* **121**, 11010 (1999).
[26]Overman, L.E., Rosen, M.D. *ACIEE* **39**, 4596 (2000).
[27]Huang, J., Nolan, S.P. *JACS* **121**, 9889 (1999).
[28]Nakamura, K., Okubo, H., Yamaguchi, M. *SL* 549 (1999).
[29]Shen, W., Thomas, S.A. *OL* **2**, 2857 (2000).
[30]Frid, M., Perez, D., Peat, A.J., Buchwald, S.L. *JACS* **121**, 9469 (1999).
[31]Dexter, C.S., Jackson, R.F.W., Elliott, J. *JOC* **64**, 7579 (1999); *T* **56**, 4539 (2000).
[32]Pena, D., Perez, D., Guitian, E., Castedo, L. *OL* **1**, 1555 (1999).
[33]Yamamoto, Y., Nagata, A., Itoh, K. *TL* **40**, 5035 (1999).
[34]Park, S.-B., Cha, J.K. *OL* **2**, 147 (2000).
[35]Oh, C.H., Han, J.W., Kim, J.S., Um, S.Y., Jung, H.H., Jang, W.H., Won, H.S. *TL* **41**, 8365 (2000).

Tris(dibenzylideneacetone)dipalladium–chloroform. 19, 390–392; **20,** 420–422

Fragmentative elimination. Cyclobutoxime benzoates are converted to unsaturated nitriles (except for the 3,3-disubstituted members whose products cannot accommodate a double bond).[1]

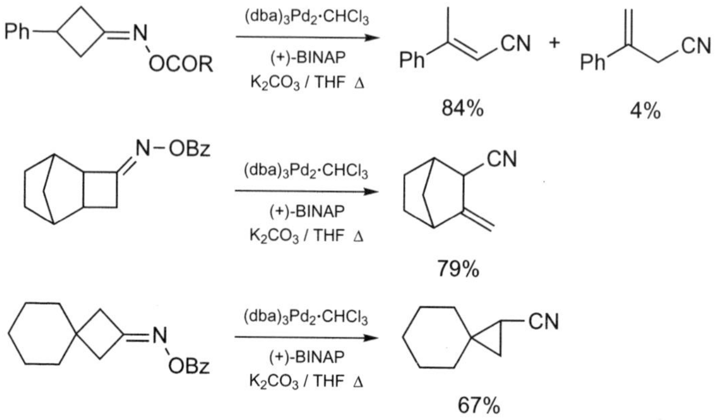

Hydration of vinyl epoxides. 1,2-Diols are formed from the Pd(0)-catalyzed ring opening with bicarbonate ion,[2] in contrast to conventional allylic substitutions of such substrates.

Coupling reactions. Scopes of many well-established reactions continue to be explored. The Heck reaction of alkenyl(2-pyridyl)dimethylsilanes is benefited by the direction of the heteroaromatic group during carbopalladation and the expediency in product purification and catalyst recovery.[3]

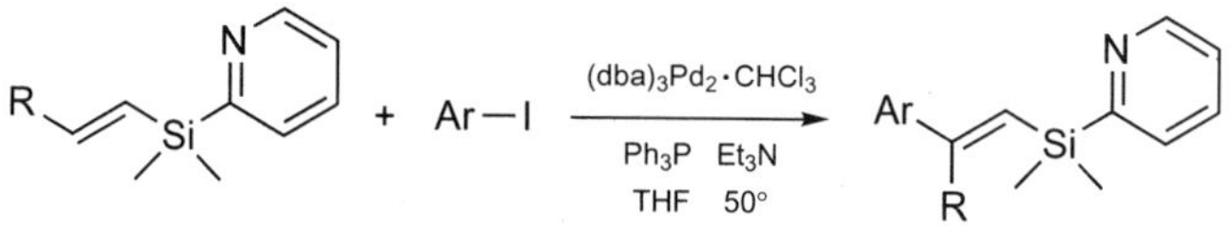

Useful structures arise from regioselective coupling of alkenylstannanes with benzylic bromides,[4] that of alkynylstannanes with allenyl bromides,[5] and hydrosilanes with alkenyl iodides.[6]

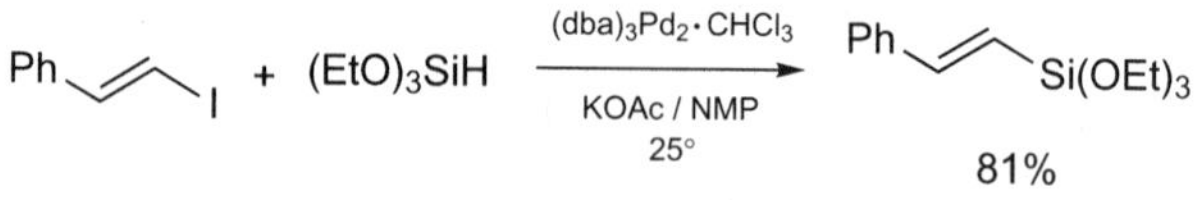

An intermolecular coupling followed by an intramolecular cyclization in tandem provides a key intermediate for the synthesis of ellipticine.[7]

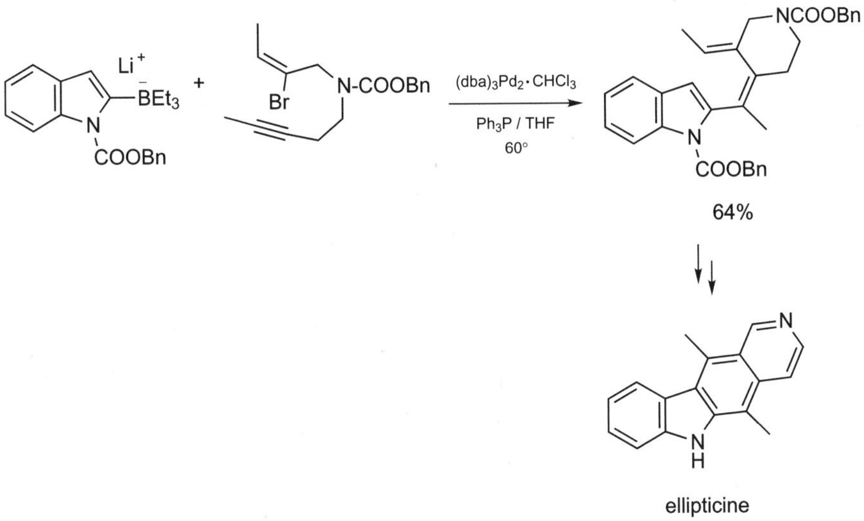

o-Diallylbenzenes are generated from a formal addition–alkylation reaction sequence involving benzyne intermediates.[8] The reagents are allylstannanes and allyl halides.

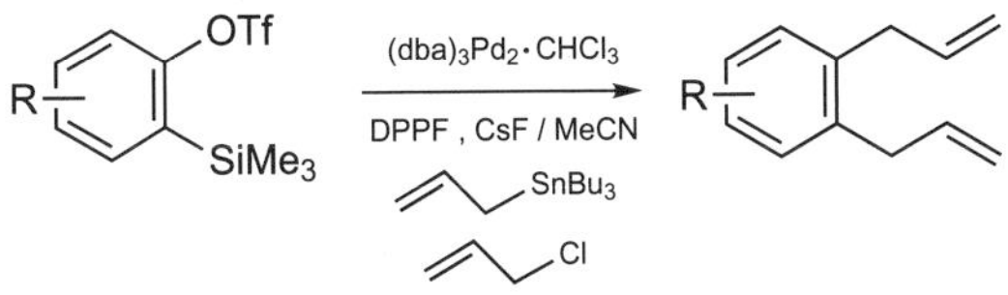

Cycloadditions. When only the allyl halide is present, the preceding reaction results in the formation of 9-substituted phenanthrenes.[9] Polysubstituted benzenes can be synthesized using allyl tosylates and stable alkynes. Furthermore, using $(PhO)_3P$ as a ligand allows 1-alkynes to participate in the reaction.[10]

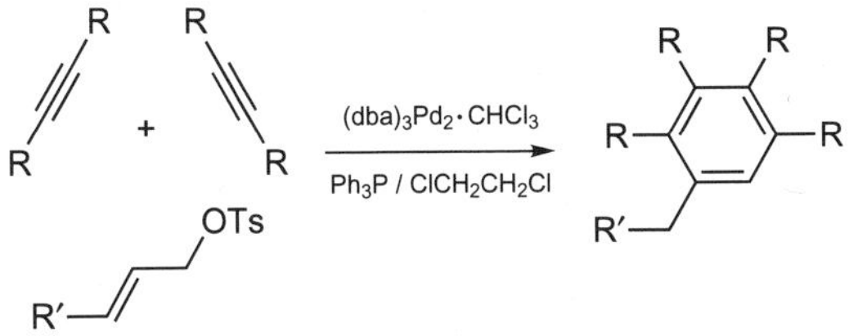

N-Protected 2-amino-1-en-3-ynes undergo [4 + 2]cycloaddition with 1,3-diynes regioselectively, thereby making many 3-substituted 4-alkynylaniline derivatives available.[11]

1,3-Oxazin-2-imines are prepared by trapping the zwitterionic species generated from vinyloxetanes and $(dba)_3Pd_2{\cdot}CHCl_3$ with carbodiimides.[12] The benzoannulated oxo-analogues are derived from carbonylated 2-iodophenols.[13]

(dba)3Pd2·CHCl3
DPPE / THF
98%

A regioselective Diels–Alder reaction between 1,2,4-alkaltrienes and conjugated dienes occurs in the presence of $(dba)_3Pd_2{\cdot}CHCl_3$.[14]

(dba)3Pd2·CHCl3
CH2Cl2

Allylic displacements. 1,1′-Bis(diphenylphosphino)ferrocene renders the Pd(0)-catalyzed alkylation of enolates usually highly diastereoselective.[15]

i-PrMgCl
(dba)3Pd2·CHCl3
ferrocenyl(PPh2)2
67%

[1]Nishimura, T., Uemura, S. *JACS* **122**, 12049 (2000).
[2]Trost, B.M., McEachern, E.J. *JACS* **121**, 8649 (1999).
[3]Itami, K., Mitsudo, K., Kamei, T., Koike, T., Nokami, T., Yoshida, J. *JACS* **122**, 12013 (2000).
[4]Kamlage, S., Sefkow, M., Peter, M.G. *JOC* **64**, 2938 (1999).
[5]Kamlage, S., Sefkow, M., Peter, M.G. *EJOC* 2367 (1999).
[6]Murata, M., Watanabe, S., Masuda, Y. *TL* **40**, 9255 (1999).
[7]Ishikura, M., Hino, A., Katagiri, N. *H* **53**, 11 (2000).
[8]Yoshikawa, E., Radhakrishnan, K.V., Yamamoto, Y. *TL* **41**, 729 (2000).

[9]Yoshikawa, E., Yamamoto, Y. *ACIEE* **39**, 173 (2000).
[10]Tsukada, N., Sugawara, S., Inoue, Y. *OL* **2**, 655 (2000).
[11]Saito, S., Uchiyama, N., Gevorgyan, V., Yamamoto, Y. *JOC* **65**, 4338 (2000).
[12]Larksarp, C., Alper, H. *JOC* **64**, 4152 (1999).
[13]Larksarp, C., Alper, H. *JOC* **64**, 9194 (1999).
[14]Murakami, M., Minamida, R., Itami, K., Sawamura, M. *CC* 2293 (2000).
[15]Braun, M., Laicher, F., Meier, T. *ACIEE* **39**, 3494 (2000).

Tris(dimethylamino)phosphoranes.

As bases.[1] The ylides $(Me_2N)_3P{=}CR_2$ serve as bases in alkylation reactions, nitroaldol condensation, and the Horner reaction.

[1]Palacios, F., Aparicio, D., de los Santos, J.M., Baceiredo, A., Bertrand, G. *T* **56**, 663 (2000).

Tris(dimethylamino)sulfonium difluorotrimethylsilicate.

Cleavage of N-(2-trimethylsilyl)ethanesulfonyl group.[1] SES-protected aziridines decompose on treatment with the title reagent.

[1]Dauben, P., Dodd, R.H. *JOC* **64**, 5304 (1999).

Tris(dimethylphosphinito)platinum hydride.

Amides from nitriles.[1] Nitriles and amines react in the presence of $[Me_2(HO)P]_3PtH$ at 160° to afford amides in moderate yields.

$$R{-}CN + R'R''N{-}H \xrightarrow[\text{DME}\ 160^\circ]{H{-}Pt[P(OH)Me_2]_3} R{-}C(=O){-}NR'R''$$

$$R{-}CN + HOCH_2CH_2NH_2 \xrightarrow[\text{DME}\ 160^\circ]{H{-}Pt[P(OH)Me_2]_3} \text{2-R-oxazoline}$$

[1]Cobley, C.J., van den Heuvel, M., Abbadi, A., de Vries, J.G. *TL* **41**, 2467 (2000).

Tris-(2,6-diphenylbenzyl)silyl bromide.

Carboxyl protection.[1] Highly hindered silyl carboxylates are formed by the AgOTf-mediated esterification. The silyl esters are stable to aq HOAc, aq NaOH, and $LiAlH_4$, but they are cleaved by *t*-BuOK–DMSO, or HF–pyridine in THF at 50°, and reduced by *i*-Bu_2AlH to alcohols.

[1]Iwasaki, A., Kondo, Y., Maruoka, K. *JACS* **122**, 10238 (2000).

Tris(pentafluorophenyl)borane. 20, 422

Imine reduction.[1] In the presence of $(C_6F_5)_3B$, imines undergo hydrosilylation with $PhMe_2SiH$. The reagent is probably $[PhMe_2Si]^+ [(C_6F_5)_3BH]^-$.

Silyl ethers.[2] A general method for *O*-silylation at room temperature in >85% yield employs Ph_3SiH and $(C_6F_5)_3B$. Under these conditions, epoxides are cleaved but ethers, esters, lactones, nitro compounds, alkenes, and alkynes are preserved.

[1]Blackwell, J.M., Sonmor, E.R., Scoccitti, T., Piers, W.E. *OL* **2**, 3921 (2000).
[2]Blackwell, J.M., Foster, K.L., Beck, V.H, Piers, W.E. *JOC* **64**, 4887 (1999).

Tris(trimethylsilyl)silane. 19, 393; **20,** 423

Elimination. Radical formation from 2-(benzenesulfinyl)allyl bromides on treatment with $(Me_3Si)_3SiH$–AIBN is immediately followed by elimination to afford allenes.[1]

$(Me_3Si)_3SiH$; AIBN / PhH; 80°

80%

Hydrodehalogenation. A convenient access to β-lactams[2] consists of desilylative cyclization and removal of the bromine atom from the α-position with $(Me_3Si)_3SiH$–AIBN.

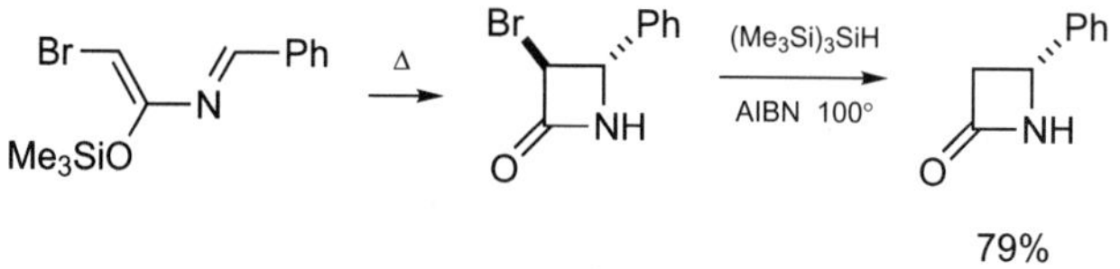

Δ; $(Me_3Si)_3SiH$; AIBN 100°

79%

α-Boryl radicals. Radicals are derived from halomethylboronates. Addition of the radicals to multiple bonds accomplishes a carbon chain homologation.[3]

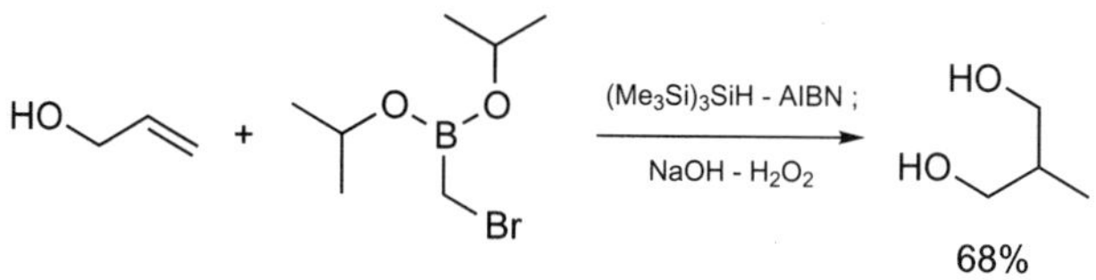

$(Me_3Si)_3SiH$ - AIBN ; NaOH - H_2O_2

68%

Cyclization. Construction of a tetracyclic intermediate of aspidospermidine is based on an intramolecular radical addition to a double bond.[4] The transposed radical interacts with an azide group to complete the process.

(Me3Si)3SiH, AIBN / PhH, 80°

40%

aspidospermidine

[1]Delouvrie, B., Lacote, E., Fensterbank, L., Malacria, M. *TL* **40**, 3565 (1999).
[2]Bandini, E., Favi, G., Martelli, G., Panunzio, M., Piersanti, G. *OL* **2**, 1077 (2000).
[3]Batey, R.A., Smil, D.V. *TL* **40**, 9183 (1999).
[4]Patro, B., Murphy, J.A. *OL* **2**, 3599 (2000).

Tris(trimethylsilyl)silyllithium.

α-(Trimethylsiloxy)alkenyl bis(dimethylsilyl)silanes.[1] These novel compounds are products from reaction of $(Me_3Si)_3SiLi$ with ketenes (after hydrolytic workup). Transsilylation of the Si→O version occurs during the process.

(Me3Si)3SiLi

78%

[1]Naka, A., Ohshita, J., Kunai, A., Lee, M.E., Ishikawa, M. *JOMC* **574**, 50 (1999).

Trityl chloride.

Hydroxyl protection.[1] Useful variations in chemoselectivity for the etherification of compounds containing different types of hydroxyl groups use TrCl and t-$BuMe_2SiCl$. Thus, *o*-hydroxybenzyl alcohol is silylated at the aliphatic moiety but tritylated at the phenolic oxygen.

[1]Sefkow, M., Kaatz, H.*TL* **40**, 6561 (1999).

Trityl tetrakis(pentafluorophenyl)borate.

Glycosylation.[1] Pyranosyl phenyl carbonates serve as glycosyl donors in their reaction with sugar derivatives containing a free hydroxyl group. The reaction is catalyzed by $Tr[B(C_6F_5)_4]$. An anomeric thioether is not affected and may be present in a coupling component. However, it can be transformed into a glycosyl donor afterward by merely performing an oxidation in situ (adding $NaIO_4$).

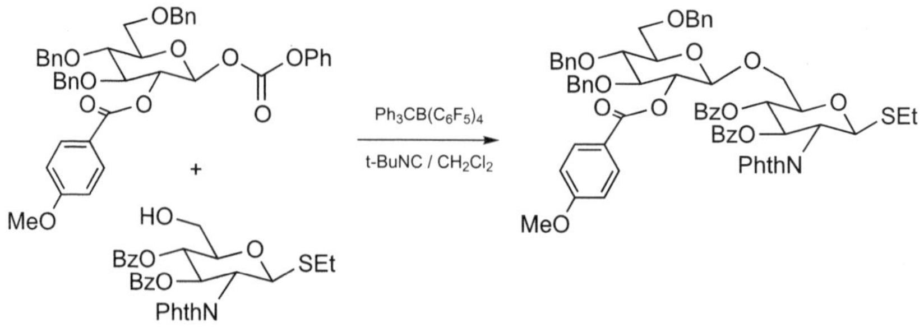

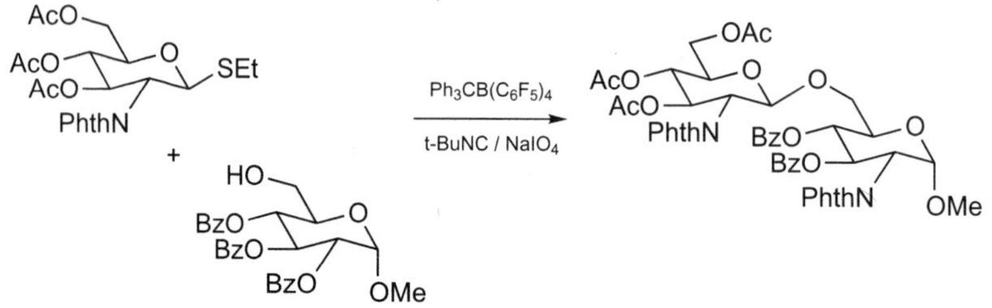

Alkylmetallation.[2] The title compound effectively promotes the reaction of alkynes with alkylzirconium species derived from hydrozirconation of alkenes.

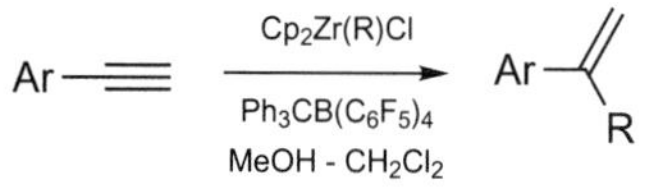

[1]Takeuchi, K., Tamura, T., Mukaiyama, T. *CL* 122, 124 (2000).
[2]Yamanoi, S., Ohrui, H., Seki, K., Matsumoto, T., Suzuki, K. *TL* **40**, 8407 (1999).

Tungsten carbene and carbyne complexes. 20, 424–425

Conjugate allylation.[1] The reaction of Fischer-type conjugated carbene complexes of tungsten with allylic alcohols under basic conditions proceeds via an oxygen attack on the tungsten atom to initiate a rearrangement. At some stage, transesterification also occurs.

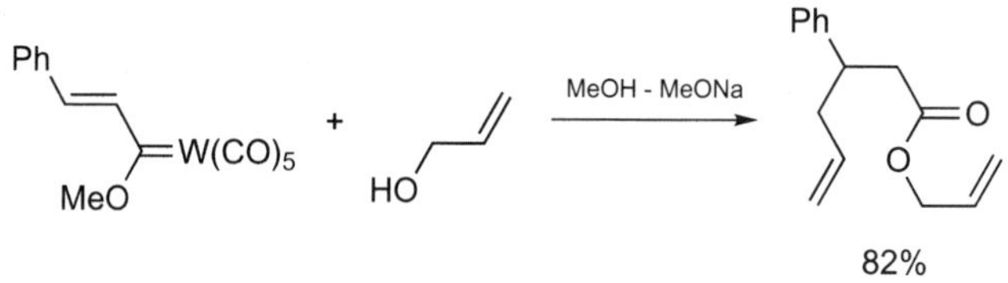

Cycloadditions. The same type of Fischer-type complexes undergo cycloaddition with enamines. Interestingly, acyclic[2] and cyclic enamines[3] afford products of different skeletons.

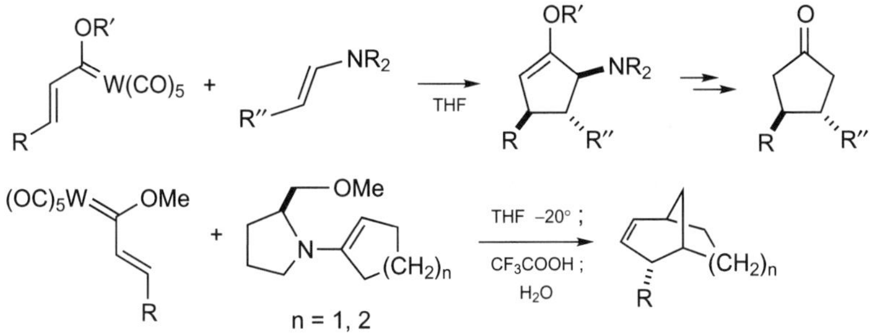

[1]Barluenga, J., Rubio, E., Lopez-Pelegrin, J.A., Tomas, M. *ACIEE* **38**, 1091 (1999).
[2]Barluenga, J., Tomas, M., Ballesteros, A., Santamaria, J., Brillet, C., Garcia-Granda, S., Pinera-Nicolas, A., Vazquez, J.T.*JACS* **121**, 4516 (1999).
[3]Barluenga, J., Ballesteros, A., Santamaria, J., de la Rua, R.B., Rubio, E., Tomas, M.*JACS* **122**, 12874 (2000).

Tungsten(VI) chloride. 19, 395; 20, 425

Reduction.[1] Sulfoxides are reduced to sulfides and sulfonyl chlorides to disulfides by the combination of WCl_6 with Zn or NaI.

Acetalization.[2] Admixture of carbonyl compounds with a trialkyl orthoformate and WCl_6 accomplishes the derivatization, usually giving the diethylacetals in >85% yield.

Ring expansion.[3] Cyclic dithioacetals (five-, six-membered) undergo rearrangement to the chlorinated 1,4-dithiacycloalkenes on treatment with WCl_6 in DMSO at room temperature.

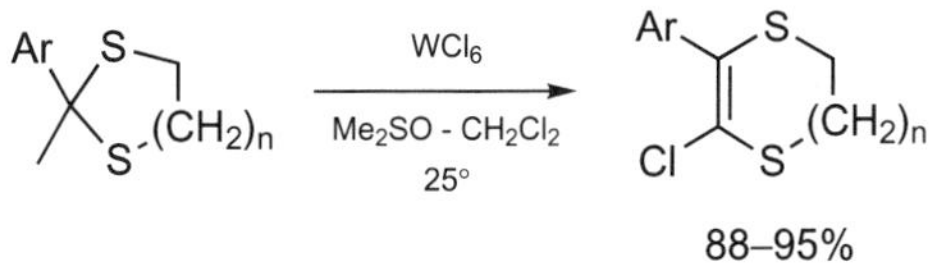

[1]Firouzabadi, H., Karimi, B. *S* 500 (1999).
[2]Firouzabadi, H., Iranpoor, N., Karimi, B. *SC* **29**, 2255 (1999).
[3]Firouzabadi, H., Iranpoor, N., Karimi, B. *SL* 413 (1999).

Tungsten pentacarbonyl tetrahydrofuran.

Cycloaddition.[1] Cyclic Fischer carbene complexes are generated from *o*-ethynylaryl carbonyl compounds with the title reagent at room temperature. When enol ethers or enamines are present, a Diels–Alder reaction–retro-Diels–Alder sequence occurs to furnish substituted naphthalenes (7 examples, 54–84%).

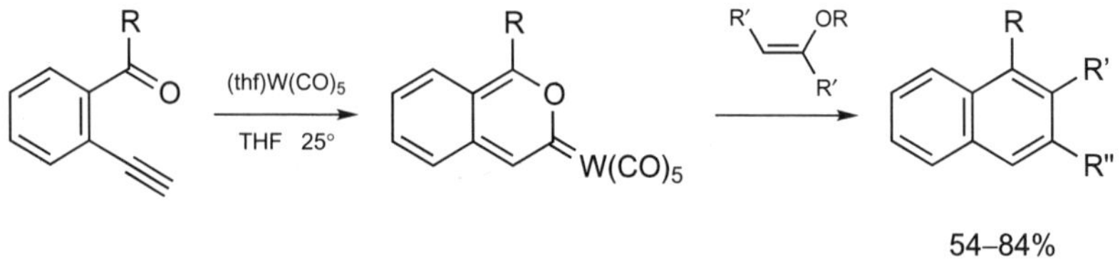

R O (thf)W(CO)5 THF 25° R O W(CO)5 R′ OR R′ R R' R" 54–84%

[1] Iwasawa, N., Shido, M., Maeyama, K., Kusama, H. *JACS* **122**, 10226 (2000).

U

Ultrasound. 15, 363; **16,** 377–379; **18,** 395; **19,** 396; **20,** 426

Selective cleavage of t-butyldimethylsilyl ethers.[1] Siyl ethers such as those situated at an ortho-position of an aryl carbonyl group is subject to rapid removal by ultrasonication in MeOH–CCl_4.

OTBS, CHO, TBSO, OTBS → (MeOH - CCl_4,)))) 50°) → OTBS, CHO, TBSO, OH; 97%

[1]de Groot, A., Dommisse, R.A., Lemiere, G.L. *T* **56**, 1541 (2000).

Urea–hydrogen peroxide–trifluoroacetic anhydride.

Oxidations. This system oxidizes electron-deficient pyridines to the *N*-oxides.[1] In this regard, it is more effective than MTO–H_2O_2. Sulfides are converted to sulfones.[2]

Saturated hydrocarbons are functionalized by this oxidation system. For example, cyclohexane gives cyclohexyl trifluoroacetate in 80% yield.[3]

[1]Caron, S., Do, N.M., Sieser, J.E. *TL* **41**, 2299 (2000).
[2]Balicki, R . *SC* **29**, 2235 (1999).
[3]Moody, C.J., O'Connell, J.L. *CC* 1311 (2000).

Urea nitrate.

Nitration.[1] A special application of this method is a regioselective synthesis of 4-halo-1,2-dinitroarene derivatives from *m*-nitrohaloarenes. The dinitroarenes are important precursors of benzimidazoles, quinoxalines, and related compounds.

X, NO_2 → (urea nitrate, H_2SO_4) → X, NO_2, NO_2

X = F, Cl, Br

[1]Mundla, S.R. *TL* **41**, 4277 (2000).

Vanadyl ethoxydichloride. 20, 428

Reductive elimination.[1] Cross-coupling of two ligands of organozincs is a reductive elimination process, it occurs when the latter species are brought into contact with $VO(OEt)Cl_2$.

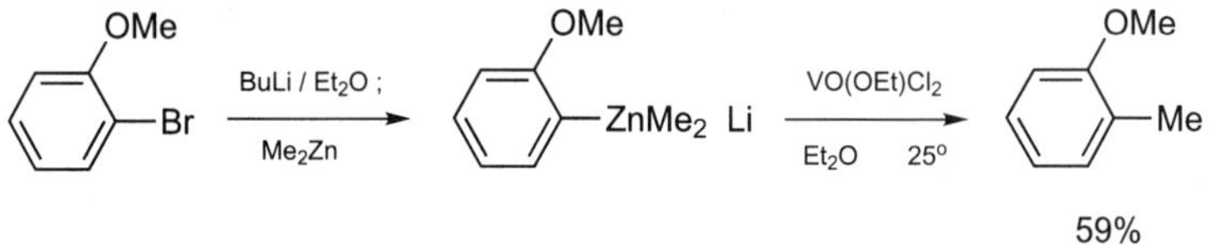

Vicinal dialkylation.[2] The oxovanadium-induced reaction of cyclic enones with dialkylzinc reagents is followed by oxidation of the alkylzinc enolates and another C—C bond formation process, resulting in the generation of vicinal dialkylation products.

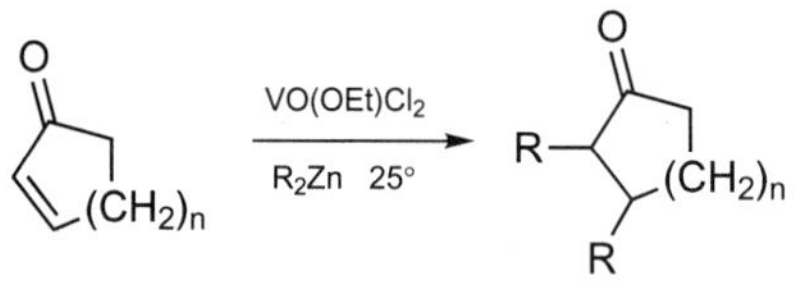

Isomerization of epoxides.[3] This vandyl compound is effective for promoting the rearrangment of epoxides to carbonyl compounds.

α-Diketones.[4] α-Ketols undergo aerial oxidation to α-diketones and α-hydroxy esters to α-ketoesters in MeCN at room temperature. $VO(OEt)Cl_2$ as well as $VOCl_3$ are equally effective as catalyst.

[1]Hirao, T., Takada, T., Ogawa, A. *JOC* **65**, 1511 (2000).
[2]Hirao, T., Takada, T., Sakurai, H. *OL* **2**, 3659 (2000).
[3]Martinez, F., del Campo, C., Llama, E.F. *JCS(P1)* 1749 (2000).
[4]Kirihara, M., Ochiai, Y., Takizawa, S., Takahata, H., Nemoto, H. *CC* 1387 (1999).

Vinamidinium hexafluorophosphate.

Pyridines.[1] Reaction with ketones leads to annulation that provides trisubstituted pyridines.

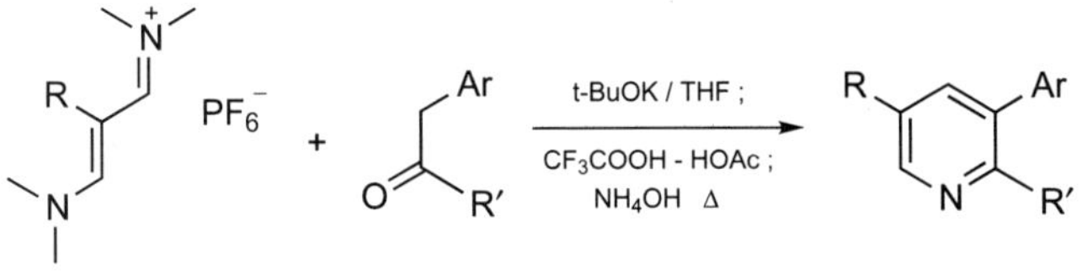

[1]Marcoux, J.-F., Corley, E.G., Rossen, K., Pye, P., Wu, J., Robbins, M.A., Davies, I.W., Larsen, R.D., Reider, P.J. *OL* **2**, 2339 (2000).

Xenon(II) fluoride. 13, 345; **19,** 399; **20,** 430

α-Fluoroketones.[1] The reaction of trimethylsilyl enol ethers with XeF_2 in MeCN involves radical cation intermediates.

[1]Ramsden, C.A., Smith, R.G. *OL* **1**, 1591 (1999).

Y

Ytterbium. 14, 348; **15,** 366; **16,** 384; **18,** 401; **19,** 400; **20,** 431

Silanes. Phenylsilanes of the type $Ph(R)SiH_2$ are arylated to provide Ph(R)ArSiH on mediation by Yb.[1]

[1]Jin, W.-S., Makioka, Y., Kitamura, T., Fujiwara, Y. *CC* 955 (1999).

Ytterbium(III) chloride.

Acetylation.[1] Monoacetylation of symmetrical 1,2-diols with acetic anhydride in the presence of $YbCl_3$ or $DyCl_3$ is possible.

Allylation.[2] Various aldehydes react with allylsilanes on catalysis by $YbCl_3$ in MeCN. *o*-Anisaldehyde is converted to 2-(1,6-heptadien-4-yl)anisole under these reaction conditions.

Diels–Alder reaction. $YbCl_3$ is effective as a catalyst for the Diels–Alder reaction involving unactivated dienes.[3]

α-Hydroxyarylacetic esters.[4] Arylglyoxals are transformed into the esters when they are treated with $YbCl_3$ in an alcoholic solvent. The presence of a base (e.g., Et_3N) is crucial.

[1]Clarke, P.A., Holton, R.A., Kayaleh, N.E. *TL* **41**, 2687 (2000).
[2]Fang, X., Watkin, J.G., Warner, B.P. *TL* **41**, 447 (2000).
[3]Fang, X., Warner, B.P., Watkin, J.G. *SC* **30**, 2669 (2000).
[4]Likhar, P.R., Bandyopadhyay, A.K. *SL* 538 (2000).

Ytterbium(III) triflate. 18, 402–403; **19,** 401–402; **20,** 431–433

Alcohols and derivatives. $Yb(OTf)_3$ serves as a catalyst for etherification of alcohols with 4-methoxybenzyl alcohols.[1] Deacetylation of esters (particularly ArOAc) by transesterification[2] to isopropanol is observed. A preparation of alkyl glycosides[3] involves reaction of glycosyl esters with trialkyl borates in the presence of $Yb(OTf)_3$.

Benzyl alcohols are oxidized to aromatic aldehydes by nitric acids using $Yb(OTf)_3$ as catalyst.[4] Byproducts from this reaction are water and nitrogen oxides only.

Detritylation.[5] $Yb(OTf)_3$ catalyzes cleavage of trityl ethers and *N*-tritylamines in THF (containing one equiv of water) at room temperature.

Conjugate additions. The $Yb(OTf)_3$-catalyzed addition of a chiral *N*-amino-2-methoxymethylpyrrolidine to alkenyl sulfones is crucial to a process for the synthesis of optically active β-sulfonylamines.[6]

Allylation.[7] A free radical addition of *N*-bromoacetyloxazolidinones to branched allylsilanes with subsequent elimination effects homologation of the *N*-acyl group. Both the addition and the elimination steps are promoted by $Yb(OTf)_3$.

PhMe$_2$Si + Br... $\xrightarrow[\text{THF } -78°]{Yb(OTf)_3}$

81% (*Z* : *E* 91 : 1)

Reactions of imines. Silyl alkynyl ethers and aldimines undergo catalyzed condensation, the strained heterocycles that ensue open to afford conjugated amides.[8] $Yb(OTf)_3$ is also a useful catalyst for the reaction of organocuprates with *N*-sulfonylimines.[9]

Friedel–Crafts reactions.[10] A synthesis of α-hydroxyarylacetic esters is by reaction of arenes with glyoxalic esters. Such esters are amenable to resolution with lipase.

Alkylations. Under high pressure, cyclopropane-1,1-dicarboxylic esters react with nucleophiles such as β-ketoesters.[11] A polyfunctional carbon chain can be built from an activated diene by sulfonylation at one end.[12]

COOMe + COOMe COOMe $\xrightarrow[\text{MeCN } 70° \text{ 8 kbar}]{Yb(OTf)_3}$ COOMe COOMe COOMe

OSiMe$_3$ Ph + O Mes $\xrightarrow[\text{CH}_2\text{Cl}_2 \; -90°; \; \text{Bu}_4\text{NF - MeI}]{SO_2 / Yb(OTf)_3}$ Ph O OR* SO$_2$Me

79%

$Yb(OTf)_3$ catalyzes an ene-type reaction of *N*-tosylimines and α-methylstyrene, that is dramatically enhanced by small amount of Me_3SiCN.[13]

Oxymercuration. Homoallylic alcohols apparently form hemiacetals with acetone and benzaldehyde in the presence of $Yb(OTf)_3$. Reaction of the hemiacetals with HgClOAc completes a diastereoselective oxymercuration,[14] a key process for generating (1,3,5, . . .)polyols. Success of the reaction depends on $Yb(OTf)_3$.

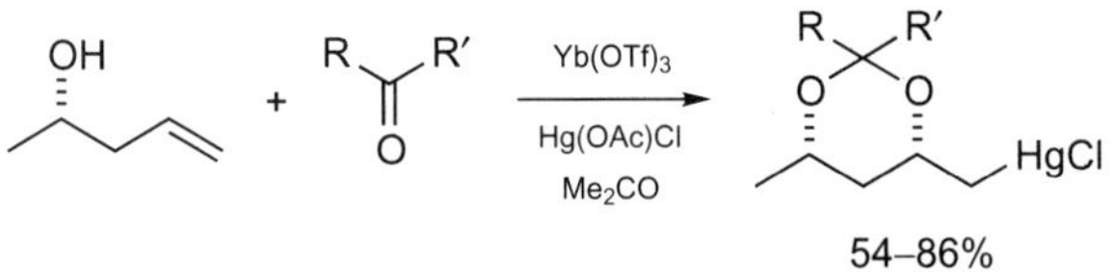

Diels–Alder reactions. The cycloaddition of 1,3-cyclohexadiene with electron-deficient dienophiles is promoted by the dihydrate of $Yb(OTf)_3$ under high pressure.[15]

Radical cyclization.[16] The dehydroabietane skeleton is readily formed by an oxidative radical cyclization reaction to which a catalytic amount of $Yb(OTf)_3$ dihydrate is also added.

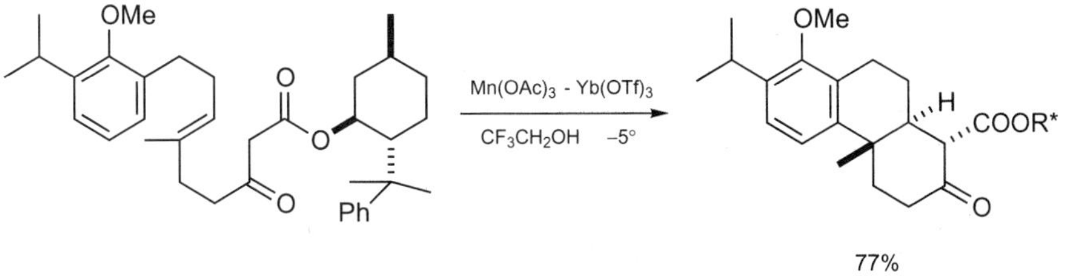

[1]Sharma, G.V.M., Mahalingam, A.K. *JOC* **64**, 8943 (1999).
[2]Sharma, G.V.M., Ilangovan, A. *SL* 1963 (1999).
[3]Yamanoi, T., Iwai, Y., Inazu, T. *H* **53**, 1263 (2000).
[4]Barrett, A.G.M., Braddock, D.C., McKinnell, R.M., Waller, F.J. *SL* 1489 (1999).
[5]Lu, R.J., Liu, D., Giese, R.W. *TL* **41**, 2817 (2000).
[6]Enders, D., Muller, S.F., Raabe, G., Runsink, J. *EJOC* 879 (2000).
[7]Porter, N.A., Zhang, G., Reed, A.D. *TL* **41**, 5773 (2000).
[8]Shindo, M., Oya, S., Sato, Y., Shishido, K. *H* **52**, 545 (2000).
[9]Li, G., Wei, H.-X., Hook, J.D. *TL* **40**, 4611 (1999).
[10]Zhang, W., Wang, P.G. *JOC* **65**, 4732 (2000).
[11]Kotsuki, H., Arimura, K., Muruzawa, R., Ohshima, R. *SL* 650 (1999).
[12]Narkevitch, V., Schenk, K., Vogel, P. *ACIEE* **39**, 1806 (2000).
[13]Yamanaka, M., Nishida, A., Nakagawa, M. *OL* **2**, 159 (2000).
[14]Dreher, S.D., Hornberger, K.R., Sarraf, S.T., Leighton, J.L. *OL* **2**, 3197 (2000).
[15]Kinsman, A.C., Kerr, M.A. *OL* **2**, 3517 (2000).
[16]Yang, D., Ye, X.-Y., Gu, S., Xu, M. *JACS* **121**, 5579 (1999).

Ytterbium(III) tris[(perfluorobutanesulfonyl)methide].

Meerwein–Ponndorf–Verley reduction.[1] This catalyst shows high activity under conditions that $AlCl_3$, $(i\text{-}PrO)_3Al$, and $Yb(OTf)_3$ are ineffective. Thus, PhCHO is reduced in isopropanol at 50° in 85% yield.

Friedel–Crafts acylation. With this catalyst acylation in fluorous biphasic system is achieved.[2] Note that the triflamide $Yb(NTf_2)_3$ is also useful for the same purpose (not specified in fluorous solvent).[3]

[1]Nishikido, J., Yamamoto, F., Nakajima, H., Mikami, Y., Matsumoto, Y., Mikami, K. *SL* 1990 (1999).
[2]Barrett, A.G.M., Braddock, D.C., Catterick, D., Chadwick, D., Henschke, J.P., McKinnell, R.M. *SL* 847 (2000).
[3]Nie, B., Xu, J., Zhou, G. *JCR(S)* 446 (1999).

Yttryl isopropoxide

Transacylation.[1] This compound, $(i\text{-PrO})_{13}Y_5O$, catalyzes acyl transfer from enol carboxylates to alcohols, and in some cases (e.g., piperidine-2-methanol), the selective *O*-acylation of amino alcohols.

[1]Lin, M.-H., RajanBabu, T.V. *OL* **2**, 997 (2000).

Z

Zeolites. 15, 367; **18,** 405–406; **19,** 403–404; **20,** 434

Reactions. Various types of zeolites continue to be exploited as catalysts or supports for common transformations: acetylation of alcohols with HOAc,[1] reduction of 2,3-epoxy alcohols and alkenyl epoxides with sodium cyanoborohydride,[2] acetalization[3] and thioacetalization[4] of carbonyl compounds. A synthesis of symmetrical dialkylureas[5] consists of heating amines with acetoacetanilide with HSZ-360 at 180°.

[1]Narender, N., Srinivasu, P., Kulkarni, S.J., Raghavan, K.V. *SC* **30**, 1887 (2000).
[2]Gupta, A., Vankar, Y.D. *TL* **40**, 1369 (1999).
[3]Tajbakhsh, M., Mohajerani, B., Heravi, M. M. *SC* **29**, 135 (1999).
[4]Ballini, R., Barboni, L., Maggi, R., Sartori, G. *SC* **29**, 767 (1999).
[5]Bigi, F., Frullanti, B., Maggi, R., Sartori, G., Zambonin, E. *JOC* **64**, 1004 (1999).

Zinc. 13, 346–347; **14,** 349–350; **16,** 386–387; **17,** 406–407; **18,** 406–408; **19,** 404–405; **20,** 435–436

Reduction. A selective reduction of nitroarenes to arylamines occurs with Zn in near-critical water (at 250°).[1] Cyclization that follows reduction is expected in the case of 2-nitroaryl compounds containing a leaving group at proper distance,[2] $ArNO_2$ is converted to $ArNR_2$ by a reductive alkylation process in protic solvents (promoted by Zn or Sn).[3]

A general method for accessing allylamines from nitrones[4] is by a Grignard reaction followed by reduction of the resulting hydroxylamines with Zn.

Zn–$AlCl_3$ in MeCN is reported for reductive coupling of carbonyl compounds to form alkenes.[5]

Nonaqueous reduction conditions compatible with $ArSO_2Cl$ employs Zn–Me_2SiCl_2 in dimethylacetamide (or dimethylimidazolone). Thiols are obtained as products.[6]

trans-1,2-Cycloalkanediols (five-, six-membered) are synthesized by reductive coupling of the dialdehydes using Zn, $Cp_2Ti(Ph)Cl$, Me_3SiCl in THF.[7] $Cp_2Ti(Ph)Cl$ is formed in situ by treating Cp_2TiCl_2 sequentially with *i*-PrMgCl and PhMgBr.

Organozinc reactions. Barbier alkylation of aldehydes with 2-arenesulfinylallyl halides is highly diastereoselective.[8] Practical conditions for the Reformatsky reaction of aldehydes are further defined.[9]

The selectivity of Rieke zinc in its reaction with organic halides is demonstrated in coupling of a benzylic bromide.[10]

Br Ph Br → Zn* / THF, CuCN-LiBr, $CH_2=C(Cl)CH_2Cl$ → Cl Ph Br 64%

[1]Boix, C., Poliakoff, M. *JCS(P1)* 1487 (1999).
[2]Le Gall, E., Malassene, R., Toupet, L., Hurvois, J.-P., Moinet, C. *SL* 1383 (1999).
[3]Bieber, L.W., da Costa, R.C., da Silva, M.F. *TL* **41**, 4827 (2000).
[4]Merino, R., Anoro, S., Franco, S., Gascon, J.M., Martin, V., Merchan, F.L., Revuelta, J., Tejero, T., Tunon, V. *SC* **30**, 2989 (2000).
[5]Dutta, D.K., Konwar, D. *TL* **41**, 6227 (2000).
[6]Uchiro, H., Kobayashi, S. *TL* **40**, 3179 (1999).
[7]Yamamoto, Y., Hattori, R., Itoh, K. *CC* 825 (1999).
[8]Marquez, F., Llebaria, A., Delgado, A. *OL* **2**, 547 (2000).
[9]Chattopadhyay, A., Salaskar, A. *S* 561 (2000).
[10]Guijarro, A., Rosenberg, D.M., Rieke, R.D. *JACS* **121**, 4155 (1999).

Zinc–copper couple.

β-Amino esters. Substitution of *N*-protected benzenesulfonylamines by Reformatsky reagents that are generated by treating α-bromoesters with a Zn–Cu couple in dichloromethane represents a new entry to the β-amino esters.[1] A second route involves addition of the Reformatsky reagents to aldimines derived from *o*-anisidine.[2] No β-lactams are formed due to steric hindrance.

OMe N R + Br COOMe → Zn-Cu, CH_2Cl_2 → OMe NH MeOOC R 51–92%

[1]Mecozzi, T., Petrini, M. *TL* **41**, 2709 (2000).
[2]Adrian, Jr., J.C., Barkin, J.L., Hassib, L. *TL* **40**, 2457 (1999).

Zinc bromide. 13, 349; **15,** 368; **16,** 389–391; **18,** 409; **19,** 409; **20,** 438–439

Cleavage of t-butyl ethers and esters.[1] A mild procedure for the cleavage of these ethers and esters employs $ZnBr_2$ in dichloromethane.

[1]Wu, Y.-q., Limburg, D.C., Wilkinson, D.E., Vaal, M.J., Hamilton, G.S. *TL* **41**, 2847 (2000).

Zinc chloride. 13, 349–350; **15,** 368–371; **16,** 391–392; **18,** 410–411; **19,** 409–410; **20,** 439

Amides from oximes.[1] Microwave irradiation of aldoximes with $ZnCl_2$ (solvent free) accomplishes the transformation into primary amides.

1,4-Diketones.[2] Methyl ketones are alkylated with α-bromoketones in the presence of $ZnCl_2$, Et_2NH, and *t*-BuOH.

Heterocycles. A route to 2,3-disubstituted furans takes advantage of the Cu–Zn-transmetallation (with $ZnCl_2$) from enolates derived from conjugate organocuprate addition to enones, and aldol reaction of zinc enolates to an alkoxyacetaldehyde.[3]

R″ O + OHC OTHP; R_2CuLi ; $ZnCl_2$; TsOH / THF; R′; R″ O R′ R

The addition of organozinc reagents to vinylstannanes afford *gem*-diorganometallic species. When the zinc reagents are derived from α-lithio-*N,N*-dimethylhydrazones, the adducts can be converted to *N*-dimethylaminopyrroles by oxygen.[4]

$NNMe_2$ R R′; t-BuLi ; BuZnI ; $SnBu_3$ $ZnCl_2$; [$NNMe_2$ R R′ ZnBu $SnBu_3$]; O_2; Me_2N N R R′

[1]Loupy, A., Regnier, S. *TL* **40**, 6221 (1999).
[2]Nevar, N.M., Kel'in, A.V., Kulinkovich, O.G. *S* 1259 (2000).
[3]Mendez-Andino, J., Paquette, L.A. *OL* **2**, 4095 (2000).
[4]Nakamura, M., Hara, K., Sakata, G., Nakamura, E. *OL* **1**, 1505 (1999).

Zinc iodide.

2,3-Dihydroisoxazoles.[1] Propargylic *N*-hydroxylamines undergo cyclization on treatment with catalytic amounts of ZnI_2 and DMAP at room temperature.

Cyclopropanation.[2] 2-Triethylsilylvinyl phenyl selenide behaves as a (2-phenylseleno-2-triethylsilyl)ethylidenating agent for conjugated double bonds in the presence of ZnI_2. Chiral adducts are obtained from reaction of di-(−)-menthyl methylenemalonate.

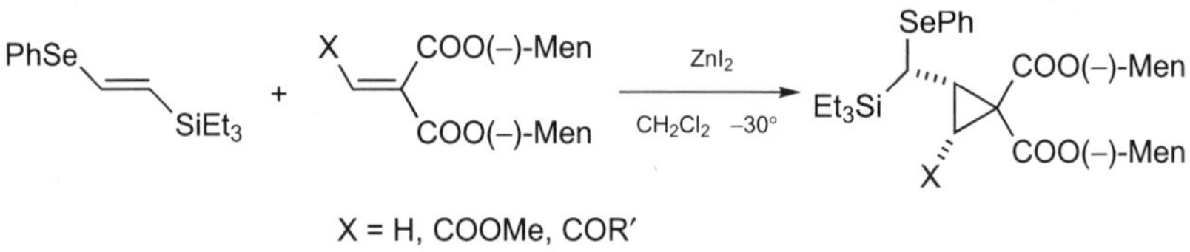

X = H, COOMe, COR′

[1]Aschwanden, P., Frantz, D.E., Carreira, E.M. *OL* **2**, 2331 (2000).
[2]Yamazaki, S., Kataoka, H., Yamabe, S. *JOC* **64**, 2367 (1999).

Zinc nitrate.

Hydrolysis of acetonides.[1] In MeCN, terminal acetonides can be selectively hydrolyzed by $Zn(NO_3)_2 \cdot 6H_2O$ to afford the diols.

[1]Vijayasaradhi, S., Singh, J., Aidhen, I.S. *SL* 110 (2000).

Zinc tetrafluoroborate.

Cleavage of TBS ethers.[1] Trimethylsilyl ethers are cleaved on exposure to $Zn(BF_4)_2$ in water.[1]

[1]Ranu, B. C., Jana, U., Majee, A. *TL* **40**, 1985 (1999).

Zinc triflate–tertiary amine.

Alkynylzinc reagents. The direct alkynylation of aldehydes is subject to asymmetric induction in the presence of a chiral base such as (+)-*N*-methylephedrine.[1] Addition to the C═N bond of *N*-tosylimines and nitrones by this procedure is also successful.[2]

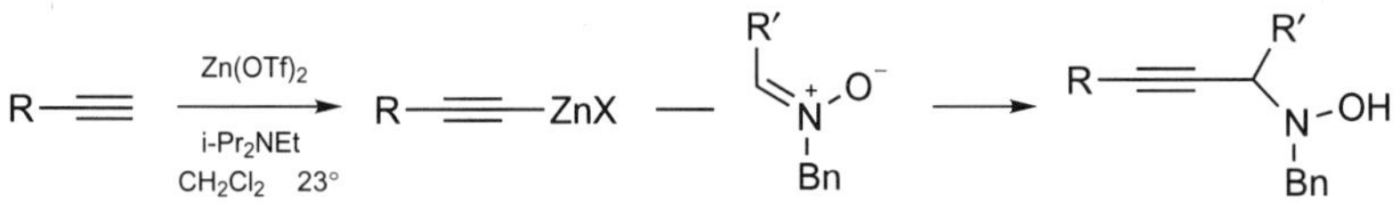

[1]Frantz, D.E., Fassler, R., Carreira, E.M. *JACS* **122**, 1806 (2000).
[2]Frantz, D.E., Fassler, R., Carreira, E.M. *JACS* **121**, 11245 (1999).

Zirconacycles.

Cyclopentenes. Zirconacyclopentadienes prepared from two alkynes react with CO in the presence of BuLi to give substituted cyclopentenones.[1] With CuCl as catalyst, the reaction of zirconacyclopentadienes with β-iodo-α,β-unsaturated carbonyl compounds provides cyclopentadienes.[2] In the cases of 3-iodo-2-cycloalkenones, the products are spirocycles.

$\xrightarrow[\text{THF } 25^\circ]{\text{CuCl}}$

48%

1,5-Dienes and 1,5-enynes.[3] Regioselective insertion of zirconacyclopentenes by alkynylmetals leads, after hydrolysis, to acyclic products.

Cp_2Zr + Li—≡—R $\xrightarrow[H_3O^+]{23^\circ\,;}$

[1]Takahashi, T., Huo, S., Hara, R., Noguchi, Y., Nakajima, K., Sun, W.-H. *JACS* **121**, 1094 (1999).
[2]Xi, C., Kotora, M., Nakajima, K., Takahashi, T. *JOC* **65**, 945 (2000).
[3]Dumond, Y., Negishi, E. *JACS* **121**, 11223 (1999).

Zirconia, sulfated.

Koch carbonylation.[1] Tertiary alcohols give carboxylic acids on treatment with the sulfated zirconia (a solid superacid) under carbon monoxide at 150°.

Cyclodehydration.[2] Heating 1,*n*-diols such as 1,4-butanediols with the zirconia readily furnishes cyclic ethers.

Glycosylation.[3] The stereoselectivity of the zirconia-catalyzed glycosylation of 2-deoxyglucopyranosyl fluoride is found to be highly solvent dependent.

$\xrightarrow{ZrO_2\,/\,SO_4]}$

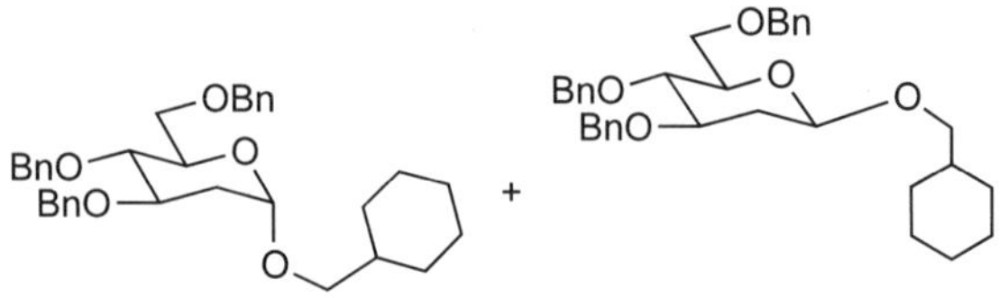

MeCN 25°	88	:	12
Et_2O / MS-5A 0°	19	:	81

[1]Mori, H., Wada, A., Xu, Q., Souma, Y. *CL* 136 (2000).
[2]Wali, A., Pillai, S.M. *JCR(S)* 326 (1999).
[3]Toshima, K., Kasumi, K., Matsumura, S. *SL* 813 (1999).

Zirconocene. 20, 441–442

α-Silylamides.[1] "Cp_2Zr" induces a retro-Brook rearrangement of 3-siloxy-2-aza-1,3-dienes. The products can be alkylated.

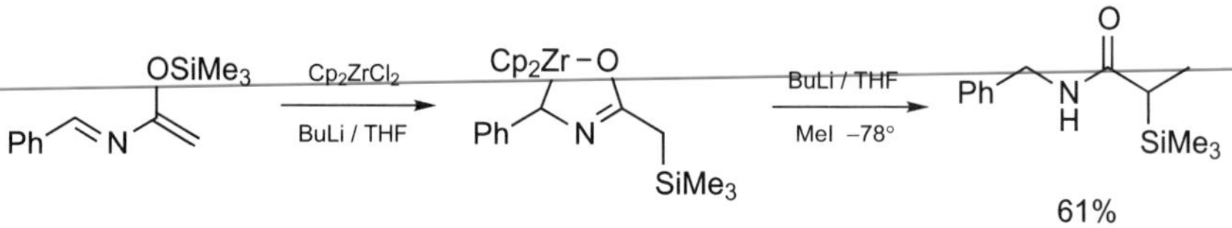

Alkenylcyclopropanes.[2] The Cp_2Zr–ethylene complex behaves as a 1,3-dipole toward the carbonyl group. The reaction products decompose on exposure to sulfuric acid are alkenylcyclopropanes, which are different from those arising from treatment with HCl.

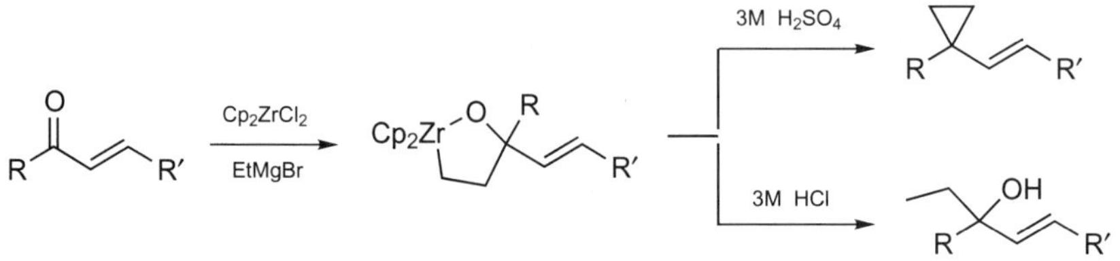

[1]Gandon, V., Bertus, V., Szymoniak, J. *T* **56**, 4467 (2000).
[2]Bertus, V., Gandon, V., Szymoniak, J. *CC* 171 (2000).

Zirconocene, Zr-alkylated. 15, 81; **18,** 414; **19,** 412–414; **20,** 442–443

α,β-Unsaturated esters.[1] Zirconacyclopropenes derived from alkynes undergo alkoxycarbonylation. The resulting alkenylzirconocene derivatives can be functionalized with various electrophiles.

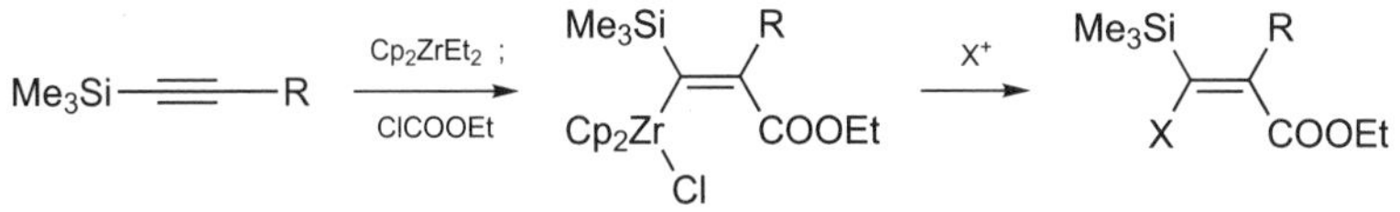

Pyridines.[2] A [2 + 2 + 2]cycloaddition is accomplished via azazirconacyclopentadiene formation from an alkyne and a nitrile and subsequent reaction with another alkyne molecule in a nickel(II)-catalyzed process.

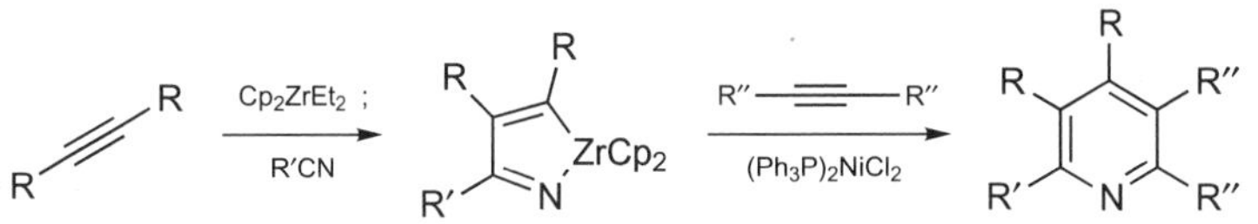

[1]Takahashi, T., Xi, C., Ura, Y., Nakajima, K. *JACS* **122**, 3228 (2000).
[2]Takahashi, T., Tsai, F.-Y., Kotora, M. *JACS* **122**, 4994 (2000).

Zirconocene dichloride. 14, 122; **15,** 120–121; **18,** 415; **19,** 414; **20,** 443–444

Aldol reaction.[1] Zirconium enolates are employed in conducting aldol reactions of amides derived from (+)-pseudoephedrine. These species are prepared from the corresponding lithium enolates on addition of Cp_2ZrCl_2.

Cyclopentanones.[2] Cyclozirconation of dienes followed by carbonylation is a valuable method for the preparation of 3,4-cyclocondensed cyclopentanones. A new application is found in a synthesis of (−)-androst-4-en-3,16-dione.

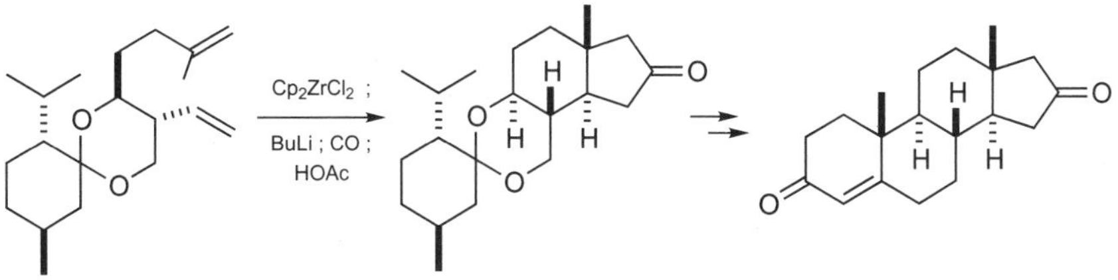

[1]Vicario, J.L., Badia, D., Dominguez, E., Rodriguez, M., Carrillo, L. *JOC* **65**, 3754 (2000).
[2]Taber, D.F., Zhang, W., Campbell, C.L., Rheingold, A.L., Incarvito, C.D. *JACS* **122**, 4813 (2000).

Zirconocene hydrochloride. 14, 81; **15,** 80–81; **18,** 416–417; **19,** 415–416; **20,** 445–446

Aldehydes from amides.[1] Tertiary amides are reduced to aldehydes at room temperature with $Cp_2Zr(H)Cl$ in THF (16 examples, 75–99%).

[1]White, J.M., Tunoori, A.R., Georg, G.I. *JACS* **122**, 11995 (2000).

AUTHOR INDEX

SUBJECT INDEX

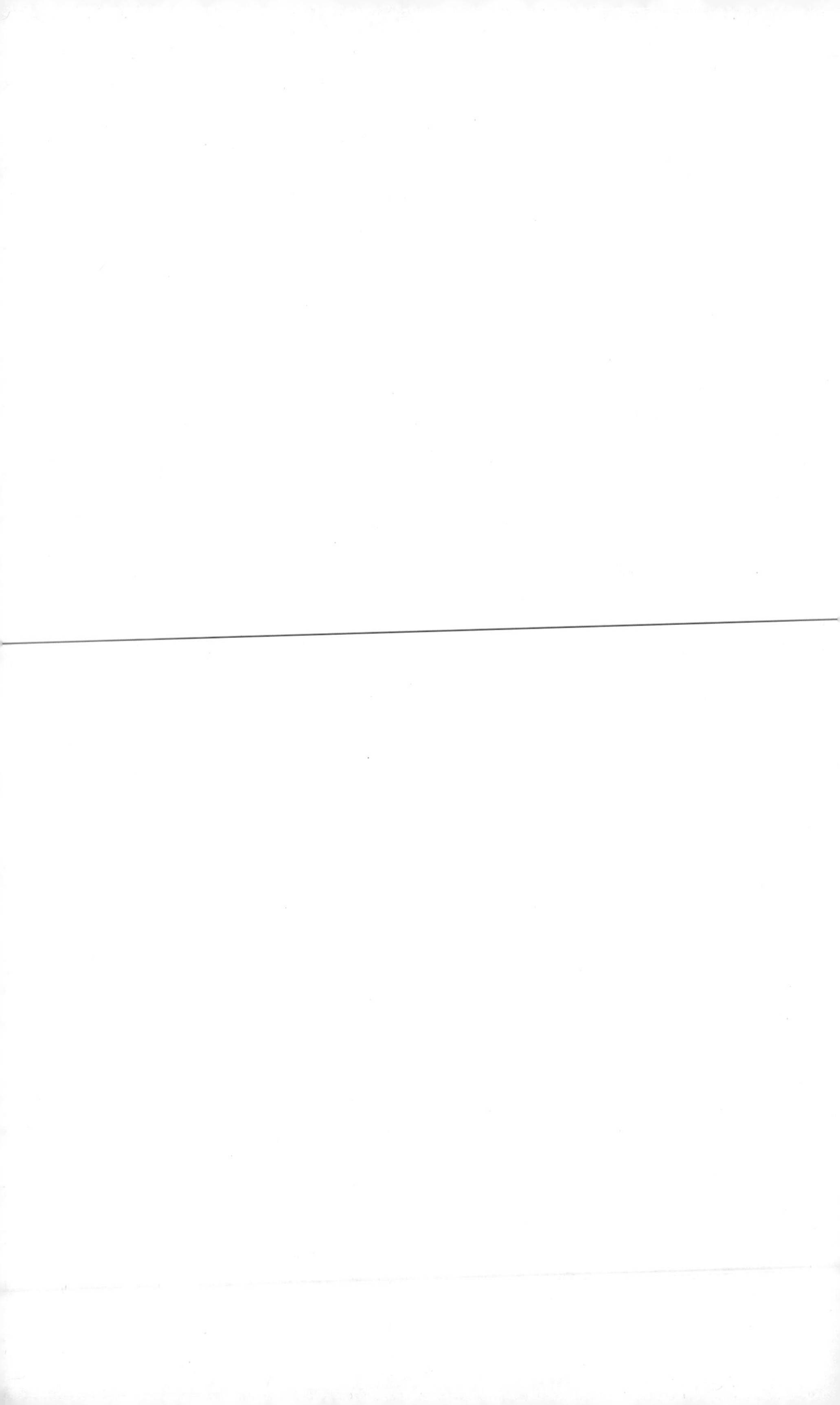

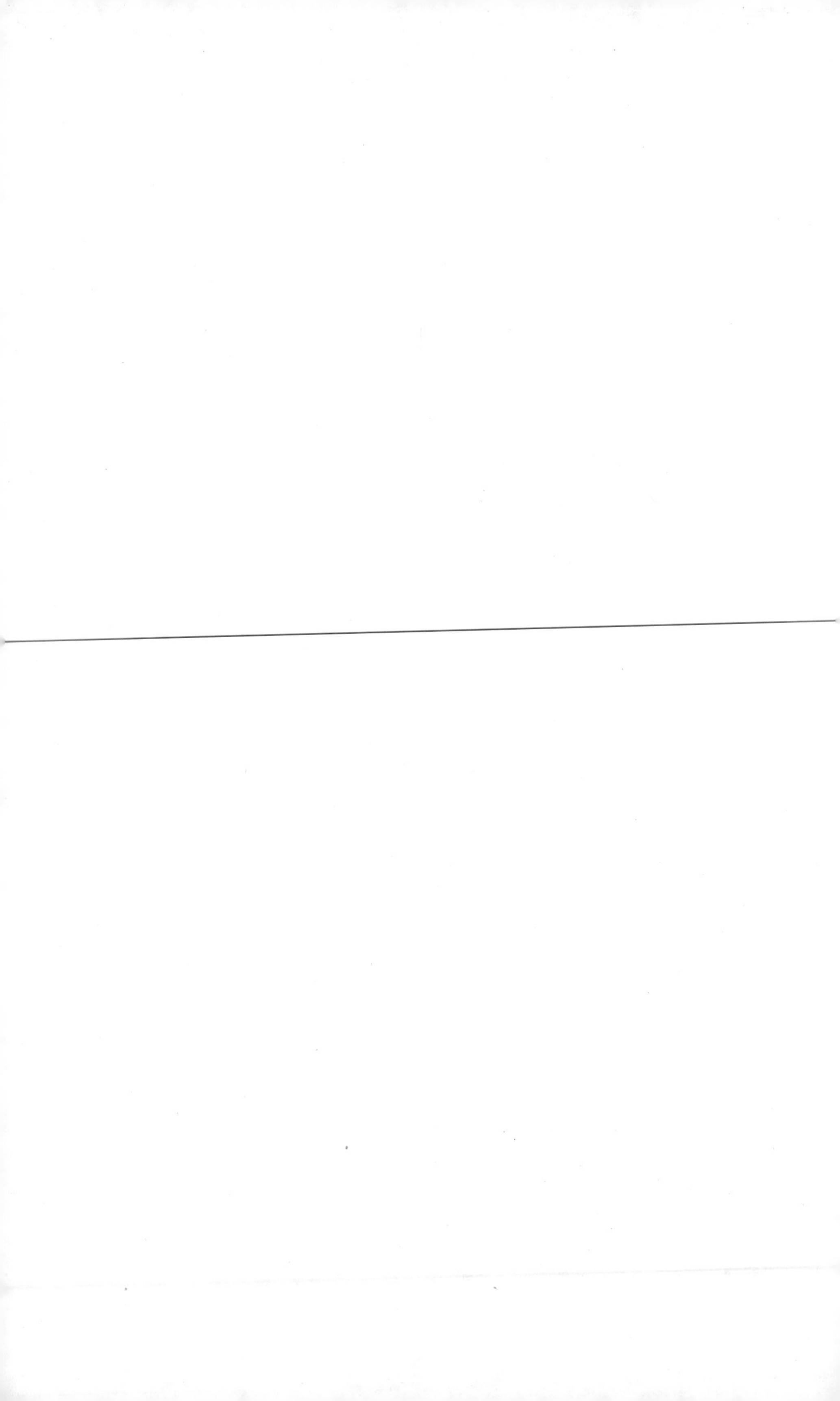

JUNIATA COLLEGE
2820 9100 067 036 4